Advanced Mechanics of Materials and Applied Elasticity

Fifth Edition

Advanced Mechanics of Materials and Applied Elasticity

Fifth Edition

ANSEL C. UGURAL

SAUL K. FENSTER

PRENTICE
HALL

Upper Saddle River, NJ • Boston • Indianapolis • San Francisco
New York • Toronto • Montreal • London • Munich • Paris • Madrid
Capetown • Sydney • Tokyo • Singapore • Mexico City

Many of the designations used by manufacturers and sellers to distinguish their products are claimed as trademarks. Where those designations appear in this book, and the publisher was aware of a trademark claim, the designations have been printed with initial capital letters or in all capitals.

The authors and publisher have taken care in the preparation of this book, but make no expressed or implied warranty of any kind and assume no responsibility for errors or omissions. No liability is assumed for incidental or consequential damages in connection with or arising out of the use of the information or programs contained herein.

The publisher offers excellent discounts on this book when ordered in quantity for bulk purchases or special sales, which may include electronic versions and/or custom covers and content particular to your business, training goals, marketing focus, and branding interests. For more information, please contact:

U.S. Corporate and Government Sales
(800) 382-3419
corpsales@pearsontechgroup.com

For sales outside the United States please contact:

International Sales
international@pearson.com

Visit us on the Web: informit.com/ph

Library of Congress Cataloging-in-Publication Data
Ugural, A. C.
    Advanced mechanics of materials and elasticity / Ansel C. Ugural, Saul K.
    Fenster. — 5th ed.
       p. cm.
    Rev. ed. of: Advanced strength and applied elasticity. 4th ed. c2003.
    Includes bibliographical references and index.
    ISBN 0-13-707920-6 (hardcover : alk. paper)
1. Strength of materials. 2. Elasticity. 3. Materials—Mechanical
properties. I. Fenster, Saul K., 1933- II. Ugural, A. C. Advanced strength
and applied elasticity. III. Title.
    TA405.U42 2011
    620.1'12—dc23   2011012705

ISBN-13: 978-0-13-707920-9
ISBN-10:   0-13-707920-6

Text printed in the United States on recycled paper at Courier in Westford, Massachusetts.
First printing, June 2011

Contents

Preface

INTRODUCTION

This text is a development of classroom notes prepared in connection with advanced undergraduate and first-year graduate courses in elasticity and the mechanics of solids. It is designed to satisfy the requirements of courses subsequent to an elementary treatment of the strength of materials. In addition to its applicability to aeronautical, civil, and mechanical engineering and to engineering mechanics curricula, the text is useful to practicing engineers. Emphasis is given to *numerical techniques* (which lend themselves to computerization) in the solution of problems resisting *analytical treatment*. The stress placed on numerical solutions is not intended to deny the value of classical analysis, which is given a rather full treatment. It instead attempts to fill what the authors believe to be a void in the world of textbooks.

An effort has been made to present a balance between the theory necessary to gain insight into the mechanics, but which can often offer no more than crude approximations to real problems because of simplifications related to geometry and conditions of loading, and numerical solutions, which are so useful in presenting stress analysis in a more realistic setting. This text emphasizes those aspects of theory and application that prepare a student for more advanced study or for professional practice in design and analysis.

The theory of elasticity plays three important roles in the text: it provides exact solutions where the configurations of loading and boundary are relatively simple; it provides a check on the limitations of the mechanics of materials approach; and it serves as the basis of approximate solutions employing numerical analysis.

To make the text as clear as possible, attention is given to the presentation of the fundamentals of the mechanics of materials. The physical significance of the solutions and practical applications are given emphasis. A special effort was made to illustrate important principles and applications with numerical examples. Consistent with announced national policy, problems are included in the text in which the physical quantities are expressed in the International System of Units (SI). All important quantities are defined in both SI and U.S. Customary System of units. A sign convention, consistent with vector mechanics, is employed throughout for loads, internal forces, and stresses. This convention conforms to that used in most classical strength of materials and elasticity texts, as well as to that most often employed in the numerical analysis of complex structures.

TEXT ARRANGEMENT

Because of the extensive subdivision into a variety of topics and the employment of alternative methods of analysis, the text should provide flexibility in the choice of assignments to cover courses of varying length and content. Most chapters are substantially self-contained. Hence, the order of presentation can be smoothly altered to meet an instructor's preference. It is suggested, however, that Chapters 1 and 2, which address the analysis of basic concepts, should be studied first. The emphasis placed on the treatment of two-dimensional problems in elasticity (Chapter 3) may differ according to the scope of the course.

This fifth edition of *Advanced Mechanics of Materials and Applied Elasticity* seeks to preserve the objectives and emphases of the previous editions. Every effort has been made to provide a more complete and current text through the inclusion of new material dealing with the fundamental principles of stress analysis and design: stress concentrations, contact stresses, failure criteria, fracture mechanics, compound cylinders, finite element analysis (FEA), energy and variational methods, buckling of stepped columns, and common shell types. The entire text has been reexamined and many improvements have been made throughout by a process of elimination and rearrangement. Some sections have been expanded to improve on previous expositions.

The references, provided as an aid to the student who wishes to further pursue certain aspects of a subject, have been updated and identified at the end of each chapter. We have resisted the temptation to increase the material covered except where absolutely necessary. However, it was considered desirable to add a number of illustrative examples and a large number of problems important in engineering practice and design. Extra care has been taken in the presentation and solution of the sample problems. All the problem sets have been reviewed and checked to ensure both their clarity and numerical accuracy. Most changes in subject-matter coverage were prompted by the suggestions of faculty familiar with earlier editions.

It is hoped that we have maintained clarity of presentation, simplicity as the subject permits, unpretentious depth, an effort to encourage intuitive understanding, and a shunning of the irrelevant. In this context, as throughout, emphasis is placed on the use of fundamentals in order to build student understanding and an ability to solve the more complex problems.

SUPPLEMENTS

The book is accompanied by a comprehensive *Solutions Manual* available to instructors. It features complete solutions to all problems in the text. Answers to selected problems are given at the end of the book. PowerPoint slides of figures and tables and a password-protected *Solutions Manual* are available for instructors at the Pearson Instructor Resource Center, pearsonhighered.com/irc.

Acknowledgments

It is a particular pleasure to acknowledge the contributions of those who assisted in the evolution of the text. Thanks, of course, are due to the many readers who have contributed general ideas and to reviewers who have made detailed comments on previous editions. These notably include the following: F. Freudenstein, Columbia University; R. A. Scott, University of Michigan; M. W. Wilcox and Y. Chan Jian, Southern Methodist University; C. T. Sun, University of Florida; B. Koplik, H. Kountouras, K. A. Narh, R. Sodhi, and C. E. Wilson, New Jersey Institute of Technology; H. Smith, Jr., South Dakota School of Mines and Technology; B. P. Gupta, Gannon University; S. Bang, University of Notre Dame; B. Koo, University of Toledo; J. T. Easley, University of Kansas; J. A. Bailey, North Carolina State University; W. F. Wright, Vanderbilt University; R. Burks, SUNY Maritime College; G. E. O. Widera, University of Illinois; R. H. Koebke, University of South Carolina; B. M. Kwak, University of Iowa; G. Nadig, Widener University; R. L. Brown, Montana State University; S. H. Advani, West Virginia University; E. Nassat, Illinois Institute of Technology; R. I. Sann, Stevens Institute of Technology; C. O. Smith, University of Nebraska; J. Kempner, Polytechnic University of New York; and P. C. Prister, North Dakota State University; R. Wetherhold, University of Buffalo, SUNY; and Shaofan Li, University of California at Berkeley.

Accuracy checking of the problems and typing of *Solutions Manual* were done expertly by my former student, Dr. Youngjin Chung. Also contributing considerably to this volume with typing new inserts, assiting with some figures, limited proofreading, and cover design was Errol A. Ugural. Their hard work is much appreciated. I am deeply indebted to my colleagues who have found the text useful through the years and to Bernard Goodwin, publisher at Prentice Hall PTR, who encouraged development of this edition. Copy editing and production were handled skillfully by Carol Lallier and Elizabeth Ryan. Their professional help is greatly appreciated. Lastly, I am very thankful for the support and understanding of my wife Nora, daughter Aileen, and son Errol during preparation of this book.

Ansel C. Ugural

About the Authors

Ansel C. Ugural, Ph.D., is visiting professor at New Jersey Institute of Technology. He has held various faculty and administrative positions at Fairleigh Dickinson University, and he taught at the University of Wisconsin. Ugural has considerable industrial experience in both full-time and consulting capacities. A member of several professional societies, he is the author of the books *Mechanics of Materials; Stresses in Beams, Plates and Shells*; and *Mechanical Design: An Integrated Approach.*

Saul K. Fenster, Ph.D., served as president and tenured professor at New Jersey Institute of Technology for more than two decades. In addition, he has held varied positions at Fairleigh Dickinson University and taught at the City University of New York. His experience includes membership on a number of corporate boards and economic development commissions. Fenster is a fellow of the American Society of Mechanical Engineers, the American Society for Engineering Education, and the American Society for Manufacturing Engineers. He is coauthor of a text on mechanics.

List of Symbols

A	area
b	width
C	carryover factor, torsional rigidity
c	distance from neutral axis to outer fiber
D	distribution factor, flexural rigidity of plate
$[D]$	elasticity matrix
d	diameter, distance
E	modulus of elasticity in tension or compression
E_s	modulus of plasticity or secant modulus
E_t	tangent modulus
e	dilatation, distance, eccentricity
$\{F\}$	nodal force matrix of bar and beam finite elements
F	body force per unit volume, concentrated force
f	coefficient of friction
$\{f\}$	displacement function of finite element
G	modulus of elasticity in shear
g	acceleration of gravity ($\approx$9.81 m/s^2)
h	depth of beam, height, membrane deflection, mesh width
I	moment of inertia of area, stress invariant
J	polar moment of inertia of area, strain invariant
K	bulk modulus, spring constant of an elastic support, stiffness factor, thermal conductivity, fatigue factor, strength coefficient, stress concentration factor
$[K]$	stiffness matrix of whole structure
k	constant, modulus of elastic foundation, spring constant
$[k]$	stiffness matrix of finite element
L	length, span
M	moment
M_{xy}	twisting moment in plates
m	moment caused by unit load
N	fatigue life (cycles), force
n	factor of safety, number, strain hardening index

l, m, n	direction cosines
P	concentrated force
p	distributed load per unit length or area, pressure, stress resultant
Q	first moment of area, heat flow per unit length, shearing force
$\{Q\}$	nodal force matrix of two-dimensional finite element
R	radius, reaction
S	elastic section modulus, shear center
r	radius, radius of gyration
r, θ	polar coordinates
s	distance along a line or a curve
T	temperature, twisting couple or torque
t	thickness
U	strain energy
U_o	strain energy per unit volume
U^*	complementary energy
V	shearing force, volume
v	velocity
W	weight, work
u, v, w	components of displacement
Z	plastic section modulus
x, y, z	rectangular coordinates
α	angle, coefficient of thermal expansion, form factor for shear
β	numerical factor, angle
γ	shear strain, weight per unit volume or specific weight, angle
δ	deflection, finite difference operator, variational symbol, displacement
$\{\delta\}$	nodal displacement matrix of finite element
Δ	change of a function
ϵ	normal strain
θ	angle, angle of twist per unit length, slope
ν	Poisson's ratio
λ	axial load factor, Lamé constant
Π	potential energy
ρ	density (mass per unit volume), radius
σ	normal stress
τ	shear stress
ϕ	total angle of twist
Φ	stress function
ω	angular velocity
ψ	stream function

Analysis of Stress

1.1 INTRODUCTION

There are two major parts to this chapter. Review of some important fundamentals of statics and mechanics of solids, the concept of stress, modes of load transmission, general sign convention for stress and force resultants that will be used throughout the book, and analysis and design principles are provided first. This is followed with treatment for changing the components of the state of stress given in one set of coordinate axes to any other set of rotated axes, as well as variation of stress within and on the boundaries of a load-carrying member. Plane stress and its transformation are of basic importance, since these conditions are most common in engineering practice. The chapter is thus also a brief guide and introduction to the remainder of the text.

Mechanics of Materials and Theory of Elasticity

The basic structure of matter is characterized by nonuniformity and discontinuity attributable to its various subdivisions: molecules, atoms, and subatomic particles. Our concern in this text is not with the particulate structure, however, and it will be assumed that the matter with which we are concerned is *homogeneous* and *continuously* distributed over its volume. There is the clear implication in such an approach that the smallest element cut from the body possesses the same properties as the body. Random fluctuations in the properties of the material are thus of no consequence. This approach is that of *continuum mechanics*, in which solid elastic materials are treated as though they are continuous media rather than composed of discrete molecules. Of the states of matter, we are here concerned only with the solid, with its ability to maintain its shape without the need of a container and to resist continuous shear, tension, and compression.

In contrast with rigid-body statics and dynamics, which treat the external behavior of bodies (that is, the equilibrium and motion of bodies without regard to small deformations associated with the application of load), the mechanics of solids is concerned with the relationships of external effect (forces and moments) to internal stresses and strains. Two different approaches used in solid mechanics are the *mechanics of materials* or *elementary theory* (also called the *technical theory*) and the *theory of elasticity*. The mechanics of materials focuses mainly on the more or less approximate solutions of practical problems. The theory of elasticity concerns itself largely with more mathematical analysis to determine the "exact" stress and strain distributions in a loaded body. The difference between these approaches is primarily in the nature of the simplifying assumptions used, described in Section 3.1.

External forces acting on a body may be classified as *surface forces* and *body forces*. A surface force is of the *concentrated* type when it acts at a point; a surface force may also be distributed *uniformly* or *nonuniformly* over a finite area. Body forces are associated with the mass rather than the surfaces of a body, and are distributed throughout the volume of a body. Gravitational, magnetic, and inertia forces are all body forces. They are specified in terms of force per unit volume. All forces acting on a body, including the reactive forces caused by supports and body forces, are considered to be *external forces*. *Internal forces* are the forces that hold together the particles forming the body. Unless otherwise stated, we assume in this text that body forces can be neglected and that forces are applied steadily and slowly. The latter is referred to as *static loading*.

In the International System of Units (SI), force is measured in newtons (N). Because the newton is a small quantity, the kilonewton (kN) is often used in practice. In the U.S. Customary System, force is expressed in pounds (lb) or kilopounds (kips). We define all important quantities in both systems of units. However, in numerical examples and problems, SI units are used throughout the text consistent with international convention. (Table D.2 compares the two systems.)

Historical Development

The study of the behavior of members in tension, compression, and bending began with Leonardo da Vinci (1452–1519) and Galileo Galilei (1564–1642). For a proper understanding, however, it was necessary to establish accurate experimental description of a material's properties. Robert Hooke (1615–1703) was the first to point out that a body is deformed subject to the action of a force. Sir Isaac Newton (1642–1727) developed the concepts of Newtonian mechanics that became key elements of the strength of materials.

Leonard Euler (1707–1783) presented the mathematical theory of columns in 1744. The renowned mathematician Joseph-Louis Lagrange (1736–1813) received credit in developing a partial differential equation to describe plate vibrations. Thomas Young (1773–1829) established a coefficient of elasticity, Young's modulus. The advent of railroads in the late 1800s provided the impetus for much of the basic work in this area. Many famous scientists and engineers, including Coulomb, Poisson, Navier, St. Venant, Kirchhoff, and Cauchy, were responsible for advances in mechanics of materials during the eighteenth and nineteenth centuries. The British physicist William Thomas Kelvin (1824–1907), better known by his knighted name, Sir Lord Kelvin, first demonstrated that torsional moments acting at the edges of plates could be decomposed into shearing forces. The prominent English mathematician Augustus Edward Hough Love (1863–1940) introduced simple analysis of shells, known as Love's approximate theory.

Over the years, most basic problems of solid mechanics had been solved. Stephan P. Timoshenko (1878–1972) made numerous original contributions to the field of applied mechanics and wrote pioneering textbooks on the mechanics of materials, theory of elasticity, and theory of elastic stability. The theoretical base for modern strength of materials had been developed by the end of the nineteenth century. Following this, problems associated with the design of aircraft, space vehicles, and nuclear reactors have led to many studies of the more advanced phases of the subject. Consequently, the mechanics of materials is being expanded into the theories of elasticity and plasticity.

In 1956, Turner, Clough, Martin, and Topp introduced the *finite element method*, which permits the numerical solution of complex problems in solid mechanics in an economical way. Many contributions in this area are owing to Argyris and Zienkiewicz. The recent trend in the development is characterized by heavy reliance on high-speed computers and by the introduction of more rigorous theories. Numerical methods presented in Chapter 7 and applied in the chapters following have clear application to computation by means of electronic digital computers. Research in the foregoing areas is ongoing, not only to meet demands for treating complex problems but to justify further use and limitations on which the theory of solid mechanics is based. Although a widespread body of knowledge exists at present, mechanics of materials and elasticity remain fascinating subjects as their areas of application are continuously expanded.* The literature dealing with various aspects of solid mechanics is voluminous. For those seeking more thorough treatment, selected references are identified in brackets and compiled at the end of each chapter.

1.2 SCOPE OF TREATMENT

As stated in the preface, this book is intended for advanced undergraduate and graduate engineering students as well as engineering professionals. To make the text as clear as possible, attention is given to the fundamentals of solid mechanics and chapter objectives. A special effort has been made to illustrate important principles and applications with numerical examples. Emphasis is placed on a thorough presentation of several classical topics in advanced mechanics of materials and applied elasticity and of selected advanced topics. Understanding is based on the explanation of the physical behavior of members and then modeling this behavior to develop the theory.

---

*Historical reviews of mechanics of materials and the theory of elasticity are given in Refs. 1.1 through 1.5.

The usual objective of mechanics of material and theory of elasticity is the examination of the load-carrying capacity of a body from three standpoints: *strength*, *stiffness*, and *stability*. Recall that these quantities relate, respectively, to the ability of a member to resist permanent deformation or fracture, to resist deflection, and to retain its equilibrium configuration. For instance, when loading produces an abrupt shape change of a member, instability occurs; similarly, an inelastic deformation or an excessive magnitude of deflection in a member will cause malfunction in normal service. The foregoing matters, by using the *fundamental principles* (Sec. 1.3), are discussed in later chapters for various types of structural members. *Failure* by yielding and fracture of the materials under combined loading is taken up in detail in Chapter 4.

Our main concern is the analysis of stress and deformation within a loaded body, which is accomplished by application of one of the methods described in the next section. For this purpose, the analysis of loads is essential. A structure or machine cannot be satisfactory unless its design is based on realistic operating loads. The principal topics under the heading of *mechanics of solids* may be summarized as follows:

1. Analysis of the stresses and deformations within a body subject to a prescribed system of forces. This is accomplished by solving the governing equations that describe the stress and strain fields (theoretical stress analysis). It is often advantageous, where the shape of the structure or conditions of loading preclude a theoretical solution or where verification is required, to apply the laboratory techniques of experimental stress analysis.
2. Determination by theoretical analysis or by experiment of the limiting values of load that a structural element can sustain without suffering damage, failure, or compromise of function.
3. Determination of the body shape and selection of the materials that are most efficient for resisting a prescribed system of forces under specified conditions of operation such as temperature, humidity, vibration, and ambient pressure. This is the design function.

The design function, item 3, clearly relies on the performance of the theoretical analyses under items 1 and 2, and it is to these that this text is directed. Particularly, emphasis is placed on the development of the equations and methods by which detailed analysis can be accomplished.

The ever-increasing industrial demand for more sophisticated structures and machines calls for a good grasp of the concepts of stress and strain and the behavior of materials — and a considerable degree of ingenuity. This text, at the very least, provides the student with the ideas and information necessary for an understanding of the advanced mechanics of solids and encourages the creative process on the basis of that understanding. Complete, carefully drawn free-body diagrams facilitate visualization, and these we have provided, all the while knowing that the subject matter can be learned best only by solving problems of practical importance. A thorough grasp of fundamentals will prove of great value in attacking new and unfamiliar problems.

1.3 ANALYSIS AND DESIGN

Throughout this text, a fundamental procedure for analysis in solving mechanics of solids problems is used repeatedly. The complete analysis of load-carrying structural members by the **method of equilibrium** requires consideration of three conditions relating to certain laws of forces, laws of material deformation, and geometric compatibility. These essential relationships, called the *basic principles of analysis*, are:

1. **Equilibrium Conditions.** The equations of *equilibrium* of forces must be satisfied throughout the member.
2. **Material Behavior.** The stress–strain or *force-deformation relations* (for example, Hooke's law) must apply to the material behavior of which the member is constructed.
3. **Geometry of Deformation.** The *compatibility conditions* of deformations must be satisfied: that is, each deformed portion of the member must fit together with adjacent portions. (Matter of compatibility is not always broached in mechanics of materials analysis.)

The stress and deformation obtained through the use of the three principles must conform to the conditions of loading imposed at the boundaries of a member. This is known as satisfying the **boundary conditions**. Applications of the preceding procedure are illustrated in the problems presented as the subject unfolds. Note, however, that it is not always necessary to execute an analysis in the exact order of steps listed previously.

As an alternative to the equilibrium methods, the analysis of stress and deformation can be accomplished by employing **energy methods** (Chap. 10), which are based on the concept of strain energy. The aspect of both the equilibrium and the energy approaches is twofold. These methods can provide solutions of acceptable accuracy where configurations of loading and member shape are regular, and they can be used as the basis of **numerical methods** (Chap. 7) in the solution of more realistic problems.

Engineering design is the process of applying science and engineering techniques to define a structure or system in detail to allow its realization. The objective of a *mechanical design* procedure includes finding of proper materials, dimensions, and shapes of the members of a structure or machine so that they will support prescribed loads and perform without failure. *Machine design* is creating new or improved machines to accomplish specific purposes. Usually, *structural design* deals with any engineering discipline that requires a structural member or system.

Design is the essence, art, and intent of engineering. A good design satisfies performance, cost, and safety requirements. An *optimum design* is the best solution to a design problem within given restrictions. Efficiency of the optimization may be gaged by such criteria as minimum weight or volume, optimum cost, and/or any other standard deemed appropriate. For a design problem with many choices, a designer may often make decisions on the basis of experience, to reduce the problem to a single variable. A solution to determine the optimum result becomes straightforward in such a situation.

A plan for satisfying a need usually includes preparation of individual preliminary design. Each *preliminary design* involves a thorough consideration of the loads and actions that the structure or machine has to support. For each situation, an analysis is necessary. Design decisions, or choosing reasonable values of the safety factors and material properties, are significant in the preliminary design process.

The **role of analysis in design** may be observed best in examining the phases of a design process. This text provides an elementary treatment of the concept of "design to meet strength requirements" as those requirements relate to individual machine or structural components. That is, the geometrical configuration and material of a component are preselected and the applied loads are specified. Then, the basic formulas for stress are employed to select members of adequate size in each case. The following is *rational procedure in the design* of a load-carrying member:

1. *Evaluate the most likely modes of failure of the member.* Failure criteria that predict the various modes of failure under anticipated conditions of service are discussed in Chapter 4.
2. *Determine the expressions relating applied loading to such effects as stress, strain, and deformation.* Often, the member under consideration and conditions of loading are so significant or so amenable to solution as to have been the subject of prior analysis. For these situations, textbooks, handbooks, journal articles, and technical papers are good sources of information. Where the situation is unique, a mathematical derivation specific to the case at hand is required.
3. *Determine the maximum usable value of stress, strain, or energy.* This value is obtained either by reference to compilations of material properties or by experimental means such as simple tension test and is used in connection with the relationship derived in step 2.
4. *Select a design factor of safety.* This is to account for uncertainties in a number of aspects of the design, including those related to the actual service loads, material properties, or environmental factors. An important area of uncertainty is connected with the assumptions made in the analysis of stress and deformation. Also, we are not likely to have a secure knowledge of the stresses that may be introduced during machining, assembly, and shipment of the element.

   The design factor of safety also reflects the consequences of failure; for example, the possibility that failure will result in loss of human life or injury or in costly repairs or danger to other components of the overall system. For these reasons, the design factor of safety is also sometimes called the *factor of ignorance.* The uncertainties encountered during the design phase may be of such magnitude as to lead to a design carrying extreme weight, volume, or cost penalties. It may then be advantageous to perform thorough tests or more exacting analysis rather to rely on overly large design factors of safety.

   The *true factor of safety*, usually referred to simply as the factor of safety, can be determined only after the member is constructed and tested. This factor is the ratio of the maximum load the member *can sustain* under severe testing without failure to the maximum load *actually* carried under normal service conditions, the working load. When a linear relationship exists between the load and the stress produced by the load, the *factor of safety n* may be expressed as

$$n = \frac{\text{maximum usable stress}}{\text{allowable stress}} \tag{1.1}$$

Maximum usable stress represents either the yield stress or the ultimate stress. The allowable stress is the working stress. The factor of safety must be greater than 1.0 if failure is to be avoided. Values for factor of safety, selected by the designer on the basis of experience and judgment, are about 1.5 or greater. For most applications, appropriate factors of safety are found in various construction and manufacturing codes.

The foregoing procedure is not always conducted in as formal a fashion as may be implied. In some design procedures, one or more steps may be regarded as unnecessary or obvious on the basis of previous experience. Suffice it to say that complete design solutions are not unique, involve a consideration of many factors, and often require a trial-and-error process [Ref. 1.6]. Stress is only one consideration in design. Other phases of the design of components are the prediction of the deformation of a given component under given loading and the consideration of buckling (Chap. 11). The methods of determining deformation are discussed in later chapters. Note that there is a very close relationship between analysis and design, and the examples and problems that appear throughout this book illustrate that connection.

We conclude this section with an appeal for the reader to exercise a degree of skepticism with regard to the application of formulas for which there is uncertainty as to the limitations of use or the areas of applicability. The relatively simple form of many formulas usually results from rather severe restrictions in its derivation. These relate to simplified boundary conditions and shapes, limitations on stress and strain, and the neglect of certain complicating factors. Designers and stress analysts must be aware of such restrictions lest their work be of no value or, worse, lead to dangerous inadequacies.

In this chapter, we are concerned with the state of *stress at a point* and the *variation of stress* throughout an elastic body. The latter is dealt with in Sections 1.8 and 1.16 and the former in the balance of the chapter.

1.4 CONDITIONS OF EQUILIBRIUM

A *structure* is a unit consisting of interconnected members supported in such a way that it is capable of carrying loads in static equilibrium. Structures are of four general types: frames, trusses, machines, and thin-walled (plate and shell) structures. *Frames* and *machines* are structures containing multiforce members. The former support loads and are usually stationary, fully restrained structures. The latter transmit and modify forces (or power) and always contain moving parts. The *truss* provides both a practical and economical solution, particularly in the design of bridges and buildings. When the truss is loaded at its joints, the only force in each member is an axial force, either tensile or compressive.

The analysis and design of structural and machine components require a knowledge of the distribution of forces within such members. Fundamental concepts and conditions of static equilibrium provide the necessary background for the determination of internal as well as external forces. In Section 1.6, we shall see that components of internal-forces resultants have special meaning in terms of the type of deformations they cause, as applied, for example, to slender members. We note that surface forces that develop at support points of a structure are called *reactions*. They equilibrate the effects of the applied loads on the structures.

The **equilibrium** of forces is the state in which the forces applied on a body are in balance. Newton's first law states that if the resultant force acting on a particle (the simplest body) is zero, the particle will remain at rest or will move with constant velocity. Statics is concerned essentially with the case where the particle or body remains at rest. A complete free-body diagram is essential in the solution of problems concerning the equilibrium.

Let us consider the equilibrium of a body in space. In this three-dimensional case, the **conditions of equilibrium** require the satisfaction of the following **equations of statics:**

$$\Sigma F_x = 0 \qquad \Sigma F_y = 0 \qquad \Sigma F_z = 0$$
$$\Sigma M_x = 0 \qquad \Sigma M_y = 0 \qquad \Sigma M_z = 0 \tag{1.2}$$

The foregoing state that the sum of all forces acting on a body in any direction must be zero; the sum of all moments about any axis must be zero.

In a *planar problem*, where all forces act in a single (xy) plane, there are only three independent equations of statics:

$$\Sigma F_x = 0 \qquad \Sigma F_y = 0 \qquad \Sigma M_A = 0 \tag{1.3}$$

That is, the sum of all forces in any (x, y) directions must be zero, and the resultant moment about axis z or any point A in the plane must be zero. By replacing a force summation with an equivalent moment summation in Eqs. (1.3), the following *alternative* sets of conditions are obtained:

$$\Sigma F_x = 0 \qquad \Sigma M_A = 0 \qquad \Sigma M_B = 0 \tag{1.4a}$$

provided that the line connecting the points A and B *is not* perpendicular to the x axis, or

$$\Sigma M_A = 0 \qquad \Sigma M_B = 0 \qquad \Sigma M_C = 0 \tag{1.4b}$$

Here points A, B, and C are *not* collinear. Clearly, the judicious selection of points for taking moments can often simplify the algebraic computations.

A structure is *statically determinate* when all forces on its members can be found by using only the conditions of equilibrium. If there are more unknowns than available equations of statics, the problem is called *statically indeterminate*. The degree of *static indeterminacy* is equal to the difference between the number of unknown forces and the number of relevant equilibrium conditions. Any reaction that is in excess of those that can be obtained by statics alone is termed *redundant*. The number of redundants is therefore the same as the degree of indeterminacy.

1.5 DEFINITION AND COMPONENTS OF STRESS

Stress and strain are most important concepts for a comprehension of the mechanics of solids. They permit the mechanical behavior of load-carrying components to be described in terms fundamental to the engineer. Both the analysis and design of a given machine or structural element involve the determination of stress and material stress–strain relationships. The latter is taken up in Chapter 2.

Consider a body in equilibrium subject to a system of external forces, as shown in Fig. 1.1a. Under the action of these forces, internal forces are developed within the body. To examine the latter at some interior point Q, we use an imaginary plane to cut the body at a section a–a through Q, dividing the body into two parts. As the forces acting on the entire body are in equilibrium, the forces acting on one part alone must be in equilibrium: this requires the presence of forces on plane a–a. These internal forces, applied to both parts, are distributed continuously over the cut surface. This process, referred to as the **method of sections** (Fig. 1.1), is relied on as a first step in solving all problems involving the investigation of internal forces.

A **free-body diagram** is simply a sketch of a body with all the appropriate forces, both known and unknown, acting on it. Figure 1.1b shows such a plot of the isolated left part of the body. An element of area ΔA, located at point Q on the cut surface, is acted on by force $\Delta \boldsymbol{F}$. Let the origin of coordinates be placed at point Q, with x normal and y, z tangent to ΔA. In general, $\Delta \boldsymbol{F}$ does not lie along x, y, or z.

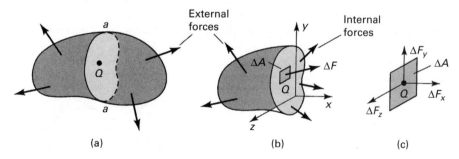

(a)   (b)   (c)

FIGURE 1.1.   *Method of sections: (a) Sectioning of a loaded body; (b) free body with external and internal forces; (c) enlarged area ΔA with components of the force $\Delta \boldsymbol{F}$.*

Decomposing $\Delta \mathbf{F}$ into components parallel to x, y, and z (Fig. 1.1c), we define the *normal stress* σ_x and the *shearing stresses* τ_{xy} and τ_{xz}:

$$\sigma_x = \lim_{\Delta A \to 0} \frac{\Delta F_x}{\Delta A} = \frac{dF_x}{dA}$$

$$\tau_{xy} = \lim_{\Delta A \to 0} \frac{\Delta F_y}{\Delta A} = \frac{dF_y}{dA}, \qquad \tau_{xz} = \lim_{\Delta A \to 0} \frac{\Delta F_z}{\Delta A} = \frac{dF_z}{dA}$$

$$(1.5)$$

These definitions provide the stress components at a point Q to which the area ΔA is reduced in the limit. Clearly, the expression $\Delta A \to 0$ depends on the idealization discussed in Section 1.1. Our consideration is with the average stress on areas, which, while small as compared with the size of the body, is large compared with interatomic distances in the solid. Stress is thus defined adequately for engineering purposes. As shown in Eq. (1.5), the intensity of force *perpendicular*, or *normal*, to the surface is termed the normal stress at a point, while the intensity of force *parallel* to the surface is the shearing stress at a point.

The values obtained in the limiting process of Eq. (1.5) differ from point to point on the surface as $\Delta \mathbf{F}$ varies. The stress components depend not only on $\Delta \mathbf{F}$, however, but also on the orientation of the plane on which it acts at point Q. Even at a given point, therefore, the stresses will differ as different planes are considered. The complete description of stress at a point thus requires the specification of the stress on all planes passing through the point.

Because the stress (σ or τ) is obtained by dividing the force by area, it has *units* of force per unit area. In SI units, stress is measured in *newtons per square meter* (N/m^2), or *pascals* (Pa). As the pascal is a very small quantity, the megapascal (MPa) is commonly used. When U.S. Customary System units are used, stress is expressed in pounds per square inch (psi) or kips per square inch (ksi).

It is verified in Section 1.12 that in order to enable the determination of the stresses on an infinite number of planes passing through a point Q, thus defining the stresses at that point, we need only specify the stress components on three mutually perpendicular planes passing through the point. These three planes, perpendicular to the coordinate axes, contain three hidden sides of an infinitesimal cube (Fig. 1.2). We emphasize that when we move from point Q to point Q' the values of stress will, in general, change. Also, body forces can exist. However, these

FIGURE 1.2. *Element subjected to three-dimensional stress. All stresses have positive sense.*

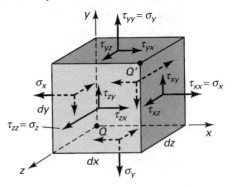

cases are not discussed here (see Sec. 1.8), as we are now merely interested in establishing the terminology necessary to specify a stress component.

The general case of a *three-dimensional state of stress* is shown in Fig. 1.2. Consider the stresses to be identical at points Q and Q' and uniformly distributed on each face, represented by a single vector acting at the center of each face. In accordance with the foregoing, a total of nine scalar *stress components* defines the state of stress at a point. The stress components can be assembled in the following *matrix form*, wherein each row represents the group of stresses acting on a plane passing through $Q(x, y, z)$:

$$[\tau_{ij}] = \begin{bmatrix} \tau_{xx} & \tau_{xy} & \tau_{xz} \\ \tau_{yx} & \tau_{yy} & \tau_{yz} \\ \tau_{zx} & \tau_{zy} & \tau_{zz} \end{bmatrix} = \begin{bmatrix} \sigma_x & \tau_{xy} & \tau_{xz} \\ \tau_{yx} & \sigma_y & \tau_{yz} \\ \tau_{zx} & \tau_{zy} & \sigma_z \end{bmatrix} \tag{1.6}$$

We note that in *indicial notation* (refer to Sec. 1.17), a stress component is written as τ_{ij}, where the subscripts i and j each assume the values of x, y, and z as required by the foregoing equation. The *double subscript notation* is interpreted as follows: The first subscript indicates the direction of a normal to the plane or face on which the stress component acts; the second subscript relates to the direction of the stress itself. Repetitive subscripts are avoided in this text, so the normal stresses τ_{xx}, τ_{yy}, and τ_{zz} are designated σ_x, σ_y, and σ_z, as indicated in Eq. (1.6). *A face or plane is usually identified by the axis normal to it*; for example, the x faces are perpendicular to the x axis.

Sign Convention

Referring again to Fig. 1.2, we observe that *both* stresses labeled τ_{yx} tend to twist the element in a clockwise direction. It would be convenient, therefore, if a sign convention were adopted under which these stresses carried the same sign. Applying a convention relying solely on the coordinate direction of the stresses would clearly not produce the desired result, inasmuch as the τ_{yx} stress acting on the upper surface is directed in the positive x direction, while τ_{yx} acting on the lower surface is directed in the negative x direction. The following *sign convention*, which applies to both normal and shear stresses, is related to the deformational influence of a stress and is based on the relationship between the direction of an outward normal drawn to a particular surface and the directions of the stress components on the same surface.

When *both* the outer normal and the stress component face in a positive direction relative to the coordinate axes, the stress is positive. When *both* the outer normal and the stress component face in a negative direction relative to the coordinate axes, the stress is positive. When the normal points in a positive direction while the stress points in a negative direction (or vice versa), the stress is negative. In accordance with this sign convention, tensile stresses are always positive and compressive stresses always negative. Figure 1.2 depicts a system of positive normal and shear stresses.

Equality of Shearing Stresses

We now examine properties of shearing stress by studying the equilibrium of forces (see Sec. 1.4) acting on the cubic element shown in Fig. 1.2. As the stresses acting on opposite faces (which are of equal area) are equal in magnitude but opposite in direction, translational equilibrium in all directions is assured; that is, $\Sigma F_x = 0$, $\Sigma F_y = 0$, and $\Sigma F_z = 0$. Rotational equilibrium is established by taking moments of the x-, y-, and z-directed forces about point Q, for example. From $\Sigma M_z = 0$,

$$(-\tau_{xy}\,dy\,dz)dx + (\tau_{yx}\,dx\,dz)dy = 0$$

Simplifying,

$$\tau_{xy} = \tau_{yx} \tag{1.7a}$$

Likewise, from $\Sigma M_y = 0$ and $\Sigma M_x = 0$, we have

$$\tau_{xz} = \tau_{zx}, \qquad \tau_{yz} = \tau_{zy}. \tag{1.7b}$$

Hence, the subscripts for the shearing stresses are commutative, and the stress tensor is symmetric. This means that *shearing stresses on mutually perpendicular planes* of the element *are equal*. Therefore, no distinction will hereafter be made between the stress components τ_{xy} and τ_{yx}, τ_{xz} and τ_{zx}, or τ_{yz} and τ_{zy}. In Section 1.8, it is shown rigorously that the foregoing is valid even when stress components vary from one point to another.

Some Special Cases of Stress

Under particular circumstances, the general state of stress (Fig. 1.2) reduces to simpler stress states, as briefly described here. These stresses, which are commonly encountered in practice, are given detailed consideration throughout the text.

a. *Triaxial Stress.* We shall observe in Section 1.13 that an element subjected to only stresses σ_1, σ_2, and σ_3 acting in mutually perpendicular directions is said to be in a state of triaxial stress. Such a state of stress can be written as

$$\begin{bmatrix} \sigma_1 & 0 & 0 \\ 0 & \sigma_2 & 0 \\ 0 & 0 & \sigma_3 \end{bmatrix} \tag{a}$$

The absence of shearing stresses indicates that the preceding stresses are the *principal stresses* for the element. A special case of triaxial stress, known as *spherical* or *dilatational stress*, occurs if all principal stresses are equal (see Sec. 1.14). Equal triaxial tension is sometimes called hydrostatic tension. An example of equal triaxial compression is found in a small element of liquid under static pressure.

b. *Two-Dimensional or Plane Stress.* In this case, only the x and y faces of the element are subjected to stress, and all the stresses act parallel to the x and y axes, as shown in Fig. 1.3a. The plane stress matrix is written

$$\begin{bmatrix} \sigma_x & \tau_{xy} \\ \tau_{xy} & \sigma_y \end{bmatrix} \tag{1.8}$$

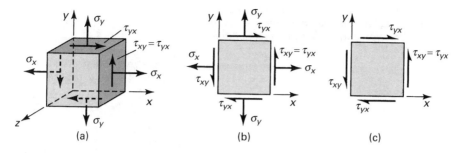

FIGURE 1.3. (a) Element in plane stress; (b) two-dimensional presentation of plane stress; (c) element in pure shear.

Although the three-dimensional nature of the element under stress should not be forgotten, for the sake of convenience we usually draw only a *two-dimensional view* of the plane stress element (Fig. 1.3b). When only two normal stresses are present, the state of stress is called *biaxial*. These stresses occur in thin plates stressed in two mutually perpendicular directions.

c. *Pure Shear.* In this case, the element is subjected to plane shearing stresses only, for example, τ_{xy} and τ_{yx} (Fig. 1.3c). Typical pure shear occurs over the cross sections and on longitudinal planes of a circular shaft subjected to torsion.

d. *Uniaxial Stress.* When normal stresses act along one direction only, the one-dimensional state of stress is referred to as a uniaxial tension or compression.

1.6 INTERNAL FORCE-RESULTANT AND STRESS RELATIONS

Distributed forces within a load-carrying member can be represented by a statically equivalent system consisting of a force and a moment vector acting at any arbitrary point (usually the centroid) of a section. These *internal force resultants*, also called *stress resultants*, exposed by an imaginary cutting plane containing the point through the member, are usually resolved into components normal and tangent to the cut section (Fig. 1.4). The sense of moments follows the right-hand screw rule, often represented by double-headed vectors, as shown in the figure. Each component can be associated with one of four modes of force transmission:

1. The *axial force P* or *N* tends to lengthen or shorten the member.
2. The *shear forces* V_y and V_z tend to shear one part of the member relative to the adjacent part and are often designated by the letter *V*.
3. The *torque* or *twisting moment T* is responsible for twisting the member.
4. The *bending moments* M_y and M_z cause the member to bend and are often identified by the letter *M*.

FIGURE 1.4. *Positive forces and moments on a cut
section of a body and components of
the force dF on an infinitesimal area
dA.*

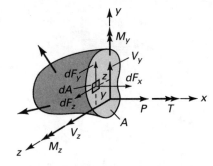

A member may be subject to any or all of the modes simultaneously. Note that the same *sign convention* is used for the force and moment components that is used for stress; a positive force (or moment) component acts on the positive face in the positive coordinate direction or on a negative face in the negative coordinate direction.

A typical infinitesimal area dA of the cut section shown in Fig. 1.4 is acted on by the components of an arbitrarily directed force dF, expressed using Eq. (1.5) as $dF_x = \sigma_x \, dA$, $dF_y = \tau_{xy} \, dA$, and $dF_z = \tau_{xz} \, dA$. Clearly, the stress components on the cut section cause the internal force resultants on that section. Thus, the incremental forces are summed in the x, y, and z directions to give

$$P = \int \sigma_x \, dA, \qquad V_y = \int \tau_{xy} \, dA, \qquad V_z = \int \tau_{xz} \, dA \qquad \textbf{(1.9a)}$$

In a like manner, the sums of the moments of the same forces about the x, y, and z axes lead to

$$T = \int (\tau_{xz}y - \tau_{xy}z) \, dA, \qquad M_y = \int \sigma_x z \, dA, \qquad M_z = -\int \sigma_x y \, dA \qquad \textbf{(1.9b)}$$

where the integrations proceed over area A of the cut section. Equations (1.9) represent the *relations between the internal force resultants and the stresses*. In the next paragraph, we illustrate the fundamental concept of stress and observe how Eqs. (1.9) connect internal force resultants and the state of stress in a specific case.

Consider a homogeneous prismatic *bar loaded by axial forces P at the ends* (Fig. 1.5a). A *prismatic* bar is a straight member having constant cross-sectional area throughout its length. To obtain an expression for the normal stress, we make an imaginary cut (section a–a) through the member at right angles to its axis. A free-body diagram of the isolated part is shown in Fig. 1.5b, wherein the stress is substituted on the cut section as a replacement for the effect of the removed part. Equilibrium of axial forces requires that $P = \int \sigma_x \, dA$ or $P = A\sigma_x$. The normal stress is therefore

$$\sigma_x = \frac{P}{A} \qquad \textbf{(1.10)}$$

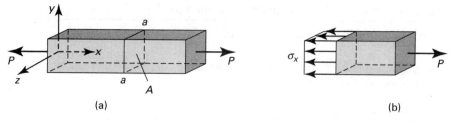

FIGURE 1.5. (a) Prismatic bar in tension; (b) Stress distribution across cross section.

where A is the cross-sectional area of the bar. Because V_y, V_z, and T all are equal to zero, the second and third of Eqs. (1.9a) and the first of Eqs. (1.9b) are satisfied by $\tau_{xy} = \tau_{xz} = 0$. Also, $M_y = M_z = 0$ in Eqs. (1.9b) requires only that σ_x be symmetrically distributed about the y and z axes, as depicted in Fig. 1.5b. When the member is being extended as in the figure, the resulting stress is a *uniaxial tensile stress*; if the direction of forces were reversed, the bar would be in compression under *uniaxial compressive stress*. In the latter case, Eq. (1.10) is applicable only to chunky or short members owing to other effects that take place in longer members.*

Similarly, application of Eqs. (1.9) to torsion members, beams, plates, and shells is presented as the subject unfolds, following the derivation of stress–strain relations and examination of the geometric behavior of a particular member. Applying the method of mechanics of materials, we develop other *elementary formulas* for stress and deformation. These, also called the *basic formulas of mechanics of materials*, are often used and extended for application to more complex problems in advanced mechanics of materials and the theory of elasticity. For reference purposes to preliminary discussions, Table 1.1 lists some commonly encountered cases. Note that in *thin-walled* vessels ($r/t \leq 10$) there is often no distinction made between the inner and outer radii because they are nearly equal. In mechanics of materials, r denotes the *inner* radius. However, the more accurate shell theory (Sec. 13.11) is based on the *average* radius, which we use throughout this text. Each equation presented in the table describes a state of stress associated with a single force, torque, moment component, or pressure at a section of a typical homogeneous and elastic structural member [Ref. 1.7]. When a member is acted on simultaneously by two or more load types, causing various internal force resultants on a section, it is assumed that each load produces the stress as if it were the only load acting on the member. The final or *combined stress* is then determined by superposition of the several states of stress, as discussed in Section 2.2.

The mechanics of materials theory is based on the simplifying assumptions related to the pattern of deformation so that the strain distributions for a cross section of the member can be determined. It is a basic assumption that *plane sections before loading remain plane after loading*. The assumption can be shown to be exact for axially loaded prismatic bars, for prismatic circular torsion members, and for

---

*Further discussion of uniaxial compression stress is found in Section 11.6, where we take up the classification of columns.

TABLE 1.1. *Commonly Used Elementary Formulas for Stress*[a]

1. Prismatic Bars of Linearly Elastic Material

Axial loading: $\sigma_x = \dfrac{P}{A}$ (a)

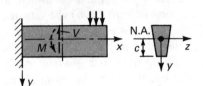

Torsion: $\tau = \dfrac{T\rho}{J}, \qquad \tau_{max} = \dfrac{Tr}{J}$ (b)

Bending: $\sigma_x = -\dfrac{My}{I}, \qquad \sigma_{max} = \dfrac{Mc}{I}$ (c)

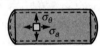

Shear: $\tau_{xy} = \dfrac{VQ}{Ib}$ (d)

where

σ_x = normal axial stress
τ = shearing stress due to torque
τ_{xy} = shearing stress due to vertical shear force
P = axial force
T = torque
V = vertical shear force
M = bending moment about z axis
A = cross-sectional area
y, z = centroidal principal axes of the area

I = moment of inertia about neutral axis (N.A.)
J = polar moment of inertia of circular cross section
b = width of bar at which τ_{xy} is calculated
r = radius
Q = first moment about N.A. of the area beyond the point at which τ_{xy} is calculated

2. Thin-Walled Pressure Vessels

Cylinder: $\sigma_\theta = \dfrac{pr}{t}, \qquad \sigma_a = \dfrac{pr}{2t}$ (e)

Sphere: $\sigma = \dfrac{pr}{2t}$ (f)

where

σ_θ = tangential stress in cylinder wall
σ_a = axial stress in cylinder wall
σ = membrane stress in sphere wall

p = internal pressure
t = wall thickness
r = mean radius

[a]Detailed derivations and limitations of the use of these formulas are discussed in Sections 1.6, 5.7, 6.2, and 13.13.

prismatic beams subjected to pure bending. The assumption is approximate for other beam situations. However, it is emphasized that there is an extraordinarily large variety of cases in which applications of the basic formulas of mechanics of materials lead to useful results. In this text we hope to provide greater insight into the meaning and limitations of stress analysis by solving problems using both the elementary and exact methods of analysis.

1.7 STRESSES ON INCLINED SECTIONS

The stresses in bars, shafts, beams, and other structural members can be obtained by using the basic formulas, such as those listed in Table 1.1. The values found by these equations are for stresses that occur on cross sections of the members. Recall that *all* of the *formulas* for stress are limited to isotropic, homogeneous, and elastic materials that behave linearly. This section deals with the states of stress at points located on *inclined sections* or *planes* under axial loading. As before, we use *stress elements* to represent the state of stress at a point in a member. However, we now wish to find normal and shear stresses acting on the sides of an element in any direction.

The directional nature of more general states of stress and finding maximum and minimum values of stress are discussed in Sections 1.10 and 1.13. Usually, the failure of a member may be brought about by a certain magnitude of stress in a certain direction. For proper design, it is necessary to determine where and in what direction the largest stress occurs. The equations derived and the graphical technique introduced here and in the sections to follow are helpful in analyzing the stress at a point under various types of loading. Note that the transformation equations for stress are developed on the basis of *equilibrium conditions* only and do not depend on material properties or on the geometry of deformation.

Axially Loaded Members

We now consider the *stresses on an inclined plane a–a* of the bar in uniaxial tension shown in Fig. 1.6a, where the normal x' to the plane forms an angle θ with the axial direction. On an isolated part of the bar to the left of section $a–a$, the resultant P may be resolved into two components: the normal force $P_{x'} = P \cos \theta$ and the shear force $P_{y'} = -P \sin \theta$, as indicated in Fig. 1.6b. Thus, the normal and shearing stresses, uniformly distributed over the area $A_{x'} = A/\cos \theta$ of the inclined plane (Fig. 1.6c), are given by

$$\sigma_{x'} = \frac{P \cos \theta}{A_{x'}} = \sigma_x \cos^2 \theta \qquad \text{(1.11a)}$$

$$\tau_{x'y'} = -\frac{P \sin \theta}{A_{x'}} = -\sigma_x \sin \theta \cos \theta \qquad \text{(1.11b)}$$

where $\sigma_x = P/A$. The negative sign in Eq. (1.11b) agrees with the sign convention for shearing stresses described in Section 1.5. The foregoing process of determining the stress in proceeding from one set of coordinate axes to another is called *stress transformation*.

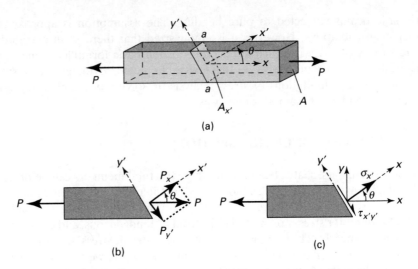

FIGURE 1.6. *(a) Prismatic bar in tension; (b, c) side views of a part cut from the bar.*

Equations (1.11) indicate how the stresses vary as the inclined plane is cut at various angles. As expected, $\sigma_{x'}$ is a maximum (σ_{max}) when θ is $0°$ or $180°$, and $\tau_{x'y'}$ is maximum (τ_{max}) when θ is $45°$ or $135°$. Also, $\tau_{max} = \pm \frac{1}{2}\sigma_{max}$. The maximum stresses are thus

$$\sigma_{max} = \sigma_x, \qquad \tau_{max} = \pm \tfrac{1}{2}\sigma_x \qquad (1.12)$$

Observe that the normal stress is either maximum or a minimum on planes for which the shearing stress is zero.

Figure 1.7 shows the manner in which the stresses vary as the section is cut at angles varying from $\theta = 0°$ to $180°$. Clearly, when $\theta > 90°$, the sign of $\tau_{x'y'}$ in Eq. (1.11b) changes; the shearing stress *changes sense*. However, the *magnitude* of the shearing stress for any angle θ determined from Eq. (1.11b) is equal to that for $\theta + 90°$. This agrees with the general conclusion reached in the preceding section: shearing stresses on mutually perpendicular planes must be equal.

We note that Eqs. (1.11) can also be used for uniaxial compression by assigning to P a negative value. The sense of each stress direction is then reversed in Fig. 1.6c.

EXAMPLE 1.1 State of Stress in a Tensile Bar

Compute the stresses on the inclined plane with $\theta = 35°$ for a prismatic bar of a cross-sectional area 800 mm^2, subjected to a tensile load of 60 kN (Fig. 1.6a). Then determine the state of stress for $\theta = 35°$ by calculating the stresses on an adjoining face of a stress element. Sketch the stress configuration.

Solution The normal stress on a cross section is

$$\sigma_x = \frac{P}{A} = \frac{60(10^3)}{800(10^{-6})} = 75 \text{ MPa}$$

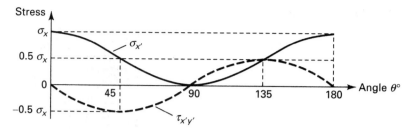

FIGURE 1.7. *Example 1.1. Variation of stress at a point with the inclined section in the bar shown in Fig. 1.6a.*

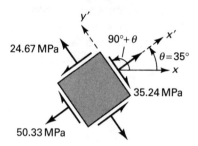

24.67 MPa

$90°+\theta$

$\theta=35°$

35.24 MPa

50.33 MPa

FIGURE 1.8. *Example 1.1. Stress element for $\theta = 35°$.*

Introducing this value in Eqs. (1.11) and using $\theta = 35°$, we have

$$\sigma_{x'} = \sigma_x \cos^2 \theta = 75(\cos 35°)^2 = 50.33 \text{ MPa}$$

$$\tau_{x'y'} = -\sigma_x \sin \theta \cos \theta = -75(\sin 35°)(\cos 35°) = -35.24 \text{ MPa}$$

The normal and shearing stresses acting on the adjoining y' face are, respectively, 24.67 MPa and 35.24 MPa, as calculated from Eqs. (1.11) by substituting the angle $\theta + 90° = 125°$. The values of $\sigma_{x'}$ and $\tau_{x'y'}$ are the same on opposite sides of the element. On the basis of the established sign convention for stress, the required sketch is shown in Fig. 1.8.

1.8 VARIATION OF STRESS WITHIN A BODY

As pointed out in Section 1.5, the components of stress generally vary from point to point in a stressed body. These variations are governed by the conditions of equilibrium of *statics*. Fulfillment of these conditions establishes certain relationships, known as the *differential equations of equilibrium*, which involve the derivatives of the stress components.

Consider a thin element of sides dx and dy (Fig. 1.9), and assume that $\sigma_x, \sigma_y, \tau_{xy}$, and τ_{yx} are functions of x, y but do not vary throughout the thickness (are independent of z) and that the other stress components are zero. Also assume that the x and y components of the body forces per unit volume, F_x and F_y, are independent of z and that the z component of the body force $F_z = 0$. This combination of stresses, satisfying the conditions described, is the plane stress. Note that because the element

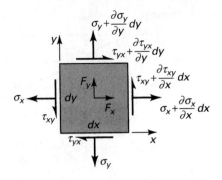

FIGURE 1.9. *Element with stresses and body forces.*

is very small, for the sake of simplicity, the stress components may be considered to be distributed uniformly over each face. In the figure they are shown by a single vector representing the mean values applied at the center of each face.

As we move from one point to another, for example, from the lower-left corner to the upper-right corner of the element, one stress component, say σ_x, acting on the negative x face, changes in value on the positive x face. The stresses σ_y, τ_{xy}, and τ_{yx} similarly change. The variation of stress with position may be expressed by a truncated Taylor's expansion:

$$\sigma_x + \frac{\partial \sigma_x}{\partial x}\, dx \tag{a}$$

The partial derivative is used because σ_x is a function of x and y. Treating all the components similarly, the state of stress shown in Fig. 1.9 is obtained.

We consider now the equilibrium of an element of unit thickness, taking moments of force about the lower-left corner. Thus, $\Sigma M_z = 0$ yields

$$\left(\frac{\partial \sigma_y}{\partial y}dx\,dy\right)\frac{dx}{2} - \left(\frac{\partial \sigma_x}{\partial x}\,dx\,dy\right)\frac{dy}{2} + \left(\tau_{xy} + \frac{\partial \tau_{xy}}{\partial x}\,dx\right)dx\,dy$$

$$-\left(\tau_{yx} + \frac{\partial \tau_{yx}}{\partial y}\,dy\right)dx\,dy + F_y dx\,dy\,\frac{dx}{2} - F_x\,dx\,dy\,\frac{dy}{2} = 0$$

Neglecting the triple products involving dx and dy, this reduces to $\tau_{xy} = \tau_{yx}$. In a like manner, it may be shown that $\tau_{yz} = \tau_{zy}$ and $\tau_{xz} = \tau_{zx}$, as already obtained in Section 1.5. From the equilibrium of x forces, $\Sigma F_x = 0$, we have

$$\left(\sigma_x + \frac{\partial \sigma_x}{\partial x}\,dx\right)dy - \sigma_x\,dy + \left(\tau_{xy} + \frac{\partial \tau_{xy}}{\partial y}\,dy\right)dx - \tau_{xy}\,dx + F_x\,dx\,dy = 0 \tag{b}$$

Upon simplification, Eq. (b) becomes

$$\left(\frac{\partial \sigma_x}{\partial x} + \frac{\partial \tau_{xy}}{\partial y} + F_x\right)dx\,dy = 0 \tag{c}$$

Inasmuch as $dx\,dy$ is nonzero, the quantity in the parentheses must vanish. A similar expression is written to describe the equilibrium of y forces. The x and y

equations yield the following differential equations of equilibrium for *two-dimensional stress*:

$$\frac{\partial \sigma_x}{\partial x} + \frac{\partial \tau_{xy}}{\partial y} + F_x = 0$$

$$\frac{\partial \sigma_y}{\partial y} + \frac{\partial \tau_{xy}}{\partial x} + F_y = 0$$

(1.13)

The differential equations of equilibrium for the case of *three-dimensional stress* may be generalized from the preceding expressions as follows:

$$\frac{\partial \sigma_x}{\partial x} + \frac{\partial \tau_{xy}}{\partial y} + \frac{\partial \tau_{xz}}{\partial z} + F_x = 0$$

$$\frac{\partial \sigma_y}{\partial y} + \frac{\partial \tau_{xy}}{\partial x} + \frac{\partial \tau_{yz}}{\partial z} + F_y = 0$$

$$\frac{\partial \sigma_z}{\partial z} + \frac{\partial \tau_{xz}}{\partial x} + \frac{\partial \tau_{yz}}{\partial y} + F_z = 0$$

(1.14)

A succinct representation of these expressions, on the basis of the range and summation conventions (Sec. 1.17), may be written as

$$\frac{\partial \tau_{ij}}{\partial x_j} + F_i = 0, \quad i, j = x, y, z$$

(1.15a)

where $x_x = x$, $x_y = y$, and $x_z = z$. The repeated subscript is j, indicating summation. The unrepeated subscript is i. Here i is termed the *free* index, and j, the *dummy* index.

If in the foregoing expression the symbol $\partial/\partial x$ is replaced by a comma, we have

$$\tau_{ij,j} + F_i = 0$$

(1.15b)

where the subscript after the comma denotes the coordinate with respect to which differentiation is performed. If no body forces exist, Eq. (1.15b) reduces to $\tau_{ij,j} = 0$, indicating that the *sum of the three stress derivatives is zero*. As the two equilibrium relations of Eqs. (1.13) contain *three* unknowns ($\sigma_x, \sigma_y, \tau_{xy}$) and the *three* expressions of Eqs. (1.14) involve the *six* unknown stress components, problems in stress analysis are *internally statically indeterminate*.

In a number of practical applications, the weight of the member is the *only* body force. If we take the y axis as upward and designate by ρ the mass density per unit volume of the member and by g, the gravitational acceleration, then $F_x = F_z = 0$ and $F_y = -\rho g$ in Eqs. (1.13) and (1.14). The resultant of this force over the volume of the member is usually so small compared with the surface forces that it can be ignored, as stated in Section 1.1. However, in dynamic systems, the stresses caused by body forces may far exceed those associated with surface forces so as to be the principal influence on the stress field.*

---

*In this case, the body is *not* in static equilibrium, and the inertia force terms $-\rho a_x$, $-\rho a_y$, and $-\rho a_z$ (where a_x, a_y, and a_z are the components of acceleration) must be included in the body force components F_x, F_y, and F_z, respectively, in Eqs. (1.14).

Application of Eqs. (1.13) and (1.14) to a variety of loaded members is presented in sections employing the approach of the theory of elasticity, beginning with Chapter 3. The following sample problem shows the pattern of the body force distribution for an arbitrary state of stress in equilibrium.

EXAMPLE 1.2   The Body Forces in a Structure

The stress field within an elastic structural member is expressed as follows:

$$\sigma_x = -x^3 + y^2, \qquad \tau_{xy} = 5z + 2y^2, \qquad \tau_{xz} = xz^3 + x^2y$$

$$\sigma_y = 2x^3 + \tfrac{1}{2}y^2, \qquad \tau_{yz} = 0, \qquad \sigma_z = 4y^2 - z^3 \tag{d}$$

Determine the body force distribution required for equilibrium.

Solution   Substitution of the given stresses into Eq. (1.14) yields

$$(-3x^2) + (4y) + (3xz^2) + F_x = 0$$

$$(y) + (0) + (0) + F_y = 0$$

$$(-3z^2) + (z^3 + 2xy) + (0) + F_z = 0$$

The body force distribution, as obtained from these expressions, is therefore

$$F_x = 3x^2 - 4y - 3xz^2, \qquad F_y = -y, \qquad F_z = -2xy + 3z^2 - z^3 \tag{e}$$

The state of stress and body force at any specific point within the member may be obtained by substituting the specific values of x, y, and z into Eqs. (d) and (e), respectively.

1.9   PLANE-STRESS TRANSFORMATION

A two-dimensional state of stress exists when the stresses and body forces are independent of one of the coordinates, here taken as z. Such a state is described by stresses σ_x, σ_y, and τ_{xy} and the x and y body forces. Two-dimensional problems are of two classes: *plane stress* and *plane strain*. In the case of plane stress, as described in the previous section, the stresses σ_z, τ_{xz}, and τ_{yz}, and the z-directed body forces are assumed to be zero. The condition that occurs in a thin plate subjected to loading uniformly distributed over the thickness and parallel to the plane of the plate typifies the state of plane stress (Fig. 1.10). In the case of plane strain, the stresses τ_{xz} and τ_{yz} and the body force F_z are likewise taken to be zero, but σ_z does not vanish* and can be determined from stresses σ_x and σ_y.

---

*More details and illustrations of these assumptions are given in Chapter 3.

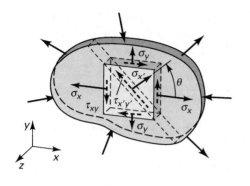

FIGURE 1.10. *Thin Plate in-plane loads.*

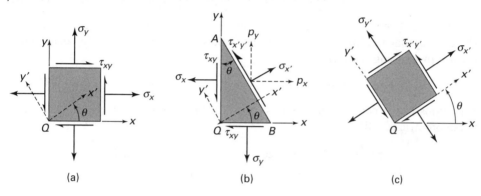

(a)   (b)   (c)

FIGURE 1.11. *Elements in plane stress.*

We shall now determine the equations for transformation of the stress components σ_x, σ_y, and τ_{xy} at any point of a body represented by an *infinitesimal* element, isolated from the plate illustrated in Fig. 1.10. The z-directed normal stress σ_z, even if it is nonzero, need not be considered here. In the following derivations, the angle θ locating the x' axis is assumed *positive when measured from the x axis in a counterclockwise direction.* Note that, according to our *sign convention* (see Sec. 1.5), the stresses are indicated as positive values.

Consider an infinitesimal wedge cut from the loaded body shown in Fig. 1.11a, b. It is required to determine the stresses $\sigma_{x'}$ and $\tau_{x'y'}$, which refer to axes x', y' making an angle θ with axes x, y, as shown in the figure. Let side AB be normal to the x' axis. Note that in accordance with the sign convention, $\sigma_{x'}$ and $\tau_{x'y'}$ are positive stresses, as shown in the figure. If the area of side AB is taken as unity, then sides QA and QB have area $\cos \theta$ and $\sin \theta$, respectively.

Equilibrium of forces in the x and y directions requires that

$$p_x = \sigma_x \cos \theta + \tau_{xy} \sin \theta$$
$$p_y = \tau_{xy} \cos \theta + \sigma_y \sin \theta \qquad \textbf{(1.16)}$$

where p_x and p_y are the *components of stress resultant* acting on AB in the x and y directions, respectively. The normal and shear stresses on the x' plane (AB plane) are obtained by projecting p_x and p_y in the x' and y' directions:

$$\sigma_{x'} = p_x \cos\theta + p_y \sin\theta$$
$$\tau_{x'y'} = p_y \cos\theta - p_x \sin\theta \tag{a}$$

From the foregoing it is clear that $\sigma_{x'}^2 + \tau_{x'y'}^2 = p_x^2 + p_y^2$. Upon substitution of the stress resultants from Eq. (1.16), Eqs. (a) become

$$\sigma_{x'} = \sigma_x \cos^2\theta + \sigma_y \sin^2\theta + 2\tau_{xy}\sin\theta\cos\theta \tag{1.17a}$$

$$\tau_{x'y'} = \tau_{xy}(\cos^2\theta - \sin^2\theta) + (\sigma_y - \sigma_x)\sin\theta\cos\theta \tag{1.17b}$$

Note that the normal stress $\sigma_{y'}$ acting on the y' face of an inclined element (Fig. 1.11c) may readily be obtained by substituting $\theta + \pi/2$ for θ in the expression for $\sigma_{x'}$. In so doing, we have

$$\sigma_{y'} = \sigma_x \sin^2\theta + \sigma_y \cos^2\theta - 2\tau_{xy}\sin\theta\cos\theta \tag{1.17c}$$

Equations (1.17) can be converted to a useful form by introducing the following trigonometric identities:

$$\cos^2\theta = \tfrac{1}{2}(1 + \cos 2\theta), \qquad \sin\theta\cos\theta = \tfrac{1}{2}\sin 2\theta,$$
$$\sin^2\theta = \tfrac{1}{2}(1 - \cos 2\theta)$$

The *transformation equations for plane stress* now become

$$\sigma_{x'} = \tfrac{1}{2}(\sigma_x + \sigma_y) + \tfrac{1}{2}(\sigma_x - \sigma_y)\cos 2\theta + \tau_{xy}\sin 2\theta \tag{1.18a}$$

$$\tau_{x'y'} = -\tfrac{1}{2}(\sigma_x - \sigma_y)\sin 2\theta + \tau_{xy}\cos 2\theta \tag{1.18b}$$

$$\sigma_{y'} = \tfrac{1}{2}(\sigma_x + \sigma_y) - \tfrac{1}{2}(\sigma_x - \sigma_y)\cos 2\theta - \tau_{xy}\sin 2\theta \tag{1.18c}$$

The foregoing expressions permit the computation of stresses acting on all possible planes AB (the *state of stress* at a point) provided that three stress components on a set of orthogonal faces are known.

Stress tensor. It is important to note that addition of Eqs. (1.17a) and (1.17c) gives the relationships

$$\sigma_x + \sigma_y = \sigma_{x'} + \sigma_{y'} = \text{constant}$$

In words then, the sum of the normal stresses on two perpendicular planes is *invariant*— that is, independent of θ. This conclusion is also valid in the case of a three-dimensional state of stress, as shown in Section 1.13. In mathematical terms, the *stress* whose components transform in the preceding way by rotation of axes is termed *tensor*. Some examples of other quantities are *strain* and *moment of inertia*. The similarities between the transformation equations for these quantities are observed in Sections 2.5 and C.4. Mohr's circle (Sec. 1.11) is a *graphical representation* of a stress tensor transformation.

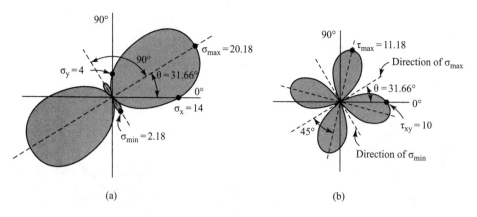

(a)   (b)

FIGURE 1.12.   *Polar representations of $\sigma_{x'}$ and $\tau_{x'y'}$ (in megapascals) versus θ.*

Polar Representations of State of Plane Stress

Consider, for example, the possible states of stress corresponding to $\sigma_x = 14$ MPa, $\sigma_y = 4$ MPa, and $\tau_{xy} = 10$ MPa. Substituting these values into Eq. (1.18) and permitting θ to vary from $0°$ to $360°$ yields the data upon which the curves shown in Fig. 1.12 are based. The plots shown, called *stress trajectories*, are *polar representations*: $\sigma_{x'}$ versus θ (Fig. 1.12a) and $\tau_{x'y'}$ versus θ (Fig. 1.12b). It is observed that the direction of each maximum shear stress bisects the angle between the maximum and minimum normal stresses. Note that the normal stress is either a maximum or a minimum on planes at $\theta = 31.66°$ and $\theta = 31.66° + 90°$, respectively, for which the shearing stress is zero. The conclusions drawn from this example are valid for any two-dimensional (or three-dimensional) state of stress and are observed in the sections to follow.

Cartesian Representation of State of Plane Stress

Now let us examine a two-dimensional condition of stress at a point in a loaded machine component on an element illustrated in Fig. 1.13a. Introducing the given values into the first two of Eqs. (1.18), gives

$$\sigma_{x'} = 4.5 + 2.5 \cos 2\theta + 5 \sin 2\theta$$

$$\tau_{x'y'} = -2.5 \sin 2\theta + 5 \cos 2\theta$$

In the foregoing, permitting θ to vary from $0°$ to $180°$ in increments of $15°$ leads to the data from which the graphs illustrated in Fig. 1.13b are obtained [Ref. 1.7]. This Cartesian representation demonstrates the variation of the normal and shearing stresses versus $\theta \leq 180°$. Observe that the direction of maximum (and minimum) shear stress *bisects* the angle between the maximum and minimum normal stresses. Moreover, the normal stress is either a maximum or a minimum on planes $\theta = 31.7°$ and $\theta = 31.7° + 90°$, respectively, for which the shear stress is zero. Note as a check that $\sigma_x + \sigma_y = \sigma_{max} + \sigma_{min} = 9$ MPa, as expected.

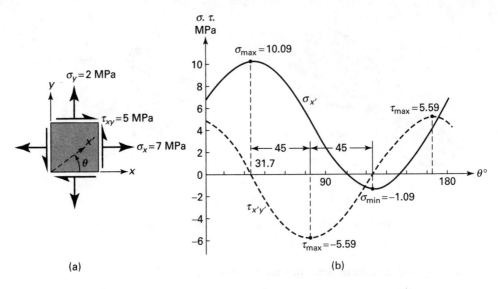

FIGURE 1.13. *Graph of normal stress $\sigma_{x'}$ and shearing stress $\tau_{x'y'}$ with angle θ (for $\theta \leq 180°$).*

The conclusions drawn from the foregoing polar and Cartesian representations are valid for *any* state of stress, as will be seen in the next section. A more convenient approach to the graphical transformation for stress is considered in Sections 1.11 and 1.15. The manner in which the three-dimensional normal and shearing stresses vary is discussed in Sections 1.12 through 1.14.

1.10 PRINCIPAL STRESSES AND MAXIMUM IN-PLANE SHEAR STRESS

The transformation equations for two-dimensional stress indicate that the normal stress $\sigma_{x'}$ and shearing stress $\tau_{x'y'}$ vary continuously as the axes are rotated through the angle θ. To ascertain the orientation of $x'y'$ corresponding to maximum or minimum $\sigma_{x'}$, the necessary condition $d\sigma_{x'}/d\theta = 0$ is applied to Eq. (1.18a). In so doing, we have

$$-(\sigma_x - \sigma_y) \sin 2\theta + 2\tau_{xy} \cos 2\theta = 0 \qquad \textbf{(a)}$$

This yields

$$\tan 2\theta_p = \frac{2\tau_{xy}}{\sigma_x - \sigma_y} \qquad \textbf{(1.19)}$$

Inasmuch as $\tan 2\theta = \tan(\pi + 2\theta)$, two directions, mutually perpendicular, are found to satisfy Eq. (1.19). These are the *principal directions*, along which the principal or maximum and minimum normal stresses act. Two values of θ_p, corresponding to the σ_1 and σ_2 planes, are represented by θ_p' and θ_p'', respectively.

When Eq. (1.18b) is compared with Eq. (a), it becomes clear that $\tau_{x'y'} = 0$ on a principal plane. A principal plane is thus a plane of zero shear. The *principal stresses* are determined by substituting Eq. (1.19) into Eq. (1.18a):

$$\sigma_{\text{max, min}} = \sigma_{1,2} = \frac{\sigma_x + \sigma_y}{2} \pm \sqrt{\left(\frac{\sigma_x - \sigma_y}{2}\right)^2 + \tau_{xy}^2} \qquad (1.20)$$

Note that the *algebraically* larger stress given here is the maximum principal stress, denoted by σ_1. The minimum principal stress is represented by σ_2. It is necessary to substitute one of the values θ_p into Eq. (1.18a) to determine which of the two corresponds to σ_1.

Similarly, employing the preceding approach and Eq. (1.18b), we determine the planes of maximum shearing stress. Thus, setting $d\tau_{x'y'}/d\theta = 0$, we now have $(\sigma_x - \sigma_y)\cos 2\theta + 2\tau_{xy} \sin 2\theta = 0$ or

$$\tan 2\theta_s = -\frac{\sigma_x - \sigma_y}{2\tau_{xy}} \qquad (1.21)$$

The foregoing expression defines two values of θ_s that are 90° apart. These directions may again be denoted by attaching a prime or a double prime notation to θ_s. Comparing Eqs. (1.19) and (1.21), we also observe that the planes of maximum shearing stress are *inclined at* 45° with respect to the planes of principal stress. Now, from Eqs. (1.21) and (1.18b), we obtain the extreme values of shearing stress as follows:

$$\tau_{\text{max}} = \pm \sqrt{\left(\frac{\sigma_x - \sigma_y}{2}\right)^2 + \tau_{xy}^2} = \pm \tfrac{1}{2}(\sigma_1 - \sigma_2) \qquad (1.22)$$

Here the largest shearing stress, regardless of sign, is referred to as the *maximum shearing stress*, designated τ_{max}. Normal stresses acting on the planes of maximum shearing stress can be determined by substituting the values of $2\theta_s$ from Eq. (1.21) into Eqs. (1.18a) and (1.18c):

$$\sigma' = \sigma_{\text{ave}} = \tfrac{1}{2}(\sigma_x + \sigma_y) \qquad (1.23)$$

The results are illustrated in Fig. 1.14. Note that the diagonal of a stress element toward which the shearing stresses act is called the *shear diagonal*. The shear diagonal of the element on which the maximum shearing stresses act lies in the direction of the algebraically larger principal stress as shown in the figure. This assists in *predicting the proper direction* of the maximum shearing stress.

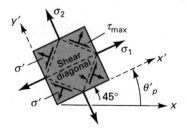

FIGURE 1.14. *Planes of principal and maximum shearing stresses.*

1.11 MOHR'S CIRCLE FOR TWO-DIMENSIONAL STRESS

A graphical technique, predicated on Eq. (1.18), permits the rapid transformation of stress from one plane to another and leads also to the determination of the maximum normal and shear stresses. In this approach, Eqs. (1.18) are depicted by a stress circle, called Mohr's circle.* In the Mohr representation, the normal stresses obey the sign convention of Section 1.5. However, for the purposes only of *constructing and reading values of stress from Mohr's circle*, the sign convention for shear stress is as follows: If the shearing stresses on opposite faces of an element would produce shearing forces that result in a *clockwise* couple, as shown in Fig. 1.15c, these stresses are regarded as *positive*. Accordingly, the shearing stresses on the y faces of the element in Fig. 1.15a are taken as positive (as before), but those on the x faces are now negative.

Given σ_x, σ_y, and τ_{xy} with algebraic sign in accordance with the foregoing sign convention, the procedure for obtaining Mohr's circle (Fig. 1.15b) is as follows:

1. Establish a rectangular coordinate system, indicating $+\tau$ and $+\sigma$. Both stress scales must be identical.
2. Locate the center C of the circle on the horizontal axis a distance $\frac{1}{2}(\sigma_x + \sigma_y)$ from the origin.
3. Locate point A by coordinates σ_x and $-\tau_{xy}$. These stresses may correspond to any face of an element such as in Fig. 1.15a. It is usual to specify the stresses on the positive x face, however.
4. Draw a circle with center at C and of radius equal to CA.
5. Draw line AB through C.

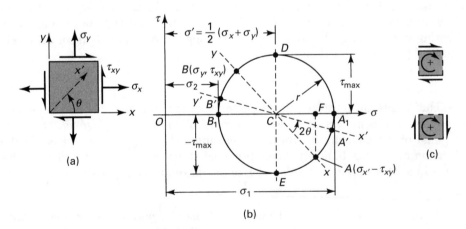

FIGURE 1.15. *(a) Stress element; (b) Mohr's circle of stress; (c) interpretation of positive shearing stresses.*

---

*After Otto Mohr (1835–1918), professor at Dresden Polytechnic. For further details, see Ref. 1.7, for example.

The angles on the circle are measured in the same direction as θ is measured in Fig. 1.15a. An angle of 2θ on the circle corresponds to an angle of θ on the element. The state of stress associated with the original x and y planes corresponds to points A and B on the circle, respectively. Points lying on diameters other than AB, such as A' and B', define states of stress with respect to any other set of x' and y' planes rotated relative to the original set through an angle θ.

It is clear that points A_1 and B_1 on the circle locate the principal stresses and provide their magnitudes as defined by Eqs. (1.19) and (1.20), while D and E represent the maximum shearing stresses, defined by Eqs. (1.21) and (1.22). The radius of the circle is

$$CA = \sqrt{CF^2 + AF^2} \qquad \text{(a)}$$

where

$$CF = \tfrac{1}{2}(\sigma_x - \sigma_y), \qquad AF = \tau_{xy}$$

Thus, the radius equals the magnitude of the maximum shearing stress. Mohr's circle shows that the planes of maximum shear are always located at 45° from planes of principal stress, as already indicated in Fig. 1.14. The use of Mohr's circle is illustrated in the first two of the following examples.

EXAMPLE 1.3   Principal Stresses in a Member

At a point in the structural member, the stresses are represented as in Fig. 1.16a. Employ Mohr's circle to determine (a) the magnitude and orientation of the principal stresses and (b) the magnitude and orientation of the maximum shearing stresses and associated normal stresses. In each case, show the results on a properly oriented element; represent the stress tensor in matrix form.

Solution   Mohr's circle, constructed in accordance with the procedure outlined, is shown in Fig. 1.16b. The center of the circle is at $(40 + 80)/2 = 60$ MPa on the σ axis.

a. The principal stresses are represented by points A_1 and B_1. Hence, the maximum and minimum principal stresses, referring to the circle, are

$$\sigma_{1,2} = 60 \pm \sqrt{\tfrac{1}{4}(80 - 40)^2 + (30)^2}$$

or

$$\sigma_1 = 96.05 \text{ MPa} \quad \text{and} \quad \sigma_2 = 23.95 \text{ MPa}$$

The planes on which the principal stresses act are given by

$$2\theta_p' = \tan^{-1}\frac{30}{20} = 56.30° \quad \text{and} \quad 2\theta_p'' = 56.30° + 180° = 236.30°$$

Hence

$$\theta_p' = 28.15° \quad \text{and} \quad \theta_p'' = 118.15°$$

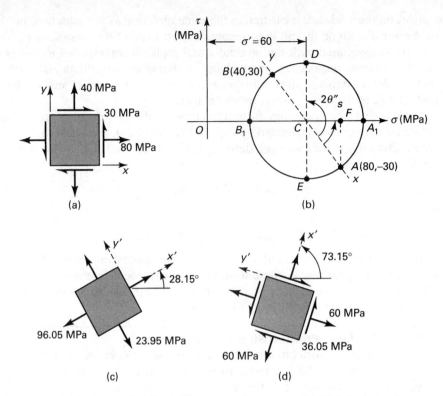

FIGURE 1.16. *Example 1.3. (a) Element in plane stress; (b) Mohr's circle of stress; (c) principal stresses; (d) maximum shear stress.*

Mohr's circle clearly indicates that θ_p' locates the σ_1 plane. The results may readily be checked by substituting the two values of θ_p into Eq. (1.18a). The state of principal stress is shown in Fig. 1.16c.

b. The maximum shearing stresses are given by points D and E. Thus,

$$\tau_{max} = \pm \sqrt{\tfrac{1}{4}(80 - 40)^2 + (30)^2} = \pm 36.05 \text{ MPa}$$

It is seen that $(\sigma_1 - \sigma_2)/2$ yields the same result. The planes on which these stresses act are represented by

$$\theta_s'' = 28.15° + 45° = 73.15° \quad \text{and} \quad \theta_s' = 163.15°$$

As Mohr's circle indicates, the positive maximum shearing stress acts on a plane whose normal x' makes an angle θ_s'' with the normal to the original plane (x plane). Thus, $+\tau_{max}$ on two opposite x' faces of the element will be directed so that a clockwise couple results. The normal stresses acting on maximum shear planes are represented by $OC, \sigma' = 60$ MPa on each face. The state of maximum shearing stress is shown in Fig. 1.16d. The direction of the τ_{max}'s may also be readily predicted by recalling that they act toward the shear diagonal. We note that, according to the general sign convention (Sec. 1.5), the shearing stress acting on the x' plane in Fig. 1.16d is negative. As a check, if $2\theta_s'' = 146.30°$ and the given initial data are substituted into Eq. (1.18b), we obtain $\tau_{x'y'} = -36.05$ MPa, as already found.

We may now describe the state of stress at the point in the following matrix forms:

$$\begin{bmatrix} 80 & 30 \\ 30 & 40 \end{bmatrix}, \quad \begin{bmatrix} 96.05 & 0 \\ 0 & 23.95 \end{bmatrix}, \quad \begin{bmatrix} 60 & -36.05 \\ -36.05 & 60 \end{bmatrix}$$

These three representations, associated with the $\theta = 0°$, $\theta = 28.15°$, and $\theta = 73.15°$ planes passing through the point, are equivalent.

Note that if we assume $\sigma_z = 0$ in this example, a much *higher* shearing stress is obtained in the planes bisecting the x' and z planes (Problem 1.56). Thus, three-dimensional analysis, Section 1.15, should be considered for determining the *true maximum shearing stress* at a point.

EXAMPLE 1.4   Stresses in a Frame

The stresses acting on an element of a loaded frame are shown in Fig. 1.17a. Apply Mohr's circle to determine the normal and shear stresses acting on a plane defined by $\theta = 30°$.

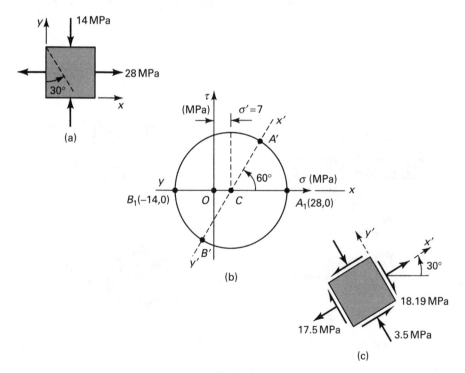

FIGURE 1.17.   *Example 1.4. (a) Element in biaxial stresses; (b) Mohr's circle of stress; (c) stress element for $\theta = 30°$.*

Solution Mohr's circle of Fig. 1.17b describes the state of stress given in Fig. 1.17a. Points A_1 and B_1 represent the stress components on the x and y faces, respectively. The radius of the circle is $(14 + 28)/2 = 21$. Corresponding to the 30° plane within the element, it is necessary to rotate through 60° counterclockwise on the circle to locate point A'. A 240° counterclockwise rotation locates point B'. Referring to the circle,

$$\sigma_{x'} = 7 + 21 \cos 60° = 17.5 \text{ MPa}$$

$$\sigma_{y'} = -3.5 \text{ MPa}$$

and $\qquad \tau_{x'y'} = \pm 21 \sin 60° = \pm 18.19 \text{ MPa}$

Figure 1.17c indicates the orientation of the stresses. The results can be checked by applying Eq. (1.18), using the initial data.

EXAMPLE 1.5 Cylindrical Vessel Under Combined Loads

A thin-walled cylindrical pressure vessel of 250-mm diameter and 5-mm wall thickness is rigidly attached to a wall, forming a cantilever (Fig. 1.18a). Determine the maximum shearing stresses and the associated normal stresses at point A of the cylindrical wall. The following loads are applied: internal pressure $p = 1.2$ MPa, torque $T = 3$ kN·m, and direct force $P = 20$ kN. Show the results on a properly oriented element.

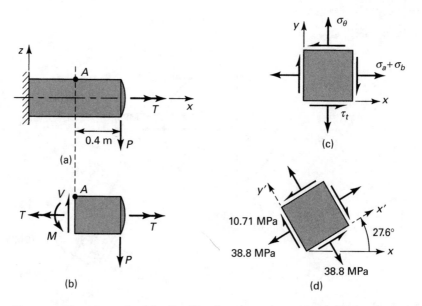

FIGURE 1.18. *Example 1.5. Combined stresses in a thin-walled cylindrical pressure vessel: (a) side view; (b) free body of a segment; (c) and (d) element A (viewed from top).*

Solution The internal force resultants on a transverse section through point A are found from the equilibrium conditions of the free-body diagram of Fig. 1.18b. They are $V = 20\,\text{kN}$, $M = 8\,\text{kN}\cdot\text{m}$, and $T = 3\,\text{kN}\cdot\text{m}$. In Fig. 1.18c, the combined axial, tangential, and shearing stresses are shown acting on a small element at point A. These stresses are (Tables 1.1 and C.1)

$$\sigma_b = \frac{Mr}{I} = \frac{8(10^3)r}{\pi r^3 t} = \frac{8(10^3)}{\pi(0.125^2)(0.005)} = 32.6\,\text{MPa}$$

$$\tau_t = -\frac{Tr}{J} = -\frac{3(10^3)r}{2\pi r^3 t} = -\frac{3(10^3)}{2\pi(0.125^2)(0.005)} = -6.112\,\text{MPa}$$

$$\sigma_a = \frac{pr}{2t} = \frac{1.2(10^6)(125)}{2(5)} = 15\,\text{MPa}, \qquad \sigma_\theta = 2\sigma_a = 30\,\text{MPa}$$

We thus have $\sigma_x = 47.6\,\text{MPa}$, $\sigma_y = 30\,\text{MPa}$, and $\tau_{xy} = -6.112\,\text{MPa}$. Note that for element $A, Q = 0$; hence, the direct shearing stress $\tau_d = \tau_{xz} = VQ/Ib = 0$.
The maximum shearing stresses are from Eq. (1.22):

$$\tau_{\max} = \pm\sqrt{\left(\frac{47.6 - 30}{2}\right)^2 + (-6.112)^2} = \pm 10.71\,\text{MPa}$$

Equation (1.23) yields

$$\sigma' = \tfrac{1}{2}(47.6 + 30) = 38.8\,\text{MPa}$$

To locate the maximum shear planes, we use Eq. (1.21):

$$\theta_s = \tfrac{1}{2}\tan^{-1}\left[-\frac{47.6 - 30}{2(-6.112)}\right] = 27.6^\circ \quad \text{and} \quad 117.6^\circ$$

Applying Eq. (1.18b) with the given data and $2\theta_s = 55.2^\circ$, $\tau_{x'y'} = -10.71\,\text{MPa}$. Hence, $\theta_s'' = 27.6^\circ$, and the stresses are shown in their proper directions in Fig. 1.18d.

1.12 THREE-DIMENSIONAL STRESS TRANSFORMATION

The physical elements studied are always three dimensional, and hence it is desirable to consider three planes and their associated stresses, as illustrated in Fig. 1.2. We note that equations governing the transformation of stress in the three-dimensional case may be obtained by the use of a similar approach to that used for the two-dimensional state of stress.

Consider a small tetrahedron isolated from a continuous medium (Fig. 1.19a), subject to a general state of stress. The body forces are taken to be negligible. In the figure, p_x, p_y, and p_z are the Cartesian components of stress resultant p acting on oblique plane ABC. It is required to relate the stresses on the perpendicular planes intersecting at the origin to the normal and shear stresses on ABC.

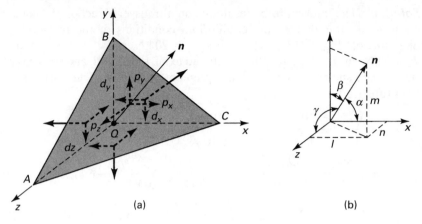

FIGURE 1.19. *Stress components on a tetrahedron.*

The orientation of plane ABC may be defined in terms of the angles between a unit normal $\boldsymbol{n}$ to the plane and the x, y, and z directions (Fig. 1.19b). The direction cosines associated with these angles are

$$\cos \alpha = \cos(\boldsymbol{n}, x) = l$$
$$\cos \beta = \cos(\boldsymbol{n}, y) = m \qquad \textbf{(1.24)}$$
$$\cos \gamma = \cos(\boldsymbol{n}, z) = n$$

The three direction cosines for the $\boldsymbol{n}$ direction are related by

$$l^2 + m^2 + n^2 = 1 \qquad \textbf{(1.25)}$$

The area of the perpendicular plane QAB, QAC, QBC may now be expressed in terms of A, the area of ABC, and the direction cosines:

$$A_{QAB} = A_x = \boldsymbol{A} \cdot \boldsymbol{i} = A(l\boldsymbol{i} + m\boldsymbol{j} + n\boldsymbol{k}) \cdot \boldsymbol{i} = Al$$

The other two areas are similarly obtained. In so doing, we have altogether

$$A_{QAB} = Al, \qquad A_{QAC} = Am, \qquad A_{QBC} = An \qquad \textbf{(a)}$$

Here $\boldsymbol{i}, \boldsymbol{j}$, and $\boldsymbol{k}$ are unit vectors in the x, y, and z directions, respectively.

Next, from the equilibrium of x, y, z-directed forces together with Eq. (a), we obtain, after canceling A,

$$p_x = \sigma_x l + \tau_{xy} m + \tau_{xz} n$$
$$p_y = \tau_{xy} l + \sigma_y m + \tau_{yz} n \qquad \textbf{(1.26)}$$
$$p_z = \tau_{xz} l + \tau_{yz} m + \sigma_z n$$

The stress resultant on A is thus determined on the basis of known stresses $\sigma_x, \sigma_y, \sigma_z, \tau_{xy}, \tau_{xz}$, and τ_{yz} and a knowledge of the orientation of A. In the limit as the sides of the tetrahedron approach zero, plane A contains point Q. It is thus demonstrated that the stress resultant at a point is specified. This in turn gives the

TABLE 1.2. *Notation for Direction Cosines*

	x	y	z
x'	l_1	m_1	n_1
y'	l_2	m_2	n_2
z'	l_3	m_3	n_3

stress components acting on any three mutually perpendicular planes passing through Q as shown next. Although perpendicular planes have been used there for convenience, these planes need not be perpendicular to define the stress at a point.

Consider now a Cartesian coordinate system x', y', z', wherein x' *coincides* with n and y', z' lie on an oblique plane. The $x'y'z'$ and xyz systems are related by the direction cosines: $l_1 = \cos(x', x)$, $m_1 = \cos(x', y)$, and so on. The notation corresponding to a complete set of direction cosines is shown in Table 1.2. The normal stress $\sigma_{x'}$ is found by projecting p_x, p_y, and p_z in the x' direction and adding

$$\sigma_{x'} = p_x l_1 + p_y m_1 + p_z n_1 \tag{1.27}$$

Equations (1.26) and (1.27) are combined to yield

$$\sigma_{x'} = \sigma_x l_1^2 + \sigma_y m_1^2 + \sigma_z n_1^2 + 2(\tau_{xy} l_1 m_1 + \tau_{yz} m_1 n_1 + \tau_{xz} l_1 n_1) \tag{1.28a}$$

Similarly, by projecting p_x, p_y, and p_z in the y' and z' directions, we obtain, respectively,

$$\tau_{x'y'} = \sigma_x l_1 l_2 + \sigma_y m_1 m_2 + \sigma_z n_1 n_2 + \tau_{xy}(l_1 m_2 + m_1 l_2)$$
$$+ \tau_{yz}(m_1 n_2 + n_1 m_2) + \tau_{xz}(n_1 l_2 + l_1 n_2) \tag{1.28b}$$

$$\tau_{x'z'} = \sigma_x l_1 l_3 + \sigma_y m_1 m_3 + \sigma_z n_1 n_3 + \tau_{xy}(l_1 m_3 + m_1 l_3)$$
$$+ \tau_{yz}(m_1 n_3 + n_1 m_3) + \tau_{xz}(n_1 l_3 + l_1 n_3) \tag{1.28c}$$

Recalling that the stresses on three mutually perpendicular planes are required to specify the stress at a point (one of these planes being the oblique plane in question), the remaining components are found by considering those planes perpendicular to the oblique plane. For one such plane, n would now coincide with the y' direction, and expressions for the stresses $\sigma_{y'}$, $\tau_{y'x'}$, and $\tau_{y'z'}$ would be derived. In a similar manner, the stresses $\sigma_{z'}$, $\tau_{z'x'}$, and $\tau_{z'y'}$ are determined when n coincides with the z' direction. Owing to the symmetry of the stress tensor, only six of the nine stress components thus developed are unique. The remaining stress components are as follows:

$$\sigma_{y'} = \sigma_x l_2^2 + \sigma_y m_2^2 + \sigma_z n_2^2 + 2(\tau_{xy} l_2 m_2 + \tau_{yz} m_2 n_2 + \tau_{xz} l_2 n_2) \tag{1.28d}$$

$$\sigma_{z'} = \sigma_x l_3^2 + \sigma_y m_3^2 + \sigma_z n_3^2 + 2(\tau_{xy} l_3 m_3 + \tau_{yz} m_3 n_3 + \tau_{xz} l_3 n_3) \tag{1.28e}$$

$$\tau_{y'z'} = \sigma_x l_2 l_3 + \sigma_y m_2 m_3 + \sigma_z n_2 n_3 + \tau_{xy}(m_2 l_3 + l_2 m_3)$$
$$+ \tau_{yz}(n_2 m_3 + m_2 n_3) + \tau_{xz}(l_2 n_3 + n_2 l_3) \tag{1.28f}$$

Equations (1.28) represent expressions *transforming* the quantities σ_x, σ_y, σ_z, τ_{xy}, τ_{xz}, and τ_{yz}, which, as we have noted, completely define the state of stress. Quantities such as stress (and moment of inertia, Appendix C), which are subject to such transformations, are tensors of second rank (see Sec. 1.9).

The equations of transformation of the components of a stress tensor, in indicial notation, are represented by

$$\tau'_{rs} = l_{ir}l_{js}\tau_{ij} \tag{1.29a}$$

Alternatively,

$$\tau_{rs} = l_{ri}l_{sj}\tau'_{ij} \tag{1.29b}$$

The repeated subscripts i and j imply the double summation in Eq. (1.29a), which, upon expansion, yields

$$\tau'_{rs} = l_{xr}l_{xs}\tau_{xx} + l_{xr}l_{ys}\tau_{xy} + l_{xr}l_{zs}\tau_{xz}$$

$$+ l_{yr}l_{xs}\tau_{xy} + l_{yr}l_{ys}\tau_{yy} + l_{ys}l_{zr}\tau_{yz}$$

$$+ l_{zr}l_{xs}\tau_{xz'} + l_{zr}l_{ys}\tau_{yz} + l_{zr}l_{zs}\tau_{zz} \tag{1.29c}$$

By assigning $r, s = x, y, z$ and noting that $\tau_{rs} = \tau_{sr}$, the foregoing leads to the six expressions of Eq. (1.28).

It is interesting to note that, because x', y', and z' are orthogonal, the nine direction cosines must satisfy trigonometric relations of the following form:

$$l_i^2 + m_i^2 + n_i^2 = 1, \qquad i = 1, 2, 3 \tag{1.30a}$$

and

$$l_1 l_2 + m_1 m_2 + n_1 n_2 = 0$$
$$l_2 l_3 + m_2 m_3 + n_2 n_3 = 0 \tag{1.30b}$$
$$l_1 l_3 + m_1 m_3 + n_1 n_3 = 0$$

From Table 1.2, observe that Eqs. (1.30a) are the sums of the squares of the cosines in each row, and Eqs. (1.30b) are the sums of the products of the adjacent cosines in any two rows.

1.13 PRINCIPAL STRESSES IN THREE DIMENSIONS

For the three-dimensional case, it is now demonstrated that three planes of zero shear stress exist, that these planes are mutually perpendicular, and that on these planes the normal stresses have maximum or minimum values. As has been discussed, these normal stresses are referred to as *principal stresses*, usually denoted σ_1, σ_2, and σ_3. The *algebraically* largest stress is represented by σ_1, and the smallest by σ_3: $\sigma_1 > \sigma_2 > \sigma_3$.

We begin by again considering an oblique x' plane. The normal stress acting on this plane is given by Eq. (1.28a):

$$\sigma_{x'} = \sigma_x l^2 + \sigma_y m^2 + \sigma_z n^2 + 2(\tau_{xy}lm + \tau_{yz}mn + \tau_{xz}ln) \tag{a}$$

The problem at hand is the determination of *extreme* or *stationary values* of $\sigma_{x'}$. To accomplish this, we examine the variation of $\sigma_{x'}$ relative to the direction cosines. Inasmuch as l, m, and n are not independent, but connected by $l^2 + m^2 + n^2 = 1$, only l and m may be regarded as independent variables. Thus,

$$\frac{\partial \sigma_{x'}}{\partial l} = 0, \qquad \frac{\partial \sigma_{x'}}{\partial m} = 0 \tag{b}$$

Differentiating Eq. (a) as indicated by Eqs. (b) in terms of the quantities in Eq. (1.26), we obtain

$$p_x + p_z \frac{\partial n}{\partial l} = 0, \qquad p_y + p_z \frac{\partial n}{\partial m} = 0 \tag{c}$$

From $n^2 = 1 - l^2 - m^2$, we have $\partial n/\partial l = -l/n$ and $\partial n/\partial m = -m/n$. Introducing these into Eq. (c), the following relationships between the components of p and n are determined:

$$\frac{p_x}{l} = \frac{p_y}{m} = \frac{p_z}{n} \tag{d}$$

These proportionalities indicate that the stress resultant must be *parallel* to the unit normal and therefore contains no shear component. It is concluded that, on a plane for which $\sigma_{x'}$ has an extreme or principal value, a principal plane, the shearing stress vanishes.

It is now shown that three principal stresses and three principal planes exist. Denoting the principal stresses by σ_p, Eq. (d) may be written as

$$p_x = \sigma_p l, \qquad p_y = \sigma_p m, \qquad p_z = \sigma_p n \tag{e}$$

These expressions, together with Eq. (1.26), lead to

$$\begin{aligned}
(\sigma_x - \sigma_p)l + \tau_{xy}m + \tau_{xz}n &= 0 \\
\tau_{xy}l + (\sigma_y - \sigma_p)m + \tau_{yz}n &= 0 \\
\tau_{xz}l + \tau_{yz}m + (\sigma_z - \sigma_p)n &= 0
\end{aligned} \tag{1.31}$$

A nontrivial solution for the direction cosines requires that the characteristic determinant vanish:

$$\begin{vmatrix}
\sigma_x - \sigma_p & \tau_{xy} & \tau_{xz} \\
\tau_{xy} & \sigma_y - \sigma_p & \tau_{yz} \\
\tau_{xz} & \tau_{yz} & \sigma_z - \sigma_p
\end{vmatrix} = 0 \tag{1.32}$$

Expanding Eq. (1.32) leads to

$$\sigma_p^3 - I_1\sigma_p^2 + I_2\sigma_p - I_3 = 0 \tag{1.33}$$

where

$$I_1 = \sigma_x + \sigma_y + \sigma_z \tag{1.34a}$$

$$I_2 = \sigma_x\sigma_y + \sigma_x\sigma_z + \sigma_y\sigma_z - \tau_{xy}^2 - \tau_{yz}^2 - \tau_{xz}^2 \tag{1.34b}$$

$$I_3 = \begin{vmatrix} \sigma_x & \tau_{xy} & \tau_{xz} \\ \tau_{xy} & \sigma_y & \tau_{yz} \\ \tau_{xz} & \tau_{yz} & \sigma_z \end{vmatrix} \tag{1.34c}$$

The three roots of the *stress cubic equation* (1.33) are the principal stresses, corresponding to which are three sets of direction cosines, which establish the relationship of the principal planes to the origin of the nonprincipal axes. The principal stresses are the characteristic values or *eigenvalues* of the stress tensor τ_{ij}. Since the stress tensor is a symmetric tensor whose elements are all real, it has real eigenvalues. That is, the three principal stresses are *real* [Refs. 1.8 and 1.9]. The direction cosines l, m, and n are the *eigenvectors* of τ_{ij}.

It is clear that the principal stresses are independent of the orientation of the original coordinate system. It follows from Eq. (1.33) that the coefficients I_1, I_2, and I_3 must likewise be independent of x, y, and z, since otherwise the principal stresses would change. For example, we can demonstrate that adding the expressions for $\sigma_{x'}$, $\sigma_{y'}$, and $\sigma_{z'}$ given by Eq. (1.28) and making use of Eq. (1.30a) leads to $I_1 = \sigma_{x'} + \sigma_{y'} + \sigma_{z'} = \sigma_x + \sigma_y + \sigma_z$. Thus, the coefficients I_1, I_2, and I_3 represent three invariants of the stress tensor in three dimensions or, briefly, the *stress invariants*. For *plane stress*, it is a simple matter to show that the following quantities are invariant (Prob. 1.27):

$$\begin{aligned} I_1 &= \sigma_x + \sigma_y = \sigma_{x'} + \sigma_{y'} \\ I_2 &= I_3 = \sigma_x\sigma_y - \tau_{xy}^2 = \sigma_{x'}\sigma_{y'} - \tau_{x'y'}^2 \end{aligned} \tag{1.35}$$

Equations (1.34) and (1.35) are particularly helpful in checking the results of a stress transformation, as illustrated in Example 1.7.

If now one of the principal stresses, say σ_1 obtained from Eq. (1.33), is substituted into Eq. (1.31), the resulting expressions, together with $l^2 + m^2 + n^2 = 1$, provide enough information to solve for the direction cosines, thus specifying the orientation of σ_1 relative to the xyz system. The direction cosines of σ_2 and σ_3 are similarly obtained. A convenient way of determining the roots of the stress cubic equation and solving for the direction cosines is presented in Appendix B, where a related computer program is also included (see Table B.1).

EXAMPLE 1.6   Three-Dimensional Stress in a Hub

A steel shaft is to be force fitted into a fixed-ended cast-iron hub. The shaft is subjected to a bending moment M, a torque T, and a vertical force P, Fig. 1.20a. Suppose that at a point Q in the hub, the stress field is as shown in Fig. 1.20b, represented by the matrix

$$\begin{bmatrix} -19 & -4.7 & 6.45 \\ -4.7 & 4.6 & 11.8 \\ 6.45 & 11.8 & -8.3 \end{bmatrix} \text{MPa}$$

Determine the principal stresses and their orientation with respect to the original coordinate system.

Solution   Substituting the given stresses into Eq. (1.33) we obtain from Eqs. (B.2)

$$\sigma_1 = 11.618 \text{ MPa}, \qquad \sigma_2 = -9.001 \text{ MPa}, \qquad \sigma_3 = -25.316 \text{ MPa}$$

Successive introduction of these values into Eq. (1.31), together with Eq. (1.30a), or application of Eqs. (B.6) yields the direction cosines that define the orientation of the planes on which σ_1, σ_2, and σ_3 act:

$$l_1 = 0.0266, \qquad l_2 = -0.6209, \qquad l_3 = 0.7834$$

$$m_1 = -0.8638, \qquad m_2 = 0.3802, \qquad m_3 = 0.3306$$

$$n_1 = -0.5031, \qquad n_2 = -0.6855, \qquad n_3 = -0.5262$$

Note that the directions of the principal stresses are seldom required for purposes of predicting the behavior of structural members.

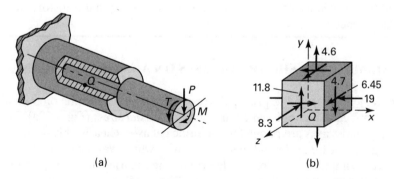

(a)   (b)

FIGURE 1.20.   *Example 1.6. (a) Hub-shaft assembly. (b) Element in three-dimensional stress.*

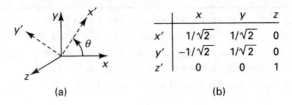

	x	y	z
x'	$1/\sqrt{2}$	$1/\sqrt{2}$	0
y'	$-1/\sqrt{2}$	$1/\sqrt{2}$	0
z'	0	0	1

(a)   (b)

FIGURE 1.21.   *Example 1.7. Direction cosines for $\theta = 45°$.*

EXAMPLE 1.7   Three-Dimensional Stress in a Machine Component

The stress tensor at a point in a machine element with respect to a Cartesian coordinate system is given by the following array:

$$[\tau_{ij}] = \begin{bmatrix} 50 & 10 & 0 \\ 10 & 20 & 40 \\ 0 & 40 & 30 \end{bmatrix} \text{MPa} \qquad \text{(f)}$$

Determine the state of stress and $I_1, I_2,$ and I_3 for an x', y', z' coordinate system defined by rotating x, y through an angle of $\theta = 45°$ counterclockwise about the z axis (Fig. 1.21a).

Solution   The direction cosines corresponding to the prescribed rotation of axes are given in Fig. 1.21b. Thus, through the use of Eq. (1.28) we obtain

$$[\tau_{i'j'}] = \begin{bmatrix} 45 & -15 & 28.28 \\ -15 & 25 & 28.28 \\ 28.28 & 28.28 & 30 \end{bmatrix} \text{MPa} \qquad \text{(g)}$$

It is seen that the arrays (f) and (g), when substituted into Eq. (1.34), both yield $I_1 = 100$ MPa, $I_2 = 1400$ (MPa)^2, and $I_3 = -53,000$ (MPa)^3, and the invariance of $I_1, I_2,$ and I_3 under the orthogonal transformation is confirmed.

1.14   NORMAL AND SHEAR STRESSES ON AN OBLIQUE PLANE

A cubic element subjected to principal stresses $\sigma_1, \sigma_2,$ and σ_3 acting on mutually perpendicular principal planes is called in a state of *triaxial stress* (Fig. 1.22a). In the figure, the $x, y,$ and z axes are parallel to the principal axes. Clearly, this stress condition is not the general case of three-dimensional stress, which was taken up in the last two sections. It is sometimes required to determine the shearing and normal stresses acting on an arbitrary oblique plane of a tetrahedron, as in Fig. 1.22b, *given the principal stresses* or triaxial stresses acting on perpendicular planes. In the figure, the $x, y,$ and z axes are parallel to the principal axes. Denoting the direction cosines of plane ABC by $l, m,$ and n, Eqs. (1.26) with $\sigma_x = \sigma_1, \tau_{xy} = \tau_{xz} = 0$, and so on, reduce to

$$p_x = \sigma_1 l, \qquad p_y = \sigma_2 m, \qquad p_z = \sigma_3 n \qquad \text{(a)}$$

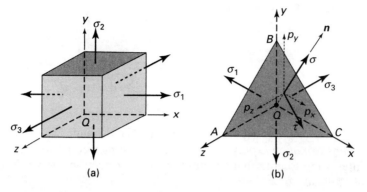

FIGURE 1.22. *Elements in triaxial stress.*

Referring to Fig. 1.22a and definitions (a), the stress resultant **p** is related to the principal stresses and the stress components on the oblique plane by the expression

$$p^2 = \sigma_1^2 l^2 + \sigma_2^2 m^2 + \sigma_3^2 n^2 = \sigma^2 + \tau^2 \qquad (1.36)$$

The normal stress σ on this plane, from Eq. (1.28a), is found as

$$\sigma = \sigma_1 l^2 + \sigma_2 m^2 + \sigma_3 n^2 \qquad (1.37)$$

Substitution of this expression into Eq. (1.36) leads to

$$\tau^2 = \sigma_1^2 l^2 + \sigma_2^2 m^2 + \sigma_3^2 n^2 - \sigma^2 \qquad (1.38a)$$

or

$$\tau^2 = \sigma_1^2 l^2 + \sigma_2^2 m^2 + \sigma_3^2 n^2 - (\sigma_1 l^2 + \sigma_2 m^2 + \sigma_3 n^2)^2 \qquad (1.38b)$$

Expanding and using the expressions $1 - l^2 = m^2 + n^2, 1 - n^2 = l^2 + m^2$, and so on, the following result is obtained for the shearing stress τ on the oblique plane:

$$\tau = \left[(\sigma_1 - \sigma_2)^2 l^2 m^2 + (\sigma_2 - \sigma_3)^2 m^2 n^2 + (\sigma_3 - \sigma_1)^2 n^2 l^2 \right]^{1/2} \qquad (1.39)$$

This clearly indicates that if the principal stresses are all equal, the shear stress vanishes, regardless of the choices of the direction cosines.

For situations in which *shear as well as normal stresses* act on perpendicular planes (Fig. 1.22b), we have p_x, p_y, and p_z defined by Eqs. (1.26). Then, Eq. (1.37) becomes

$$\sigma = \sigma_x l^2 + \sigma_y m^2 + \sigma_z n^2 + 2(\tau_{xy} lm + \tau_{yz} mn + \tau_{xz} ln) \qquad (1.40)$$

Hence,

$$\tau = \left[(\sigma_x l + \tau_{xy} m + \tau_{xz} n)^2 + (\tau_{xy} l + \sigma_y m + \tau_{yz} n)^2 \right.$$

$$\left. + (\tau_{xz} l + \tau_{yz} m + \sigma_z n)^2 - \sigma^2 \right]^{1/2} \qquad (1.41)$$

1.14 Normal and Shear Stresses on an Oblique Plane

FIGURE 1.23. *Stress ellipsoid.*

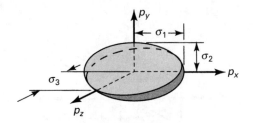

where σ is given by Eq. (1.40). Formulas (1.37) through (1.41) represent the *simplified transformation expressions* for the three-dimensional stress.

It is interesting to note that substitution of the direction cosines from Eqs. (a) into Eq. (1.25) leads to

$$\left(\frac{p_x}{\sigma_1}\right)^2 + \left(\frac{p_y}{\sigma_2}\right)^2 + \left(\frac{p_z}{\sigma_3}\right)^2 = 1 \qquad \textbf{(1.42)}$$

which is a *stress ellipsoid* having its three semiaxes as the principal stresses (Fig. 1.23). This geometrical interpretation helps to explain the earlier conclusion that the principal stresses are the extreme values of the normal stress. In the event that $\sigma_1 = \sigma_2 = \sigma_3$, a state of hydrostatic stress exists, and the stress ellipsoid becomes a *sphere*. In this case, note again that *any* three mutually perpendicular axes can be taken as the principal axes.

Octahedral Stresses

The stresses acting on an octahedral plane is represented by face ABC in Fig. 1.22b with $QA = QB = QC$. The normal to this oblique face thus has equal direction cosines relative to the principal axes. Since $l^2 + m^2 + n^2 = 1$, we have

$$l = m = n = \frac{1}{\sqrt{3}} \qquad \textbf{(b)}$$

Plane ABC is clearly one of eight such faces of a regular octahedron (Fig. 1.24). Equations (1.39) and (b) are now applied to provide an expression for the *octahedral shearing stress*, which may be rearranged to the form

$$\tau_{\text{oct}} = \tfrac{1}{3}[(\sigma_1 - \sigma_2)^2 + (\sigma_2 - \sigma_3)^2 + (\sigma_3 - \sigma_1)^2]^{1/2} \qquad \textbf{(1.43)}$$

Through the use of Eqs. (1.37) and (b), we obtain the *octahedral normal stress*:

$$\sigma_{\text{oct}} = \tfrac{1}{3}(\sigma_1 + \sigma_2 + \sigma_3) \qquad \textbf{(1.44)}$$

The normal stress acting on an octahedral plane is thus the average of the principal stresses, the *mean stress*. The orientations of σ_{oct} and τ_{oct} are indicated in Fig. 1.24. That the normal and shear stresses are the same for the eight planes is a powerful tool for failure analysis of ductile materials (see Sec. 4.8). Another useful form of Eq. (1.43) is developed in Section 2.15.

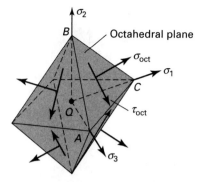

FIGURE 1.24. *Stresses on an octahedron.*

1.15 MOHR'S CIRCLES IN THREE DIMENSIONS

Consider a wedge shown in Fig. 1.25a, cut from the cubic element subjected to triaxial stresses (Fig. 1.22a). The only stresses on the inclined x' face (parallel to the z axis) are the normal stress $\sigma_{x'}$ and the shear stress $\tau_{x'y'}$ acting in the $x'y'$ plane. Inasmuch as the foregoing stresses are determined from force equilibrium equations in the $x'y'$ plane, they are independent of the stress σ_3. Thus, the transformation equations of plane stress (Sec. 1.9) and Mohr's circle can be employed to obtain the stresses $\sigma_{x'}$ and $\tau_{x'y'}$. The foregoing conclusion is also valid for normal and shear stresses acting on inclined faces cut through the element parallel to the x and y axes.

The stresses acting on elements oriented at various angles to the principal axes can be visualized with the aid of Mohr's circle. The cubic element (Fig. 1.22a) viewed from three different directions is sketched in Figs. 1.26a to c. A Mohr's circle is drawn corresponding to each projection of an element. The *cluster of three circles* represents Mohr's circles for triaxial stress (Fig. 1.26d). The radii of the circles are equal to the maximum shear stresses, as indicated in the figure. The normal stresses acting on the planes of maximum shear stresses have the magnitudes given by the abscissa as of the centers of the circles.

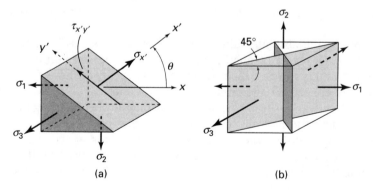

(a)   (b)

FIGURE 1.25. *Triaxial state of stress: (a) wedge; (b) planes of maximum shear stress.*

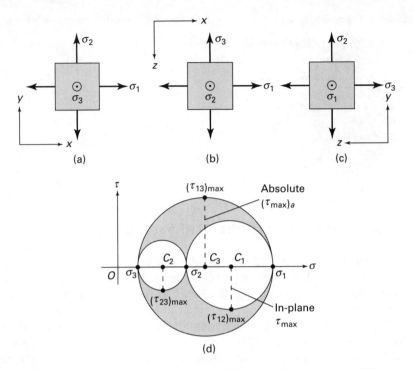

FIGURE 1.26. *(a–c) Views of elements in triaxial stresses on different principal axes; (d) Mohr's circles for three-dimensional stress.*

The largest shear stresses occur on planes oriented at 45° to the principal planes. The shear stress is a maximum located as the highest point on the outer circle. The value of the *absolute maximum shearing stress* is therefore

$$(\tau_{\max})_a = (\tau_{13})_a = \tfrac{1}{2}(\sigma_1 - \sigma_3) \tag{1.45}$$

acting on the planes that *bisect* the planes of the maximum and minimum principal stresses, as shown in Fig. 1.25b. It is noted that the planes of maximum shear stress may also be ascertained by substituting $n^2 = 1 - l^2 - m^2$ into Eq. (1.38b), differentiating with respect to l and m, and equating the resulting expressions to zero (Prob. 1.80).

Determining the absolute value of maximum shear stress is significant when designing members made of ductile materials, since the strength of the material depends on its ability to resist shear stress (Sec. 4.6). Obviously, as far as the stress magnitudes are concerned, the largest circle is the most significant one. However, all stresses in their various transformations may play a role in causing failure, and it is usually instructive to plot all three principal circles of stress, as depicted in the figure. An example of this type occurs in thin-walled pressurized cylinders, where $\sigma_\theta = \sigma_1$, $\sigma_a = \sigma_2$, and $\sigma_r = \sigma_3 = 0$ at the outer surface (Table 1.1). It is also interesting to note that, in special cases, where two or all principal stresses are equal, a Mohr's circle becomes a point.

Equations of Three Mohr's Circles for Stress

It has been demonstrated that, given the values of the principal stresses and of the direction cosines for any oblique plane (Fig. 1.22b), the normal and shear stresses on the plane may be ascertained through the application of Eqs. (1.37) and (1.38). This may also be accomplished by means of a graphical technique due to Mohr [Refs. 1.10 through 1.12]. The latter procedure was used in the early history of stress analysis, but today it is employed only as a *heuristic* device.

In the following discussion, we demonstrate that the aforementioned equations together with the relation $l^2 + m^2 + n^2 = 1$ are represented by three circles of stress, and the coordinates (σ, τ) locate a point in the shaded area of Fig. 1.26d [Ref. 1.13]. These simultaneous equations are

$$
\begin{aligned}
1 &= l^2 + m^2 + n^2 \\
\sigma &= \sigma_1 l^2 + \sigma_2 m^2 + \sigma_3 n^2 \\
\tau^2 &= \sigma_1^2 l^2 + \sigma_2^2 m^2 + \sigma_3^2 n^2 - \sigma^2
\end{aligned}
\tag{a}
$$

where $l^2 \geq 0$, $m^2 \geq 0$, and $n^2 \geq 0$. Solving for the direction cosines, results in

$$
l^2 = \frac{\sigma^2 + (\sigma - \sigma_2)(\sigma - \sigma_3)}{(\sigma_1 - \sigma_2)(\sigma_1 - \sigma_3)} \geq 0
$$

$$
m^2 = \frac{\sigma^2 + (\sigma - \sigma_3)(\sigma - \sigma_1)}{(\sigma_2 - \sigma_3)(\sigma_2 - \sigma_1)} \geq 0
\tag{1.46}
$$

$$
n^2 = \frac{\sigma^2 + (\sigma - \sigma_1)(\sigma - \sigma_2)}{(\sigma_3 - \sigma_1)(\sigma_3 - \sigma_2)} \geq 0
$$

Inasmuch as $\sigma_1 > \sigma_2 > \sigma_3$, the numerators of Eqs. (1.46) satisfy

$$
\sigma^2 + (\sigma - \sigma_2)(\sigma - \sigma_3) \geq 0
$$

$$
\sigma^2 + (\sigma - \sigma_3)(\sigma - \sigma_1) \leq 0
\tag{b}
$$

$$
\sigma^2 + (\sigma - \sigma_1)(\sigma - \sigma_2) \geq 0
$$

as the denominators of Eqs. (1.46) are $(\sigma_1 - \sigma_2) > 0$ and $(\sigma_1 - \sigma_3) > 0$, $(\sigma_2 - \sigma_3) > 0$ and $(\sigma_2 - \sigma_1) < 0$, $(\sigma_3 - \sigma_1) < 0$ and $(\sigma_3 - \sigma_2) < 0$, respectively.

Finally, the preceding inequalities may be expressed as follows

$$
\sigma^2 + [\sigma - \tfrac{1}{2}(\sigma_2 + \sigma_3)]^2 \geq \tfrac{1}{4}(\sigma_2 - \sigma_3)^2 = (\tau_{23})^2_{\max}
$$

$$
\sigma^2 + [\sigma - \tfrac{1}{2}(\sigma_1 + \sigma_3)]^2 \leq \tfrac{1}{4}(\sigma_1 - \sigma_3)^2 = (\tau_{13})^2_{\max}
\tag{1.47}
$$

$$
\sigma^2 + [\sigma - \tfrac{1}{2}(\sigma_1 + \sigma_2)]^2 \geq \tfrac{1}{4}(\sigma_1 - \sigma_2)^2 = (\tau_{12})^2_{\max}
$$

Equations (1.47) represent the formulas of the three Mohr's circles for stress, shown in Fig. 1.26d. Stress points (σ, τ) satisfying the equations for circles centered at C_1 and C_2 lie on or *outside* circles, but for the circle centered at C_3 lie on or *inside* circle. We conclude therefore that an admissible state of stress must lie on Mohr's circles or within the *shaded area* enclosed by these circles.

EXAMPLE 1.8 Analysis of Three-Dimensional Stresses in a Member

The state of stress on an element of a structure is illustrated in Fig. 1.27a. Using Mohr's circle, determine (a) the principal stresses and (b) the maximum shearing stresses. Show results on a properly oriented element. Also, (c) apply the equations developed in Section 1.14 to calculate the octahedral stresses.

Solution

a. First, Mohr's circle for the transformation of stress in the xy plane is sketched in the usual manner as shown, centered at C_2 with diameter A_2A_3 (Fig. 1.27b). Next, we complete the three-dimensional Mohr's circle by drawing two additional circles of diameters A_1A_2 and A_1A_3 in the figure. Referring to the circle, the principal stresses are $\sigma_1 = 100$ MPa, $\sigma_2 = 40$ MPa, and $\sigma_3 = -60$ MPa. Angle $\theta_p''' = 26.56°$, as $\tan 2\theta_p''' = 4/3$. The results are sketched on a properly oriented element in Fig. 1.27c.

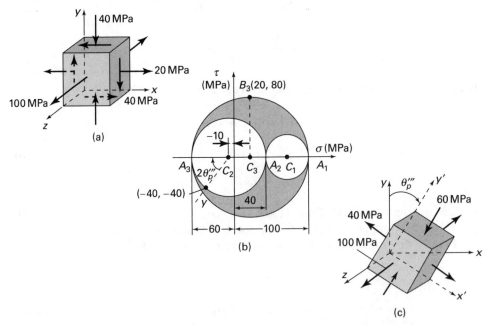

FIGURE 1.27. *Example 1.8. (a) Element in three-dimensional stress; (b) Mohr's circles of stress; (c) stress element for $\theta_p''' = 26.56°$.*

b. The absolute maximum shearing stress, point B_3, equals the radius of the circle centered at C_3 of diameter $A_1 A_3$. Thus,

$$(\tau_{13})_{max} = (\tau_{max})_a = \tfrac{1}{2}[100 - (-60)] = 80 \text{ MPa}$$

The maximum shearing stress occurs on the planes 45° from the y' and z faces of the element of Fig. 1.27c.

c. The octahedral normal stress, from Eq. (1.44), is

$$\sigma_{oct} = \tfrac{1}{3}(100 + 40 - 60) = 26.7 \text{ MPa}$$

The octahedral shearing stress, using Eq. (1.43), is

$$\tau_{oct} = \tfrac{1}{3}[(100 - 40)^2 + (40 + 60)^2 + (-60 - 100)^2]^{\frac{1}{2}} = 66 \text{ MPa}$$

Comments A comparison of the results (see Fig. 1.27b) shows that

$$\sigma_{oct} < \sigma_1 \quad \text{and} \quad \tau_{oct} < (\tau_{max})_a$$

That is, the maximum principal stress and absolute maximum shear stress are greater than their octahedral counterparts.

1.16 BOUNDARY CONDITIONS IN TERMS OF SURFACE FORCES

We now consider the relationship between the stress components and the given surface forces acting on the boundary of a body. The equations of equilibrium that must be satisfied within a body are derived in Section 1.8. The distribution of stress in a body must also be such as to accommodate the conditions of equilibrium with respect to externally applied forces. The external forces may thus be regarded as a continuation of the internal stress distribution.

Consider the equilibrium of the forces acting on the tetrahedron shown in Fig. 1.19b, and assume that oblique face ABC is coincident with the surface of the body. The components of the stress resultant p are thus now the *surface forces* per unit area, or *surface tractions*, p_x, p_y, and p_z. The equations of equilibrium for this element, representing *boundary conditions*, are, from Eqs. (1.26),

$$\begin{aligned}
p_x &= \sigma_x l + \tau_{xy} m + \tau_{xz} n \\
p_y &= \tau_{xy} l + \sigma_y m + \tau_{yz} n \\
p_z &= \tau_{xz} l + \tau_{yz} m + \sigma_z n
\end{aligned} \qquad (1.48)$$

For example, if the boundary is a plane with an x-directed surface normal, Eqs. (1.48) give $p_x = \sigma_x$, $p_y = \tau_{xy}$, and $p_z = \tau_{xz}$; under these circumstances, the applied surface force components p_x, p_y, and p_z are balanced by σ_x, τ_{xy}, and τ_{xz}, respectively.

It is of interest to note that, instead of prescribing the distribution of surface forces on the boundary, the boundary conditions of a body may also be given in terms of displacement components. Furthermore, we may be given boundary conditions that prescribe surface forces on one part of the boundary and displacements on another. When displacement boundary conditions are given, the equations of equilibrium express the situation in terms of strain, through the use of Hooke's law and subsequently in terms of the displacements by means of strain–displacement relations (Sec. 2.3). It is usual in engineering problems, however, to specify the boundary conditions in terms of surface forces, as in Eq. (1.48), rather than surface displacements. This practice is adhered to in this text.

1.17 INDICIAL NOTATION

A system of symbols, called *indicial notation*, *index notation*, also known as *tensor notation*, to represent components of force, stress, displacement, and strain is used throughout this text. Note that a particular class of tensor, a *vector*, requires only a single subscript to describe each of its components. Often the components of a tensor require more than a single subscript for definition. For example, second-*order* or second-*rank* tensors, such as those of stress or inertia, require double subscripting: τ_{ij}, I_{ij}. Quantities such as temperature and mass are scalars, classified as tensors of zero rank.

Tensor or *indicial* notation, here briefly explored, offers the advantage of succinct representation of lengthy equations through the minimization of symbols. In addition, physical laws expressed in tensor form are independent of the choice of coordinate system, and therefore similarities in seemingly different physical systems are often made more apparent. That is, indicial notation generally provides insight and understanding not readily apparent to the relative newcomer to the field. It results in a saving of space and serves as an aid in nonnumerical computation.

The displacement components u, v, and w, for instance, are written u_1, u_2, u_3 (or u_x, u_y, u_z) and collectively as u_i, with the understanding that the subscript i can be 1, 2, and 3 (or x, y, z). Similarly, the coordinates themselves are represented by x_1, x_2, x_3, or simply $x_i (i = 1, 2, 3)$, and x_x, x_y, x_z, or $x_i (i = x, y, z)$. Many equations of elasticity become unwieldy when written in full, unabbreviated term; see, for example, Eqs. (1.28). As the complexity of the situation described increases, so does that of the formulations, tending to obscure the fundamentals in a mass of symbols. For this reason, the more compact indicial notation is sometimes found in publications.

Two simple conventions enable us to write most equations developed in this text in indicial notation. These conventions, relative to range and summation, are as follows:

Range convention: When a lowercase alphabetic subscript is *unrepeated*, it takes on all values indicated.

Summation convention: When a lowercase alphabetic subscript is *repeated* in a term, then summation over the range of that subscript is indicated, making unnecessary the use of the summation symbol.

The introduction of the summation convention is attributable to A. Einstein (1879–1955). This notation, in conjunction with the tensor concept, has far-reaching consequences not restricted to its notational convenience [Refs. 1.14 and 1.15].

REFERENCES

1.1. TIMOSHENKO, S. P., *History of Strength of Materials*, Dover, New York, 1983.

1.2. TODHUNTER, L., and PEARSON, K., *History of the Theory of Elasticity and the Strength of Materials*, Vols. I and II, Dover, New York, 1960.

1.3. LOVE, A. E. H., *A Treatise on the Mathematical Theory of Elasticity*, 4th ed., Dover, New York, 1944.

1.4. UGURAL, A. C., *Stresses in Beams, Plates, and Shells*, 3rd ed., CRC Press, Taylor and Francis Group, Boca Raton, Fla., 2010, Sec. 3.2.

1.5. GERE, J., and TIMOSHENKO, S. P., *Mechanics of Materials*, 3rd ed., PWS-Kent, Boston, 1990.

1.6. UGURAL, A. C., *Mechanical Design: An Integrated Approach*, McGraw-Hill, New York, 2004, Sec. 1.6.

1.7. UGURAL, A. C., *Mechanics of Materials*, Wiley, Hoboken, N.J., 2008.

1.8. BORESI, A. P., and CHONG, K. P., *Elasticity in Engineering Mechanics*, 2nd ed., Wiley, Hoboken, N.J., 2000.

1.9. SOKOLNIKOFF, I. S., *Mathematical Theory of Elasticity*, 2nd ed., McGraw-Hill, New York, 1956.

1.10. UGURAL, A. C., and FENSTER, S. K., *Advanced Strength and Applied Elasticity*, 4th ed., Prentice Hall, Englewood Cliffs, N.J., 2003, Sec. 1.15.

1.11. SHAMES, I. H., and COZZARELLI, F. A., *Elastic and Inelastic Stress Analysis*, Prentice Hall, Englewood Cliffs, N.J., 1992.

1.12. FORD, H., *Advanced Mechanics of Materials*, 2nd ed., Ellis Horwood, Chichester, England, 1977, Chap. 4.

1.13. HOFFMAN, O., and SACHS, G., *Introduction to the Theory of Plasticity for Engineers*, McGraw-Hill, New York, 1953.

1.14. REISMANN, H., and PAWLIK, P. S., *Elasticity: Theory and Applications*, Wiley, Hoboken, N.J., 1980, Chap. 1.

1.15. CHOU, P. C., and PAGANO, N. J., *Elasticity*, Dover, New York, 1992, Chaps. 8 and 9.

PROBLEMS

Sections 1.1 through 1.8

1.1. Two prismatic bars of a by b rectangular cross section are glued as shown in Fig. P1.1. The allowable normal and shearing stresses for the glued joint are 700 and 560 kPa, respectively. Assuming that the strength of the joint controls the design, what is the largest axial load P that may be applied? Use $\phi = 40°$, $a = 50$ mm, and $b = 75$ mm.

1.2. A prismatic steel bar of $a = b = 50$-mm square cross section is subjected to an axial tensile load $P = 125$ kN (Fig. P1.1). Calculate the normal and shearing stresses on all faces of an element oriented at (a) $\phi = 70°$, and (b) $\phi = 45°$.

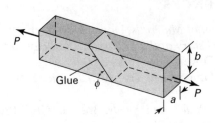

FIGURE P1.1.   FIGURE P1.3.

1.3. A prismatic bar is under an axial load, producing a compressive stress of 75 MPa on a plane at an angle $\theta = 30°$ (Fig. P1.3). Determine the normal and shearing stresses on all faces of an element at an angle of $\theta = 50°$.

1.4. A square prismatic bar of 1300-mm^2 cross-sectional area is composed of two pieces of wood glued together along the x' plane, which makes an angle θ with the axial direction (Fig. 1.6a). The normal and shearing stresses acting simultaneously on the joint are limited to 20 and 10 MPa, respectively, and on the bar itself, to 56 and 28 MPa, respectively. Determine the maximum allowable axial load that the bar can carry and the corresponding value of the angle θ.

1.5. Calculate the maximum normal and shearing stresses in a circular bar of diameter $d = 50$ mm subjected to an axial compression load of $P = 150$ kN through rigid end plates at its ends.

1.6. A frame is formed by two metallic rectangular cross sectional parts soldered along their inclined planes as illustrated in Fig. P1.6. What is the permissible axial load P_{all} that can be applied to the frame, without exceeding a normal stress of σ_{all} or a shearing stress of τ_{all} on the inclined plane? *Given:* $a = 10$ mm, $b = 75$ mm, $t = 20$ mm, $\theta = 55°$, $\sigma_{all} = 25$ MPa, and $\tau_{all} = 12$ MPa. *Assumption*: Material strength in tension is 90 MPa.

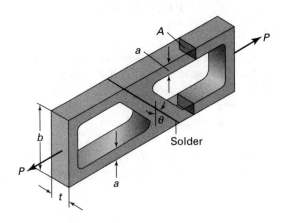

FIGURE P1.6.

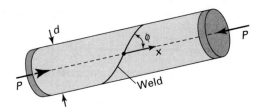

FIGURE P1.10.

1.7. Redo Prob. 1.6 for the case in which $\sigma_{all} = 20\ \text{MPa}$, $\tau_{all} = 8\ \text{MPa}$, and $\theta = 40°$.

1.8. Determine the normal and shearing stresses on an inclined plane at an angle ϕ through the bar subjected to an axial tensile force of P (Fig. P1.1). Given: $a = 15\ \text{mm}$, $b = 30\ \text{mm}$, $\phi = 50°$, $P = 120\ \text{kN}$.

1.9. Redo Prob. 1.8, for an angle of $\phi = 30°$ and $P = -100\ \text{kN}$.

1.10. A cylindrical pipe of 160-mm outside diameter and 10-mm thickness, spirally welded at an angle of $\phi = 40°$ with the axial (x) direction, is subjected to an axial compressive load of $P = 150\ \text{kN}$ through the rigid end plates (Fig. P1.10). Determine the normal $\sigma_{x'}$ and shearing stresses $\tau_{x'y'}$ acting simultaneously in the plane of the weld.

1.11. The following describes the stress distribution in a body (in megapascals):

$$\sigma_x = x^2 + 2y, \quad \sigma_y = xy - y^2z, \quad \tau_{xy} = -xy^2 + 1$$
$$\tau_{yz} = 0, \quad \tau_{xz} = xz - 2x^2y, \quad \sigma_z = x^2 - z^2$$

Determine the body force distribution required for equilibrium and the magnitude of its resultant at the point $x = -10\ \text{mm}$, $y = 30\ \text{mm}$, $z = 60\ \text{mm}$.

1.12. Given zero body forces, determine whether the following stress distribution can exist for a body in equilibrium:

$$\sigma_x = -2c_1xy, \qquad \sigma_y = c_2z^2, \qquad \sigma_z = 0$$
$$\tau_{xy} = c_1(c_2 - y^2) + c_3xz, \quad \tau_{xz} = -c_3y, \quad \tau_{yz} = 0$$

Here the c's are constants.

1.13. Determine whether the following stress fields are possible within an elastic structural member in equilibrium:

(a) $\begin{bmatrix} c_1x + c_2y & c_5x - c_1y \\ c_5x - c_1y & c_3x + c_4 \end{bmatrix}$, (b) $\begin{bmatrix} -\frac{3}{2}x^2y^2 & xy^3 \\ xy^3 & -\frac{1}{4}y^4 \end{bmatrix}$

The c's are constant, and it is assumed that the body forces are negligible.

1.14. For what body forces will the following stress field describe a state of equilibrium?

$$\sigma_x = -2x^2 + 3y^2 - 5z, \quad \tau_{xy} = z + 4xy - 7$$
$$\sigma_y = -2y^2, \quad \tau_{xz} = -3x + y + 1$$
$$\sigma_z = 3x + y + 3z - 5, \quad \tau_{yz} = 0$$

Sections 1.9 through 1.11

1.15. and 1.16. The states of stress at two points in a loaded body are represented in Figs. P1.15 and P1.16. Calculate for each point the normal and shearing stresses acting on the indicated inclined plane. As is done in the derivations given in Section 1.9, *use an approach based on the equilibrium equations* applied to the wedge-shaped element shown.

1.17. and 1.18. Resolve Probs. 1.15 and 1.16 using Eqs. (1.18).

1.19. At a point in a loaded machine, the normal and shear stresses have the magnitudes and directions acting on the inclined element shown in Fig. P1.19. What are the stresses σ_x, σ_y, and τ_{xy} on an element whose sides are parallel to the xy axes?

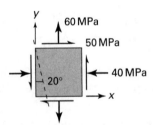

FIGURE P1.15.

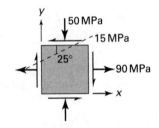

FIGURE P1.16.

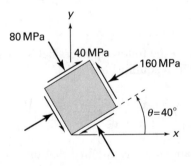

FIGURE P1.19.

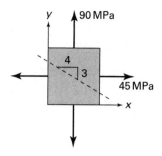

FIGURE P1.20.

1.20. The stresses at a point in the enclosure plate of a tank are as depicted in the element of Fig. P1.20. Find the normal and shear stresses at the point on the indicated inclined plane. Show the results on a sketch of properly oriented element.

1.21. A welded plate carries the uniform biaxial tension illustrated in Fig. P1.21. Determine the maximum stress σ for two cases: (a) The weld has an allowable shear stress of 30 MPa. (b) The weld has an allowable normal stress of 80 MPa.

1.22. Using Mohr's circle, solve Prob. 1.15.

1.23. Using Mohr's circle, solve Prob. 1.16.

1.24. Using Mohr's circle, solve Prob. 1.20.

1.25. Using Mohr's circle, solve Prob. 1.21.

1.26. The states of stress at two points in a loaded beam are represented in Fig. P1.26a and b. Determine the following for each point: (a) The magnitude of the maximum and minimum principal stresses and the maximum shearing stress; use Mohr's circle. (b) The orientation of the principal and maximum shear planes; use Mohr's circle. (c) Sketch the results on properly oriented elements. Check the values found in (a) and (b) by applying the appropriate equations.

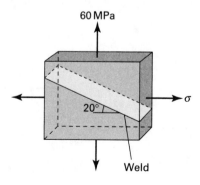

FIGURE P1.21.

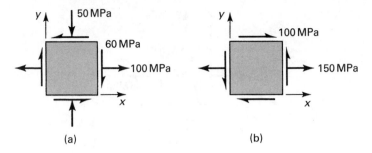

FIGURE P1.26.

1.27. By means of Mohr's circle, verify the results given by Eqs. (1.35).

1.28. An element in plane stress (Fig. 1.3b) is subjected to stresses $\sigma_x = 50$ MPa, $\sigma_y = -190$ MPa, and $\tau_{xy} = -70$ MPa. Determine the principal stresses and show them on a sketch of a properly oriented element.

1.29. For an element in plane stress (Fig. 1.3b), the normal stresses are $\sigma_x = 60$ MPa and $\sigma_y = -100$ MPa. What is the maximum permissible value of shearing stress τ_{xy} if the shearing stress in the material is not to exceed 140 MPa?

1.30. The state of stress on an element oriented at $\theta = 60°$ is shown in Fig. P1.30. Calculate the normal and shearing stresses on an element oriented at $\theta = 0°$.

1.31. A thin skewed plate is subjected to a uniform distribution of stress along its sides, as shown in Fig. P1.31. Calculate (a) the stresses $\sigma_x, \sigma_y, \tau_{xy}$, and (b) the principal stresses and their orientations.

1.32. The stress acting uniformly over the sides of a rectangular block is shown in Fig. P1.32. Calculate the stress components on planes parallel and perpendicular to *mn*. Show the results on a properly oriented element.

1.33. Redo Prob. 1.31 for the stress distribution shown in Fig. P1.33.

1.34. A thin-walled cylindrical tank of radius *r* is subjected simultaneously to internal pressure *p* and a compressive force *P* through rigid end plates. Determine the magnitude of force *P* to produce pure shear in the cylindrical wall.

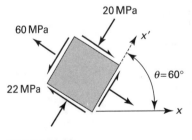

FIGURE P1.30.

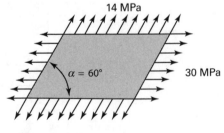

FIGURE P1.31.

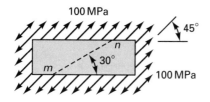

FIGURE P1.32.

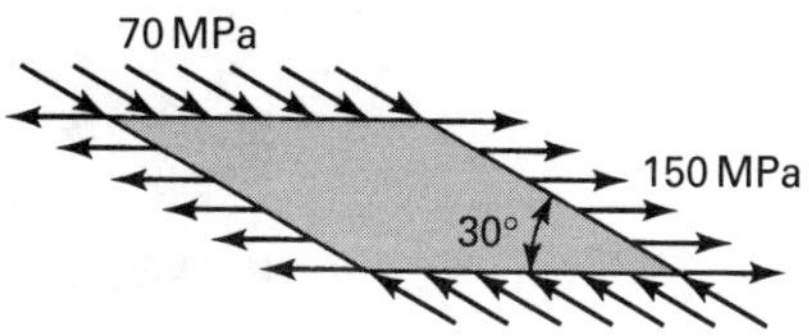

70 MPa

150 MPa

30°

FIGURE P1.33.

1.35. A thin-walled cylindrical pressure vessel of radius 120 mm and a wall thickness of 5 mm is subjected to an internal pressure of $p = 4$ MPa. In addition, an axial compression load of $P = 30\pi$ kN and a torque of $T = 10\pi$ kN·m are applied to the vessel through the rigid end plates (Fig. P1.35). Determine the maximum shearing stresses and associated normal stresses in the cylindrical wall. Show the results on a properly oriented element.

1.36. A pressurized thin-walled cylindrical tank of radius $r = 60$ mm and wall thickness $t = 4$ mm is acted on by end torques $T = 600$ N·m and tensile forces P (Fig. P1.35 with sense of P reversed). The internal pressure is $p = 5$ MPa. Calculate the maximum permissible value of P if the allowable tensile stress in the cylinder wall is 80 MPa.

1.37. A shaft of diameter d carries an axial compressive load P and two torques T_1, T_2 (Fig. P1.37). Determine the maximum shear stress at a point A on the surface of the shaft. *Given*: $d = 100$ mm, $P = 400$ kN, $T_1 = 10$ kN·m, and $T_2 = 2$ kN·m.

FIGURE P1.35.

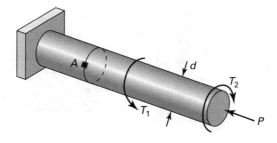

FIGURE P1.37.

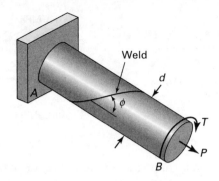

FIGURE P1.38.

1.38. What are the normal and shearing stresses on the spiral weld of the aluminum shaft of diameter d subjected to an axial load P and a torque T (Fig. P1.38)? *Given*: $P = 120$ kN, $T = 1.5$ kN·m, $d = 40$ mm, and $\phi = 50°$.

1.39. A hollow generator shaft of 180-mm outer diameter and 120-mm inner diameter carries simultaneously a torque $T = 20$ kN·m and axial compressive load $P = 700$ kN. What is the maximum tensile stress?

1.40. A cantilever beam of thickness t is subjected to a constant traction τ_0 (force per unit area) at its upper surface, as shown in Fig. P1.40. Determine, in terms of τ_0, h, and L, the principal stresses and the maximum shearing stress at the corner points A and B.

1.41. A hollow shaft of 60-mm outer diameter and 30-mm inner diameter is acted on by an axial tensile load of 50 kN, a torque of 500 N·m, and a bending moment of 200 N·m. Use Mohr's circle to determine the principal stresses and their directions.

1.42. Given the stress acting uniformly over the sides of a thin, flat plate (Fig. P1.42), determine (a) the stresses on planes inclined at 20° to the horizontal and (b) the principal stresses and their orientations.

1.43. A steel shaft of radius $r = 75$ mm is subjected to an axial compression $P = 81$ kN, a twisting couple $T = 15.6$ kN·m, and a bending moment $M = 13$ kN·m at both ends. Calculate the magnitude of the principal stresses, the maximum shear stress, and the planes on which they act in the shaft.

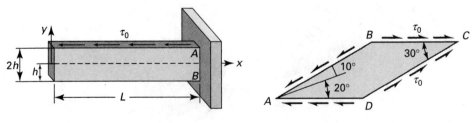

FIGURE P1.40.   FIGURE P1.42.

1.44. A structural member is subjected to a set of forces and moments. Each separately produces the stress conditions at a point shown in Fig. P1.44. Determine the principal stresses and their orientations at the point under the effect of combined loading.

1.45. Redo Prob. 1.44 for the case shown in Fig. P1.45.

1.46. Redo Prob. 1.44 for the case shown in Fig. P1.46.

1.47. The shearing stress at a point in a loaded structure is $\tau_{xy} = 40$ MPa. Also, it is known that the principal stresses at this point are $\sigma_1 = 40$ MPa and $\sigma_2 = -60$ MPa. Determine σ_x (compression) and σ_y and indicate the principal and maximum shearing stresses on an appropriate sketch.

1.48. The state of stress at a point in a structure is depicted in Fig. P1.48. Calculate the normal stress σ and the angle θ.

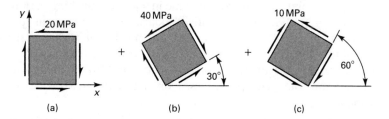

FIGURE P1.44.

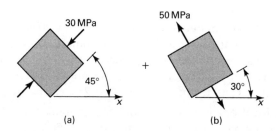

FIGURE P1.45.

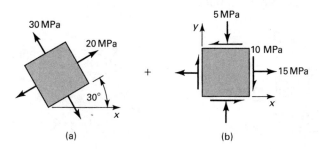

FIGURE P1.46.

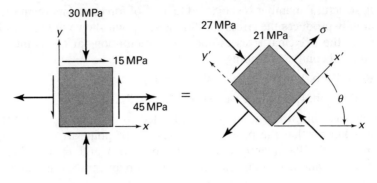

FIGURE P1.48.

1.49. Acting at a point on a horizontal plane in a loaded machine part are normal stress $\sigma_y = 20$ MPa and a (negative) shearing stress. One principal stress at the point is 10 MPa (tensile), and the maximum shearing stress is of magnitude 50 MPa. Find, by the use of Mohr's circle, (a) the unknown stresses on the horizontal and vertical planes and (b) the unknown principal stress. Show the principal stresses on a sketch of a properly oriented element.

1.50. For a state of stress at a point in a structure, certain stress components are given for each of the two orientations (Fig. P1.50). Applying transformation equations, calculate stress components $\sigma_{y'}$ and $\tau_{x'y'}$ and the angle θ_1 between zero and 90°.

1.51. A solid shaft 200 mm in diameter rotates at $f = 20$ rps and is subjected to a bending moment of 21π kN·m. Determine the torque T and power P that can also act simultaneously on the shaft without exceeding a resultant shearing stress of 56 MPa and a resultant normal stress of 98 MPa (with f expressed in rps and torque in N·m, $P = 2\pi f \cdot T$ in watts).

1.52. The cylindrical portion of a compressed-air tank is made of 5-mm-thick plate welded along a helix at an angle of $\phi = 60°$ with the axial direction (Fig. P1.52). The radius of the tank is 250 mm. If the allowable shearing stress parallel to the weld is 30 MPa, calculate the largest internal pressure p that may be applied.

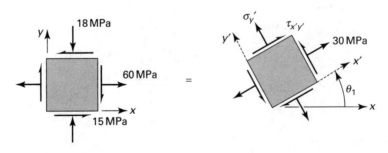

FIGURE P1.50.

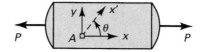

FIGURE P1.52.　　　　　　　　FIGURE P1.53.

1.53. A thin-walled cylindrical tank is subjected to an internal pressure p and uniform axial tensile load P (Fig. P1.53). The radius and thickness of the tank are $r = 0.45$ m and $t = 5$ mm. The normal stresses at a point A on the surface of the tank are restricted to $\sigma_{x'} = 84$ MPa and $\sigma_{y'} = 56$ MPa, while shearing stress $\tau_{x'y'}$ is not specified. Determine the values of p and P. Use $\theta = 30°$.

1.54. For a given state of stress at a point in a frame, certain stress components are known for each of the two orientations shown in Fig. P1.54. Using Mohr's circle, determine the following stress components: (a) τ_{xy} and (b) $\tau_{x'y'}$ and $\sigma_{y'}$.

1.55. The state of stress at a point in a machine member is shown in Fig. P1.55. The allowable compression stress at the point is 14 MPa. Determine (a) the tensile stress σ_x and (b) the maximum principal and maximum shearing stresses in the member. Sketch the results on properly oriented elements.

1.56. In Example 1.3, taking $\sigma_z = 0$, investigate the maximum shearing stresses on all possible (three-dimensional) planes.

1.57. A thin-walled pressure vessel of 60-mm radius and 4-mm thickness is made from spirally welded pipe and fitted with two rigid end plates (Fig. P1.57). The vessel is subjected to an internal pressure of $p = 2$ MPa and a $P = 50$ kN axial load. Calculate (a) the normal stress perpendicular to the weld; (b) the shearing stress parallel to the weld.

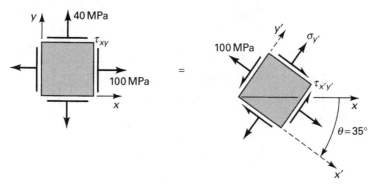

FIGURE P1.54.

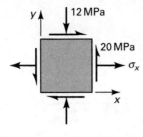

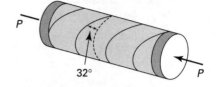

FIGURE P1.55.

FIGURE P1.57.

1.58. A thin-walled cylindrical pressure vessel of 0.3-m radius and 6-mm wall thickness has a welded spiral seam at an angle of $\phi = 30°$ with the axial direction (Fig. P1.10). The vessel is subjected to an internal gage pressure of p Pa and an axial compressive load of $P = 9\pi$ kN applied through rigid end plates. Find the allowable value of p if the normal and shearing stresses acting simultaneously in the plane of welding are limited to 21 and 7 MPa, respectively.

Sections 1.12 and 1.13

1.59. The state of stress at a point in an x, y, z coordinate system is

$$\begin{bmatrix} 20 & 12 & -15 \\ 12 & 0 & 10 \\ -15 & 10 & 6 \end{bmatrix} \text{MPa}$$

Determine the stresses and stress invariants relative to the x', y', z' coordinate system defined by rotating x, y through an angle of 30° counterclockwise about the z axis.

1.60. Redo Prob. 1.59 for the case in which the state of stress at a point in an x, y, z coordinate system is

$$\begin{bmatrix} 60 & 40 & -40 \\ 40 & 0 & -20 \\ -40 & -20 & 20 \end{bmatrix} \text{MPa}$$

1.61. The state of stress at a point relative to an x, y, z coordinate system is given by

$$\begin{bmatrix} 12 & 4 & 2 \\ 4 & -8 & -1 \\ 2 & -1 & 6 \end{bmatrix} \text{MPa}$$

Calculate the maximum shearing stress at the point.

1.62. At a point in a loaded member, the stresses relative to an x, y, z coordinate system are given by

$$\begin{bmatrix} 60 & 20 & 10 \\ 20 & -40 & -5 \\ 10 & -5 & 30 \end{bmatrix} \text{MPa}$$

Calculate the magnitude and direction of maximum principal stress.

1.63. For the stresses given in Prob. 1.59, calculate the maximum shearing stress.

1.64. At a specified point in a member, the state of stress with respect to a Cartesian coordinate system is given by

$$\begin{bmatrix} 12 & 6 & 9 \\ 6 & 10 & 3 \\ 9 & 3 & 14 \end{bmatrix} \text{MPa}$$

Calculate the magnitude and direction of the maximum principal stress.

1.65. At a point in a loaded structure, the stresses relative to an x, y, z coordinate system are given by

$$\begin{bmatrix} 30 & 0 & 20 \\ 0 & 0 & 0 \\ 20 & 0 & 0 \end{bmatrix} \text{MPa}$$

Determine by expanding the characteristic stress determinant: (a) the principal stresses; (b) the direction cosines of the maximum principal stress.

1.66. The stresses (in megapascals) with respect to an x, y, z coordinate system are described by

$$\sigma_x = x^2 + y, \qquad \sigma_z = -x + 6y + z$$
$$\sigma_y = y^2 - 5, \qquad \tau_{xy} = \tau_{xz} = \tau_{yz} = 0$$

At point $(3, 1, 5)$, determine (a) the stress components with respect to x', y', z' if

$$l_1 = 1, \qquad m_2 = \tfrac{1}{2}, \qquad n_2 = \frac{\sqrt{3}}{2}, \qquad n_3 = \tfrac{1}{2}, \qquad m_3 = -\frac{\sqrt{3}}{2}$$

and (b) the stress components with respect to x'', y'', z'' if $l_1 = 2/\sqrt{5}$, $m_1 = -1/\sqrt{5}$, and $n_3 = 1$. Show that the quantities given by Eq. (1.34) are invariant under the transformations (a) and (b).

1.67. Determine the stresses with respect to the x', y', z' axes in the element of Prob. 1.64 if

$$l_1 = \tfrac{1}{2}, \qquad l_2 = -\frac{\sqrt{3}}{2}, \qquad l_3 = 0$$
$$m_1 = \frac{\sqrt{3}}{2}, \qquad m_2 = \tfrac{1}{2}, \qquad m_3 = 0$$
$$n_1 = 0, \qquad n_2 = 0, \qquad n_3 = 1$$

1.68. For the case of plane stress, verify that Eq. (1.33) reduces to Eq. (1.20).

1.69. Obtain the principal stresses and the related direction cosines for the following cases:

$$\text{(a)} \quad \begin{bmatrix} 3 & 4 & 6 \\ 4 & 2 & 5 \\ 6 & 5 & 1 \end{bmatrix} \text{MPa}, \qquad \text{(b)} \quad \begin{bmatrix} 14.32 & 0.8 & 1.55 \\ 0.8 & 6.97 & 5.2 \\ 1.55 & 5.2 & 16.3 \end{bmatrix} \text{MPa}$$

1.70. The stress at a point in a machine component relative to an x, y, z coordinate system is given by

$$\begin{bmatrix} 100 & 40 & 0 \\ 40 & 60 & 80 \\ 0 & 80 & 20 \end{bmatrix} \text{MPa}$$

Referring to the parallelepiped shown in Fig. P1.70, calculate the normal stress σ and the shear stress τ at point Q for the surface parallel to the following planes: (a) $CEBG$, (b) $ABEF$, (c) AEG. [Hint: The position vectors of points G, E, A and any point on plane AEG are, respectively, $r_g = 3i$, $r_e = 4j$, $r_a = 2k$, $r = xi + yj + zk$. The equation of the plane is given by

$$(r - r_g) \cdot (r_e - r_g) \times (r_a - r_g) = 0 \tag{P1.70}$$

from which

$$\begin{bmatrix} x - 3 & y & z \\ -3 & 4 & 0 \\ -3 & 0 & 2 \end{bmatrix} = 0 \quad \text{or} \quad 4x + 3y + 6z = 12$$

The *direction cosines* are then

$$l = \frac{4}{\sqrt{4^2 + 3^2 + 6^2}} = \frac{4}{\sqrt{61}}, \qquad m = \frac{3}{\sqrt{61}}, \qquad n = \frac{6}{\sqrt{61}}$$

1.71. Re-solve Prob. 1.70 for the case in which the dimensions of the parallelepiped are as shown in Fig. P1.71.

1.72. The state of stress at a point in a member relative to an x, y, z coordinate system is

$$\begin{bmatrix} 20 & 10 & -10 \\ 10 & 30 & 0 \\ -10 & 0 & 50 \end{bmatrix} \text{MPa}$$

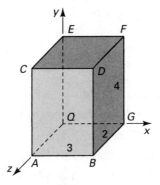

FIGURE P1.70.

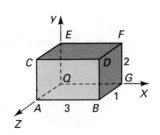

FIGURE P1.71.

Determine the normal stress σ and the shearing stress τ on the surface intersecting the point and parallel to the plane: $2x + y - 3z = 9$.

1.73. For the stresses given in Prob. 1.62, calculate the normal stress σ and the shearing stress τ on a plane whose outward normal is oriented at angles $35°, 60°$, and $73.6°$ with the x, y, and z axes, respectively.

1.74. At a point in a loaded body, the stresses relative to an x, y, z coordinate system are

$$\begin{bmatrix} 40 & 40 & 30 \\ 40 & 20 & 0 \\ 30 & 0 & 20 \end{bmatrix} \text{MPa}$$

Determine the normal stress σ and the shearing stress τ on a plane whose outward normal is oriented at angles of $40°, 75°$, and $54°$ with the x, y, and z axes, respectively.

1.75. Determine the magnitude and direction of the maximum shearing stress for the cases given in Prob. 1.69.

1.76. The stresses at a point in a loaded machine bracket with respect to the x, y, z axes are given as

$$\begin{bmatrix} 36 & 0 & 0 \\ 0 & 48 & 0 \\ 0 & 0 & -72 \end{bmatrix} \text{MPa}$$

Determine (a) the octahedral stresses; (b) the maximum shearing stresses.

1.77. The state of stress at a point in a member relative to an x, y, z coordinate system is given by

$$\begin{bmatrix} -100 & 0 & -80 \\ 0 & 20 & 0 \\ -80 & 0 & 20 \end{bmatrix} \text{MPa}$$

Calculate (a) the principal stresses by expansion of the characteristic stress determinant; (b) the octahedral stresses and the maximum shearing stress.

1.78. Given the principal stresses σ_1, σ_2, and σ_3 at a point in an elastic solid, prove that the maximum shearing stress at the point always exceeds the octahedral shearing stress.

1.79. Determine the value of the octahedral stresses of Prob. 1.64.

1.80. By using Eq. (1.38b), verify that the planes of maximum shearing stress in three dimensions bisect the planes of maximum and minimum principal stresses. Also find the normal stresses associated with the shearing plane by applying Eq. (1.37).

1.81. A point in a structural member is under three-dimensional stress with $\sigma_x = 100$ MPa, $\sigma_y = 20$ MPa, $\tau_{xy} = 60$ MPa, and σ_z, as shown in Fig. P1.81. Calculate (a) the absolute maximum shear stress for $\sigma_z = 30$ MPa; (b) the absolute maximum shear stress for $\sigma_z = -30$ MPa.

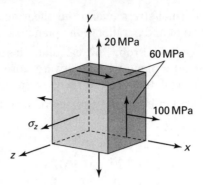

FIGURE P1.81.

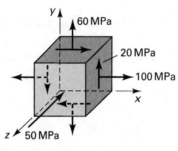

FIGURE P1.82.

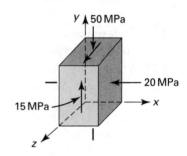

FIGURE P1.84.

1.82. Consider a point in a loaded body subjected to the stress field represented in Fig. P1.82. Determine, using *only* Mohr's circle, the principal stresses and their orientation with respect to the original system.

1.83. Re-solve Prob. 1.82 for the case of a point in a loaded body subjected to the following nonzero stress components: $\sigma_x = 80$ MPa, $\sigma_z = -60$ MPa, and $\tau_{xy} = 40$ MPa.

1.84. The state of stress at a point in a loaded structure is represented in Fig. P1.84. Determine (a) the principal stresses; (b) the octahedral stresses and maximum shearing stress.

1.85. Find the normal and shearing stresses on an oblique plane defined by
$$l = \sqrt{\frac{3}{13}}, m = \sqrt{\frac{1}{13}} \text{ and } n = \sqrt{\frac{9}{13}}.$$
The principal stresses are $\sigma_1 = 35$ MPa, $\sigma_2 = -14$ MPa, and $\sigma_3 = -28$ MPa. If this plane is on the boundary of a structural member, what should be the values of surface forces p_x, p_y, and p_x on the plane?

1.86. Redo Prob. 1.85 for an octahedral plane, $\sigma_1 = 40$ MPa, $\sigma_2 = 15$ MPa, and $\sigma_3 = 25$ MPa.

CHAPTER 2

Strain and Material Properties

2.1 INTRODUCTION

In Chapter 1, our concern was with the stresses within a body subject to a system of external forces. We now turn to the deformations caused by these forces and to a measure of deformational intensity called *strain*, discussed in Sections 2.3 through 2.5. Deformations and strains, which are necessary to an analysis of stress, are also important quantities in themselves, for they relate to changes in the size and shape of a body.

Recall that the state of stress at a point can be determined if the stress components on mutually perpendicular planes are given. A similar operation applies to the state of strain to develop the transformation relations that give two-dimensional and three-dimensional strains in inclined directions in terms of the strains in the coordinate directions. The plane strain transformation equations are especially important in experimental investigations, where normal strains are measured with strain gages. It is usually necessary to use some combination of strain gages or a strain rosette, with each gage measuring the strain in a different direction.

The mechanical properties of engineering materials, as determined from tension test, are considered in Sections 2.6 through 2.8. Material selection and stress–strain curves in tension, compression, and shear are also briefly discussed. Following this, there is a discussion of the relationship between strain and stress under uniaxial, shear, and multiaxial loading conditions. The measurement of strain and the concept of strain energy are taken up in Sections 2.12 through 2.15. Finally, Saint-Venant's principle, which is extremely useful in the solution of practical problems, is introduced in Section 2.16.

2.2 DEFORMATION

Let us consider a body subjected to external loading that causes it to take up the position pictured by the dashed lines in Fig. 2.1, in which A is displaced to A', B to B', and so on, until all the points in the body are displaced to new positions. The displacements of *any two* points such as A and B are simply AA' and BB', respectively, and may be a consequence of deformation (straining), rigid-body motion (translation and rotation), or some combination. The body is said to be *strained* if the *relative positions* of points in the body are *altered*. If no straining has taken place, displacements AA' and BB' are attributable to rigid-body motion. In the latter case, the distance between A and B remains fixed; that is, $L_o = L$. Such displacements are not discussed in this chapter.

To describe the magnitude and direction of the displacements, points within the body are located with respect to an appropriate coordinate reference as, for example, the xyz system. Therefore, in the two-dimensional case shown in Fig. 2.1, the components of displacement of point A to A' can be represented by u and v in the x and y coordinate directions, respectively. In general, the components of displacement at a point, occurring in the x, y, and z directions, are denoted by u, v, and w, respectively. The displacement at *every* point within the body constitutes the *displacement field*, $u = u(x, y, z), v = v(x, y, z)$, and $w = w(x, y, z)$. In this text, mainly *small displacements* are considered, a simplification consistent with the magnitude of deformation commonly found in engineering structures. The strains produced by small deformations are small compared to unity, and their products (higher-order terms) are neglected. For purposes of clarity, small displacements with which we are concerned will be shown highly exaggerated on all diagrams.

Superposition

The small displacement assumption leads to one of the basic fundamentals of solid mechanics, called the *principle of superposition*. This principle is valid whenever the quantity (stress or displacement) to be determined is a *linear* function of the loads that produce it. For the foregoing condition to exist, material must be linearly elastic. In such situations, the total quantity owing to the *combined loads* acting simultaneously on a member may be obtained by *determining separately the quantity attributable to each load* and combining the individual results.

For example, normal stresses caused by axial forces and bending simultaneously (see Table 1.1) may be obtained by superposition, provided that the combined stresses do not exceed the proportional limit of the material. Likewise,

FIGURE 2.1. *Plane displacement and strain in a body.*

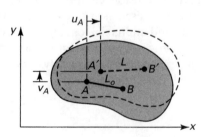

shearing stresses caused by a torque and a vertical shear force acting simultaneously in a beam may be treated by superposition. Clearly, superposition cannot be applied to plastic deformations. The principle of superposition is employed repeatedly in this text. The motivation for superposition is the replacement of a complex load configuration by two or more simpler loads.

2.3 STRAIN DEFINED

For purposes of defining normal strain, refer to Fig. 2.2, where line AB of an axially loaded member has suffered deformation to become $A'B'$. The length of AB is Δx (Fig. 2.2a). As shown in Fig. 2.2b, points A and B have each been displaced: A an amount u, and $B, u + \Delta u$. Stated differently, point B has been displaced by an amount Δu in addition to displacement of point A, and the length Δx has been increased by Δu. Normal strain, the unit change in length, is defined as

$$\varepsilon_x = \lim_{\Delta x \to 0} \frac{\Delta u}{\Delta x} = \frac{du}{dx} \tag{2.1}$$

In view of the limiting process, Eq. (2.1) represents the strain at a point, the point to which Δx shrinks.

If the deformation is distributed *uniformly* over the original length, the normal strain may be written

$$\varepsilon_o = \frac{L - L_o}{L_o} = \frac{\delta}{L_o} \tag{2.2}$$

where L, L_o, and δ are the final length, the original length, and the change of length of the member, respectively. When uniform deformation does not occur, the aforementioned is the average strain.

Plane Strain

We now investigate the case of two-dimensional or *plane strain*, wherein all points in the body, before and after application of load, remain in the same plane. Two-dimensional views of an element with edges of *unit* lengths subjected to plane strain are shown in three parts in Fig. 2.3. We note that this element has no normal strain ε_z and no shearing strains γ_{xz} and γ_{yz} in the xz and yz planes, respectively.

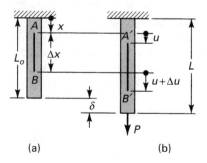

(a)   (b)

FIGURE 2.2.   *Normal strain in a prismatic bar: (a) undeformed state; (b) deformed state.*

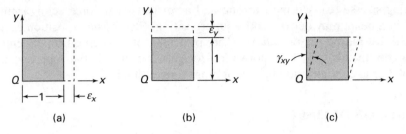

FIGURE 2.3. *Strain components ε_x, ε_y, and γ_{xy} in the xy plane.*

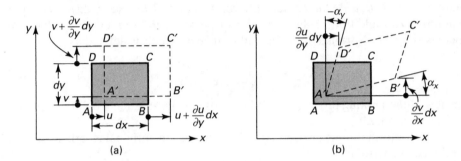

FIGURE 2.4. *Deformations of an element: (a) normal strain; (b) shearing strain.*

Referring to Fig. 2.4, consider an element with dimensions dx, dy and of unit thickness. The total deformation may be regarded as possessing the following features: a change in length experienced by the sides (Fig. 2.4a) and a relative rotation without accompanying changes of length (Fig. 2.4b).

Recalling the basis of Eq. (2.1), two normal or longitudinal strains are apparent upon examination of Fig. 2.4a:

$$\varepsilon_x = \frac{\partial u}{\partial x}, \qquad \varepsilon_y = \frac{\partial v}{\partial y} \tag{2.3a}$$

A *positive* sign is applied to *elongation*; a *negative* sign, to *contraction*.

Now consider the change experienced by right angle DAB (Fig. 2.4b). We shall assume the angle α_x between AB and $A'B'$ to be so small as to permit the approximation $\alpha_x \approx \tan\alpha_x$. Also, in view of the smallness of α_x, the normal strain is small, so $AB \approx A'B'$. As a consequence of the aforementioned considerations, $\alpha_x \approx \partial v/\partial x$, where the *counterclockwise rotation* is defined as *positive*. Similar analysis leads to $-\alpha_y \approx \partial u/\partial y$. The total angular change of angle DAB, the angular change between lines in the x and y directions, is defined as the shearing strain and denoted by γ_{xy}:

$$\gamma_{xy} = \alpha_x - \alpha_y = \frac{\partial u}{\partial y} + \frac{\partial v}{\partial x} \tag{2.3b}$$

The shear strain is *positive* when the *right angle* between two positive (or negative) axes *decreases*. That is, if the angle between $+x$ and $+y$ or $-x$ and $-y$ decreases, we have positive γ_{xy}; otherwise the shear strain is negative.

Three-Dimensional Strain

In the case of a three-dimensional element, a rectangular prism with sides dx, dy, dz, an essentially identical analysis leads to the following normal and shearing strains:

$$\varepsilon_x = \frac{\partial u}{\partial x}, \qquad \varepsilon_y = \frac{\partial v}{\partial y}, \qquad \varepsilon_z = \frac{\partial w}{\partial z}$$

$$\gamma_{xy} = \frac{\partial u}{\partial y} + \frac{\partial v}{\partial x}, \qquad \gamma_{yz} = \frac{\partial v}{\partial z} + \frac{\partial w}{\partial y}, \qquad \gamma_{zx} = \frac{\partial w}{\partial x} + \frac{\partial u}{\partial z} \tag{2.4}$$

Clearly, the angular change is not different if it is said to occur between the x and y directions or between the y and x directions; $\gamma_{xy} = \gamma_{yx}$. The remaining components of shearing strain are similarly related:

$$\gamma_{xy} = \gamma_{yx}, \qquad \gamma_{yz} = \gamma_{zy}, \qquad \gamma_{zx} = \gamma_{xz}$$

The symmetry of shearing strains may also be deduced from an examination of Eq. (2.4). The expressions (2.4) are the *strain–displacement* relations of continuum mechanics. They are also referred to as the *kinematic* relations, treating the *geometry* of strain rather than the matter of cause and effect.

A succinct statement of Eq. (2.3) is made possible by tensor notation:

$$\varepsilon_{ij} = \frac{1}{2}\left(\frac{\partial u_i}{\partial x_j} + \frac{\partial u_j}{\partial x_i}\right), \qquad i, j = x, y, z \tag{2.5a}$$

or expressed more concisely by using commas,

$$\varepsilon_{ij} = \tfrac{1}{2}(u_{i,j} + u_{j,i}) \tag{2.5b}$$

where $u_x = u$, $u_y = v$, $x_x = x$, and so on. The factor $\frac{1}{2}$ in Eq. (2.5) facilitates the representation of the strain transformation equations in indicial notation. The longitudinal strains are obtained when $i = j$; the shearing strains are found when $i \neq j$ and $\varepsilon_{ij} = \varepsilon_{ji}$. It is apparent from Eqs. (2.4) and (2.5) that

$$\varepsilon_{xy} = \tfrac{1}{2}\gamma_{xy}, \qquad \varepsilon_{yz} = \tfrac{1}{2}\gamma_{yz}, \qquad \varepsilon_{xz} = \tfrac{1}{2}\gamma_{xz} \tag{2.6}$$

Just as the state of stress at a point is described by a nine-term array, so Eq. (2.5) represents nine strains composing the symmetric strain tensor ($\varepsilon_{ij} = \varepsilon_{ji}$):

$$[\varepsilon_{ij}] = \begin{bmatrix} \varepsilon_x & \frac{1}{2}\gamma_{xy} & \frac{1}{2}\gamma_{xz} \\ \frac{1}{2}\gamma_{yx} & \varepsilon_y & \frac{1}{2}\gamma_{yz} \\ \frac{1}{2}\gamma_{zx} & \frac{1}{2}\gamma_{zy} & \varepsilon_z \end{bmatrix} \tag{2.7}$$

It is interesting to observe that the Cartesian coordinate systems of Chapters 1 and 2 are not identical. In Chapter 1, the equations of statics pertain to the deformed state, and the coordinate set is thus established in a *deformed* body; *xyz* is, in this instance, a *Eulerian coordinate system*. In discussing the kinematics of deformation in this chapter, recall that the *xyz* set is established in the *undeformed* body. In this case, *xyz* is referred to as a *Lagrangian coordinate system*. Although these systems are clearly not the same, the assumption of small deformation permits us to regard *x*, *y*, and *z*, the coordinates in the undeformed body, as applicable to equations of stress *or* strain. Choice of the Lagrangian system should lead to no errors of consequence unless applications in finite elasticity or large deformation theory are attempted. Under such circumstances, the approximation discussed is not valid, and the resulting equations are more difficult to formulate [Refs. 2.1 and 2.2].

Throughout the text, strains are indicated as *dimensionless* quantities. The normal and shearing strains are also frequently described in terms of units such as inches per inch or micrometers per meter and radians or microradians, respectively. The strains for engineering materials in ordinary use seldom exceed 0.002, which is equivalent to 2000×10^{-6} or $2000\,\mu$. We read this as "2000 micros."

EXAMPLE 2.1   Plane Strains in a Plate

A 0.8-m by 0.6-m rectangle *ABCD* is drawn on a thin plate prior to loading. Subsequent to loading, the deformed geometry is shown by the dashed lines in Fig. 2.5. Determine the components of plane strain at point *A*.

Solution   The following *approximate version* of the strain–displacement relations of Eqs. (2.3) must be used:

$$\varepsilon_x = \frac{\Delta u}{\Delta x}, \qquad \varepsilon_y = \frac{\Delta v}{\Delta y}, \qquad \gamma_{xy} = \frac{\Delta u}{\Delta y} + \frac{\Delta v}{\Delta x} \qquad \textbf{(2.8)}$$

Thus, by setting $\Delta x = 800$ mm and $\Delta y = 600$ mm, the normal strains are calculated as follows:

$$\varepsilon_x = \frac{u_B - u_A}{\Delta x} = \frac{1.6 - 0.6}{800} = 1250\,\mu$$

$$\varepsilon_y = \frac{v_D - v_A}{\Delta y} = \frac{-1.2 - 0}{600} = -2000\,\mu$$

FIGURE 2.5.   *Example 2.1. Deformation of a thin plate.*

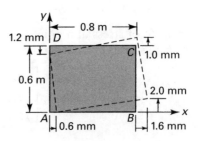

In a like manner, we obtain the shearing strain:

$$\gamma_{xy} = \frac{u_D - u_A}{\Delta y} + \frac{v_B - v_A}{\Delta x} = \frac{0 - 0.6}{600} + \frac{2 - 0}{800} = 1500\,\mu$$

The positive sign indicates that angle BAD has decreased.

Large Strains

As pointed out in Section 2.1, the small deformations or deflections are considered in most applications of this book. The preceding is consistent with the magnitude of deformations usually found in engineering practice. The following more general *large* or *finite strain–displacement relationships* are included here so that the reader may better understand the approximations resulting in the relations of small-deformation theory.

When displacements are relatively large, the strain components are given in terms of the square of the element length instead of the length itself. Therefore, with reference to Fig. 2.4b, we write

$$\varepsilon_x = \frac{(A'D')^2 - (AD)^2}{2(AD)^2} \tag{2.9}$$

in which

$$\begin{aligned} (A'D')^2 &= \left(dx + \frac{\partial u}{\partial x}dx \right)^2 + \left(\frac{\partial v}{\partial x}dx \right)^2 \\ &= \left[1 + 2\frac{\partial u}{\partial x} + \left(\frac{\partial u}{\partial x} \right)^2 + \left(\frac{\partial v}{\partial x} \right)^2 \right](dx)^2 \end{aligned}$$

and $AD = dx$.

Carrying the foregoing terms into Eq. (2.9) leads to a two-dimensional *finite normal strain–displacement* relationship:

$$\varepsilon_x = \frac{\partial u}{\partial x} + \frac{1}{2}\left[\left(\frac{\partial u}{\partial x} \right)^2 + \left(\frac{\partial v}{\partial x} \right)^2 \right] \tag{2.10a}$$

Likewise, we have

$$\varepsilon_y = \frac{\partial v}{\partial y} + \frac{1}{2}\left[\left(\frac{\partial u}{\partial y} \right)^2 + \left(\frac{\partial v}{\partial y} \right)^2 \right] \tag{2.10b}$$

It can also be verified [Refs. 2.3 and 2.4], that the *finite shearing strain–displacement relation* is

$$\gamma_{xy} = \frac{\partial v}{\partial x} + \frac{\partial u}{\partial y} + \frac{\partial u}{\partial x}\frac{\partial u}{\partial y} + \frac{\partial v}{\partial x}\frac{\partial v}{\partial y} \tag{2.10c}$$

In small displacement theory, the higher-order terms in Eqs. (2.10) are omitted. In so doing, these equations reduce to Eqs. (2.4), as expected. The expressions for three-dimensional state of strain may readily be generalized from the preceding equations.

2.4 EQUATIONS OF COMPATIBILITY

The concept of compatibility has both mathematical and physical significance. From a mathematical point of view, it asserts that the displacements u, v, and w match the geometrical boundary conditions and are single-valued and continuous functions of position with which the strain components are associated [Refs. 2.1 and 2.2]. Physically, this means that the body must be pieced together; no voids are created in the deformed body.

Recall, for instance, the uniform state of stress at a section a–a of an axially loaded member as shown in Fig. 1.5b (Sec. 1.6). This, as well as any other stress distribution symmetric with respect to the centroidal axis, such as a parabolic distribution, can ensure equilibrium provided that $\int \sigma_x \, dA = P$. However, the reason the uniform distribution is the acceptable or correct one is that it also ensures a piecewise-continuous strain and displacement field consistent with the boundary conditions of the axially loaded member, the essential characteristic of compatibility.

We now develop the equations of compatibility, which establish the geometrically possible form of variation of strains from point to point within a body. The kinematic relations, Eqs. (2.4), connect six components of strain to only three components of displacement. We cannot therefore arbitrarily specify all the strains as functions of x, y, and z. As the strains are evidently not independent of one another, in what way are they related? In two-dimensional strain, differentiation of ε_x twice with respect to y, ε_y twice with respect to x, and γ_{xy} with respect to x and y results in

$$\frac{\partial^2 \varepsilon_x}{\partial y^2} = \frac{\partial^3 u}{\partial x \, \partial y^2}, \qquad \frac{\partial^2 \varepsilon_y}{\partial x^2} = \frac{\partial^3 v}{\partial x^2 \, \partial y}, \qquad \frac{\partial^2 \gamma_{xy}}{\partial x \, \partial y} = \frac{\partial^3 u}{\partial x \, \partial y^2} + \frac{\partial^3 v}{\partial x^2 \, \partial y}$$

or

$$\frac{\partial^2 \varepsilon_x}{\partial y^2} + \frac{\partial^2 \varepsilon_y}{\partial x^2} = \frac{\partial^2 \gamma_{xy}}{\partial x \, \partial y} \tag{2.11}$$

This is the *condition of compatibility* of the two-dimensional problem, expressed in terms of strain. The three-dimensional *equations of compatibility* are obtained in a like manner:

$$\frac{\partial^2 \varepsilon_x}{\partial y^2} + \frac{\partial^2 \varepsilon_y}{\partial x^2} = \frac{\partial^2 \gamma_{xy}}{\partial x \, \partial y}, \qquad 2\frac{\partial^2 \varepsilon_x}{\partial y \, \partial z} = \frac{\partial}{\partial x}\left(-\frac{\partial \gamma_{yz}}{\partial x} + \frac{\partial \gamma_{xz}}{\partial y} + \frac{\partial \gamma_{xy}}{\partial z} \right)$$

$$\frac{\partial^2 \varepsilon_y}{\partial z^2} + \frac{\partial^2 \varepsilon_z}{\partial y^2} = \frac{\partial^2 \gamma_{yz}}{\partial y \, \partial z}, \qquad 2\frac{\partial^2 \varepsilon_y}{\partial z \, \partial x} = \frac{\partial}{\partial y}\left(\frac{\partial \gamma_{yz}}{\partial x} - \frac{\partial \gamma_{xz}}{\partial y} + \frac{\partial \gamma_{xy}}{\partial z} \right) \tag{2.12}$$

$$\frac{\partial^2 \varepsilon_z}{\partial x^2} + \frac{\partial^2 \varepsilon_x}{\partial z^2} = \frac{\partial^2 \gamma_{xz}}{\partial z \, \partial x}, \qquad 2\frac{\partial^2 \varepsilon_z}{\partial x \, \partial y} = \frac{\partial}{\partial z}\left(\frac{\partial \gamma_{yz}}{\partial x} + \frac{\partial \gamma_{xz}}{\partial y} - \frac{\partial \gamma_{xy}}{\partial z} \right)$$

These equations were first derived by Saint-Venant in 1860. The application of the equations of compatibility is illustrated in Example 2.2(a) and in various sections that use the method of the theory of elasticity.

To gain further insight into the meaning of compatibility, imagine an elastic body subdivided into a number of small cubic elements prior to deformation. These cubes may, upon loading, be deformed into a system of parallelepipeds. The deformed system will, in general, be impossible to arrange in such a way as to compose a continuous body unless the components of strain satisfy the equations of compatibility.

2.5 STATE OF STRAIN AT A POINT

Recall from Chapter 1 that, given the components of stress at a point, it is possible to determine the stresses on any plane passing through the point. A similar operation pertains to the strains at a point.

Consider a small linear element AB of length ds is an unstrained body (Fig. 2.6a). The projections of the element on the coordinate axes are dx and dy. After straining, AB is displaced to position $A'B'$ and is now ds' long. The x and y displacements are $u + du$ and $v + dv$, respectively. The variation with position of the displacement is expressed by a truncated Taylor's expansion as follows:

$$du = \frac{\partial u}{\partial x} dx + \frac{\partial u}{\partial y} dy, \qquad dv = \frac{\partial v}{\partial x} dx + \frac{\partial v}{\partial y} dy \qquad \textbf{(a)}$$

Figure 2.6b shows the relative displacement of B with respect to A, the straining of AB. It is observed that AB has been translated so that A coincides with A'; it is now in the position $A'B''$. Here $B''D = du$ and $DB' = dv$ are the components of displacement.

Transformation of Two-Dimensional Strain

We now choose a new coordinate system $x'y'$, as shown in Fig. 2.6, and examine the components of strain with respect to it: $\varepsilon_{x'}, \varepsilon_{y'}, \gamma_{x'y'}$. First we determine the

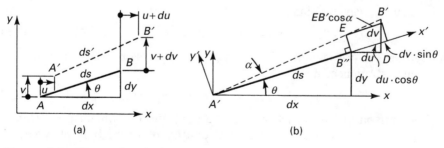

FIGURE 2.6. *Plane straining of an element.*

unit elongation of ds', $\varepsilon_{x'}$. The projections of du and dv on the x' axis, after taking $EB' \cos \alpha = EB'(1)$ by virtue of the small angle approximation, lead to the approximation (Fig. 2.6b)

$$EB' = du \cos \theta + dv \sin \theta \tag{b}$$

By definition, $\varepsilon_{x'}$ is found from EB'/ds. Thus, applying Eq. (b) together with Eqs. (a), we obtain

$$\varepsilon_{x'} = \left(\frac{\partial u}{\partial x} \frac{dx}{ds} + \frac{\partial u}{\partial y} \frac{dy}{ds} \right) \cos \theta + \left(\frac{\partial v}{\partial y} \frac{dy}{ds} + \frac{\partial v}{\partial x} \frac{dx}{ds} \right) \sin \theta$$

Substituting $\cos \theta$ for dx/ds, $\sin \theta$ for dy/ds, and Eq. (2.3) into this equation, we have

$$\varepsilon_{x'} = \varepsilon_x \cos^2 \theta + \varepsilon_y \sin^2 \theta + \gamma_{xy} \sin \theta \cos \theta \tag{2.13a}$$

This represents the transformation equation for the x-directed normal strain, which, through the use of trigonometric identities, may be converted to the form

$$\varepsilon_{x'} = \frac{\varepsilon_x + \varepsilon_y}{2} + \frac{\varepsilon_x - \varepsilon_y}{2} \cos 2\theta + \frac{\gamma_{xy}}{2} \sin 2\theta \tag{2.14a}$$

The normal strain $\varepsilon_{y'}$ is determined by replacing θ by $\theta + \pi/2$ in Eq. (2.14a).

To derive an expression for the shearing strain $\gamma_{x'y'}$, we first determine the angle α through which AB (the x' axis) is rotated. Referring again to Fig. 2.6b, $\tan \alpha = B''E/ds$, where $B''E = dv \cos \theta - du \sin \theta - EB' \sin \alpha$. By letting $\sin \alpha = \tan \alpha = \alpha$, we have $EB' \sin \alpha = \varepsilon_{x'} ds \, \alpha = 0$. The latter is a consequence of the smallness of both $\varepsilon_{x'}$ and α. Substituting Eqs. (a) and (2.3) into $B''E$, $\alpha = B''E/ds$ may be written as follows:

$$\alpha = -(\varepsilon_x - \varepsilon_y) \sin \theta \cos \theta + \frac{\partial v}{\partial x} \cos^2 \theta - \frac{\partial u}{\partial y} \sin^2 \theta \tag{c}$$

Next, the angular displacement of y' is readily derived by replacing θ by $\theta + \pi/2$ in Eq. (c):

$$\alpha_{\theta + \pi/2} = -(\varepsilon_y - \varepsilon_x) \sin \theta \cos \theta + \frac{\partial v}{\partial x} \sin^2 \theta - \frac{\partial u}{\partial y} \cos^2 \theta$$

Now, taking counterclockwise rotations to be positive (see Fig. 2.4b), it is necessary, in finding the shear strain $\gamma_{x'y'}$, to add α and $-\alpha_{\theta + \pi/2}$. By so doing and substituting $\gamma_{xy} = \partial v/\partial x + \partial u/\partial y$, we obtain

$$\gamma_{x'y'} = 2(\varepsilon_y - \varepsilon_x) \sin \theta \cos \theta + \gamma_{xy} (\cos^2 \theta - \sin^2 \theta) \tag{2.13b}$$

Through the use of trigonometric identities, this expression for the transformation of the shear strain becomes

$$\gamma_{x'y'} = -(\varepsilon_x - \varepsilon_y) \sin 2\theta + \gamma_{xy} \cos 2\theta \tag{2.14b}$$

Comparison of Eqs. (1.18) with Eqs. (2.14), the two-dimensional transformation equations of strain, reveals an identity of form. It is observed that transformations expressions for stress are converted into strain relationships by *replacing*

$$\sigma \text{ with } \varepsilon \quad \text{and} \quad \tau \text{ with } \tfrac{1}{2}\gamma$$

These substitutions can be made in *all the analogous relations*. For instance, the principal strain directions (where $\gamma_{x'y'} = 0$) are found from Eq. (1.19):

$$\tan 2\theta_p = \frac{\gamma_{xy}}{\varepsilon_x - \varepsilon_y} \tag{2.15}$$

Similarly, the magnitudes of the principal strains are

$$\varepsilon_{1,2} = \frac{\varepsilon_x + \varepsilon_y}{2} \pm \sqrt{\left(\frac{\varepsilon_x - \varepsilon_y}{2}\right)^2 + \left(\frac{\gamma_{xy}}{2}\right)^2} \tag{2.16}$$

The maximum shearing strains are found on planes 45° relative to the principal planes and are given by

$$\gamma_{max} = \pm 2\sqrt{\left(\frac{\varepsilon_x - \varepsilon_y}{2}\right)^2 + \left(\frac{\gamma_{xy}}{2}\right)^2} = \pm(\varepsilon_1 - \varepsilon_2) \tag{2.17}$$

Transformation of Three-Dimensional Strain

This case may also proceed from the corresponding stress relations by replacing σ by ε and τ by $\gamma/2$. Therefore, using Eqs. (1.28), we have

$$\varepsilon_{x'} = \varepsilon_x l_1^2 + \varepsilon_y m_1^2 + \varepsilon_z n_1^2 + \gamma_{xy} l_1 m_1 + \gamma_{yz} m_1 n_1 + \gamma_{xz} l_1 n_1 \tag{2.18a}$$

$$\begin{aligned}\gamma_{x'y'} = {}& 2(\varepsilon_x l_1 l_2 + \varepsilon_y m_1 m_2 + \varepsilon_z n_1 n_2) + \gamma_{xy}(l_1 m_2 + m_1 l_2) \\ & + \gamma_{yz}(m_1 n_2 + n_1 m_2) + \gamma_{xz}(n_1 l_2 + l_1 n_2)\end{aligned} \tag{2.18b}$$

$$\begin{aligned}\gamma_{x'z'} = {}& 2(\varepsilon_x l_1 l_3 + \varepsilon_y m_1 m_3 + \varepsilon_z n_1 n_3) + \gamma_{xy}(l_1 m_3 + m_1 l_3) \\ & + \gamma_{yz}(m_1 n_3 + n_1 m_3) + \gamma_{xz}(n_1 l_3 + l_1 n_3)\end{aligned} \tag{2.18c}$$

$$\varepsilon_{y'} = \varepsilon_x l_2^2 + \varepsilon_y m_2^2 + \varepsilon_z n_2^2 + \gamma_{xy} l_2 m_2 + \gamma_{yz} m_2 n_2 + \gamma_{xz} l_2 n_2 \tag{2.18d}$$

$$\varepsilon_{z'} = \varepsilon_x l_3^2 + \varepsilon_y m_3^2 + \varepsilon_z n_3^2 + \gamma_{xy} l_3 m_3 + \gamma_{yz} m_3 n_3 + \gamma_{xz} l_3 n_3 \tag{2.18e}$$

$$\begin{aligned}\gamma_{y'z'} = {}& 2(\varepsilon_x l_2 l_3 + \varepsilon_y m_2 m_3 + \varepsilon_z n_2 n_3) + \gamma_{xy}(m_2 l_3 + l_2 m_3) \\ & + \gamma_{yz}(n_2 m_3 + m_2 n_3) + \gamma_{xz}(l_2 n_3 + n_2 l_3)\end{aligned} \tag{2.18f}$$

where l_1 is the cosine of the angle between x and x', m_1 is the cosine of the angle between y and x', and so on (see Table 1.2). The foregoing equations are succinctly expressed, referring to Eqs. (1.29), as follow:

$$\varepsilon'_{rs} = l_{ir} l_{js} \varepsilon_{ij} \tag{2.19a}$$

Conversely,

$$\varepsilon_{rs} = l_{ri} l_{sj} \varepsilon'_{ij} \tag{2.19b}$$

These equations represent the law of transformation for a strain tensor of rank 2.

Also, referring to Eqs. (1.33) and (1.34), the principal strains in three dimensions are the roots of the following cubic equation:

$$\varepsilon_p^3 - J_1 \varepsilon_p^2 + J_2 \varepsilon_p - J_3 = 0 \qquad (2.20)$$

The *strain invariants* are

$$J_1 = \varepsilon_x + \varepsilon_y + \varepsilon_z$$
$$J_2 = \varepsilon_x\varepsilon_y + \varepsilon_x\varepsilon_z + \varepsilon_y\varepsilon_z - \tfrac{1}{4}(\gamma_{xy}^2 + \gamma_{yz}^2 + \gamma_{xz}^2)$$

$$J_3 = \begin{vmatrix} \varepsilon_x & \tfrac{1}{2}\gamma_{xy} & \tfrac{1}{2}\gamma_{xz} \\ \tfrac{1}{2}\gamma_{xy} & \varepsilon_y & \tfrac{1}{2}\gamma_{yz} \\ \tfrac{1}{2}\gamma_{xz} & \tfrac{1}{2}\gamma_{yz} & \varepsilon_z \end{vmatrix} \qquad (2.21)$$

For a given state of strain, the three roots ε_1, ε_2, and ε_3 of Eqs. (2.20) and the corresponding direction cosines may conveniently be computed using Table B.1 with some notation modification.

EXAMPLE 2.2 Three-Dimensional Strain in a Block

A 2-m by 1.5-m by 1-m parallelepiped is deformed by movement of corner point A to A' (1.9985, 1.4988, 1.0009), as shown by the dashed lines in Fig. 2.7. Calculate the following quantities at point A: (a) the strain components; (b) the normal strain in the direction of line AB; and (c) the shearing strain for perpendicular lines AB and AC.

Solution The components of displacement of point A are given by

$$u_A = -1.5 \text{ mm}, \qquad v_A = -1.2 \text{ mm}, \qquad w_A = 0.9 \text{ mm} \qquad \textbf{(d)}$$

a. We can readily obtain the strain components, by using an approximate version of Eqs. (2.4) and Eqs. (d), as in Example 2.1. Alternatively, these strains can be determined as follows. First, referring to Fig. 2.7, we represent the displacement field in the form

$$u = c_1\, xyz, \qquad v = c_2\, xyz, \qquad w = c_3\, xyz \qquad (2.22)$$

where c_1, c_2, and c_3 are constants. From these and Eqs. (d), $-1.5(10^{-3}) = c_1(2 \times 1.5 \times 1)$ or $c_1 = -500(10^{-6})$; similarly, $c_2 = -400(10^{-6})$, and $c_3 = 300(10^{-6})$. Therefore,

FIGURE 2.7. *Example 2.2. Deformation of a parallelpiped.*

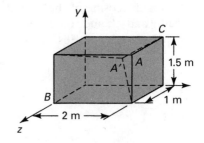

$$u = -500(10^{-6})xyz, \qquad v = -400(10^{-6})xyz, \qquad w = 300(10^{-6})xyz$$

(e)

Applying Eqs. (2.4) and substituting $10^{-6} = \mu$, we have

$$\varepsilon_x = \frac{\partial u}{\partial x} = -500\mu yz, \qquad \varepsilon_y = -400\mu xz, \qquad \varepsilon_z = 300\mu xy$$

$$\gamma_{xy} = \frac{\partial v}{\partial x} + \frac{\partial u}{\partial y} = -\mu(400yz + 500xz), \qquad \gamma_{yz} = -\mu(400xy - 300xz)$$

$$\gamma_{xz} = -\mu(500xy - 300yz)$$

(f)

By introducing the foregoing into Eqs. (2.12), we readily find that these conditions are satisfied and the strain field obtained is therefore possible. The calculations proceed as follows:

$$\varepsilon_x = -500\,\mu(1.5 \times 1) = -750\,\mu, \qquad \varepsilon_y = -800\,\mu, \qquad \varepsilon_z = 900\,\mu$$

$$\gamma_{xy} = -\mu(400 \times 1.5 \times 1 + 500 \times 2 \times 1) = -1600\,\mu,$$

$$\gamma_{yz} = -600\,\mu, \qquad \gamma_{xz} = -1050\,\mu$$

b. Let the x' axis be placed along the line from A to B. The direction cosines of AB are $l_1 = -0.8, m_1 = -0.6,$ and $n_1 = 0$. Applying Eq. (2.18a), we thus have

$$\varepsilon_{x'} = \varepsilon_x l_1^2 + \varepsilon_y m_1^2 + \gamma_{xy} l_1 m_1$$

$$= [-750(-0.8)^2 - 800(-0.6)^2 - 1600(-0.8)(-0.6)]\,\mu = -1536\,\mu$$

c. Let the y' axis be placed along the line A to C. The direction cosines of AC are $l_2 = 0, m_2 = 0,$ and $n_2 = -1$. Thus, from Eq. (2.18b),

$$\gamma_{x'y'} = \gamma_{yz} m_1 n_2 + \gamma_{xz} l_1 n_2$$

$$= [-600(-0.6)(-1) - 1050(-0.8)(-1)]\,\mu = -1200\,\mu$$

where the negative sign indicates that angle BAC has increased.

Mohr's Circle for Plane Strain

Because we have concluded that the transformation properties of stress and strain are identical, it is apparent that a *Mohr's circle for strain* may be drawn and that the construction technique does not differ from that of Mohr's circle for stress. In Mohr's circle for strain, the normal strains are plotted on the horizontal axis, positive to the right. When the shear strain is positive, the point representing the x axis strains is plotted a distance $\gamma/2$ *below* the ε line, and the y axis point a distance $\gamma/2$ *above* the ε line, and vice versa when the shear strain is negative. Note that this convention for shear strain, used *only* in constructing and reading values from Mohr's circle, agrees with the convention employed for stress in Section 1.11.

An illustration of the use of Mohr's circle of strain is given in the solution of the following numerical problem.

EXAMPLE 2.3 State of Plane Strain in a Plate

The state of strain at a point on a thin plate is given by $\varepsilon_x = 510\,\mu$, $\varepsilon_y = 120\,\mu$, and $\gamma_{xy} = 260\,\mu$. Determine, using Mohr's circle of strain, (a) the state of strain associated with axes x', y', which make an angle $\theta = 30°$ with the axes x, y (Fig. 2.8a); (b) the principal strains and directions of the principal axes; (c) the maximum shear strains and associated normal strains; display the given data and the results obtained on properly oriented elements of unit dimensions.

Solution A sketch of Mohr's circle of strain is shown in Fig. 2.8b, constructed by determining the position of point C at $\frac{1}{2}(\varepsilon_x + \varepsilon_y)$ and A at $(\varepsilon_x, \frac{1}{2}\gamma_{xy})$ from the origin O. Note that $\gamma_{xy}/2$ is positive, so point A, representing x-axis strains, is plotted below the ε axis (or B above). Carrying out calculations similar to that for Mohr's circle of stress (Sec. 1.11), the required quantities are determined. The radius of the circle is $r = (195^2 + 130^2)^{1/2}\,\mu = 234\,\mu$, and the angle $2\theta'_p = \tan^{-1}(130/195) = 33.7°$.

a. At a position $60°$ counterclockwise from the x axis lies the x' axis on Mohr's circle, corresponding to twice the angle on the plate. The angle $A'CA_1$ is $60° - 33.7° = 26.3°$. The strain components associated with $x'y'$ are therefore

$$\varepsilon_{x'} = 315\,\mu + 234\,\mu \cos 26.3° = 525\,\mu$$
$$\varepsilon_{y'} = 315\,\mu - 234\,\mu \cos 26.3° = 105\,\mu \qquad \text{(g)}$$
$$\gamma_{x'y'} = -2(234\,\mu \sin 26.3°) = -207\,\mu$$

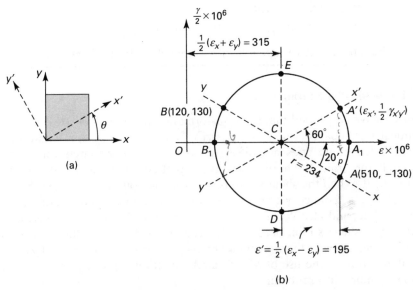

FIGURE 2.8. *Example 2.3: (a) Axes rotated for $\theta = 30°$; (b) Mohr's circle of strain.*

The shear strain is taken as negative because the point representing the x axis strains, A', is above the ε axis. The negative sign indicates that the angle between the element faces x' and y' at the origin increases (Sec. 2.3). As a check, Eq. (2.14b) is applied with the given data to obtain $-207\,\mu$ as before.

b. The principal strains, represented by points A_1 and B_1 on the circle, are found to be

$$\varepsilon_1 = 315\,\mu + 234\,\mu = 549\,\mu$$
$$\varepsilon_2 = 315\,\mu - 234\,\mu = 81\,\mu$$

The axes of ε_1 and ε_2 are directed at $16.85°$ and $106.85°$ from the x axis, respectively.

c. Points D and E represent the maximum shear strains. Thus,

$$\gamma_{max} = \pm 468\,\mu$$

Observe from the circle that the axes of maximum shear strain make an angle of $45°$ with respect to the principal axes. The normal strains associated with the axes of γ_{max} are equal, represented by OC on the circle: $315\,\mu$.

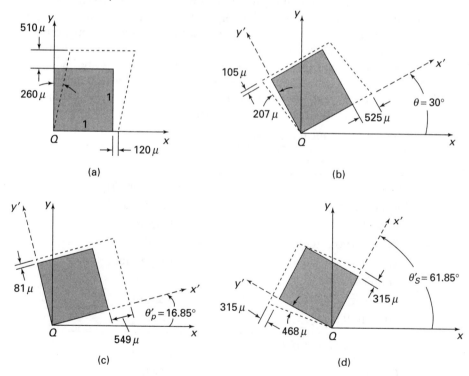

FIGURE 2.9. *Example 2.3. (a) Element with edges of unit lengths in plane strain; (b) element at $\theta = 30°$; (c) principal strains; and (d) maximum shearing strains.*

d. The given data is depicted in Fig. 2.9a. The strain components obtained, Eqs. (g), are portrayed in Fig. 2.9b for an element at $\theta = 30°$. Observe that the angle at the corner Q of the element at the origin increases because γ_{xy} is negative. The principal strains are given in Fig. 2.9c. The sketch of the maximum shearing strain element is shown in Fig. 2.9d.

2.6 ENGINEERING MATERIALS

The equations of equilibrium derived in Chapter 1 and the kinematic relations of this chapter together represent nine equations involving 15 unknowns (six stresses, six strains, and three displacements). The insufficiency noted in the number of available equations is made up for by a set of material-dependent relationships, discussed in Section 2.9, that connect stress with strain. We first define some important characteristics of engineering materials, such as those in widespread commercial usage, including a variety of metals, plastics, and concretes. Table 2.1 gives a general classification of materials commonly used in engineering [Refs. 2.5 through 2.8]. Following this, the tension test is discussed (Sec. 2.7), providing information basic to material behavior.

An *elastic material* is one that returns to its original (unloaded) shape upon the removal of applied forces. Elastic behavior thus precludes permanent or plastic deformation. In many cases, the *elastic range* includes a region throughout which stress and strain bear a linear relationship. This portion of the stress–strain variation ends at a point termed the *proportional limit*. Such materials are linearly elastic. It is not necessary for a material to possess such linearity for it to be elastic. In a *viscoelastic* material, the state of stress is a function not only of the strains but of the time rates of change of stress and strain as well.

TABLE 2.1. *Typical Engineering Materials*

Metallic Materials	
Ferrous Metals	**Nonferrous Metals**
Cast iron	Aluminum
Cast steel	Copper
Plain carbon steel	Lead
Steel alloys	Magnesium
Stainless steel	Nickel
Special steels	Platinum
Structural steel	Silver
Nonmetallic Materials	
Graphite	Plastics
Ceramics	Brick
Glass	Stone
Concrete	Wood

Combinations of elastic (springlike) and viscous (dashpotlike) elements form a *viscous–elastic* model. Glasses, ceramics, biomaterials, synthetic rubbers, and plastics may frequently be considered to be linear viscoelastic materials. Also, most rocks exhibit properties that can be represented by inclusion of viscous terms in the stress–strain relationship. Viscoelastic solids return to their original state when unloaded. A *plastically* deformed solid, on the other hand, does not return to its original shape when the load is removed; there is some *permanent* deformation. With the exception of Chapter 12, our considerations will be limited to the behavior of elastic materials.

Leaving out Section 5.9, it is also assumed in this text that the material is homogeneous and isotropic. A *homogeneous* material displays identical properties throughout. If the properties are identical in all directions at a point, the material is termed *isotropic*. A nonisotropic or *anisotropic* solid such as wood displays direction-dependent properties. An orthotropic material, such as wood, is a special case of an anisotropic material, which has greater strength in a direction parallel to the grain than perpendicular to the grain (see Sec. 2.11). Single crystals also display pronounced anisotropy, manifesting different properties along the various crystallographic directions.

Materials composed of many crystals (polycrystalline aggregates) may exhibit either isotropy or anisotropy. Isotropy results when the crystal size is small relative to the size of the sample, provided that nothing has acted to disturb the random distribution of crystal orientations within the aggregate. Mechanical processing operations such as cold rolling may contribute to minor anisotropy, which in practice is often disregarded. These processes may also result in high internal stress, termed *residual stress*. In the cases treated in this volume, materials are assumed initially *entirely free* of such stress.

General Properties of Some Common Materials

There are various engineering materials, as listed in Table 2.1. The following is a brief description of a few frequently employed materials. The common classes of materials of engineering interest are *metals, plastics, ceramics*, and *composites*. Each group generally has similar properties (such as chemical makeup and atomic structure) and applications. Selection of materials plays a significant role in mechanical design. The choice of a particular material for the members depends on the purpose and type of operation as well as on mode of failure of this component. Strength and stiffness are principal factors considered in selection of a material. But selecting a material from both its functional and economical standpoints is very important. Material properties are determined by standardized test methods outlined by the *American Society for Testing Materials* (ASTM).

Metals. Metals can be made stronger by alloying and by various mechanical and heat treatments. Most metals are ductile and good conductors of electricity and heat. Cast iron and steel are iron alloys containing over 2% carbon and less than 2% carbon, respectively. *Cast irons* constitute a whole family of materials including carbon. *Steels* can be grouped as plain carbon steels, alloy steels, high-strength steels, cast steels, and special-purpose steels. Low-carbon steels or mild steels are also known as the *structural steels*. There are many effects of adding any alloy to a basic carbon steel. *Stainless steels* (in addition to carbon) contain at least 12% chromium as the basic alloying element. *Aluminum* and *magnesium* alloys possess a high strength-to-weight ratio.

Plastics. Plastics are synthetic materials, also known as *polymers*. They are used increasingly for structural purposes, and many different types are available. Polymers are corrosion resistant and have low coefficient of friction. The mechanical characteristics of these materials vary notably, with some plastics being brittle and others ductile. The polymers are of two classes: thermoplastics and thermosets. *Thermoplastics* include acetal, acrylic, nylon, teflon, polypropylene, polystyrene, PVC, and saran. Examples of thermosets are epoxy, polyster, polyurethane, and bakelite. Thermoplastic materials repeatedly soften when heated and harden when cooled. There are also highly elastic flexible materials known as thermoplastic *elastomers*. A common elastomer is a rubber band. *Thermosets* sustain structural change during processing; they can be shaped only by cutting or machining.

Ceramics. Ceramics represent ordinary compounds of nonmetallic as well as metallic elements, mostly oxides, nitrides, and carbides. They are considered an important class of engineering materials for use in machine and structural parts. Ceramics have high hardness and brittleness, high compressive but low tensile strengths. High temperature and chemical resistance, high dielectric strength, and low weight characterize many of these materials. Glasses are also made of metallic and nonmetallic elements, just as are ceramics. But glasses and ceramics have different structural forms. Glass ceramics are widely used as electrical, electronic, and laboratory ware.

Composites. Composites are made up of two or more distinct constituents. They often consist of a high-strength material (for example, fiber made of steel, glass, graphite, or polymers) embedded in a surrounding material (such as resin or concrete), which is termed a matrix. Therefore, a composite material shows a relatively large strength-to-weight ratio compared with a homogeneous material; composite materials generally have other desirable characteristics and are widely used in various structures, pressure vessels, and machine components. A composite is designed to display a combination of the best characteristics of each component material. A *fiber-reinforced composite* is formed by imbedding fibers of a strong, stiff material into a weaker reinforcing material. A layer or lamina of a composite material consists of a variety of arbitrarily oriented bonded layers or laminas. If all fibers in all layers are given the same orientation, the laminate is orthotropic. A typical composite usually consists of bonded three-layer orthotropic material. Our discussions will include isotropic composites like reinforced-concrete beam and multilayered members, single-layer orthotropic materials, and compound cylinders.

2.7 STRESS–STRAIN DIAGRAMS

Let us now discuss briefly the nature of the typical static tensile test. In such a test, a specimen is inserted into the jaws of a machine that permits tensile straining at a relatively low rate (since material strength is strain-rate dependent). Normally, the stress–strain curve resulting from a tensile test is predicated on *engineering* (conventional) *stress* as the ordinate and *engineering* (conventional) *strain* as the abscissa. The latter is defined by Eq. (2.2). The former is the load or tensile force (P) divided by the original cross-sectional area (A_o) of the specimen and, as such, is simply a measure of load (force divided by a constant) rather than *true stress*. True stress is the load divided by the *actual instantaneous* or *current* area (A) of the specimen.

Ductile Materials in Tension

Figure 2.10a shows two stress–strain plots, one (indicated by a solid line) based on engineering stress, the other on true stress. The material tested is a relatively ductile, polycrystalline metal such as steel. A *ductile* metal is capable of substantial elongation prior to failure, as in a drawing process. The converse applies to *brittle* materials. Note that beyond the point labeled "proportional limit" is a point labeled *"yield point"* (for most cases these two points are taken as one). At the yield point, a great deal of deformation occurs while the applied loading remains essentially constant. The engineering stress curve for the material when strained beyond the yield point shows a characteristic maximum termed the *ultimate* tensile stress and a lower value, the *rupture* stress, at which failure occurs. Bearing in mind the definition of engineering stress, this decrease is indicative of a decreased load-carrying capacity of the specimen with continued straining beyond the ultimate tensile stress.

For materials such as heat-treated steel, aluminum, and copper that do not exhibit a distinctive yield point, it is usual to employ a quasi-yield point. According to the *0.2-percent offset method*, a line is drawn through a strain of 0.002, parallel to the initial straight line portion of the curve (Fig. 2.10b). The intersection of this line with the stress–strain curve defines the yield point as shown. Corresponding yield stress is commonly referred to as the *yield strength*.

Geometry Change of Specimen. In the vicinity of the ultimate stress, the reduction of the cross-sectional area becomes clearly visible, and a *necking* of the specimen occurs in the range between ultimate and rupture stresses. Figure 2.11 shows the geometric change in the portion of a ductile specimen under tensile loading. The local elongation is always greater in the necking zone than elsewhere. The standard measures of ductility of a material are expressed as follows:

$$\text{Percent elongation} = \frac{L_f - L_o}{L_o}(100) \qquad \textbf{(2.23a)}$$

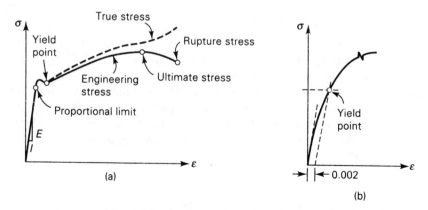

FIGURE 2.10.   (a) Stress–strain diagram of a typical ductile material; (b) determination of yield strength by the offset method.

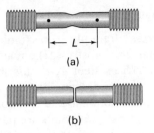

(a)

(b)

FIGURE 2.11. *A typical round specimen of ductile material in tension: (a) neck-ing; (b) fractured.*

$$\text{Percent reduction in area} = \frac{A_o - A_f}{A_o}(100) \qquad \textbf{(2.23b)}$$

Here A_o and L_o designate, respectively, the initial cross-sectional area and gage length between two punch marks of the specimen. The ruptured bar must be pieced together in order to measure the final gage length L_f. Similarly, the final area A_f is measured at the fracture site where the cross section is minimum. The elongation is not uniform over the length of the specimen but concentrated on the region of neck-ing. Percentage of elongation thus depends on the gage length. For structural steel, about 25 percent elongation (for a 50-mm gage length) and 50 percent reduction in area usually occur.

True Stress and True Strain

The large disparity between the engineering stress and true stress curves in the region of a large strain is attributable to the significant localized decrease in area (necking down) prior to fracture. In the area of large strain, particularly that occur-ring in the *plastic range*, the engineering strain, based on small deformation, is clearly inadequate. It is thus convenient to introduce true or logarithmic strain. The *true strain*, denoted by ε, is defined by

$$\varepsilon = \int_{L_o}^{L} \frac{dL}{L} = \ln\frac{L}{L_o} = \ln(1 + \varepsilon_o) \qquad \textbf{(2.24)}$$

This strain is observed to represent the sum of the increments of deformation divided by the length L corresponding to a particular increment of length dL. Here L_o is the original length and ε_o is the engineering strain.

For small strains, Eqs. (2.2) and (2.24) yield approximately the same results. Note that the curve of true stress versus true strain is more informative in examin-ing plastic behavior and will be discussed in detail in Chapter 12. In the plastic range, the material is assumed to be *incompressible* and the *volume constant* (Sec. 2.10). Hence,

$$A_o L_o = AL \qquad \textbf{(a)}$$

where the left and right sides of this equation represent the original and the current volume, respectively. If P is the current load, then

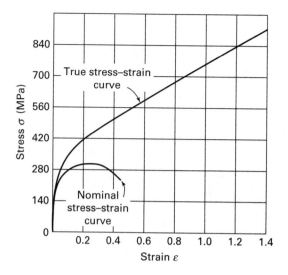

FIGURE 2.12. *Stress–strain curves for a low-carbon (0.05%) steel in tension.*

$$\sigma = \frac{P}{A} = \frac{P}{A_o} \frac{L}{L_o} = \sigma_o \frac{L}{L_o}$$

But, from Eq. (2.2), we have $L/L_o = 1 + \varepsilon_o$. The *true stress* is thus defined by

$$\sigma = \sigma_o(1 + \varepsilon_o) \tag{2.25}$$

That is, the *true stress is equal to the engineering stress multiplied by 1 plus the engineering strain.*

A comparison of a true and nominal stress–strain plot is given in Fig. 2.12 [Ref. 2.8]. The true $\sigma - \varepsilon$ curve shows that as straining progresses, more and more stress develops. On the contrary, in the nominal $\sigma - \varepsilon$ curve, beyond the ultimate strength the stress decreases with the increase in strain. This is particularly important for large deformations involved in metal-forming operations [Ref. 2.7].

Brittle Materials in Tension

Brittle materials are characterized by the fact that rupture occurs with little deformation. The behavior of typical brittle materials, such as magnesium alloy and cast iron, under axial tensile loading is shown in Figs. 2.13a. Observe from the diagrams that there is no well-defined linear region, rupture takes place with no noticeable prior change in the rate of elongation, there is no difference between the ultimate stress and the fracture stress, and the strain at rupture is much smaller than in ductile materials. The fracture of a brittle material is associated with the tensile stress. A brittle material thus breaks normal to the axis of the specimen, as depicted in Fig. 2.13b, because this is the plane of maximum tensile stress.

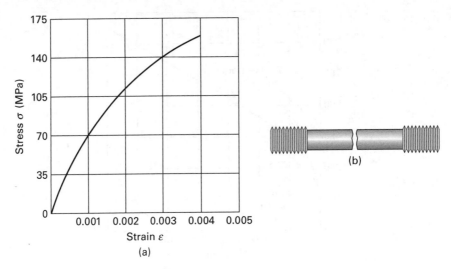

FIGURE 2.13. *Cast iron in tension: (a) Stress–strain diagram; (b) fractured specimen.*

Materials in Compression

Diagrams analogous to those in tension may also be obtained for various materials in compression. Most *ductile materials* behave approximately the same in tension and compression over the elastic range, and the yield-point stress is about the same in tension as in compression. But, in plastic range, the behavior is notably different. Many *brittle materials* have ultimate stresses in compression that are much greater than in tension. Their entire compression stress–strain diagram has a form similar to the form of the tensile diagram. In compression, as the load increases, the brittle material, such as gray cast iron, will generally bulge out or become barrel shaped.

Materials in Shear

Shear stress–strain diagrams can be determined from the results of direct-shear or torsion tests [Ref. 2.9]. These diagrams of torque (T) versus shear strain (γ) are analogous to those seen in Fig. 2.10 for the same materials. But properties such as yield stress and ultimate stress are often half as large in shear as they are in tension. For ductile materials, yield stress in shear is about 0.5 to 0.6 times the yield stress in tension.

2.8 ELASTIC VERSUS PLASTIC BEHAVIOR

The preceding section dealt with the behavior of a variety of materials as they are **loaded** statically under tension, compression, or shear. We now discuss what happens when the load is slowly removed and the material is **unloaded**. Let us consider the stress–stain curve in Fig. 2.14, where E and F represent the *elastic limit* and point of fracture, respectively. The elastic strain is designated by ε_e. It is seen from Fig. 2.14a that when the load is removed at (or under) point E, the material follows exactly same curve back to the origin O. This elastic characteristic of a material, by which it returns to its original size and shape during unloading, is called the *elasticity*. Inasmuch as the stress–strain curve from O to E is not a straight line, the material is *nonlinearly elastic*.

When unloaded at a point A beyond E, the material follows the line AB on the curve (Fig. 2.14b). The slope of this line is *parallel* to the tangent to the stress–strain curve at the origin. Note that ε does not return to zero after the load has been removed. This means that a *residual strain* or *permanent strain* remains in the material. The corresponding elongation of the specimen is called *permanent set*. The property of a material that experiences strains beyond those at the elastic limit is called the *plasticity*. On the stress–strain curve, an elastic range is therefore followed by a plastic region (Fig. 2.14a), in which total recovery of the size and shape of a material does not occur.

Upon reloading (BA), the unloading path is retracted and further loading results in a continuation of the original stress–strain curve. It is seen that the material behaves in a linearly elastic manner in this second loading. There is now proportional limit (A) that is higher than before but reduced ductility, inasmuch as the amount of yielding from E to F is less than from A to F. This process can be repeated until the material becomes brittle and fractures. A significant implication of the preceding is that the strength and ductility characteristics of metals change considerably during fabrication process involving cold working.

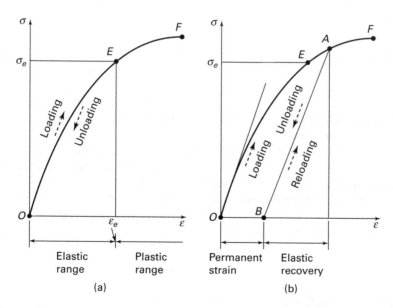

FIGURE 2.14. *Stress–strain diagrams showing (a) elastic behavior; (b) partially elastic behavior.*

A final point to be noted is that, so far, we discussed the behavior of a test specimen subjected to only static loading; passage of time and change in temperatures did not enter into our considerations. However, under certain circumstances, some materials may continue to deform permanently. On the contrary, a loss of stress is observed with time though strain level remains constant in a load-carrying member. The study of material behavior under various loading and environmental conditions is taken up in Chapters 4 and 12.

2.9 HOOKE'S LAW AND POISSON'S RATIO

Most structural materials exhibit an initial region of the stress–strain diagram in which the material behaves both elastically and linearly. This *linear elasticity* is extremely important in engineering because many structures and machines are designed to experience relatively small deformations. For that straight-line portion of the diagram (Fig. 2.10a), stress is directly proportional to strain. If the normal stress acts in the x direction,

$$\sigma_x = E\varepsilon_x \tag{2.26}$$

This relationship is known as *Hooke's law*, after Robert Hooke (1635–1703). The constant E is called the *modulus of elasticity*, or *Young's modulus*, in honor of Thomas Young (1773–1829). As ε is a dimensionless quantity, E has the units of σ. Thus, E is expressed in pascals (or gigapascals) in SI units and in pounds (or kilopounds) per square inch in the U.S. Customary System. Graphically, E is the *slope* of the stress–strain diagram in the linearly elastic region, as shown Fig. 2.10a. It differs from material to material. For most materials, E in compression is the same as that in tension (Table D.1).

Elasticity can similarly be measured in two-dimensional *pure shear* (Fig. 1.3c). It is found experimentally that, in the linearly elastic range, stress and strain are related by *Hooke's law in shear:*

$$\tau_{xy} = G\gamma_{xy} \tag{2.27}$$

Here G is the *shear modulus of elasticity* or *modulus of rigidity*. Like E, G is a constant for a given material.

It was stated in Section 2.7 that axial tensile loading induces a reduction or lateral contraction of a specimen's cross-sectional area. Similarly, a contraction owing to an axial compressive load is accompanied by a lateral extension. In the linearly elastic region, it is found experimentally that lateral strains, say in the y and z directions, are related by a constant of proportionality, ν, to the axial strain caused by *uniaxial stress only* $\varepsilon_x = \sigma_x/E$, in the x direction:

$$\varepsilon_y = \varepsilon_z = -\nu\frac{\sigma_x}{E} \tag{a}$$

Alternatively, the definition of ν may be stated as

$$\nu = -\frac{\text{lateral strain}}{\text{axial strain}} \tag{2.28}$$

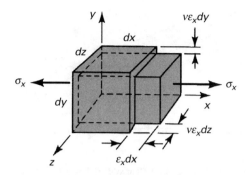

FIGURE 2.15. *Lateral contraction of an element in tension.*

Here v is known as *Poisson's ratio*, after S. D. Poisson (1781–1840), who calculated v to be $\frac{1}{4}$ for isotropic materials employing molecular theory. Note that more recent calculations based on a model of atomic structure yield $v = \frac{1}{3}$. Both values given here are close to the actual measured values, 0.25 to 0.35 for most metals. Extreme cases range from a low of 0.1 (for some concretes) to a high of 0.5 (for rubber).

Volume Change

The lateral contraction of a cubic *element* from a bar *in tension* is illustrated in Fig. 2.15, where it is assumed that the faces of the element at the origin are fixed in position. From the figure, subsequent to straining, the final volume is

$$V_f = (1 + \varepsilon_x)\, dx(1 - v\varepsilon_x)\, dy(1 - v\varepsilon_x)\, dz \qquad \textbf{(b)}$$

Expanding the right side and neglecting higher-order terms involving ε_x^2 and ε_x^3, we have

$$V_f = [1 + (\varepsilon_x - 2v\varepsilon_x)]dx\, dy\, dz = V_o + \Delta V$$

where V_o is the initial volume $dx\, dy\, dz$ and ΔV is the change in volume. The *unit volume change e*, also referred to as the *dilatation*, may now be expressed in the form

$$e = \frac{\Delta V}{V_o} = (1 - 2v)\varepsilon_x = \frac{1 - 2v}{E}\sigma_x \qquad \textbf{(2.29)}$$

Observe from this equation that a tensile force *increases* and a compressive force *decreases* the volume of the element.

EXAMPLE 2.4   Deformation of a Tension Bar

An aluminum alloy bar of circular cross-sectional area A and length L is subjected to an axial tensile force P (Fig. 2.16). The modulus of elasticity and Poisson's ratio of the material are E and v, respectively. For the bar, determine (a) the axial deformation; (b) the change in diameter d; and (c) the change in volume ΔV. (d) Evaluate the numerical values of the quantities obtained in (a) through (c) for the case in which $P = 60\,\text{kN}$,

FIGURE 2.16. *Example 2.4. A bar under tensile forces.*

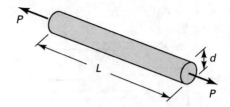

$d = 25$ mm, $L = 3$ m, $E = 70$ GPa, $\nu = 0.3$, and the yield strength $\sigma_{yp} = 260$ MPa.

Solution If the resulting axial stress $\sigma = P/A$ does not exceed the proportional limit of the material, we may apply Hooke's law and write $\sigma = E\varepsilon$. Also, the axial strain is defined by $\varepsilon = \delta/L$.

a. The preceding expressions can be combined to yield the *axial deformation*,

$$\delta = \frac{PL}{AE} \tag{2.30}$$

where the product AE is known as the *axial rigidity* of the bar.

b. The *change in diameter* equals the product of transverse or lateral strain and diameter: $\varepsilon_t d = -\nu\varepsilon d$. Thus,

$$\Delta d = -\frac{\nu P d}{AE} \tag{2.31a}$$

c. The *change in volume*, substituting $V_o = AL$ and $\varepsilon_x = P/AE$ into Eq. (2.29), is

$$\Delta V = \frac{PL}{E}(1 - 2\nu) \tag{2.31b}$$

d. For $A = (\pi/4)\,(25^2) = 490.9(10^{-6})$ m², the axial stress σ in the bar is obtained from

$$\sigma = \frac{P}{A} = \frac{60(10^3)}{490.9(10^{-6})} = 122.2 \text{ MPa}$$

which is well below the yield strength of 260 MPa. Thus, introducing the given data into the preceding equations, we have

$$\delta = \frac{60(10^3)3(10^3)}{490.9(70)10^3} = 5.24 \text{ mm}$$

$$\Delta d = -\frac{0.3(60 \times 10^3)25}{490.9(70)10^3} = -0.0131 \text{ mm}$$

$$\Delta V = \frac{60(10^3)3(10^3)}{70(10^3)}(1 - 2 \times 0.3) = 1029 \text{ mm}^3$$

Comment A positive sign indicates an increase in length and volume; the negative sign means that the diameter has decreased.

2.10 GENERALIZED HOOKE'S LAW

For a three-dimensional state of stress, each of the six stress components is expressed as a linear function of six components of strain within the linear elastic range, and vice versa. We thus express the *generalized Hooke's law* for any homogeneous elastic material as follows:

$$
\begin{Bmatrix} \sigma_x \\ \sigma_y \\ \sigma_z \\ \tau_{xy} \\ \tau_{yz} \\ \tau_{xz} \end{Bmatrix} = \begin{bmatrix} c_{11} & c_{12} & c_{13} & c_{14} & c_{15} & c_{16} \\ c_{21} & c_{22} & c_{23} & c_{24} & c_{25} & c_{26} \\ c_{31} & c_{32} & c_{33} & c_{34} & c_{35} & c_{36} \\ c_{41} & c_{42} & c_{43} & c_{44} & c_{45} & c_{46} \\ c_{51} & c_{52} & c_{53} & c_{54} & c_{55} & c_{56} \\ c_{61} & c_{62} & c_{63} & c_{64} & c_{65} & c_{66} \end{bmatrix} \begin{Bmatrix} \varepsilon_x \\ \varepsilon_y \\ \varepsilon_z \\ \gamma_{xy} \\ \gamma_{yz} \\ \gamma_{xz} \end{Bmatrix}
\tag{2.32}
$$

The coefficients $c_{ij}(i, j = 1, 2, 3 \ldots, 6)$ are the material-dependent *elastic constants*. A succinct representation of the preceding stress–strain relationships are given in the following form:

$$
\tau_{ij} = c_{mnij}\varepsilon_{mn}
\tag{2.33}
$$

which is valid in *all* coordinate systems. Thus, it follows that the c_{mnij}, requiring four subscripts for definition, are components of a tensor of rank 4. We note that, to avoid repetitive subscripts, the material constants $c_{1111}, c_{1122}, \ldots, c_{6666}$ are denoted $c_{11}, c_{12}, \ldots, c_{66}$, as indicated in Eqs. (2.32).

In a homogeneous body, each of the 36 constants c_{ij} has the same value at all points. A material without any planes of symmetry is fully *anisotropic*. Strain energy considerations can be used to show that for such materials, $c_{ij} = c_{ji}$; thus the number of independent material constants can be as large as 21 (see Sec. 2.13). In case of a general orthotropic material, the number of constants reduces to nine, as shown in Section 2.11. For a homogeneous isotropic material, the constants must be identical in *all* directions at any point. An isotropic material has every plane as a plane of symmetry. Next, it is observed that if the material is *isotropic*, the number of essential elastic constants reduces to two.

In the following derivation, we rely on certain experimental evidence: a normal stress (σ_x) creates no shear strain whatsoever, and a shear stress (τ_{xy}) creates only a shear strain (γ_{xy}). Also, according to the small deformation assumption, the principle of superposition applies under multiaxial stressing. Consider now a two-dimensional homogeneous isotropic rectangular element of unit thickness, subjected to a biaxial state of stress (Fig. 2.17). Were σ_x to act, not only would the direct strain σ_x/E but a y contraction would take place as well, $-\nu\sigma_x/E$. Application of σ_y alone would result in an x contraction $-\nu\sigma_y/E$ and a y strain σ_y/E. The simultaneous action of σ_x and σ_y, applying the principle of superposition, leads to the following strains:

$$
\varepsilon_x = \frac{\sigma_x}{E} - \nu\frac{\sigma_y}{E}, \qquad \varepsilon_y = \frac{\sigma_y}{E} - \nu\frac{\sigma_x}{E}
\tag{a}
$$

FIGURE 2.17. *Element deformations caused by biaxial stresses.*

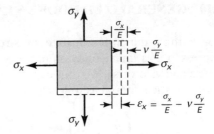

For pure shear (Fig. 1.3c), it is noted in Section 2.9 that, in the linearly elastic range, stress and strain are related by

$$\gamma_{xy} = \frac{\tau_{xy}}{G}$$

Similar analysis enables us to express the components ε_z, γ_{yz}, and γ_{xz} of strain in terms of stress and material properties. In the case of a three-dimensional state of stress, this procedure leads to the *generalized Hooke's law*, valid for an isotropic homogeneous material:

$$\varepsilon_x = \frac{1}{E}[\sigma_x - \nu(\sigma_y + \sigma_z)], \qquad \gamma_{xy} = \frac{\tau_{xy}}{G}$$

$$\varepsilon_y = \frac{1}{E}[\sigma_y - \nu(\sigma_x + \sigma_z)], \qquad \gamma_{yz} = \frac{\tau_{yz}}{G} \qquad \textbf{(2.34)}$$

$$\varepsilon_z = \frac{1}{E}[\sigma_z - \nu(\sigma_x + \sigma_y)], \qquad \gamma_{xz} = \frac{\tau_{xz}}{G}$$

It is demonstrated next that the elastic constants E, ν, and G are related, serving to reduce the number of independent constants in Eq. (2.34) to two. For this purpose, refer again to the element subjected to pure shear (Fig. 1.3c). In accordance with Section 1.9, a pure shearing stress τ_{xy} can be expressed in terms of the principal stresses acting on planes (in the x' and y' directions) making an angle of 45° with the shear planes: $\sigma_{x'} = \tau_{xy}$ and $\sigma_{y'} = -\tau_{xy}$. Then, applying Hooke's law, we find that

$$\varepsilon_{x'} = \frac{\sigma_{x'}}{E} - \nu\frac{\sigma_{y'}}{E} = \frac{\tau_{xy}}{E}(1 + \nu) \qquad \textbf{(b)}$$

On the other hand, because $\varepsilon_x = \varepsilon_y = 0$ for pure shear, Eq. (2.13a) yields, for $\theta = 45°$, $\varepsilon_{x'} = \gamma_{xy}/2$, or

$$\varepsilon_{x'} = \frac{\tau_{xy}}{2G} \qquad \textbf{(c)}$$

Equating the alternative relations for $\varepsilon_{x'}$ in Eqs. (b) and (c), we find that

$$G = \frac{E}{2(1 + \nu)} \qquad \textbf{(2.35)}$$

It is seen that, when any two of the constants v, E, and G are determined experimentally, the third may be found from Eq. (2.35). From Eq. (2.34) together with Eq. (2.35), we obtain the following stress–strain relationships:

$$\begin{aligned}
\sigma_x &= 2G\varepsilon_x + \lambda e, & \tau_{xy} &= G\gamma_{xy} \\
\sigma_y &= 2G\varepsilon_y + \lambda e, & \tau_{yz} &= G\gamma_{yz} \\
\sigma_z &= 2G\varepsilon_z + \lambda e, & \tau_{xz} &= G\gamma_{xz}
\end{aligned} \tag{2.36}$$

Here

$$e = \varepsilon_x + \varepsilon_y + \varepsilon_z = \frac{1 - 2v}{E}(\sigma_x + \sigma_y + \sigma_z) \tag{2.37}$$

and

$$\lambda = \frac{vE}{(1 + v)(1 - 2v)} \tag{2.38}$$

The shear modulus G and the quantity λ are referred to as the Lamé constants. Following a procedure similar to that used for axial stress in Section 2.9, it can be shown that Eq. (2.37) represents the unit volume change or *dilatation* of an element in *triaxial stress*.

The bulk modulus of elasticity is another important constant. The physical significance of this quantity is observed by considering, for example, the case of a cubic element subjected to hydrostatic pressure p. Because the stress field is described by $\sigma_x = \sigma_y = \sigma_z = -p$ and $\tau_{xy} = \tau_{yz} = \tau_{xz} = 0$, Eq. (2.37) reduces to $e = -3(1 - 2v)p/E$. The foregoing may be written in the form

$$K = -\frac{p}{e} = \frac{E}{3(1 - 2v)} \tag{2.39}$$

where K is the modulus of volumetric expansion or *bulk modulus of elasticity*. It is seen that the unit volume contraction is proportional to the pressure and inversely proportional to K. Equation (2.39) also indicates that for incompressible materials, for which $e = 0$, Poisson's ratio is 1/2. For all common materials, however, $v < 1/2$, since they demonstrate some change in volume, $e \neq 0$. Table D.1 lists average mechanical properties for a number of common materials. The relationships connecting the elastic constants introduced in this section are given by Eqs. (P2.51) in Prob. 2.51.

EXAMPLE 2.5   Volume Change of a Metal Block

Calculate the volumetric change of the metal block shown in Fig. 2.18 subjected to uniform pressure $p = 160$ MPa acting on all faces. Use $E = 210$ GPa and $v = 0.3$.

Solution   The bulk modulus of elasticity of the material, using Eq. (2.39), is

$$K = \frac{E}{3(1 - 2v)} = \frac{210(10^9)}{3(1 - 2 \times 0.3)} = 175 \text{ GPa}$$

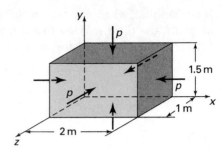

FIGURE 2.18. *Example 2.5. A parallelpiped under pressure.*

and the dilatation is

$$e = -\frac{p}{K} = -\frac{160}{175(10^3)} = -9.14 \times 10^{-4}$$

Since the initial volume of the block (Fig. 2.18) is $V_o = 2 \times 1.5 \times 1 = 3\,\mathrm{m}^3$, Eq. (2.29) yields

$$\Delta V = eV_o = (-9.14 \times 10^{-4})(3 \times 10^9) = -2.74 \times 10^6\,\mathrm{mm}^3$$

where a minus sign means that the block experiences a decrease in the volume, as expected intuitively.

2.11 HOOKE'S LAW FOR ORTHOTROPIC MATERIALS

A general orthotropic material has three *planes of symmetry* and three corresponding orthogonal axes called the *orthotropic axes*. Within each plane of symmetry, material properties may be different and independent of direction. A familiar example of such an orthotropic material is wood. Strength and stiffness of wood along its grain and in each of the two perpendicular directions vary. These properties are greater in a direction parallel to the fibers than in the transverse direction. A polymer reinforced by parallel glass or graphite fibers represents a typical orthotropic material with two axes of symmetry.

Materials such as corrugated and rolled metal sheet, reinforced concrete, various composites, gridwork, and particularly laminates can also be treated as orthotropic [Refs. 2.8 and 2.10]. We note that a gridwork consists of two systems of equally spaced parallel ribs (beams), mutually perpendicular and attached rigidly at the points of intersection. For an elastic orthotropic material, the elastic *coefficients c_{ij} remain invariant* at a point under a *rotation of 180°* about any of the orthotropic axes. In the following derivations, we shall assume that the directions of orthotropic axes are *parallel* to the directions of the x, y, and z coordinates.*

---

*For a detailed discussion of the elastic properties of various classes of materials, see, for example, Refs. 2.1 and 2.11.

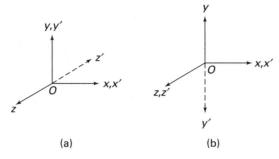

(a)   (b)

FIGURE 2.19. *Orthotropic coordinates x, y, z: (a) with Oz rotated 180°;*
(b) with Oy rotated 180°.

First, let the *xy plane* be *a plane material symmetry* and rotate Oz through 180°
(Fig. 2.19a). Accordingly, under the coordinate transformation $x: \ x', y: \ y'$, and
$z: \ -z'$, the direction cosines (see Table 1.1) are

$$l_1 = m_2 = 1, n_3 = -1, \qquad l_2 = l_3 = m_1 = m_3 = n_1 = n_2 = 0 \qquad \textbf{(a)}$$

Carrying Eqs. (a) into Eqs. (1.28) and (2.18), we have

$$\sigma_{x'} = \sigma_x, \ \sigma_{y'} = \sigma_y, \ \sigma_{z'} = \sigma_z, \qquad \tau_{x'y'} = \tau_{xy}, \tau_{y'z'} = -\tau_{yz}, \tau_{x'z'} = -\tau_{xz} \qquad \textbf{(b)}$$

and

$$\varepsilon_{x'} = \varepsilon_x, \varepsilon_{y'} = \varepsilon_y, \varepsilon_{z'} = \varepsilon_z, \qquad \gamma_{x'y'} = \gamma_{xy}, \gamma_{y'z'} = -\gamma_{yz}, \gamma_{x'z'} = -\gamma_{xz} \qquad \textbf{(c)}$$

Inasmuch as the c_{ij} remain the same, the first of Eqs. (2.28) may be written as

$$\sigma_{x'} = c_{11}\varepsilon_{x'} + c_{12}\varepsilon_{y'} + c_{13}\varepsilon_{z'} + c_{14}\gamma_{x'y'} + c_{15}\gamma_{y'z'} + c_{16}\gamma_{x'z'} \qquad \textbf{(d)}$$

Inserting Eqs. (b) and (c) into Eq. (d) gives

$$\sigma_x = \sigma_{x'} = c_{11}\varepsilon_x + c_{12}\varepsilon_y + c_{13}\varepsilon_z + c_{14}\gamma_{xy} - c_{15}\gamma_{yz} - c_{16}\gamma_{xz} \qquad \textbf{(e)}$$

Comparing Eq. (e) with the first of Eqs. (2.28) shows that $c_{15} = -c_{15}, c_{16} = -c_{16}$, im-
plying that $c_{15} = c_{16} = 0$. Likewise, considering $\sigma_{y'}, \sigma_{z'}, \tau_{x'y'}, \tau_{y'z'}, \tau_{x'z'}$, we obtain that
$c_{25} = c_{26} = c_{35} = c_{36} = c_{45} = c_{46} = 0$. The elastic coefficient matrix $[c_{ij}]$ is therefore

$$
\begin{bmatrix}
c_{11} & c_{12} & c_{13} & c_{14} & 0 & 0 \\
c_{12} & c_{22} & c_{23} & c_{24} & 0 & 0 \\
c_{13} & c_{23} & c_{33} & c_{34} & 0 & 0 \\
c_{14} & c_{24} & c_{34} & c_{44} & 0 & 0 \\
0 & 0 & 0 & 0 & c_{55} & c_{56} \\
0 & 0 & 0 & 0 & c_{56} & c_{66}
\end{bmatrix}
\qquad \textbf{(f)}
$$

Next, consider the *xz plane of elastic symmetry* by rotating Oy through a
180° angle (Fig. 2.19b). This gives $l_1 = n_3 = 1, m_2 = -1$ and $l_2 = l_3 = m_1 = m_3 = n_1 =
n_2 = 0$. Upon following a procedure similar to that in the preceding, we now
obtain $c_{14} = c_{24} = c_{34} = c_{56} = 0$. The matrix of elastic coefficients, Eqs. (f),

become then

$$
\begin{bmatrix}
c_{11} & c_{12} & c_{13} & 0 & 0 & 0 \\
c_{12} & c_{22} & c_{23} & 0 & 0 & 0 \\
c_{13} & c_{23} & c_{33} & 0 & 0 & 0 \\
0 & 0 & 0 & c_{44} & 0 & 0 \\
0 & 0 & 0 & 0 & c_{55} & 0 \\
0 & 0 & 0 & 0 & 0 & c_{66}
\end{bmatrix}
\tag{2.40}
$$

Finally, letting yz be the *plane of elastic symmetry* and repeating the foregoing procedure do *not* lead to further reduction in the number of nine elastic coefficients of Eqs. (2.40). Hence, the generalized Hooke's law for the most general orthotropic elastic material is given by

$$\sigma_x = c_{11}\varepsilon_x + c_{12}\varepsilon_y + c_{13}\varepsilon_z$$

$$\sigma_y = c_{12}\varepsilon_x + c_{22}\varepsilon_y + c_{23}\varepsilon_z$$

$$\sigma_z = c_{13}\varepsilon_x + c_{23}\varepsilon_y + c_{33}\varepsilon_z \tag{2.41}$$

$$\tau_{xy} = c_{44}\,\gamma_{xy}$$

$$\tau_{yz} = c_{55}\,\gamma_{yz}$$

$$\tau_{xz} = c_{66}\,\gamma_{xz}$$

The inversed form of Eqs. (2.41), referring to Eqs. (2.34), may be expressed in terms of orthotropic moduli and orthotropic Poisson's ratios as follows:

$$\varepsilon_x = \frac{1}{E_x}\sigma_x - \frac{\nu_{yx}}{E_y}\sigma_y - \frac{\nu_{zx}}{E_z}\sigma_z$$

$$\varepsilon_y = -\frac{\nu_{xy}}{E_x}\sigma_x - \frac{1}{E_y}\sigma_y - \frac{\nu_{yz}}{E_z}\sigma_z$$

$$\varepsilon_z = -\frac{\nu_{xz}}{E_x}\sigma_x - \frac{\nu_{yz}}{E_y}\sigma_y - \frac{1}{E_z}\sigma_z \tag{2.42}$$

$$\gamma_{xy} = \frac{1}{G_{xy}}\tau_{xy}$$

$$\gamma_{yz} = \frac{1}{G_{yz}}\tau_{yz}$$

$$\gamma_{xz} = \frac{1}{G_{xz}}\tau_{xz}$$

Because of symmetry in the material constants (Sec. 2.13), we have

$$\frac{\nu_{xy}}{E_x} = \frac{\nu_{yx}}{E_y}, \qquad \frac{\nu_{yz}}{E_y} = \frac{\nu_{zy}}{E_z}, \qquad \frac{\nu_{xz}}{E_x} = \frac{\nu_{zx}}{E_z}, \tag{2.43}$$

In the foregoing, the quantities E_x, E_y, E_z designate the orthotropic moduli of elasticity, and G_{xy}, G_{yz}, G_{xz} are the orthotropic shear moduli in the orthotropic coordinate system. Poisson's ratio ν_{xy} indicates the strain in the y direction produced by the stress in the x direction. The remaining Poisson's ratios $\nu_{xz}, \nu_{yz}, ..., \nu_{xz}$ are interpreted in a like manner. We observe from Eqs. (2.42) that, in an orthotropic material, there is *no* interaction between the normal stresses and the shearing strains.

2.12 MEASUREMENT OF STRAIN: STRAIN ROSETTE

A wide variety of mechanical, electrical, and optical systems has been developed for measuring the *normal strain* at a point on a *free surface* of a member [Ref. 2.12]. The method in widest use employs the bonded electric wire or foil resistance strain gages. The *bonded wire gage* consists of a grid of fine wire filament cemented between two sheets of treated paper or plastic backing (Fig. 2.20a). The backing insulates the grid from the metal surface on which it is to be bonded and functions as a carrier so that the filament may be conveniently handled. Generally, 0.025-mm diameter wire is used. The grid in the case of *bonded foil gages* is constructed of very thin metal foil (approximately 0.0025 mm) rather than wire. Because the filament cross section of a foil gage is rectangular, the ratio of surface area to cross-sectional area is higher than that of a round wire. This results in increased heat dissipation and improved adhesion between the grid and the backing material. Foil gages are readily manufactured in a variety of configurations. In general, the selection of a particular bonded gage depends on the specific service application.

The ratio of the unit change in the resistance of the gage to the unit change in length (strain) of the gage is called the *gage factor*. The metal of which the filament element is made is the principal factor determining the magnitude of this factor. *Constantan*, an alloy composed of 60% copper and 40% nickel, produces wire or foil gages with a gage factor of approximately 2.

The operation of the bonded strain gage is based on the change in electrical resistance of the filament that accompanies a change in the strain. Deformation of the surface on which the gage is bonded results in a deformation of the backing and

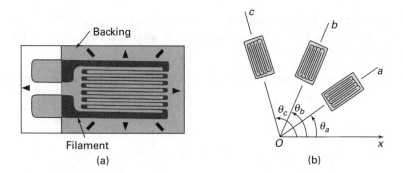

FIGURE 2.20. *(a) Strain gage (courtesy of Micro-Measurements Division, Vishay Intertechnology, Inc.) and (b) schematic representation of a strain rosette.*

the grid as well. Thus, with straining, a variation in the resistance of the grid will manifest itself as a change in the voltage across the grid. An electrical bridge circuit, attached to the gage by means of lead wires, is then used to translate electrical changes into strains. The *Wheatstone bridge*, one of the most accurate and convenient systems of this type employed, is capable of measuring strains as small as $1 \, \mu$.

Strain Rosette

Special combination gages are available for the measurement of the state of strain at a point on a surface simultaneously in three or more directions. It is usual to cluster together three gages to form a **strain rosette**, which may be cemented on the surface of a member. Generally, these consist of three gages whose axes are either 45° or 60° apart. Consider three strain gages located at angles θ_a, θ_b, and θ_c with respect to reference axis x (Fig. 2.20b). The a-, b-, and c-directed normal strains are, from Eq. (2.13a),

$$\varepsilon_a = \varepsilon_x \cos^2 \theta_a + \varepsilon_y \sin^2 \theta_a + \gamma_{xy} \sin \theta_a \cos \theta_a$$
$$\varepsilon_b = \varepsilon_x \cos^2 \theta_b + \varepsilon_y \sin^2 \theta_b + \gamma_{xy} \sin \theta_b \cos \theta_b \qquad \textbf{(2.44)}$$
$$\varepsilon_c = \varepsilon_x \cos^2 \theta_c + \varepsilon_y \sin^2 \theta_c + \gamma_{xy} \sin \theta_c \cos \theta_c$$

When the values of ε_a, ε_b, and ε_c are measured for given *gage orientations* θ_a, θ_b, and θ_c, the values of ε_x, ε_y, and γ_{xy} can be obtained by simultaneous solution of Eqs. (2.44). The arrangement of gages employed for this kind of measurement is called a *strain rosette*.

Once strain components are known, we can apply Eq. (3.11b) of Section 3.4 to determine the *out-of-plane principal strain* ε_z. The in-plane principal strains and their orientations may be obtained readily using Eqs. (2.15) and (2.16), as illustrated next, or Mohr's circle for strain.

EXAMPLE 2.6 Principal Strains on Surface of a Steel Frame

Strain rosette readings are made at a critical point on the free surface in a structural steel member. The 60° rosette contains three wire gages positioned at 0°, 60°, and 120° (Fig. 2.20b). The readings are

$$\varepsilon_a = 190 \, \mu, \qquad \varepsilon_b = 200 \, \mu, \qquad \varepsilon_c = -300 \, \mu \qquad \textbf{(a)}$$

Determine (a) the in-plane principal strains and stresses and their directions, and (b) the true maximum shearing strain. The material properties are $E = 200$ GPa and $\nu = 0.3$.

Solution For the situation described, Eq. (2.44) provides three simultaneous expressions:

$$\varepsilon_a = \varepsilon_x$$

$$\varepsilon_b = \tfrac{1}{4}\varepsilon_x + \tfrac{3}{4}\varepsilon_y + \frac{\sqrt{3}}{4}\gamma_{xy}$$

$$\varepsilon_c = \tfrac{1}{4}\varepsilon_x + \tfrac{3}{4}\varepsilon_y - \frac{\sqrt{3}}{4}\gamma_{xy}$$

From these,

$$\varepsilon_x = \varepsilon_a$$

$$\varepsilon_y = \tfrac{1}{3}[2(\varepsilon_b + \varepsilon_c) - \varepsilon_a]$$

$$\gamma_{xy} = \frac{2}{\sqrt{3}}(\varepsilon_b - \varepsilon_c)$$

(b)

Note that the relationships among ε_a, ε_b, and ε_c may be observed from a Mohr's circle construction corresponding to the state of strain $\varepsilon_x, \varepsilon_y$, and γ_{xy} at the point under consideration.

a. Upon substituting numerical values, we obtain $\varepsilon_x = 190\,\mu$, $\varepsilon_y = -130\,\mu$, and $\gamma_{xy} = 577\,\mu$. Then, from Eq. (2.16), the principal strains are

$$\varepsilon_{1,2} = \frac{190 - 130}{2}\,\mu \pm \mu\left[\left(\frac{190 + 130}{2}\right)^2 + \left(\frac{577}{2}\right)^2\right]^{1/2}$$

$$= 30\,\mu \pm 330\,\mu$$

or

$$\varepsilon_1 = 360\,\mu, \qquad \varepsilon_2 = -300\,\mu \qquad\qquad \textbf{(c)}$$

The maximum shear strain is found from

$$\gamma_{max} = \pm(\varepsilon_1 - \varepsilon_2) = \pm[360 - (-300)]\,\mu = \pm660\,\mu$$

The orientations of the principal axes are given by Eq. (2.15):

$$2\theta_p = \tan^{-1}\frac{577}{320} = 61° \quad \text{or} \quad \theta'_p = 30.5°, \qquad \theta''_p = 120.5° \quad \textbf{(d)}$$

When θ'_p is substituted into Eq. (2.14) together with Eq. (b), we obtain $360\,\mu$. Therefore, 30.5° and 120.5° are the respective directions of ε_1 and ε_2, measured from the horizontal axis in a counterclockwise direction. The principal stresses may now be found from the generalized Hooke's law. Thus, the first two equations of (2.34) for plane stress, letting $\sigma_z = 0$, $\sigma_x = \sigma_1$, and $\sigma_y = \sigma_2$, together with Eqs. (c), yield

$$\sigma_1 = \frac{200 \times 10^9}{1 - 0.09}[360 + 0.3(-300)](10^{-6}) = 59.34 \text{ MPa}$$

$$\sigma_2 = \frac{200 \times 10^3}{0.91}[-300 + 0.3(360)] = -42.2 \text{ MPa}$$

The directions of σ_1 and σ_2 are given by Eq. (d). From Eq. (2.36), the maximum shear stress is

$$\tau_{max} = \frac{200 \times 10^9}{2(1 + 0.3)}\,660 \times 10^{-6} = 50.77 \text{ MPa}$$

Note as a check that $(\sigma_1 - \sigma_2)/2$ yields the same result.

b. Applying Eq. (3.11b), the out-of-plane principal strain is

$$\varepsilon_z = -\frac{\nu}{1-\nu}(\varepsilon_x + \varepsilon_y) = -\frac{0.3}{1-0.3}(190 - 130)\,\mu = -26\,\mu$$

The principal strain ε_2 found in part (a) is redesignated $\varepsilon_3 = -300\,\mu$ so that algebraically $\varepsilon_2 > \varepsilon_3$, where $\varepsilon_2 = -26\,\mu$. The *true or absolute maximum shearing strain*

$$(\gamma_{\max})_t = \pm(\varepsilon_1 - \varepsilon_3) \tag{2.45}$$

is therefore $\pm 660\,\mu$, as already calculated in part (a).

Employing a procedure similar to that used in the preceding numerical example, it is possible to develop expressions relating three-element gage outputs of various rosettes to principal strains and stresses. Table 2.2 provides two typical cases: equations for the *rectangular rosette* ($\theta_a = 0°$, $\theta_b = 45°$, and $\theta_c = 90°$, Fig. 2.20b)

TABLE 2.2. *Strain Rosette Equations*

1. *Rectangular rosette or 45° strain rosette*

   Principal strains:

$$\varepsilon_{1,2} = \tfrac{1}{2}\left[\varepsilon_a + \varepsilon_c \pm \sqrt{(\varepsilon_a - \varepsilon_c)^2 + (2\varepsilon_b - \varepsilon_a - \varepsilon_c)^2}\right] \tag{2.46a}$$

   Principal stresses:

$$\sigma_{1,2} = \frac{E}{2}\left[\frac{\varepsilon_a + \varepsilon_c}{1-\nu} \pm \frac{1}{1+\nu}\sqrt{(\varepsilon_a - \varepsilon_c)^2 + (2\varepsilon_b - \varepsilon_a - \varepsilon_c)^2}\right] \tag{2.46b}$$

   Directions of principal planes:

$$\tan 2\theta_p = \frac{2\varepsilon_b - \varepsilon_a - \varepsilon_c}{\varepsilon_a - \varepsilon_c} \tag{2.46c}$$

2. *Delta rosette or 60° strain rosette*

   Principal strains:

$$\varepsilon_{1,2} = \tfrac{1}{3}\left[\varepsilon_a + \varepsilon_b + \varepsilon_c \pm \sqrt{2}\sqrt{(\varepsilon_a - \varepsilon_b)^2 + (\varepsilon_b - \varepsilon_c)^2 + (\varepsilon_c - \varepsilon_a)^2}\right] \tag{2.47a}$$

   Principal stresses:

$$\sigma_{1,2} = \frac{E}{3}\left[\frac{\varepsilon_a + \varepsilon_b + \varepsilon_c}{1-\nu} \pm \frac{\sqrt{2}}{1+\nu}\sqrt{(\varepsilon_a - \varepsilon_b)^2 + (\varepsilon_b - \varepsilon_c)^2 + (\varepsilon_c - \varepsilon_a)^2}\right] \tag{2.47b}$$

   Directions of principal planes:

$$\tan 2\theta_p = \frac{\sqrt{3}(\varepsilon_b - \varepsilon_c)}{2\varepsilon_a - \varepsilon_b - \varepsilon_c} \tag{2.47c}$$

and the *delta rosette* ($\theta_a = 0°$, $\theta_b = 60°$, and $\theta_c = 120°$, Fig. 2.20b). Experimental stress analysis is facilitated by this kind of compilation.

2.13 STRAIN ENERGY

The work done by external forces in causing deformation is stored within the body in the form of *strain energy*. In an ideal elastic process, no dissipation of energy takes place, and all the stored energy is recoverable upon unloading. The concept of elastic strain energy, introduced in this section, is useful as applied to the solution of problems involving both static and dynamic loads. It is particularly significant for predicting failure in members under combined loading.

Strain Energy Density for Normal and Shear Stresses

We begin our analysis by considering a rectangular prism of dimensions dx, dy, dz subjected to *uniaxial tension*. The front view of the prism is represented in Fig. 2.21a. If the stress is applied very slowly, as is generally the case in this text, it is reasonable to assume that equilibrium is maintained at all times. In evaluating the work done by stresses σ_x on either side of the element, it is noted that each stress acts through a different displacement. Clearly, the work done by oppositely directed forces ($\sigma_x \, dy \, dz$) through positive displacement (u) cancel one another. The *net* work done on the element by force ($\sigma_x \, dy \, dz$) is therefore

$$dW = dU = \int_0^{\varepsilon_x} \sigma_x \, d\left(\frac{\partial u}{\partial x} \, dx\right) dy \, dz = \int_0^{\varepsilon_x} \sigma_x \, d\varepsilon_x (dx \, dy \, dz)$$

where $\partial u / \partial x = \varepsilon_x$. Note that dW is the work done on $dx \, dy \, dz$, and dU is the corresponding increase in strain energy. Designating the *strain energy per unit volume* (strain energy density) as U_o, for a linearly elastic material we have

$$U_o = \int_0^{\varepsilon_x} \sigma_x \, d\varepsilon_x = \int_0^{\varepsilon_x} E\varepsilon_x \, d\varepsilon_x \qquad \text{(a)}$$

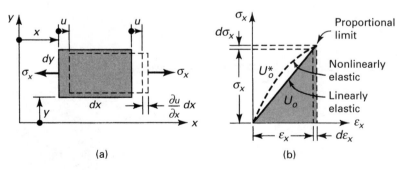

(a)   (b)

FIGURE 2.21.   *(a) Displacement under uniaxial stress; (b) work done by uniaxial stress.*

After integration, Eq. (a) yields

$$U_o = \tfrac{1}{2}E\varepsilon_x^2 = \tfrac{1}{2}\sigma_x\varepsilon_x = \frac{1}{2E}\sigma_x^2 \qquad (2.48)$$

This quantity represents the shaded area in Fig. 2.21b. The area above the stress–strain curve, termed the complementary energy density, may be determined from

$$U_o^* = \int_0^{\sigma_x} \varepsilon_x\, d\sigma_x \qquad (2.49)$$

For a linearly elastic material, $U_o = U_o^*$, but for a nonlinearly elastic material, U_o and U_o^* will differ, as seen in the figure. The unit of strain energy density in SI units is the joules per cubic meter (J/m^3), or pascals; in U.S. Customary Units, it is expressed in inch-pounds per cubic inch (in. $\cdot$ lb/in.^{3}), or pounds per square inch (psi).

When the material is stressed to the proportional limit, the strain energy density is referred to as the *modulus of resilience*. It is equal to the area under the straight-line portion of the stress–strain diagram (Fig. 2.10a) and represents a measure of the material's ability to store or absorb energy without permanent deformation. Similarly, the area under an entire stress–strain diagram provides a measure of a material's ability to absorb energy up to the point of fracture; it is called the *modulus of toughness*. The greater the total area under a stress–strain diagram, the tougher the material.

In the case in which σ_x, σ_y, and σ_z act simultaneously, the total work done by these normal stresses is simply the sum of expressions similar to Eq. (2.48) for each direction. This is because an x-directed stress does no work in the y or z directions. The total strain energy per volume is thus

$$U_o = \tfrac{1}{2}(\sigma_x\varepsilon_x + \sigma_y\varepsilon_y + \sigma_z\varepsilon_z) \qquad \textbf{(b)}$$

The elastic strain energy associated with *shear deformation* is now analyzed by considering an element of thickness dz subject only to shearing stresses τ_{xy} (Fig. 2.22). From the figure, we note that shearing force $\tau_{xy}\, dxdz$ causes a displacement of $\gamma_{xy}\, dy$. The strain energy due to shear is $\tfrac{1}{2}(\tau_{xy}\, dxdz)(\gamma_{xy}\, dy)$, where the factor $\tfrac{1}{2}$

FIGURE 2.22. *Deformation due to pure shear.*

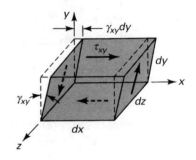

arises because the stress varies linearly with strain from zero to its final value, as before. The strain energy density is therefore

$$U_o = \tfrac{1}{2}\tau_{xy}\gamma_{xy} = \frac{1}{2G}\tau_{xy}^2 = \tfrac{1}{2}G\gamma_{xy}^2 \qquad (2.50)$$

Because the work done by τ_{xy} accompanying perpendicular strains γ_{yz} and γ_{xz} is zero, the total strain energy density attributable to shear alone is found by superposition of three terms identical in form with Eq. (2.50):

$$U_o = \tfrac{1}{2}(\tau_{xy}\gamma_{xy} + \tau_{yz}\gamma_{yz} + \tau_{xz}\gamma_{xz}) \qquad \text{(c)}$$

Strain Energy Density for Three-Dimensional Stresses

Given a *general state of stress*, the strain energy density is found by adding Eqs. (b) and (c):

$$U_o = \tfrac{1}{2}(\sigma_x\varepsilon_x + \sigma_y\varepsilon_y + \sigma_z\varepsilon_z + \tau_{xy}\gamma_{xy} + \tau_{yz}\gamma_{yz} + \tau_{xz}\gamma_{xz}) \qquad (2.51)$$

Introducing Hooke's law into Eq. (2.51) leads to the following form involving only stresses and elastic constants:

$$U_o = \frac{1}{2E}(\sigma_x^2 + \sigma_y^2 + \sigma_z^2) - \frac{\nu}{E}(\sigma_x\sigma_y + \sigma_y\sigma_z + \sigma_x\sigma_z)$$
$$+ \frac{1}{2G}(\tau_{xy}^2 + \tau_{yz}^2 + \tau_{xz}^2) \qquad (2.52)$$

An alternative form of Eq. (2.51), written in terms of strains, is

$$U_o = \tfrac{1}{2}[\lambda e^2 + 2G(\varepsilon_x^2 + \varepsilon_y^2 + \varepsilon_z^2) + G(\gamma_{xy}^2 + \gamma_{yz}^2 + \gamma_{xz}^2)] \qquad (2.53)$$

The quantities λ and e are defined by Eqs. (2.38) and (2.37), respectively.
It is interesting to observe that we have the relationships

$$\frac{\partial U_o(\tau)}{\partial \tau_{ij}} = \varepsilon_{ij}, \qquad \frac{\partial U_o(\varepsilon)}{\partial \varepsilon_{ij}} = \tau_{ij}, \qquad i, j = x, y, z \qquad (2.54)$$

Here $U_o(\tau)$ and $U_o(\varepsilon)$ designate the strain energy densities expressed in terms of stress and strain, respectively [Eqs. (2.52) and (2.53)]. Derivatives of this type will be discussed again in connection with energy methods in Chapter 10. We note that Eqs. (2.54) and (2.32) give

$$\frac{\partial U_o}{\partial \varepsilon_x} = \sigma_x = c_{11}\varepsilon_x + c_{12}\varepsilon_y + c_{13}\varepsilon_z + c_{14}\gamma_{xy} + c_{15}\gamma_{yz} + c_{16}\gamma_{xz}$$

$$\cdots\cdots\cdots\cdots\cdots\cdots\cdots\cdots\cdots\cdots\cdots\cdots\cdots\cdots\cdots\cdots \qquad (2.55)$$
$$\cdots\cdots\cdots\cdots\cdots\cdots\cdots\cdots\cdots\cdots\cdots\cdots\cdots\cdots\cdots\cdots$$

$$\frac{\partial U_o}{\partial \gamma_{xz}} = \tau_{xz} = c_{61}\varepsilon_x + c_{62}\varepsilon_y + c_{63}\varepsilon_z + c_{64}\gamma_{xy} + c_{65}\gamma_{yz} + c_{66}\gamma_{xz}$$

Differentiations of these equations as indicated result in

$$\frac{\partial^2 U_o}{\partial \varepsilon_x \, \partial \varepsilon_y} = c_{12} = c_{21}, \qquad \frac{\partial^2 U_o}{\partial \varepsilon_x \, \partial \varepsilon_z} = c_{13} = c_{31}, \ldots,$$

$$\frac{\partial^2 U_o}{\partial \gamma_{xz} \, \partial \gamma_{xy}} = c_{46} = c_{64}, \qquad \frac{\partial^2 U_o}{\partial \gamma_{xz} \, \partial \gamma_{yz}} = c_{56} = c_{65}, \ldots, \tag{2.56}$$

We are led to conclude from these results that $c_{ij} = c_{ji}(i, j = 1, 2, \ldots, 6)$. Because of this symmetry of elastic constants, there can be at most $[(36 - 6)/2] + 6 = 21$ independent elastic constants for an anisotropic elastic body.

2.14 STRAIN ENERGY IN COMMON STRUCTURAL MEMBERS

To determine the *elastic strain energy* stored within an entire body, the elastic energy density is integrated over the original or undeformed volume V. Therefore,

$$U = \int_V U_o \, dV = \int \int \int U_o \, dx \, dy \, dz \tag{2.57}$$

The foregoing shows that the energy-absorbing capacity of a body (that is, the failure resistance), which is critical when loads are dynamic in character, is a function of material *volume*. This contrasts with the resistance to failure under static loading, which depends on the cross-sectional area or the section modulus.

Equation (2.57) permits the strain energy to be readily evaluated for a number of commonly encountered geometries and loadings. Note especially that the strain energy is a nonlinear (quadratic) function of load or deformation. The principle of superposition is thus *not* valid for the strain energy. That is, the effects of several forces (or moments) on strain energy are not simply additive, as demonstrated in Example 2.7. Some special cases of Eq. (2.57) follow.

Strain Energy for Axially Loaded Bars

The normal stress at any given transverse section through a nonprismatic bar subjected to an axial force P is $\sigma_x = P/A$, where A represents the cross-sectional area (Fig. 2.23). Substituting this and Eq. (2.48) into Eq. (2.57) and setting $dV = A \, dx$, we have

$$U = \int_V \frac{\sigma_x^2}{2E} \, dV = \int_0^L \frac{P^2}{2AE} \, dx \tag{2.58}$$

FIGURE 2.23. *Nonprismatic bar with varying axial loading.*

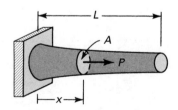

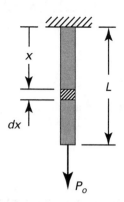

FIGURE 2.24. *Example 2.7. A prismatic bar loaded by its weight and load P_o.*

When a *prismatic bar* is subjected at its ends to equal and opposite forces of magnitude P, the foregoing becomes

$$U = \frac{P^2 L}{2AE} \tag{2.59}$$

where L is the length of the bar.

EXAMPLE 2.7   Strain Energy in a Bar under Combined Loading

A prismatic bar suspended from one end carries, in addition to its own weight, an axial load P_o (Fig. 2.24). Determine the strain energy U stored in the bar.

Solution The axial force P acting on the shaded element indicated is expressed

$$P = \gamma A(L - x) + P_o \tag{a}$$

where γ is the specific weight of the material and A, the cross-sectional area of the bar. Inserting Eq. (a) into Eq. (2.58), we have

$$U = \int_0^L \frac{[\gamma A(L - x) + P_o]^2}{2\,AE}\,dx = \frac{\gamma^2 A L^3}{6\,E} + \frac{\gamma P_o L^2}{2\,E} + \frac{P_o^2 L}{2\,AE} \tag{2.60}$$

The first and the third terms on the right side represent the strain energy of the bar subjected to its own weight and the strain energy of a bar supporting only axial force P_o, respectively. The presence of the middle term indicates that the strain energy produced by the two loads acting simultaneously is not simply equal to the sum of the strain energies associated with the loads acting separately.

Strain Energy of Circular Bars in Torsion

Consider a circular bar of varying cross section and varying torque along its axis (Fig. 2.23, with double-headed torque vector T replacing force vector P). The state of stress is pure shear. The torsion formula (Table 1.1) for an arbitrary distance ρ from the centroid of the cross section results in $\tau = T\rho/J$. The strain energy density,

Eq. (2.50), becomes then $U_o = T^2\rho^2/2J^2G$. When this is introduced into Eq. (2.57), we obtain

$$U = \int_0^L \frac{T^2}{2J^2G} \left(\int \rho^2 \, dA \right) dx \qquad \textbf{(b)}$$

where $dV = dA\,dx$; dA represents the cross-sectional area of an element. By definition, the term in parentheses is the polar moment of inertia J of the cross-sectional area. The strain energy is therefore

$$U = \int_0^L \frac{T^2}{2JG} \, dx \qquad \textbf{(2.61)}$$

In the case of a *prismatic shaft* subjected at its ends to equal and opposite torques T, Eq. (2.61) yields

$$U = \frac{T^2 L}{2JG} \qquad \textbf{(2.62)}$$

where L is the length of the bar.

Strain Energy for Beams in Bending

For the case of a beam in *pure bending*, the flexure formula gives us the axial normal stress $\sigma_x = -My/I$ (see Table 1.1). From Eq. (2.48), the strain energy density is $U_o = M^2 y^2/2EI^2$. Upon substituting this into Eq. (2.57) and noting that $M^2/2EI^2$ is a function of x alone, we have

$$U = \int_0^L \frac{M^2}{2EI^2} \left(\int y^2 \, dA \right) dx \qquad \textbf{(c)}$$

Here, as before, $dV = dA\,dx$, and dA represents an element of the cross-sectional area. Recalling that the integral in parentheses defines the moment of inertia I of the cross-sectional area about the neutral axis, the strain energy is expressed as

$$U = \int_0^L \frac{M^2}{2EI} dx \qquad \textbf{(2.63)}$$

where integration along beam length L gives the required quantity.

2.15 COMPONENTS OF STRAIN ENERGY

A new perspective on strain energy may be gained by viewing the general state of stress (Fig. 2.25a) in terms of the superposition shown in Fig. 2.25. The state of stress in Fig. 2.25b, represented by

$$\begin{bmatrix} \sigma_m & 0 & 0 \\ 0 & \sigma_m & 0 \\ 0 & 0 & \sigma_m \end{bmatrix} \qquad \textbf{(a)}$$

results in volume change without distortion and is termed the *dilatational* stress tensor. Here $\sigma_m = \frac{1}{3}(\sigma_x + \sigma_y + \sigma_z)$ is the mean stress defined by Eq. (1.44). Associated with σ_m is the *mean strain*, $\varepsilon_m = \frac{1}{3}(\varepsilon_x + \varepsilon_y + \varepsilon_z)$. The sum of the normal strains accompanying the application of the dilatational stress tensor is the dilatation $e = \varepsilon_x + \varepsilon_y + \varepsilon_z$, representing a change in volume only. Thus, the dilatational strain energy absorbed per unit volume is given by

$$U_{ov} = \tfrac{3}{2}\sigma_m \varepsilon_m = \frac{\sigma_m^2}{2K} = \frac{1}{18K}(\sigma_x + \sigma_y + \sigma_z)^2 \tag{2.64}$$

where K is defined by Eq. (2.39).

The state of stress in Fig. 2.25c, represented by

$$\begin{bmatrix} \sigma_x - \sigma_m & \tau_{xy} & \tau_{xz} \\ \tau_{xy} & \sigma_y - \sigma_m & \tau_{yz} \\ \tau_{xz} & \tau_{yz} & \sigma_z - \sigma_m \end{bmatrix} \tag{b}$$

is called the *deviator* or *distortional* stress tensor. This produces deviator strains or distortion without change in volume because the sum of the normal strains is $(\varepsilon_x - \varepsilon_m) + (\varepsilon_y - \varepsilon_m) + (\varepsilon_z - \varepsilon_m) = 0$. The distortional energy per unit volume, U_{od}, associated with the deviator stress tensor is attributable to the change of shape of the unit volume, while the volume remains constant. Since U_{ov} and U_{od} are the only components of the strain energy, we have $U_o = U_{ov} + U_{od}$. By subtracting Eq. (2.64) from Eq. (2.52), the distortional energy is readily found to be

$$U_{od} = \frac{3}{4G}\tau_{oct}^2 \tag{2.65}$$

This is the elastic strain energy absorbed by the unit volume as a result of its change in shape (distortion). In the preceding, the octahedral shearing stress τ_{oct} is given by

$$\tau_{oct} = \tfrac{1}{3}[(\sigma_x - \sigma_y)^2 + (\sigma_y - \sigma_z)^2 + (\sigma_z - \sigma_x)^2 + 6(\tau_{xy}^2 + \tau_{xz}^2 + \tau_{yz}^2)]^{1/2} \tag{2.66}$$

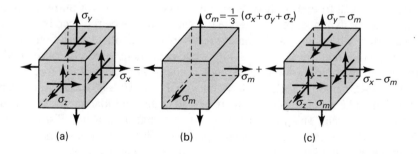

FIGURE 2.25.   *Resolution of (a) state of stress into (b) dilatational stresses and (c) distortional stresses.*

The planes where the τ_{oct} acts are shown in Fig. 1.24 of Section 1.14. The strain energy of distortion plays an important role in the theory of failure of a ductile metal under any condition of stress. This is discussed further in Chapter 4. The stresses and strains associated with both components of the strain energy are also very useful in describing the plastic deformation (Chap. 12).

EXAMPLE 2.8   Strain Energy Components in a Tensile Bar

A mild steel bar of uniform cross section A is subjected to an axial tensile load P. Derive an expression for the strain energy density, its components, and the total strain energy stored in the bar. Let $\nu = 0.25$.

Solution   The state of stress at any point in the bar is axial tension, $\tau_{xy} = \tau_{xz} = \tau_{yz} = \sigma_y = \sigma_z = 0$, $\sigma_x = \sigma = P/A$ (Fig. 2.25a). We therefore have the stresses associated with volume change $\sigma_m = \sigma/3$ and shape change $\sigma_x - \sigma_m = 2\sigma/3$, $\sigma_y - \sigma_m = \sigma_z - \sigma_m = -\sigma/3$ (Fig. 2.25b, c). The strain energy densities for the state of stress in cases a, b, and c are found, respectively, as follows:

$$U_o = \frac{\sigma^2}{2E}$$

$$U_{ov} = \frac{(1 - 2\nu)\sigma^2}{6E} = \frac{\sigma^2}{12E} \tag{c}$$

$$U_{od} = \frac{(1 + \nu)\sigma^2}{3E} = \frac{5\sigma^2}{12E}$$

Observe from these expressions that $U_o = U_{ov} + U_{od}$ and that $5U_{ov} = U_{od}$. Thus, we see that in changing the shape of a unit volume element under uniaxial stressing, five times more energy is absorbed than in changing the volume.

2.16 SAINT-VENANT'S PRINCIPLE

The reader will recall from a study of Newtonian mechanics that, for purposes of analyzing the statics or dynamics of a body, one force system may be replaced by an equivalent force system whose force and moment resultants are identical. It is often added in discussing this point that the force resultants, while equivalent, need not cause an identical distribution of strain, owing to difference in the *arrangement* of the forces. *Saint-Venant's principle*, named for Barré de Saint-Venant (1797–1886), a famous French mathematician and elastician, permits the use of an equivalent loading for the calculation of stress and strain. This principle or *rule* states that if an actual distribution of forces is replaced by a statically equivalent system, the distribution of stress and strain throughout the body is altered only near the regions of load application.*

The contribution of Saint-Venant's principle to the solution of engineering problems is very important, for it often frees the analyst of the burden of prescribing

---

*See, for example, Refs. 2.13 and 2.14.

the boundary conditions very precisely when it is difficult to do so. Furthermore, where a certain solution is predicated on a particular boundary loading, the solution can serve equally for another type of statically equivalent boundary loading, not quite the same as the first. That is, when an analytical solution calls for a certain distribution of stress on a boundary (such as σ_x in Sec. 5.5), we need not discard the solution merely because the boundary distribution is not quite the same as that required by the solution. The value of existing solutions is thus greatly extended.

Saint-Venant's principle is confirmed in Fig. 2.26, which shows the stress distribution, obtained using the methods of the theory of elasticity, across three sections of a rectangular elastic plate of width b subjected to a concentrated load [Ref. 2.15]. The *average stress* σ_{avg} as given by Eq. (1.10) is also sketched in the diagrams. From these, note that the maximum stress σ_{max} greatly exceeds the average stress near the point of application of the load and diminishes as we move along the vertical center axis of the plate away from an end. At a distance equal to the width of the plate, the *stress is nearly uniform*. With rare exceptions, this rule applies to members made of *linearly elastic* materials.

The foregoing observation also holds true for most stress concentrations and practically any type of loading. Thus, the basic formulas of the mechanics of materials give the stress in a member with high accuracy, provided that the cross section in question is at *least a distance b (or h) away* from any concentrated load or discontinuity of shape. Here, *b (or h)* denotes the largest lateral dimension of a member. We note that within this distance the stresses depend on the details of loading, boundary conditions, and geometry of the stress concentrations, as is seen in Chapter 3.

Consider, for example, the substitution of a uniform distribution of stress at the ends of a tensile test specimen for the actual irregular distribution that results from end clamping. If we require the stress in a region away from the ends, the stress variation at the ends need not be of concern, since it does not lead to significant variation in the region of interest. As a further example, according to Saint-Venant's principle, the complex distribution of force supplied by the wall to a cantilever beam (Fig. 2.27a) may be replaced by vertical and horizontal forces and a

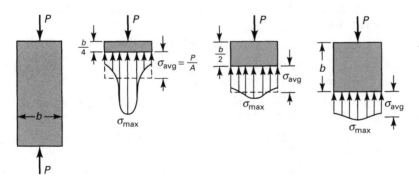

FIGURE 2.26.   *Stress distribution due to a concentrated load in a rectangular elastic plate, confirming the Saint-Venant's principle.*

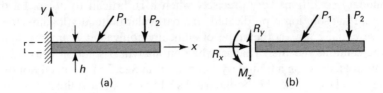

FIGURE 2.27.   *Cantilever beam illustrating use of Saint-Venant's principle: (a) actual support; (b) statical equivalent.*

moment (Fig. 2.27b) for purposes of determining the stresses acting at a distance from the wall equal to or greater than the depth h of the beam.

REFERENCES

2.1. SOKOLNIKOFF, I. S., *Mathematical Theory of Elasticity*, 2nd ed., Krieger, Melbourne, Fla., 1986.

2.2. CHOU, P. C., and PAGANO, N. J., *Elasticity*, Dover, New York, 1992.

2.3. FLUGGE, W., ed., *Handbook of Engineering Mechanics*, McGraw-Hill, New York, 1962, Chap. 2.

2.4. BORESI, A. P., and SCHMIDT, R. J., *Advanced Mechanics of Materials*, 6th ed., Wiley, Hoboken, N.J., 2003, Chap. 2.

2.5. AVALLONE, E. A., and BAUMEISTER III, T., eds., *Mark's Standard Handbook for Mechanical Engineers,* 10th ed., McGraw-Hill, New York, 1997.

2.6. American Society of Testing and Materials, *Annual Book of ASTM*, Philadelphia, Pa, 2009.

2.7. UGURAL, A. C., *Mechanical Design: An Integrated Approach*, McGraw-Hill, New York, 2004.

2.8. FAUPEL, J. H., and FISHER, F. E., *Engineering Design*, 2nd ed., Wiley, Hoboken, N.J., 1981.

2.9. UGURAL, A. C., *Mechanics of Materials*, Wiley, Hoboken, N.J., 2008.

2.10. UGURAL, A. C., *Stresses in Beams, Plates, and Shells*, CRC Press, Taylor & Francis Group, Boca Raton, Fla., 2010.

2.11. SHAMES, I. H., and COZZARELLI, F. A., *Elastic and Inelastic Stress Analysis*, Prentice Hall, Englewood Cliffs, N.J., 1992.

2.12. HETENYI, M., ed., *Handbook of Experimental Stress Analysis*, 2nd ed., Wiley, Hoboken, N.J., 1987.

2.13. STERNBERG, E., On St. Venant's principle. *Quart. Appl. Math.* 11:393, 1954.

2.14. STERNBERG, E., and KOTTER, T., The wedge under a concentrated couple, *J. Appl., Mech.* 25-575-581, 1958.

2.15. TIMOSHENKO, S. P., and GOODIER, J. N., *Theory of Elasticity*, 3rd ed., McGraw Hill, New York, 1970, p. 60.

PROBLEMS

Sections 2.1 through 2.8

2.1. Determine whether the following strain fields are possible in a continuous material:

$$\text{(a)} \quad \begin{bmatrix} c(x^2 + y^2) & cxy \\ cxy & y^2 \end{bmatrix}, \qquad \text{(b)} \quad \begin{bmatrix} cz(x^2 + y^2) & cxyz \\ cxyz & y^2z \end{bmatrix}$$

Here c is a small constant, and it is assumed that $\varepsilon_z = \gamma_{xz} = \gamma_{yz} = 0$.

2.2. Rectangle $ABCD$ is scribed on the surface of a member prior to loading (Fig. P2.2). Following the application of the load, the displacement field is expressed by

$$u = c(2x + y^2), \qquad v = c(x^2 - 3y^2)$$

where $c = 10^{-4}$. Subsequent to the loading, determine (a) the length of the sides AB and AD; (b) the change in the angle between sides AB and AD; and (c) the coordinates of point A.

2.3. A displacement field in a body is given by

$$u = c(x^2 + 10)$$
$$v = 2cyz$$
$$w = c(-xy + z^2)$$

where $c = 10^{-4}$. Determine the state of strain on an element positioned at $(0, 2, 1)$.

2.4. The displacement field and strain distribution in a member have the form

$$u = a_0 x^2 y^2 + a_1 xy^2 + a_2 x^2 y$$
$$v = b_0 x^2 y + b_1 xy$$
$$\gamma_{xy} = c_0 x^2 y + c_1 xy + c_2 x^2 + c_3 y^2$$

What relationships connecting the constants (a's, b's, and c's) make the foregoing expressions possible?

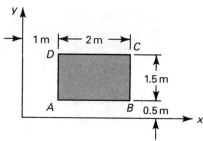

FIGURE P2.2.

2.5. Redo Prob. 2.4 for the following system of strains:

$$\varepsilon_x = a_0 + a_1 y^2 + y^4$$
$$\varepsilon_y = b_0 + b_1 x^2 + x^4$$
$$\gamma_{xy} = c_0 + c_1 xy(x^2 + y^2 + c_2)$$

2.6. A *rigid* horizontal bar BE is supported as illustrated in Fig. P2.6. After the load P is applied, point E moves 3 mm down and the axial strain in the bar AB is $-500\ \mu$. Calculate the axial strain in the bar CD.

2.7. Find the normal strain in the members AB and CB of the pin-connected plane structure (Fig. P2.7) if point B is moved leftward 2.5 mm. Assume that axial deformation is uniform throughout the length of each member.

2.8. The thin, triangular plate ABC is uniformly deformed into a shape ABC as depicted by the dashed lines in Fig. P2.8. Determine (a) the plane stress components $\varepsilon_x, \varepsilon_y,$ and γ_{xy}; (b) the shearing strain between edges AC and BC.

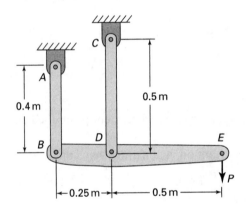

FIGURE P2.6.

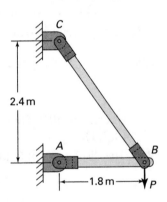

FIGURE P2.7.

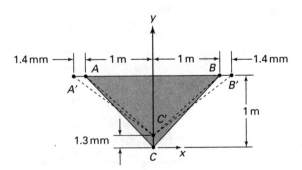

FIGURE P2.8.

2.9. A 100-mm by 150-mm rectangular plate $QABC$ is deformed into the shape shown by the dashed lines in Fig. P2.9. All dimensions shown in the figure are in millimeters. Determine at point Q (a) the strain components ε_x, ε_y, γ_{xy}, and (b) the principal strains and the direction of the principal axes.

2.10. Calculate the principal strains and their orientations at point A of the deformed rectangular plate shown in Fig. 2.5 of Example 2.1.

2.11. As a result of loading, the rectangle shown in Fig. P2.11 deforms into a parallelogram in which sides QA and BC shorten 0.003 mm and rotate $500\,\mu$ radian counterclockwise while sides AB and QC elongate 0.004 mm and rotate $1000\,\mu$ radian clockwise. Determine the principal strains and the direction of the principal axes at point Q. Take $a = 20$ mm and $b = 12$ mm.

2.12. A thin rectangular plate $a = 20$ mm $\times$ $b = 12$ mm (Fig. P2.11) is acted upon by a stress distribution resulting in the uniform strains $\varepsilon_x = 300\,\mu$, $\varepsilon_y = 500\,\mu$, and $\gamma_{xy} = 200\,\mu$. Determine the changes in length of diagonals QB and AC.

2.13. Redo Prob. 2.12 using the following information: $a = 30$ mm, $b = 15$ mm, $\varepsilon_x = 400\,\mu$, $\varepsilon_y = 200\,\mu$, and $\gamma_{xy} = -300\,\mu$.

2.14. A thin plate is subjected to uniform shear stress $\tau_o = 70$ MPa (Fig. P1.42 of Chap. 1). Let $E = 200$ GPa, $\nu = 0.3$, $AB = 40$ mm, and $BC = 60$ mm. Determine (a) the change in length AB, (b) the changes in length of diagonals AC and BD, and (c) the principal strains and their directions at point A.

2.15. The principal strains at a point are $\varepsilon_1 = 400\,\mu$ and $\varepsilon_2 = 200\,\mu$. Determine (a) the maximum shear strain and the direction along which it occurs and (b) the strains in the directions at $\theta = 30°$ from the principal axes. Solve the problem by using the formulas developed and check the results by employing Mohr's circle.

2.16. A 3-m by 2-m rectangular thin plate is deformed by the movement of point B to B' as shown by the dashed lines in Fig. P2.16. Assuming a displacement field of the form $u = c_1 xy$ and $v = c_2 xy$, wherein c_1 and c_2 are constants, determine (a) expressions for displacements u and v; (b) strain

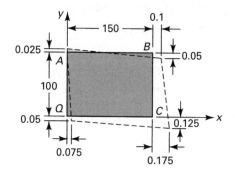

FIGURE P2.9.

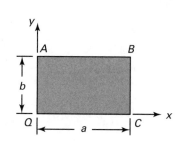

FIGURE P2.11.

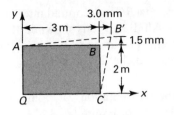

FIGURE P2.16.

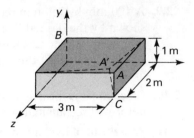

FIGURE P2.19.

components ε_x, ε_y, and γ_{xy} at point B; and (c) the normal strain $\varepsilon_{x'}$ in the direction of line QB. Verify that the strain field is possible.

2.17. If the strains at a point are $\varepsilon_x = -900\,\mu$, $\varepsilon_y = -300\,\mu$, and $\gamma_{xy} = 900\,\mu$, what are the principal strains, and in what direction do they occur? Use Mohr's circle of strain.

2.18. Solve Prob. 2.17 for $\varepsilon_x = 300\,\mu$, $\varepsilon_y = 900\,\mu$, and $\gamma_{xy} = -900\,\mu$.

2.19. A 3-m by 1-m by 2-m parallelepiped is deformed by movement of corner A to A' (2.9995, 1.0003, 1.9994), as shown in Fig. P2.19. Assuming that the displacement field is given by Eqs. (2.22), calculate at point A (a) the strain components and ascertain whether this strain distribution is possible; (b) the normal strain in the direction of line AB; and (c) the shearing strain for the perpendicular lines AB and AC.

2.20. Redo Prob. 2.19 for the case in which corner point A is moved to A' (3.0006, 0.9997, 1.9996).

2.21. At a point in a stressed body, the strains, related to the coordinate set xyz, are given by

$$\begin{bmatrix} 200 & 300 & 200 \\ 300 & -100 & 500 \\ 200 & 500 & -400 \end{bmatrix}\mu$$

Determine (a) the strain invariants; (b) the normal strain in the x' direction, which is directed at an angle $\theta = 30°$ from the x axis; (c) the principal strains ε_1, ε_2, and ε_3; and (d) the maximum shear strain.

2.22. Solve Prob. 2.21 for a state of strain given by

$$\begin{bmatrix} 400 & 100 & 0 \\ 100 & 0 & -200 \\ 0 & -200 & 600 \end{bmatrix}\mu$$

2.23. The following describes the state of strain at a point in a structural member:

$$\begin{bmatrix} 450 & 600 & 900 \\ 600 & 300 & 750 \\ 900 & 750 & 150 \end{bmatrix}\mu$$

Determine the magnitudes and directions of the principal strains.

2.24. A tensile test is performed on a 12-mm-diameter aluminum alloy specimen ($\nu = 0.33$) using a 50-mm gage length. When an axial tensile load reaches a value of 16 kN, the gage length has increased by 0.10 mm. Determine (a) the modulus of elasticity; (b) the decrease Δd in diameter and the dilatation e of the bar.

2.25. A 12-mm-diameter specimen is subjected to tensile loading. The increase in length resulting from a load of 9 kN is 0.025 mm for an original length L_o of 75 mm. What are the true and conventional strains and stresses? Calculate the modulus of elasticity.

Sections 2.9 through 2.12

2.26. Find the smallest diameter and shortest length that may be selected for a steel control rod of a machine under an axial load of 5 kN if the rod must stretch 2 mm. Use $E = 210$ GPa and $\sigma_{all} = 160$ MPa.

2.27. A 40-mm diameter bar ABC is composed of an aluminum part AB and a steel part BC (Fig. P2.27). After axial force P is applied, a strain gage attached to the steel measures normal strain at the longitudinal direction as $\varepsilon_s = 600 \ \mu$. Determine (a) the magnitude of the applied force P; (b) the total elongation of the bar if each material behaves elastically. Take $E_a = 70$ GPa and $E_s = 210$ GPa.

2.28. A 5-m-long truss member is made of two 40-mm-diameter steel bars. For a tensile load of 600 kN, find (a) the change in the length of the member; (b) the change in the diameter of the member. Use $E = 200$ GPa, $\sigma_{yp} = 250$ MPa, and $\nu = 0.3$.

2.29. The cast-iron pipe of length L, outer diameter D, and thickness t is subjected to an axial compressive P. Calculate (a) the change in length ΔL; (b) the change in outer diameter D; (c) the change in thickness Δt. *Given:* $D = 100$ mm, $t = 10$ mm, $L = 0.4$ m, $P = 150$ kN, $E = 70$ GPa, and $\nu = 0.3$.

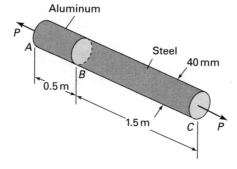

FIGURE P2.27.

2.30. A typical vibration isolation device consists of rubber cylinder of diameter d compressed inside of a steel cylinder by a force Q applied to a steel rod, as schematically depicted in Fig. P2.30. Find, in terms of d, Q, and Poisson's ratio ν for the rubber, as needed: (a) an expression for the lateral pressure p between the rubber and the steel cylinder; (b) the lateral pressure p between the rubber and the steel cylinder for $d = 50$ mm, $\nu = 0.3$, and $Q = 5$ kN. *Assumptions:* 1. Friction between the rubber and steel can be neglected; 2. Steel cylinder and rod are rigid.

2.31. A solid sphere of diameter d experiences a uniform pressure of p. Determine (a) the decrease in circumference of the sphere; (b) the decrease in volume of the sphere ΔV. *Given:* $d = 250$ mm, $p = 160$ MPa, $E = 70$ GPa, and $\nu = 0.3$. Note: Volume of a sphere is $V_0 = \frac{4}{3}\pi r^3$, where $r = d/2$.

2.32. The state of strain at a point in a *thin* steel plate is $\varepsilon_x = 500\,\mu$, $\varepsilon_y = -100\,\mu$, and $\gamma_{xy} = 150\,\mu$. Determine (a) the in-plane principal strains and the maximum in-plane shear strain; (b) true maximum shearing strain $\nu = 0.3$. Sketch the results found in part (a) on properly oriented deformed elements.

2.33. An element at a point on a loaded frame has strains as follows: $\varepsilon_x = 480\,\mu$, $\varepsilon_y = 800\,\mu$, and $\gamma_{xy} = -1120\,\mu$. Determine (a) the principal strains; (b) the maximum shear strain; (c) the true maximum shearing strain.

2.34. A metallic plate of width w and thickness t is subjected to a uniform axial force P as shown in Fig. P2.34. Two strain gages placed at point A measure the strains $\varepsilon_{x'}$ and at $30°$ and $60°$, respectively, to the axis of the plate. Calculate (a) the normal strains ε_x and ε_y; (b) the normal strains $\varepsilon_{x'}$ and $\varepsilon_{y'}$; (c) the shearing strain $\gamma_{x'y'}$. *Given:* $w = 60$ mm, $t = 6$ mm, $E = 200$ GPa, $\nu = 0.3$, and $P = 25$ kN.

2.35. During the static test of a panel, a $45°$ rosette reads the following normal strains on the free surface (Fig. P2.35): $\varepsilon_a = -800\,\mu$, $\varepsilon_b = -1000\,\mu$, and $\varepsilon_c = 400\,\mu$. Find the principal strains and show the results on a properly oriented deformed element.

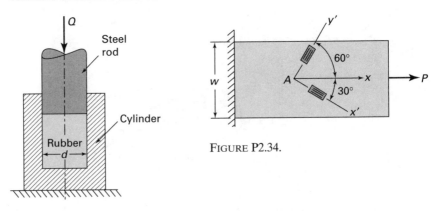

FIGURE P2.30.

FIGURE P2.34.

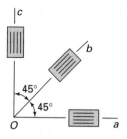

FIGURE P2.35.

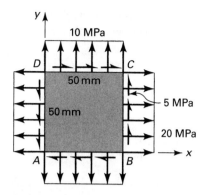

FIGURE P2.36.

2.36. A 50-mm-square plate is subjected to the stresses shown in Fig. P2.36. What deformation is experienced by diagonal BD? Express the solution, in terms of E, for $\nu = 0.3$ using two approaches: (a) determine the components of strain along the x and y directions and then employ the equations governing the transformation of strain; (b) determine the stress on planes perpendicular and parallel to BD and then employ the generalized Hooke's law.

2.37. A uniform pressure p acts over the entire straight edge of a large plate (Fig. P2.37). What are normal stress components σ_x and σ_z acting on a volumetric element at some distance from the loading in terms of Poisson's ratio ν and p, as required? Assume that $\varepsilon_x = \varepsilon_z = 0$ and $\sigma_y = -p$ everywhere.

2.38. A 45° rosette is used to measure strain at a critical point on the surface of a loaded beam. The readings are $\varepsilon_a = -100\,\mu$, $\varepsilon_b = 50\,\mu$, $\varepsilon_c = 100\,\mu$ for $\theta_a = 0°$, $\theta_b = 45°$, and $\theta_c = 90°$ (Fig. 2.20b). Calculate the principal strains and stresses and their directions. Use $E = 200$ GPa and $\nu = 0.3$.

2.39. The following state of strain has been measured at a point on the surface of a crane hook: $\varepsilon_a = 1000\,\mu$, $\varepsilon_b = -250\,\mu$, and $\varepsilon_c = 200\,\mu$ for $\theta_a = -15°$, $\theta_b = 30°$, and $\theta_c = 75°$ (Fig. 2.20b). Determine strain components ε_x, ε_y, and γ_{xy}.

2.40. The strains measured at a point on the surface of a machine element are $\varepsilon_a = 400\,\mu$, $\varepsilon_b = 300\,\mu$, and $\varepsilon_c = -50\,\mu$ for $\theta_a = 30°$, $\theta_b = -30°$, and $\theta_c = 90°$ (Fig. 2.20b). Calculate (a) the in-plane maximum shearing strain, and (b) the true maximum shearing strain. Use $\nu = \frac{1}{3}$.

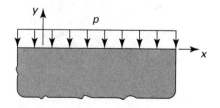

FIGURE P2.37.

2.41. For a given steel, $E = 200$ GPa and $G = 80$ GPa. If the state of strain at a point within this material is given by

$$\begin{bmatrix} 200 & 100 & 0 \\ 100 & 300 & 400 \\ 0 & 400 & 0 \end{bmatrix} \mu$$

ascertain the corresponding components of the stress tensor.

2.42. For a material with $G = 80$ GPa and $E = 200$ GPa, determine the strain tensor for a state of stress given by

$$\begin{bmatrix} 20 & -4 & 5 \\ -4 & 0 & 10 \\ 5 & 10 & 15 \end{bmatrix} \text{MPa}$$

2.43. The distribution of stress in an aluminum machine component is given (in megapascals) by

$$\sigma_x = y + 2z^2, \qquad \tau_{xy} = 3z^2$$
$$\sigma_y = x + z, \qquad \tau_{yz} = x^2$$
$$\sigma_z = 3x + y, \qquad \tau_{xz} = 2y^2$$

Calculate the state of strain of a point positioned at (1, 2, 4). Use $E = 70$ GPa and $\nu = 0.3$.

2.44. The distribution of stress in a structural member is given (in megapascals) by Eqs. (d) of Example 1.2 of Chapter 1. Calculate the strains at the specified point $Q(\frac{3}{4}, \frac{1}{4}, \frac{1}{2})$ for $E = 200$ GPa and $\nu = 0.25$.

2.45. An aluminum alloy plate ($E = 70$ GPa, $\nu = 1/3$) of dimensions $a = 300$ mm, $b = 400$ mm, and thickness $t = 10$ mm is subjected to biaxial stresses as shown in Fig. P2.45. Calculate the change in (a) the length AB; (b) the volume of the plate.

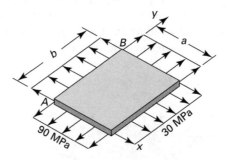

FIGURE P2.45.

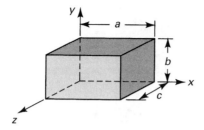

FIGURE P2.46.

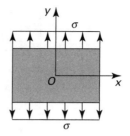

FIGURE P2.49.

2.46. The steel rectangular parallelepiped ($E = 200$ GPa and $\nu = 0.3$) shown in Fig. P2.46 has dimensions $a = 250$ mm, $b = 200$ mm, and $c = 150$ mm. It is subjected to triaxial stresses $\sigma_x = -60$ MPa, $\sigma_y = -50$ MPa, and $\sigma_z = -40$ MPa acting on the x, y, and z faces, respectively. Determine (a) the changes Δa, Δb, and Δc in the dimensions of the block, and (b) the change ΔV in the volume.

2.47. Redo Prob. 2.46 for an aluminum block ($E = 70$ GPa and $\nu = \frac{1}{3}$) for which $a = 150$ mm, $b = 100$ mm, and $c = 75$ mm, subjected to stresses $\sigma_x = 70$ MPa, $\sigma_y = -30$ MPa, and $\sigma_z = -15$ MPa.

2.48. At a point in an elastic body, the principal strains ε_3, ε_2, ε_1 are in the ratio 3: 4: 5; the largest principal stress is $\sigma_1 = 140$ MPa. Determine the ratio σ_3: σ_2: σ_1 and the values of σ_2 and σ_3. Take $\nu = 0.3$ and $E = 200$ GPa.

2.49. A rectangular plate is subjected to uniform tensile stress σ along its upper and lower edges, as shown in Fig. P2.49. Determine the displacements u and v in terms of x, y, and material properties (E, ν): (a) using Eqs. (2.3) and the appropriate conditions at the origin; (b) by the mechanics of materials approach.

2.50. The stress field in an elastic body is given by

$$\begin{bmatrix} cy^2 & 0 \\ 0 & -cx^2 \end{bmatrix}$$

where c is a constant. Derive expressions for the displacement components $u(x, y)$ and $v(x, y)$ in the body.

2.51. Derive the following relations involving the elastic constants:

$$K = \lambda + \tfrac{2}{3}G = \frac{2G(1 + \nu)}{3(1 - 2\nu)} = \frac{E}{3(1 - 2\nu)}$$

$$G = \frac{\lambda(1 - 2\nu)}{2\nu} = \frac{3K(1 - 2\nu)}{2(1 + \nu)} = \frac{3KE}{9K - E}$$

$$E = \frac{G(3\lambda + 2G)}{\lambda + G} = 2G(1 + \nu) = 3K(1 - 2\nu) = \frac{9KG}{3K + G} \quad \textbf{(P2.51)}$$

$$\nu = \frac{\lambda}{2(\lambda + G)} = \frac{E}{2G} - 1 = \frac{3K - 2G}{2(3K + G)} = \frac{3K - E}{6K}$$

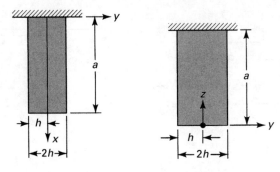

FIGURE P2.52.   FIGURE P2.53.

2.52. As shown in Fig. P2.52, a thin prismatical bar of specific weight γ and constant cross section hangs in the vertical plane. Under the effect of its own weight, the displacement field is described by

$$u = \frac{\gamma}{2E}(2xa - x^2 - vy^2), \qquad v = -\frac{v\gamma}{E}(a - x)y$$

The z displacement and stresses may be neglected. Find the strain and stress components in the bar. Check to see whether the boundary conditions [Eq. (1.48)] are satisfied by the stresses found.

2.53. A uniform bar of rectangular cross section $2h \times b$ and specific weight γ hangs in the vertical plane (Fig. P2.53). Its weight results in displacements

$$u = -\frac{v\gamma}{E}xz$$

$$v = -\frac{v\gamma}{E}yz$$

$$w = \frac{\gamma}{2E}[(z^2 - a^2) + v(x^2 + y^2)]$$

Demonstrate whether this solution satisfies the 15 equations of elasticity and the boundary conditions.

Sections 2.13 through 2.16

2.54. A bar of uniform cross-sectional area A, modulus of elasticity E, and length L is fixed at its right end and subjected to axial forces P_1 and P_2 at its free end. Verify that the total strain energy stored in the bar is given by

$$U = \frac{P_1^2 L}{2AE} + \frac{P_2^2 L}{2AE} + \frac{P_1 P_2 L}{AE} \qquad \text{(P2.54)}$$

Note that U is *not* the sum of the strain energies due to P_1 and P_2 acting separately. Find the components of the energy for $P_1 = P_2 = P$ and $v = 0.25$.

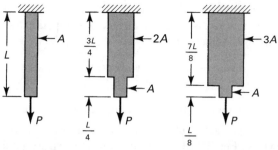

FIGURE P2.55.

2.55. Three bars of successively larger volume are to support the same load P (Fig. P2.55). Note that the first bar has a uniform cross-sectional area A over its length L. Neglecting stress concentrations, compare the strain energy stored in the three bars.

2.56. A solid bronze sphere ($E = 110$ GPa, $\nu = \frac{1}{3}, r = 150$ mm) is subjected to hydrostatic pressure p so that its volume is reduced by 0.5%. Determine (a) the pressure p, and (b) the strain energy U stored in the sphere. (Note: volume of a sphere $V = \frac{4}{3}\pi r^3$.)

2.57. Calculate the total strain energy U stored in the block described in Prob. 2.46.

2.58. A round bar is composed of three segments of the same material (Fig. P2.58). The diameter is d for the lengths BC and DE and nd for length CD, where n is the ratio of the two diameters. Neglecting the stress concentrations, verify that the strain energy of the bar when subjected to axial load P is

$$U = \frac{1 + 3n^2}{4n^2} \frac{P^2 L}{2AE} \qquad \textbf{(P2.58)}$$

where $A = \pi d^2/4$. Compare the result for $n = 1$ with those for $n = \frac{1}{2}$ and $n = 2$.

2.59. (a) Taking into account only the effect of normal stress, determine the strain energy of prismatic beam AB due to the axial force P and moment M_o acting simultaneously (Fig. P2.59). (b) Evaluate the strain energy for the case in which the beam is rectangular, 100-mm deep by 75-mm wide, $P = 8$ kN, $M_o = 2$ kN·m, $L = 1.2$ m, $a = 0.3$ m, $b = 0.9$ m, and $E = 70$ GPa.

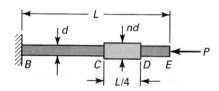

FIGURE P2.58.   FIGURE P2.59.

2.60. A stepped shaft is subjected to pure torsion, as shown in Fig. P2.60. Neglecting the stress concentrations, develop the following equation for energy stored in the shaft:

$$U = \frac{\pi G \phi^2}{32\,L}\left(\frac{d_1^4\,d_2^4}{d_1^4 + d_2^4}\right) \tag{P2.60}$$

Here ϕ is the angle of twist and G represents the modulus of rigidity.

2.61. (a) Determine the strain energy of a solid brass circular shaft ABC loaded as shown in Fig. P2.61, assuming that the stress concentrations may be omitted. (b) Calculate the strain energy for $T = 1.4$ kN·m, $a = 500$ mm, $d = 20$ mm, and $G = 42$ GPa.

2.62. Consider a simply supported rectangular beam of depth h, width b, and length L subjected to a uniform load of intensity p. Verify that the maximum strain energy density equals

$$U_o = \frac{45}{8}\frac{U}{V} \tag{P2.62}$$

in which U is the strain energy of the beam and V its volume.

2.63. Consider a beam with simple supports at B and C and an overhang AB (Fig. P2.63). What is the strain energy in the beam due to the load P?

2.64. A simply supported beam carries a concentrated force P and a moment M_o as shown in Fig. P2.64. How much strain energy is stored in the beam owing to the loads acting simultaneously?

2.65. Consider the state of stress given in Fig. 1.20b. Determine how many times more energy is absorbed in changing the shape than in changing the volume of a unit element. Let $E = 200$ GPa and $\nu = 0.3$.

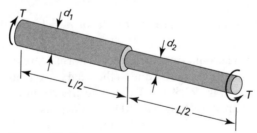

FIGURE P2.60.

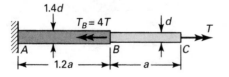

FIGURE P2.61.

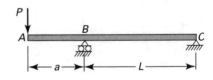

FIGURE P2.63.

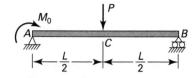

FIGURE P2.64.

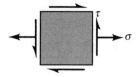

FIGURE P2.68.

2.66. The state of stress at a point is

$$\begin{bmatrix} 200 & 20 & 10 \\ 20 & -50 & 0 \\ 10 & 0 & 40 \end{bmatrix} \text{MPa}$$

Decompose this array into a set of dilatational stresses and a set of deviator stresses. Determine the values of *principal deviator stress*.

2.67. Calculate the strain energy per unit volume in changing the volume and in changing the shape of the material at any point on the surface of a steel shaft 120 mm in diameter subjected to torques of $20 \text{ kN} \cdot \text{m}$ and moments of $15 \text{ kN} \cdot \text{m}$ at its ends. Use $E = 200 \text{ GPa}$ and $\nu = 0.25$.

2.68. The state of stress at a point in a loaded member is represented in Fig. P2.68. Express the dilatational energy density and the distortional energy density in terms of the given stresses (σ, τ) at the point and the material properties (E, ν).

2.69. A circular prismatic cantilever is subjected to a torque T and an axial force P at its free end. The length of the bar is L, the radius is r, and the modulus of elasticity of the material is E. Determine the total strain energy stored in the bar and its components. Assume $\nu = \frac{1}{4}$ for the material.

CHAPTER 3

Problems in Elasticity

3.1 INTRODUCTION

As pointed out in Section 1.1, the approaches in widespread use for determining the influence of applied loads on elastic bodies are the *mechanics of materials* or *elementary theory* (also known as *technical theory*) and the *theory of elasticity*. Both must rely on the conditions of equilibrium and make use of a relationship between stress and strain that is usually considered to be associated with elastic materials. The essential difference between these methods lies in the extent to which the strain is described and in the types of simplifications employed.

The mechanics of materials approach uses an assumed deformation mode or strain distribution in the body as a whole and hence yields the *average* stress at a section under a given loading. Moreover, it usually treats separately each simple type of complex loading, for example, axial centric, bending, or torsion. Although of practical importance, the formulas of the mechanics of materials are best suited for relatively slender members and are derived on the basis of very restrictive conditions. On the other hand, the method of elasticity does not rely on a prescribed deformation mode and deals with the general equations to be satisfied by a body in equilibrium under any external force system.

The theory of elasticity is preferred when critical design constraints such as minimum weight, minimum cost, or high reliability dictate more exact treatment or when prior experience is limited and intuition does not serve adequately to supply the needed simplifications with any degree of assurance. If properly applied, the theory of elasticity should yield solutions more closely approximating the actual distribution of strain, stress, and displacement. Thus, elasticity theory provides a check on the limitations of the mechanics of materials solutions. We emphasize, however, that both techniques cited are approximations of nature, each of considerable value and each supplementing the other. The influences of material anisotropy, the extent to which boundary conditions depart from reality, and numerous other factors all contribute to error.

In this chapter, we present the *applied theory of elasticity*, emphasizing physical significance and employing engineering notations in Cartesian and polar coordinates. The main purpose here is to give the reader a clear and basic understanding of methods for solving typical problems in elasticity. The high localized stresses created by abrupt changes in cross section, as well as the high stresses produced by concentrated loads and reactions, are treated. Our analysis is carried out for isotropic and linearly elastic materials under ordinary and elevated temperatures. The later chapters include solutions by the theory of elasticity on bending of beams, noncircular torsion, and various axisymmetrically loaded structural and machine components, plates, and shells.

3.2 FUNDAMENTAL PRINCIPLES OF ANALYSIS

To ascertain the distribution of stress, strain, and displacement within an elastic body subject to a prescribed system of forces requires consideration of a number of conditions relating to certain physical laws, material properties, and geometry. These fundamental principles of analysis, also called the *three aspects of solid mechanics problems*, summarized in Section 1.3, are *conditions of equilibrium*, *material behavior* or stress–strain relations, and *conditions of compatibility*.

In addition, the stress, strain, and displacement fields must be such as to satisfy the *boundary conditions* for a particular problem. If the problem is *dynamic*, the equations of equilibrium become the more general conservation of momentum; conservation of energy may be a further requirement.

Three-Dimensional Problems

The conditions described, and stated mathematically in the previous chapters, are used to derive the equations of elasticity. In the case of a *three-dimensional* problem in elasticity, it is required that the following 15 quantities be ascertained: six stress components, six strain components, and three displacement components. These components must satisfy 15 governing equations throughout the body in addition to the boundary conditions: three equations of equilibrium, six stress–strain relations, and six strain–displacement relations. Note that the equations of compatibility are derived from the strain–displacement relations, which are already included in the preceding description. Thus, if the 15 expressions are satisfied, the equations of compatibility will also be satisfied. Three-dimensional problems in elasticity are often very complex. It may not always be possible to use the direct method of solution in treating the general equations and given boundary conditions. Only a useful indirect method of solution will be presented in Sections 6.4 and 6.5.

Two-Dimensional Problems

In many engineering applications, ample justification may be found for simplifying assumptions with respect to the state of strain and stress. Of special importance, because of the resulting decrease in complexity, are those reducing a three-dimensional problem to one involving only two dimensions. In this regard, we discuss throughout the text various plane strain and plane stress problems.

This chapter is subdivided into two parts. In Part A, derivations of the governing differential equations and various approaches for solution of two-dimensional problems in Cartesian and polar coordinates are considered. Part B treats stress concentrations in members whose cross sections manifest pronounced changes and cases of load application over small areas.

Part A—Formulation and Methods of Solution

3.3 PLANE STRAIN PROBLEMS

Consider a long prismatic member subject to lateral loading (for example, a cylinder under pressure), held between *fixed*, smooth, rigid planes (Fig. 3.1). Assume the external force to be functions of the x and y coordinates only. As a consequence, we expect all cross sections to experience identical deformation, including those sections near the ends. The frictionless nature of the end constraint permits x, y deformation but precludes z displacement; that is, $w = 0$ at $z = \pm L/2$. Considerations of symmetry dictate that w must also be zero at midspan. Symmetry arguments can again be used to infer that $w = 0$ at $\pm L/4$, and so on, until every cross section is taken into account. For the case described, the strain depends on x and y only:

$$\varepsilon_x = \frac{\partial u}{\partial x}, \qquad \varepsilon_y = \frac{\partial v}{\partial y}, \qquad \gamma_{xy} = \frac{\partial u}{\partial y} + \frac{\partial v}{\partial x} \tag{3.1}$$

$$\varepsilon_z = \frac{\partial w}{\partial z} = 0, \qquad \gamma_{xz} = \frac{\partial w}{\partial x} + \frac{\partial u}{\partial z} = 0, \qquad \gamma_{yz} = \frac{\partial w}{\partial y} + \frac{\partial v}{\partial z} = 0 \tag{3.2}$$

The latter expressions depend on $\partial u/\partial z$ and $\partial v/\partial z$ vanishing, since w and its derivatives are zero. A state of plane strain has thus been described wherein each point remains within its transverse plane, following application of the load. We next develop the equations governing the behavior of bodies under plane strain.

FIGURE 3.1. *Plane strain in a cylindrical body.*

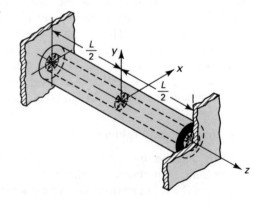

Substitution of $\varepsilon_z = \gamma_{yz} = \gamma_{xz} = 0$ into Eq. (2.36) provides the following *stress–strain relationships*:

$$
\begin{aligned}
\sigma_x &= 2\,G\varepsilon_x + \lambda(\varepsilon_x + \varepsilon_y) \\
\sigma_y &= 2\,G\varepsilon_y + \lambda(\varepsilon_x + \varepsilon_y) \\
\tau_{xy} &= G\gamma_{xy}
\end{aligned}
\tag{3.3}
$$

and

$$
\tau_{xz} = \tau_{yz} = 0, \qquad \sigma_z = \lambda(\varepsilon_x + \varepsilon_y) = \nu(\sigma_x + \sigma_y)
\tag{3.4}
$$

Because σ_z is not contained in the other governing expressions for plane strain, it is determined independently by applying Eq. (3.4). The strain–stress relations, Eqs. (2.34), for this case become

$$
\varepsilon_x = \frac{1 - \nu^2}{E}\left(\sigma_x - \frac{\nu}{1 - \nu}\sigma_y\right)
$$

$$
\varepsilon_y = \frac{1 - \nu^2}{E}\left(\sigma_y - \frac{\nu}{1 - \nu}\sigma_x\right)
\tag{3.5}
$$

$$
\gamma_{xy} = \frac{\tau_{xy}}{G}
$$

Inasmuch as these stress components are functions of x and y only, the first two equations of (1.14) yield the following *equations of equilibrium* of plane strain:

$$
\begin{aligned}
\frac{\partial\sigma_x}{\partial x} + \frac{\partial\tau_{xy}}{\partial y} + F_x &= 0 \\
\frac{\partial\sigma_y}{\partial y} + \frac{\partial\tau_{xy}}{\partial x} + F_y &= 0
\end{aligned}
\tag{3.6}
$$

The third equation of (1.14) is satisfied if $F_z = 0$. In the case of plane strain, therefore, no body force in the axial direction can exist.

A similar restriction is imposed on the surface forces. That is, plane strain will result in a prismatic body if the surface forces p_x and p_y are each functions of x and y and $p_z = 0$. On the lateral surface, $n = 0$ (Fig. 3.2). The *boundary conditions*, from the first two equations of (1.41), are thus given by

$$
\begin{aligned}
p_x &= \sigma_x l + \tau_{xy} m \\
p_y &= \tau_{xy} l + \sigma_y m
\end{aligned}
\tag{3.7}
$$

Clearly, the last equation of (1.48) is also satisfied.

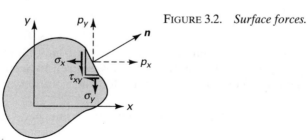

FIGURE 3.2.   *Surface forces.*

In the case of a plane strain problem, therefore, *eight quantities*, σ_x, σ_y, τ_{xy}, ε_x, ε_y, γ_{xy}, u, and v, must be determined so as to satisfy Eqs. (3.1), (3.3), and (3.6) and the boundary conditions (3.7). How eight governing equations, (3.1), (3.3), and (3.6), may be reduced to three is now discussed.

Three expressions for two-dimensional strain at a point [Eq. (3.1)] are functions of only *two* displacements, u and v, and therefore a compatibility relationship exists among the strains [Eq. (2.11)]:

$$\frac{\partial^2 \varepsilon_x}{\partial y^2} + \frac{\partial^2 \varepsilon_y}{\partial x^2} = \frac{\partial^2 \gamma_{xy}}{\partial x\, \partial y} \tag{3.8}$$

This equation must be satisfied for the strain components to be related to the displacements as in Eqs. (3.1). The condition as expressed by Eq. (3.8) may be transformed into one involving components of stress by substituting the strain–stress relations and employing the equations of equilibrium. Performing the operations indicated, using Eqs. (3.5) and (3.8), we have

$$\frac{\partial^2}{\partial y^2} [(1 - \nu)\sigma_x - \nu\sigma_y] + \frac{\partial^2}{\partial x^2} [(1 - \nu)\sigma_y - \nu\sigma_x] = 2 \frac{\partial^2 \tau_{xy}}{\partial x\, \partial y} \tag{a}$$

Next, the first and second equations of (3.6) are differentiated with respect to x and y, respectively, and added to yield

$$2\frac{\partial^2 \tau_{xy}}{\partial x\, \partial y} = -\left(\frac{\partial^2 \sigma_x}{\partial x^2} + \frac{\partial^2 \sigma_y}{\partial y^2} \right) - \left(\frac{\partial F_x}{\partial x} + \frac{\partial F_y}{\partial y} \right)$$

Finally, substitution of this into Eq. (a) results in

$$\left(\frac{\partial^2}{\partial x^2} + \frac{\partial^2}{\partial y^2} \right)(\sigma_x + \sigma_y) = -\frac{1}{1 - \nu}\left(\frac{\partial F_x}{\partial x} + \frac{\partial F_y}{\partial y} \right) \tag{3.9}$$

This is the *equation of compatibility* in terms of stress.

We now have three expressions, Eqs. (3.6) and (3.9), in terms of *three unknown quantities*: σ_x, σ_y, and τ_{xy}. This set of equations, together with the boundary conditions (3.7), is used in the solution of plane strain problems. For a given situation, after determining the stress, Eqs. (3.5) and (3.1) yield the strain and displacement, respectively. In Section 3.6, Eqs. (3.6) and (3.9) will further be reduced to one equation containing a single variable.

3.4 PLANE STRESS PROBLEMS

In many problems of practical importance, the stress condition is one of *plane stress*. The basic definition of this state of stress was given in Section 1.8. In this section we present the governing equations for the solution of plane stress problems.

To exemplify the case of plane stress, consider a thin plate, as in Fig. 3.3, wherein the loading is uniformly distributed over the thickness, parallel to the

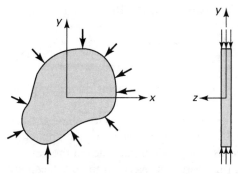

FIGURE 3.3. *Thin plate under plane stress.*

plane of the plate. This geometry contrasts with that of the long prism previously discussed, which is in a state of plane strain. To arrive at tentative conclusions with regard to the stress within the plate, consider the fact that σ_z, τ_{xz}, and τ_{yz} are zero on both faces of the plate. Because the plate is thin, the stress distribution may be very closely approximated by assuming that the foregoing is likewise true throughout the plate.

We shall, as a condition of the problem, take the body force $F_z = 0$ and F_x and F_y each to be functions of x and y only. As a consequence of the preceding, the stress is specified by

$$\sigma_x, \qquad \sigma_y, \qquad \tau_{xy}$$

$$\sigma_z = \tau_{xz} = \tau_{yz} = 0 \tag{a}$$

The nonzero stress components remain constant over the thickness of the plate and are functions of x and y only. This situation describes a state of plane stress. Equations (1.14) and (1.48), together with this combination of stress, again reduce to the forms found in Section 3.3. Thus, Eqs. (3.6) and (3.7) describe the equations of equilibrium and the boundary conditions in this case, as in the case of plane strain.

Substitution of Eq. (a) into Eq. (2.34) yields the following *stress–strain relations* for plane stress:

$$\varepsilon_x = \frac{1}{E}(\sigma_x - \nu\sigma_y)$$

$$\varepsilon_y = \frac{1}{E}(\sigma_y - \nu\sigma_x) \tag{3.10}$$

$$\gamma_{xy} = \frac{\tau_{xy}}{G}$$

and

$$\gamma_{xz} = \gamma_{yz} = 0, \qquad \varepsilon_z = -\frac{\nu}{E}(\sigma_x + \sigma_y) \tag{3.11a}$$

Solving for $\sigma_x + \sigma_y$ from the sum of the first two of Eqs. (3.10) and inserting the result into Eq. (3.11a), we obtain

$$\varepsilon_z = -\frac{\nu}{1-\nu}(\varepsilon_x + \varepsilon_y) \tag{3.11b}$$

Equations (3.11) define the *out-of-plane principal strain* in terms of the in-plane stresses (σ_x, σ_y) or strains $(\varepsilon_x, \varepsilon_y)$.

Because ε_z is not contained in the other governing expressions for plane stress, it can be obtained independently from Eqs. (3.11); then $\varepsilon_z = \partial w/\partial z$ may be applied to yield w. That is, only u and v are considered as independent variables in the governing equations. In the case of plane stress, therefore, the basic strain–displacement relations are again given by Eqs. (3.1). Exclusion from Eq. (2.4) of $\varepsilon_z = \partial w/\partial z$ makes the plane stress equations approximate, as is demonstrated in the section that follows.

The governing equations of plane stress will now be reduced, as in the case of plane strain, to three equations involving stress components only. Since Eqs. (3.1) apply to plane strain and plane stress, the compatibility condition represented by Eq. (3.8) applies in both cases. The latter expression may be written as follows, substituting strains from Eqs. (3.10) and employing Eqs. (3.6):

$$\left(\frac{\partial^2}{\partial x^2} + \frac{\partial^2}{\partial y^2}\right)(\sigma_x + \sigma_y) = -(1+\nu)\left(\frac{\partial F_x}{\partial x} + \frac{\partial F_y}{\partial y}\right) \tag{3.12}$$

This *equation of compatibility*, together with the equations of equilibrium, represents a useful form of the governing equations for problems of plane stress.

Stress–Strain Relations for Orthotropic Materials

Three-dimensional stress–strain relations for orthotropic materials in terms of orthotropic moduli of elasticity and orthotropic Poisson's ratios were developed in Section 2.11. Now consider an orthotropic member with orthotropic axes x, y, z subjected to state of stress relative to the xy plane. Thus, for the orthotropic material in state of plane stress, introducing Eqs. (a) into Eqs. (2.42), we obtain the *strain–stress relations*:

$$\varepsilon_x = \frac{\sigma_x}{E_x} - \frac{\nu_{yx}\sigma_y}{E_y}$$

$$\varepsilon_y = -\frac{\nu_{xy}\sigma_x}{E_x} + \frac{\sigma_y}{E_y} \tag{3.13}$$

$$\gamma_{xy} = \frac{\tau_{xy}}{G_{xy}}$$

and

$$\gamma_{yz} = \gamma_{xz} = 0, \quad \varepsilon_z = -\frac{\nu_{xz}}{E_x}\sigma_x - \frac{\nu_{yz}}{E_y}\sigma_y \tag{3.14}$$

It is recalled that E_x, E_y, denote the orthotropic moduli of elasticity, G_{xy} the orthotropic shear modulus of elasticity, and ν_{xy}, ν_{xz}, ν_{yz} are the orthotropic

Poisson ratios. Through the inversion of Eqs. (3.13), the *stress–strain relations* are found as

$$\sigma_x = \frac{E_x}{1 - \nu_{xy}\nu_{yx}}\left(\varepsilon_x + \nu_{yx}\,\varepsilon_y\right)$$

$$\sigma_y = \frac{E_y}{1 - \nu_{xy}\nu_{yx}}\left(\nu_{xy}\,\varepsilon_x + \varepsilon_y\right) \tag{3.15}$$

$$\tau_{xy} = G_{xy}\,\gamma_{xy}$$

Having the preceding equations, the plane stress orthotropic problems are treated similarly to plane stress problems for isotropic materials.

3.5 COMPARISON OF TWO-DIMENSIONAL ISOTROPIC PROBLEMS

To summarize the two-dimensional situations discussed, the equations of equilibrium [Eqs. (3.6)], together with those of compatibility [Eq. (3.9) for plane strain and Eq. (3.12) for plane stress] and the boundary conditions [Eqs. (3.7)], provide a system of equations sufficient for determination of the complete stress distribution. It can be shown that a solution satisfying all these equations is, for a given problem, unique [Ref. 3.1]. That is, it is the *only* solution to the problem.

   In the absence of body forces or in the case of constant body forces, the compatibility equations for plane strain and plane stress are the same. In these cases, the equations governing the distribution of stress do not contain the elastic constants. Given identical geometry and loading, a bar of steel and one of Lucite should thus display identical stress distributions. This characteristic is important in that any convenient isotropic material may be used to substitute for the actual material, as, for example, in *photoelastic* studies.

   It is of interest to note that by comparing Eqs. (3.5) with Eqs. (3.10) we can form Table 3.1, which facilitates the conversion of a plane stress solution into a plane strain solution, and vice versa. For instance, conditions of plane stress and plane strain prevail in a *narrow* beam and a *very wide* beam, respectively. Hence, in a result pertaining to a thin beam, EI would become $EI/(1 - \nu^2)$ for the case of a wide beam. The stiffness in the latter case is, for $\nu = 0.3$, about 10% greater owing to the prevention of sidewise displacement (Secs. 5.2 and 13.4).

TABLE 3.1.   *Conversion between Plane Stress and Plane Strain Solutions*

Solution	To Convert to:	E is Replaced by:	ν is Replaced by:
Plane stress	Plane strain	$\dfrac{E}{1 - \nu^2}$	$\dfrac{\nu}{1 - \nu}$
Plane strain	Plane stress	$\dfrac{1 + 2\nu}{(1 + \nu)^2}E$	$\dfrac{\nu}{1 + \nu}$

3.6 AIRY'S STRESS FUNCTION

The preceding sections demonstrated that the solution of two-dimensional problems in elasticity requires integration of the differential equations of equilibrium [Eqs. (3.6)], together with the compatibility equation [Eq. (3.9) or (3.12)] and the boundary conditions [Eqs. (3.7)]. In the event that the body forces F_x and F_y are negligible, these equations reduce to

$$\frac{\partial \sigma_x}{\partial x} + \frac{\partial \tau_{xy}}{\partial y} = 0, \qquad \frac{\partial \sigma_y}{\partial y} + \frac{\partial \tau_{xy}}{\partial x} = 0 \tag{a}$$

$$\left(\frac{\partial^2}{\partial x^2} + \frac{\partial^2}{\partial y^2} \right)(\sigma_x + \sigma_y) = 0 \tag{b}$$

together with the boundary conditions (3.7). The equations of equilibrium are identically satisfied by the *stress function*, $\Phi(x, y)$, introduced by G. B. Airy, related to the stresses as follows:

$$\sigma_x = \frac{\partial^2 \Phi}{\partial y^2}, \qquad \sigma_y = \frac{\partial^2 \Phi}{\partial x^2}, \qquad \tau_{xy} = -\frac{\partial^2 \Phi}{\partial x \, \partial y} \tag{3.16}$$

Substitution of (3.16) into the compatibility equation, Eq. (b), yields

$$\frac{\partial^4 \Phi}{\partial x^4} + 2 \frac{\partial^4 \Phi}{\partial x^2 \, \partial y^2} + \frac{\partial^4 \Phi}{\partial y^4} = \nabla^4 \Phi = 0 \tag{3.17}$$

What has been accomplished is the formulation of a two-dimensional problem in which body forces are absent, in such a way as to require the solution of a single *biharmonic equation*, which must of course satisfy the boundary conditions.

It should be noted that in the case of plane stress we have $\sigma_z = \tau_{xz} = \tau_{yz} = 0$ and σ_x, σ_y, and τ_{xy} independent of z. As a consequence, $\gamma_{xz} = \gamma_{yz} = 0$, and ε_x, ε_y, ε_z, and γ_{xy} are independent of z. In accordance with the foregoing, from Eq. (2.12), it is seen that in addition to Eq. (3.17), the following compatibility equations also hold:

$$\frac{\partial^2 \varepsilon_z}{\partial x^2} = 0, \qquad \frac{\partial^2 \varepsilon_z}{\partial y^2} = 0, \qquad \frac{\partial^2 \varepsilon_z}{\partial x \partial y} = 0 \tag{c}$$

Clearly, these additional conditions will not be satisfied in a case of plane stress by a solution of Eq. (3.17) alone. Therefore, such a solution of a plane stress problem has an approximate character. However, it can be shown that for thin plates the error introduced is negligibly small.

Generalized Plane Strain Problems

It is also important to note that if the ends of the cylinder shown in Fig. 3.1 are *free* to expand, we may assume the longitudinal strain ε_z to be a constant. Such a state may be called that of *generalized plane strain*. Therefore, we now have

$$\varepsilon_x = \frac{1 - \nu^2}{E}\left(\sigma_x - \frac{\nu}{1 - \nu}\sigma_y\right) - \nu\varepsilon_z$$

$$\varepsilon_y = \frac{1 - \nu^2}{E}\left(\sigma_y - \frac{\nu}{1 - \nu}\sigma_x\right) - \nu\varepsilon_z \qquad \textbf{(3.18)}$$

$$\gamma_{xy} = \frac{\tau_{xy}}{G}$$

and

$$\sigma_z = \nu(\sigma_x + \sigma_y) + E\varepsilon_z \qquad \textbf{(3.19)}$$

Introducing Eqs. (3.18) into Eq. (3.8) and simplifying, we again obtain Eq. (3.17) as the governing differential equation. Having determined σ_x and σ_y, the constant value of ε_z can be found from the condition that the resultant force in the z direction acting on the ends of the cylinder is zero. That is,

$$\iint \sigma_z \, dx \, dy = 0 \qquad \textbf{(d)}$$

where σ_z is given by Eq. (3.19). A detailed discussion of pressured thick-cylinders is given in Section 8.2.

3.7 SOLUTION OF ELASTICITY PROBLEMS

Unfortunately, solving directly the equations of elasticity derived may be a formidable task, and it is often advisable to attempt a solution by an *indirect method*: the *inverse* or *semi-inverse method*. The inverse method requires examination of the assumed solutions with a view toward finding one that will satisfy the governing equations and boundary conditions. The semi-inverse method requires the assumption of a partial solution formed by expressing stress, strain, displacement, or stress function in terms of known or undetermined coefficients. The governing equations are thus rendered more manageable.

It is important to note that the preceding assumptions, based on the mechanics of a particular problem, are subject to later verification. This is in contrast with the mechanics of materials approach, in which analytical verification does not occur. The applications of indirect and *direct* methods are found in examples to follow and in Chapters 5, 6, and 8.

A number of problems may be solved by using a linear combination of *polynomials* in x and y and undetermined coefficients of the stress function Φ. Clearly, an assumed polynomial form must satisfy the biharmonic equation and must be of second degree or higher in order to yield a nonzero stress solution of Eq. (3.16), as described in the following paragraphs. In general, finding the desirable polynomial form is laborious and requires a systematic approach [Refs. 3.2 and 3.3]. The *Fourier series*, indispensible in the analytical treatment of many problems in the field of applied mechanics, is also often employed (Secs. 10.10 and 13.7).

Another way to overcome the difficulty involved in the solution of Eq. (3.17) is to use the method of finite differences. Here the governing equation is replaced by series of finite difference equations (Sec. 7.3), which relate the stress function at stations that are removed from one another by finite distances. These equations, although not exact, frequently lead to solutions that are close to the exact solution. The results obtained are, however, applicable only to specific numerical problems.

Polynomial Solutions

An elementary approach to obtaining solutions of the biharmonic equation uses polynomial functions of various degree with their coefficients adjusted so that $\nabla^4 \Phi = 0$ is satisfied. A brief discussion of this procedure follows.

A polynomial of the *second* degree,

$$\Phi_2 = \frac{a_2}{2} x^2 + b_2 xy + \frac{c_2}{2} y^2 \tag{3.20}$$

satisfies Eq. (3.14). The associated stresses are

$$\sigma_x = c_2, \qquad \sigma_y = a_2, \qquad \tau_{xy} = -b_2$$

All three stress components are *constant* throughout the body. For a rectangular plate (Fig. 3.4a), it is apparent that the foregoing may be adapted to represent *simple tension* $(c_2 \neq 0)$, *double tension* $(c_2 \neq 0, a_2 \neq 0)$, or *pure shear* $(b_2 \neq 0)$.

A polynomial of the *third* degree

$$\Phi_3 = \frac{a_3}{6} x^3 + \frac{b_3}{2} x^2 y + \frac{c_3}{2} xy^2 + \frac{d_3}{6} y^3 \tag{3.21}$$

fulfills Eq. (3.17). It leads to stresses

$$\sigma_x = c_3 x + d_3 y, \qquad \sigma_y = a_3 x + b_3 y, \qquad \tau_{xy} = -b_3 x - c_3 y$$

For $a_3 = b_3 = c_3 = 0$, these expressions reduce to

$$\sigma_x = d_3 y, \qquad \sigma_y = \tau_{xy} = 0$$

representing the case of *pure bending* of the rectangular plate (Fig. 3.4b).

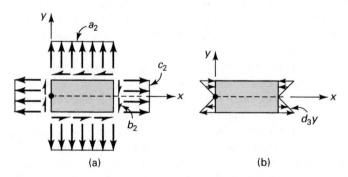

(a)   (b)

FIGURE 3.4. *Stress fields of (a) Eq. (3.20) and (b) Eq. (3.21).*

A polynomial of the *fourth* degree,

$$\Phi_4 = \frac{a_4}{12}x^4 + \frac{b_4}{6}x^3y + \frac{c_4}{2}x^2y^2 + \frac{d_4}{6}xy^3 + \frac{e_4}{12}y^4 \tag{3.22}$$

satisfies Eq. (3.17) if $e_4 = -(2c_4 + a_4)$. The corresponding stresses are

$$\sigma_x = c_4x^2 + d_4\,xy - (2c_4 + a_4)y^2$$
$$\sigma_y = a_4x^2 + b_4xy + c_4y^2$$
$$\tau_{xy} = -\frac{b_4}{2}x^2 - 2c_4xy - \frac{d_4}{2}y^2$$

A polynomial of the *fifth* degree

$$\Phi_5 = \frac{a_5}{20}x^5 + \frac{b_5}{12}x^4y + \frac{c_5}{6}x^3y^2 + \frac{d_5}{6}x^2y^3 + \frac{e_5}{12}xy^4 + \frac{f_5}{20}y^5 \tag{3.23}$$

fulfills Eq. (3.17) provided that

$$(3a_5 + 2c_5 + e_5)x + (b_5 + 2d_5 + 3f_5)y = 0$$

It follows that

$$e_5 = -3a_5 - 2c_5, \qquad b_5 = -2d_5 - 3f_5$$

The components of stress are then

$$\sigma_x = \frac{c_5}{3}x^3 + d_5x^2y - (3a_5 + 2c_5)xy^2 + f_5y^3$$

$$\sigma_y = a_5x^3 - (3f_5 + 2d_5)x^2y + c_5xy^2 + \frac{d_5}{3}y^3$$

$$\tau_{xy} = \tfrac{1}{3}(3f_5 + 2d_5)x^3 - c_5x^2y - d_5xy^2 + \tfrac{1}{3}(3d_5 + 2c_5)y^3$$

Problems of practical importance may be solved by combining functions (3.20) through (3.23), as required. With experience, the analyst begins to understand the types of stress distributions arising from a variety of polynomials.

EXAMPLE 3.1   Stress Distribution in a Cantilever Beam

A narrow cantilever of rectangular cross section is loaded by a concentrated force at its free end of such magnitude that the beam weight may be neglected (Fig. 3.5a). Determine the stress distribution in the beam.

Solution   The situation described may be regarded as a case of *plane stress* provided that the beam thickness t is small relative to the beam depth $2h$.

The following boundary conditions are consistent with the coordinate system in Fig. 3.5a:

$$(\tau_{xy})_{y=\pm h} = 0, \qquad (\sigma_y)_{y=\pm h} = 0 \tag{a}$$

These conditions simply express the fact that the top and bottom edges of the beam are not loaded. In addition to Eq. (a), it is necessary, on the

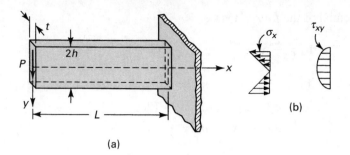

(a)

(b)

FIGURE 3.5.   *Example 3.1. (a) End-loaded cantilever beam; (b) stress distribution away from ends.*

basis of zero external loading in the x direction at $x = 0$, that $\sigma_x = 0$ along the vertical surface at $x = 0$. Finally, the applied load P must be equal to the resultant of the shearing forces distributed across the free end:

$$P = -\int_{-h}^{+h} \tau_{xy} t \, dy \qquad \textbf{(b)}$$

The negative sign agrees with the convention for stress discussed in Section 1.5.

For purposes of illustration, three approaches are employed to determine the distribution of stress within the beam.

Method 1.   Inasmuch as the bending moment varies linearly with x and σ_x at any section depends on y, it is reasonable to assume a general expression of the form

$$\sigma_x = \frac{\partial^2 \Phi}{\partial y^2} = c_1 xy \qquad \textbf{(c)}$$

in which c_1 represents a constant. Integrating twice with respect to y,

$$\Phi = \tfrac{1}{6} c_1 xy^3 + yf_1(x) + f_2(x) \qquad \textbf{(d)}$$

where $f_1(x)$ and $f_2(x)$ are functions of x to be determined. Introducing the Φ thus obtained into Eq. (3.17), we have

$$y \frac{d^4 f_1}{dx^4} + \frac{d^4 f_2}{dx^4} = 0$$

Since the second term is independent of y, a solution exists for all x and y provided that $d^4 f_1/dx^4 = 0$ and $d^4 f_2/dx^4 = 0$, which, upon integrating, leads to

$$f_1(x) = c_2 x^3 + c_3 x^2 + c_4 x + c_5$$
$$f_2(x) = c_6 x^3 + c_7 x^2 + c_8 x + c_9$$

where $c_2, c_3, \ldots$, are constants of integration. Substitution of $f_1(x)$ and $f_2(x)$ into Eq. (d) gives

$$\Phi = \tfrac{1}{6}c_1 xy^3 + (c_2 x^3 + c_3 x^2 + c_4 x + c_5)y$$
$$+ c_6 x^3 + c_7 x^2 + c_8 x + c_9$$

Expressions for σ_y and τ_{xy} follow from Eq. (3.16):

$$\sigma_y = \frac{\partial^2 \Phi}{\partial x^2} = 6(c_2 y + c_6)x + 2(c_3 y + c_7)$$

$$\tau_{xy} = -\frac{\partial^2 \Phi}{\partial x \, \partial y} = -\tfrac{1}{2}c_1 y^2 - 3c_2 x^2 - 2c_3 x - c_4 \qquad \text{(e)}$$

At this point, we are prepared to apply the boundary conditions. Substituting Eqs. (a) into (e), we obtain $c_2 = c_3 = c_6 = c_7 = 0$ and $c_4 = -\tfrac{1}{2}c_1 h^2$. The final condition, Eq. (b), may now be written as

$$-\int_{-h}^{h} \tau_{xy} t \, dy = \int_{-h}^{h} \tfrac{1}{2}c_1 t(y^2 - h^2) dy = P$$

from which

$$c_1 = -\frac{3P}{2th^3} = -\frac{P}{I}$$

where $I = \tfrac{2}{3}th^3$ is the moment of inertia of the cross section about the neutral axis. From Eqs. (c) and (e), together with the values of the constants, the stresses are found to be

$$\sigma_x = -\frac{Pxy}{I}, \qquad \sigma_y = 0, \qquad \tau_{xy} = -\frac{P}{2I}(h^2 - y^2) \qquad \textbf{(3.24)}$$

The distribution of these stresses at sections away from the ends is shown in Fig. 3.5b.

Method 2. Beginning with bending moments $M_z = Px$, we may assume a stress field similar to that for the case of pure bending:

$$\sigma_x = -\left(\frac{Px}{I}\right)y, \qquad \tau_{xy} = \tau_{xy}(x, y), \qquad \sigma_y = \sigma_z = \tau_{xz} = \tau_{yz} = 0 \qquad \textbf{(f)}$$

Equation of compatibility (3.12) is satisfied by these stresses. On the basis of Eqs. (f), the equations of equilibrium lead to

$$\frac{\partial \sigma_x}{\partial x} + \frac{\partial \tau_{xy}}{\partial y} = 0, \qquad \frac{\partial \tau_{xy}}{\partial x} = 0 \qquad \textbf{(g)}$$

From the second expression, τ_{xy} can depend only on y. The first equation of (g) together with Eqs. (f) yields

$$\frac{d\tau_{xy}}{dy} = \frac{Py}{I}$$

from which

$$\tau_{xy} = \frac{Py^2}{2I} + c$$

Here c is determined on the basis of $(\tau_{xy})_{y+\pm h} = 0: c = -Ph^2/2I$. The resulting expression for τ_{xy} satisfies Eq. (b) and is identical with the result previously obtained.

Method 3. The problem may be treated by superimposing the polynomials Φ_2 and Φ_4,

$$a_2 = c_2 = a_4 = b_4 = c_4 = e_4 = 0$$

Thus,

$$\Phi = \Phi_2 + \Phi_4 = b_2 xy + \frac{d_4}{6} xy^3$$

The corresponding stress components are

$$\sigma_x = d_4 xy, \qquad \sigma_y = 0, \qquad \tau_{xy} = -b_2 - \frac{d_4}{2} y^2$$

It is seen that the foregoing satisfies the second condition of Eqs. (a). The first of Eqs. (a) leads to $d_4 = -2b_2/h^2$. We then obtain

$$\tau_{xy} = -b_2 \left(1 - \frac{y^2}{h^2} \right)$$

which when substituted into condition (b) results in $b_2 = -3P/4ht = Ph^2/2I$. As before, τ_{xy} is as given in Eqs. (3.24).

Comments Observe that the stress distribution obtained is the same as that found by employing the elementary theory. If the boundary forces result in a stress distribution as indicated in Fig. 3.5b, the solution is exact. Otherwise, the solution is not exact. In any case, however, recall that Saint-Venant's principle permits us to regard the result as quite accurate for sections away from the ends.

Section 5.4 illustrates the determination of the displacement field after derivation of the curvature–moment relation.

3.8 THERMAL STRESSES

Consider the consequences of increasing or decreasing the *uniform* temperature of an entirely unconstrained elastic body. The resultant expansion or contraction occurs in such a way as to cause a cubic element of the solid to remain cubic, while experiencing changes of length on each of its sides. Normal strains occur in each direction unaccompanied by normal stresses. In addition, there are neither shear strains nor shear stresses. If the body is heated in such a way as to produce a nonuniform temperature field, or if the thermal expansions are prohibited from taking place freely because of restrictions placed on the boundary even if the temperature is uniform, or if the material exhibits anisotropy in a uniform temperature field, thermal stresses will occur. The effects of such stresses can be severe, especially since the most adverse thermal environments are often associated with design requirements involving unusually stringent constraints as to weight and volume. This is especially true in aerospace applications but is of considerable importance, too, in many everyday machine design applications.

Equations of Thermoelasticity

Solution of thermal stress problems requires reformulation of the stress–strain relationships accomplished by superposition of the strain attributable to stress and that due to temperature. For a change in temperature $T(x, y)$, the change of length, δL, of a small linear element of length L in an unconstrained body is $\delta L = \alpha L T$. Here α, usually a positive number, is termed the coefficient of linear thermal expansion. The thermal strain ε_t associated with the free expansion at a point is then

$$\varepsilon_t = \alpha T \tag{3.25}$$

The total x and y strains, ε_x and ε_y, are obtained by adding to the thermal strains of the type described, the strains due to stress resulting from external forces:

$$\varepsilon_x = \frac{1}{E}(\sigma_x - \nu\sigma_y) + \alpha T$$

$$\varepsilon_y = \frac{1}{E}(\sigma_y - \nu\sigma_x) + \alpha T \tag{3.26a}$$

$$\gamma_{xy} = \frac{\tau_{xy}}{G}$$

In terms of strain components, these expressions become

$$\sigma_x = \frac{E}{1 - \nu^2}(\varepsilon_x + \nu\varepsilon_y) - \frac{E\alpha T}{1 - \nu}$$

$$\sigma_y = \frac{E}{1 - \nu^2}(\varepsilon_y + \nu\varepsilon_x) - \frac{E\alpha T}{1 - \nu} \tag{3.26b}$$

$$\tau_{xy} = G\gamma_{xy}$$

Because free thermal expansion results in no angular distortion in an isotropic material, the shearing strain is unaffected, as indicated. Equations (3.26) represent modified strain–stress relations for *plane stress*. Similar expressions may be written for the case of plane strain. The differential equations of equilibrium (3.6) are based on purely mechanical considerations and are unchanged for *thermoelasticity*. The same is true of the strain–displacement relations (2.3) and the compatibility equation (3.8), which are geometrical in character. Thus, for given boundary conditions (expressed either as surface forces or displacements) and temperature distribution, thermoelasticity and ordinary elasticity *differ only* to the extent of the strain–stress relationship.

By substituting the strains given by Eq. (3.26a) into the equation of compatibility (3.8), employing Eq. (3.6) as well, and neglecting body forces, a compatibility equation is derived in terms of stress:

$$\left(\frac{\partial^2}{\partial x^2} + \frac{\partial^2}{\partial y^2}\right)(\sigma_x + \sigma_y + \alpha ET) = 0 \tag{3.27}$$

Introducing Eq. (3.16), we now have

$$\nabla^4\Phi + \alpha E \nabla^2 T = 0 \tag{3.28}$$

This expression is valid for plane strain or plane stress provided that the body forces are negligible.

It has been implicit in treating the matter of thermoelasticity as a superposition problem that the distribution of stress or strain plays a negligible role in influencing the temperature field [Refs. 3.4 and 3.5]. This lack of coupling enables the temperature field to be determined independently of any consideration of stress or strain. If the effect of the temperature distribution on material properties cannot be disregarded, the equations become coupled and analytical solutions are significantly more complex, occupying an area of considerable interest and importance. Numerical solutions can, however, be obtained in a relatively simple manner through the use of finite difference methods.

EXAMPLE 3.2   Thermal Stress and Strain in a Beam
A rectangular beam of small thickness t, depth $2h$, and length $2L$ is subjected to an arbitrary variation of temperature throughout its depth, $T = T(y)$. Determine the distribution of stress and strain for the case in which (a) the beam is entirely free of surface forces (Fig. 3.6a) and (b) the beam is held by rigid walls that prevent the x-directed displacement only (Fig. 3.6b).

Solution   The beam geometry indicates a problem of plane stress. We begin with the assumptions

$$\sigma_x = \sigma_x(y), \qquad \sigma_y = \tau_{xy} = 0 \tag{a}$$

Direct substitution of Eqs. (a) into Eqs. (3.6) indicates that the equations of equilibrium are satisfied. Equations (a) reduce the compatibility equation (3.27) to the form

$$\frac{d^2}{dy^2}(\sigma_x + \alpha ET) = 0 \tag{b}$$

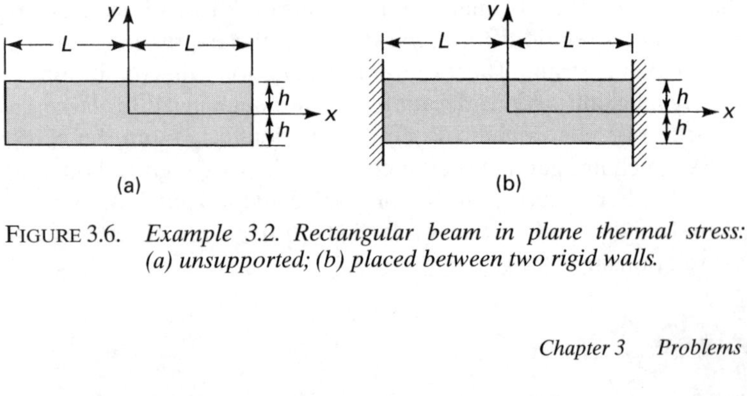

FIGURE 3.6.   *Example 3.2. Rectangular beam in plane thermal stress: (a) unsupported; (b) placed between two rigid walls.*

from which

$$\sigma_x = -\alpha ET + c_1 y + c_2 \tag{c}$$

where c_1 and c_2 are constants of integration. The requirement that faces $y = \pm h$ be free of surface forces is obviously fulfilled by Eq. (b).

a. The boundary conditions at the end faces are satisfied by determining the constants that assume zero resultant force and moment at $x = \pm L$:

$$\int_{-h}^{h} \sigma_x t \, dy = 0, \qquad \int_{-h}^{h} \sigma_x y t \, dy = 0 \tag{d}$$

Substituting Eq. (c) into Eqs. (d), it is found that $c_1 = (3/2h^3) \int_{-h}^{h} \alpha ET y \, dy$ and $c_2 = (1/2h) \int_{-h}^{h} \alpha ET \, dy$. The normal stress, upon substituting the values of the constants obtained, together with the moment of inertia $I = 2h^3 t/3$ and area $A = 2ht$, into Eq. (c), is thus

$$\sigma_x = E\alpha \left[-T + \frac{t}{A} \int_{-h}^{h} T \, dy + \frac{yt}{I} \int_{-h}^{h} Ty \, dy \right] \tag{3.29}$$

The corresponding strains are

$$\varepsilon_x = \frac{\sigma_x}{E} + \alpha T, \qquad \varepsilon_y = -\frac{\nu \sigma_x}{E} + \alpha T, \qquad \gamma_{xy} = 0 \tag{e}$$

The displacements can readily be determined from Eqs. (3.1).

From Eq. (3.29), observe that the temperature distribution for $T = $ constant results in zero stress, as expected. Of course, the strains (e) and the displacements will, in this case, not be zero. It is also noted that when the temperature is symmetrical about the midsurface $(y = 0)$, that is, $T(y) = T(-y)$, the final integral in Eq. (3.29) vanishes. For an antisymmetrical temperature distribution about the midsurface, $T(y) = -T(-y)$, and the first integral in Eq. (3.29) is zero.

b. For the situation described, $\varepsilon_x = 0$ for all y. With $\sigma_y = \tau_{xy} = 0$ and Eq. (c), Eqs. (3.26a) lead to $c_1 = c_2 = 0$, regardless of how T varies with y. Thus,

$$\sigma_x = -E\alpha T \tag{3.30}$$

and

$$\varepsilon_x = \gamma_{xy} = 0, \qquad \varepsilon_y = (1 + \nu)\alpha T \tag{f}$$

Comment Note that the axial stress obtained here can be large even for modest temperature changes, as can be verified by substituting properties of a given material.

3.9 BASIC RELATIONS IN POLAR COORDINATES

Geometrical considerations related either to the loading or to the boundary of a loaded system often make it preferable to employ polar coordinates rather than the Cartesian system used exclusively thus far. In general, polar coordinates are used advantageously where a degree of axial symmetry exists. Examples include a cylinder, a disk, a wedge, a curved beam, and a large thin plate containing a circular hole.

The polar coordinate system (r, θ) and the Cartesian system (x, y) are related by the following expressions (Fig. 3.7a):

$$x = r \cos \theta, \qquad r^2 = x^2 + y^2$$
$$y = r \sin \theta, \qquad \theta = \tan^{-1} \frac{y}{x}$$

(a)

These equations yield

$$\frac{\partial r}{\partial x} = \frac{x}{r} = \cos \theta, \qquad \frac{\partial r}{\partial y} = \frac{y}{r} = \sin \theta$$

$$\frac{\partial \theta}{\partial x} = -\frac{y}{r^2} = -\frac{\sin \theta}{r}, \qquad \frac{\partial \theta}{\partial y} = \frac{x}{r^2} = \frac{\cos \theta}{r}$$

(b)

Any derivatives with respect to x and y in the Cartesian system may be transformed into derivatives with respect to r and θ by applying the *chain rule*:

$$\frac{\partial}{\partial x} = \frac{\partial r}{\partial x} \frac{\partial}{\partial r} + \frac{\partial \theta}{\partial x} \frac{\partial}{\partial \theta} = \cos \theta \frac{\partial}{\partial r} - \frac{\sin \theta}{r} \frac{\partial}{\partial \theta}$$

$$\frac{\partial}{\partial y} = \frac{\partial r}{\partial y} \frac{\partial}{\partial r} + \frac{\partial \theta}{\partial y} \frac{\partial}{\partial \theta} = \sin \theta \frac{\partial}{\partial r} + \frac{\cos \theta}{r} \frac{\partial}{\partial \theta}$$

(c)

Relations governing properties at a point not containing any derivatives are *not* affected by the curvilinear nature of the coordinates, as is observed next.

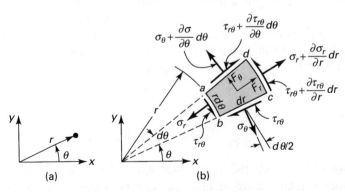

FIGURE 3.7. *(a) Polar coordinates; (b) stress element in polar coordinates.*

Equations of Equilibrium

Consider the state of stress on an infinitesimal element *abcd* of unit thickness described by polar coordinates (Fig. 3.7b). The r and θ-directed body forces are denoted by F_r and F_θ. Equilibrium of radial forces requires that

$$\left(\sigma_r + \frac{\partial \sigma_r}{\partial r}\, dr\right)(r + dr)d\theta - \sigma_r r\, d\theta - \left(\sigma_\theta + \frac{\partial \sigma_\theta}{\partial \theta}\, d\theta\right)dr \sin\frac{d\theta}{2}$$

$$- \sigma_\theta\, dr \sin\frac{d\theta}{2} + \left(\tau_{r\theta} + \frac{\partial \tau_{r\theta}}{\partial \theta}\, d\theta\right)dr \cos\frac{d\theta}{2} - \tau_{r\theta}\, dr \cos\frac{d\theta}{2} + F_r r\, dr\, d\theta = 0$$

Inasmuch as $d\theta$ is small, $\sin(d\theta/2)$ may be replaced by $d\theta/2$ and $\cos(d\theta/2)$ by 1. Additional simplication is achieved by dropping terms containing higher-order infinitesimals. A similar analysis may be performed for the tangential direction. When both equilibrium equations are divided by $r\, dr\, d\theta$, the results are

$$\frac{\partial \sigma_r}{\partial r} + \frac{1}{r}\frac{\partial \tau_{r\theta}}{\partial \theta} + \frac{\sigma_r - \sigma_\theta}{r} + F_r = 0$$

$$\frac{1}{r}\frac{\partial \sigma_\theta}{\partial \theta} + \frac{\partial \tau_{r\theta}}{\partial r} + \frac{2\tau_{r\theta}}{r} + F_\theta = 0$$

(3.31)

In the absence of body forces, Eqs. (3.31) are satisfied by a stress function $\Phi(r, \theta)$ for which the stress components in the radial and tangential directions are given by

$$\sigma_r = \frac{1}{r}\frac{\partial \Phi}{\partial r} + \frac{1}{r^2}\frac{\partial^2 \Phi}{\partial \theta^2}$$

$$\sigma_\theta = \frac{\partial^2 \Phi}{\partial r^2}$$

(3.32)

$$\tau_{r\theta} = \frac{1}{r^2}\frac{\partial \Phi}{\partial \theta} - \frac{1}{r}\frac{\partial^2 \Phi}{\partial r\, \partial \theta} = -\frac{\partial}{\partial r}\left(\frac{1}{r}\frac{\partial \Phi}{\partial \theta}\right)$$

Strain–Displacement Relations

Consider now the deformation of the infinitesimal element *abcd*, denoting the r and θ displacements by u and v, respectively. The general deformation experienced by an element may be regarded as composed of (1) a change in length of the sides, as in Figs. 3.8a and b, and (2) rotation of the sides, as in Figs. 3.8c and d.

In the analysis that follows, the small angle approximation $\sin \theta \approx \theta$ is employed, and arcs *ab* and *cd* are regarded as straight lines. Referring to Fig. 3.8a, it is observed that a u displacement of side *ab* results in both radial and tangential strain. The radial strain ε_r, the deformation per unit length of side *ad*, is associated only with the u displacement:

$$\varepsilon_r = \frac{\partial u}{\partial r}$$

(3.33a)

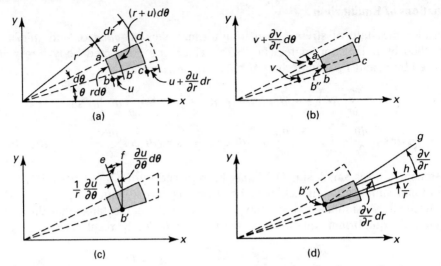

FIGURE 3.8. *Deformation and displacement of an element in polar coordinates.*

The tangential strain owing to u, the deformation per unit length of ab, is

$$(\varepsilon_\theta)_u = \frac{(r + u)\, d\theta - r\, d\theta}{r\, d\theta} = \frac{u}{r} \qquad \textbf{(d)}$$

Clearly, a v displacement of element $abcd$ (Fig. 3.8b) also produces a tangential strain,

$$(\varepsilon_\theta)_v = \frac{(\partial v/\partial\theta)\, d\theta}{r\, d\theta} = \frac{1}{r}\frac{\partial v}{\partial\theta} \qquad \textbf{(e)}$$

since the increase in length of ab is $(\partial v/\partial\theta)d\theta$. The resultant tangential strain, combining Eqs. (d) and (e), is

$$\varepsilon_\theta = \frac{1}{r}\frac{\partial v}{\partial\theta} + \frac{u}{r} \qquad \textbf{(3.33b)}$$

Figure 3.8c shows the angle of rotation $eb'f$ of side $a'b'$ due to a u displacement. The associated strain is

$$(\gamma_{r\theta})_u = \frac{(\partial u/\partial\theta)\, d\theta}{r\, d\theta} = \frac{1}{r}\frac{\partial u}{\partial\theta} \qquad \textbf{(f)}$$

The rotation of side bc associated with a v displacement alone is shown in Fig. 3.8d. Since an initial rotation of b'' through an angle v/r has occurred, the relative rotation $gb''h$ of side bc is

$$(\gamma_{r\theta})_v = \frac{\partial v}{\partial r} - \frac{v}{r} \qquad \textbf{(g)}$$

The sum of Eqs. (f) and (g) provides the total shearing strain

$$\gamma_{r\theta} = \frac{\partial v}{\partial r} + \frac{1}{r}\frac{\partial u}{\partial \theta} - \frac{v}{r} \tag{3.33c}$$

The strain–displacement relationships in polar coordinates are thus given by Eqs. (3.33).

Hooke's Law

To write Hooke's law in polar coordinates, we need only replace subscripts x with r and y with θ in the appropriate Cartesian equations. In the case of plane stress, from Eqs. (3.10) we have

$$\varepsilon_r = \frac{1}{E}(\sigma_r - \nu\sigma_\theta)$$

$$\varepsilon_\theta = \frac{1}{E}(\sigma_\theta - \nu\sigma_r) \tag{3.34}$$

$$\gamma_{r\theta} = \frac{1}{G}\tau_{r\theta}$$

For plane strain, Eqs. (3.5) lead to

$$\varepsilon_r = \frac{1+\nu}{E}[(1-\nu)\sigma_r - \nu\sigma_\theta]$$

$$\varepsilon_\theta = \frac{1+\nu}{E}[(1-\nu)\sigma_\theta - \nu\sigma_r] \tag{3.35}$$

$$\gamma_{r\theta} = \frac{1}{G}\tau_{r\theta}$$

Transformation Equations

Replacement of the subscripts x' with r and y' with θ in Eqs. (1.17) results in

$$\sigma_r = \sigma_x \cos^2\theta + \sigma_y \sin^2\theta + 2\tau_{xy}\sin\theta\cos\theta$$
$$\tau_{r\theta} = (\sigma_y - \sigma_x)\sin\theta\cos\theta + \tau_{xy}(\cos^2\theta - \sin^2\theta) \tag{3.36}$$
$$\sigma_\theta = \sigma_x \sin^2\theta + \sigma_y \cos^2\theta - 2\tau_{xy}\sin\theta\cos\theta$$

We can also express σ_x, τ_{xy}, and σ_y in terms of σ_r, $\tau_{r\theta}$, and σ_θ (Problem 3.26) by replacing θ with $-\theta$ in Eqs. (1.17). Thus,

$$\sigma_x = \sigma_r \cos^2\theta + \sigma_\theta \sin^2\theta - 2\tau_{r\theta}\sin\theta\cos\theta$$
$$\tau_{xy} = (\sigma_r - \sigma_\theta)\sin\theta\cos\theta + \tau_{r\theta}(\cos^2\theta - \sin^2\theta) \tag{3.37}$$
$$\sigma_y = \sigma_r \sin^2\theta + \sigma_\theta \cos^2\theta + 2\tau_{r\theta}\sin\theta\cos\theta$$

Similar transformation equations may also be written for the *strains* ε_r, $\gamma_{r\theta}$, and ε_θ.

Compatibility Equation

It can be shown that Eqs. (3.33) result in the following form of the equation of compatibility:

$$\frac{\partial^2 \varepsilon_\theta}{\partial r^2} + \frac{1}{r^2}\frac{\partial^2 \varepsilon_r}{\partial \theta^2} + \frac{2}{r}\frac{\partial \varepsilon_\theta}{\partial r} - \frac{1}{r}\frac{\partial \varepsilon_r}{\partial r} = \frac{1}{r}\frac{\partial^2 \gamma_{r\theta}}{\partial r \partial \theta} + \frac{1}{r^2}\frac{\partial \gamma_{r\theta}}{\partial \theta} \qquad (3.38)$$

To arrive at a compatibility equation expressed in terms of the stress function Φ, it is necessary to evaluate the partial derivatives $\partial^2 \Phi / \partial x^2$ and $\partial^2 \Phi / \partial y^2$ in terms of r and θ by means of the chain rule together with Eqs. (a). These derivatives lead to the Laplacian operator:

$$\nabla^2 \Phi = \frac{\partial^2 \Phi}{\partial x^2} + \frac{\partial^2 \Phi}{\partial y^2} = \frac{\partial^2 \Phi}{\partial r^2} + \frac{1}{r}\frac{\partial \Phi}{\partial r} + \frac{1}{r^2}\frac{\partial^2 \Phi}{\partial \theta^2} \qquad (3.39)$$

The equation of compatibility in alternative form is thus

$$\nabla^4 \Phi = \left(\frac{\partial^2}{\partial r^2} + \frac{1}{r}\frac{\partial}{\partial r} + \frac{1}{r^2}\frac{\partial^2}{\partial \theta^2}\right)(\nabla^2 \Phi) = 0 \qquad (3.40)$$

For the axisymmetrical, zero body force case, the compatibility equation is, from Eq. (3.9) [referring to (3.39)],

$$\nabla^2 (\sigma_r + \sigma_\theta) = \frac{d^2(\sigma_r + \sigma_\theta)}{dr^2} + \frac{1}{r}\frac{d(\sigma_r + \sigma_\theta)}{dr} = 0 \qquad (3.41)$$

The remaining relationships appropriate to two-dimensional elasticity are found in a manner similar to that outlined in the foregoing discussion.

EXAMPLE 3.3   State of Stress in a Plate in Tension

A large thin plate is subjected to uniform tensile stress σ_o at its ends, as shown in Fig. 3.9. Determine the field of stress existing within the plate.

Solution For purposes of this analysis, it will prove convenient to locate the origin of coordinate axes at the center of the plate as shown. The state of stress in the plate is expressed by

$$\sigma_x = \sigma_o, \qquad \sigma_y = \tau_{xy} = 0$$

The stress function, $\Phi = \sigma_o y^2 / 2$, satisfies the biharmonic equation, Eq. (3.17). The geometry suggests polar form. The stress function Φ may be transformed by substituting $y = r \sin \theta$, with the following result:

$$\Phi = \tfrac{1}{4}\sigma_o r^2 (1 - \cos 2\theta) \qquad \textbf{(h)}$$

FIGURE 3.9.   *Example 3.3. A plate in uniaxial tension.*

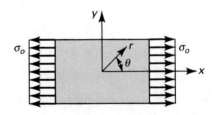

The stresses in the plate now follow from Eqs. (h) and (3.32):

$$\sigma_r = \tfrac{1}{2}\sigma_o(1 + \cos 2\theta)$$
$$\sigma_\theta = \tfrac{1}{2}\sigma_o(1 - \cos 2\theta) \qquad\qquad \textbf{(3.42)}$$
$$\tau_{r\theta} = -\tfrac{1}{2}\sigma_o \sin 2\theta$$

Clearly, substitution of $\sigma_y = \tau_{xy} = 0$ could have led directly to the foregoing result, using the transformation expressions of stress, Eqs. (3.36).

Part B—Stress Concentrations

3.10 STRESSES DUE TO CONCENTRATED LOADS

Let us now consider a *concentrated force P or F* acting at the vertex of a very large or *semi-infinite wedge* (Fig. 3.10). The load distribution along the thickness (*z* direction) is uniform. The thickness of the wedge is taken as unity, so *P* or *F* is the *load per unit thickness*. In such situations, it is convenient to use polar coordinates and the semi-inverse method.

In actuality, the concentrated load is assumed to be a theoretical *line load* and will be spread over an area of small finite width. Plastic deformation may occur locally. Thus, the solutions that follow are *not* valid in the immediate vicinity of the application of load.

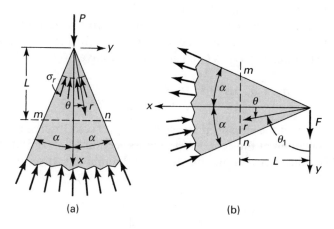

FIGURE 3.10. *Wedge of unit thickness subjected to a concentrated load per unit thickness: (a) knife edge or pivot; (b) wedge cantilever.*

Compression of a Wedge (Fig. 3.10a)

Assume the stress function

$$\Phi = cPr\theta \sin \theta \tag{a}$$

where c is a constant. It can be verified that Eq. (a) satisfies Eq. (3.40) and compatibility is ensured. For equilibrium, the stresses from Eqs. (3.32) are

$$\sigma_r = 2cP \frac{\cos \theta}{r}, \qquad \sigma_\theta = 0, \qquad \tau_{r\theta} = 0 \tag{b}$$

The force resultant acting on a cylindrical surface of small radius, shown by the dashed lines in Fig. 3.10a, must balance P. The boundary conditions are therefore expressed by

$$\sigma_\theta = \tau_{r\theta} = 0, \qquad \theta = \pm\alpha \tag{c}$$

$$2 \int_0^\alpha (\sigma_r \cos \theta) r d\theta = -P \tag{d}$$

Conditions (c) are fulfilled by the last two of Eqs. (b). Substituting the first of Eqs. (b) into condition (d) results in

$$4cP \int_0^\alpha \cos^2 \theta \, d\theta = -P$$

Integrating and solving for c: $c = -1/(2\alpha + \sin 2\alpha)$. The stress distribution in the knife edge is therefore

$$\sigma_r = -\frac{P \cos \theta}{r(\alpha + \frac{1}{2} \sin 2\alpha)}, \qquad \sigma_\theta = 0, \qquad \tau_{r\theta} = 0 \tag{3.43}$$

This solution is due to J. H. Mitchell [Ref. 3.6].

The distribution of the normal stresses σ_x over any cross section $m-n$ perpendicular to the axis of symmetry of the wedge is *not* uniform (Fig. 3.10a). Applying Eq. (3.37) and substituting $r = L/\cos \theta$ in Eq. (3.43), we have

$$\sigma_x = \sigma_r \cos^2 \theta = -\frac{P \cos^4 \theta}{L(\alpha + \frac{1}{2} \sin 2\alpha)} \tag{3.44}$$

The foregoing shows that the stresses increase as L decreases. Observe also that the normal stress is maximum at the center of the cross section ($\theta = 0$) and minimum at $\theta = \alpha$. The difference between the maximum and minimum stress, $\Delta\sigma_x$, is from Eq. (3.44),

$$\Delta\sigma_x = -\frac{P(1 - \cos^4 \alpha)}{L(\alpha + \frac{1}{2} \sin 2\alpha)} \tag{e}$$

For instance, if $\alpha = 10°$, $\Delta\sigma_x = -0.172P/L$ is about 6% of the average normal stress calculated from the elementary formula $(\sigma_x)_{\text{elem}} = -P/A = -P/2L \tan \alpha = -2.836P/L$.

For larger angles, the difference is greater; the error in the mechanics of materials solution increases (Prob. 3.31). It may be demonstrated that the stress distribution over the cross section approaches *uniformity* as the taper of the wedge diminishes. Analogous conclusions may also be drawn for a *conical bar*. Note that Eqs. (3.43) can be applied as well for the uniaxial *tension* of tapered members by assigning σ_r a positive value.

Bending of a Wedge (Fig. 3.10b)

We now employ $\Phi = cFr\theta_1 \sin \theta_1$, with θ_1 measured from the line of action of the force. The equilibrium condition is

$$\int_{(\pi/2)-\alpha}^{(\pi/2)+\alpha} (\sigma_r \cos \theta_1) r \, d\theta_1 = 2cF \int_{(\pi/2)-\alpha}^{(\pi/2)+\alpha} \cos^2 \theta_1 \, d\theta_1 = -F$$

from which, after integration, $c = -1/(2\alpha - \sin 2\alpha)$. Thus, by replacing θ_1 with $90° - \theta$, we have

$$\sigma_r = -\frac{F \cos \theta_1}{r(\alpha - \frac{1}{2}\sin 2\alpha)} = -\frac{F \sin \theta}{r(\alpha - \frac{1}{2}\sin 2\alpha)}, \qquad \sigma_\theta = 0, \qquad \tau_{r\theta} = 0 \quad \textbf{(3.45)}$$

It is seen that if θ_1 is larger than $\pi/2$, the radial stress is positive, that is, tension exists. Because $\sin \theta = y/r$, $\cos \theta = x/r$, and $r = \sqrt{x^2 + y^2}$, the normal and shearing stresses at a point over any cross section $m - n$, using Eqs. (3.37) and (3.45), may be expressed as

$$\sigma_x = \sigma_r \cos^2 \theta = -\frac{F \sin \theta \cos^2 \theta}{r(\alpha - \frac{1}{2}\sin 2\alpha)}$$

$$= -\frac{F}{\alpha - \frac{1}{2}\sin 2\alpha} \frac{x^2 y}{(x^2 + y^2)^2}$$

$$\sigma_y = \sigma_r \sin^2 \theta = -\frac{F \sin^3 \theta}{r(\alpha - \frac{1}{2}\sin 2\alpha)}$$

$$= -\frac{F}{\alpha - \frac{1}{2}\sin 2\alpha} \frac{y^3}{(x^2 + y^2)^2} \qquad \textbf{(3.46)}$$

$$\tau_{xy} = \sigma_r \sin \theta \cos \theta = -\frac{F \sin^2 \theta \cos \theta}{r(\alpha - \frac{1}{2}\sin 2\alpha)}$$

$$= -\frac{F}{\alpha - \frac{1}{2}\sin 2\alpha} \frac{xy^2}{(x^2 + y^2)^2}$$

Using Eqs. (3.46), it can be shown that (Prob. 3.33) across a transverse section $x = L$ of the wedge: σ_x is a maximum for $\theta = \pm 30°$, σ_y is a maximum for $\theta = \pm 60°$, and τ_{xy} is a maximum for $\theta = \pm 45°$.

To compare the results given by Eqs. (3.46) with the results given by the elementary formulas for stress, consider the series

$$\sin 2\alpha = 2\alpha - \frac{(2\alpha)^3}{3!} + \frac{(2\alpha)^5}{5!}$$

It follows that, *for small angle* α, we can disregard all but the first two terms of this series to obtain

$$2\alpha = \sin 2\alpha + \frac{(2\alpha)^3}{6} \qquad \text{(f)}$$

By introducing the moment of inertia of the cross section $m - n$, $I = \frac{2}{3}y^3 = \frac{2}{3}x^3 \cdot \tan^3\alpha$, and Eq. (f), we find from Eqs. (3.46) that

$$\sigma_x = -\frac{Fxy}{I}\left[\left(\frac{\tan\alpha}{\alpha}\right)^3 \cos^4\theta\right], \qquad \tau_{xy} = -\frac{Fy^2}{I}\left[\left(\frac{\tan\alpha}{\alpha}\right)^3 \cos^4\theta\right] \qquad \text{(g)}$$

For small values of α, the factor in the bracket is approximately equal to unity. The expression for σ_x then coincides with that given by the flexure formula, $-My/I$, of the mechanics of materials. In the elementary theory, the lateral stress σ_y given by the second of Eqs. (3.46) is ignored. The maximum shearing stress τ_{xy} obtained from Eq. (g) is twice as great as the shearing stress calculated from VQ/Ib of the elementary theory and occurs at the extreme fibers (at points m and n) rather than the neutral axis of the rectangular cross section.

In the case of loading in both compression and bending, superposition of the effects of P and F results in the following expression for *combined stress* in a pivot or in a wedge–cantilever:

$$\sigma_r = -\frac{P\cos\theta}{r(\alpha + \frac{1}{2}\sin 2\alpha)} - \frac{F\cos\theta_1}{r(\alpha - \frac{1}{2}\sin 2\alpha)}, \qquad \sigma_\theta = 0, \qquad \tau_{r\theta} = 0 \quad \textbf{(3.47)}$$

The foregoing provides the local stresses at the support of a beam of narrow rectangular cross section.

Concentrated Load on a Straight Boundary (Fig. 3.11a)

By setting $\alpha = \pi/2$ in Eq. (3.43), the result

$$\sigma_r = -\frac{2P}{\pi}\frac{\cos\theta}{r}, \qquad \sigma_\theta = 0, \qquad \tau_{r\theta} = 0 \qquad \textbf{(3.48)}$$

is an expression for radial stress distribution in a very large plate (semi-infinite solid) under normal load at its horizontal surface. For a circle of *any* diameter d with center on the x axis and tangent to the y axis, as shown in Fig. 3.11b, we have, for point A of the circle, $d \cdot \cos\theta = r$. Equation (3.48) then becomes

$$\sigma_r = -\frac{2P}{\pi d} \qquad \textbf{(3.49)}$$

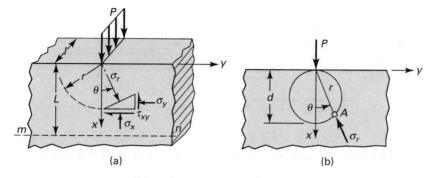

FIGURE 3.11. (a) Concentrated load on a straight boundary of a large plate;
(b) a circle of constant radial stress.

We thus observe that, except for the point of load application, the *stress is the same at all points on the circle.*

The stress components in Cartesian coordinates may be obtained readily by following a procedure similar to that described previously for a wedge:

$$\sigma_x = -\frac{2P}{\pi x} \cos^4 \theta = -\frac{2P}{\pi} \frac{x^3}{(x^2 + y^2)^2}$$

$$\sigma_y = -\frac{2P}{\pi x} \sin^2 \theta \cos^2 \theta = -\frac{2P}{\pi} \frac{xy^2}{(x^2 + y^2)^2} \tag{3.50}$$

$$\tau_{xy} = -\frac{2P}{\pi x} \sin \theta \cos^3 \theta = -\frac{2P}{\pi} \frac{x^2 y}{(x^2 + y^2)^2}$$

The state of stress is shown on a properly oriented element in Fig. 3.11a.

3.11 STRESS DISTRIBUTION NEAR CONCENTRATED LOAD ACTING ON A BEAM

The elastic flexure formula for beams gives satisfactory results only at some distance away from the point of load application. Near this point, however, there is a significant perturbation in stress distribution, which is very important. In the case of a beam of narrow rectangular cross section, these irregularities can be studied by using the equations developed in Section 3.10.

Consider the case of a simply supported beam of depth h, length L, and width b, loaded at the midspan (Fig. 3.12a). The origin of coordinates is taken to be the center of the beam, with x the axial axis as shown in the figure. Both force P and the supporting reactions are applied along lines across the width of the beam. The bending stress distribution, using the flexure formula, is expressed by

$$\sigma'_x = -\frac{My}{I} = \frac{6P}{bh^3}\left(\frac{L}{2} - x\right)y$$

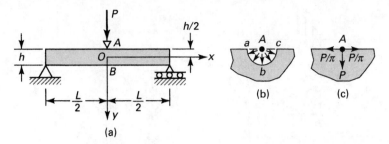

FIGURE 3.12.   *Beam subjected to a concentrated load P at the midspan.*

where $I = bh^3/12$ is the moment of inertia of the cross section. The stress at the loaded section is obtained by substituting $x = 0$ into the preceding equation:

$$\sigma_x' = \frac{3PL}{bh^3}y \tag{a}$$

To obtain the total stress along section AB, we apply the superposition of the bending stress distribution and stresses created by the line load, given by Eq. (3.48) for a semi-infinite plate. Observe that the radial pressure distribution created by a line load over quadrant ab of cylindrical surface abc at point A (Fig. 3.12b) produces a horizontal force

$$\int_0^{\pi/2} (\sigma_r \sin \theta) r\, d\theta = \int_0^{\pi/2} \frac{2P}{\pi} \sin \theta \cos \theta\, d\theta = \frac{P}{\pi} \tag{b}$$

and a vertical force

$$\int_{-\pi/2}^{\pi/2} (\sigma_r \cos \theta) r\, d\theta = \int_{-\pi/2}^{\pi/2} \frac{2P}{\pi} \cos^2 \theta\, d\theta = P \tag{c}$$

applied at A (Fig. 3.12c). In the case of a beam (Fig. 3.12a), the latter force is balanced by the supporting reactions that give rise to the bending stresses [Eq. (a)]. On the other hand, the horizontal forces create tensile stresses at the midsection of the beam of

$$\sigma_x'' = \frac{P}{\pi bh} \tag{d}$$

as well as bending stresses of

$$\sigma_x''' = -\frac{Ph}{2\pi} \frac{y}{I} = -\frac{6P}{\pi bh^2}y \tag{e}$$

Here $Ph/2\pi$ is the bending moment of forces P/π about the point 0.

Combining the stresses of Eqs. (d) and (e) with the bending stress given by Eq. (a), we obtain the axial normal stress distribution over beam cross section AB:

$$\sigma_x = \frac{3P}{bh^3}\left(L - \frac{2h}{\pi}\right)y + \frac{P}{\pi bh} \tag{3.51}$$

At point $B(0, h/2)$, the tensile stress is

$$(\sigma_x)_B = \frac{3PL}{2bh^2}\left(1 - \frac{4h}{3\pi L}\right) = \frac{3PL}{2bh^2} - 0.637\frac{P}{bh} \tag{3.52}$$

The second term represents a correction to the simple beam formula owing to the presence of the line load. It is observed that for short beams this stress is of considerable magnitude. The axial normal stresses at other points in the midsection are determined in a like manner.

The foregoing procedure leads to the poorest accuracy for point B, the point of maximum tensile stress. A better approximation [see Ref. 3.7] of this stress is given by

$$(\sigma_x)_B = \frac{3PL}{2bh^2} - 0.508\frac{P}{bh} \tag{3.53}$$

Another more detailed study demonstrates that the *local stresses decrease very rapidly* with increase of the distance (x) from the point of load application. At a *distance equal to the depth* of the beam, they are usually negligible. Furthermore, along the loaded section, the normal stress σ_x does *not* obey a linear law.

In the preceding discussion, the disturbance caused by the reactions at the ends of the beam, which are also applied as line loads, are not taken into account. To determine the radial stress distribution at the *supports* of the beam of narrow rectangular cross section, Eq. (3.47) can be used. Clearly, for the beam under consideration, we use $F = 0$ and replace P by $P/2$ in this expression.

3.12 STRESS CONCENTRATION FACTORS

The discussion in Section 3.10 shows that, for situations in which the cross section of a load-carrying member varies gradually, reasonably accurate results can be expected if we apply equations derived on the basis of constant section. On the other hand, where abrupt changes in the cross section exist, the mechanics of materials approach cannot predict the high values of stress that actually exist. The condition referred to occurs in such frequently encountered configurations as holes, notches, and fillets. While the stresses in these regions can in some cases (for example, Fig. 3.13) be analyzed by applying the theory of elasticity, it is more usual to rely on experimental techniques and, in particular, photoelastic methods. The finite element method (Chap. 7) is very efficient for this purpose.

It is to be noted that irregularities in stress distribution associated with abrupt changes in cross section are of practical importance in the design of machine

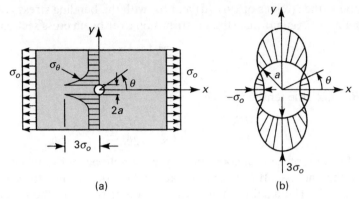

FIGURE 3.13. *Example 3.4. Circular hole in a plate subjected to uniaxial tension: (a) tangential stress distribution for $\theta = \pm\pi/2$; (b) tangential stress distribution along periphery of the hole.*

elements subject to variable external forces and stress reversal. Under the action of stress reversal, progressive cracks (Sec. 4.4) are likely to start at certain points at which the stress is far above the average value. The majority of fractures in machine elements in service can be attributed to such progressive cracks.

A geometric or theoretical *stress concentration factor K* is used to relate the maximum stress at the discontinuity to the nominal stress. The factor is defined by

$$K = \frac{\text{maximum stress}}{\text{nominal stress}} = \frac{\sigma_{\max}}{\sigma_{\text{nom}}} \tag{3.54}$$

In the foregoing, the nominal stress is the stress that occurs if the abrupt change in the cross section did not exist or had no influence on stress distribution. It is important that a stress concentration factor is applied to the stress computed for the *net* or reduced cross section. Equation (3.54) is valid as long as computed values of maximum stress do *not exceed* the *proportional limit* of the material.

We note that for ductile materials statically loaded beyond the yield point, the stress concentration factors decrease to a value approaching unity because of the redistribution of stress around a discontinuity. That is, the effect of an abrupt change in geometry is nullified, and $\sigma_{\max} = \sigma_{\text{nom}}$, or $K = 1$; a nearly uniform stress distribution exists across the net section. This is referred to as a fully plastic condition (Chap. 12). Therefore, the stress-concentration factor is of *no* significance in design of a ductile material under *static* loading. However, for *dynamic* (such as repeated, impact, or thermal) loading, even a ductile material may fail as a result of propagation of cracks originating at points of high stress. So, the presence of stress concentration in the case of dynamic loading must *not* be ignored, regardless of whether the material is brittle or ductile. More is said about this in Chapter 4.

Circular Hole in a Large Plate in Simple Tension

The theory of elasticity can be applied to evaluate the stress concentration associated with some incomplex geometric configurations under static loadings. One solution is that of a large, thin plate containing a small circular hole of radius a subjected to a tension (Fig. 3.13a). In the following, we determine the field of stress and compare it with those of Example 3.3.

The *boundary conditions* appropriate to the circumference of the hole are

$$\sigma_r = \tau_{r\theta} = 0, \qquad r = a \tag{a}$$

For large distances away from the origin, we set σ_r, σ_θ, and $\tau_{r\theta}$ equal to the values found for a solid plate in Example 3.3. Thus, from Eq. (3.42), for $r = \infty$,

$$\sigma_r = \tfrac{1}{2}\sigma_o (1 + \cos 2\theta)$$

$$\sigma_\theta = \tfrac{1}{2}\sigma_o(1 - \cos 2\theta), \qquad \tau_{r\theta} = -\tfrac{1}{2}\sigma_o \sin 2\theta \tag{b}$$

For this case, we assume a stress function analogous to Eq. (h) of Example 3.3,

$$\Phi = f_1(r) + f_2(r) \cos 2\theta \tag{c}$$

in which f_1 and f_2 are yet to be determined. Substituting Eq. (c) into the biharmonic equation (3.40) and noting the validity of the resulting expression for all θ, we have

$$\left(\frac{d^2}{dr^2} + \frac{1}{r}\frac{d}{dr}\right)\left(\frac{d^2 f_1}{dr^2} + \frac{1}{r}\frac{df_1}{dr}\right) = 0 \tag{d}$$

$$\left(\frac{d^2}{dr^2} + \frac{1}{r}\frac{d}{dr} - \frac{4}{r^2}\right)\left(\frac{d^2 f_2}{dr^2} + \frac{1}{r}\frac{df_2}{dr} - \frac{4f_2}{r^2}\right) = 0 \tag{e}$$

The solutions of Eqs. (d) and (e) are (Prob. 3.35)

$$f_1 = c_1 r^2 \ln r + c_2 r^2 + c_3 \ln r + c_4 \tag{f}$$

$$f_2 = c_5 r^2 + c_6 r^4 + \frac{c_7}{r^2} + c_8 \tag{g}$$

where the c's are the constants of integration. The stress function is then obtained by introducing Eqs. (f) and (g) into (c). By substituting Φ into Eq. (3.32), the stresses are found to be

$$\sigma_r = c_1(1 + 2\ln r) + 2c_2 + \frac{c_3}{r^2} - \left(2c_5 + \frac{6c_7}{r^4} + \frac{4c_8}{r^2}\right)\cos 2\theta$$

$$\sigma_\theta = c_1(3 + 2\ln r) + 2c_2 - \frac{c_3}{r^2} + \left(2c_5 + 12c_6 r^2 + \frac{6c_7}{r^4}\right)\cos 2\theta \tag{h}$$

$$\tau_{r\theta} = \left(2c_5 + 6c_6 r^2 - \frac{6c_7}{r^4} - \frac{2c_8}{r^2}\right)\sin 2\theta$$

The absence of c_4 indicates that it has no influence on the solution.

According to the boundary conditions (b), $c_1 = c_6 = 0$ in Eq. (h), because as $r \to \infty$, the stresses must assume finite values. Then, according to the conditions (a), the equations (h) yield

$$2c_2 + \frac{c_3}{a^2} = 0, \qquad 2c_5 + \frac{6c_7}{a^4} + \frac{4c_8}{a^2} = 0, \qquad 2c_5 - \frac{6c_7}{a^4} - \frac{2c_8}{a^2} = 0$$

Also, from Eqs. (b) and (h) we have

$$\sigma_o = -4c_5, \qquad \sigma_o = 4c_2$$

Solving the preceding five expressions, we obtain $c_2 = \sigma_o/4$, $c_3 = -a^2\sigma_o/2$, $c_5 = -\sigma_o/4$, $c_7 = -a^4\sigma_o/4$, and $c_8 = a^2\sigma_o/2$. The determination of the *stress distribution* in a large plate containing a small circular hole is completed by substituting these constants into Eq. (h):

$$\sigma_r = \tfrac{1}{2}\sigma_o\left[\left(1 - \frac{a^2}{r^2}\right) + \left(1 + \frac{3a^4}{r^4} - \frac{4a^2}{r^2}\right)\cos 2\theta\right] \qquad \text{(3.55a)}$$

$$\sigma_\theta = \tfrac{1}{2}\sigma_o\left[\left(1 + \frac{a^2}{r^2}\right) - \left(1 + \frac{3a^4}{r^4}\right)\cos 2\theta\right] \qquad \text{(3.55b)}$$

$$\tau_{r\theta} = -\tfrac{1}{2}\sigma_o\left(1 - \frac{3a^4}{r^4} + \frac{2a^2}{r^2}\right)\sin 2\theta \qquad \text{(3.55c)}$$

The tangential stress distribution along the edge of the hole, $r = a$, is shown in Fig. 3.13b using Eq. (3.55b). We observe from the figure that

$$(\sigma_\theta)_{\max} = 3\sigma_o, \qquad \theta = \pm\pi/2$$
$$(\sigma_\theta)_{\min} = -\sigma_o, \qquad \theta = 0, \quad \theta = \pm\pi$$

The latter indicates that there exists a small area experiencing compressive stress. On the other hand, from Eq. (3.42), for $\theta = \pm\pi/2, (\sigma_\theta)_{\max} = \sigma_o$. The stress concentration factor, defined as the ratio of the maximum stress at the hole to the nominal stress σ_o, is therefore $K = 3\sigma_o/\sigma_o = 3$.

To depict the variation of $\sigma_r(r, \pi/2)$ and $\sigma_\theta(r, \pi/2)$ over the distance from the origin, dimensionless stresses are plotted against the dimensionless radius in Fig. 3.14. The shearing stress $\tau_{r\theta}(r, \pi/2) = 0$. At a distance of *twice the diameter of the hole*, that is, $r = 4a$, we obtain $\sigma_\theta \approx 1.037\sigma_o$ and $\sigma_r \approx 0.088\sigma_o$. Similarly, at a distance $r = 9a$, we have $\sigma_\theta \approx 1.006\sigma_o$ and $\sigma_r \approx 0.018\sigma_o$, as is observed in the figure. Thus, simple tension prevails at a distance of approximately nine radii; the hole has a *local effect* on the distribution of stress. This is a verification of Saint-Venant's principle.

Circular Hole in a Large Plate in Biaxial Tension

The results expressed by Eqs. (3.55) are applied, together with the method of superposition, to the case of biaxial loading. Distributions of maximum stress $\sigma_\theta(r, \pi/2)$, obtained in this way (Prob. 3.36), are given in Fig. 3.15. Such conditions of stress

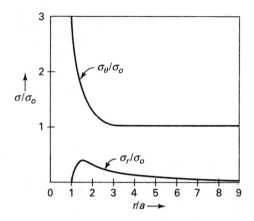

FIGURE 3.14. *Example 3.4. Graph of tangential and radial stresses for θ = π/2 versus the distance from the center of the plate shown in Fig. 3.13a.*

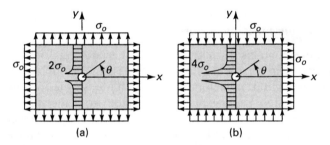

(a) (b)

FIGURE 3.15. *Tangential stress distribution for θ = ±π/2 in the plate with the circular hole subject to biaxial stresses: (a) uniform tension; (b) pure shear.*

concentration occur in a thin-walled spherical pressure vessel with a small circular hole (Fig. 3.15a) and in the torsion of a thin-walled circular tube with a small circular hole (Fig. 3.15b).

Elliptic Hole in a Large Plate in Tension

Several simple geometries of practical importance were the subject of stress concentration determination by Inglis and Neuber on the basis of mathematical analysis discussed in the preceding. We note that a similar concentration of stress is caused by a small elliptic hole in a thin, large plate (Fig. 3.16). It can be shown that stress concentration factor at the ends of the major axis of the hole is given by

$$K = 1 + 2\left(\frac{b}{a}\right) \tag{3.56}$$

FIGURE 3.16. *Elliptical hole in a plate under uniaxial tension.*

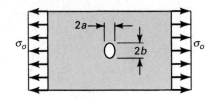

The maximum tensile stress is thus

$$\sigma_{max} = \sigma_o \left(1 + 2\frac{b}{a} \right) \tag{3.57}$$

where a is the half-width of the ellipse and b is the half-height.

Clearly, the stress increases with the ratio b/a. In the limit, as $a \to 0$, the ellipse becomes a *narrow crack* of length $2b$, and a very high stress concentration is produced; material will yield plastically around the ends of the crack or the crack will propagate. To prevent such spreading, holes may be drilled at the ends of the crack to effectively increase the radii to correspond to a smaller b/a. Thus, a high stress concentration is replaced by a relatively smaller one. When the hole is a circle, $a = b$ and $K = 3$.

Graphs for Stress Concentration Factors

Technical literature contains an abundance of specialized information on stress concentration factors in the form of graphs, tables, and formulas.* Values of the calculated stress concentration factors for bars with fillets, holes, and grooves under axial, bending, or torsion loading may be obtained from the *diagram* (a nomograph) given by *Neuber* (1958). The *best source book* on stress-concentration factors is Peterson [Ref. 3.8], which compiles the theoretical and experimental results of many researchers into useful design charts. Some examples of most commonly used graphs for stress concentration factors for a variety of geometries are provided in Appendix D. Observe that these charts indicate the advisability of streamlining junctures and transitions of portions that make up a member; that is, stress concentration can be reduced in intensity by properly proportioning the parts. Large fillet radii help at reentrant corners. There are many other well-established techniques for smoothing out the stress distribution in a part and thus reducing the stress-concentration factor.

EXAMPLE 3.4   Stresses at the Groove in a Circular Shaft

A circular shaft of diameter D with a circumferential circular groove (of diameter d and radius r) is subjected to axial force P, bending moment M, and torque T (Fig. 3.17). Determine the maximum principal stress.

---

*See, for example, Refs. 3.8 through 3.14.

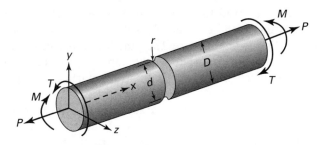

FIGURE 3.17. *Example 3.4. A grooved circular shaft with combined loadings.*

Solution For the loading described, the principal stresses occur at a point at the root of the notch, which, from Eq. (1.20), are given by

$$\sigma_{1,2} = \frac{\sigma_x}{2} \pm \sqrt{\left(\frac{\sigma_x}{2}\right)^2 + \tau_{xy}^2}, \qquad \sigma_3 = 0 \qquad \textbf{(i)}$$

where σ_x and τ_{xy} represent the normal and shear stresses in the *reduced* cross section of the shaft, respectively. We have

$$\sigma_x = K_a \frac{P}{A} + K_b \frac{My}{I}, \qquad \tau_{xy} = K_t \frac{Tr}{J}$$

or

$$\sigma_x = K_a \frac{P}{\pi b^2} + K_b \frac{4M}{\pi b^3}, \qquad \tau_{xy} = K_t \frac{2T}{\pi b^3} \qquad \textbf{(j)}$$

Here K_a, K_b, and K_t denote the stress concentration factors for axial force, bending moment, and torque, respectively. These factors are determined from Figures D.5, D.7, and D.6, respectively. Thus, given a set of shaft dimensions and the loading, formulas (a) and (b) lead to the value of the maximum principal stress σ_1 (see Problem 3.46).

In addition, note that a shear force V may also act on the shaft, as in Fig. 5.11 (Chap. 5). For slender members, however, this shear contributes very little to the deflection (Sec. 5.4) and to the maximum stress.

3.13 CONTACT STRESSES

Application of a load over a small area of contact results in unusually high stresses. Situations of this nature are found on a microscopic scale whenever force is transmitted through bodies in contact. There are important practical cases when the geometry of the contacting bodies results in large stresses, disregarding the stresses associated with the asperities found on any nominally smooth surface. The original analysis of elastic contact stresses, by H. Hertz, was published in 1881. In his honor, the stresses at the mating surfaces of curved bodies in compression are called *Hertz contact stresses*, or simply referred to as the *contact stresses*. The Hertz problem relates to the stresses owing to the contact of a sphere on a plane, a sphere on a sphere, a cylinder on a cylinder, and the like. The practical implications with respect to ball and roller bearings, locomotive wheels, valve tappets, gear teeth, pin joints in linkages, cams, push rod mechanisms, and numerous machine components are apparent.

Consider, in this regard, the contact without deformation of two bodies having spherical surfaces of radii r_1 and r_2, in the vicinity of contact. If now a collinear pair of forces F acts to press the bodies together, deformation will occur, and the point of contact will be replaced by a small area of contact. The first steps taken toward the solution of this problem are the determination of the size and shape of the contact area as well as the distribution of normal pressure acting on the area. The stresses and deformations resulting from the interfacial pressure are then evaluated.

The following *basic assumptions* are generally made in the solution of the contact problem:

1. The contacting bodies are isotropic and elastic.
2. The contact areas are essentially flat and small relative to the radii of curvature of the undeformed bodies in the vicinity of the interface.
3. The contacting bodies are perfectly smooth, and therefore only normal pressures need be taken into account.

The foregoing set of assumptions enables an elastic analysis to be conducted. Without going into the derivations, we shall, in the following sections, introduce some of the results.* It is important to note that, in all instances, the *contact pressure* varies from zero at the side of the contact area to a *maximum* value p_o *at its center*.

3.14 SPHERICAL AND CYLINDRICAL CONTACTS

In this section, maximum contact pressure and deflection of two bodies held in contact by normal forces to the area of contact will be discussed. The *deflection* is the *relative displacement* δ of centers of the two bodies. It represents the *sum* of the deflections of the two bodies as they approach each other.

Two Spheres in Contact

The contact area and corresponding stress distribution between two spheres, loaded with a force F, is illustrated in Fig. 3.18. Observe that the contact pressure within each sphere has a semi-elliptical distribution. It varies from 0 at the side of

---

*A summary and complete list of publications dealing with contact stress problems are given by Refs. 3.13 through 3.19.

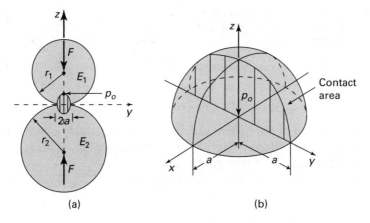

FIGURE 3.18. (a) Spherical surfaces of two members held in contact by force F; (b) contact stress distribution. Note: The contact area is a circle of radius a.

the contact area to a maximum value p_o on the load axis z at its center. Here a is the radius of the *circular contact area* (πa^2). Because of forces F, the contact pressure is distributed over a small *circular area* of radius a given by

$$a = \left[\frac{3F}{4} \frac{(1 - \nu_1^2)/E_1 + (1 - \nu_2^2)/E_2}{1/r_1 + 1/r_2} \right]^{1/3} \tag{3.58}$$

where E_i, r_i, and ν_i (with $i = 1, 2$) are respective moduli of elasticity, radii, and Poisson's ratios of the spheres.

For simplicity, Poisson's ratios ν_1 and ν_2 will be taken as 0.3 in the following equations. In so doing, Eq. (3.58) becomes

$$a = 0.88 \left[\frac{F(E_1 + E_2)r_1 r_2}{E_1 E_2 (r_1 + r_2)} \right]^{1/3} \tag{3.59}$$

The force F causing the contact pressure acts in the direction of the normal axis, perpendicular to the tangent plane passing through the contact area. The *maximum contact pressure* is found to be

$$p_o = 1.5 \frac{F}{\pi a^2} \tag{3.60}$$

This is the maximum principal stress owing to the fact that at the center of the contact area, material is compressed not only in the normal direction but also in the lateral directions. The relationship between the force of contact F and the relative *displacement* of the centers of the two elastic spheres, owing to local deformation, is

$$\delta = 0.77 \left[F^2 \left(\frac{1}{E_1} + \frac{1}{E_2} \right)^2 \left(\frac{1}{r_1} + \frac{1}{r_2} \right) \right]^{1/3} \tag{3.61}$$

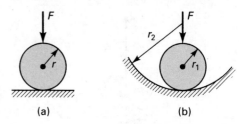

(a)   (b)

In the special case of a *sphere* of radius r contacting a body of the same material but having a *flat surface* (Fig. 3.19a), substitution of $r_1 = r, r_2 = \infty$ and $E_1 = E_2 = E$ into Eqs. (3.59) through (3.61) leads to

$$a = 0.88\left(\frac{2Fr}{E}\right)^{1/3}, \qquad p_o = 0.62\left(\frac{FE^2}{4r^2}\right)^{1/3}, \qquad \delta = 1.54\left(\frac{F^2}{2E^2r}\right)^{1/3} \quad \textbf{(3.62)}$$

For the case of a *sphere* in a *spherical seat* of the same material (Fig. 3.19b), substituting $r_2 = -r_2$ and $E_1 = E_2 = E$ in Eqs. (3.59) through (3.61), we obtain

$$a = 0.88\left[\frac{2Fr_1r_2}{E(r_2 - r_1)}\right]^{1/3}, \qquad p_o = 0.623\left[FE^2\left(\frac{r_2 - r_1}{2r_1r_2}\right)^2\right]^{1/3}$$

$$\delta = 1.54\left[\frac{F^2(r_2 - r_1)}{2E^2r_1r_2}\right]^{1/3}$$

$$\textbf{(3.63)}$$

Two Parallel Cylinders in Contact

Figure 3.20a shows the contact area and corresponding stress distribution between two spheres, loaded with a force F. It is seen from the figure that the contact pressure within each cylinder has a semi-elliptical distribution; it varies from 0 at the side of the contact area to the largest value p_o at its center. The quantity a represents the half-width of a *narrow* rectangular contact area ($2aL$). Note that the maximum contact pressure p_o occurs on the load axis z.

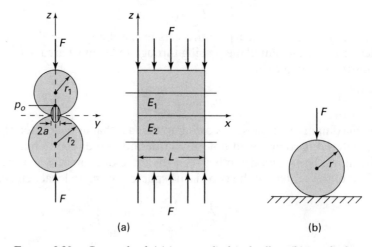

(a)   (b)

FIGURE 3.20. *Contact load: (a) in two cylindrical rollers; (b) in cylinder on a plane.*

The *maximum contact pressure* is given by

$$p_o = \frac{2}{\pi} \frac{F}{aL} \tag{3.64}$$

where

$$a = \left[\frac{4Fr_1r_2}{\pi L(r_1 + r_2)} \left(\frac{1 - \nu_1^2}{E_1} + \frac{1 - \nu_2^2}{E_2} \right) \right]^{1/2} \tag{3.65}$$

In this expression, $E_i(\nu_i)$ and r_i, with $i = 1, 2$, are the moduli of elasticity (Poisson's ratio) of the two rollers and the corresponding radii, respectively. If the cylinders have the *same* elastic modulus E and Poisson's ratio $\nu = 0.3$, these expressions reduce to

$$p_o = 0.418 \sqrt{\frac{FE}{L} \frac{r_1 + r_2}{r_1 r_2}}, \qquad a = 1.52 \sqrt{\frac{F}{EL} \frac{r_1 r_2}{r_1 + r_2}} \tag{3.66}$$

Figure 3.20b depicts the special case of contact between a circular *cylinder* of radius r and a *flat surface*, both bodies of the same material. After rearranging the terms and taking $r_1 = r$ and $r_2 = \infty$ in Eqs. (3.66), we have

$$p_o = 0.418 \sqrt{\frac{FE}{Lr}}, \qquad a = 1.52 \sqrt{\frac{Fr}{EL}} \tag{3.67}$$

In Table 3.2, the preceding and some additional results are presented as an aid in solving problems.

3.15   CONTACT STRESS DISTRIBUTION

The material along the axis compressed in the z direction tends to expand in the x and y directions. But, the surrounding material does not permit this expansion. Thus, the compressive stresses are produced in the x and y directions. The maximum stresses occur along the load axis z, and they are the principal stresses, as shown in Fig. 3.21. These and the maximum shearing stresses are given in terms of the maximum contact pressure p_o by the following equations [Ref. 3.14].

Two Spheres in Contact (Figure 3.18a)

$$\sigma_x = \sigma_y = -p_o \left\{ \left(1 - \frac{z}{a} \tan^{-1} \frac{1}{z/a} \right) (1 + \nu) - \frac{1}{2[1 + (z/a)^2]} \right\} \tag{3.68a}$$

$$\sigma_z = -\frac{p_o}{1 + (z/a)^2} \tag{3.68b}$$

We have $\tau_{xy} = 0$ and

$$\tau_{yz} = \tau_{xz} = \tfrac{1}{2}(\sigma_x - \sigma_z) \tag{3.68c}$$

Figure 3.22a shows a plot of the preceding equations.

TABLE 3.2.   *Maximum Pressure p_o and Deflection δ of Two Bodies in Contact*

Configuration	*Spheres:* $p_o = 1.5\dfrac{F}{\pi a^2}$	*Cylinders:* $p_o = \dfrac{2}{\pi}\dfrac{F}{aL}$
A	Sphere on a Flat Surface	Cylinder on a Flat Surface
	$a = 0.880\,\sqrt[3]{Fr_1\Delta}$	$a = 1.076\,\sqrt{\dfrac{F}{L}r_l\Delta}$
		For $E_1 = E_2 = E$:
	$\delta = 0.775\,\sqrt[3]{F^2\dfrac{\Delta^2}{r_1}}$	$\delta = \dfrac{0.579F}{EL}\left(\dfrac{1}{3} + \ln\dfrac{2r_1}{a}\right)$
B	Two Spherical Balls	Two Cylindrical Rollers
	$a = 0.880\,\sqrt[3]{F\dfrac{\Delta}{m}}$	$a = 1.076\,\sqrt{\dfrac{F\Delta}{Lm}}$
	$\delta = 0.775\,\sqrt[3]{F^2\Delta^2 n}$	
C	Sphere on a Spherical Seat	Cylinder on a Cylindrical Seat
	$a = 0.880\,\sqrt[3]{F\dfrac{\Delta}{n}}$	$a = 1.076\,\sqrt{\dfrac{F\Delta}{Ln}}$
	$\delta = 0.775\,\sqrt[3]{F^2\Delta^2 n}$	

Note: $\Delta = \dfrac{1}{E_1} + \dfrac{1}{E_2}$, $m = \dfrac{1}{r_1} + \dfrac{1}{r_2}$, $n = \dfrac{1}{r_1} - \dfrac{1}{r_2}$

where the modulus of elasticity (E) and radius (r) are for the contacting members, 1 and 2. The L represents the length of the cylinder. The total force pressing two spheres of cylinders is F. Poisson's ratio ν in the formulas is taken as 0.3. Source: Ref. 3.13.

FIGURE 3.21.   *Principal stresses below the surface along the load axis z.*

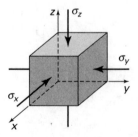

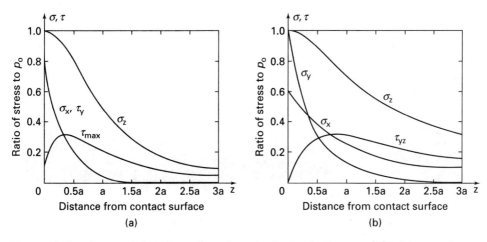

FIGURE 3.22. *Stresses below the surface along the load axis (for $v = 0.3$): (a) two spheres; (b) two parallel cylinders. Note: All normal stresses are compressive stresses.*

Two Parallel Cylinders in Contact (Figure 3.20a)

$$\sigma_x = -2\nu p_o \left[\sqrt{1+\left(\frac{z}{a}\right)^2} - \frac{z}{a} \right] \tag{3.69a}$$

$$\sigma_y = -p_o \left\{ \left[2 - \frac{1}{1+(z/a)^2} \right] \sqrt{1+\left(\frac{z}{a}\right)^2} - 2\frac{z}{a} \right\} \tag{3.69b}$$

$$\sigma_z = -\frac{p_o}{\sqrt{1+(z/a)^2}} \tag{3.69c}$$

$$\tau_{xy} = \tfrac{1}{2}(\sigma_x - \sigma_y), \quad \tau_{yz} = \tfrac{1}{2}(\sigma_y - \sigma_z), \quad \tau_{xz} = \tfrac{1}{2}(\sigma_x - \sigma_z) \tag{3.69d}$$

A plot of Eqs. (3.69a–c) and the second of Eqs. (3.69d) is given in Fig. 3.22b.

For each case, observe how principal stress decreases below the surface. Figure 3.22 also illustrates how the shearing stress reaches a maximum value slightly below the surface and decreases. The maximum shear stresses act on the planes bisecting the planes of maximum and minimum principal stresses (Sec. 1.15). As already pointed out, all stresses considered in this section exist along the load axis z. The states of stress off the z axis are not required for design purposes, because the maxima occur on the z axis.

EXAMPLE 3.5 Cam and Follower

A camshaft and follower of an intermittent motion mechanism is illustrated in Fig. 3.23. For the position depicted, the cam exerts a force F_{max} on the follower. Determine (a) the maximum stress at the contact

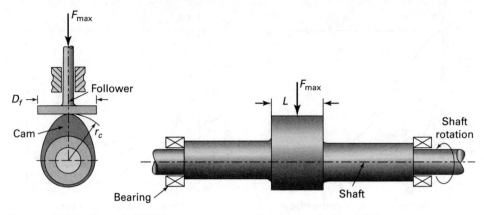

FIGURE 3.23.   *Example 3.5. Schematic representation of camshaft and follower.*

line between the cam and follower; (b) the deflection. *Data:* F_{max} = 8 kN, r_c = 40 mm, $D_f = L$ = 35 mm, E = 200 GPa, and σ_{yp} = 510 MPa. *Assumptions:* The material of all parts is hardened on the surface. Frictional forces can be omitted. The rotational speed is slow so that the loading is considered static.

Solution   Formulas on the second column of case A of Table 3.2 apply. We begin by calculating the half-width a of the contact patch. Inasmuch as $E_1 = E_2 = E$ and $\Delta = 2/E$, hence,

$$a = 1.076 \sqrt{\frac{F_{max}}{L} r_c \Delta}$$

Introducing the given numerical values results in

$$a = 1.076 \left[\frac{800}{35} (40) \left(\frac{2}{200 \times 10^9} \right) \right]^{1/2}$$

$$= 0.325 \, (10^{-3}) \, \text{m} = 0.325 \, \text{mm}$$

a.   The largest contact pressure is therefore

$$p_o = \frac{2}{\pi} \frac{F_{max}}{aL}$$

$$= \frac{2}{\pi} \frac{8000}{(0.325)(35)} = 448 \, \text{MPa}$$

b.   The deflection δ of the cam and follower at the line of contact is given by

$$\delta = \frac{0.579 F_{max}}{EL_4} \left(\frac{1}{3} + \ln \frac{2r_c}{a} \right)$$

Inserting the given data,

$$\delta = \frac{0.579(8000)}{(2000 \times 10^9)(35 \times 10^{-3})} \left(\frac{1}{3} + \ln \frac{2 \times 40}{0.325} \right)$$

$$= 3.9 \, (10^{-6}) \, \text{m} = 0.0039 \, \text{mm}$$

Comments The maximum contact stress is calculated to be smaller than the yield strength of 510 MPa; the design is satisfactory. Deflection obtained between the cam and the follower is very small and does not affect the performance of the mechanism.

3.16 GENERAL CONTACT

Consider now two rigid bodies of equal elastic moduli E, compressed by force F (Fig. 3.24). The load lies along the axis passing through the centers of the bodies and through the point of contact and is perpendicular to the plane tangent to both bodies at the point of contact. The minimum and maximum radii of curvature of the surface of the upper body are r_1 and r_1'; those of the lower body are r_2 and r_2' at the point of contact. Thus, $1/r_1$, $1/r_1'$, $1/r_2$, and $1/r_2'$ are the principal curvatures. The *sign convention* of the *curvature* is such that it is *positive* if the corresponding center of curvature is *inside* the body. If the center of the curvature is *outside* the body, the curvature is *negative*. (For example, in Fig. 3.25a, r_1, r_1' are positive, while r_2, r_2' are negative.)

Let θ be the angle between the normal planes in which radii r_1 and r_2 lie. Subsequent to loading, the area of contact will be an *ellipse* with semiaxes a and b (Table C.1). The *maximum contact pressure* is

$$p_o = 1.5 \frac{F}{\pi ab} \tag{3.70}$$

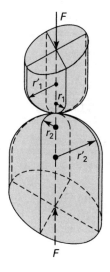

FIGURE 3.24. *Curved surfaces of different radii of two bodies compressed by forces F.*

In this expression the semiaxes are given by

$$a = c_a \sqrt[3]{\frac{Fm}{n}}, \qquad b = c_b \sqrt[3]{\frac{Fm}{n}} \tag{3.71}$$

Here

$$m = \frac{4}{\dfrac{1}{r_1} + \dfrac{1}{r'_1} + \dfrac{1}{r_2} + \dfrac{1}{r'_2}}, \qquad n = \frac{4E}{3(1 - \nu^2)} \tag{3.72}$$

The constants c_a and c_b are read in Table 3.3. The first column of the table lists values of α, calculated from

$$\cos \alpha = \frac{B}{A} \tag{3.73}$$

where

$$A = \frac{2}{m}, \qquad B = \frac{1}{2}\left[\left(\frac{1}{r_1} - \frac{1}{r'_1}\right)^2 + \left(\frac{1}{r_2} - \frac{1}{r'_2}\right)^2 \right.$$

$$\left. + 2\left(\frac{1}{r_1} - \frac{1}{r'_1}\right)\left(\frac{1}{r_2} - \frac{1}{r'_2}\right)\cos 2\theta\right]^{1/2} \tag{3.74}$$

By applying Eq. (3.70), many problems of practical importance may be treated, for example, contact stresses in ball bearings (Fig. 3.25a), contact stresses between a cylindrical wheel and a rail (Fig. 3.25b), and contact stresses in cam and pushrod mechanisms.

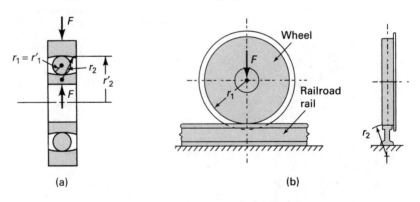

(a) (b)

FIGURE 3.25. *Contact load: (a) in a single-row ball bearing; (b) in a cylindrical wheel and rail.*

TABLE 3.3.   *Factors for Use in Eqs. (3.71)*

α (degrees)	c_a	c_b
20	3.778	0.408
30	2.731	0.493
35	2.397	0.530
40	2.136	0.567
45	1.926	0.604
50	1.754	0.641
55	1.611	0.678
60	1.486	0.717
65	1.378	0.759
70	1.284	0.802
75	1.202	0.846
80	1.128	0.893
85	1.061	0.944
90	1.000	1.000

EXAMPLE 3.6   Steel Railway Car Wheel

A railway car wheel rolls on a rail. Both rail and wheel are made of steel for which $E = 210$ GPa and $\nu = 0.3$. The wheel has a radius of $r_1 = 0.4$ m, and the cross radius of the rail top surface is $r_2 = 0.3$ m (Fig. 3.25b). Determine the size of the contact area and the maximum contact pressure, given a compression load of $F = 90$ kN.

Solution   For the situation described, $1/r'_1 = 1/r'_2 = 0$, and, because the axes of the members are mutually perpendicular, $\theta = \pi/2$. The first of Eqs. (3.72) and Eqs. (3.74) reduce to

$$m = \frac{4}{1/r_1 + 1/r_2}, \qquad A = \frac{1}{2}\left(\frac{1}{r_1} + \frac{1}{r_2}\right), \qquad B = \pm\frac{1}{2}\left(\frac{1}{r_1} - \frac{1}{r_2}\right) \quad (3.75)$$

The proper sign in B must be chosen so that its values are positive. Now Eq. (3.73) has the form

$$\cos\alpha = \pm\frac{1/r_1 - 1/r_2}{1/r_1 + 1/r_2} \qquad (3.76)$$

Substituting the given numerical values into Eqs. (3.75), (3.76), and the second of (3.72), we obtain

$$m = \frac{4}{1/0.4 + 1/0.3} = 0.6857, \qquad n = \frac{4(210 \times 10^9)}{3(0.91)} = 3.07692 \times 10^{11}$$

$$\cos \alpha = \pm \frac{1/0.4 - 1/0.3}{1/0.4 + 1/0.3} = 0.1428 \quad \text{or} \quad \alpha = 81.79°$$

Corresponding to this value of α, interpolating in Table 3.3, we have

$$c_a = 1.1040, \qquad c_b = 0.9113$$

The semiaxes of the elliptical contact are found by applying Eqs. (3.71):

$$a = 1.1040 \left[\frac{90,000 \times 0.6857}{3.07692 \times 10^{11}} \right]^{1/3} = 0.00646 \text{ m}$$

$$b = 0.9113 \left[\frac{90,000 \times 0.6857}{3.07692 \times 10^{11}} \right]^{1/3} = 0.00533 \text{ m}$$

The maximum contact pressure, or maximum principal stress, is thus

$$p_o = 1.5 \frac{90,000}{\pi(0.00646 \times 0.00533)} = 1248 \text{ MPa}$$

Comment A hardened steel material is capable of resisting this or somewhat higher stress levels for the body geometries and loading conditions described in this section.

REFERENCES

3.1. SOKOLNIKOFF, I. S., *Mathematical Theory of Elasticity*, 2nd ed., Krieger, Melbourne, Fla., 1986.

3.2. TIMOSHENKO, S. P., and GOODIER, J. N., *Theory of Elasticity*, 3rd ed., McGraw-Hill, New York, 1970, Chaps. 3 and 4.

3.3. NEOU, C. Y., Direct method of determining Airy polynomial stress functions, J. Appl. Mech. 24/3, 387, 1957.

3.4. BOLEY, B. A., and WEINER, J. H., *Theory of Thermal Stresses*, Wiley, Hoboken, N. J., 1960, Chap. 2: reprinted, R. E. Krieger, Melbourne, Fla., 1985.

3.5. NOWACKI, W., *Thermoelasticity*, Addison-Wesley, Reading, Mass., 1963.

3.6. MITCHELL, J. N., *The Collected Mathematical Works of J. H. and A. G. M. Mitchell*, P. Nordhoff, Ltd., Groningen, Holland, 1964.

3.7. TIMOSHENKO, S. P., and GOODIER, J. N., *Theory of Elasticity*, 3rd ed., McGraw-Hill, New York, 1970, Sec. 40.

3.8. PETERSON, R. E., *Stress Concentration Factors*, Wiley, Hoboken, N. J., 1974.

3.9. INGLIS, C. E., Stresses in the plate due to presence of cracks and sharp corners, Trans. Inst. Naval Arch., 60, 219, 1913.

3.10. NEUBER, H. P., *Kerbspannungslehre*, 2nd ed., Springer, New York, 1958.

3.11. NEUBER, H. P., Research on the distribution of tension in notched construction parts. WADD Rept. 60-906, Jan. 1961.

3.12. TIMOSHENKO, S. P., and GOODIER, J. N., *Theory of Elasticity*, 3rd ed., McGraw-Hill, New York, 1970, Sec. 35.

3.13. YOUNG, W. C., *Roark's Formulas for Stress and Strain*, 6th ed., McGraw-Hill, New York, 1989, Sec. 2.10 and Table 37.

3.14. UGURAL, A. C., *Mechanical Design: An Integrated Approach*, McGraw-Hill, New York, 2004, Chap. 3.

3.15. FLUGGE, W., ed., *Handbook of Engineering Mechanics*, McGraw-Hill, New York, 1908, Chap. 42.

3.16. HERTZ, H., Contact of elastic solids; in *Miscellaneous Papers*, P. Lenard, ed., MacMillan, London, pp. 146–162, 1892.

3.17. FAUPEL, J. H., and FISHER, F. E., *Engineering Design*, 2nd ed., Wiley, Hoboken, N. J., 1981, Chap. 11.

3.18. BORESI, A. P., and SCHMIDT, R. J., *Advanced Mechanics of Materials*, 6th ed., Wiley, Hoboken, N. J., 2003, Chap. 18.

3.19. TIMOSHENKO, S. P., and GOODIER, J. N., *Theory of Elasticity*, 3rd ed., McGraw-Hill, New York, 1970, Chap. 12.

PROBLEMS

Sections 3.1 through 3.8

3.1. A stress distribution is given by

$$\sigma_x = pyx^3 - 2c_1xy + c_2y$$
$$\sigma_y = pxy^3 - 2px^3y \tag{a}$$
$$\tau_{xy} = -\tfrac{3}{2}px^2y^2 + c_1y^2 + \tfrac{1}{2}px^4 + c_3$$

where the p and c's are constants. (a) Verify that this field represents a solution for a thin plate of thickness t (Fig. P3.1); (b) obtain the corresponding stress function; (c) find the *resultant* normal and shearing boundary forces (P_y and V_x) along edges $y = 0$ and $y = b$ of the plate.

3.2. If the stress field given by Eq. (a) of Prob. 3.1 acts in the thin plate shown in Fig. P3.1 and p is a known constant, determine the c's so that edges $x = \pm a$ are free of shearing stress and no normal stress acts on edge $x = a$.

3.3. In bending of a rectangular plate (Fig. P3.3), the state of stress is expressed by

$$\sigma_x = c_1y + c_2xy \qquad \tau_{xy} = c_3(b^2 - y^2)$$

(a) What conditions among the constants (the c's) make the preceding expressions possible? Body forces may be neglected. (b) Draw a sketch showing the boundary stresses on the plate.

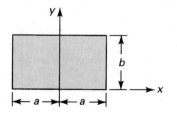

FIGURE P3.1.

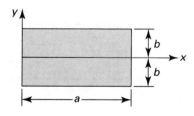

FIGURE P3.3.

3.4. Given the following stress field within a structural member,

$$\sigma_x = a[y^2 + b(x^2 - y^2)] \qquad \tau_{xy} = -2abxy$$
$$\sigma_y = a[x^2 + b(y^2 - x^2)] \qquad \tau_{yz} = \tau_{xz} = 0$$
$$\sigma_z = ab(x^2 + y^2)$$

where a and b are constants, determine whether this stress distribution represents a solution for a plane strain problem. The body forces are omitted.

3.5. Determine whether the following stress functions satisfy the conditions of compatibility for a two-dimensional problem:

$$\Phi_1 = ax^2 + bxy + cy^2 \qquad \textbf{(a)}$$
$$\Phi_2 = ax^3 + bx^2y + cxy^2 + dy^3 \qquad \textbf{(b)}$$

Here $a, b, c,$ and d are constants. Also obtain the stress fields that arise from Φ_1 and Φ_2.

3.6. Figure P3.6 shows a long, thin steel plate of thickness t, width $2h$, and length $2a$. The plate is subjected to loads that produce the uniform stresses σ_o at the ends. The edges at $y = \pm h$ are placed between the two rigid walls. Show that, by using an inverse method, the displacements are expressed by

$$u = -\frac{1 - \nu^2}{E}\sigma_o x, \qquad v = 0, \qquad w = \frac{\nu(1 + \nu)}{E}\sigma_o z$$

3.7. Determine whether the following stress distribution is a valid solution for a two-dimensional problem:

$$\sigma_x = -ax^2y \qquad \sigma_y = -\frac{1}{3}ay^3 \qquad \tau_{xy} = axy^2$$

where a is a constant. Body forces may be neglected.

3.8. The strain distribution in a thin plate has the form

$$\begin{bmatrix} ax^3 & axy^2 \\ axy^2 & ax^2y \end{bmatrix}$$

in which a is a small constant. Show whether this strain field is a valid solution of an elasticity problem. Body forces may be disregarded.

3.9. The components of the displacement of a thin plate (Fig. P3.9) are given by

$$u = -c(y^2 + \nu x) \qquad v = 2cxy$$

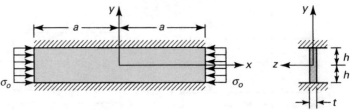

FIGURE P3.6.

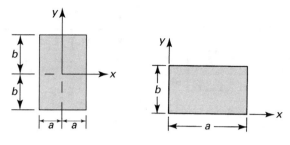

FIGURE P3.9. FIGURE P3.10.

Here c is a constant and v represents Poisson's ratio. Determine the stresses σ_x, σ_y, and τ_{xy}. Draw a sketch showing the boundary stresses on the plate.

3.10. Consider a rectangular plate with sides a and b of thickness t (Fig. P3.10). (a) Determine the stresses σ_x, σ_y, and τ_{xy} for the stress function $\Phi = px^3y$, where p is a constant. (b) Draw a sketch showing the boundary stresses on the plate. (c) Find the *resultant* normal and shearing boundary forces (P_x, P_y, V_x, and V_y) along all edges of the plate.

3.11. Redo Prob. 3.10 for the case of a square plate of side dimensions a and

$$\Phi = \frac{p}{2a^2}(x^2y^2 + \tfrac{1}{3}xy^3)$$

where p is a constant.

3.12. Resolve Prob. 3.10 a and b for the stress function of the form

$$\Phi = -\frac{p}{b^3}xy^2(3b - 2y)$$

where p represents a constant.

3.13. A vertical force P per unit thickness is applied on the horizontal boundary of a semi-infinite solid plate of unit thickness (Fig. 3.11a). Show that the stress function $\Phi = -(P/\pi)y\tan^{-1}(y/x)$ results in the following stress field within the plate:

$$\sigma_x = -\frac{2P}{\pi}\frac{x^3}{(x^2 + y^2)^2}, \qquad \sigma_y = -\frac{2P}{\pi}\frac{xy^2}{(x^2 + y^2)^2}, \qquad \tau_{xy} = -\frac{2P}{\pi}\frac{yx^2}{(x^2 + y^2)^2}$$

Also plot the resulting stress distribution for σ_x and τ_{xy} at a constant depth L below the boundary.

3.14. The thin cantilever shown in Fig. P3.14 is subjected to uniform shearing stress τ_o along its upper surface $(y = +h)$, while surfaces $y = -h$ and $x = L$ are free of stress. Determine whether the Airy stress function

$$\Phi = \tfrac{1}{4}\tau_o\left(xy - \frac{xy^2}{h} - \frac{xy^3}{h^2} + \frac{Ly^2}{h} + \frac{Ly^3}{h^2}\right)$$

satisfies the required conditions for this problem.

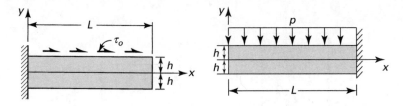

FIGURE P3.14.   FIGURE P3.15.

3.15. Figure P3.15 shows a thin cantilever beam of unit thickness carrying a uniform load of intensity p per unit length. Assume that the stress function is expressed by

$$\Phi = ax^2 + bx^2y + cy^3 + dy^5 + ex^2y^3$$

in which $a, \ldots, e$ are constants. Determine (a) the requirements on $a, \ldots, e$ so that Φ is biharmonic; (b) the stresses σ_x, σ_y, and τ_{xy}.

3.16. Consider a thin square plate with sides a. For a stress function $\Phi = (p/a^2)(\frac{1}{2}x^2y^2 - \frac{1}{6}y^4)$, determine the stress field and sketch it along the boundaries of the plate. Here p represents a uniformly distributed loading per unit length. Note that the origin of the x, y coordinate system is located at the lower-left corner of the plate.

3.17. Consider a thin cantilever loaded as shown in Fig. P3.17. Assume that the bending stress is given by

$$\sigma_x = -\frac{M_z y}{I} = -\frac{p}{2I}x^2y \tag{P3.17}$$

and $\sigma_z = \tau_{xz} = \tau_{yz} = 0$. Determine the stress components σ_y and τ_{xy} as functions of x and y.

3.18. Show that for the case of plane stress, in the absence of body forces, the equations of equilibrium may be expressed in terms of displacements u and v as follows:

$$\frac{\partial^2 u}{\partial x^2} + \frac{\partial^2 u}{\partial y^2} + \frac{1+v}{1-v}\frac{\partial}{\partial x}\left(\frac{\partial u}{\partial x} + \frac{\partial v}{\partial y}\right) = 0$$

$$\frac{\partial^2 v}{\partial y^2} + \frac{\partial^2 v}{\partial x^2} + \frac{1+v}{1-v}\frac{\partial}{\partial y}\left(\frac{\partial v}{\partial y} + \frac{\partial v}{\partial x}\right) = 0 \tag{P3.18}$$

[Hint: Substitute Eqs. (3.10) together with (2.3) into (3.6).]

3.19. Determine whether the following compatible stress field is possible within an elastic uniformly loaded cantilever beam (Fig. P3.17):

$$\sigma_x = -\frac{p}{10I}(5x^2 + 2h^2)y + \frac{p}{3I}y^3$$

$$\tau_{xy} = -\frac{px}{2I}(h^2 - y^2) \tag{P3.19}$$

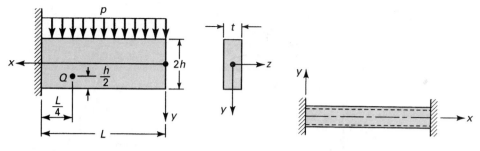

FIGURE P3.17.   FIGURE P3.24.

$$\sigma_y = -\frac{p}{6I}(2h^3 - 3h^2y + y^3)$$

Here $I = 2th^3/3$ and the body forces are omitted. Given $p = 10$ kN/m, $L = 2$ m, $h = 100$ mm, $t = 40$ mm, $v = 0.3$, and $E = 200$ GPa, calculate the magnitude and direction of the maximum principal strain at point Q.

3.20. A prismatic bar is restrained in the x (axial) and y directions but free to expand in z direction. Determine the stresses and strains in the bar for a temperature rise of T_1 degrees.

3.21. Under free thermal expansion, the strain components within a given elastic solid are $\varepsilon_x = \varepsilon_y = \varepsilon_z = \alpha T$ and $\gamma_{xy} = \gamma_{yz} = \gamma_{xz} = 0$. Show that the temperature field associated with this condition is of the form

$$\alpha T = c_1 x + c_2 y + c_3 z + c_4$$

in which the c's are constants.

3.22. Redo Prob. 3.6 adding a temperature change T_1, with all other conditions remaining unchanged.

3.23. Determine the axial force P_x and moment M_z that the walls in Fig. 3.6b apply to the beam for $T = a_1 y + a_2$, where a_1 and a_2 are constant.

3.24. A copper tube of 800-mm^2 cross-sectional area is held at both ends as in Fig. P3.24. If at 20°C no axial force P_x exists in the tube, what will P_x be when the temperature rises to 120°C? Let $E = 120$ GPa and $\alpha = 16.8 \times 10^{-6}$ per °C.

Sections 3.9 through 3.11

3.25. Show that the case of a concentrated load on a straight boundary (Fig. 3.11a) is represented by the stress function

$$\Phi = -\frac{P}{\pi}r\theta \sin \theta$$

and derive Eqs. (3.48) from the result.

3.26. Verify that Eqs. (3.37) are determined from the equilibrium of forces acting on the elements shown in Fig. P3.26.

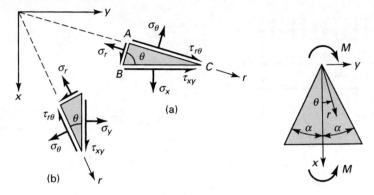

FIGURE P3.26. FIGURE P3.29.

3.27. Demonstrate that the biharmonic equation $\nabla^4\Phi = 0$ in polar coordinates can be written as

$$\left(\frac{\partial^2}{\partial r^2} + \frac{1}{r}\frac{\partial}{\partial r} + \frac{1}{r^2}\frac{\partial^2}{\partial\theta^2}\right)\left(\frac{\partial^2\Phi}{\partial r^2} + \frac{1}{r}\frac{\partial\Phi}{\partial r} + \frac{1}{r^2}\frac{\partial^2\Phi}{\partial\theta^2}\right) = 0$$

3.28. Show that the compatibility equation in polar coordinates, for the axisymmetrical problem of thermal elasticity, is given by

$$\frac{1}{r}\frac{d}{dr}\left(r\frac{d\Phi}{dr}\right) + E\alpha T = 0 \qquad \textbf{(P3.28)}$$

3.29. Assume that moment M acts in the plane and at the vertex of the wedge–cantilever shown in Fig. P3.29. Given a stress function

$$\Phi = -\frac{M(\sin 2\theta - 2\theta\cos 2\alpha)}{2(\sin 2\alpha - 2\alpha\cos 2\alpha)} \qquad \textbf{(P3.29a)}$$

determine (a) whether Φ satisfies the condition of compatibility; (b) the stress components σ_r, σ_θ, and $\tau_{r\theta}$; and (c) whether the expressions

$$\sigma_r = -\frac{2M\sin 2\theta}{\pi r^2}, \qquad \sigma_\theta = 0, \qquad \tau_{r\theta} = \frac{2M\cos^2\theta}{\pi r^2} \qquad \textbf{(P3.29b)}$$

represent the stress field in a semi-infinite plate (that is, for $\alpha = \pi/2$).

3.30. Referring to Fig. P3.30, verify the results given by Eqs. (b) and (c) of Section 3.11.

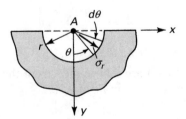

FIGURE P3.30.

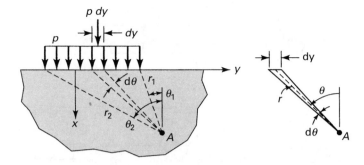

FIGURE P3.34.

3.31. Consider the pivot of unit thickness subject to force P per unit thickness at its vertex (Fig. 3.10a). Determine the maximum values of σ_x and τ_{xy} on a plane a distance L from the apex through the use of σ_r given by Eq. (3.43) and the formulas of the elementary theory: (a) take $\alpha = 15°$; (b) take $\alpha = 60°$. Compare the results given by the two approaches.

3.32. Solve Prob. 3.31 for $\alpha = 30°$.

3.33. Redo Prob. 3.31 in its entirety for the wedge–cantilever shown in Fig. 3.10b.

3.34. A uniformly distributed load of intensity p is applied over a short distance on the straight edge of a large plate (Fig. P3.34). Determine stresses σ_x, σ_y, and τ_{xy} in terms of p, θ_1, and θ_2, as required. [Hint: Let $dP = pdy$ denote the load acting on an infinitesimal length $dy = rd\,\theta/\cos\theta$ (from geometry) and hence $dP = prd\,\theta/\cos\theta$. Substitute this into Eqs. (3.50) and integrate the resulting expressions.]

Sections 3.12 through 3.16

3.35. Verify the result given by Eqs. (f) and (g) of Section 3.12 (a) by rewriting Eqs. (d) and (e) in the following forms, respectively,

$$\frac{1}{r}\frac{d}{dr}\left\{r\frac{d}{dr}\left[\frac{1}{r}\frac{d}{dr}\left(r\frac{df_1}{dr}\right)\right]\right\} = 0$$

$$r\frac{d}{dr}\left(\frac{1}{r^3}\frac{d}{dr}\left\{r^3\frac{d}{dr}\left[\frac{1}{r^3}\frac{d}{dr}(r^2f_2)\right]\right\}\right) = 0 \qquad \textbf{(P3.35)}$$

and by integrating (P3.35); (b) by expanding Eqs. (d) and (e), setting $t = \ln r$, and thereby transforming the resulting expressions into two ordinary differential equations with constant coefficients.

3.36. Verify the results given in Fig. 3.15 by employing Eq. (3.55b) and the method of superposition.

3.37. A 20-mm-thick steel bar with a slot (25-mm radii at ends) is subjected to an axial load P, as shown in Fig. P3.37. What is the maximum stress for $P = 180$ kN? Use Fig. D.8B to estimate the value of the K.

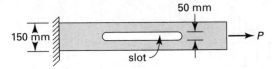

FIGURE P3.37.

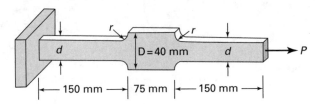

FIGURE P3.38.

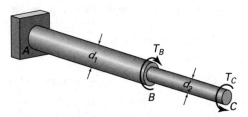

FIGURE P3.40.

3.38. What is the full-fillet radius r and width d of the steel plate with $D/d = 1.5$ in tension (Fig. P3.38)? Use a maximum allowable stress of 130 MPa and an allowable nominal stress in the reduced section of 80 MPa.

3.39. For the 20-mm-thick full-filleted steel bar ($\sigma_{yp} = 250$ MPa) shown in Fig. P3.38, given the ratio of $r/d = 0.15$, find the maximum axial load P that can be applied without causing permanent deformation.

3.40. As seen in Fig. P3.40, a stepped shaft ABC with built-in end at A carries the torques T_B and T_C sections B and C. Based on a stress concentration factor $K = 1.6$, determine the maximum shearing stress in the shaft. *Given:* $d_1 = 50$ mm, $d_2 = 40$ mm, $T_B = 3$ kN · m, and $T_C = 1$ kN · m.

3.41. Figure 3.17 illustrates a circular shaft consisting of diameters D and d and a groove of radius r carries a torque T with $M = 0$ and $P = 0$. What is the minimum yield strength in shear required for the shaft material? *Given:* $D = 40$ mm, $d = 35$ mm, $r = 2$ mm, and $T = 100$ N · m.

3.42. A circular shaft having diameters D and d and a groove of radius r (see Fig. 3.17, with $M = 0$ and $P = 0$) is made of steel with the allowable shear stress τ_{all}. Find the maximum torque T that can be transmitted by the shaft. *Given:* $D = 40$ mm, $d = 16$ mm, $r = 8$ mm, and $\tau_{all} = 250$ MPa.

3.43. For a flat bar consisting of two portions, both 10-mm thick, and respectively 25-mm and 37.5-mm wide, connected by fillets of radius $r = 5$ mm

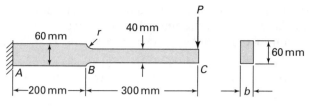

FIGURE P3.44.

(see Fig. D.1), determine the largest axial load P that can be supported by the bar. *Given:* $\sigma_{yp} = 210$ MPa and a factor of safety of $n = 1.4$.

3.44. Figure P3.44 depicts a filleted cantilever spring. Find the largest bending stress for two cases: (a) the fillet radius is $r = 5$ mm; (b) the fillet radius is $r = 10$ mm. *Given:* $b = 12$ mm and $P = 400$ N.

3.45. A thin-walled circular cylindrical vessel of diameter d and wall thickness t is subjected to internal pressure p (see Table 1.1). Given a small circular hole in the vessel wall, show that the maximum tangential and axial stresses at the hole are $\sigma_\theta = 5pd/4t$ and $\sigma_a = pd/4t$, respectively.

3.46. The shaft shown in Fig. 3.17 has the following dimensions: $r = 20$ mm, $d = 400$ mm, and $D = 440$ mm. The shaft is subjected simultaneously to a torque $T = 20$ kN·m, a bending moment $M = 10$ kN·m, and an axial force $P = 50$ kN. Calculate at the root of the notch (a) the maximum principal stress, (b) the maximum shear stress, and (c) the octahedral stresses.

3.47. Redo Prob. 3.46 for $r = 10$ mm, $d = 250$ mm, $D = 500$ mm, $T = 5$ kN·m, $M = 20$ kN·m, and $P = 0$.

3.48. A 50-mm-diameter ball is pressed into a spherical seat of diameter 75 mm by a force of 500 N. The material is steel ($E = 200$ GPa, $\nu = 0.3$). Calculate (a) the radius of the contact area, (b) the maximum contact pressure, and (c) the relative displacement of the centers of the ball and seat.

3.49. Calculate the maximum contact pressure p_o in Prob. 3.48 for the cases when the 50-mm-diameter ball is pressed against (a) a flat surface and (b) an identical ball.

3.50. Calculate the maximum pressure between a steel wheel of radius $r_1 = 400$ mm and a steel rail of crown radius of the head $r_2 = 250$ mm (Fig. 3.25b) for $P = 4$ kN. Use $E = 200$ GPa and $\nu = 0.3$.

3.51. A concentrated load of 2.5 kN at the center of a deep steel beam is applied through a 10-mm-diameter steel rod laid across the 100-mm beam width. Compute the maximum contact pressure and the width of the contact between rod and beam surface. Use $E = 200$ GPa and $\nu = 0.3$.

3.52. Two identical 400-mm-diameter steel rollers of a rolling mill are pressed together with a force of 2 MN/m. Using $E = 200$ GPa and $\nu = 0.25$, compute the maximum contact pressure and width of contact.

3.53. Determine the size of the contact area and the maximum pressure between two circular cylinders with mutually perpendicular axes. Denote by r_1 and r_2 the radii of the cylinders. Use $r_1 = 500$ mm, $r_2 = 200$ mm, $F = 5$ kN, $E = 210$ GPa, and $\nu = 0.25$.

3.54. Solve Prob. 3.53 for the case of two cylinders of equal radii, $r_1 = r_2 = 200$ mm.

3.55. Two 340-mm-diameter balls of a rolling mill are pressed together with a force of 400 N. Calculate (a) the half-width of contact, (b) the maximum contact pressure, (c) the maximum principal stresses and shear stress in the center of the contact area. *Assumption*: Both balls are made of steel of $E = 210$ GPa and $\nu = 0.3$.

3.56. A 16-mm-diameter cylindrical roller runs on the inside of a ring of inner diameter 100 mm. Determine (a) the width a of the contact area, (b) the maximum contact pressure. *Given:* The roller load is $F = 240$ kN per meter of axial length. *Assumption:* Both roller and ring are made of steel having $E = 210$ GPa and $\nu = 0.3$.

3.57. It is seen in Fig. P3.25b, a wheel of radius $r_1 = 480$ mm and a rail of crown radius of the head $r_2 = 340$ mm. Calculate the maximum contact pressure p_o between the members. *Given:* Contact force $F = 4$ kN. *Assumption:* Both roller and ring are made of steel having $E = 210$ GPa and $\nu = 0.3$.

3.58. Determine the maximum pressure at the contact point between the outer race and a ball in the single-row ball bearing assembly shown in Fig. 3.25a. The ball diameter is 50 mm; the radius of the grooves, 30 mm; the diameter of the outer race, 250 mm; and the highest compressive force on the ball, $F = 1.8$ kN. Take $E = 200$ GPa and $\nu = 0.3$.

3.59. Redo Prob. 3.58 for a ball diameter of 40 mm and a groove radius of 22 mm. Assume the remaining data to be unchanged.

CHAPTER 4

Failure Criteria

4.1 INTRODUCTION

The efficiency of design relies in great measure on an ability to predict the circumstances under which failure is likely to occur. The important variables connected with structural failure include the nature of the material; the load configuration; the rate of loading; the shape, surface peculiarities, and temperature of the member; and the characteristics of the medium surrounding the member (environmental conditions). Exact quantitative formulation of the problem of failure and accurate means for predicting failure represent areas of current research.

In Chapter 2 the stress–strain properties and characteristics of engineering materials were presented. We now discuss the mechanical behavior of materials associated with failure. The relations introduced for each theory are represented in a graphical form, which are extremely useful in visualizing impending failure in a stressed member. Note that a *yield* criterion is a part of plasticity theory (see Sec. 12.1). An introduction to fracture mechanics theory that provides a means to predict a sudden failure as the basis of a computed stress-intensity factor compared to a tested toughness criterion for the material is given in Section 4.13. Theories of failure for repeated loading and response of materials to dynamic loading and temperature change are taken up in the remaining sections.

4.2 FAILURE

In the most general terms, *failure* refers to any action leading to an inability on the part of the structure or machine to function in the manner intended. It follows that permanent deformation, fracture, or even excessive linear elastic deflection may be regarded as modes of failure, the last being the most easily predicted. Another way in which a member may fail is through *instability*, by undergoing large displacements from its design configuration when the applied load reaches a critical value, the buckling load (Chap. 11). In this chapter, the failure of homogeneous materials by yielding or permanent deformation and by fracture are given particular emphasis.*

Among the variables cited, one of the most important factors in regard to influencing the threshold of failure is the rate at which the load is applied. Loading at high rate, that is, dynamic loading, may lead to a variety of adverse phenomena associated with impact, acceleration, and vibration, and with the concomitant high levels of stress and strain, as well as rapid reversal of stress. In a conventional tension test, the rate referred to may relate to either the application of load or changes in strain. Ordinarily, strain rates on the order of 10^{-4} s^{-1} are regarded as "static" loading.

Our primary concern in this chapter and in this text is with *polycrystalline structural metals* or alloys, which are composed of crystals or grains built up of atoms. It is reasonable to expect that very small volumes of a given metal will not exhibit isotropy in such properties as elastic modulus. Nevertheless, we adhere to the basic assumption of isotrophy and homogeneity, because we deal primarily with an entire body or a large enough segment of the body to contain many randomly distributed crystals, which behave as an isotropic material would.

The brittle or ductile character of a metal has relevance to the mechanism of failure. If a metal is capable of undergoing an appreciable amount of yielding or permanent deformation, it is regarded as *ductile*. Such materials include mild steel, aluminum and some of its alloys, copper, magnesium, lead, Teflon, and many others. If, prior to fracture, the material can suffer only small yielding (less than 5%), the material is classified as *brittle*. Examples are concrete, stone, cast iron, glass, ceramic materials, and many common metallic alloys. The distinction between ductile and brittle materials is not as simple as might be inferred from this discussion. The nature of the stress, the temperature, and the material itself all play a role, as discussed in Section 4.17, in defining the boundary between ductility and brittleness.

4.3 FAILURE BY YIELDING

Whether because of material inhomogeneity or nonuniformity of loading, regions of high stress may be present in which *localized* yielding occurs. As the load increases, the inelastic action becomes more widespread, resulting eventually in a state of *general* yielding. The rapidity with which the transition from localized to general yielding occurs depends on the service conditions as well as the distribution of stress and the properties of the materials. Among the various service conditions, temperature represents a particularly significant factor.

The relative motion or *slip* between two planes of atoms (and the relative displacement of two sections of a crystal that results) represents the most common mechanism of yielding. Slip occurs most readily along certain crystallographic

---

*For further details, see Ref. 4.1 and texts on material science, for example, Refs. 4.2 through 4.4.

planes, termed *slip* or *shear planes*. The planes along which slip takes place easily are generally those containing the largest number of atoms per unit area. Inasmuch as the gross yielding of material represents the total effect of slip occurring along many randomly oriented planes, the yield strength is clearly a statistical quantity, as are other material properties such as the modulus of elasticity. If a metal fails by yielding, one can, on the basis of the preceding considerations, expect the shearing stress to play an important role.

It is characteristic of most ductile materials that after yielding has taken place, the load must be increased to produce further deformation. In other words, the material exhibits a strengthening termed *strain hardening* or *cold working*, as shown in Section 2.8. The slip occurring on intersecting planes of randomly oriented crystals and their resulting interaction is believed to be a factor of prime importance in strain hardening.

Creep

The deformation of a material under short-time loading (as occurs in a simple tension test) is simultaneous with the increase in load. Under certain circumstances, deformation may continue with time while the load remains constant. This deformation, beyond that experienced as the material is initially loaded, is termed *creep*. Turbine disks and reinforced concrete floors offer examples in which creep may be a problem. In materials such as lead, rubber, and certain plastics, creep may occur at ordinary temperatures. Most metals, on the other hand, begin to evidence a loss of strain hardening and manifest appreciable creep only when the absolute temperature is roughly 35 to 50% of the melting temperature. The rate at which creep proceeds in a given material depends on the stress, temperature, and history of loading.

A deformation time curve (creep curve), as in Fig. 4.1, typically displays a segment of decelerating *creep rate* (stage 0 to 1), a segment of essentially constant deformation or minimum creep rate (stage 1 to 2), and finally a segment of accelerating creep rate (stage 2 to 3). In the figure, curve *A* might correspond to a condition of either higher stress or higher temperature than curve *B*. Both curves terminate in fracture at point 3. The *creep strength* refers to the maximum employable strength of the material at a prescribed elevated temperature. This value of stress corresponds to a given rate of creep in the second stage (2 to 3), for example, 1% creep in 10,000 hours. Inasmuch as the creep stress and creep strain are not linearly related, calculations involving such material behavior are generally not routine.

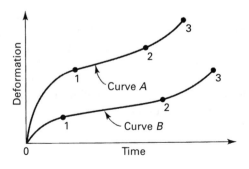

FIGURE 4.1. *Typical creep curves for a bar in tension.*

Stress relaxation refers to a loss of stress with time at a constant strain or deformation level. It is essentially a relief of stress through the mechanism of internal creep. Bolted flange connections and assemblies with shrink or press fits operating at high temperatures are examples of this variable stress condition. Insight into the behavior of viscoelastic models, briefly described in Section 2.6, can be achieved by subjecting models to standard creep and relaxation tests [Ref. 4.5]. In any event, allowable stresses should be kept low in order to prevent intolerable deformations caused by creep.

4.4 FAILURE BY FRACTURE

Separation of a material under stress into two or more parts (thereby creating new surface area) is referred to as *fracture*. The determination of the conditions of combined stress that lead to either elastic or inelastic termination of deformation, that is, predicting the failure strength of a material, is difficult. In 1920, A. A. Griffith was the first to equate the strain energy associated with material failure to that required for the formation of new surfaces. He also concluded that, with respect to its capacity to cause failure, tensile stress represents a more important influence than does compressive stress. The Griffith theory assumes the presence in brittle materials of minute cracks, which as a result of applied stress are caused to grow to macroscopic size, leading eventually to failure.

Although Griffith's experiments dealt primarily with glass, his results have been widely applied to other materials. Application to metals requires modification of the theory, however, because failure does not occur in an entirely brittle manner. Due to major catastrophic failures of ships, buildings, trains, airplanes, pressure vessels, and bridges in the 1940s and 1950s, increasing attention has been given by design engineers to the conditions of the growth of a crack. Griffith's concept has been considerably expanded by G. R. Irwin [Ref. 4.6].

Brittle materials most commonly fracture through the grains in what is termed a transcrystalline failure. Here the tensile stress is usually regarded as playing the most significant role. Examination of the failed material reveals very little deformation prior to fracture.

Types of Fracture in Tension

There are two types of fractures to be considered in tensile tests of *polycrystalline* specimens: *brittle fracture*, as in the case of cast iron, and *shear fracture*, as in the case of mild steel, aluminum, and other metals. In the former case, fracture occurs essentially without yielding over a cross section perpendicular to the axis of the specimen. In the latter case, fracture occurs only after considerable plastic stretching and subsequent local reduction of the cross-sectional area (necking) of the specimen, and the familiar *cup-and-cone* formation is observed.

At the narrowest neck section in cup-and-cone fracture, the tensile forces in the longitudinal fibers exhibit directions, as shown in Fig. 4.2a. The horizontal components of these forces produce radial tangential stresses, so each infinitesimal element is in a condition of *three-dimensional stress* (Fig. 4.2b). Based on the assumption that plastic flow requires a constant maximum shearing stress, we conclude that the axial tensile stresses σ are *nonuniformly* distributed over the minimum cross section of the specimen. These stresses have a maximum value σ_{max} at the center of the minimum cross section, where σ_r and σ_θ are also maximum, and a minimum value σ_{min} at the surface (Fig. 4.2a). The magnitudes of the maximum and minimum axial stresses depend on the radius a of the minimum cross section and the radius of the curvature r of the neck. The following relationships are used to calculate σ_{max} and σ_{min} [Ref. 4.7]:

$$\sigma_{max} = \frac{1 + (a/2r)}{1 + (a/4r)} \sigma_{avg}, \qquad \sigma_{min} = \frac{1}{1 + (a/4r)} \sigma_{avg} \tag{a}$$

Here $\sigma_{avg} = P/\pi a^2$ and represents the average stress.

Note that, owing to the condition of three-dimensional stress, the material near the center of the minimum cross section of the tensile specimen has its ductility reduced. During stretching, therefore, the crack begins in that region, while the material near the surface continues to stretch plastically. This explains why the central portion of a cup-and-cone fracture is of brittle character, while near the surface a ductile type of failure is observed.

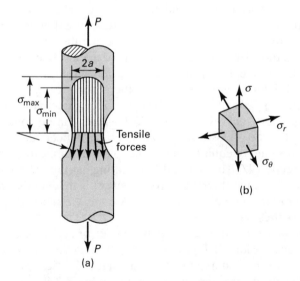

FIGURE 4.2.   *Necking of a bar in tension: (a) the distribution of the axial stresses; (b) stress element in the plane of the minimum cross section.*

Progressive Fracture: Fatigue

Multiple application and removal of load, usually measured in thousands of episodes or more, are referred to as *repeated loading*. Machine and structural members subjected to repeated, fluctuating, or alternating stresses, which are below the ultimate tensile strength or even the yield strength, may nevertheless manifest diminished strength and ductility. Since the phenomenon described, termed *fatigue*, is difficult to predict and is often influenced by factors eluding recognition, increased uncertainty in strength and in service life must be dealt with [Ref. 4.8]. As is true for brittle behavior in general, fatigue is importantly influenced by minor structural discontinuities, the quality of surface finish, and the chemical nature of the environment.

The *types of fracture* produced in ductile metals subjected to repeated loading differs greatly from that of fracture under static loading discussed in Section 2.7. In fatigue fractures, two zones of failure can be obtained: the beachmarks (so called because the resemble ripples left on sand by retracting waves) region produced by the gradual development of a crack and the sudden fracture region. As the name suggest, the fracture region is the portion that fails suddenly when the crack reaches its size limit. The appearance of the surfaces of fracture greatly helps in identifying the cause of crack initiation to be corrected in redesign.

A fatigue crack is generally observed to have, as its origin, a point of high stress concentration, for example, the corner of a keyway or a groove. This failure, through the involvement of slip planes and spreading cracks, is progressive in nature. For this reason, *progressive fracture* is probably a more appropriate term than fatigue failure. Tensile stress, and to a lesser degree shearing stress, lead to fatigue crack propagation, while compressive stress probably does not. The *fatigue life* or *endurance* of a material is defined as the number of stress repetitions or *cycles* prior to fracture. The fatigue life of a given material depends on the magnitudes (and the algebraic signs) of the stresses at the extremes of a stress cycle.

Experimental determination is made of the number of cycles (N) required to break a specimen at a particular stress level (S) under a fluctuating load. From such tests, called *fatigue tests*, curves termed *S–N diagrams* can be constructed. Various types of simple fatigue stress-testing machines have been developed. Detailed information on this kind of equipment may be found in publications such as cited in the footnote of Section 4.2. The simplest is a rotating bar fatigue-testing machine on which a specimen (usually of circular cross section) is held so that it rotates while under condition of alternating *pure bending*. A *complete reversal* (tension to compression) of stress thus results.

It is usual practice to plot stress versus number of cycles with semilogarithmic scales (that is, σ against $\log N$). For most steels, the *S–N* diagram obtained in a simple fatigue test performed on a number of nominally identical specimens loaded at different stress levels has the appearance shown in Fig. 4.3. The stress at which the curve levels off is called the *fatigue* or *endurance limit* σ_e. Beyond the point (σ_e, N_e) failure *will not* take place no matter how great the number of cycles. For a lower number of cycles $N < N_f$, the loading is regarded as *static*. At $N = N_f$ cycles, failure occurs at static tensile *fracture stress* σ_f. The *fatigue strength* for *complete stress*

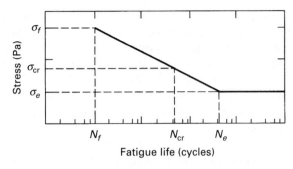

FIGURE 4.3. *Typical S–N diagram for steel.*

reversal at a specified number of cycles N_{cr}, is designated σ_{cr} on the diagram. The S–N curve relationships are utilized in Section 4.15, in which combined stress fatigue properties are discussed.

While yielding and fracture may well depend on the rate of load application or the rate at which the small permanent strains form, we shall, with the exception of Sections 4.16, 4.17, and 12.13, assume that yielding and fracture in solids are functions solely of the states of stress or strain.

4.5 YIELD AND FRACTURE CRITERIA

As the tensile loading of a ductile member is increased, a point is eventually reached at which changes in geometry are no longer entirely reversible. The beginning of inelastic behavior (yield) is thus marked. The extent of the inelastic deformation preceding fracture very much depends on the material involved.

Consider an element subjected to a general state of stress, where $\sigma_1 > \sigma_2 > \sigma_3$. Recall that subscripts 1, 2, and 3 refer to the principal directions. The state of stress in uniaxial loading is described by σ_1, equal to the normal force divided by the cross-sectional area, and $\sigma_2 = \sigma_3 = 0$. Corresponding to the start of the yielding event in this *simple tension test* are the quantities pertinent to stress and strain energy shown in the second column of Table 4.1. Note that the items listed in this column, expressed in terms of the uniaxial yield point stress σ_{yp}, have special significance in predicting failure involving multiaxial states of stress. In the case of a material in *simple torsion*, the state of stress is given by $\tau = \sigma_1 = -\sigma_3$ and $\sigma_2 = 0$.

TABLE 4.1. *Shear Stress and Strain Energy at the Start of Yielding*

Quantity	Tension test	Torsion test
Maximum shear stress	$\tau_{yp} = \sigma_{yp}/2$	τ_{yp}
Maximum energy of distortion	$U_{od} = [(1 + \nu)/3E]\sigma_{yp}^2$	$U_{od} = \tau_{yp}^2(1 + \nu)/E$
Maximum octahedral shear stress	$\tau_{oct} = (\sqrt{2}/3)\sigma_{yp}$	$\tau_{oct} = (\sqrt{2}/\sqrt{3})\tau_{yp}$

In the foregoing, τ is calculated using the standard torsion formula. Corresponding to this case of pure shear, at the onset of yielding, are the quantities shown in the third column of the table, expressed in terms of yield point stress in torsion, τ_{yp}.

The behavior of materials subjected to uniaxial normal stresses or pure shearing stresses is readily presented on stress–strain diagrams. The onset of yielding or fracture in these cases is considerably more apparent than in situations involving combined stress. From the viewpoint of mechanical design, it is imperative that some practical guides be available to predict yielding or fracture under the conditions of stress as they are likely to exist in service. To meet this need and to understand the basis of material failure, a number of failure criteria have been developed. In this chapter we discuss only the classical idealizations of yield and fracture criteria of materials. These strength theories are structured to apply to particular classes of materials. The three most widely accepted theories to predict the onset of inelastic behavior for ductile materials under combined stress are described first in Sections 4.6 through 4.9. This is followed by a presentation of three fracture theories pertaining to brittle materials under combined stress (Secs. 4.10 through 4.12).

In addition to the failure theories, failure is sometimes predicted conveniently using the interaction curves discussed in Section 12.7. Experimentally obtained curves of this kind, unless complicated by a buckling phenomenon, are equivalent to the strength criteria considered here.

4.6   MAXIMUM SHEARING STRESS THEORY

The maximum shearing stress theory is an outgrowth of the experimental observation that a ductile material yields as a result of slip or shear along crystalline planes. Proposed by C. A. Coulomb (1736–1806), it is also referred to as the *Tresca yield criterion* in recognition of the contribution of H. E. Tresca (1814–1885) to its application. This theory predicts that yielding will start when the maximum shearing stress in the material equals the maximum shearing stress at yielding in a simple tension test. Thus, by applying Eq. (1.45) and Table 4.1, we obtain

$$\tfrac{1}{2}|\sigma_1 - \sigma_3| = \tau_{yp} = \tfrac{1}{2}\sigma_{yp}$$

or

$$|\sigma_1 - \sigma_3| = \sigma_{yp} \tag{4.1}$$

In the case of plane stress, $\sigma_3 = 0$, there are two combinations of stresses to be considered. When σ_1 and σ_2 are of *opposite sign*, that is, one tensile and the other compressive, the maximum shearing stress is $(\sigma_1 - \sigma_2)/2$. Thus, the yield condition is given by

$$|\sigma_1 - \sigma_2| = \sigma_{yp} \tag{4.2a}$$

which may be restated as

$$\frac{\sigma_1}{\sigma_{yp}} - \frac{\sigma_2}{\sigma_{yp}} = \pm 1 \tag{4.2b}$$

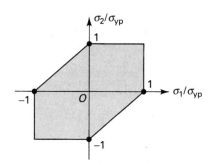

FIGURE 4.4. *Yield criterion based on maximum shearing stress.*

When σ_1 and σ_2 carry the *same sign*, the maximum shearing stress equals $(\sigma_1 - \sigma_3)/2 = \sigma_1/2$. Then, for $|\sigma_1| > |\sigma_2|$ and $|\sigma_2| > |\sigma_1|$, we have the following yield conditions, respectively:

$$|\sigma_1| = \sigma_{yp} \quad \text{and} \quad |\sigma_2| = \sigma_{yp} \tag{4.3}$$

Figure 4.4 is a plot of Eqs. (4.2) and (4.3). Note that Eq. (4.2) applies to the second and fourth quadrants, while Eq. (4.3) applies to the first and third quadrants. The boundary of the hexagon thus marks the onset of yielding, with points outside the shaded region representing a yielded state. The foregoing describes the Tresca yield condition. Good agreement with experiment has been realized for *ductile* materials. The theory offers an additional advantage in its ease of application.

4.7 MAXIMUM DISTORTION ENERGY THEORY

The maximum distortion energy theory, also known as the *von Mises theory*, was proposed by M. T. Huber in 1904 and further developed by R. von Mises (1913) and H. Hencky (1925). In this theory, failure by yielding occurs when, at any point in the body, the distortion energy per unit volume in a state of combined stress becomes equal to that associated with yielding in a simple tension test. Equation (2.65) and Table 4.1 thus lead to

$$(\sigma_x - \sigma_y)^2 + (\sigma_y - \sigma_z)^2 + (\sigma_z - \sigma_x)^2 + 6(\tau_{xy}^2 + \tau_{yz}^2 + \tau_{xz}^2) = 2\,\sigma_{yp}^2 \tag{4.4a}$$

or, in terms of principal stresses,

$$(\sigma_1 - \sigma_2)^2 + (\sigma_2 - \sigma_3)^2 + (\sigma_3 - \sigma_1)^2 = 2\,\sigma_{yp}^2 \tag{4.4b}$$

For plane stress $\sigma_3 = 0$, and the criterion for yielding becomes

$$\sigma_1^2 - \sigma_1\sigma_2 + \sigma_2^2 = \sigma_{yp}^2 \tag{4.5a}$$

or, alternatively,

$$\left(\frac{\sigma_1}{\sigma_{yp}}\right)^2 - \left(\frac{\sigma_1}{\sigma_{yp}}\right)\left(\frac{\sigma_2}{\sigma_{yp}}\right) + \left(\frac{\sigma_2}{\sigma_{yp}}\right)^2 = 1 \tag{4.5b}$$

Expression (4.5b) defines the ellipse shown in Fig. 4.5a. We note that, for simplification, Eq. (4.4b) or (4.5a) may be written $\sigma_e = \sigma_{yp}$, where σ_e is known as the *von*

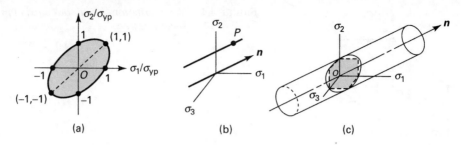

FIGURE 4.5. *Yield criterion based on distortion energy: (a) plane stress yield ellipse; (b) a state of stress defined by position; (c) yield surface for triaxial state of stress.*

Mises stress or the *effective stress* (Sec. 12.12). For example, in the latter case we have $\sigma_e = (\sigma_1^2 - \sigma_1\sigma_2 + \sigma_2^2)^{1/2}$.

Returning to Eq. (4.4b), it is observed that only the *differences* of the principal stresses are involved. Consequently, addition of an equal amount to each stress does not affect the conclusion with respect to whether or not yielding will occur. In other words, yielding does not depend on hydrostatic tensile or compressive stresses. Now consider Fig. 4.5b, in which a state of stress is defined by the position $P(\sigma_1, \sigma_2, \sigma_3)$ in a principal stress coordinate system as shown. It is clear that a hydrostatic alteration of the stress at point P requires shifting of this point along a direction parallel to direction n, making equal angles with coordinate axes. This is because changes in hydrostatic stress involve changes of the normal stresses by equal amounts. On the basis of the foregoing, it is concluded that the yield criterion is properly described by the cylinder shown in Fig. 4.5c and that the surface of the cylinder is the yield surface. Points within the surface represent states of nonyielding. The ellipse of Fig. 4.5a is defined by the intersection of the cylinder with the σ_1, σ_2 plane. Note that the yield surface appropriate to the maximum shearing stress criterion (shown by the dashed lines for plane stress) is described by a hexagonal surface placed within the cylinder.

The maximum distortion energy theory of failure finds considerable experimental support in situations involving *ductile* materials and plane stress. For this reason, it is in common use in design.

4.8 OCTAHEDRAL SHEARING STRESS THEORY

The octahedral shearing stress theory (also referred to as the Mises–Hencky or simply the von Mises criterion) predicts failure by yielding when the octahedral shearing stress at a point achieves a particular value. This value is determined by the relationship of τ_{oct} to σ_{yp} in a simple tension test. Referring to Table 4.1, we obtain

$$\tau_{oct} = 0.47\sigma_{yp} \tag{4.6}$$

where τ_{oct} for a general state of stress is given by Eq. (2.66).

The Mises–Hencky criterion may also be viewed in terms of distortion energy [Eq. (2.65)]:

$$U_{od} = \frac{3}{2}\frac{1+\nu}{E}\tau_{oct}^2 \qquad\qquad \textbf{(a)}$$

If it is now asserted that yielding will, in a general state of stress, occur when U_{od} defined by Eq. (a) is equal to the value given in Table 4.1, then Eq. (4.6) will again be obtained. We conclude, therefore, that the octahedral shearing stress theory enables us to apply the distortion energy theory while dealing with stress rather than energy.

EXAMPLE 4.1   Circular Shaft under Combined Loads

A circular shaft of tensile strength $\sigma_{yp} = 350$ MPa is subjected to a combined state of loading defined by bending moment $M = 8$ kN·m and torque $T = 24$ kN·m (Fig. 4.6a). Calculate the required shaft diameter d in order to achieve a factor of safety $n = 2$. Apply (a) the maximum shearing stress theory and (b) the maximum distortion energy theory.

Solution   For the situation described, the principal stresses are

$$\sigma_{1,2} = \frac{\sigma_x}{2} \pm \tfrac{1}{2}\sqrt{\sigma_x^2 + 4\tau_{xy}^2}, \quad \sigma_3 = 0 \qquad\qquad \textbf{(b)}$$

where

$$\sigma_x = \frac{My}{I} = \frac{32M}{\pi d^3}, \qquad \tau_{xy} = \frac{Tr}{J} = \frac{16T}{\pi d^3}$$

Therefore

$$\sigma_{1,2} = \frac{16}{\pi d^3}(M \pm \sqrt{M^2 + T^2}) \qquad\qquad \textbf{(4.7)}$$

a. *Maximum shearing stress theory:*   For the state of stress under consideration, it may be observed from Mohr's circle, shown in Fig. 4.6b,

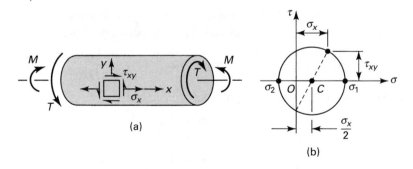

(a)

(b)

FIGURE 4.6.   *Example 4.1. (a) Torsion–flexure of a shaft; (b) Mohr's circle for torsion–flexure loading.*

that σ_1 is tensile and σ_2 is compressive. Thus, through the use of Eqs. (b) and (4.2a),

$$\frac{\sigma_{yp}}{n} = \sqrt{\sigma_x^2 + 4\tau_{xy}^2} \qquad (4.8a)$$

or

$$\frac{\sigma_{yp}}{n} = \frac{32}{\pi d^3}\sqrt{M^2 + T^2} \qquad (4.8b)$$

After substitution of the numerical values, Eq. (4.8b) gives $d = 113.8$ mm.

b. *Maximum distortion energy theory*: From Eqs. (4.5a), (b), and (4.7),

$$\frac{\sigma_{yp}}{n} = \sqrt{\sigma_x^2 + 3\tau_{xy}^2} \qquad (4.9a)$$

or

$$\frac{\sigma_{yp}}{n} = \frac{16}{\pi d^3}\sqrt{4M^2 + 3T^2} \qquad (4.9b)$$

This result may also be obtained from the octahedral shearing stress theory by applying Eqs. (4.6) and (4.7). Substituting the data into Eq. (4.9b) and solving for d, we have $d = 109$ mm.

Comments The diameter based on the shearing stress theory is thus 4.4% larger than that based on the maximum energy of distortion theory. A 114-mm shaft should be used to prevent initiation of yielding.

EXAMPLE 4.2   Conical Tank filled with Liquid
A steel conical tank, supported at its edges, is filled with a liquid of density γ (Fig. P13.32.). The yield point stress (σ_{yp}) of the material is known. The cone angle is 2α. Determine the required wall thickness t of the tank based on a factor of safety n. Apply (a) the maximum shear stress theory and (b) the maximum energy of distortion theory.

Solution The variations of the circumferential and longitudinal principal, stresses $\sigma_\theta = \sigma_1$ and $\sigma_\phi = \sigma_2$, in the tank are, respectively (Prob. 13.32),

$$\phi_1 = \gamma(a - y)y\frac{\tan \alpha}{t \cos \alpha}, \qquad \phi_2 = \gamma(a - \tfrac{2}{3}y)y\frac{\tan \alpha}{2t \cos \alpha} \qquad (c)$$

These stresses have the largest magnitude:

$$\sigma_{1,\,max} = \frac{\gamma a^2}{4t}\frac{\tan \alpha}{\cos \alpha}, \quad \sigma_2 = \frac{\gamma a^2}{12t}\frac{\tan \alpha}{\cos \alpha}, \quad at \ y = \frac{a}{2}$$

$$\sigma_{2,\,max} = \frac{3\gamma a^2}{16t}\frac{\tan \alpha}{\cos \alpha}, \quad \sigma_1 = \frac{\gamma a^2}{4t}\frac{\tan \alpha}{\cos \alpha}, \quad at \ y = \frac{3a}{4} \qquad (d)$$

a. *Maximum shear stress theory*: Because σ_1 and σ_2 are of the same sign and $|\sigma_1| > |\sigma_2|$, we have, from the first equations of (4.3) and (d),

$$\frac{\sigma_{yp}}{n} = \frac{\gamma a^2 \tan \alpha}{4t \cos \alpha}$$

The thickness of the tank is found from this equation to be

$$t = 0.250 \frac{\gamma a^2 n}{\sigma_{yp}} \frac{\tan \alpha}{\cos \alpha} \tag{e}$$

b. *Maximum distortion energy theory*: It is observed in Eq. (d) that the largest values of principal stress are found at different locations. We shall therefore first locate the section at which the combined principal stresses are at a critical value. For this purpose, we insert Eq. (c) into Eq. (4.5a):

$$\frac{\sigma_{yp}^2}{n^2} = \left[\gamma(a - y)y \frac{\tan \alpha}{t \cos \alpha} \right]^2 + \left[\gamma(a - \tfrac{2}{3}y)y \frac{\tan \alpha}{2t \cos \alpha} \right]^2$$
$$- \left[\gamma(a - y)y \frac{\tan \alpha}{t \cos \alpha} \right]\left[\gamma(a - \tfrac{2}{3}y)y \frac{\tan \alpha}{2t \cos \alpha} \right] \tag{f}$$

Upon differentiating Eq. (f) with respect to the variable y and equating the result to zero, we obtain

$$y = 0.52a$$

Upon substitution of this value of y into Eq. (f), the thickness of the tank is determined:

$$t = 0.225 \frac{\gamma a^2 n}{\sigma_{yp}} \frac{\tan \alpha}{\cos \alpha} \tag{g}$$

Comments   The thickness based on the maximum shear stress theory is thus 10% larger than that based on the maximum energy of distortion theory.

4.9   COMPARISON OF THE YIELDING THEORIES

Two approaches may be employed to compare the theories of yielding heretofore discussed. The first comparison equates, for each theory, the critical values corresponding to uniaxial loading and torsion. Referring to Table 4.1, we have

Maximum shearing stress theory: $\tau_{yp} = 0.50\sigma_{yp}$
Maximum energy of distortion theory, or its
equivalent, the octahedral shearing stress theory: $\tau_{yp} = 0.577\sigma_{yp}$

Observe that the difference in strength predicted by these theories is not substantial. A second comparison may be made by means of a superposition of Figs. 4.4 and 4.5a. This is left as an exercise for the reader.

Experiment shows that, for ductile materials, the yield stress obtained from a torsion test is 0.5 to 0.6 times that determined from a simple tension test. We conclude, therefore, that the energy of distortion theory, or the octahedral shearing stress theory, is most suitable for ductile materials. However, the shearing stress theory, which gives $\tau_{yp} = 0.50\sigma_{yp}$, is in widespread use because it is simple to apply and offers a conservative result in design.

Consider, as an example, a solid shaft of diameter d and tensile yield strength σ_{yp}, subjected to combined loading consisting of tension P and torque T. The yield criteria based on the maximum shearing stress and energy of distortion theories, for $n = 1$, are given by Eqs. (4.8a) and (4.9a):

$$\sigma_{yp} = \sqrt{\sigma_x^2 + 4\tau_{xy}^2}, \qquad \sigma_{yp} = \sqrt{\sigma_x^2 + 3\tau_{xy}^2} \tag{a}$$

where

$$\sigma_x = \frac{4P}{\pi d^2}, \qquad \tau_{xy} = \frac{16T}{\pi d^3}$$

A comparison of a dimensionless plot of Eqs. (a) with some experimental results is shown in Fig. 4.7. Note again the particularly good agreement between the maximum distortion energy theory and experimental data for ductile materials.

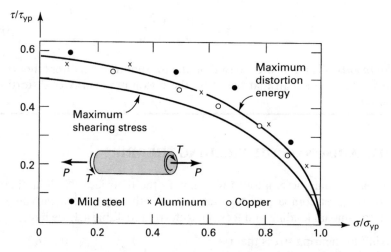

FIGURE 4.7. *Yield curves for torsion–tension shaft. The points indicated in this figure are based on experimental data obtained by G. I. Taylor and H. Quinney [Ref. 4.9].*

4.10 MAXIMUM PRINCIPAL STRESS THEORY

According to the maximum principal stress theory, credited to W. J. M. Rankine (1820–1872), a material fails by fracturing when the largest principal stress exceeds the ultimate strength σ_u in a simple tension test. That is, at the onset of fracture,

$$|\sigma_1| = \sigma_u \quad \text{or} \quad |\sigma_3| = \sigma_u \tag{4.10}$$

Thus, a crack will start at the most highly stressed point in a brittle material when the largest principal stress at that point reaches σ_u.

Note that, while a material may be weak in simple compression, it may nevertheless sustain very high hydrostatic pressure without fracturing. Furthermore, brittle materials are much stronger in compression than in tension, while the maximum principal stress criterion is based on the assumption that the ultimate strength of a material is the same in tension and compression. Clearly, these are inconsistent with the theory. Moreover, the theory makes no allowance for influences on the failure mechanism other than those of normal stresses. However, for brittle materials in all stress ranges, the maximum principal stress theory has good experimental verification, provided that there exists a tensile principal stress.

In the case of plane stress ($\sigma_3 = 0$), Eq. (4.10) becomes

$$|\sigma_1| = \sigma_u \quad \text{or} \quad |\sigma_2| = \sigma_u \tag{4.11a}$$

This may be rewritten as

$$\frac{\sigma_1}{\sigma_u} = \pm 1 \quad \text{or} \quad \frac{\sigma_2}{\sigma_u} = \pm 1 \tag{4.11b}$$

The foregoing is depicted in Fig. 4.8 with points a, b, and c, d indicating the tensile and compressive principal stresses, respectively. For this case, the boundaries represent the onset of failure due to fracture. The area within the boundary of the figure is thus a region of no failure.

4.11 MOHR'S THEORY

The Mohr theory of failure is used to predict the fracture of a material having different properties in tension and compression when results of various types of tests are available for that material. This criterion makes use of the well-known Mohr's

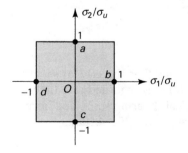

FIGURE 4.8. *Fracture criterion based on maximum principal stress.*

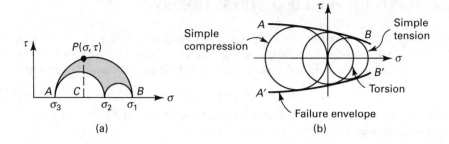

FIGURE 4.9. (a) Mohr's circles of stress; (b) Mohr's envelopes.

circles of stress. As discussed in Section 1.15, in a Mohr's circle representation, the shear and normal components of stress acting on a particular plane are specified by the coordinates of a point within the shaded area of Fig. 4.9a. Note that τ depends on σ; that is, $|\tau| = f(\sigma)$.

The figure indicates that a vertical line such as PC represents the states of stress on planes with the same σ but with differing τ. It follows that the weakest of all these planes is the one on which the maximum shearing stress acts, designated P. The same conclusion can be drawn regardless of the position of the vertical line between A and B; the points on the outer circle correspond to the weakest planes. On these planes, the *maximum* and *minimum* principal stresses alone are sufficient to decide whether or not failure will occur, because these stresses determine the outer circle shown in Fig. 4.9a. Using these extreme values of principal stress thus enables us to apply the Mohr approach to either two- or three-dimensional situations.

The foregoing provides a background for the Mohr theory of failure, which relies on stress plots in σ, τ coordinates. The particulars of the Mohr approach are presented next.

Experiments are performed on a given material to determine the states of stress that result in failure. Each such stress state defines a Mohr's circle. If the data describing states of limiting stress are derived from only simple tension, simple compression, and pure shear tests, the three resulting circles are adequate to construct the *envelope*, denoted by lines AB and $A'B'$ in Fig. 4.9b. The Mohr envelope thus represents the locus of all possible failure states. Many solids, particularly those that are brittle, exhibit greater resistance to compression than to tension. As a consequence, higher limiting shear stresses will, for these materials, be found to the left of the origin, as shown in the figure.

4.12 COULOMB–MOHR THEORY

The *Coulomb–Mohr* or *internal friction* theory assumes that the critical shearing stress is related to internal friction. If the frictional force is regarded as a function of the normal stress acting on a shear plane, the critical shearing stress and normal stress can be connected by an equation of the following form (Fig. 4.10a):

$$\tau = a\sigma + b \tag{a}$$

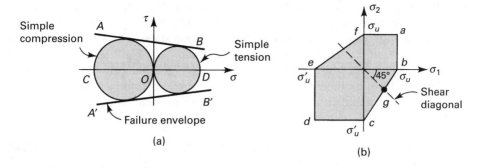

FIGURE 4.10. *(a) Straight-line Mohr's envelopes; (b) Coulomb–Mohr fracture criterion.*

The constants a and b represent properties of the particular material. This expression may also be viewed as a *straight-line version of the Mohr envelope*.

For the case of plane stress, $\sigma_3 = 0$ when σ_1 is tensile and σ_2 is compressive. The maximum shearing stress τ and the normal stress σ acting on the shear plane are, from Eqs. (1.22) and (1.23), given by

$$\tau = \frac{\sigma_1 - \sigma_2}{2}, \qquad \sigma = \frac{\sigma_1 + \sigma_2}{2} \tag{b}$$

Introducing these expressions into Eq. (a), we obtain

$$\sigma_1(1 - a) - \sigma_2(1 + a) = 2b \tag{c}$$

To evaluate the material constants, the following conditions are applied:

$$\begin{aligned} \sigma_1 &= \sigma_u \quad \text{when} \quad \sigma_2 = 0 \\ \sigma_2 &= -\sigma'_u \quad \text{when} \quad \sigma_1 = 0 \end{aligned} \tag{d}$$

Here σ_u and σ'_u represent the ultimate strength of the material in tension and compression, respectively. If now Eqs. (d) are inserted into Eq. (c), the results are

$$\sigma_u(1 - a) = 2b \quad \text{and} \quad \sigma'_u(1 + a) = 2b$$

from which

$$a = \frac{\sigma_u - \sigma'_u}{\sigma_u + \sigma'_u}, \qquad b = \frac{\sigma_u \sigma'_u}{\sigma_u + \sigma'_u} \tag{e}$$

These constants are now introduced into Eq. (c) to complete the equation of the envelope of failure by fracturing. When this is done, the following expression is obtained, applicable for $\sigma_1 > 0, \sigma_2 < 0$:

$$\frac{\sigma_1}{\sigma_u} - \frac{\sigma_2}{\sigma'_u} = 1 \tag{4.12a}$$

For any given ratio σ_1/σ_2, the individual stresses at fracture, σ_1 and σ_2, can be calculated by applying expression (4.12a) (Prob. 4.22).

Relationships for the case where the principal stresses have the same sign ($\sigma_1 > 0, \sigma_2 > 0$ or $\sigma_1 < 0, \sigma_2 < 0$) may be deduced from Fig. 4.10a without resort to the preceding procedure. In the case of *biaxial tension* (now $\sigma_{min} = \sigma_3 = 0$, σ_1 and σ_2 are tensile), the corresponding Mohr's circle is represented by diameter OD. Therefore, fracture occurs if either of the two tensile stresses achieves the value σ_u. That is,

$$\sigma_1 = \sigma_u, \qquad \sigma_2 = \sigma_u \qquad\qquad \textbf{(4.12b)}$$

For biaxial compression (now $\sigma_{max} = \sigma_3 = 0$, σ_1 and σ_2 are compressive), a Mohr's circle of diameter OC is obtained. Failure by fracture occurs if either of the compressive stresses attains the value σ_u':

$$\sigma_2 = -\sigma_u', \qquad \sigma_1 = -\sigma_u' \qquad\qquad \textbf{(4.12c)}$$

Figure 4.10b is a graphical representation of the Coulomb–Mohr theory plotted in the σ_1, σ_2 plane. Lines ab and af represent Eq. (4.12b), and lines dc and de, Eq. (4.12c). The boundary bc is obtained through the application of Eq. (4.12a). Line ef completes the hexagon in a manner analogous to Fig. 4.4. Points lying within the shaded area should not represent failure, according to the theory. In the case of pure shear, the corresponding limiting point is g. The magnitude of the limiting shear stress may be graphically determined from the figure or calculated from Eq. (4.12a) by letting $\sigma_1 = -\sigma_2$:

$$\sigma_1 = \tau_u = \frac{\sigma_u \, \sigma_u'}{\sigma_u + \sigma_u'} \qquad\qquad \textbf{(4.13)}$$

EXAMPLE 4.3 Tube Torque Requirement

A thin-walled tube is fabricated of a brittle material having ultimate tensile and compressive strengths $\sigma_u = 300$ MPa and $\sigma_u' = 700$ MPa. The radius and thickness of the tube are $r = 100$ mm and $t = 5$ mm. Calculate the limiting torque that can be applied without causing failure by fracture. Apply (a) the maximum principal stress theory and (b) the Coulomb–Mohr theory.

Solution The torque and maximum shearing stress are related by the torsion formula:

$$T = \frac{J}{r}\tau = 2\pi r^2 t\tau = 2\pi(0.1)^2(0.005)\tau = 314 \times 10^{-6}\tau \qquad\qquad \textbf{(f)}$$

The state of stress is described by $\sigma_1 = -\sigma_2 = \tau, \sigma_3 = 0$.

a. *Maximum principal stress theory:* Equations (4.10) are applied with σ_3 replaced by σ_2 because the latter is negative: $|\sigma_1| = |\sigma_2| = \sigma_u$. Because we have $\sigma_1 = \sigma_u = 300 \times 10^6 = \tau$, from Eq. (f),

$$T = 314 \times 10^{-6}(300 \times 10^6) = 94.2 \text{ kN} \cdot \text{m}$$

b. *Coulomb–Mohr theory*: Applying Eq. (4.12a),

$$\frac{\tau}{300 \times 10^6} - \frac{-\tau}{700 \times 10^6} = 1$$

from which $\tau = 210$ MPa. Equation (f) gives $T = 314 \times 10^{-6}$ $(210 \times 10^6) = 65.9$ kN·m.

Based on the maximum principal stress theory, the torque that can be applied to the tube is thus 30% larger than that based on the Coulomb–Mohr theory. To prevent fracture, the torque should not exceed 65.9 kN·m.

EXAMPLE 4.4 Design of a Cast Iron Torsion Bar

A torsion-bar spring made of ASTM grade A-48 cast iron is loaded as shown in Fig. 4.11. The stress concentration factors are 1.7 for bending and 1.4 for torsion. Determine the diameter d to resist loads $P = 25$ N and $T = 10$ N·m, using a factor of safety $n = 2.5$. Apply (a) the maximum principal stress theory and (b) the Coulomb–Mohr theory.

Solution The stresses produced by bending moment $M = 0.1P$ and torque T at the shoulder are

$$\sigma_x = K_b \frac{32M}{\pi d^3}, \qquad \tau_{xy} = K_t \frac{16T}{\pi d^3} \qquad (4.14)$$

The principal stresses, using Eq. (4.7), are then

$$\sigma_{1,2} = \frac{16}{\pi d^3}(K_b M \pm \sqrt{K_b^2 M^2 + K_t^2 T^2}) \qquad (4.15)$$

Substituting the given data, we have

$$\sigma_{1,2} = \frac{16}{\pi d^3}\left[1.7(0.1 \times 25) \pm \sqrt{(1.7 \times 0.1 \times 25)^2 + (1.4 \times 10)^2}\right]$$

The foregoing results in

$$\sigma_1 = \frac{96.16}{d^3}, \qquad \sigma_2 = -\frac{52.87}{d^3} \qquad (g)$$

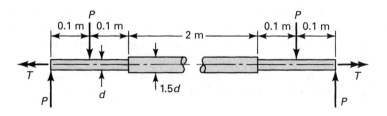

FIGURE 4.11. *Example 4.4. A torsion-bar spring.*

The allowable ultimate strengths of the material in tension and compression are $170/2.5 = 68$ MPa and $650/2.5 = 260$ MPa, respectively (see Table D.1).

a. *Maximum principal stress theory*: On the basis of Eqs. (g) and (4.10),

$$\frac{96.16}{d^3} = 68 \times 10^6 \quad \text{or} \quad d = 11.2 \text{ mm}$$

Similarly, $52.87/d^3 = 68 \times 10^6$ gives $d = 9.2$ mm.

b. *Coulomb–Mohr theory*: Using Eqs. (g) and (4.12a),

$$\frac{96.16}{68(10^6)d^3} - \frac{-52.87}{260(10^6)d^3} = 1 \quad \text{or} \quad d = 11.7 \text{ mm}$$

Comments The diameter of the spring based on the Coulomb–Mohr theory is therefore about 4.5% larger than that based on the maximum principal stress theory. A 12-mm-diameter bar, a commercial size, should be used to prevent fracture.

4.13 FRACTURE MECHANICS

As noted in Section 4.4, *fracture* is defined as the separation of a part into two or more pieces. It normally constitutes a "pulling apart" associated with the tensile stress. This type of failure often occurs in some materials in an instant. The mechanisms of brittle fracture are the concern of *fracture mechanics*, which is based on a stress analysis in the vicinity of a crack or defect of unknown small radius in a part. A *crack* is a microscopic flaw that may exist under normal conditions on the surface or within the material. These may vary from nonmetallic inclusions and microvoids to weld defects, grinding cracks, and so on. Scratches in the surface due to mishandling can also serve as incipient cracks.

Recall that the stress concentration factors are limited to elastic structures for which all dimensions are precisely known, particularly the radius of the curvature in regions of high stress concentrations. When exists a crack, the stress concentration factor approaches infinity as the root radius approaches 0. Therefore, analysis from the viewpoint of stress concentration factors is inadequate when cracks are present. Space limitations preclude our including more detailed treatment of the subject of fracture mechanics. However, the basic principles and some important results are briefly stated.

The fracture mechanics approach starts with an assumed initial minute crack (or cracks), for which the size, shape, and location can be defined. If brittle failure occurs, it is because the conditions of loading and environment are such that they cause an almost sudden propagation to failure of the original crack. When there is fatigue loading, the initial crack may grow slowly until it reaches a critical size at which the rapid fracture occurs [Ref. 4.8].

Stress-Intensity Factors

In the fracture mechanics approach, a stress-intensity factor, K, is evaluated. This can be thought of as a measure of the effective local stress at the crack root. The three modes of crack deformation of a plate are shown in Fig. 4.12. The most currently available values of K are for tensile loading normal to the crack, which is called mode I (Fig. 4.12a) and denoted as K_I. Modes II and III are essentially associated with the in-plane and out-of-plane loads, respectively (Figs. 4.12b and 4.12c). The treatment here is concerned only with mode I. We eliminate the subscript and let $K = K_I$.

Solutions for numerous configurations, specific initial crack shapes, and orientations have been developed analytically and by computational techniques, including finite element analysis (FEA) [Ref. 4.10 and 4.11]. For plates and beams, the *stress-intensity factor* is defined as

$$K = \lambda \sigma \sqrt{\pi a} \tag{4.16}$$

In the foregoing, we have σ = normal stress; λ = geometry factor, depends on a/w, listed in Table 4.2; a = crack length (or half crack length); w = member width (or half width of member). It is seen from Eq. (4.16) and Table 4.2 that the stress-intensity factor depends on the applied load and geometry of the specimen as well as on the size and shape of the crack. The units of the stress-intensity factors are commonly MPa $\sqrt{\text{m}}$ in SI and ksi $\sqrt{\text{in}}$. in U.S. customary system.

It is obvious that most cracks may not be as basic as shown in Table 4.2. They may be at an angle embedded in a member or sunken into surface. A shallow surface crack in a component may be considered semi-elliptical. A circular or elliptical form has proven to be adequate for many studies. Publications on fracture mechanics provide methods of analysis, applications, and extensive references [Ref. 4.8 through 4.12].

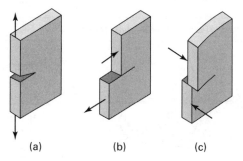

(a)   (b)   (c)

FIGURE 4.12. *Crack deformation types: (a) mode I, opening; (b) mode II, sliding; (c) mode III, tearing.*

TABLE 4.2. *Geometry Factors λ for Initial Crack Shapes*

Case A   *Tension of a long plate with central crack*

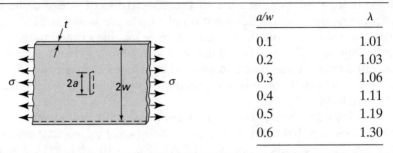

a/w	λ
0.1	1.01
0.2	1.03
0.3	1.06
0.4	1.11
0.5	1.19
0.6	1.30

Case B   *Tension of a long plate with edge crack*

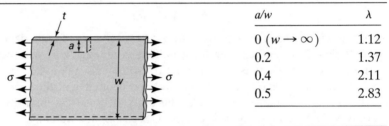

a/w	λ
0 ($w \rightarrow \infty$)	1.12
0.2	1.37
0.4	2.11
0.5	2.83

Case C   *Tension of a long plate with double edge cracks*

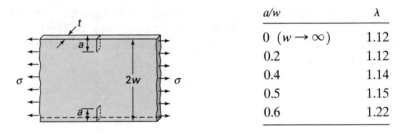

a/w	λ
0 ($w \rightarrow \infty$)	1.12
0.2	1.12
0.4	1.14
0.5	1.15
0.6	1.22

Case D   *Pure bending of a beam with edge crack*

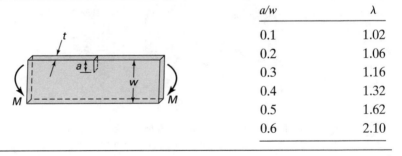

a/w	λ
0.1	1.02
0.2	1.06
0.3	1.16
0.4	1.32
0.5	1.62
0.6	2.10

Interestingly, crack propagation occurring after an increase in load may be interrupted if a small zone forms ahead of the crack. However, stress intensity has risen with the increase in crack length and, in time, the crack may advance again a short amount. When stress continues to increase owing to the reduced load-carrying area or different manner, the crack may grow, leading to failure. A final point to be noted is that the stress-intensity factors are also used to predict the rate of growth of a fatigue crack.

4.14 FRACTURE TOUGHNESS

In a toughness test of a given material, the stress-intensity factor at which a crack will propagate is measured. This is the critical stress-intensity factor, referred to as the *fracture toughness* and denoted by the symbol K_c. Ordinarily, testing is done on an ASTM standard specimen, either a beam or tension member with an edge crack at the root of a notch. Loading is increased slowly, and a record is made of load versus notch opening. The data are interpreted for the value of fracture toughness [Ref. 4.13].

For a known applied stress acting on a member of known or assumed crack length, when the magnitude of stress-intensity factor K reaches fracture toughness K_c, the crack will propagate, leading to rupture in an instant. The *factor of safety* for fracture mechanics n, strength-to-stress ratio, is thus

$$n = \frac{K_c}{K} \tag{4.17}$$

Introducing the stress-intensity factor from Eq. (4.16), this becomes

$$n = \frac{K_c}{\lambda \sigma \sqrt{\pi a}} \tag{4.18}$$

Table 4.3 furnishes the values of the yield strength and fracture toughness for some metal alloys, measured at room temperature in a single edge-notch test specimen.

For consistency of results, the ASTM specifications require a crack length a or member thickness t are defined as

$$a, t \geq 2.5 \left(\frac{K_c}{\sigma_{yp}} \right) \tag{4.19}$$

This ensures plane strain and flat crack surfaces. The values of a and t found by Eq. (4.19) are also included in Table 4.3.

Application of the foregoing equations is demonstrated in the solution of the following numerical problems.

TABLE 4.3.   *Yield Strength σ_{yp} and Fatigue Toughness K_c for Some Materials*

	σ_{yp}		K_c		Minimum Values of a and t	
Metals	*MPa*	*(ksi)*	*MPa$\sqrt{m}$*	*(ksi$\sqrt{in.}$)*	*mm*	*(in.)*
Steel						
AISI 4340	1503	(218)	59	(53.7)	3.9	(0.15)
Stainless steel						
AISI 403	690	(100)	77	(70.1)	31.1	(1.22)
Aluminum						
2024-T851	444	(64.4)	23	(20.9)	6.7	(0.26)
7075-T7351	392	(56.9)	31	(28.2)	15.6	(0.61)
Titanium						
Ti-6AI-6V	1149	(167)	66	(60.1)	8.2	(0.32)
Ti-6AI-4V	798	(116)	111	(101)	48.4	(1.91)

EXAMPLE 4.5   Aluminum Bracket with an Edge Crack

A 2024-T851 aluminum alloy frame with an edge crack supports a concentrated load (Fig. 4.13a). Determine the magnitude of the fracture load P based on a safety factor of $n = 1.5$ for crack length of $a = 4$ mm. The dimensions are $w = 50$ mm, $d = 125$ mm, and $t = 25$ mm.

Solution   From Table 4.3, we have

$$K_c = 23 \text{ MPa}\sqrt{m} \qquad \sigma_{yp} = 444 \text{ MPa}$$

Note that that values of a and t both satisfy the table. At the section through the point B (Fig. 4.13b), the bending moment equals $M = Pd = 0.125P$. Nominal stress, by superposition of two states of stress for axial force P and moment M, is $\lambda\sigma = \lambda_a\sigma_a + \lambda_b\sigma_b$. Thus

$$\lambda\sigma = \lambda_a \frac{P}{wt} + \lambda_b \frac{6M}{tw^2} \qquad \textbf{(a)}$$

FIGURE 4.13.   *Example 4.5. Aluminum bracket with an edge crack under a concentrated load.*

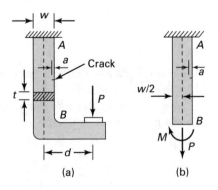

(a)   (b)

in which w and t represent the width and thickness of the member, respectively.

The ratio of crack length to bracket width is $a/w = 0.08$. For cases of B and C of Table 4.2, $\lambda_a = 1.12$ and $\lambda_b = 1.02$, respectively. Substitution of the numerical values into Eq. (4.19) results in

$$\lambda\sigma = 1.12 \frac{P}{(0.05)(0.025)} + 1.02 \frac{6(0.125P)}{(0.025)(0.05)^2} \tag{b}$$
$$= 896P + 12{,}000P = 12{,}896P$$

Therefore, by Eq. (4.17):

$$\lambda\sigma = \frac{K_C}{n\sqrt{\pi a}}; \qquad 12{,}896P = \frac{23(10^6)}{1.5\sqrt{\pi(0.004)}}$$

The foregoing gives $P = 10.61$ kN. Note that the normal stress at fracture, $10.61/1(0.05 - 0.004) = 9.226$ MPa is well below the yield strength of the material.

EXAMPLE 4.6 Titanium Panel with a Central Crack

A long plate of width $2w$ is subjected to a tensile force P in longitudinal direction with a safety factor of n (see case A, Table 4.2). Determine the thickness t required (a) to resist yielding, (b) to prevent a central crack from growing to a length of $2a$. *Given*: $w = 50$ mm, $P = 50$ kN, $n = 3$, and $a = 10$ mm. *Assumption*: The plate will be made of Ti-6AI-6V alloy.

Solution Through the use of Table 4.2, we have $K = 66\sqrt{1000}$ MPa$\sqrt{mm}$ and $\sigma_{yp} = 1149$ MPa.

a. The permissible tensile stress on the basis of the net area is

$$\sigma_{all} = \frac{\sigma_{yp}}{n} = \frac{P}{2(w - a)t}$$

Therefore

$$t = \frac{Pn}{2(w - a)\sigma_{yp}} = \frac{150(10^3)(3)}{2(50 - 6)1149} = 4.45 \text{ mm}$$

b. From the case A of Table 4.2,

$$\frac{a}{w} = \frac{10}{50} = 0.2, \qquad\qquad \lambda = 1.03$$

Applying Eq. (4.18), the stress at fracture is

$$\sigma = \frac{K_c}{\lambda n\sqrt{\pi a}} = \frac{66\sqrt{1000}}{1.03(3)\sqrt{\pi(10)}} = 120.5 \text{ MPa}$$

Inasmuch as this stress is smaller than the yield strength, the *fracture governs the design*; $\sigma_{all} = 120.5$ MPa. Hence,

$$t_{reg} = \frac{P}{2w\sigma_{all}} = \frac{150(10^3)}{2(50)(120.5)} = 12.45 \text{ mm}$$

Comment   Use a thickness of 13 mm. Both values of a and t satisfy Table 4.3.

4.15   FAILURE CRITERIA FOR METAL FATIGUE

A very common type of fatigue loading consists of an alternating sinusoidal stress superimposed on a uniform stress (Fig. 4.14). Such variation of stress with time occurs, for example, if a forced vibration of constant amplitude is applied to a structural member already transmitting a constant load. Referring to the figure, we define the *mean* stress and the *alternating* or *range* stress as follows:

$$\sigma_m = \tfrac{1}{2}(\sigma_{max} + \sigma_{min})$$
$$\sigma_a = \tfrac{1}{2}(\sigma_{max} - \sigma_{min})$$

(4.20)

In the case of complete stress reversal, it is clear that the average stress equals zero. The alternating stress component is the most important factor in determining the number of cycles of load the material can withstand before fracture; the mean stress level is less important, particularly if σ_m is negative (compressive).

As mentioned in Section 4.4, the local character of fatigue phenomena makes it necessary to analyze carefully the stress field within an element. A fatigue crack can start in one small region of high alternating stress and propagate, producing complete failure regardless of how adequately proportioned the remainder of the member may be. To predict whether the state of stress at a critical point will result in failure, a criterion is employed on the basis of the mean and alternating stresses and utilizing the simple S–N curve relationships.

FIGURE 4.14.   *Typical stress–time variation in fatigue.*

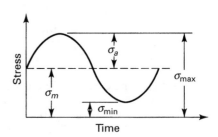

TABLE 4.4.[a] *Failure Criteria for Fatigue*

Fatigue criterion	Modified Goodman	Soderberg	Gerber	SAE
Equation	$\dfrac{\sigma_a}{\sigma_{cr}} + \dfrac{\sigma_m}{\sigma_u} = 1$	$\dfrac{\sigma_a}{\sigma_{cr}} + \dfrac{\sigma_m}{\sigma_{yp}} = 1$	$\dfrac{\sigma_a}{\sigma_{cr}} + \left(\dfrac{\sigma_m}{\sigma_u}\right)^2 = 1$	$\dfrac{\sigma_a}{\sigma_{cr}} + \dfrac{\sigma_m}{\sigma_f} = 1$

[a] Subscript notations: a = alternating, yp = static tensile yield, m = mean, u = static tensile ultimate, cr = completely reversed, f = fracture.

Single Loading

Many approaches have been suggested for interpreting fatigue data. Table 4.4 lists commonly employed *criteria*, also referred to as *mean stress–alternating stress relations*. In each case, fatigue strength for complete stress reversal at a specified number of cycles may have a value between the fracture stress and endurance stress; that is, $\sigma_e \leq \sigma_{cr} \leq \sigma_f$ (Fig. 4.3).

Experience has shown that for steel, the *Soderberg* or *modified Goodman* relations are the most reliable for predicting fatigue failure. The *Gerber* criterion leads to more liberal results and, hence, is less safe to use. For hard steels, the SAE and modified Goodman relations result in identical solutions, since for brittle materials $\sigma_u = \sigma_f$.

Relationships presented in Table 4.4 together with specified material properties form the basis for practical fatigue calculations for members under single loading.

EXAMPLE 4.7 Fatigue Load of Tension-Bending Bar

A square prismatic bar of sides 0.05 m is subjected to an axial thrust (tension) $F_m = 90$ kN (Fig 4.15). The fatigue strength for completely reversed stress at 10^6 cycles is 210 MPa, and the static tensile yield strength is 280 MPa. Apply the Soderberg criterion to determine the limiting value of completely reversed axial load F_a that can be superimposed to F_m at the midpoint of a side of the cross section without causing fatigue failure at 10^6 cycles.

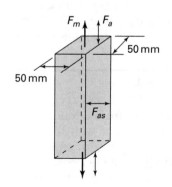

FIGURE 4.15. *Example 4.7. Bar subjected to axial tension F_m and eccentric alternating F_a loads.*

Solution The alternating and mean stresses are given by

$$\sigma_a = \frac{M_a c}{I} = \frac{0.025 F_a (0.025)}{(0.05)^4/12} = 1200 F_a$$

$$\sigma_m = \frac{F_m}{A} = \frac{90,000}{0.05 \times 0.05} = 36 \text{ MPa}$$

Applying the Soderberg criterion,

$$\frac{1200 F_a}{210 \times 10^6} + \frac{36 \times 10^6}{280 \times 10^6} = 1$$

we obtain $F_a = 152.5$ kN.

Combined Loading

Often structural and machine elements are subjected to *combined* fluctuating bending, torsion, and axial loading. Examples include crankshafts, propeller shafts, and aircraft wings. Under cyclic conditions of a general state of stress, it is common practice to modify the static failure theories for purposes of analysis, substituting the subscripts *a, m,* and *e* in the expressions developed in the preceding sections. In so doing, the maximum distortion energy theory, for example, is expressed as

$$(\sigma_{xa} - \sigma_{ya})^2 + (\sigma_{ya} - \sigma_{za})^2 + (\sigma_{za} - \sigma_{xa})^2 + 6(\tau_{xya}^2 + \tau_{xza}^2 + \tau_{yza}^2) = 2\sigma_{ea}^2$$
$$(\sigma_{xm} - \sigma_{ym})^2 + (\sigma_{ym} - \sigma_{zm})^2 + (\sigma_{zm} - \sigma_{xm})^2 + 6(\tau_{xym}^2 + \tau_{xym}^2 + \tau_{yzm}^2) = 2\sigma_{em}^2$$

$$(4.21)$$

or

$$(\sigma_{1a} - \sigma_{2a})^2 + (\sigma_{2a} - \sigma_{3a})^2 + (\sigma_{3a} - \sigma_{1a})^2 = 2\sigma_{ea}^2$$
$$(\sigma_{1m} - \sigma_{2m})^2 + (\sigma_{2m} - \sigma_{3m})^2 + (\sigma_{3m} - \sigma_{1m})^2 = 2\sigma_{em}^2$$

$$(4.22)$$

Here σ_{ea} and σ_{em}, the *equivalent alternating* stress and *equivalent mean* stress, respectively, replace the quantity σ_{yp} (or σ_u) used thus far. Relations for other failure theories can be written in a like manner.

The equivalent mean stress–equivalent alternating stress fatigue failure relations are represented in Table 4.4, replacing σ_a and σ_m with σ_{ea} and σ_{em}. These criteria together with modified static failure theories are used to compute fatigue strength under combined loading.

EXAMPLE 4.8 Fatigue Pressure of a Cylindrical Tank

Consider a thin-walled cylindrical tank of radius $r = 120$ mm and thickness $t = 5$ mm, subject to an internal pressure varying from a value of $-p/4$ to p. Employ the octahedral shear theory together with the Soderberg criterion to compute the value of p producing failure after 10^8 cycles. The material tensile yield strength is 300 MPa and the fatigue strength is $\sigma_{cr} = 250$ MPa at 10^8 cycles.

Solution The maximum and minimum values of the tangential and axial principal stresses are given by

$$\sigma_{\theta, \max} = \frac{pr}{t} = 24p, \qquad \sigma_{\theta, \min} = \frac{(-p/4)r}{t} = -6p$$

$$\sigma_{z, \max} = \frac{pr}{2t} = 12p, \qquad \sigma_{z, \min} = \frac{(-p/4)r}{2t} = -3p$$

The alternating and mean stresses are therefore

$$\sigma_{\theta a} = \tfrac{1}{2}(\sigma_{\theta, \max} - \sigma_{\theta, \min}) = 15p, \qquad \sigma_{\theta m} = \tfrac{1}{2}(\sigma_{\theta, \max} + \sigma_{\theta, \min}) = 9p$$

$$\sigma_{za} = \tfrac{1}{2}(\sigma_{z, \max} - \sigma_{z, \min}) = 7.5p, \qquad \sigma_{zm} = \tfrac{1}{2}(\sigma_{z, \max} + \sigma_{z, \min}) = 4.5p$$

The octahedral shearing stress theory, Eq. (4.6), for cyclic combined stress is expressed as

$$
\begin{aligned}
(\sigma_{\theta a}^2 - \sigma_{\theta a}\sigma_{za} + \sigma_{za}^2)^{1/2} &= \sigma_{ea} \\
(\sigma_{\theta m}^2 - \sigma_{\theta m}\sigma_{zm} + \sigma_{zm}^2)^{1/2} &= \sigma_{em}
\end{aligned}
\tag{4.23}
$$

In terms of computed alternating and mean stresses, Eqs. (4.23) appear as

$$(225p^2 - 112.5p^2 + 56.25p^2)^{1/2} = \sigma_{ea}$$

$$(81p^2 - 40.5p^2 + 20.25p^2)^{1/2} = \sigma_{em}$$

from which $\sigma_{ea} = 12.99p$ and $\sigma_{em} = 7.794p$.

The Soderberg relation then leads to

$$\frac{12.99p}{250 \times 10^6} + \frac{7.794p}{300 \times 10^6} = 1$$

Solving this equation, $p = 12.82$ MPa.

Fatigue Life

Combined stress conditions can lower fatigue life appreciably. The approach described here predicts the durability of a structural or machine element loaded in fatigue. The procedure applies to any uniaxial, biaxial, or general state of stress. According to the method, *fatigue life N_{cr}* (Fig. 4.3) is defined by the formula*

$$N_{cr} = N_f \left(\frac{\sigma_{cr}}{\sigma_f} \right)^{1/b} \tag{4.24}$$

---

*From Ref. 4.14. For further details related to failure criteria for metal fatigue, see Refs. 4.8 and 4.15 through 4.17.

TABLE 4.5. *Fracture Stress σ_f (Fracture Cycles N_f) and Fatigue Strength σ_e (Fatigue Life N_e) for Steels*

	End points for S–N diagram (Fig. 4.3)							
	Ductile steels $(\sigma_u \leq 1750\ MPa)$				Brittle (hard) steels $(\sigma_\mu > 1750\ MPa)$			
Fatigue Criterion	σ_f	N_f	σ_e	N_e	σ_f	N_f	σ_e	N_e
Modified Goodman	$0.9\sigma_u$	10^3	$\frac{1}{2}K\sigma_u$	10^6	$0.9\sigma_u$	10^3	$\frac{1}{3}K\sigma_u$	10^8
Soderberg	$0.9\sigma_u$	10^3	$\frac{1}{2}K\sigma_u$	10^6	$0.9\sigma_u$	10^3	$\frac{1}{3}K\sigma_u$	10^8
Gerber	$0.9\sigma_u$	10^3	$\frac{1}{2}K\sigma_u$	10^6	$0.9\sigma_u$	10^3	$\frac{1}{3}K\sigma_u$	10^8
SAE	$\sigma_u + 350 \times 10^6$	1	$\frac{1}{2}K\sigma_u$	10^6	σ_u	1	$\frac{1}{3}K\sigma_u$	10^8

in which

$$b = \frac{\ln(\sigma_f/\sigma_e)}{\ln(N_f/N_e)} \qquad (4.25)$$

Here the values of σ_e and σ_f are specified in terms of material static tensile strengths, while N_e and N_f are given in cycles (Table 4.5). The *fatigue-strength reduction factor K*, listed in the table, can be ascertained on the basis of tests or from finite element analysis. The data will be scattered (in general, $K > 0.3$), and considerable variance requires the stress analyst to use a statistically acceptable value. The reversed stress σ_{cr} is computed applying the relations of Table 4.4, as required.

Alternatively, the fatigue life may be determined graphically from the S–N diagram (Fig. 4.3) constructed by connecting points with coordinates (σ_f, N_f) and (σ_e, N_e). Interestingly, b represents the *slope* of the diagram. Following is a solution of a triaxial stress problem illustrating the use of the preceding approach.

EXAMPLE 4.9   Fatigue Life of an Assembly

A rotating hub and shaft assembly is subjected to bending moment, axial thrust, *bidirectional* torque, and a uniform shrink fit pressure so that the following stress levels (in megapascals) occur at an outer critical point of the shaft:

$$\begin{bmatrix} 700 & 14 & 0 \\ 14 & -350 & 0 \\ 0 & 0 & -350 \end{bmatrix}, \quad \begin{bmatrix} -660 & -7 & 0 \\ -7 & -350 & 0 \\ 0 & 0 & -350 \end{bmatrix}$$

These matrices represent the maximum and minimum stress components, respectively. Determine the fatigue life, using the maximum energy of distortion theory of failure together with (a) the SAE fatigue criterion and (b) the Gerber criterion. The material properties are $\sigma_u = 2400$ MPa and $K = 1$.

Solution From Table 4.5, we have $\sigma_e = 1(2400 \times 10^6)/3 = 800$ MPa, $N_f = 1$ cycle for SAE, $N_f = 10^3$ cycles for Gerber, and $N_e = 10^8$ cycles. The alternating and mean values of the stress components are

$$\sigma_{xa} = \tfrac{1}{2}(700 + 660) = 680 \text{ MPa}$$
$$\sigma_{xm} = \tfrac{1}{2}(700 - 660) = 20 \text{ MPa}$$
$$\sigma_{ya} = \sigma_{za} = \tfrac{1}{2}(-350 + 350) = 0$$
$$\sigma_{ym} = \sigma_{zm} = \tfrac{1}{2}(-350 - 350) = -350 \text{ MPa}$$
$$\tau_{xya} = \tfrac{1}{2}(14 + 7) = 10.5 \text{ MPa}$$
$$\tau_{xym} = \tfrac{1}{2}(14 - 7) = 3.5 \text{ MPa}$$

Upon application of Eq. (4.21), the equivalent alternating and mean stresses are found to be

$$\sigma_{ea} = \{\tfrac{1}{2}[(680 - 0)^2 + (0 - 680)^2 + 6(10.5)^2]\}^{1/2} = 680.24 \text{ MPa}$$
$$\sigma_{em} = \{\tfrac{1}{2}[(20 + 350)^2 + (-350 - 20)^2 + 6(3.5^2)]\}^{1/2} = 370.05 \text{ MPa}$$

a. The fatigue strength for complete reversal of stress, referring to Table 4.4, is

$$\sigma_{cr} = \frac{\sigma_{ea}}{1 - (\sigma_{em}/\sigma_f)} = \frac{680.24 \times 10^6}{1 - (370.05 \times 10^6/2400 \times 10^6)} = 804.24 \text{ MPa}$$

Equation (4.25) yields

$$b = \frac{\ln(2400/800)}{\ln(1/10^8)} = -0.0596$$

The fatigue life, from Eq. (4.24), is thus

$$N_{cr} = 1\left(\frac{804.24}{2400}\right)^{-(1/0.0596)} = 92.6 \times 10^6 \text{ cycles}$$

b. From Table 4.4, we now apply

$$\sigma_{cr} = \frac{\sigma_{ea}}{1 - (\sigma_{em}/\sigma_u)^2} = \frac{680.24 \times 10^6}{1 - (370.05/2400)^2} = 696.81 \text{ MPa}$$

and

$$b = \frac{\ln(0.9 \times 2400/800)}{\ln(10^3/10^8)} = -0.0863$$

It follows that

$$N_{cr} = 10^3 \left(\frac{696.81}{0.9 \times 2400} \right)^{-11.587} = 493.34 \times 10^6 \text{ cycles}$$

Comment Upon comparison of the results of (a) and (b), we observe that the Gerber criterion overestimates the fatigue life.

4.16 IMPACT OR DYNAMIC LOADS

Forces suddenly applied to structures and machines are termed *shock* or *impact* loads and result in *dynamic loading*. Examples include *rapidly* moving loads, such as those caused by a railroad train passing over a bridge or a high-speed rocket-propelled test sled moving on a track, or direct *impact* loads, such as result from a drop hammer. In machine service, impact loads are due to gradually increasing clearances that develop between mating parts with progressive wear, for example, steering gears and axle journals of automobiles.

A *dynamic force* acts to *modify* the static stress and strain fields as well as the resistance properties of a material. Shock loading is usually produced by a sudden application of force or motion to a member, whereas impact loading results from the collision of bodies. When the time of application of a load is *equal to* or *smaller* than the *largest natural period* of vibration of the structural element, shock or impact loading is produced.

Although following a shock or impact loading, vibrations commence, our concern here is only with the influence of impact forces on the maximum stress and deformation of the body. It is important to observe that the design of engineering structures subject to suddenly applied loads is complicated by a number of factors, and theoretical considerations generally serve only qualitatively to guide the design [Ref. 4.18]. Note that the effect of shock loading on members has been neglected in the preceding sections. For example, various failure criteria for metal fatigue result in shaft design equations that include stresses due to fluctuating loads but ignore shock loads. To take into account shock conditions, correction factors should be used in the design equations. The use of static material properties in the design of members under impact loading is regarded as conservative and satisfactory. Details concerning the behavior of materials under impact loading are presented in the next section.

The impact problem is analyzed using the elementary theory together with the following assumptions:

1. The displacement is proportional to the applied forces, static and dynamic.
2. The inertia of a member subjected to impact loading may be neglected.
3. The material behaves elastically. In addition, it is assumed that there is no energy loss associated with the local inelastic deformation occurring at the point of impact or at the supports. Energy is thus conserved within the system.

To idealize an elastic system subjected to an impact force, consider Fig. 4.16, in which is shown a weight W, which falls through a distance h, striking the end of a

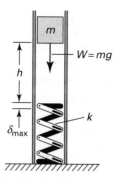

FIGURE 4.16. *A falling weight W striking a spring.*

freestanding spring. As the velocity of the weight is zero initially and is again zero at the *instant* of maximum deflection of the spring (δ_{max}), the change in kinetic energy of the system is zero, and likewise the work done on the system. The total work consists of the work done by gravity on the mass as it falls and the resisting work done by the spring:

$$W(h + \delta_{max}) - \tfrac{1}{2}k\delta_{max}^2 = 0 \tag{a}$$

where k is known as the *spring constant*.

Note that the weight is assumed to remain in contact with the spring. The deflection corresponding to a static force equal to the weight of the body is simply W/k. This is termed the *static deflection*, δ_{st}. Then the general expression of maximum dynamic deflection is, from Eq. (a),

$$\delta_{max} = \delta_{st} + \sqrt{(\delta_{st})^2 + 2\delta_{st}h} \tag{4.26}$$

or, by rearrangement,

$$\delta_{max} = \delta_{st}\left(1 + \sqrt{1 + \frac{2h}{\delta_{st}}}\right) = K\delta_{st} \tag{4.27}$$

The *impact factor* K, the ratio of the maximum dynamic deflection to the static deflection, δ_{max}/δ_{st}, is given by

$$K = 1 + \sqrt{1 + \frac{2h}{\delta_{st}}} \tag{4.28}$$

Multiplication of the impact factor by W yields an *equivalent static* or *dynamic load*:

$$P_{dyn} = KW \tag{4.29}$$

To compute the maximum stress and deflection resulting from impact loading, the preceding load may be used in the relationships derived for static loading.

Two extreme cases are clearly of particular interest. When $h \gg \delta_{max}$, the work term, $W\delta_{max}$, in Eq. (a) may be neglected, reducing the expression to $\delta_{max} = \sqrt{2\delta_{st}h}$. On the other hand, when $h = 0$, the load is suddenly applied, and Eq. (a) becomes $\delta_{max} = 2\delta_{st}$.

The expressions derived may readily be applied to analyze the dynamic effects produced by a falling weight causing axial, flexural, or torsional loading. Where bending is concerned, the results obtained are acceptable for the deflections but poor in accuracy for predictions of maximum stress, with the error increasing as h/δ_{st} becomes larger or $h \gg \delta_{st}$. This departure is attributable to the variation in the shape of the actual static deflection curve. Thus, the curvature of the beam axis and, in turn, the maximum stress at the location of the impact may differ considerably from that obtained through application of the strength of materials approach.

An analysis similar to the preceding may be employed to derive expressions for the case of a weight W in *horizontal* motion with a velocity v, arrested by an elastic body. In this instance, the kinetic energy $Wv^2/2g$ replaces $W(h + \delta_{max})$, the work done by W, in Eq. (a). Here g is the gravitational acceleration. By so doing, the maximum dynamic load and deflection are found to be, respectively,

$$P_{dyn} = W\sqrt{\frac{v^2}{g\delta_{st}}}, \qquad \delta_{max} = \delta_{st}\sqrt{\frac{v^2}{g\delta_{st}}} \qquad (4.30)$$

where δ_{st} is the static deflection caused by a horizontal force W.

EXAMPLE 4.10   Dynamic Stress and Deflection of a Metal Beam
A weight $W = 180$ N is dropped a height $h = 0.1$ m, striking at midspan a simply supported beam of length $L = 1.16$ m. The beam is of rectangular cross section: $a = 25$ mm width and $b = 75$ mm depth. For a material with modulus of elasticity $E = 200$ GPa, determine the instantaneous maximum deflection and maximum stress for the following cases: (a) the beam is rigidly supported (Fig. 4.17); (b) the beam is supported at each end by springs of stiffness $k = 180$ kN/m.

Solution   The deflection of a point at midspan, owing to a statically applied load, is

$$\delta_{st} = \frac{WL^3}{48EI} = \frac{180(1.16)^3(12)}{48(200 \times 10^9)(0.025)(0.075)^3} = 0.033 \times 10^{-3} \text{ m}$$

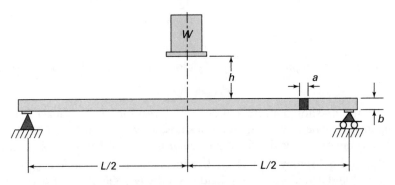

FIGURE 4.17.   *Example 4.10. A simple beam under center impact due to a falling weight W.*

The maximum static stress, also occurring at midspan, is calculated from

$$\sigma_{st,\,max} = \frac{Mc}{I} = \frac{180(1.16)(0.0375)(12)}{4(0.025)(0.075)^3} = 2.23 \text{ MPa}$$

a. The impact factor is, from Eq. (4.28),

$$K = 1 + \sqrt{1 + \frac{2(0.1)}{0.033 \times 10^{-3}}} = 78.86$$

We thus have

$$\delta_{max} = 0.033 \times 78.86 = 2.602 \text{ mm}$$

$$\sigma_{max} = 2.23 \times 78.86 = 175.86 \text{ MPa}$$

b. The static deflection of the beam due to its own bending and the deformation of the spring is

$$\delta_{st} = 0.033 \times 10^{-3} + \frac{90}{180,000} = 0.533 \times 10^{-3} \text{ m}$$

The impact factor is thus

$$K = 1 + \sqrt{1 + \frac{2(0.1)}{0.533 \times 10^{-3}}} = 20.40$$

Hence,

$$\delta_{max} = 0.533 \times 20.40 = 10.87 \text{ mm}$$

$$\sigma_{max} = 2.23 \times 20.40 = 45.49 \text{ MPa}$$

Comments It is observed from a comparison of the results that dynamic loading increases the value of deflection and stress considerably. Also noted is a reduction in stress with increased flexibility attributable to the springs added to the supports. The values calculated for the dynamic stress are probably somewhat high, because $h \gg \delta_{st}$ in both cases.

4.17 DYNAMIC AND THERMAL EFFECTS

We now explore the conditions under which metals may manifest a change from ductile to brittle behavior, and vice versa. The matter of ductile–brittle transition has important application where the operating environment includes a wide variation in temperature or when the rate of loading changes.

Let us, to begin with, identify two tensile stresses. The first, σ_f, leads to brittle fracture, that is, failure by cleavage or separation. The second, σ_y, corresponds to failure by yielding or permanent deformation. These stresses are shown in Fig. 4.18a as functions of material temperature. Referring to the figure, the point of intersection of the two stress curves defines the critical temperature, T_{cr}. If, at a given temperature

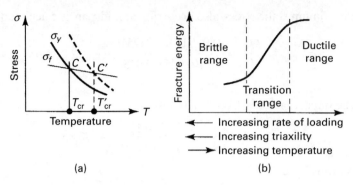

FIGURE 4.18. *Typical transition curves for metals.*

above T_{cr}, the stress is progressively increased, failure will occur by yielding, and the fracture curve will never be encountered. Similarly, for a test conducted at $T < T_{cr}$, the yield curve is not intercepted, inasmuch as failure occurs by fracture. The *principal factors* governing whether failure will occur by fracture or yielding are summarized as follows:

Temperature. If the temperature of the specimen exceeds T_{cr}, resistance to yielding is less than resistance to fracture $(\sigma_y < \sigma_f)$, and the specimen yields. If the temperature is less than T_{cr}, then $\sigma_f < \sigma_y$, and the specimen fractures without yielding. Note that σ_f exhibits only a small decrease with increasing temperature.

Loading Rate. Increasing the rate at which the load is applied increases a material's ability to resist yielding, while leaving comparatively unaffected its resistance to fracture. The increased loading rate thus results in a shift to the position occupied by the dashed curve. Point C moves to C', meaning that accompanying the increasing loading rate an increase occurs in the critical temperature. In impact tests, brittle fractures are thus observed to occur at higher temperatures than in static tests.

Triaxiality. The effect on the transition of a three-dimensional stress condition, or *triaxiality*, is similar to that of loading rate. This phenomenon may be illustrated by comparing the tendency to yield in a uniform cylindrical tensile specimen with that of a specimen containing a circumferential groove. The unstressed region above and below the groove tends to resist the deformation associated with the tensile loading of the central region, therefore contributing to a radial stress field in addition to the longitudinal stress. This state of triaxial stress is thus indicative of a tendency to resist yielding (become less ductile), the material behaving in a more brittle fashion.

Referring once more to Fig. 4.18b, in the region to the right of T_{cr} the material behaves in a ductile manner, while to the left of T_{cr} it is brittle. At temperatures close to T_{cr}, the material generally exhibits some yielding prior to a partially brittle fracture. The width of the temperature range over which the transition from brittle to ductile failure occurs is material dependent.

Transition phenomena may also be examined from the viewpoint of the energy required to fracture the material, the *toughness* rather than the stress (Fig. 4.18b). Notches and grooves serve to reduce the energy required to cause fracture and to shift the transition temperature, normally very low, to the range of normal temperatures. This is one reason that experiments are normally performed on notched specimens.

REFERENCES

4.1. NADAI, A., *Theory of Flow and Fracture of Solids*, McGraw-Hill, New York, 1950.

4.2. MARIN, J., *Mechanical Behavior of Materials*, Prentice Hall, Englewood Cliffs, N.J., 1962.

4.3. VAN VLACK, L. H., *Elements of Material Science and Engineering*, 6th ed., Addison-Wesley, Reading, Mass., 1989.

4.4. American Society of Metals (ASM), *Metals Handbook*, Metals Park, Ohio, 1985.

4.5. SHAMES, I. H., and COZZARELLI, F. A., *Elastic and Inelastic Stress Analysis*, Prentice Hall, Englewood Cliffs, N.J., 1992, Chap. 7.

4.6. IRWIN, G. R., "Fracture Mechanics," *First Symposium on Naval Structural Mechanics*, Pergamon, Elmsford, N.Y., 1958, p. 557.

4.7. TIMOSHENKO, S. P., *Strength of Materials*, Part II, 3rd ed., Van Nostrand, New York, 1956, Chap. 10.

4.8. UGURAL, A. C., *Mechanical Design: An Integrated Approach*, McGraw-Hill, New York, 2004, Chaps. 7 and 8.

4.9. TAYLOR, G. L., and QUINNEY, H., *Philosophical Transactions of the Royal Society*, Section A, No. 230, 1931, p. 323.

4.10. KNOTT, J. F., *Fundamentals of Fracture Mechanics*, Wiley, Hoboken, N.J., 1973.

4.11. BROEK, D., *The Practical Use of Fracture Mechanics*, Kluwer, Dordrecht, the Netherlands, 1988.

4.12. MEGUID, S. A., *Engineering Fracture Mechanics*, Elsevier, London, 1989.

4.13. DIETER, G. E., *Mechanical Metallurgy*, 3rd ed., McGraw-Hill, New York, 1986.

4.14. SULLIVAN, J. L., Fatigue life under combined stress, Mach. Des., January 25, 1979.

4.15. JUVINAL, R. C., and MARSHEK, K. M., *Fundamentals of Machine Component Design*, 3rd ed., Wiley, Hoboken, N.J., 2000.

4.16. DEUTSCMAN, A. D., MICHELS, W. J., and WILSON, C. E., *Machine Design, Theory and Practice,* Macmillan, New York, 1975.

4.17. JUVINALL, R. C., *Stress, Strain, and Strength,* McGraw-Hill, New York, 1967.

4.18. HARRIS, C. M., *Shock and Vibration Handbook,* McGraw-Hill, New York. 1988.

PROBLEMS

Sections 4.1 through 4.9

4.1. A steel circular bar ($\sigma_{yp} = 250$ MPa) of $d = 60$-mm diameter is acted upon by combined moments M and axial compressive loads P at its ends. Taking $M = 1.5$ kN · m, determine, based on the maximum energy of distortion theory of failure, the largest allowable value of P.

4.2. A 5-m-long steel shaft of allowable strength ($\sigma_{all} = 100$ MPa) supports a torque of $T = 325$ N · m and its own weight. Find the required shaft diameter d applying the von Mises theory of failure. *Assumptions:* Use $\rho = 7.86$ Mg/m^3 as the mass per unit volume for steel (see Table D.1). The shaft is supported by frictionless bearings that act as simple supports at its ends.

4.3. At a critical point in a loaded ASTM-A36 structural steel bracket, the plane stresses have the magnitudes and directions depicted on element A in Fig. P4.3. Calculate whether the loadings will cause the shaft to fail, based on a safety factor of $n = 1.5$, applying (a) the maximum shear stress theory; (b) the maximum energy of distortion theory.

4.4. A steel circular cylindrical bar of 0.1-m diameter is subject to compound bending and tension at its ends. The material yield strength is 221 MPa. Assume failure to occur by yielding and take the value of the applied moment to be $M = 17$ kN · m. Using the octahedral shear stress theory, determine the limiting value of P that can be applied to the bar without causing permanent deformation.

4.5. The state of stress at a critical point in a ASTM-A36 steel member is shown in Fig. P4.5. Determine the factor of safety using (a) the maximum shearing stress criterion; (b) the maximum energy of distortion criterion.

4.6. At a point in a structural member, yielding occurred under a state of stress given by

$$\begin{bmatrix} 0 & 40 & 0 \\ 40 & 50 & -60 \\ 0 & -60 & 0 \end{bmatrix} \text{MPa}$$

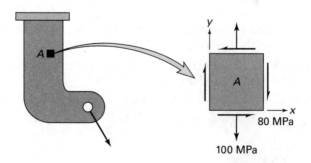

FIGURE P4.3.

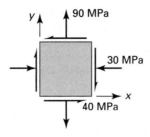

FIGURE P4.5.

Determine the uniaxial tensile yield strength of the material according to (a) maximum shearing stress theory, and (b) octahedral shear stress theory.

4.7. A circular shaft of 120-mm diameter is subjected to end loads $P = 45$ kN, $M = 4$ kN·m, and $T = 11.2$ kN·m. Let $\sigma_{yp} = 280$ MPa. What is the factor of safety, assuming failure to occur in accordance with the octahedral shear stress theory?

4.8. Determine the width t of the cantilever of height $2t$ and length 0.25 m subjected to a 450-N concentrated force at its free end. Apply the maximum energy of distortion theory. The tensile and compressive strengths of the material are both 280 MPa.

4.9. Determine the required diameter of a steel transmission shaft 10 m in length and of yield strength 350 MPa in order to resist a torque of up to 500 N·m. The shaft is supported by frictionless bearings at its ends. Design the shaft according to the maximum shear stress theory, selecting a factor of safety of 1.5, (a) neglecting the shaft weight, and (b) including the effect of shaft weight. Use $\gamma = 77$ kN/m³ as the weight per unit volume of steel.

4.10. The state of stress at a point is described by

$$\begin{bmatrix} 63.4 & 0 & 0 \\ 0 & 0.53 & 0 \\ 0 & 0 & -12.2 \end{bmatrix} \text{MPa}$$

Using $\sigma_{yp} = 90$ MPa, $\nu = 0.3$, and a factor of safety of 1.2, determine whether failure occurs at the point for (a) the maximum shearing stress theory, and (b) the maximum distortion energy theory.

4.11. A solid cylinder of radius 50 mm is subjected to a twisting moment T and an axial load P. Assume that the energy of distortion theory governs and that the yield strength of the material is $\sigma_{yp} = 280$ MPa. Determine the maximum twisting moment consistent with elastic behavior of the bar for (a) $P = 0$ and (b) $P = 400\pi$ kN.

4.12. A simply supported nonmetallic beam of 0.25-m height, 0.1-m width, and 1.5-m span is subjected to a uniform loading of 6 kN/m. Determine the factor of safety for this loading according to (a) the maximum distortion energy theory, and (b) the maximum shearing stress theory. Use $\sigma_{yp} = 28$ MPa.

4.13. The state of stress at a point in a machine element of irregular shape, subjected to combined loading, is given by

$$\begin{bmatrix} 3 & 4 & 6 \\ 4 & 2 & 5 \\ 6 & 5 & 1 \end{bmatrix} \text{MPa}$$

A torsion test performed on a specimen made of the same material shows that yielding occurs at a shearing stress of 9 MPa. Assuming the same ratios are maintained between the stress components, predict the values of the normal stresses σ_y and σ_x at which yielding occurs at the point. Use maximum distortion energy theory.

4.14. A steel rod of diameter $d = 50$ mm ($\sigma_{yp} = 260$ MPa) supports an axial load $P = 50R$ and vertical load R acting at the end of an 0.8-m-long arm (Fig. P4.14). Given a factor of safety $n = 2$, compute the largest permissible value of R using the following criteria: (a) maximum shearing stress and (b) maximum energy of distortion.

4.15. Redo Prob. 4.13 for the case in which the stresses at a point in the member are described by

$$\begin{bmatrix} 50 & 60 & 100 \\ 60 & 40 & 80 \\ 100 & 80 & 20 \end{bmatrix} \text{MPa}$$

and yielding occurs at a shearing stress of 140 MPa.

4.16. A thin-walled cylindrical pressure vessel of diameter $d = 0.5$ m and wall thickness $t = 5$ mm is fabricated of a material with 280-MPa tensile yield strength. Determine the internal pressure p required according to the following theories of failure: (a) maximum distortion energy and (b) maximum shear stress.

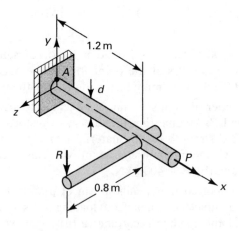

FIGURE P4.14.

4.17. The state of stress at a point is given by

$$\begin{bmatrix} 2.8 & 7 & 14 \\ 7 & 21 & 35 \\ 14 & 35 & 28 \end{bmatrix} \text{MPa}$$

Taking $\sigma_{yp} = 82$ MPa and a factor of safety of 1.2, determine whether failure takes place at the point, using (a) the maximum shearing stress theory and (b) the maximum distortion energy theory.

4.18. A structural member is subjected to combined loading so that the following stress occur at a critical point:

$$\begin{bmatrix} 120 & 50 & 30 \\ 50 & 80 & 20 \\ 30 & 20 & 10 \end{bmatrix} \text{MPa}$$

The tensile yield strength of the material is 300 MPa. Determine the factor of safety n according to (a) maximum shearing stress theory and (b) maximum energy of distortion theory.

4.19. Solve Prob. 4.18 assuming that the state of stress at a critical point in the member is given by

$$\begin{bmatrix} 100 & 50 & 40 \\ 50 & 80 & 20 \\ 40 & 20 & 30 \end{bmatrix} \text{MPa}$$

The yield strength is $\sigma_{yp} = 220$ MPa.

Sections 4.10 through 4.12

4.20. A thin-walled, closed-ended metal tube with ultimate strengths in tension σ_u and compression σ'_u, outer and inner diameters D and d, respectively, is under an internal pressure of p and a torque of T. Calculate the factor of safety n according to the maximum principal stress theory. *Given:* $\sigma_u = 250$ MPa, $\sigma'_u = 520$ MPa, $D = 210$ mm, $d = 200$ mm, $p = 5$ MPa, and $T = 50$ kN· m.

4.21. Design the cross section of a rectangular beam b meters wide by $2b$ meter deep, supported and uniformly loaded as illustrated in Fig. P4.21. *Assumptions:* $\sigma_{all} = 120$ MPa and $w = 150$ kN/m. Apply the maximum principal stress theory of failure.

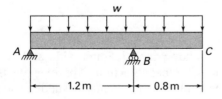

FIGURE P4.21.

4.22. Simple tension and compression tests on a brittle material reveal that failure occurs by fracture at $\sigma_u = 260$ MPa and $\sigma_u' = 420$ MPa, respectively. In an actual application, the material is subjected to perpendicular tensile and compressive stresses, σ_1 and σ_2, respectively, such that $\sigma_1/\sigma_2 = -\frac{1}{4}$. Determine the limiting values of σ_1 and σ_2 according to (a) the Mohr theory for an ultimate stress in torsion of $\tau_u = 175$ MPa and (b) the Coulomb–Mohr theory. [Hint: For case (a), the circle representing the given loading is drawn by a trial-and-error procedure.]

4.23. The state of stress at a point in a cast-iron structure ($\sigma_u = 290$ MPa, $\sigma_u' = 650$ MPa) is described by $\sigma_x = 0, \sigma_y = -180$ MPa, and $\tau_{xy} = 200$ MPa. Determine whether failure occurs at the point according to (a) the maximum principal stress criterion and (b) the Coulomb–Mohr criterion.

4.24. A thin-walled cylindrical pressure vessel of 250-mm diameter and 5-mm thickness is subjected to an internal pressure $p_i = 2.8$ MPa, a twisting moment of 31.36 kN·m, and an axial end thrust (tension) $P = 45$ kN. The ultimate strengths in tension and compression are 210 and 500 MPa, respectively. Apply the following theories to evaluate the ability of the tube to resist failure by fracture: (a) Coulomb–Mohr and (b) maximum principal stress.

4.25. A piece of chalk of ultimate strength σ_u is subjected to an axial force producing a tensile stress of $3\sigma_u/4$. Applying the principal stress theory of failure, determine the shear stress produced by a torque that acts simultaneously on the chalk and the orientation of the fracture surface.

4.26. The ultimate strengths in tension and compression of a material are 420 and 900 MPa, respectively. If the stress at a point within a member made of this material is

$$\begin{bmatrix} 200 & 150 \\ 150 & 20 \end{bmatrix} \text{MPa}$$

determine the factor of safety according to the following theories of failure: (a) maximum principal stress and (b) Coulomb–Mohr.

4.27. A plate, t meters thick, is fabricated of a material having ultimate strengths in tension and compression of σ_u and σ_u' Pa, respectively. Calculate the force P required to punch a hole of d meters in diameter through the plate (Fig. P4.27). Employ (a) the maximum principal stress theory and (b) the Mohr–Coulomb theory. Assume that the shear force is uniformly distributed through the thickness of the plate.

Sections 4.13 through 4.17

4.28. A 2024-T851 aluminum alloy panel, 125-mm wide and 20-mm thick, is loaded in tension in longitudinal direction. Approximate the maximum axial load P that can be applied without causing sudden fracture when an edge crack grows to a 25-mm length (Case B, Table 4.2).

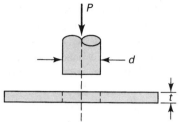

FIGURE P4.27.

FIGURE P4.33.

4.29. An AISI 4340 steel ship deck, 10-mm wide and 5-mm thick, is subjected to longitudinal tensile stress of 100 MPa. If a 60-mm-long central transverse crack is present (Case A, Table 4.2), estimate (a) the factor of safety against crack; (b) tensile stress at fracture.

4.30. A long Ti-6Al-6V alloy plate of 130-mm width is loaded by a 200-kN tensile force in longitudinal direction with a safety factor of 2.2. Determine the thickness t required to prevent a central crack to grow to a length of 200 mm (Case A, Table 4.2).

4.31. Resolve Example 4.5 if the frame is made of AISI 4340 steel. Use $a = 8$ mm, $d = 170$ mm, $w = 40$ mm, $t = 10$ mm, and $n = 1.8$.

4.32. A 2024-T851 aluminum-alloy plate, $w = 150$ mm wide and $t = 30$ mm thick, is under a tensile loading. It has a 24-mm-long transverse crack on one edge (Case B, Table 4.2). Determine the maximum allowable axial load P when the plate will undergo sudden fracture. Also find the nominal stress at fracture.

4.33. An AISI-4340 steel pressure vessel of 60-mm diameter and 5-mm wall thickness contains a 12-mm-long crack (Fig. P4.33). Calculate the pressure that will cause fracture when (a) the crack is longitudinal; (b) the crack is circumferential. *Assumption*: A factor of safety $n = 2$ and geometry factor $\lambda = 1.01$ are used (Table 4.2).

4.34. A 7075-T7351 aluminum alloy beam with $a = 48$-mm-long edge crack is in pure bending (see Case D, Table 4.2). Using $w = 120$ mm and $t = 30$ mm, find the maximum moment M that can be applied without causing sudden fracture.

4.35. Redo Example 4.7 using the SAE criterion. Take $\sigma_f = 700$ MPa, $\sigma_{cr} = 240$ MPa, and $F_m = 120$ kN.

4.36. Redo Example 4.8 employing the maximum shear stress theory together with the Soderberg criterion.

4.37. A bolt is subjected to an alternating axial load of maximum and minimum values F_{max} and F_{min}. The static tensile ultimate and fatigue strength for completely reversed stress of the material are σ_u and σ_{cr}. Verify that, according to the modified Goodman relation, the expression

$$A = \frac{1}{\sigma_{cr}}\left[F_{max} - \tfrac{1}{2}(F_{max} + F_{min})\left(1 - \frac{\sigma_{cr}}{\sigma_u}\right)\right] \qquad \textbf{(P4.37)}$$

represents the required cross-sectional area of the bolt.

4.38. An electrical contact contains a flat spring in the form of a cantilever beam, $b = 5$ mm wide by $L = 50$ mm long by t mm thick, is subjected at its free end to a load P that varies continuously from 0 to 10 N (Fig. P4.38). Employ the Soderberg criterion to calculate the value of t based on yield strength $\sigma_{yp} = 1050$ MPa, fatigue strength $\sigma_{cr} = 510$ MPa, and a factor of safety $n = 1.5$.

4.39. A small leaf spring $b = 10$ mm wide by $L = 125$ mm long by t mm thick is simply supported at its ends and subjected to a center load P that varies continuously from 0 to 20 N (Fig. P4.39). Using the modified Goodman criterion, determine the value of t, given a fatigue strength $\sigma_{cr} = 740$ MPa, ultimate tensile strength $\sigma_u = 1500$ MPa, and safety factor of $n = 2.5$.

4.40. A circular rotating shaft is subjected to a static bending moment M and a torque that varies from a value of zero to T. Apply the energy of distortion theory together with Soderberg's relation to obtain the following expression for the required shaft radius:

$$ r = \left[\frac{T}{\pi \sigma_{cr}} \left(\frac{\sigma_{cr}}{\sigma_{yp}} \sqrt{16\left(\frac{M}{T}\right)^2 + 3} + \sqrt{3} \right) \right]^{1/3} \tag{P4.40} $$

4.41. Compute the fatigue life of the rotating hub and shaft assembly described in Example 4.9 if at a critical point in the shaft the state of stress is described by $\sigma_{x,\,max} = 1000$ MPa, $\sigma_{x,\,min} = -800$ MPa, $\tau_{xy,\,max} = 300$ MPa, $\tau_{xy,\,min} = -100$ MPa, and $\sigma_y = \sigma_z = \tau_{xz} = \tau_{yz} = 0$. Employ the maximum shear stress theory of failure together with the four criteria given in Table 4.4. Take $\sigma_{yp} = 1600$ MPa, $\sigma_u = 2400$ MPa, and $K = 1$.

4.42. Determine the fatigue life of a machine element subjected to the following respective maximum and minimum stresses (in megapascals):

$$ \begin{bmatrix} 800 & 200 \\ 200 & 500 \end{bmatrix}, \quad \begin{bmatrix} -600 & -150 \\ -150 & -300 \end{bmatrix} $$

Use the maximum energy of distortion theory of failure together with the (a) modified Goodman criterion and (b) Soderberg criterion. Let $\sigma_u = 1600$ MPa, $\sigma_{yp} = 1000$ MPa, and $K = 1$.

4.43. A steel cantilever of width $t = 0.05$ m, height $2c = 0.1$ m, and length $L = 1.2$ m is subjected to a downward-acting alternating end load of maximum and minimum values P_{max} and $P_{min} = 10$ kN. The static tensile yield

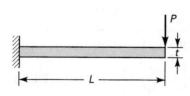

FIGURE P4.38.

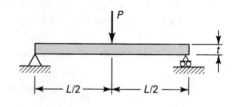

FIGURE P4.39.

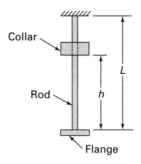

FIGURE P4.44.

and fatigue strengths for the completely reversed stress of the material are $\sigma_{yp} = 300$ MPa and $\sigma_{cr} = 200$ MPa. Use the Soderberg criterion to determine the value of P_{max} that will result in failure.

4.44. A sliding collar of $m = 80$ kg falls onto a flange at the bottom of a vertical rod (Fig. P4.44). Calculate the height h through which the mass m should drop to produce a maximum stress in the rod of 350 MPa. The rod has length $L = 2$ m, cross-sectional area $A = 250$ mm^2, and modulus of elasticity $E = 105$ GPa.

4.45. A sliding collar W is dropped from a distance h onto a flange at the bottom of the rod of length L, cross-sectional area A, and modulus of elasticity E (Fig. 4.44). Verify that the weight may be expressed in the form

$$W = \frac{\sigma_{max}^2 A}{2\left(\dfrac{hE}{L} + \sigma_{max}\right)} \qquad \textbf{(P4.45)}$$

in which σ_{max} represents the maximum stress in the rod.

4.46. A 0.125-m-diameter and 1.5-m-long circular shaft has a flywheel at one end and rotates at 240 rpm. The shaft is suddenly stopped at the free end. Determine the maximum shear stress and the maximum angle of twist produced by the impact. It is given that the shaft is made of steel with $G = 80.5$ GPa, $\nu = 0.3$, the weight of the flywheel is $W = 1.09$ kN, and the flywheel's radius of gyration is $r = 0.35$ m. [Note that kinetic energy $E_k = W\omega^2 r^2/2g = T\phi/2$. Here ω, g, T, ϕ represent the angular velocity, acceleration of gravity, torque, and angle of twist, respectively.]

4.47. A weight W is dropped from a height $h = 0.75$ m onto the free end of a cantilever beam of length $L = 1.2$ m. The beam is of 50-mm by 50-mm square cross section. Determine the value of W required to result in yielding. Omit the weight of the beam. Let $\sigma_{yp} = 280$ MPa and $E = 200$ GPa.

CHAPTER 5

Bending of Beams

5.1 INTRODUCTION

In this chapter we are concerned with the bending of straight as well as curved *beams*—that is, structural elements possessing one dimension significantly greater than the other two, usually loaded in a direction normal to the longitudinal axis. The elasticity or "exact" solutions of beams that are straight and made of homogeneous, linearly elastic materials are taken up first. Then, solutions for straight beams using mechanics of materials or elementary theory, special cases involving members made of composite materials, and the shear center are considered. The deflections and stresses in beams caused by pure bending as well as those due to lateral loading are discussed. We analyze stresses in curved beams using both exact and elementary methods, and compare the results of the various theories.

Except in the case of very simple shapes and loading systems, the theory of elasticity yields beam solutions only with considerable difficulty. Practical considerations often lead to assumptions with regard to stress and deformation that result in mechanics of materials or elementary theory solutions. The theory of elasticity can sometimes be applied to test the validity of such assumptions. The role of the theory of elasticity is then threefold. It can serve to place limitations on the use of the elementary theory, it can be used as the basis of approximate solutions through numerical analysis, and it can provide exact solutions when configurations of loading and shape are simple.

Part A—Exact Solutions

5.2 PURE BENDING OF BEAMS OF SYMMETRICAL CROSS SECTION

The simplest case of *pure bending* is that of a beam possessing a vertical axis of symmetry, subjected to equal and opposite end couples (Fig. 5.1a). The semi-inverse method is now applied to analyze this problem. The *moment* M_z shown in the figure is defined as *positive*, because it acts on a positive (negative) face with its vector in the positive (negative) coordinate direction. This *sign convention* agrees with that of stress (Sec. 1.5). We shall assume that the normal stress over the cross section varies linearly with y and that the remaining stress components are zero:

$$\sigma_x = ky, \qquad \sigma_y = \sigma_z = \tau_{xy} = \tau_{xz} = \tau_{yz} = 0 \tag{5.1}$$

Here k is a constant, and $y = 0$ contains the *neutral* surface, that is, the surface along which $\sigma_x = 0$. The intersection of the neutral surface and the cross section locates the neutral axis (abbreviated N.A.). Figure 5.1b shows the linear stress field in a section located an arbitrary distance a from the left end.

Since Eqs. (5.1) indicate that the lateral surfaces are free of stress, we need only be assured that the stresses are consistent with the boundary conditions at the ends. These *conditions of equilibrium* require that the resultant of the internal forces be zero and that the moments of the internal forces about the neutral axis equal the applied moment M_z:

$$\int_A \sigma_x \, dA = 0, \qquad -\int_A y\sigma_x \, dA = M_z \tag{5.2}$$

where A is the cross-sectional area. It should be noted that the zero stress components τ_{xy}, τ_{xz} in Eqs. (5.1) satisfy the conditions that no y- and z-directed forces exist at the end faces, and because of the y symmetry of the section, $\sigma_x = ky$ produces no moment about the y axis. The negative sign in the second expression implies that a *positive* moment M_z is one that results in *compressive* (negative) stress at points of *positive* y. Substitution of Eqs. (5.1) into Eqs. (5.2) yields

$$k \int_A y \, dA = 0, \qquad -k \int_A y^2 \, dA = M_z \tag{5.3a, b}$$

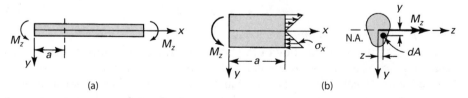

(a)   (b)

FIGURE 5.1.   *(a) Beam of singly symmetric cross section in pure bending; (b) stress distribution across cross section of the beam.*

Inasmuch as $k \neq 0$, Eq. (5.3a) indicates that the first moment of cross-sectional area about the neutral axis is zero. This requires that *the neutral and centroidal axes of the cross section coincide.* Neglecting body forces, it is clear that the equations of equilibrium (3.4), are satisfied by Eqs. (5.1). It may readily be verified also that Eqs. (5.1) together with Hooke's law fulfill the compatibility conditions, Eq. (2.12). Thus, Eqs. (5.1) represent an exact solution.

The integral in Eq. (5.3b) defines the moment of inertia I_z of the cross section about the z axis of the beam cross section (Appendix C); therefore

$$k = -\frac{M_z}{I_z} \tag{a}$$

An expression for normal stress can now be written by combining Eqs. (5.1) and (a):

$$\sigma_x = -\frac{M_z y}{I_z} \tag{5.4}$$

This is the familiar elastic *flexure formula* applicable to straight beams.

Since, at a given section, M and I are constant, the maximum stress is obtained from Eq. (5.4) by taking $|y|_{max} = c$:

$$\sigma_{max} = \frac{Mc}{I} = \frac{M}{I/c} = \frac{M}{S} \tag{5.5}$$

Here S is the *elastic section modulus.* Formula (5.5) is widely employed in practice because of its simplicity. To facilitate its use, section moduli for numerous common sections are tabulated in various handbooks. A fictitious stress in extreme fibers, computed from Eq. (5.5) for experimentally obtained *ultimate* bending moment (Sec. 12.7), is termed the *modulus of rupture* of the material in bending. This, $\sigma_{max} = M_u/S$, is frequently used as a *measure of the bending strength* of materials.

Kinematic Relationships

To gain further insight into the beam problem, consideration is now given to the *geometry of deformation,* that is, beam kinematics. Fundamental to this discussion is the hypothesis that sections originally plane remain so subsequent to bending. For a beam of symmetrical cross section, *Hooke's law* and Eq. (5.4) lead to

$$\varepsilon_x = -\frac{M_z y}{EI_z}, \qquad \varepsilon_y = \varepsilon_z = \nu \frac{M_z y}{EI_z}$$

$$\gamma_{xy} = \gamma_{xz} = \gamma_{yz} = 0 \tag{5.6}$$

where EI_z is the *flexural rigidity.*

Let us examine the deflection of the beam axis, the axial deformation of which is zero. Figure 5.2a shows an element of an initially straight beam, now in a deformed state. Because the beam is subjected to pure bending, uniform throughout, each element of infinitesimal length experiences identical deformation, with the result that the beam curvature is everywhere the same. The deflected axis of the

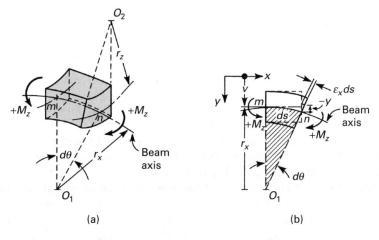

FIGURE 5.2. *(a) Segment of a bent beam; (b) geometry of deformation.*

beam or the deflection curve is thus shown deformed, with *radius of curvature* r_x. The *curvature of the beam* axis in the xy plane in terms of the y deflection v is

$$\frac{1}{r_x} = \frac{d^2v/dx^2}{[1 + (dv/dx)^2]^{3/2}} \approx \frac{d^2v}{dx^2} \tag{5.7}$$

where the approximate form is valid for small deformations $(dv/dx \ll 1)$. The sign convention for curvature of the beam axis is such that it is *positive* when the beam is bent concave *downward* as shown in the figure. From the geometry of Fig. 5.2b, the shaded sectors are similar. Hence, the radius of curvature and the strain are related as follows:

$$d\theta = \frac{ds}{r_x} = -\frac{\varepsilon_x\,ds}{y} \tag{5.8}$$

Here ds is the arc length mn along the longitudinal axis of the beam. For small displacement, $ds \approx dx$, and θ represents the slope dv/dx of the beam axis. Clearly, for the positive curvature shown, θ increases as we move from left to right along the beam axis. On the basis of Eqs. (5.6) and (5.8),

$$\frac{1}{r_x} = -\frac{\varepsilon_x}{y} = \frac{M_z}{EI_z} \tag{5.9a}$$

Following a similar procedure and noting that $\varepsilon_z \approx -\nu\varepsilon_x$, we may also obtain the curvature in the yz plane as

$$\frac{1}{r_z} = -\frac{\varepsilon_z}{y} = -\frac{\nu M_z}{EI_z} \tag{5.9b}$$

The basic *equation of the deflection curve* of a beam is obtained by combining Eqs. (5.7) and (5.9a) as follows:

$$\frac{d^2v}{dx^2} = \frac{M_z}{EI_z} \tag{5.10}$$

This expression, relating the beam curvature to the bending moment, is known as the Bernoulli–Euler law of *elementary bending theory*. It is observed from Fig. 5.2 and Eq. (5.10) that a positive moment produces positive curvature. If the sign convention adopted in this section for either moment or deflection (and curvature) should be reversed, the plus sign in Eq. (5.10) should likewise be reversed.

Reference to Fig. 5.2a reveals that the top and bottom lateral surfaces have been deformed into saddle-shaped or *anticlastic* surfaces of curvature $1/r_z$. The vertical sides have been simultaneously rotated as a result of bending. Examining Eq. (5.9b) suggests a method for determining Poisson's ratio [Ref. 5.1, Sec. 102]. For a given beam and bending moment, a measurement of $1/r_z$ leads directly to v. The effect of anticlastic curvature is small when the beam depth is comparable to its width.

Timoshenko Beam Theory

The Timoshenko theory of beams, developed by S. P. Timoshenko in the beginning of the 20th century, constitutes an improvement over the Euler–Bernoulli theory. In *static* case, the difference between the two hypotheses is that the former includes the effect of shear stresses on the deformation by assuming a constant shear over the beam height. The latter ignores the influence of transverse shear on beam deformation. The Timoshenko theory is also said to be an extension of the ordinary beam theory to allow for the effect of the transverse *shear deformation* while relaxing the assumption that plane sections remain plane and normal to the deformed beam axis.

The Timoshenko beam theory is highly suited for describing the behavior of short beams and sandwich composite beams. In *dynamic* case, the theory incorporates *shear deformation* as well as *rotational inertia* effects, and it will be more accurate for not very slender beams. By taking into account mechanism of deformation effectively, Timoshenko's theory lowers the stiffness of the beam, while the result is a larger deflection under static load and lower predicted fundamental frequencies of vibration for a prescribed set of boundary conditions.

5.3 PURE BENDING OF BEAMS OF ASYMMETRICAL CROSS SECTION

The development of Section 5.2 is now extended to the more general case in which a beam of arbitrary cross section is subjected to end couples M_y and M_z about the y and z axes, respectively (Fig. 5.3). Following a procedure similar to that of Section 5.2, plane sections are again taken to remain plane. Assume that the normal stress σ_x

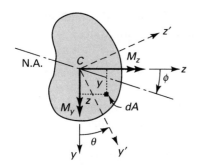

FIGURE 5.3. *Pure bending of beams of asymmetrical cross section.*

acting at a point within dA is a linear function of the y and z coordinates of the point; assume further that the remaining stresses are zero. The stress field is thus

$$\sigma_x = c_1 + c_2 y + c_3 z$$
$$\sigma_y = \sigma_z = \tau_{xy} = \tau_{xz} = \tau_{yz} = 0 \tag{5.11}$$

where c_1, c_2, c_3 are constants to be evaluated.

The *equilibrium conditions* at the beam ends, as before, relate to the force and bending moment:

$$\int_A \sigma_x \, dA = 0 \tag{a}$$

$$\int_A z\sigma_x \, dA = M_y, \qquad -\int_A y\sigma_x \, dA = M_z \tag{b, c}$$

Substitution of σ_x, as given by Eq. (5.11), into Eqs. (a), (b), and (c) results in the following expressions:

$$c_1 \int_A dA + c_2 \int_A y \, dA + c_3 \int_A z \, dA = 0 \tag{d}$$

$$c_1 \int_A z \, dA + c_2 \int_A yz \, dA + c_3 \int_A z^2 \, dA = M_y \tag{e}$$

$$c_1 \int_A y \, dA + c_2 \int_A y^2 \, dA + c_3 \int_A yz \, dA = -M_z \tag{f}$$

For the origin of the y and z axes to be coincident with the centroid of the section, it is required that

$$\int_A y \, dA = \int_A z \, dA = 0 \tag{g}$$

We conclude, therefore, from Eq. (d) that $c_1 = 0$, and from Eqs. (5.11) that $\sigma_x = 0$ at the origin. The neutral axis is thus observed to pass through the centroid, as in the beam of symmetrical section. It may be verified that the field of stress by Eqs. (5.11) satisfies the equations of *equilibrium* and *compatibility* and that the lateral surfaces are free of stress. Now consider the defining relationships

$$I_y = \int_A z^2\, dA, \qquad I_z = \int_A y^2\, dA, \qquad I_{yz} = \int_A yz\, dA \qquad (5.12)$$

where I_y and I_z are the moments of inertia about the y and z axes, respectively, and I_{yz} is the product of inertia about the y and z axes. From Eqs. (e) and (f), together with Eqs. (5.12), we obtain expressions for c_2 and c_3.

Substitution of the constants into Eqs. (5.11) results in the following *generalized flexure formula*:

$$\sigma_x = \frac{(M_y I_z + M_z I_{yz})z - (M_y I_{yz} + M_z I_y)y}{I_y I_z - I_{yz}^2} \qquad (5.13)$$

The *equation of the neutral axis* is found by equating this expression to zero:

$$(M_y I_z + M_z I_{yz})z - (M_y I_{yz} + M_z I_y)y = 0 \qquad (5.14)$$

This is an inclined line through the centroid C. The *angle ϕ* between the neutral axis and the z axis is determined as follows:

$$\tan \phi = \frac{y}{z} = \frac{M_y I_z + M_z I_{yz}}{M_y I_{yz} + M_z I_y} \qquad (5.15)$$

The angle ϕ (measured from the z axis) is positive in the *clockwise* direction, as shown in Fig. 5.3. The highest bending stress occurs at a point located *farthest* from the neutral axis.

There is a specific orientation of the y and z axes for which the product of inertia I_{yz} vanishes. Labeling the axes so oriented y' and z', we have $I_{y'z'} = 0$. The flexure formula under these circumstances becomes

$$\sigma_x = \frac{M_{y'}\, z'}{I_{y'}} - \frac{M_{z'}\, y'}{I_{z'}} \qquad (5.16)$$

The y' and z' axes now coincide with the *principal* axes of inertia of the cross section. The stresses at any point can now be ascertained by applying Eq. (5.13) or (5.16).

The kinematic relationships discussed in Section 5.2 are valid for beams of asymmetrical section provided that y and z represent principal axes.

Recall that the two-dimensional stress (or strain) and the moment of inertia of an area are second-order tensors (Sec. 1.17). Thus, *the transformation equations for stress and moment of inertia are analogous* (Sec. C.4). The Mohr's circle analysis and all conclusions drawn for stress therefore apply to the moment of inertia. With reference to the coordinate axes shown in Fig. 5.3, applying Eq. (C.12a), the moment of inertia about the y' axis is found to be

$$I_{y'} = \frac{I_y + I_z}{2} + \frac{I_y - I_z}{2} \cos 2\theta - I_{yz} \sin 2\theta \qquad (5.17)$$

From Eq. (C.13) the orientation of the principal axes is given by

$$\tan 2\theta_p = -\frac{2I_{yz}}{I_y - I_z}$$ (5.18)

The principal moment of inertia, I_1 and I_2, from Eq. (C.14) are

$$I_{1,2} = \frac{I_y + I_z}{2} \pm \sqrt{\left(\frac{I_y - I_z}{2}\right)^2 + I_{yz}^2}$$ (5.19)

Subscripts 1 and 2 refer to the maximum and minimum values, respectively.

Determination of the moments of inertia and stresses in an asymmetrical section will now be illustrated.

EXAMPLE 5.1   Analysis of an Angle in Pure Bending

A 150- by 150-mm slender angle of 20-mm thickness is subjected to oppositely directed end couples $M_z = 11$ kN·m at the centroid of the cross section. What bending stresses exist at points A and B on a section away from the ends (Fig. 5.4a)? Determine the orientation of the neutral axis.

Solution   Equations (5.13) and (5.16) are applied to ascertain the normal stress. This requires first the determination of a number of section properties through the use of familiar expressions of mechanics given in Appendix C. Note that the FORTRAN computer program presented in Table C.2 provides a check of the numerical values obtained here for the area characteristics and may easily be extended to compute the stresses.

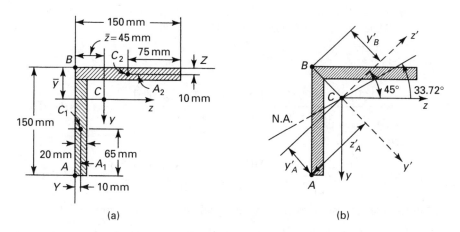

(a)   (b)

FIGURE 5.4.   *Example 5.1. An equal leg angle cross section of beam.*

Location of the Centroid C. Let $\bar{y}$ and $\bar{z}$ represent the distances from C to arbitrary reference lines (denoted Z and Y):

$$\bar{z} = \frac{\Sigma A_i \bar{z}_i}{\Sigma A_i} = \frac{A_1 \bar{z}_1 + A_2 \bar{z}_2}{A_1 + A_2} = \frac{130 \times 20 \times 10 + 150 \times 20 \times 75}{130 \times 20 + 150 \times 20} = 45 \text{ mm}$$

Here $\bar{z}_i$ represents the z distance from the Y reference line to the centroid of each subarea (A_1 and A_2) composing the total cross section. Since the section is symmetrical, $\bar{z} = \bar{y}$.

Moments and Products of Inertia. For a rectangular section of depth h and width b, the moment of inertia about the neutral $\bar{z}$ axis is $I_{\bar{z}} = bh^3/12$ (Table C.1). We now use yz axes as reference axes through C. Representing the distances from C to the centroids of each subarea by $d_{y1}, d_{y2}, d_{z1},$ and d_{z2}, we obtain the moments of inertia with respect to these axes using the parallel-axis theorem. Applying Eq. (C.9),

$$I_z = \Sigma(I_{\bar{z}} + Ad_y^2) = I_{\bar{z}1} + A_1 d_{y1}^2 + I_{\bar{z}2} + A_2 d_{y2}^2$$

Thus, referring to Fig. 5.4a,

$$I_z = I_y = \tfrac{1}{12} \times 20 \times (130)^3 + 130 \times 20 \times (40)^2$$
$$+ \tfrac{1}{12} \times 150 \times (20)^3 + 150 \times 20 \times (35)^2$$
$$= 11.596 \times 10^6 \text{ mm}^4$$

The transfer formula (C.11) for a product of inertia yields

$$I_{yz} = \Sigma(I_{\bar{y}\bar{z}} + Ad_y d_z)$$
$$= 0 + 130 \times 20 \times 40 \times (-35) + 0 + 150 \times 20 \times (-35) \times 30$$
$$= -6.79 \times 10^6 \text{ mm}^4$$

Stresses Using Formula (5.13). We have $y_A = 0.105$ m, $y_B = -0.045$ m, $z_A = -0.045$ m, $z_B = -0.045$ m, and $M_y = 0$. Thus,

$$(\sigma_x)_A = \frac{M_z(I_{yz}z_A - I_y y_A)}{I_y I_z - I_{yz}^2}$$

$$= \frac{11(10^3)[(-6.79)(-0.045) - (11.596)(0.105)]}{[(11.596)^2 - (-6.79)^2]10^{-6}} = -114 \text{ MPa} \quad \textbf{(h)}$$

Similarly,

$$(\sigma_x)_B = \frac{11(10^3)[(-6.79)(-0.045) - (11.596)(-0.045)]}{[(11.596)^2 - (-6.79)^2]10^{-6}} = 103 \text{ MPa}$$

Alternatively, these stresses may be calculated by proceeding as follows.

Directions of the Principal Axes and the Principal Moments of Inertia. Employing Eq. (5.18), we have

$$\tan 2\theta_p = \frac{-2(-6.79)}{11.596 - 11.596} = \infty, \qquad 2\theta_p = 90° \quad \text{or} \quad 270°$$

Therefore, the two values of θ_p are 45° and 135°. Substituting the first of these values into Eq. (5.17), we obtain $I_{y'} = [11.596 + 6.79 \sin 90°]$ $10^6 = 18.386 \times 10^6$ mm^4. Since the principal moments of inertia are, by application of Eq. (5.19),

$$I_{1,2} = [11.596 \pm \sqrt{0 + 6.79^2}]10^6 = [11.596 \pm 6.79]10^6$$

it is observed that $I_1 = I_{y'} = 18.386 \times 10^6$ mm^4 and $I_2 = I_{z'} = 4.806 \times 10^6$ mm^4. The principal axes are indicated in Fig. 5.4b as the $y'z'$ axes.

Stresses Using Formula (5.16). The components of bending moment about the principal axes are

$$M_{y'} = 11(10^3) \sin 45° = 7778 \text{ N} \cdot \text{m}$$
$$M_{z'} = 11(10^3) \cos 45° = 7778 \text{ N} \cdot \text{m}$$

Equation (5.16) is now applied, referring to Fig. 5.4b, with $y'_A = 0.043$ m, $z'_A = -0.106$ m, $y'_B = -0.0636$ m, and $z'_B = 0$, determined from geometrical considerations:

$$(\sigma_x)_A = \frac{7778(-0.106)}{18.386 \times 10^{-6}} - \frac{7778(0.043)}{4.806 \times 10^{-6}} = -114 \text{ MPa}$$

$$(\sigma_x)_B = 0 - \frac{7778(-0.0636)}{4.806 \times 10^{-6}} = 103 \text{ MPa}$$

as before.

Direction of the Neutral Axis. From Eq. (5.15), with $M_y = 0$,

$$\tan \phi = \frac{I_{yz}}{I_y} \quad \text{or} \quad \phi = \arctan \frac{-6.79}{11.596} = -33.72°$$

The negative sign indicates that the neutral is located counterclockwise from the z axis (Fig. 5.4b).

5.4 BENDING OF A CANTILEVER OF NARROW SECTION

Consider a narrow cantilever beam of rectangular cross section, loaded at its free end by a concentrated force of such magnitude that the beam weight may be neglected (Fig. 5.5). The situation described may be regarded as a case of plane stress provided that the beam thickness t is small relative to beam depth $2h$. The distribution of stress in the beam, as we have already found in Example 3.1, is given by

$$\sigma_x = -\left(\frac{Px}{I}\right)y, \qquad \sigma_y = 0, \qquad \tau_{xy} = -\frac{P}{2I}(h^2 - y^2) \qquad \text{(3.21)}$$

FIGURE 5.5. *Deflections of an end-loaded cantilever beam.*

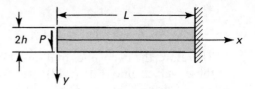

To derive expressions for the beam displacement, it is necessary to relate stress, described by Eq. (3.21), to strain. This is accomplished through the use of the strain–displacement relations and Hooke's law:

$$\frac{\partial u}{\partial x} = -\frac{Pxy}{EI}, \qquad \frac{\partial v}{\partial y} = \frac{\nu Pxy}{EI} \tag{a, b}$$

$$\frac{\partial u}{\partial y} + \frac{\partial v}{\partial x} = \frac{2(1+\nu)\tau_{xy}}{E} = -\frac{(1+\nu)P}{EI}(h^2 - y^2) \tag{c}$$

Integration of Eqs. (a) and (b) yields

$$u = -\frac{Px^2 y}{2EI} + u_1(y) \tag{d}$$

$$v = \frac{\nu Pxy^2}{2EI} + v_1(x) \tag{e}$$

Differentiating Eqs. (d) and (e) with respect to y and x, respectively, and substituting into Eq. (c), we have

$$\frac{du_1}{dy} - \frac{P}{2EI}(2+\nu)y^2 = -\frac{dv_1}{dx} + \frac{P}{2EI}x^2 - \frac{(1+\nu)Ph^2}{EI}$$

In this expression, note that the left and right sides depend only on y and x, respectively. These variables are independent of one another, and it is therefore concluded that the equation can be valid only if each side is equal to the same constant:

$$\frac{du_1}{dy} - \frac{P}{2EI}(2+\nu)y^2 = a_1, \qquad \frac{dv_1}{dx} - \frac{Px^2}{2EI} + \frac{(1+\nu)Ph^2}{EI} = -a_1$$

These are integrated to yield

$$u_1(y) = \frac{P}{6EI}(2+\nu)y^3 + a_1 y + a_2$$

$$v_1(x) = \frac{Px^3}{6EI} - \frac{(1+\nu)Pxh^2}{EI} - a_1 x + a_3$$

in which a_2 and a_3 are constants of integration. The displacements may now be written

$$u = -\frac{Px^2 y}{2EI} + \frac{P}{6EI}(2+\nu)y^3 + a_1 y + a_2 \tag{5.20a}$$

$$v = \frac{\nu P x y^2}{2EI} + \frac{P x^3}{6EI} - \frac{(1 + \nu)P x h^2}{EI} - a_1 x + a_3 \qquad \textbf{(5.20b)}$$

The constants a_1, a_2, and a_3 depend on known conditions. If, for example, the situation at the fixed end is such that

$$\frac{\partial u}{\partial y} = 0, \qquad v = u = 0, \qquad x = L, y = 0$$

then, from Eqs. (5.20),

$$a_1 = \frac{PL^2}{2EI}, \qquad a_2 = 0, \qquad a_3 = \frac{PL^3}{3EI} + \frac{PLh^2(1 + \nu)}{EI}$$

The beam displacement is therefore

$$u = \frac{P}{2EI}(L^2 - x^2)y + \frac{(2 + \nu)P y^3}{6EI} \qquad \textbf{(5.21)}$$

$$v = \frac{P}{EI}\left[\frac{x^3}{6} + \frac{L^3}{3} + \frac{x}{2}(\nu y^2 - L^2) + h^2(1 + \nu)(L - x) \right] \qquad \textbf{(5.22)}$$

It is clear on examining these equations that u and v do not obey a simple linear relationship with y and x. We conclude, therefore, that plane sections do not, as assumed in elementary theory, remain plane subsequent to bending.

Comparison of the Results with That of Elementary Theory

The vertical displacement of the beam axis is obtained by substituting $y = 0$ into Eq. (5.22):

$$(v)_{y=0} = \frac{P x^3}{6EI} - \frac{PL^2 x}{2EI} + \frac{PL^3}{3EI} + \frac{Ph^2(1 + \nu)}{EI}(L - x) \qquad \textbf{(5.23)}$$

Introducing the foregoing into Eq. (5.7), the radius of curvature is given by

$$\frac{1}{r_x} \approx \frac{P x}{EI} = \frac{M}{EI}$$

provided that dv/dx is a small quantity. Once again we obtain Eq. (5.9a), the beam curvature–moment relationship of elementary bending theory.

It is also a simple matter to compare the total vertical deflection at the free end $(x = 0)$ with the deflection derived in elementary theory. Substituting $x = 0$ into Eq. (5.23), the total deflection is

$$(v)_{x=y=0} = \frac{PL^3}{3EI} + \frac{Ph^2(1 + \nu)L}{EI} = \frac{PL^3}{3EI} + \frac{Ph^2 L}{2GI} \qquad \textbf{(5.24)}$$

wherein the deflection associated with shear is clearly $Ph^2L/2GI = 3PL/2GA$. The ratio of the shear deflection to the bending deflection at $x = 0$ provides a measure of beam slenderness:

$$\frac{Ph^2L/2GI}{PL^3/3EI} = \frac{3}{2}\frac{h^2E}{L^2G} = \tfrac{3}{4}(1 + \nu)\left(\frac{2h}{L}\right)^2 \approx \left(\frac{2h}{L}\right)^2$$

If, for example, $L = 10(2h)$, the preceding quotient is only $\frac{1}{100}$. For a slender beam, $2h \ll L$, and it is clear that the deflection is mainly due to bending. It should be mentioned here, however, that in vibration at higher modes, and in wave propagation, the effect of shear is of great importance in slender as well as in other beams.

In the case of *wide beams* ($t \gg 2h$), Eq. (5.24) must be modified by replacing E and ν as indicated in Table 3.1.

5.5 BENDING OF A SIMPLY SUPPORTED NARROW BEAM

Consideration is now given to the stress distribution in a narrow beam of thickness t and depth $2h$ subjected to a uniformly distributed loading (Fig. 5.6). The situation as described is one of plane stress, subject to the following boundary conditions, consistent with the origin of an x, y coordinate system located at midspan and mid-height of the beam, as shown:

$$(\tau_{xy})_{y=\pm h} = 0, \qquad (\sigma_y)_{y=+h} = 0, \qquad (\sigma_y)_{y=-h} = -p/t \qquad \textbf{(a)}$$

Since at the ends no longitudinal load is applied, it would appear reasonable to state that $\sigma_x = 0$ at $x = \pm L$. However, this boundary condition leads to a complicated solution, and a less severe statement is instead used:

$$\int_{-h}^{h} \sigma_x t\, dy = 0 \qquad \textbf{(b)}$$

The corresponding condition for bending couples at $x = \pm L$ is

$$\int_{-h}^{h} \sigma_x t y\, dy = 0 \qquad \textbf{(c)}$$

For y equilibrium, it is required that

$$\int_{-h}^{h} \tau_{xy} t\, dy = \pm pL \qquad x = \pm L \qquad \textbf{(d)}$$

FIGURE 5.6. *Bending of a simply supported beam with a uniform load.*

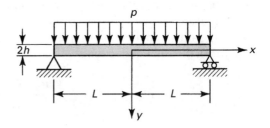

The problem is treated by superimposing the solutions Φ_2, Φ_3, and Φ_5 (Sec. 3.6), with

$$c_2 = b_2 = a_3 = c_3 = a_5 = b_5 = c_5 = e_5 = 0$$

We then have

$$\Phi = \Phi_2 + \Phi_3 + \Phi_5 = \frac{a_2}{2}x^2 + \frac{b_3}{2}x^2y + \frac{d_3}{6}y^3 + \frac{d_5}{6}x^2y^3 - \frac{2d_5}{60}y^5$$

The stresses are

$$\sigma_x = d_3y + d_5(x^2y - \tfrac{2}{3}y^3)$$
$$\sigma_y = a_2 + b_3y + \frac{d_5}{3}y^3 \qquad\qquad \text{(e)}$$
$$\tau_{xy} = -b_3x - d_5xy^2$$

The conditions (a) are

$$-b_3 - d_5h^2 = 0$$
$$a_2 + b_3h + \frac{d_5}{3}h^3 = 0$$
$$a_2 - b_3h - \frac{d_5}{3}h^3 = -\frac{p}{t}$$

and the solution is

$$a_2 = -\frac{p}{2t}, \qquad b_3 = \frac{3p}{4th}, \qquad d_5 = -\frac{3p}{4th^3}$$

The constant d_3 is obtained from condition (c) as follows:

$$\int_{-h}^{h}\left[d_3y + d_5\left(L^2y - \frac{2}{3}y^3\right)\right]yt\,dy = 0$$

or

$$d_3 = -d_5\left(L^2 - \frac{2}{5}h^2\right) = \frac{3p}{4th}\left(\frac{L^2}{h^2} - \frac{2}{5}\right)$$

Expressions (e) together with the values obtained for the constants also fulfill conditions (b) and (d).

The state of stress is thus represented by

$$\sigma_x = \frac{py}{2I}(L^2 - x^2) + \frac{py}{I}\left(\frac{y^2}{3} - \frac{h^2}{5}\right) \qquad\qquad \textbf{(5.25a)}$$

$$\sigma_y = \frac{-p}{2I}\left(\frac{y^3}{3} - h^2y + \frac{2h^3}{3}\right) \qquad\qquad \textbf{(5.25b)}$$

$$\tau_{xy} = \frac{-px}{2I}(h^2 - y^2) \qquad\qquad \textbf{(5.25c)}$$

Here $I = \frac{2}{3}th^3$ is the area moment of inertia taken about a line through the centroid, parallel to the z axis. Although the solutions given by Eqs. (5.25) satisfy the equations of elasticity and the boundary conditions, they are nevertheless not exact. This is indicated by substituting $x = \pm L$ into Eq. (5.25a) to obtain the following expression for the normal distributed forces per unit area at the ends:

$$p_x = \frac{py}{I}\left(\frac{y^2}{3} - \frac{h^2}{5}\right)$$

which cannot exist, as no forces act at the ends. From Saint-Venant's principle we may conclude, however, that the solutions do predict the correct stresses throughout the beam, except near the supports.

Comparison of the Results with That of Elementary Theory

Recall that the longitudinal normal stress derived from elementary beam theory is $\sigma_x = -My/I$; this is equivalent to the first term of Eq. (5.25a). The second term is then the difference between the longitudinal stress results given by the two approaches. To gauge the magnitude of the deviation, consider the ratio of the second term of Eq. (5.25a) to the result of elementary theory at $x = 0$. At this point, the bending moment is a maximum. Substituting $y = h$ for the condition of maximum stress, we obtain

$$\frac{\Delta\sigma_x}{(\sigma_x)_{\text{elem. theory}}} = \frac{(ph/I)(h^2/3 - h^2/5)}{phL^2/2I} = \frac{4}{15}\left(\frac{h}{L}\right)^2$$

For a beam of length 10 times its depth, the ratio is small, $\frac{1}{1500}$. For beams of ordinary proportions, we can conclude that elementary theory provides a result of sufficient accuracy for σ_x. As for σ_y, this stress is not found in the elementary theory. The result for τ_{xy} is, on the other hand, the same as that of elementary beam theory.

The displacement of the beam may be determined in a manner similar to that described for a cantilever beam (Sec. 5.4).

Part B—Approximate Solutions

5.6 ELEMENTARY THEORY OF BENDING

We may conclude, on the basis of the previous sections, that exact solutions are difficult to obtain. It was also observed that for a slender beam the results of the exact theory do not differ markedly from that of the mechanics of materials or elementary approach provided that solutions close to the ends are not required. The bending deflection was found to be very much larger than the shear deflection. Thus, the stress associated with the former predominates. We deduce therefore that the normal strain ε_y resulting from transverse loading may be neglected. Because it is more easily applied, the elementary approach is usually preferred in engineering practice. The exact and elementary theories should be regarded as complementary rather than competitive approaches, enabling the analyst to obtain the degree of accuracy required in the context of the specific problem at hand.

The basic assumptions of the elementary theory [Ref. 5.2], for a slender beam whose cross section is symmetrical about the vertical plane of loading, are

$$\varepsilon_y = \frac{\partial v}{\partial y} = 0, \qquad \gamma_{xy} = \frac{\partial u}{\partial y} + \frac{\partial v}{\partial x} = 0$$

$$\varepsilon_x = \frac{\sigma_x}{E} \qquad (\text{independent of } z)$$

$$\varepsilon_z = 0, \qquad \gamma_{yz} = \gamma_{xz} = 0 \tag{5.27}$$

(5.26)

The first equation of (5.26) is equivalent to the assertion $v = v(x)$. Thus, all points in a beam at a given longitudinal location x experience identical deformation. The second equation of (5.26), together with $v = v(x)$, yields, after integration,

$$u = -y\frac{dv}{dx} + u_0(x) \tag{a}$$

The third equation of (5.26) and Eqs. (5.27) imply that the beam is considered *narrow*, and we have a case of *plane stress*.

At $y = 0$, the bending deformation should vanish. Referring to Eq. (a), it is clear, therefore, that $u_0(x)$ must represent axial deformation. The term dv/dx is the *slope* θ of the beam axis, as shown in Fig. 5.7a, and is very much smaller than unity. Therefore,

$$u = -y\frac{dv}{dx} = -y\theta$$

The slope is *positive* when *clockwise*, provided that the x and y axes have the directions shown. Since u is a linear function of y, this equation restates the kinematic hypothesis of the elementary theory of bending: *Plane sections perpendicular to the longitudinal axis of the beam remain plane subsequent to bending.* This assumption is confirmed by the exact theory *only* in the case of *pure bending*.

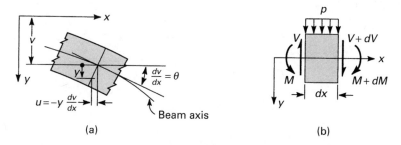

(a)   (b)

FIGURE 5.7.   (a) Longitudinal displacements in a beam due to rotation of a plane section; (b) element between adjoining sections of a beam.

Method of Integration

In the next section, we obtain the stress distribution in a beam according to the elementary theory. We now derive some useful relations involving the shear force V, the bending moment M, the load per unit length p, the slope θ, and the *deflection v*. Consider a beam element of length dx subjected to a distributed loading (Fig. 5.7b). Note that as dx is small, the variation in the load per unit length p is omitted. In the free-body diagram, all the forces and the moments are positive. The shear force obeys the sign convention discussed in Section 1.4; the bending moment is in agreement with the convention adopted in Section 5.2. In general, the shear force and bending moment vary with the distance x, and it thus follows that these quantities will have different values on each face of the element. The increments in shear force and bending moment are denoted by dV and dM, respectively. Equilibrium of forces in the vertical direction is governed by $V - (V + dV) - p\,dx = 0$, or

$$\frac{dV}{dx} = -p \tag{5.28}$$

That is, the rate of change of shear force with respect to x is equal to the algebraic value of the distributed loading. Equilibrium of the moments about a z axis through the left end of the element, neglecting the higher-order infinitesimals, leads to

$$\frac{dM}{dx} = -V \tag{5.29}$$

This relation states that the rate of change of bending moment is equal to the algebraic value of the shear force, valid only if a distributed load or no load acts on the beam segment. Combining Eqs. (5.28) and (5.29), we have

$$\frac{d^2M}{dx^2} = p \tag{5.30}$$

The basic equation of bending of a beam, Eq. (5.10), combined with Eq. (5.30), may now be written as

$$\frac{d^2}{dx^2}\left(EI\frac{d^2v}{dx^2}\right) = p \tag{5.31}$$

For a beam of *constant* flexural rigidity EI, the beam equations derived here may be expressed as

$$EI\frac{d^4v}{dx^4} = EIv^{IV} = p$$

$$EI\frac{d^3v}{dx^3} = EIv''' = -V$$

$$EI\frac{d^2v}{dx^2} = EIv'' = M \tag{5.32}$$

$$EI\frac{dv}{dx} = EIv' = \int M\,dx$$

These relationships also apply to *wide beams* provided that $E/(1 - v^2)$ is substituted for E (Table 3.1).

In many problems of practical importance, the deflection due to transverse loading of a beam may be obtained through successive integration of the beam equation:

$$EIv^{IV} = p$$

$$EIv''' = \int_0^x p \, dx + c_1$$

$$EIv'' = \int_0^x dx \int_0^x p \, dx + c_1 x + c_2 \qquad \text{(5.33)}$$

$$EIv' = \int_0^x dx \int_0^x dx \int_0^x p \, dx + \tfrac{1}{2} c_1 x^2 + c_2 x + c_3$$

$$EIv = \int_0^x dx \int_0^x dx \int_0^x dx \int_0^x p \, dx + \tfrac{1}{6} c_1 x^3 + \tfrac{1}{2} c_2 x^2 + c_3 x + c_4$$

Alternatively, we could begin with $EIv'' = M(x)$ and integrate twice to obtain

$$EIv = \int_0^x dx \int_0^x M \, dx + c_3 x + c_4 \qquad \text{(5.34)}$$

In either case, the constants, $c_1, c_2, c_3,$ and c_4, which correspond to the homogeneous solution of the differential equations, may be evaluated from the boundary conditions. The constants $c_1, c_2, c_3/EI,$ and c_4/EI represent the values at the origin of $V, M, \theta,$ and v, respectively. In the method of successive integration, there is no need to distinguish between statically determinate and statically indeterminate systems (Sec. 5.11), because the equilibrium equations represent only two of the boundary conditions (on the first two integrals), and because the *total* number of boundary conditions is always equal to the total number of unknowns.

EXAMPLE 5.2   Displacements of a Cantilever Beam

A cantilever beam AB of length L and constant flexural rigidity EI carries a moment M_o at its free end A (Fig. 5.8a). Derive the equation of the deflection curve and determine the slope and deflection at A.

Solution   From the free-body diagram of Fig. 5.8b, observe that the bending moment is $+M_o$ throughout the beam. Thus, the third of Eqs. (5.32) becomes

$$EIv'' = M_o$$

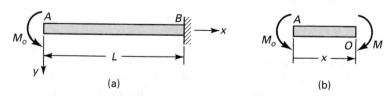

(a)   (b)

FIGURE 5.8.   *Example 5.2. (a) A cantilever beam is subjected to moment at its free end; (b) free-body diagram of part AO.*

Integrating, we obtain

$$EIv' = M_ox + c_1$$

The constant of integration c_1 can be found from the condition that the slope is zero at the support; therefore, we have $v'(L) = 0$, from which $c_1 = -M_oL$. The slope is then

$$v' = \frac{M_o}{EI}(x - L) \tag{5.35}$$

Integrating, we obtain

$$v = \frac{M_o}{2EI}(x^2 - 2Lx) + c_2$$

The boundary condition on the deflection at the support is $v(L) = 0$, which yields $c_2 = M_oL^2/2EI$. The equation of the deflection curve is thus a parabola:

$$v = \frac{M_o}{2EI}(L^2 + x^2 - 2Lx) \tag{5.36}$$

However, every element of the beam experiences equal moments and deforms alike. The deflection curve should therefore be part of a circle. This inconsistancy results from the use of an approximation for the curvature, Eq. (5.7). The error is very small, however, when the deformation v is small [Ref. 5.1].

The slope and deflection at A are readily found by letting $x = 0$ into Eqs. (5.35) and (5.36):

$$\theta_A = -\frac{M_oL}{EI}, \qquad v_A = \frac{M_oL^2}{2EI} \tag{5.37}$$

The minus sign indicates that the angle of rotation is counterclockwise (Fig. 5.8a).

5.7 NORMAL AND SHEAR STRESSES

When a beam is bent by transverse loads, usually both a bending moment M and a shear force V act on each cross section. The distribution of the normal stress associated with the bending moment is given by the *flexure formula*, Eq. (5.4):

$$\sigma_x = -\frac{My}{I} \tag{5.38}$$

where M and I are taken with respect to the z axis (Fig. 5.7).

In accordance with the assumptions of elementary bending, Eqs. (5.26) and (5.27), the contribution of the shear strains to beam deformation is omitted. However, shear stresses do exist, and the shearing forces are the resultant of the stresses. The shearing stress τ_{xy} acting at section mn, assumed uniformly distributed

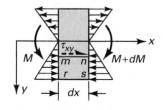
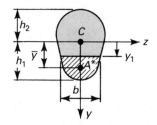

FIGURE 5.9. *(a) Beam segment for analyzing shear stress; (b) cross section of beam.*

over the area $b \cdot dx$, can be determined on the basis of equilibrium of forces acting on the shaded part of the beam element (Fig. 5.9). Here b is the width of the beam a distance y_1 from the neutral axis and dx is the length of the element. The distribution of normal stresses produced by M and $M + dM$ is indicated in the figure. The normal force distributed over the left face *mr* on the shaded area A^* is equal to

$$\int_{-b/2}^{b/2} \int_{y_1}^{h_1} \sigma_x \, dy \, dz = \int_{A^*} -\frac{My}{I} \, dA \qquad \textbf{(a)}$$

Similarly, an expression for the normal force on the right face *ns* may be written in terms of $M + dM$. The equilibrium of x-directed forces acting on the beam element is governed by

$$-\int_{A^*} \frac{(M + dM)y}{I} \, dA - \int_{A^*} -\frac{My}{I} \, dA = \tau_{xy} \, b \, dx$$

from which we have

$$\tau_{xy} = -\frac{1}{Ib} \int_{A^*} \frac{dM}{dx} y \, dA$$

Upon substitution of Eq. (5.29), we obtain the *shear formula* for beams:

$$\tau_{xy} = \frac{V}{Ib} \int_{A^*} y \, dA = \frac{VQ}{Ib} \qquad \textbf{(5.39)}$$

The integral represented by Q is the *first moment of the shaded area A^** with respect to the neutral axis z:

$$Q = \int_{A^*} y \, dA = A^* \bar{y} \qquad \textbf{(5.40)}$$

By definition, $\bar{y}$ is the distance from the neutral axis to the centroid of A^*. In the case of sections of regular geometry, $A^*\bar{y}$ provides a convenient means of calculating Q. Note that the shear force acting across the width of the beam per unit length $q = \tau_{xy} \, b = VQ/I$ is called the *shear flow*.

For example, in the case of a *rectangular cross section* of width b and depth $2h$, the shear stress at y_1 is

$$\tau_{xy} = \frac{V}{Ib} \int_{-b/2}^{b/2} \int_{y_1}^{h} y \, dy \, dz = \frac{V}{2I} (h^2 - y_1^2) \qquad \textbf{(5.41)}$$

This shows that the shear stress varies parabolically with y_1; it is zero when $y_1 = \pm h$ and has its maximum value at the neutral axis, $y_1 = 0$:

$$\tau_{max} = \frac{Vh^2}{2I} = \frac{3}{2}\frac{V}{2bh} \tag{5.42}$$

Here, $2bh$ is the area of the rectangular cross section. It is observed that the maximum shear stress (either horizontal or vertical: $\tau_{xy} = \tau_{yx}$) is 1.5 times larger than the average shear stress V/A. As observed in Section 5.4, for a *thin* rectangular beam, Eq. (5.42) is the exact distribution of shear stress. However, in general, for wide rectangular sections and for other sections, Eq. (5.39) yields only approximate values of the shearing stress.

It should be pointed out that the maximum shear stress does not always occur at the neutral axis. For instance, in the case of a cross section having nonparallel sides, such as a triangular section, the maximum value of Q/b (and thus τ_{xy}) takes place at midheight, $h/2$, while the neutral axis is located at a distance $h/3$ from the base.

The following sample problem illustrates the application of the shear stress formula.

EXAMPLE 5.3 Shear Stresses in a Flanged Beam

A cantilever wide-flange beam is loaded by a force P at the free end acting through the centroid of the section. The beam is of constant thickness t (Fig. 5.10a). Determine the shear stress distribution in the section.

Solution The vertical shear force at every section is P. It is assumed that the shear stress τ_{xy} is uniformly distributed over the web thickness. Then, in the web, for $0 \le y_1 \le h_1$, applying Eq. (5.39),

$$\tau_{xy} = \frac{V}{Ib}A^*\bar{y} = \frac{P}{It}\left[b(h - h_1)\left(h_1 + \frac{h - h_1}{2}\right) + t(h_1 - y_1)\left(y_1 + \frac{h_1 - y_1}{2}\right)\right]$$

This equation may be written as

$$\tau_{xy} = \frac{P}{2It}[b(h^2 - h_1^2) + t(h_1^2 - y_1^2)] \tag{b}$$

FIGURE 5.10. *Example 5.3. (a) Cross section of a wide-flange beam; (b) shearing stress distribution.*

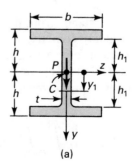
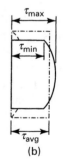

The shearing stress thus varies parabolically in the web (Fig. 5.10b). The extreme values of τ_{xy} found at $y_1 = 0$ and $y_1 = \pm h_1$, are, from Eq. (b), as follows:

$$\tau_{max} = \frac{P}{2It}(bh^2 - bh_1^2 + th_1^2), \qquad \tau_{min} = \frac{Pb}{2It}(h^2 - h_1^2)$$

Note that it is usual that $t \ll b$, and therefore the maximum and minimum stresses do not differ appreciably, as is seen in the figure. Similarly, the shear stress in the flange, for $h_1 < y_1 \le h$, is

$$\tau_{xz} = \frac{P}{Ib}\left[b(h - y_1)\left(y_1 + \frac{h - y_1}{2}\right)\right] = \frac{P}{2I}(h^2 - y_1^2) \qquad \textbf{(c)}$$

This is the parabolic equation for the variation of stress in the flange, shown by the dashed lines in the figure.

Comments Clearly, for a *thin* flange, the shear stress is very small as compared with the shear stress in the web. It is concluded that the approximate *average* value of *shear stress* in the beam may be found by dividing P by the web cross section with the web height assumed equal to the beam's overall height: $\tau_{avg} = P/2th$. The preceding is indicated by the dotted lines in the figure. The distribution of stress given by Eq. (c) is fictitious, because the inner planes of the flanges must be free of shearing stress, as they are load-free boundaries of the beam. This contradiction cannot be resolved by the elementary theory; the theory of elasticity must be applied to obtain the correct solution. Fortunately, this defect of the shearing stress formula does not lead to serious error since, as pointed out previously, the web carries almost all the shear force. To reduce the stress concentration at the juncture of the web and the flange, the sharp corners should be rounded.

EXAMPLE 5.4   Beam of Circular Cross Section
A cantilever beam of circular cross section supporting a concentrated load P at its free end (Fig. 5.11a). The shear force V in this beam is constant and equal to the magnitude of the load $P = V$. Determine the maximum shearing stresses (a) in a solid cross section; (b) in a hollow cross section. *Assumptions*: All shear stresses do not act parallel to the y. At a point such as a or b on the boundary of the cross section, the shear stress τ must act parallel to the boundary. The shear stresses at line ab across the cross section are not parallel to the y axis and cannot be determined by the shear formula, $\tau = VQ/Ib$. The maximum shear stresses occur along the neutral axis z, uniformly distributed, and act parallel to the y axis. These stresses are within about 5% of their true value [Ref. 5.1, Sec. 122].

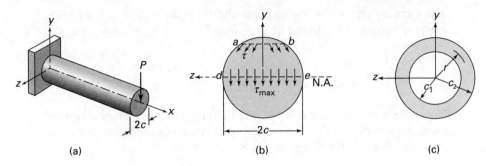

FIGURE 5.11. *Example 5.4. (a) A cantilever beam under a load P; (b) shear stress distribution on a circular cross section; (c) hollow circular cross section.*

Solution

a. *Solid Cross Section* (Fig. 5.11b). The shear formula may be used to calculate with reasonable accuracy of the shear stresses at the neutral axis. The area properties for a circular cross section of radius c (see Table C.1) are

$$I = \frac{\pi c^4}{4} \qquad Q = A^* \bar{y} = \frac{\pi c^2}{2}\left(\frac{4c}{3\pi}\right) = \frac{2c^2}{3} \qquad \textbf{(d)}$$

and $b = 2c$. The maximum shear stress is thus

$$\tau_{max} = \frac{VQ}{Ib} = \frac{4V}{3\pi c^2} = \frac{4V}{3A} \qquad \textbf{(5.43)}$$

in which A is the cross-sectional area of the beam.

Comment The result shows that the largest shear stress in a circular beam is 4/3 times the average shear stress $\tau_{avg} = V/A$.

b. *Hollow Circular Cross Section* (Fig. 5.11c). Equation (5.43) applies with equal rigor to circular tubes, since the same assumptions stated in the foregoing are valid. But in this case, by Eq. (C.3), we have

$$Q = \frac{2}{3}(c_2^3 - c_1^3) \qquad b = 2(c_2 - c_1) \qquad A = \pi(c_2^2 - c_1^2)$$

and

$$I = \frac{\pi}{4}(c_2^4 - c_1^4)$$

So, the maximum shear stress may be written in the form:

$$\tau_{max} = \frac{VQ}{Ib} = \frac{4V}{3A}\frac{c_2^2 + c_2 c_1 + c_1^2}{c_2^2 + c_1^2} \qquad \textbf{(e)}$$

Comment Observe that for $c_1 = 0$, Eq. (e) reduces to Eq. (5.43) for a solid circular beam, as expected. In the special case of a *thin-walled tube*, we have $r/t > 10$, where r and t represent the mean radius and thickness, respectively. As a theoretical limiting case, setting $c_2 = c_1 = r$, the Eq. (e) results in $\tau_{max} = 2V/A$.

5.8 EFFECT OF TRANSVERSE NORMAL STRESS

When a beam is subjected to a transverse load, there is a resulting *transverse normal stress*. According to Eq. (5.26), this stress is not related to the normal strain ε_y and thus cannot be determined from Hooke's law. However, an expression for the average transverse normal stress can be obtained from the *equilibrium requirement* of force balance along the axis of the beam. For this purpose, a procedure is used similar to that employed for determining the shear stress in Section 5.7.

Consider, for example, a rectangular cantilever beam of width b and depth $2h$ subject to a uniform load of intensity p (Fig. 5.12a). The free-body diagram of an isolated beam segment of length dx is shown in Fig. 5.12b. Passing a horizontal plane through this segment results in the free-body diagram of Fig. 5.11c, for which the condition of statics $\Sigma F_y = 0$ yields

$$\sigma_y \cdot b \, dx = \int_{-b/2}^{b/2} \int_y^h \frac{\partial \tau_{xy}}{\partial x} \, dx \cdot dy \, dz = b \int_y^h \frac{\partial \tau_{xy}}{\partial x} \, dx \, dy \qquad \textbf{(a)}$$

Here, the shear stress is defined by Eq. (5.41) as

$$\tau_{xy} = \frac{V}{2I}(h^2 - y^2) = \frac{3}{4} \frac{V}{bh}\left[1 - \left(\frac{y}{h}\right)^2\right] \qquad \textbf{(b)}$$

Upon substitution of Eqs. (5.28) and (b) into Eq. (a), we have

$$\sigma_y = -\int_y^h \frac{3p}{4bh}\left[1 - \left(\frac{y}{h}\right)^2\right] dy$$

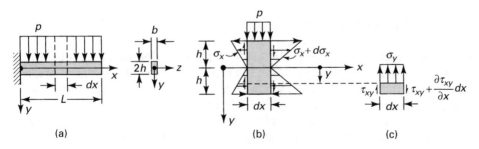

FIGURE 5.12. *(a) Uniformly loaded cantilever beam of rectangular cross section; (b) free-body diagram of a segment; (c) stresses in a beam element.*

Integration yields the transverse normal stress in the form

$$\sigma_y = -\frac{p}{b}\left[\frac{1}{2} - \frac{3}{4}\left(\frac{y}{h}\right) + \frac{1}{4}\left(\frac{y}{h}\right)^3\right]$$ (5.44a)

We see that this stress varies as a *cubic parabola* from $-p/b$ at the surface ($y = -h$), where the load acts, to zero at the opposite surface ($y = h$).

The distribution of the bending and the shear stresses in a uniformly loaded cantilever beam (Fig. 5.12a) is determined from Eqs. (5.38) and (b):

$$\sigma_x = -\frac{My}{I} = -\frac{3p}{4bh^3}(L - x)^2 y$$

$$\tau_{xy} = \frac{3p}{4bh}(L - x)\left[1 - \left(\frac{y}{h}\right)^2\right]$$ (5.44b)

The largest values of the σ_x, τ_{xy}, and σ_y given by Eqs. (5.44) are

$$\sigma_{x,max} = \pm\frac{3pL^2}{4bh^2}, \qquad \tau_{max} = \frac{3pL}{4bh}, \qquad \sigma_{y,max} = -\frac{p}{b}$$ (c)

To *compare* the magnitudes of the maximum stresses, consider the ratios

$$\frac{\tau_{max}}{\sigma_{x,max}} = \frac{h}{L}, \qquad \frac{\sigma_{y,max}}{\sigma_{x,max}} = \frac{4}{3}\left(\frac{h}{L}\right)^2$$ (d,e)

Because L is much greater than h in most beams, $L \geq 20h$, it is observed from the preceding that the shear and the transverse normal stresses will usually be orders of magnitude smaller than the bending stresses. This is justification for assuming $\gamma_{xy} = 0$ and $\varepsilon_y = 0$ in the technical theory of bending. Note that Eq. (e) results in even smaller values than Eq. (d). Therefore, in practice it is reasonable to *neglect* σ_y.

The foregoing conclusion applies, *in most cases*, to beams of a variety of cross-sectional shapes and under various load configurations. Clearly, the factor of proportionality in Eqs. (d) and (e) will differ for beams of different sectional forms and for different loadings of a given beam.

5.9 COMPOSITE BEAMS

Beams constructed of two or more materials having different moduli of elasticity are referred to as *composite beams*. Examples include multilayer beams made by bonding together multiple sheets, sandwich beams consisting of high-strength material faces separated by a relatively thick layer of low-strength material such as plastic foam, and reinforced concrete beams. The assumptions of the technical theory for a homogeneous beam (Sec. 5.6) are valid for a beam of more than one material.

We shall employ the common *transformed-section method* to analyze a composite beam. In this technique, the cross section of several materials is transformed into an *equivalent* cross section of one material on which the resisting forces and the neutral axis are the same as on the original section. The usual flexure formula is then applied to the new section. To illustrate the method, a frequently used beam with a symmetrical cross section built of two different materials is considered (Fig. 5.13a).

The *cross sections of the beam remain plane* during bending. Hence, geometric *compatibility of deformation* is satisfied. It follows that the normal strain ε_x varies linearly with the distance y from the neutral axis of the section; that is, $\varepsilon_x = ky$ (Figs. 5.13a and b). The location of the neutral axis is yet to be determined. Both materials of the beam are assumed to obey Hooke's law, and their moduli of elasticity are designated E_1 and E_2. Then, the *stress–strain relation* gives

$$\sigma_{x1} = E_1\varepsilon_x = E_1 ky, \qquad \sigma_{x2} = E_2\varepsilon_2 = E_2 ky \qquad \textbf{(5.45a, b)}$$

This result is sketched in Fig. 5.13c for the assumption that $E_2 > E_1$. We introduce the notation

$$n = \frac{E_2}{E_1} \qquad \textbf{(5.46)}$$

where n is called the *modular ratio*. Note that $n > 1$ in Eq. (5.46). However, this choice is arbitrary; the technique applies as well for $n < 1$.

Referring to the cross section (Figs. 5.13a and c), *equilibrium equations* $\Sigma F_x = 0$ and $\Sigma M_z = 0$ lead to

$$\int_A \sigma_x \, dA = \int_{A_1} \sigma_{x1} \, dA + \int_{A_2} \sigma_{x2} \, dA = 0 \qquad \textbf{(a)}$$

$$-\int_A y\sigma_x \, dA = -\int_{A_1} y\sigma_{x1} \, dA - \int_{A_2} y\sigma_{x2} \, dA = M \qquad \textbf{(b)}$$

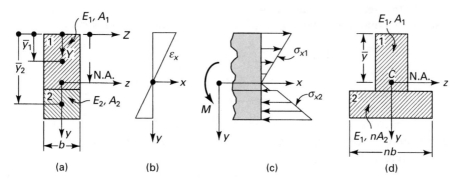

FIGURE 5.13. *Beam of two materials: (a) composite cross section; (b) strain distribution; (d) transformed cross section.*

wherein A_1 and A_2 denote the cross-sectional areas for materials 1 and 2, respectively. Substituting into Eq. (a) σ_{x1}, σ_{x2}, and n as given by Eqs. (5.45) and (5.46) results in

$$\int_{A_1} y \, dA + n \int_{A_2} y \, dA = 0 \tag{5.47}$$

Electing the top of the section as a reference (Fig. 5.13a), from Eq. (5.47) with $y = Y - \bar{y}$,

$$\int_{A_1} (Y - \bar{y}) \, dA + n \int_{A_2} (Y - \bar{y}) \, dA = 0$$

or, setting

$$\int_{A_1} Y \, dA = \bar{y}_1 A_1 \qquad \text{and} \qquad \int_{A_2} Y \, dA = \bar{y}_2 A_2$$

we have

$$A_1 \bar{y}_1 - A_1 \bar{y} + n A_2 \bar{y}_2 - n A_2 \bar{y} = 0$$

The foregoing yields an alterative form of Eq. (5.47):

$$\bar{y} = \frac{A_1 \bar{y}_1 + n A_2 \bar{y}_2}{A_1 + n A_2} \tag{5.47'}$$

Expression (5.47) or (5.47′) can be used to locate the *neutral axis* for a beam of two materials. These equations show that the transformed section will have the same neutral axis as the original beam, provided the width of area 2 is changed by a factor n and area 1 remains the same (Fig. 5.13d). Clearly, this widening must be effected in a direction *parallel* to the neutral axis, since the distance $\bar{y}_2$ to the centroid of area 2 is unchanged. The new section constructed this way represents the cross section of a beam made of a homogeneous material with a modulus of elasticity E_1, and the neutral axis passes through its *centroid*, as shown in the figure.

Similarly, condition (b) together with Eqs. (5.45) and (5.46) leads to

$$M = -k E_1 \left(\int_{A_1} y^2 \, dA + n \int_{A_2} y^2 \, dA \right)$$

or

$$M = -k E_1 (I_1 + n I_2) = -k E_1 I_t \tag{5.48}$$

where I_1 and I_2 are the moments of inertia about the neutral axis of the cross-sectional areas 1 and 2, respectively. Note that

$$I_t = I_1 + n I_2 \tag{5.49}$$

is the moment of inertia of the *entire* transformed area about the neutral axis. From Eq. (5.48), we have

$$k = -\frac{M}{E_1 I_t}$$

The *flexure formulas* for a composite beam are obtained upon introduction of this relation into Eqs. (5.45):

$$\sigma_{x1} = -\frac{My}{I_t}, \qquad \sigma_{x2} = -\frac{nMy}{I_t} \tag{5.50}$$

in which σ_{x1} and σ_{x2} are the stresses in materials 1 and 2, respectively. Note that when $E_1 = E_2 = E$, Eqs. (5.50) reduce to the flexure formula for a beam of homogeneous material, as expected.

The preceding discussion may be extended to include composite beams of *more than* two materials. It is readily shown that for m different materials, Eqs. (5.47′), (5.49), and (5.50) take the forms

$$\bar{y} = \frac{A_1 \bar{y}_1 + \Sigma n_i A_i \bar{y}_i}{A_1 + \Sigma n_i A_i}, \qquad n_i = \frac{E_i}{E_1} \tag{5.51}$$

$$I_t = I_1 + \Sigma n_i I_i \tag{5.52}$$

$$\sigma_{x1} = -\frac{My}{I_t}, \qquad \sigma_{xi} = -\frac{n_i My}{I_t} \tag{5.53}$$

where $i = 2, 3, \ldots, m$ denotes the ith material.

The use of the formulas developed in this section is demonstrated in the solutions of two numerical problems that follow.

EXAMPLE 5.5   Aluminum-Reinforced Wood Beam

A wood beam $E_w = 8.75$ GPa, 100-mm wide by 220-mm deep, has an aluminum plate $E_a = 70$ GPa with a net section 80 mm by 20 mm securely fastened to its bottom face, as shown in Fig. 5.14a. Dimensions are given in millimeters. The beam is subjected to a bending moment of

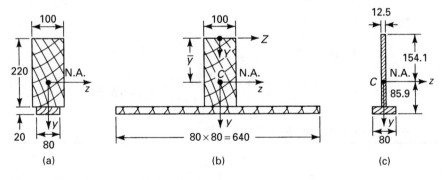

FIGURE 5.14.   *Example 5.5. (a) Composite cross section; (b) equivalent wood cross section; (c) equivalent aluminum cross section.*

20 kN · m around a horizontal axis. Calculate the maximum stresses in both materials (a) using a transformed section of wood and (b) using a transformed section of aluminum.

Solution

a. The modular ratio $n = E_a/E_w = 8$. The centroid and the moment of inertia about the neutral axis of the transformed section (Fig. 5.14b) are

$$\bar{y} = \frac{100(220)(110) + 20(640)(230)}{100(220) + 20(640)} = 154.1 \text{ mm}$$

$$I_t = \tfrac{1}{12}(100)(220)^3 + 100(220)(44.1)^2 + \tfrac{1}{12}(640)(20)^3 + 640(20)(75.9)^2$$
$$= 205.6 \times 10^6 \text{ mm}^4$$

The maximum stresses in the wood and aluminum portions are therefore

$$\sigma_{w,\,max} = \frac{Mc}{I_t} = \frac{20 \times 10^3(0.1541)}{205.6(10^{-6})} = 14.8 \text{ MPa}$$

$$\sigma_{a,\,max} = \frac{nMc}{I_t} = \frac{8(20 \times 10^3)(0.0859)}{205.6(10^{-6})} = 66.8 \text{ MPa}$$

It is noted that *at the juncture* of the two parts:

$$\sigma_{w,\,min} = \frac{Mc}{I_t} = \frac{20 \times 10^3(0.0659)}{205.6(10^{-6})} = 6.4 \text{ MPa}$$

$$\sigma_{a,\,min} = n(\sigma_{w,min}) = 8(6.4) = 51.2 \text{ MPa}$$

b. For this case, the modular ratio $n = E_w/E_a = 1/8$ and the transformed area is shown in Fig. 5.14c. We now have

$$I_t = \tfrac{1}{12}(12.5)(220)^3 + 12.5(220)(44.1)^2 + \tfrac{1}{12}(80)(20)^3 + 80(20)(75.9)^2$$
$$= 25.7 \times 10^6 \text{ mm}^4$$

Then

$$\sigma_{a,\,max} = \frac{Mc}{I_t} = \frac{20 \times 10^3(0.0859)}{25.7(10^{-6})} = 66.8 \text{ MPa}$$

$$\sigma_{w,\,max} = \frac{nMc}{I_t} = \frac{20 \times 10^3(0.1541)}{8(25.7 \times 10^{-6})} = 14.8 \text{ MPa}$$

as have already been found in part (a).

EXAMPLE 5.6   Steel-Reinforced-Concrete Beam

A concrete beam of width $b = 250$ mm and *effective depth* $d = 400$ mm is reinforced with three steel bars providing a total cross-sectional area $A_s = 1000$ mm² (Fig. 5.15a). Dimensions are given in millimeters. Note that it is usual for an approximate *allowance* $a = 50$ mm to be used to protect the steel from corrosion and fire. Let $n = E_s/E_c = 10$. Calculate the maximum stresses in the materials produced by a *negative* bending moment of $M = 60$ kN · m.

Solution   Concrete is very weak in tension but strong in compression. Thus, only the portion of the cross section located a distance kd above the neutral axis is used in the transformed section (Fig. 5.15b); *the concrete is assumed to take no tension*. Notice that the transformed area of the steel nA_s is located by a single dimension from the neutral axis to its centroid. The compressive stress in the concrete is assumed to vary linearly from the neutral axis. The steel is taken to be uniformly stressed.

The condition that the first moment of the transformed section with respect to the neutral axis be zero is satisfied by

$$b(kd)\frac{kd}{2} - nA_s(d - kd) = 0$$

or

$$(kd)^2 + (kd)\frac{2n}{b}A_s - \frac{2n}{b}dA_s = 0 \qquad \textbf{(5.54)}$$

Solving this quadratic expression for kd, the position of the neutral axis is obtained.

Introducing the data given, Eq. (5.54) reduces to

$$(kd)^2 + 80(kd) - 32 \times 10^3 = 0$$

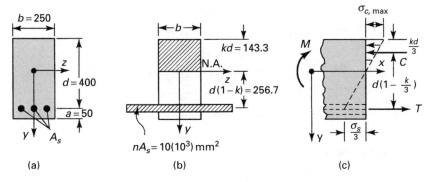

(a)   (b)   (c)

FIGURE 5.15.   *Example 5.6. (a) Reinforced-concrete cross section; (b) equivalent concrete cross section; (c) compressive force C in concrete and tensile force T in the steel rods.*

from which

$$kd = 143.3 \text{ mm} \quad \text{and hence} \quad k = 0.358 \qquad \text{(c)}$$

The moment of inertia of the transformed cross section about the neutral axis is

$$I_t = \tfrac{1}{12}(0.25)(0.1433)^3 + 0.25(0.1433)(0.0717)^2 + 0 + 10 \times 10^{-3}(0.2567)^2$$
$$= 904.4 \times 10^{-6} \text{ m}^4$$

Thus, the *peak compressive* stress in the concrete and the *tensile* stress in the steel are

$$\sigma_{c,\,max} = \frac{Mc}{I_t} = \frac{60 \times 10^3(0.1433)}{904.4 \times 10^{-6}} = 9.5 \text{ MPa}$$

$$\sigma_s = \frac{nMc}{I_t} = \frac{10(60 \times 10^3)(0.2567)}{904.4 \times 10^{-6}} = 170 \text{ MPa}$$

The stresses act as shown in Fig. 5.15c.

An *alternative method* of solution is to obtain $\sigma_{c,\,max}$ and σ_s from a free-body diagram of the portion of the beam (Fig. 5.15c) without computing I_t. The first equilibrium condition, $\Sigma F_x = 0$, gives $C = T$, where

$$C = \tfrac{1}{2}\sigma_{c,\,max}(b \cdot kd), \qquad T = \sigma_s(A_s) \qquad \text{(d)}$$

are the compressive and tensile stress resultants, respectively. From the second requirement of statics, $\Sigma M_z = 0$, we have

$$M = Cd(1 - k/3) = Td(1 - k/3) \qquad \text{(e)}$$

Equations (d) and (e) result in

$$\sigma_{c,\,max} = \frac{2M}{bd^2k(1 - k/3)}, \qquad \sigma_s = \frac{M}{A_s d(1 - k/3)} \qquad \text{(5.55)}$$

Substitution of the data given and Eq. (c) into Eq. (5.55) yields

$$\sigma_{c,\,max} = \frac{2(60 \times 10^3)}{0.25(0.4^2)(0.358)(1 - 0.358/3)} = 9.5 \text{ MPa}$$

$$\sigma_s = \frac{60 \times 10^3}{1000 \times 10^{-6}(0.4)(1 - 0.358/3)} = 170 \text{ MPa}$$

as before.

5.10 SHEAR CENTER

Given any cross-sectional configuration, one point may be found in the plane of the cross section through which passes the resultant of the transverse shearing stresses. A transverse load applied on the beam must act through this point, called the *shear center* or *flexural center*, if no twisting is to occur.* The center of shear is sometimes defined as the point in the end section of a cantilever beam at which an applied load results in bending only. When the load does not act through the shear center, in addition to bending, a twisting action results (Sec. 6.1). The location of the shear center is independent of the direction and magnitude of the transverse forces. For singly symmetrical sections, the shear center lies on the axis of symmetry, while for a beam with two axes of symmetry, the shear center coincides with their point of intersection (also the centroid). It is not necessary, in general, for the shear center to lie on a principal axis, and it may be located outside the cross section of the beam.

Thin-Walled Open Cross Sections

For thin-walled sections, the shearing stresses are taken to be distributed uniformly over the thickness of the wall and directed so as to parallel the boundary of the cross section. If the shear center S for the typical section of Fig. 5.16a is required, we begin by calculating the shear stresses by means of Eq. (5.39). The moment M_x of these stresses about arbitrary point A is then obtained. Inasmuch as the external moment attributable to V_y about A is $V_y e$, the distance between A and the shear center is given by

$$e = \frac{M_x}{V_y} \tag{5.56}$$

If the force is parallel to the z axis rather than the y axis, the position of the line of action may be established in the manner discussed previously. In the event that both V_y and V_z exist, the intersection of the two lines of action locates the shear center.

The determination of M_x is simplified by propitious selection of point A, such as in Fig. 5.16b. Here it is observed that the moment M_x of the shear forces about A is zero; point A is also the shear center. For all sections consisting of two intersecting rectangular elements, the same situation exists.

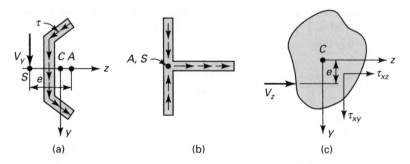

(a) (b) (c)

FIGURE 5.16. *Shear centers.*

*For a detailed discussion, see Ref. 5.3.

For the thin-walled *box-beams* (with boxlike cross section), the point or points in the wall where the shear flow $q = 0$ (or $\tau_{xy} = 0$) is unknown. Here shear flow is represented by the superposition of transverse and torsional flow (see Sec. 6.8). Hence, the unit angle of twist equation, Eq. (6.23), along with $q = VQ/I$ is required to find the shear flow for a cross section of a box beam. The analysis procedure is as follows: First, introduce a free edge by *cutting* the section *open*; second, close it again by obtaining the shear flow that makes the angle of twist in the beam zero [Refs. 5.4 through 5.6].

Arbitrary Solid Cross Sections

The preceding considerations can be extended to beams of arbitrary solid cross section, in which the shearing stress varies with both cross-sectional coordinates y and z. For these sections, the exact theory can, in some cases, be successfully applied to locate the shear center. Examine the section of Fig. 5.16c, subjected to the shear force V_z, which produces the stresses indicated. Denote y and z as the principal directions. The moment about the x axis is

$$M_x = \iint (\tau_{xy}z - \tau_{xz}y)\, dz\, dy \tag{5.57}$$

V_z must be located a distance e from the z axis, where $e = M_x/V_z$.

In the following examples, the determination of the shear center of an open, thin-walled section is illustrated in the solution for two typical situations. The first refers to a section having only one axis of symmetry, the second to an asymmetrical section.

EXAMPLE 5.7   Shearing Stress Distribution in a Channel Section

Locate the shear center of the channel section loaded as a cantilever (Fig. 5.17a). Assume that the flange thicknesses are small when compared with the depth and width of the section.

Solution   The shearing stress in the upper flange at any section *nn* will be found first. This section is located a distance s from the free edge m, as shown in the figure. At m the shearing stress is zero. The first moment of area st_1 about the z axis is $Q_z = st_1h$. The shear stress at nn, from Eq. (5.39), is thus

$$\tau_{xz} = \frac{V_y Q_z}{I_z b} = P\frac{sh}{I_z} \tag{a}$$

The direction of τ along the flange can be determined from the equilibrium of the forces acting on an element of length dx and width s (Fig. 5.17b). Here the normal force $N = t_1 s\sigma_x$, owing to the bending of the beam, increases with dx by dN. Hence, the x equilibrium of the element requires that $\tau t_1 \cdot dx$ must be directed as shown. As a consequence, this flange force is directed to the left, because the shear forces must intersect at the corner of the element.

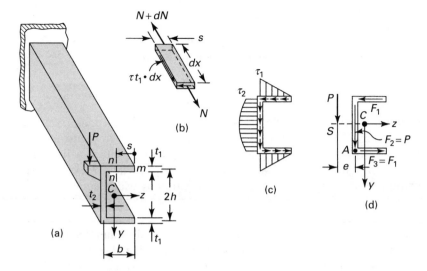

FIGURE 5.17. *Example 5.7. (a) Cantilever beam with a concentrated load at free end; (b) an element of upper flange; (c) shear distribution; (d) location of shear center S.*

The distribution of the shear stress τ_{xz} on the flange, as Eq. (a) indicates, is linear with s. Its maximum value occurs at $s = b$:

$$\tau_1 = P \frac{bh}{I_z} \qquad \textbf{(b)}$$

Similarly, the value of stress τ_{xy} at the top of the web is

$$\tau_2 = P \frac{bt_1 h}{t_2 I_2} \qquad \textbf{(c)}$$

The stress varies parabolically over the web, and its maximum value is found at the neutral axis. A sketch of the shear stress distribution in the channel is shown in Fig. 5.17c. As the shear stress is linearly distributed across the flange length, from Eq. (b), the flange force is expressed by

$$F_1 = \tfrac{1}{2}\tau_1 b t_1 = P \frac{b^2 h t_1}{2I_z} \qquad \textbf{(d)}$$

Symmetry of the section dictates that $F_1 = F_3$ (Fig. 5.17d). We shall assume that the web force $F_2 = P$, since the vertical shearing force transmitted by the flange is negligibly small, as shown in Example 5.3. The shearing force components acting in the section must be statically equivalent to the resultant shear load P. Thus, the principle of the moments for the system of forces in Fig. 5.17d or Eq. (5.56), applied at A, yields $M_x = Pe = 2F_1 h$. Upon substituting F_1 from Eq. (d) into this expression, we obtain

$$e = \frac{b^2 h^2 t_1}{I_z}$$

where

$$I_z = \tfrac{2}{3}t_2h^3 + 2bt_1h^2$$

The shear center is thus located by the expression

$$e = \frac{3}{2}\frac{b^2 t_1}{ht_2 + 3bt_1} \tag{e}$$

Comments   Note that e depends on only section dimensions. Examination reveals that e may vary from a minimum of zero to a maximum of $b/2$. A zero or near-zero value of e corresponds to either a flangeless beam $(b = 0, e = 0)$ or an especially deep beam $(h \gg b)$. The extreme case, $e = b/2$, is obtained for an infinitely wide beam.

EXAMPLE 5.8   Shear Flow in an Asymmetrical Channel Section
Locate the shear center S for the asymmetrical channel section shown in Fig. 5.18a. All dimensions are in millimeters. Assume that the beam thickness $t = 1.25$ mm is constant.

Solution   The centroid C of the section is located by $\bar{y}$ and $\bar{z}$ with respect to nonprincipal axes z and y. By performing the procedure given in Example 5.1, we obtain $\bar{y} = 15.63$ mm, $\bar{z} = 5.21$ mm, $I_y = 4765.62$ mm^4, $I_z = 21{,}054.69$ mm^4, and $I_{yz} = 3984.37$ mm^4. Equation (5.18) then yields the direction of the principal axis x', y' as $\theta_p = 13.05°$, and Eq. (5.19), the principal moments of inertia $I_{y'} = 3828.12$ mm^4, $I_{z'} = 21{,}953.12$ mm^4 (Fig. 5.18a).

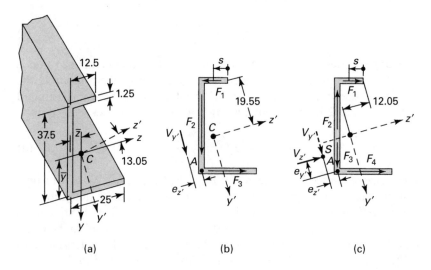

(a)   (b)   (c)

FIGURE 5.18.   *Example 5.8. (a) Portion of a beam with channel cross section; (b) shear flow; (c) location of shear center S.*

Let us now assume that a shear load $V_{y'}$ is applied in the y', z' plane (Fig. 5.18b). This force may be considered the resultant of force components $F_1, F_2,$ and F_3 acting in the flanges and web in the directions indicated in the figure. The algebra will be minimized if we choose point A, where F_2 and F_3 intersect, in finding the line of action of $V_{y'}$ by applying the principle of moments. In so doing, we need to determine the value of F_1 acting in the upper flange. The shear stress τ_{xz} in this flange, from Eq. (5.39), is

$$\tau_{xz} = \frac{V_{y'}Q_{z'}}{I_{z'}b} = \frac{V_{y'}}{I_{z'}t}[st(19.55 + \tfrac{1}{2}s \sin 13.05°)] \qquad \textbf{(f)}$$

where s is measured from right to left along the flange. Note that $Q_{z'}$, the bracketed expression, is the first moment of the shaded flange element area with respect to the z' axis. The constant 19.55 is obtained from the geometry of the section. Upon substituting the numerical values and integrating Eq. (f), the total shear force in the upper flange is found to be

$$F_1 = \int_0^s \tau_{xz}t \, ds = \frac{V_{y'}t}{I_{z'}} \int_0^{12.5} s(19.55 + \tfrac{1}{2}s \sin 13.05°) \, ds = 0.0912 V_{y'} \qquad \textbf{(g)}$$

Application of the principle of moments at A gives $V_{y'}e_{z'} = 37.5F_1$. Introducing F_1 from Eq. (g) into this equation, the distance $e_{z'}$, which locates the line of action of $V_{y'}$ from A, is

$$e_{z'} = 3.42 \text{ mm} \qquad \textbf{(h)}$$

Next, assume that the shear loading $V_{z'}$ acts on the beam (Fig. 5.18c). The distance $e_{y'}$ may be obtained as in the situation just described. Because of $V_{z'}$, the force components F_1 to F_4 will be produced in the section. The shear stress in the upper flange is given by

$$\tau_{xz} = \frac{V_{z'}Q_{y'}}{I_{y'}b} = \frac{V_{z'}}{I_{y'}t}[st(12.05 - \tfrac{1}{2}s \cos 13.05°)] \qquad \textbf{(i)}$$

Here $Q_{y'}$ represents the first moment of the flange segment area with respect to the y' axis, and 12.05 is found from the geometry of the section. The total force F_1 in the flange is

$$F_1 = \frac{V_{z'}}{I_{y'}} \int_0^{12.5} st(12.05 - \tfrac{1}{2}s \cos 13.05°) \, ds = 0.204 V_{z'}$$

The principle of moments applied at A, $V_{z'}e_{y'} = 37.5F_1 = 7.65V_{z'}$, leads to

$$e_{y'} = 7.65 \text{ mm} \qquad \textbf{(j)}$$

Thus, the intersection of the lines of action of $V_{y'}$ and $V_{z'}$, and $e_{z'}$ and $e_{y'}$, locates the shear center S of the asymmetrical channel section.

5.11 STATICALLY INDETERMINATE SYSTEMS

A large class of problems of considerable practical interest relates to structural systems for which the equations of statics are not sufficient (though are necessary) for determination of the reactions or other unknown forces. Such systems are *statically indeterminate*, requiring supplementary information for solution. Additional equations usually describe certain geometrical conditions associated with displacement or strain. These *equations of compatibility* state that the strain owing to deflection or rotation must be such as to preserve continuity. With this additional information, the solution proceeds in essentially the same manner as for statically determinate systems. The number of reactions in excess of the number of equilibrium equations is called the *degree of statical indeterminacy*. Any reaction in excess of that which can be obtained by statics alone is said to be *redundant*. Thus, the number of redundants is the same as the degree of indeterminacy.

Several methods are available to analyze statically indeterminate structures. The principle of superposition, briefly discussed next, offers for many cases an effective approach. In Section 5.6 and in Chapters 7 and 10, a number of commonly employed methods are discussed for the solution of the indeterminate beam, frame, and truss problems.

The Method of Superposition

In the event of complicated load configurations, the method of *superposition* may be used to good advantage to simplify the analysis. Consider, for example, the continuous beam of Fig. 5.19a, replaced by the beams shown in Fig. 5.19b and c. At point A, the beam now experiences the deflections $(v_A)_P$ and $(v_A)_R$ due respectively to P and R. Subject to the restrictions imposed by small deformation theory and a material obeying Hooke's law, the deflections and stresses are linear functions of transverse loadings, and superposition is valid:

$$v_A = (v_A)_P + (v_A)_R$$
$$\sigma_A = (\sigma_A)_P + (\sigma_A)_R$$

The procedure may in principle be extended to situations involving any degree of indeterminacy.

FIGURE 5.19. *Superposition of displacements in a continuous beam.*

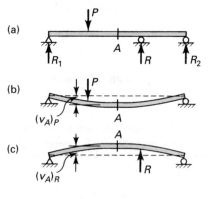

EXAMPLE 5.9 Displacements of a Propped Cantilever Beam

A propped cantilever beam AB subject to a uniform load of intensity p is shown in Fig. 5.20. Determine (a) the reactions, (b) the equation of the deflection curve, and (c) the slope at A.

Solution Reactions R_A, R_B, and M_B are statically indeterminate because there are only two equilibrium conditions ($\Sigma F_y = 0$, $\Sigma M_z = 0$); the beam is statically indeterminate to the first degree. With the origin of coordinates taken at the left support, the equation for the beam moment is

$$M = -R_A x + \tfrac{1}{2} px^2$$

The third of Eqs. (5.32) then becomes

$$EIv'' = -R_A x + \tfrac{1}{2} px^2$$

and successive integrations yield

$$EIv' = -\tfrac{1}{2} R_A x^2 + \tfrac{1}{6} px^3 + c_1$$
$$EIv = -\tfrac{1}{6} R_A x^3 + \tfrac{1}{24} px^4 + c_1 x + c_2 \qquad \textbf{(a)}$$

There are three unknown quantities in these equations (c_1, c_2, and R_A) and three boundary conditions:

$$v(0) = 0, \qquad v'(L) = 0, \qquad v(L) = 0 \qquad \textbf{(b)}$$

a. Introducing Eqs. (b) into the preceding expressions, we obtain $c_2 = 0$, $c_1 = pL^3/48$, and

$$R_A = \tfrac{3}{8} pL \qquad \textbf{(5.58a)}$$

We can now determine the remaining reactions from the equations of equilibrium:

$$R_B = \tfrac{5}{8} pL, \qquad M_B = \tfrac{1}{8} pL^2 \qquad \textbf{(5.58b, c)}$$

b. Substituting for R_A, c_1, and c_2 in Eq. (a), the equation of the deflection curve is obtained:

$$v = \frac{p}{48EI} (2x^4 - 3Lx^3 + L^3 x) \qquad \textbf{(5.59)}$$

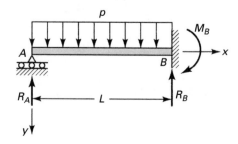

FIGURE 5.20. *Example 5.9. A propped beam under uniform load.*

c. Differentiating the foregoing with respect to x, the equation of the angle of rotation is

$$\theta = \frac{p}{48EI}(8x^3 - 9Lx^2 + L^3) \qquad (5.60)$$

Setting $x = 0$, we have the slope at A:

$$\theta_A = \frac{pL^3}{48EI} \qquad (5.61)$$

EXAMPLE 5.10   Reactions of a Propped Cantilever Beam

Consider again the statically indeterminate beam of Fig. 5.20. Determine the reactions using the method of superposition.

Solution   Reaction R_A is selected as redundant and is considered an unknown load by eliminating the support at A (Fig. 5.21a). The loading is resolved into those shown in Fig. 5.21b. The solution for each case is (see Table D.4)

$$(v_A)_P = \frac{pL^4}{8EI}, \qquad (v_A)_R = -\frac{R_A L^3}{3EI}$$

The compatibility condition for the original beam requires that

$$v_A = \frac{pL^4}{8EI} - \frac{R_A L^3}{3EI} = 0$$

from which $R_A = 3pL/8$. Reaction R_B and moment M_B can now be found from the equilibrium requirements. The results correspond to those of Example 5.9.

5.12   ENERGY METHOD FOR DEFLECTIONS

Strain energy methods are frequently employed to analyze the deflections of beams and other structural elements. Of the many approaches available, *Castigliano's second theorem* is one of the most widely used. In applying this theory,

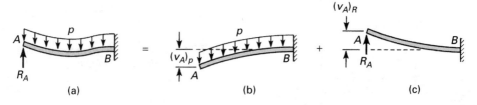

FIGURE 5.21.   *Example 5.10. Method of superposition: (a) reaction R_A is selected as redundant; (b) deflection at end A due to load P; (c) deflection at end A due to reaction R_A.*

the strain energy must be represented as a function of loading. Detailed discussions of energy techniques are found in Chapter 10. In this section we limit ourselves to a simple example to illustrate how the strain energy in a beam is evaluated and how the deflection is obtained by the use of Castigliano's theorem (Sec. 10.4).

The strain energy stored in a beam under bending stress σ_x only, substituting $M = EI(d^2v/dx^2)$ into Eq. (2.63), is expressed in the form

$$U_b = \int \frac{M^2 dx}{2EI} = \int \frac{EI}{2}\left(\frac{d^2v}{dx^2}\right)^2 dx \qquad (5.62)$$

Here the integrations are carried out over the beam length. We next determine the strain energy stored in a beam, *only* due to the *shear* loading V. As we described in Section 5.7, this force produces shear stress τ_{xy} at every point in the beam. The strain energy density is, from Eq. (2.50), $U_o = \tau_{xy}^2/2G$. Substituting τ_{xy} as expressed by Eq. (5.39), we have $U_o = V^2Q^2/2GI^2b^2$. Integrating this expression over the volume of the beam of cross-sectional area A, we obtain

$$U_s = \int \frac{V^2}{2GI^2}\left[\int \frac{Q^2}{b^2}dA\right]dx \qquad (a)$$

Let us denote

$$\alpha = \frac{A}{I^2}\int \frac{Q^2}{b^2}dA \qquad (5.63)$$

This is termed the *form factor for shear*, which when substituted in Eq. (a) yields

$$U_s = \int \frac{\alpha V^2 dx}{2AG} \qquad (5.64)$$

where the integration is carried over the beam length.

The form factor is a dimensionless quantity specific to a given cross-section geometry. For example, for a *rectangular* cross section of width b and height $2h$, the first moment Q, from Eq. (5.41), is $Q = (b/2)(h^2 - y_1^2)$. Because $A/I^2 = 9/2bh^5$, Eq. (5.63) provides the following result:

$$\alpha = \frac{9}{2bh^5}\int_{-h}^{h}\frac{1}{4}(h^2 - y_1^2)^2 b\, dy_1 = \frac{6}{5} \qquad (b)$$

In a like manner, the form factor for other cross sections can be determined. Table 5.1 lists several typical cases. Following the determination of α, the strain energy is evaluated by applying Eq. (5.64).

For a linearly elastic beam, Castigliano's theorem, from Eq. (10.3), is expressed by

$$\delta = \frac{\partial U}{\partial P} \qquad (c)$$

TABLE 5.1. *Form Factor for Shear for Various Beam Cross Sections*

Cross Section	Form Factor α
A. Rectangle	6/5
B. I-section, box section, or channels[a]	A/A_{web}
C. Circle	10/9
D. Thin-walled circular	2

[a] A = area of the entire section, A_{web} = area of the web ht, where h is the beam depth and t is the web thickness.

where P is a load acting on the beam and δ is the displacement of the point of application in the direction of P. Note that the strain energy $U = U_b + U_s$ is expressed as a function of the externally applied forces (or moments).

As an illustration, consider the bending of a cantilever beam of rectangular cross section and length L, subjected to a concentrated force P at the free end (Fig. 5.5). The bending moment at any section is $M = Px$, and the shear force V is equal in magnitude to P. Upon substituting these together with $\alpha = \frac{6}{5}$ into Eqs. (5.62) and (5.64) and integrating, the strain energy stored in the cantilever is found to be

$$U = \frac{P^2 L^3}{6EI} + \frac{3P^2 L}{5AG}$$

The displacement of the free end owing to bending and shear is, by application of Castigliano's theorem, therefore

$$\delta = v = \frac{PL^3}{3EI} + \frac{6PL}{5AG}$$

The exact solution is given by Eq. (5.24).

Part C—Curved Beams

5.13 ELASTICITY THEORY

Our treatment of stresses and deflections caused by the bending has been restricted so far to straight members. But many members, such as crane hooks, chain links, C-lamps, and punch-press frames, are curved and loaded as beams. *Part C* deals with the stresses caused by the bending of bars that are initially curved.

A curved bar or beam is a structural element for which the locus of the centroids of the cross sections is a curved line. This section concerns itself with an application of the theory of elasticity. We deal here with a bar characterized by a constant narrow rectangular cross section and a circular axis. The axis of symmetry of the cross section lies in a single plane throughout the length of the member.

Consider a beam subjected to equal end couples M such that bending takes place in the plane of curvature, as shown in Fig. 5.22a. Inasmuch as the bending

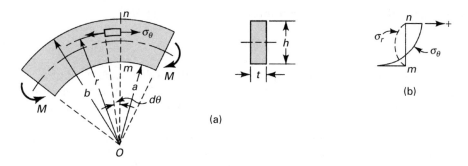

FIGURE 5.22. *Pure bending of a curved beam of rectangular cross section.*

moment is constant along the length of the bar, the stress distribution should be identical in any radial cross section. Stated differently, we seek a distribution of stress displaying θ independence. It is clear that the appropriate expression of equilibrium is Eq. (8.2),

$$\frac{d\sigma_r}{dr} + \frac{\sigma_r - \sigma_\theta}{r} = 0 \tag{a}$$

and that the condition of compatibility for plane stress, Eq. (3.41),

$$\frac{d^2(\sigma_r + \sigma_\theta)}{dr^2} + \frac{1}{r}\frac{d(\sigma_r + \sigma_\theta)}{dr} = 0$$

must also be satisfied. The latter is an equidimensional equation, reducible to a second-order equation with constant coefficients by substituting $r = e^t$ or $t = \ln r$. Direct integration then leads to $\sigma_r + \sigma_\theta = c'' + c' \ln r$, which may be written in the form $\sigma_r + \sigma_\theta = c''' + c' \ln(r/a)$. Solving this expression together with Eq. (a) results in the following equations for the radial and tangential stress:

$$\sigma_r = c_1 + c_2 \ln\frac{r}{a} + \frac{c_3}{r^2}$$

$$\sigma_\theta = c_1 + c_2\left(1 + \ln\frac{r}{a}\right) - \frac{c_3}{r^2} \tag{5.65}$$

To evaluate the constants of integration, the boundary conditions are applied as follows:

1. No normal forces act along the curved boundaries at $r = a$ and $r = b$, and therefore

$$(\sigma_r)_{r=a} = (\sigma_r)_{r=b} = 0 \tag{b}$$

2. Because there is no force acting at the ends, the normal stresses acting at the straight edges of the bar must be distributed to yield a zero resultant:

$$t\int_a^b \sigma_\theta \, dr = 0 \tag{c}$$

where t represents the beam thickness.

3. The normal stresses at the ends must produce a couple M:

$$t \int_a^b r\sigma_\theta \, dr = M \tag{d}$$

The conditions (c) and (d) apply not only at the ends, but because of σ_θ independence, at any θ. In addition, shearing stresses have been assumed zero throughout the beam, and $\tau_{r\theta} = 0$ is thus satisfied at the boundaries, where no tangential forces exist.

Combining the first equation of (5.65) with the conditions (b), we find that

$$c_3 = -a^2 c_1, \qquad c_1\left(\frac{a^2}{b^2} - 1\right) = c_2 \ln\frac{b}{a}$$

These constants together with the second of Eqs. (5.65) satisfy condition (c). Thus, we have

$$c_1 = \frac{b^2 \ln (b/a)}{a^2 - b^2} c_2, \qquad c_3 = \frac{a^2 b^2 \ln (b/a)}{b^2 - a^2} c_2 \tag{e}$$

Finally, substitution of the second of Eqs. (5.65) and (e) into (d) provides

$$c_2 = \frac{M}{N} \frac{4(b^2 - a^2)}{tb^4} \tag{f}$$

where

$$N = \left(1 - \frac{a^2}{b^2}\right)^2 - 4\frac{a^2}{b^2} \ln^2\frac{b}{a} \tag{5.66}$$

When the expressions for constants c_1, c_2, and c_3 are inserted into Eq. (5.65), the following equations are obtained for the radial stress and tangential stress:

$$\sigma_r = \frac{4M}{tb^2 N}\left[\left(1 - \frac{a^2}{b^2}\right)\ln\frac{r}{a} - \left(1 - \frac{a^2}{r^2}\right)\ln\frac{b}{a}\right]$$

$$\sigma_\theta = \frac{4M}{tb^2 N}\left[\left(1 - \frac{a^2}{b^2}\right)\left(1 + \ln\frac{r}{a}\right) - \left(1 + \frac{a^2}{r^2}\right)\ln\frac{b}{a}\right] \tag{5.67}$$

If the end moments are applied so that the force couples producing them are distributed in the manner indicated by Eq. (5.67), then these equations are applicable throughout the bar. If the distribution of applied stress (to produce M) differs from Eq. (5.67), the results may be regarded as valid in regions away from the ends, in accordance with Saint-Venant's principle. The foregoing results, when applied to a beam with radius a, large relative to its depth h, yield an interesting comparison between straight and curved beam theory. For *slender beams $h \ll a$, radial* stress σ_r in Eq. (5.67) becomes negligible, and *tangential* stress σ_θ is approximately the same as that obtained from My/I. Note that radial stresses developed in *nonslender* curved beams made of isotropic materials are small enough that they can be neglected in analysis and design.

The bending moment is taken as positive when it tends to decrease the radius of curvature of the beam, as in Fig. 5.22a. Employing this sign convention, σ_r as determined from Eq. (5.67) is always negative, indicating that it is compressive. Similarly, when σ_θ is found to be positive, it is tensile; otherwise, compressive. In Fig. 5.22b, a plot of the stresses at section mn is presented. Note that the maximum stress magnitude is found at the extreme fiber of the concave side.

Deflections

Substitution of σ_r and σ_θ from Eq. (5.67) into Hooke's law provides expressions for the strains ε_θ, ε_r, and $\gamma_{r\theta}$. The displacements u and v then follow, upon integration, from the strain–displacement relationships, Eqs. (3.33). The resulting displacements indicate that plane sections of the curved beam subjected to pure bending remain plane subsequent to bending. Castigliano's theorem (Sec. 5.12) is particularly attractive for determining the deflection of curved members.

For beams in which the *depth of the member is small relative to the radius of curvature* or, as is usually assumed, $\bar{r}/c > 4$, the initial curvature may be neglected in evaluating the strain energy. Here $\bar{r}$ represents the radius to the centroid, and c is the distance from the centroid to the extreme fiber on the concave side. Thus, the strain energy due to the bending of a straight beam [Eq. (5.62)] is a good approximation also for curved, slender beams.

5.14 CURVED BEAM FORMULA

The approach to curved beams now explored is due to E. Winkler (1835–1888). As an extension of the elementary theory of straight beams, in *Winkler's theory* it will be assumed that all conditions required to make the straight-beam formula applicable are satisfied except that the beam is initially curved. Consider the pure bending of a curved beam as in Fig. 5.23a. The distance from the center of curvature to the centroidal axis is $\bar{r}$. Observe that the *positive y* coordinate is measured *toward* the center of curvature O from the neutral axis (Fig. 5.23b). The outer and inner fibers are at distances of r_o and r_i from the center of curvature, respectively. Derivation of the stress in the beam is again based on the three principles of solid mechanics and the familiar assumptions:

1. All cross sections possess a vertical axis of symmetry lying in the plane of the centroidal axis passing through C.
2. The beam is subjected to end couples M. The bending moment vector is everywhere normal to the plane of symmetry of the beam.
3. Sections originally plane and perpendicular to the centroidal beam axis remain so subsequent to bending. (The influence of transverse shear on beam deformation is not taken into account.)

Referring to assumption (3), note the relationship in Fig. 5.23a between lines bc and ef representing plane sections before and after the bending of an initially

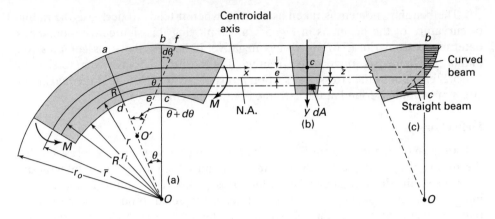

FIGURE 5.23. *(a) Curved beam in pure bending with a cross-sectional vertical (y) axis of symmetry; (b) cross section; (c) stress distributions over the cross section.*

curved beam. Note also that the initial length of a beam fiber such as *gh* depends on the distance *r* from the center of curvature *O*. On the basis of plane sections remaining plane, we can state that the total *deformation* of a beam fiber obeys a linear law, as the beam element rotates through small angle $d\theta$.

Location of the Neutral Axis

It may be seen from Fig. 5.23a that the initial length of any arbitrary fiber *gh* of the beam depends on the distance *r* from the center of curvature *O*. On this basis, we can state that the total deformation of a beam fiber obeys a linear law, as the beam element rotates through a small angle $d\theta$. However, the normal or tangential strain ε_θ does *not* follow a linear relationship. The contraction of fiber *gh* equals $-(R - r)d\theta$, in which *R* is the distance from *O* to the neutral axis (yet to be determined) and its initial length $r\theta$. So, the normal strain of this fiber is given by $\varepsilon_\theta = -(R - r)d\theta/r\theta$. For convenience, we denote $\lambda = d\theta/\theta$, which is constant for any element.

The tangential normal stress, acting on an area *dA* of the cross section can now be obtained through the use of *Hooke's law* $\sigma_\theta = E\varepsilon_\theta$. It follows that

$$\sigma_\theta = -E\lambda \frac{R - r}{r} \tag{a}$$

The *equations of equilibrium*, $\Sigma F_x = 0$ and $\Sigma M_z = 0$ are, respectively,

$$\int \sigma_\theta \, dA = 0, \tag{b}$$

$$\int \sigma_\theta (R - r) \, dA = M \tag{c}$$

When the tangential stress of Eq. (a) is inserted into Eq. (b), we obtain

$$\int_A -E\lambda \left(\frac{R-r}{r}\right) dA = 0 \qquad \textbf{(d)}$$

Inasmuch as $E\lambda$ and R are constants, they may be taken outside of the integral sign, as

$$E\lambda \left(R\int_A \frac{dA}{r} - \int_A dA\right) = 0$$

The *radius of the neutral axis* R is then written in the form

$$R = \frac{A}{\displaystyle\int_A \frac{dA}{r}} \qquad \textbf{(5.68)}$$

in which A is the cross-sectional area of the beam. The integral in Eq.(5.68) may be evaluated for various cross-sectional shapes (see Example 5.11 and Problems 5.39 to 5.41). For reference, Table 5.2 lists explicit formulas for R and A for some commonly used cases.

The *distance e* between the *centroidal axis and the neutral axis* ($y = 0$) of the cross section of a curved beam (Fig. 5.23b) is equal to

$$e = \bar{r} - R \qquad \textbf{(5.69)}$$

Thus, it is we concluded that, in a curved member, the *neutral axis does not coincide with the centriodal axis*. This differs from the situation found to be true for straight elastic beams.

Tangential Stress

Having the location of the neutral axis known, the equation for the stress distribution is found by introducing Eq. (a) into Eq. (c). Therefore,

$$M = E\lambda \int_A \frac{(R-r)^2}{r} dA$$

Expanding this equation, we have

$$M = E\lambda \left(R^2 \int_A \frac{dA}{r} - 2R\int_A dA + \int_A r\,dA\right)$$

Here, the first integral is equivalent to A/R as determined by Eq. (5.68), and the second integral equals the cross-sectional area A. The third integral, by definition, represents $\bar{r}A$ in which $\bar{r}$ is the radius of the centroidal axis. So,

$$M = E\lambda\, A(\bar{r} - R) = E\lambda Ae$$

TABLE 5.2. *Properties for Various Cross-Sectional Shapes*

Cross Section	Radius of Neutral Surface R
A. Rectangle 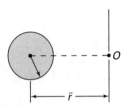	$$R = \frac{h}{\ln\dfrac{r_o}{r_1}}$$ $$A = bh$$
B. Circle	$$R = \frac{A}{2\pi(\bar{r} + \sqrt{\bar{r}^2 - c^2})}$$ $$A = \pi c^2$$
C. Ellipse	$$R = \frac{A}{\dfrac{2\pi b}{a}(\bar{r} - \sqrt{\bar{r}^2 - a^2})}$$ $$A = \pi ab$$
D. Triangle	$$R = \frac{A}{\dfrac{br_o}{h}\left(\ln\dfrac{r_o}{r_i}\right) - b}$$ $$A = \tfrac{1}{2}bh$$
E. Trapezoid	$$R = \frac{A}{\frac{1}{h}[(b_1 r_o - b_2 r_i)\ln\dfrac{r_o}{r_i} - h(b_1 - b_2)]}$$ $$A = \tfrac{1}{2}(b_1 + b_2)h$$

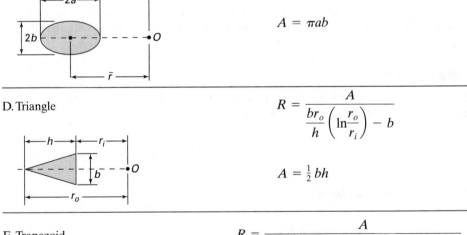

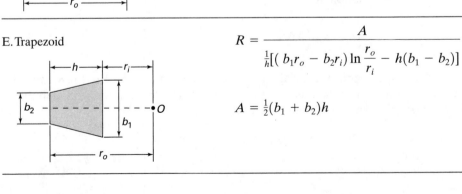

We now introduce E from Eq. (a) into the preceding and solve for σ_θ from the resulting expression. Then, the *tangential stress* in a curved beam, subject to pure bending at a distance r from the center of curvature, is expressed in the form:

$$\sigma_\theta = -\frac{M(R - r)}{Aer} \tag{5.70}$$

where e is defined by Eq. (5.69). Alternatively, substituting $y = R - r$ or $r = R - y$ (Fig. 5.23a) into Eq. (5.70) results in

$$\sigma_\theta = -\frac{My}{Ae(R - y)} \tag{5.71}$$

Equations (5.70) or (5.71) represent two forms of the so-called *curved-beam formula*. Another alternate form of these equations is often referred to as *Winkler's formula*. The variation of stress over the cross section is *hyperbolic*, as sketched in Fig. 5.23c. The *sign convention* applied to bending moment is the same as that used in Section 5.13. The bending moment is positive when directed toward the concave side of the beam, as shown in the figure. If Eq. (5.70) or (5.71) results in a positive value, it is indicative of a tensile stress.

5.15 COMPARISON OF THE RESULTS OF VARIOUS THEORIES

We now compare the solutions obtained in Sections 5.13 and 5.14 with results determined using the flexure formula for straight beams. To do this, consider a curved beam of rectangular cross section and unit thickness experiencing pure bending. The tangential stress predicted by the elementary theory (based on a linear distribution of stress) is My/I. The Winkler approach, leading to a hyperbolic distribution, is given by Eq. (5.70) or (5.71), while the exact theory results in Eqs. (5.67). In each case, the maximum and minimum values of stress are expressible by

$$\sigma_\theta = m\frac{M}{a^2} \tag{5.72}$$

In Table 5.3, values of m are listed as a function of b/a for the four cases cited [Ref. 5.1], in which $b = r_o$ and $a = r_i$; see Figs. 5.22 and 5.23. Observe that there is good agreement between the exact and Winkler results. On this basis as well as from more extensive comparisons, it may be concluded that the Winkler approach is adequate for practical applications. Its advantage lies in the relative ease with which it may be applied to *any* symmetric section.

The agreement between the Winkler and exact analyses is not as good in situations of combined loading, as for the case of pure bending. As might be expected, for beams of only slight curvature, the simple flexure formula provides good results while requiring only simple computation. The *linear and hyperbolic stress distributions are approximately the same for $b/a = 1.1$.* As the curvature of the beam increases ($b/a > 1.3$), the stress on the concave side rapidly increases over the one given by the flexure formula.

b/a	Flexure Formula	Curved Beam Formula		Elasticity Theory	
		r = a	r = b	r = a	r = b
1.3	±66.67	−72.980	61.270	−73.050	61.350
1.5	±24.00	−26.971	20.647	−27.858	21.275
2.0	±6.00	−7.725	4.863	−7.755	4.917
3.0	±1.50	−2.285	1.095	−2.292	1.130

Correction of σ_θ for Beams with Thin-Walled Cross Sections

It is noted that where I-, T-, or thin-walled tubular curved beams are involved, the stresses predicted by the approaches developed in this chapter will be in error. This is attributable to high stresses existing in certain sections such as the flanges, which cause significant beam distortion. A modified Winkler's equation finds application in such situations if more accurate results are required [Ref. 5.6]. The distortion, and thus error in σ_θ, is reduced if the flange thickness is increased. Inasmuch as material yielding is highly localized, its effect is not of concern unless the curved beam is under fatigue loading.

EXAMPLE 5.11 Maximum Stress in a Curved Rectangular Bar

A rectangular aluminum bar having mean radius $\bar{r}$ carries end moments M, as illustrated in Fig. 5.24. Calculate the stresses in the member (a) using the flexure formula; (b) by the curved beam formula. *Given*: $M = 1.2$ kN·m, $b = 30$ mm, $h = 50$ mm, and $\bar{r} = 125$ mm.

Solution The subscripts i and o refer to the quantities of the inside and outside fibers, respectively.

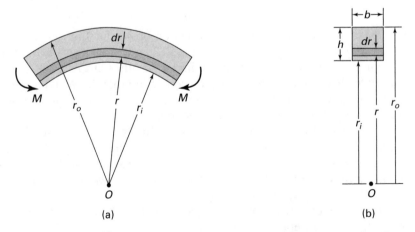

FIGURE 5.24. *Example 5.11. (a) Rectangular curved beam in pure bending; (b) cross section.*

a. Applying the flexure formula, Eq. (5.38) with $y = h/2$, we obtain

$$\sigma_o = -\sigma_i = \frac{My}{I} = \frac{1200(0.025)}{\frac{1}{12}(0.03)(0.05)^3} = 96 \text{ MPa}$$

which is the result we would get for a straight beam.

b. We first derive the expression for the radius R of the neutral axis. From Fig. 5.24: $A = bh$ and $dA = bdr$. Integration of Eq. (5.68) between the limits r_i and r_o results in

$$R = \frac{A}{\int_A \frac{dA}{r}} = \frac{bh}{\int_{r_1}^{r_o} \frac{bdr}{r}} = \frac{h}{\int_{r_1}^{r_o} \frac{dr}{r}}$$

or

$$R = \frac{h}{\ln \dfrac{r_o}{r_i}} \tag{5.73}$$

The given data leads to

$$A = bh = (30)(50) = 1500 \text{ mm}^2$$

$$r_i = \bar{r} - \tfrac{1}{2}h = 125 - 25 = 100 \text{ mm}$$

$$r_o = \bar{r} + \tfrac{1}{2}h = 125 + 25 = 150 \text{ mm}$$

Then, Eqs. (5.73) and (5.69) yield, respectively,

$$R = \frac{h}{\ln \dfrac{r_o}{r_i}} = \frac{50}{\ln \frac{3}{2}} = 123.3152 \text{ mm}$$

$$e = \bar{r} - R = 125 - 123.3152 = 1.6848 \text{ mm}$$

It is important to note that the radius of the neutral axis R must be calculated with *five significant* figures.

The maximum compressive and tensile stresses are calculated through the use of Eq. (5.70) as follows:

$$\sigma_i = -\frac{M(R - r_i)}{Aer_i} = -\frac{1.2(123.3152 - 100)}{1.5(10^{-3})(1.6848(10^{-3})(0.1))}$$

$$= -110.7 \text{ MPa}$$

$$\sigma_o = -\frac{M(R - r_o)}{Aer_o} = -\frac{1.2(123.3152 - 150)}{1.5(10^{-3})(1.6848(10^{-3})(0.15))}$$

$$= 84.5 \text{ MPa}$$

The negative sign means a compressive stress.

Comment The maximum stress 96 MPa obtained in part (a) by the flexure formula represents an error of about 13% from the more accurate value for the maximum stress 110.7 found in part (b).

5.16 COMBINED TANGENTIAL AND NORMAL STRESSES

Curved beams are often loaded so that there is an axial force as well as a moment on the cross section, as is shown in the example to follow. The tangential stress given by Eq. (5.70) may then be algebraically added to the stress due to an axial force P acting through the centroid of cross-sectional area A. For this simple case of *superposition*, the total stress at a point located at distance r from the center of curvature O may be expressed in the form

$$\sigma_\theta = \frac{P}{A} - \frac{M(R - r)}{Aer} \tag{5.74}$$

As before, a negative sign would be associated with a compressive load P. The theory developed in this section applies, of course, only to the elastic stress distribution in curved beams. Stresses in straight members under various combined loads are discussed in detail throughout this text.

The following sample problems illustrate the application of the formulas developed to statically determinate and statically indeterminate beams under combined loadings. Observe that, in the latter case, the energy method (Sec. 10.4) facilitates the determination of the unknown, redundant moment in the member.

EXAMPLE 5.12 **Stresses in a Steel Crane Hook by Various Methods**
A load P is applied to the simple steel hook having a rectangular cross section, as illustrated in Fig. 5.25a. Calculate the tangential stresses at points A

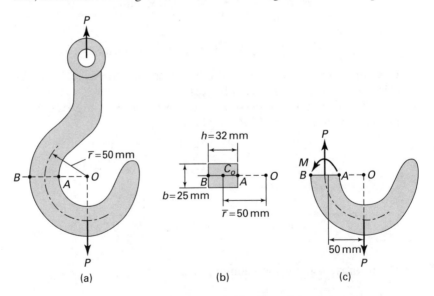

FIGURE 5.25. *Example 5.12. A crane hook of rectangular cross section.*

and B, using (a) the curved beam formula; (b) the flexure formula; (c) elasticity theory. *Given:* $P = 6$ kN, $\bar{r} = 50$ mm, $b = 25$ mm, and $h = 32$ mm.

Solution

a. *Curved Beam Formula.* For the given numerical values, we obtain (Fig. 5.25b):

$$A = bh = (25)(32) = 800 \text{ mm}^2$$

$$r_i = \bar{r} - \tfrac{1}{2}h = 50 - 16 = 34 \text{ mm}$$

$$r_o = \bar{r} + \tfrac{1}{2}h = 50 + 16 = 66 \text{ mm}$$

Then, Eqs. (5.73) and (5.69) result in

$$R = \frac{h}{\ln\dfrac{r_o}{r_i}} = \frac{32}{\ln\dfrac{66}{34}} = 48.2441 \text{ mm}$$

$$e = \bar{r} - R = 50 - 48.2441 = 1.7559 \text{ mm}$$

In order to maintain applied force P in equilibrium, there must be an axial tensile force P and a moment $M = -Pr$ at the centroid of the section (Fig. 5.25c). Thus, by Eq. (5.74), the *stress at the inner edge* $(r = r_i)$ of the section $A–B$:

$$(\sigma_\theta)_A = \frac{P}{A} - \frac{(-P\bar{r})(R - r_i)}{Aer_i} = \frac{P}{A}\left[1 + \frac{\bar{r}(R - r_i)}{er_i}\right] \quad \textbf{(5.75a)}$$

$$= \frac{6000}{0.0008}\left[1 + \frac{50(48.2441 - 34)}{(1.7559)(34)}\right] = 97 \text{ MPa}$$

Likewise, the *stress at the outer edge* $(r = r_o)$,

$$(\sigma_\theta)_B = \frac{P}{A} - \frac{(-P\bar{r})(R - r_o)}{Aer_o} = \frac{P}{A}\left[1 + \frac{\bar{r}(R - r_o)}{er_o}\right] \quad \textbf{(5.75b)}$$

$$= \frac{6000}{0.0008}\left[1 + \frac{50(48.2441 - 66)}{(1.7559)(66)}\right] = -50 \text{ MPa}$$

The negative sign of $(\sigma_\theta)_B$ means a compressive stress. The maximum tensile stress is at A and equals 97 MPa.

Comment   The stress due to the axial force,

$$\frac{P}{A} = \frac{6000}{0.0008} = 7.5 \text{ MPa}$$

which is negligibly small compared to the combined stresses at points A and B of the cross section.

b. *Flexure Formula.* Equation (5.5), with $M = P\bar{r} = 6(50) = 300 \text{ N} \cdot \text{m}$, gives

$$(\sigma_\theta)_B = -(\sigma_\theta)_A = \frac{My}{I} = \frac{300(0.016)}{\frac{1}{12}(0.025)(0.032)^3} = 70.3 \text{ MPa}$$

c. *Elasticity Theory.* Using Eq. (5.66) with $a = r_i = 34 \text{ mm}$ and $b = r_o = 66 \text{ mm}$, we find

$$N = \left[1 - \left(\frac{34}{66}\right)^2\right]^2 - 4\left(\frac{34}{66}\right)^2 \ln^2\left(\frac{66}{34}\right) = 0.0726$$

Superposition of $-P/A$ and the second of Eqs. (5.67) with $t = 25 \text{ mm}$ at $r = a$ leads to

$$(\sigma_\theta)_A = -\frac{6000}{0.0008} + \frac{4(300)}{(0.025)(0.066)^2(0.0726)}\left[\left(1 - \frac{34^2}{66^2}\right)(1 + 0) - (1 + 1)\ln\frac{66}{34}\right]$$

$$= -7.5 - 89.85 = -97.4 \text{ MPa}$$

Similarly, at $r = b$, we find $(\sigma_\theta)_B = -7.5 + 58.1 = 50.6 \text{ MPa}$

Comments The preceding indicates that the results of the curved beam formula and elasticity theory are in good agreement. But the flexure formula provides a result of *unacceptable* accuracy for the tangential stress in this nonslender curved beam.

EXAMPLE 5.13   Ring with a Diametral Bar

A steel ring of 350-mm mean diameter and of uniform rectangular section 60-mm wide and 12-mm thick is shown in Fig. 5.26a. A rigid bar is fitted across diameter AB, and a tensile force P applied to the ring as

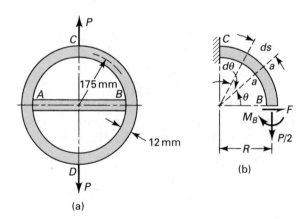

(a)

(b)

FIGURE 5.26.   *Example 5.13. (a) Ring with a bar AB is subjected to a concentrated load P; (b) moment at a section.*

shown. Assuming an allowable stress of 140 MPa, determine the maximum tensile force that can be carried by the ring.

Solution   Let the thrust induced in bar AB be denoted by $2F$. The moment at any section a–a (Fig. 5.26b) is then

$$M_\theta = -FR \sin \theta + M_B + \frac{PR}{2}(1 - \cos \theta) \tag{a}$$

Note that before and after deformation, the relative slope between B and C remains unchanged. Therefore, the relative angular rotation between B and C is zero. Applying Eq. (5.32), we therefore obtain

$$EI\theta = 0 = \int_B^C M_\theta \, dx = R \int_0^{\pi/2} M_\theta \, d\theta$$

where $dx = ds = R \, d\theta$ is the length of beam segment corresponding to $d\theta$. Upon substitution of Eq. (a), this becomes, after integrating,

$$\tfrac{1}{2}\pi M_B + \tfrac{1}{2}PR\left(\frac{\pi}{2} - 1\right) - FR = 0 \tag{b}$$

This expression involves two unknowns, M_B and F. Another expression in terms of M_B and F is found by recognizing that the deflection at B is zero. By application of Castigliano's theorem,

$$\delta_B = \frac{\partial U}{\partial F} = \frac{1}{EI} \int_0^{\pi/2} M_\theta \frac{\partial M_\theta}{\partial F}(R d\theta) = 0$$

where U is the strain energy of the segment. This expression, upon introduction of Eq. (a), takes the form

$$\int_0^{\pi/2} \left\{ -FR \sin \theta + M_B + \frac{PR}{2}(1 - \cos \theta) \right\} \sin \theta \, d\theta = 0$$

After integration,

$$-\tfrac{1}{4}\pi FR + \tfrac{1}{4}PR + M_B = 0 \tag{c}$$

Solution of Eqs. (b) and (c) yields $M_B = 0.1106PR$ and $FR = 0.4591PR$. Substituting Eq. (a) gives, for $\theta = 90°$,

$$M_C = -FR + M_B + \tfrac{1}{2}PR = 0.1515PR$$

Thus, $M_C > M_B$. Since $R/c = 0.175/0.006 = 29$, the simple flexure formula offers the most efficient means of computation. The maximum stress is found at points A and B:

$$(\sigma_\theta)_{A, B} = \frac{P/2}{A} + \frac{M_B c}{I} = 694P + 13{,}441P = 14{,}135P$$

Similarly, at C and D,

$$(\sigma_\theta)_{C, D} = \frac{M_C\, c}{I} = 18{,}411P$$

Hence $\sigma_{\theta C} > \sigma_{\theta B}$. Since $\sigma_{max} = 140$ MPa, $140 \times 10^6 = 18{,}411P$. The maximum tensile load is therefore $P = 7.604$ kN.

REFERENCES

5.1. TIMOSHENKO, S. P., and GOODIER, J. N., *Theory of Elasticity*, 3rd ed., McGraw-Hill, New York, 1970.

5.2. UGURAL, A. C., *Mechanics of Materials*, Wiley, Hoboken, N.J., 2008, Chap. 8.

5.3. SOKOLNIKOFF, I. S., *Mathematical Theory of Elasticity*, 2nd ed., Krieger, Melbourne, Fla., 1986, Sec. 53.

5.4. COOK, R. D., and YOUNG, W. C., *Advanced Mechanics of Materials*, Macmillan, New York, 1985.

5.5. BUDYNAS, R. G., *Advanced Strength and Applied Stress Analysis*, 2nd ed., McGraw-Hill, New York, 1999.

5.6. BORESI, A. P., and SCHMIDT, R. J., *Advanced Mechanics of Materials*, 6th ed., Wiley, Hoboken, N.J., 2003.

PROBLEMS

Sections 5.1 through 5.5

5.1. A simply supported beam constructed of a $0.15 \times 0.15 \times 0.015$ m angle is loaded by concentrated force $P = 22.5$ kN at its midspan (Fig. P5.1). Calculate stress σ_x at A and the orientation of the neutral axis. Neglect the effect of shear in bending and assume that beam twisting is prevented.

5.2. A wood cantilever beam with cross section as shown in Fig. P5.2 is subjected to an inclined load P at its free end. Determine (a) the orientation of the neutral axis; (b) the maximum bending stress. *Given:* $P = 1$ kN, $\alpha = 30°$, $b = 80$ mm, $h = 150$ mm, and length $L = 1.2$ m.

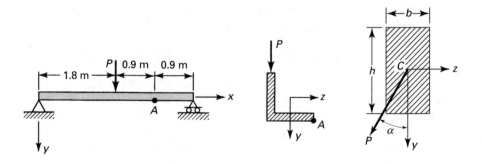

FIGURE P5.1.   FIGURE P5.2.

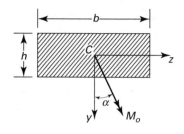

FIGURE P5.3.

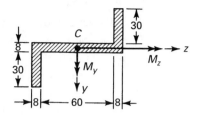

FIGURE P5.4.

5.3. A moment M_o is applied to a beam of the cross section shown in Fig. P5.3 with its vector forming an angle of α. Use $b = 100$ mm, $h = 40$ mm, $M_o = 800 \, \text{N} \cdot \text{m}$, and $\alpha = 25°$. Calculate (a) the orientation of the neutral axis; (b) the maximum bending stress.

5.4. Couples $M_y = M_o$ and $M_z = 1.5 M_o$ are applied to a beam of cross section shown in Fig. P5.4. Determine the largest allowable value of M_o for the maximum stress not to exceed 80 MPa. All dimensions are in millimeters.

5.5. For the simply supported beam of Fig. P5.5, determine the bending stress at points D and E. The cross section is a $0.15 \times 0.15 \times 0.02$ m angle (Fig. 5.4).

5.6. A concentrated load P acts on a cantilever, as shown in Fig. P5.6. The beam is constructed of a 2024-T4 aluminum alloy having a yield strength $\sigma_{yp} = 290$ MPa, $L = 1.5$ m, $t = 20$ mm, $c = 60$ mm, and $b = 80$ mm. Based on a factor of safety $n = 1.2$ against initiation of yielding, calculate the magnitude of P for (a) $\alpha = 0°$ and (b) $\alpha = 15°$. Neglect the effect of shear in bending and assume that beam twisting is prevented.

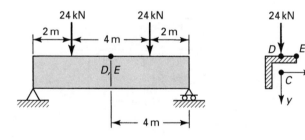

FIGURE P5.5.

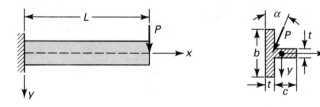

FIGURE P5.6.

5.7. Redo Prob. 5.6 for $\alpha = 30°$. Assume the remaining data to be unchanged.

5.8. A cantilever beam has a Z section of uniform thickness for which $I_y = \frac{2}{3}th^3$, $I_z = \frac{8}{3}th^3$, and $I_{yz} = -th^3$. Determine the maximum bending stress in the beam subjected to a load P at its free end (Fig. P5.8).

5.9. A beam with cross section as shown in Fig. P5.9 is acted on by a moment $M_o = 3\ kN \cdot m$ with its vector forming an angle $\alpha = 20°$. Determine (a) the orientation of the neutral axis, and (b) the maximum bending stress.

5.10. For the thin cantilever of Fig. P5.10, the stress function is given by

$$\Phi = -c_1xy + c_2\frac{x^3}{6} - c_3\frac{x^3y}{6} - c_4\frac{xy^3}{6} - c_5\frac{x^3y^3}{9} - c_6\frac{xy^5}{20}$$

a. Determine the stresses σ_x, σ_y, and τ_{xy} by using the elasticity method.
b. Determine the stress σ_x by using the elementary method.
c. Compare the values of maximum stress obtained by the preceding approaches for $L = 10h$.

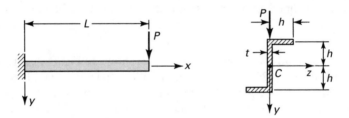

FIGURE P5.8.

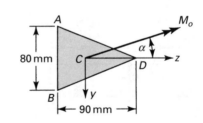

FIGURE P5.9.

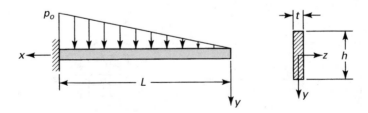

FIGURE P5.10.

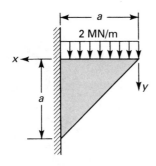

FIGURE P5.11.

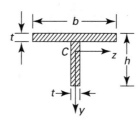

FIGURE P5.12.

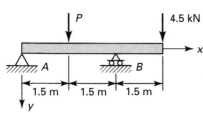

FIGURE P5.14.

5.11. Consider a cantilever beam of constant unit thickness subjected to a uniform load of $p = 2000$ kN per unit length (Fig. P5.11). Determine the maximum stress in the beam:

a. Based on a stress function

$$\Phi = \frac{p}{0.43}\left[-x^2 + xy + (x^2 + y^2)\left(0.78 - \tan^{-1}\frac{y}{x}\right)\right]$$

b. Based on the elementary theory. Compare the results of (a) and (b).

Sections 5.6 through 5.11

5.12. A bending moment acting about the z axis is applied to a T-beam shown in Fig. P5.12. Take the thickness $t = 15$ mm and depth $h = 90$ mm. Determine the width b of the flange in order that the stresses at the bottom and top of the beam will be in the ratio $3:1$, respectively.

5.13. A wooden, simply supported beam of length L is subjected to a uniform load p. Determine the beam length and the loading necessary to develop simultaneously $\sigma_{max} = 8.4$ MPa and $\tau_{max} = 0.7$ MPa. Take thickness $t = 0.05$ m and depth $h = 0.15$ m.

5.14. A box beam supports the loading shown in Fig. P5.14. Determine the maximum value of P such that a flexural stress $\sigma = 7$ MPa or a shearing stress $\tau = 0.7$ MPa will not be exceeded.

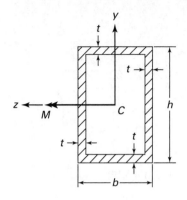

FIGURE P5.15.

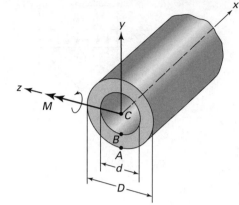

FIGURE P5.16.

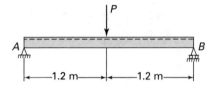

FIGURE P5.17.

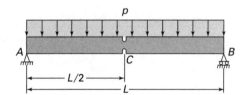

FIGURE P5.18.

5.15. A steel beam of the tubular cross section seen in Fig. P5.15 is subjected to the bending moment M about the z axis. Determine (a) the bending moment M; (b) the radius of curvature r_x of the beam. *Given:* $\sigma_{all} = 150$ MPa, $E = 70$ GPa, $b = 120$ mm, $h = 170$ mm, and $t = 10$ mm.

5.16. An aluminum alloy beam of hollow circular cross section is subjected to a bending moment M about the z axis (Fig. 5.16). Determine (a) the normal stress at point A; (b) the normal stress at point B; (c) the radius of curvature r_z of the beam of a transverse cross section. *Given:* $M = 600$ N $\cdot$ m, $D = 60$ mm, $d = 40$ mm, $E = 70$ GPa, and $\nu = 0.29$.

5.17. A simply supported beam AB of the channel cross section carries a concentrated load P at midpoint (Fig. P5.17). Find the maximum allowable load P based on an allowable normal stress of $\sigma_{all} = 60$ MPa in the beam.

5.18. A uniformly loaded, simply supported rectangular beam has two 15-mm deep vertical grooves opposite each other on the edges at midspan, as illustrated in Fig. P5.18. Find the smallest permissible radius of the grooves for the case in which the normal stress is limited to $\sigma_{max} = 95$ MPa. *Given:* $p = 12$ kN/m, $L = 3$ m, $b = 80$ mm, and $h = 120$ mm.

5.19. A simple wooden beam is under a uniform load of intensity p, as illustrated in Fig. P5.19. (a) Find the ratio of the maximum shearing stress to the largest bending stress in terms of the depth h and length L of the beam. (b) Using $\sigma_{all} = 9$ MPa, $\tau_{all} = 1.4$ MPa, $b = 50$ mm, and $h = 160$ mm, also calculate the maximum permissible length L and the largest permissible distributed load of intensity p.

5.20. A composite cantilever beam 140-mm wide, 300-mm deep, and 3-m long is fabricated by fastening two timber planks ($E_t = 10$ GPa), 60 mm × 300 mm, to the sides of a steel plate ($E_s = 200$ GPa), 20-mm wide by 300-mm deep. Note that the 300-mm dimension is vertical. The allowable stresses in bending for timber and steel are 7 and 120 MPa, respectively. Calculate the maximum vertical load P the beam can carry at its free end.

5.21. A 180-mm-wide by 300-mm-deep wood beam ($E_w = 10$ GPa) 4-m long is reinforced with 180-mm-wide and 10-mm-deep aluminum plates ($E_a = 70$ GPa) on the top and bottom faces. The beam is simply supported and subject to a uniform load of intensity 25 kN/m over its entire length. Calculate the maximum stresses in each material.

5.22. Referring to the reinforced concrete beam of Fig. 5.15a, $b = 300$ mm, $d = 450$ mm, $A_s = 1200$ mm^2, and $n = 10$. Given allowable stresses in steel and concrete of 150 and 12 MPa, respectively, calculate the maximum bending moment the section can carry.

5.23. Referring to the reinforced concrete beam of Fig. 5.15a, $b = 300$ mm, $d = 500$ mm, and $n = 8$. Given the actual maximum stresses developed to be $\sigma_s = 80$ MPa and $\sigma_c = 5$ MPa, calculate the applied bending moment and the steel area required.

5.24. A beam is constructed of half a hollow tube of mean radius R and wall thickness t (Fig. P5.24). Assuming $t \ll R$, locate the shear center S. The moment of inertia of the section about the z axis is $I_z = \pi R^3 t/2$.

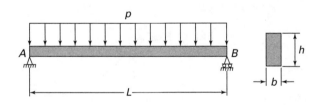

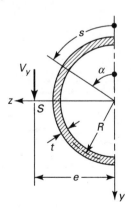

FIGURE P5.19.

FIGURE P5.24.

5.25. An H-section beam with unequal flanges is subjected to a vertical load P (Fig. P5.25). The following assumptions are applicable:
1. The total resisting shear occurs in the flanges.
2. The rotation of a plane section during bending occurs about the symmetry axis so that the radii of curvature of both flanges are equal.

Determine the location of the shear center S.

5.26. Determine the shear center S of the section shown in Fig. P5.26. All dimensions are in millimeters.

5.27. A cantilever beam AB supports a triangularly distributed load of maximum intensity p_o (Fig. P5.27). Determine (a) the equation of the deflection curve, (b) the deflection at the free end, and (c) the slope at the free end.

5.28. The slope at the wall of a built-in beam (Fig. P5.28a) is as shown in Fig. P5.28b and is given by $pL^3/96EI$. Determine the force acting at the simple support, expressed in terms of p and L.

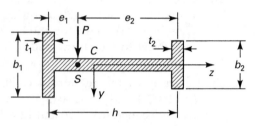

FIGURE P5.25.

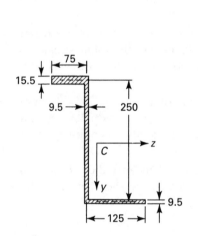

FIGURE P5.26.

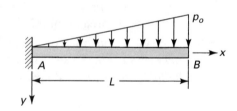

FIGURE P5.27.

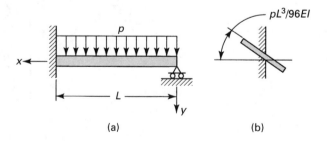

(a)   (b)

FIGURE P5.28.

5.29. A fixed-ended beam of length L is subjected to a concentrated force P at a distance c away from the left end. Derive the equations of the elastic curve.

5.30. A propped cantilever beam AB is subjected to a couple M_o acting at support B, as shown in Fig. P5.30. Derive the equation of the deflection curve and determine the reaction at the roller support.

5.31. A welded bimetallic strip (Fig. P5.31) is initially straight. A temperature increment ΔT causes the element to curve. The coefficients of thermal expansion of the constituent metals are α_1 and α_2. Assuming elastic deformation and $\alpha_2 > \alpha_1$, determine (a) the radius of curvature to which the strip bends, (b) the maximum stress occurring at the interface, and (c) the temperature increase that would result in the simultaneous yielding of both elements.

Sections 5.12 through 5.16

5.32. Verify the values of α for cases B, C, and D of Table 5.1.

5.33. Consider a curved bar subjected to pure bending (Fig. 5.22). Assume the stress function

$$\Phi = A \ln r + Br^2 \ln r + Cr^2 + D$$

to rederive the stress field in the bar given by Eqs. (5.67).

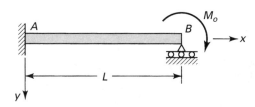

FIGURE P5.30.

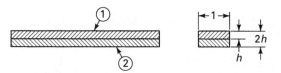

FIGURE P5.31.

5.34. The allowable stress in tension and compression for the clamp body shown in Fig. P5.34 is 80 MPa. Calculate the maximum permissible load the member can resist. Dimensions are in millimeters.

5.35. A curved frame of rectangular cross section is loaded as shown in Fig. P5.35. Determine the maximum tangential stress (a) using the second of Eqs. (5.67) together with the method of superposition; (b) applying Eq. (5.73). *Given:* $h = 100$ mm, $\bar{r} = 150$ mm, and $P = 70$ kN.

5.36. A curved frame having a channel-shaped cross section is subjected to bending by end moments M, as illustrated in Fig. P5.36. Determine the dimension b required if the tangential stresses at points A and B of the beam are equal in magnitude.

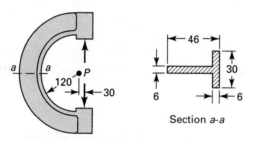

FIGURE P5.34.

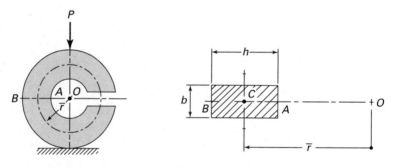

FIGURE P5.35.

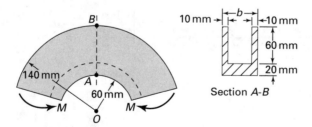

FIGURE P5.36.

5.37. A curved beam of a circular cross section of diameter d is fixed at one end and subjected to a concentrated load P at the free end (Fig. P5.37). Calculate (a) the tangential stress at point A; (b) the tangential stress at point B. Given: $P = 800$ N, $d = 20$ mm, $a = 25$ mm, and $b = 15$ mm.

5.38. The circular steel frame has a cross section approximated by the trapezoidal form shown in Fig. P5.38. Calculate (a) the tangential stress at point A; (b) the tangential stress at point B. Given: $r_i = 100$ mm, $r_o = 250$ mm, $b = 75$ mm, $b = 50$ mm, and $P = 50$ kN.

5.39. The triangular cross section of a curved beam is shown in Fig. P5.39. Derive the expression for the radius R along the neutral axis. Compare the result with that given for Fig. D in Table 5.2.

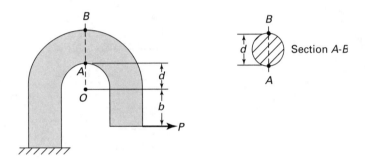

FIGURE P5.37.

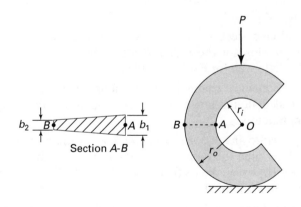

FIGURE P5.38.

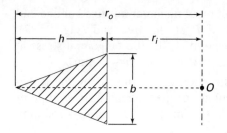

FIGURE P5.39.

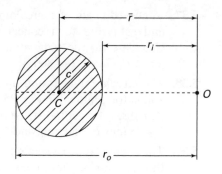

FIGURE P5.40.

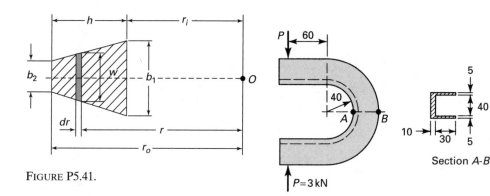

FIGURE P5.41.

FIGURE P5.42.

5.40. The circular cross section of a curved beam is illustrated in Fig. P5.40. Derive the expression for the radius R along the neutral axis. Compare the result with that given for Fig. B in Table 5.2.

5.41. The trapezoidal cross section of a curved beam is depicted in Fig. P5.41. Derive the expression for the radius R along the neutral axis. Compare the result with that given for Fig. E in Table 5.2.

5.42. A machine component of channel cross-sectional area is loaded as shown in Fig. P5.42. Calculate the tangential stress at points A and B. All dimensions are in millimeters.

5.43. A load P is applied to an *eye bar with rigid insert* for the purpose of pulling (Fig. P5.43). Determine the tangential stress at points A and B (a) by the elasticity theory, (b) by Winkler's theory, and (c) by the elementary theory. Compare the results obtained.

5.44. A ring of mean radius R and constant rectangular section is subjected to a concentrated load (Fig. P5.44). You may omit the effect of shear in

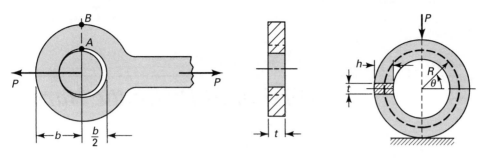

bending. Derive the following general expression for the tangential stress at any section of the ring:

$$\sigma_\theta = -\frac{(P/2)\cos\theta}{A} + \frac{M_\theta}{A}\left(\frac{R-\bar{r}}{e\bar{r}}\right)$$ (P5.44)

where

$$M_\theta = 0.182P\bar{r} - \tfrac{1}{2}P\bar{r}(1-\cos\theta)$$

Use Castigliano's theorem.

5.45. The ring shown in Fig. P5.44 has the following dimensions: $\bar{r} = 150$ mm, $t = 50$ mm, and $h = 100$ mm. Taking $E = \tfrac{5}{2}G$, determine (a) the tangential stress on the inner fiber at $\theta = \pi/4$ and (b) the deflection along the line of action of the load P, considering the effects of the normal and shear forces, as well as bending moment (Sec. 10.4).

CHAPTER 6

Torsion of Prismatic Bars

6.1 INTRODUCTION

In this chapter, consideration is given to stresses and deformations in prismatic members subject to equal and opposite end torques. In general, these bars are assumed free of end constraint. Usually, members that transmit torque, such as propeller shafts and torque tubes of power equipment, are circular or tubular in cross section. For circular cylindrical bars, the torsion formulas are readily derived employing the method of mechanics of materials, as illustrated in the next section. We shall observe that a shaft having a circular cross section is *most efficient* compared to a shaft having an arbitrary cross section.

Slender members with other than circular cross sections are also often used. In treating noncircular prismatic bars, cross sections initially plane (Fig. 6.1a) experience out-of-plane deformation or *warping* (Fig. 6.1b), and the basic kinematic assumptions of the elementary theory are no longer appropriate. Consequently, the theory of elasticity, a general analytic approach, is employed, as discussed in Section 6.4. The governing differential equations derived using this method are applicable to both the linear elastic and the fully plastic torsion problems. The latter is treated in Section 12.10.

For cases that cannot be conveniently solved analytically, the governing expressions are used in conjunction with the membrane and fluid flow analogies, as will be treated in Sections 6.6 through 6.9. Computer-oriented numerical approaches (Chap. 7) are also very efficient for such situations. The chapter concludes with discussions of warping of thin-walled open cross sections and combined torsion and bending of curved bars.

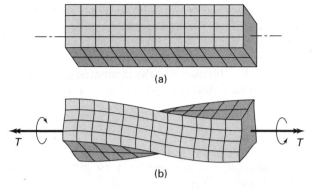

FIGURE 6.1. *Rectangular bar: (a) before loading; (b) after a torque is applied.*

6.2 ELEMENTARY THEORY OF TORSION OF CIRCULAR BARS

Consider a torsion bar or shaft of circular cross section (Fig. 6.2). Assume that the right end twists relative to the left end so that longitudinal line AB deforms to AB'. This results in a shearing stress τ and an angle of twist or angular deformation ϕ. The reader will recall from an earlier study of mechanics of materials [Ref. 6.1] the *basic assumptions* underlying the formulations for the torsional loading of *circular bars*:

1. All plane sections perpendicular to the longitudinal axis of the bar remain plane following the application of torque; that is, points in a given cross-sectional plane remain in that plane after twisting.
2. Subsequent to twisting, cross sections are undistorted in their individual planes; that is, the shearing strain γ varies linearly from zero at the center to a maximum on the outer surface.

 The preceding assumptions hold for both elastic and inelastic material behavior. In the elastic case, the following also applies:

3. The material is homogeneous and obeys Hooke's law; hence, the magnitude of the maximum shear angle γ_{max} must be less than the yield angle.

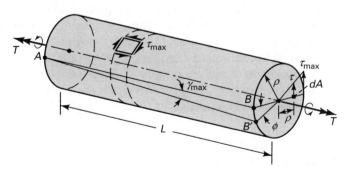

FIGURE 6.2. *Variation of stress and angular rotation of a circular member in torsion.*

We now derive the elastic stress and deformation relationships for circular bars in torsion in a manner similar to the most fundamental equations of the mechanics of materials: by employing the previously considered procedures and the foregoing assumptions. In the case of these elementary formulas, there is complete agreement between the experimentally obtained and the computed quantities. Moreover, their validity can be demonstrated through application of the theory of elasticity (see Example 6.3).

Shearing Stress

On any bar cross section, the resultant of the stress distribution must be equal to the applied torque T (Fig. 6.2). That is,

$$T = \int \rho(\tau dA) = \int \rho\left(\frac{\rho}{r}\tau_{max}\right)dA$$

where the integration proceeds over the entire area of the cross section. At any given section, the maximum shearing stress τ_{max} and the distance r from the center are constant. Hence, the foregoing can be written

$$T = \frac{\tau_{max}}{r}\int \rho^2 dA \tag{a}$$

in which $\int \rho^2 dA = J$ is the polar moment of inertia of the circular cross section (Sec. C.2). For circle of radius r, $J = \pi r^4/2$. Thus,

$$\tau_{max} = \frac{Tr}{J} \tag{6.1}$$

This is the well-known *torsion formula* for circular bars. The shearing stress at a distance ρ from the center is

$$\tau = \frac{T\rho}{J} \tag{6.2}$$

The *transverse* shearing stress obtained by Eq. (6.1) or (6.2) is accompanied by a *longitudinal* shearing stress of equal value, as shown on a surface element in the figure.

We note that, when a shaft is subjected to torques at several points along its length, the *internal torques* will vary from section to section. A graph showing the variation of torque along the axis of the shaft is called the *torque diagram*. That is, this diagram represents a plot of the internal torque T versus its position x along the shaft length. As a *sign convention*, T will be positive if its vector is in the direction of a positive coordinate axis. However, the diagram is not used commonly, because in practice, only a few variations in torque occur along the length of a given shaft.

Angle of Twist

According to Hooke's law, $\gamma_{max} = \tau_{max}/G$; introducing the torsion formula, $\gamma_{max} = Tr/JG$, where G is the modulus of elasticity in shear. For small deformations, $\tan \gamma_{max} = \gamma_{max}$, and we may write $\gamma_{max} = r\phi/L$ (Fig. 6.2). The foregoing expressions lead to the angle of twist, the angle through which one cross section of a circular bar rotates with respect to another:

$$\phi = \frac{TL}{JG} \tag{6.3}$$

Angle ϕ is measured in radians. The product JG is termed the *torsional rigidity* of the member. Note that Eqs. (6.1) through (6.3) are valid for both solid and hollow circular bars; this follows from the assumptions used in the derivations. For a circular tube of inner radius r_i and outer radius r_o, we have $J = \pi(r_o^4 - r_i^4)/2$.

EXAMPLE 6.1   Stress and Deformation in an Aluminum Shaft
A hollow aluminum alloy 6061-T6 shaft of outer radius $c = 40$ mm, inner radius $b = 30$ mm, and length $L = 1.2$ m is fixed at one end and subjected to a torque T at the other end, as shown in Fig. 6.3. If the shearing stress is limited to $\tau_{max} = 140$ MPa, determine (a) the largest value of the torque; (b) the corresponding minimum value of shear stress; (c) the angle of twist that will create a shear stress $\tau_{min} = 100$ MPa on the inner surface.

Solution   From Table D.1, we have the shear modulus of elasticity $G = 72$ GPa and $\tau_{yp} = 220$ MPa.

a. Inasmuch as $\tau_{max} < \tau_{yp}$, we can apply Eq. (6.1) with $r = c$ to obtain

$$T = \frac{J\tau_{max}}{c} \tag{b}$$

By Table C.1, the polar moment of inertia of the hollow circular tube is

$$J = \frac{\pi}{2}(c^4 - b^4) = \frac{\pi}{2}(40^4 - 30^4) = 2.749(10^6)\text{mm}^4$$

Inserting J and τ_{max} into Eq. (b) results in

$$T = \frac{2.749(10^{-6})(140 \times 10^6)}{0.04} = 9.62 \text{ MPa}$$

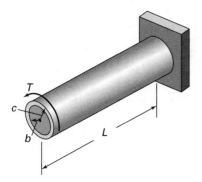

FIGURE 6.3.   *Example 6.1. A tubular circular bar in torsion.*

b. The smallest value of the shear stress takes place on the inner surface of the shaft, and τ_{min} and τ_{max} are respectively proportional to b and c. Therefore,

$$\tau_{min} = \frac{b}{c}\,\tau_{max} = \frac{30}{40}\,(140) = 105 \text{ MPa}$$

c. Through the use of Eq. (2.27), the shear strain on the inner surface of the shaft is equal to

$$\gamma_{min} = \frac{\tau_{min}}{G} = \frac{100 \times 10^6}{72 \times 10^9} = 1389\,\mu$$

Referring to Fig. 6.2,

$$\phi = \frac{L\gamma_{min}}{b} = \frac{1200}{30}\,(1389 \times 10^{-6}) = 55.6 \times 10^{-3} \text{ rad}$$

or

$$\phi = 3.19°$$

EXAMPLE 6.2   Redundantly Supported Shaft

A solid circular shaft AB is fixed to rigid walls at both ends and subjected to a torque T at section C, as shown in Fig. 6.4a. The shaft diameters are d_a and d_b for segments AC and CB, respectively. Determine the lengths a and b if the maximum shearing stress in both shaft segments is to be the same for $d_a = 20$ mm, $d_b = 12$ mm, and $L = 600$ mm.

Solution   From the free-body diagram of Fig. 6.4a, we observe that the problem is *statically indeterminate* to the first degree; the one equation of equilibrium available is not sufficient to obtain the two unknown reactions T_A and T_B.

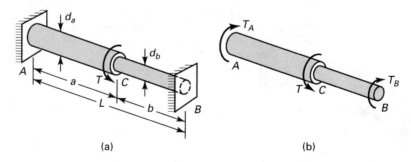

(a)   (b)

FIGURE 6.4.   *Example 6.2. (a) A fixed-ended circular shaft in torsion; (b) free-body diagram of the entire shaft.*

Condition of Equilibrium.   Using the free-body diagram of Fig. 6.4b,

$$\Sigma T = 0, \qquad T_A + T_B = T \tag{c}$$

Torque-Displacement Relations.   The angle of twist at section C is expressed in terms of the left and right segments of the solid shaft, respectively, as

$$\phi_a = \frac{T_A a}{J_a G}, \qquad \phi_b = \frac{T_B b}{J_b G} \tag{d}$$

Here, the polar moments of inertia are $J_a = \pi d_a^4/32$ and $J_b = \pi d_b^4/32$.

Condition of Compatibility.   The two segments must have the same angle of twist where they join. Thus,

$$\phi_a = \phi_b \quad \text{or} \quad \frac{T_A a}{J_a G} = \frac{T_B b}{J_b G} \tag{e}$$

Equations (c) and (e) can be solved simultaneously to yield the reactions

$$T_A = \frac{T}{1 + (aJ_b/bJ_a)}, \qquad T_B = \frac{T}{1 + (bJ_a/aJ_b)} \tag{f}$$

The maximum shearing stresses in each segment of the shaft are obtained from the torsion formula:

$$\tau_a = \frac{T_A d_a}{2J_a}, \qquad \tau_b = \frac{T_B d_b}{2J_b} \tag{g}$$

For the case under consideration, $\tau_a = \tau_b$, or

$$\frac{T_A d_a}{J_a} = \frac{T_B d_b}{J_b}$$

Introducing Eqs. (f) into the foregoing and simplifying, we obtain

$$\frac{a}{b} = \frac{d_a}{d_b}$$

from which

$$a = \frac{d_a L}{d_a + d_b}, \qquad b = \frac{d_b L}{d_a + d_b} \tag{h}$$

where $L = a + b$. Insertion of the given data results in

$$a = \frac{20(600)}{20 + 12} = 375 \text{ mm}, \qquad b = \frac{12(600)}{20 + 12} = 225 \text{ mm}$$

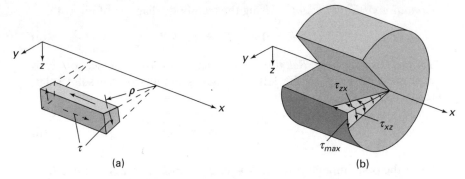

FIGURE 6.5. *Shearing stresses on transverse (xz) and axial (zx) planes of a shaft (Fig. 6.2): (a) shaft element in pure shear; (b) shaft segment in pure shear.*

For a prescribed value of torque T, we can now compute the reactions, angle of twist at section C, and maximum shearing stress from Eqs. (f), (d), and (g), respectively.

Axial and Transverse Shear Stresses

In the foregoing, we considered shear stresses acting in the plane of a cut perpendicular to the axis of the shaft, defined by Eqs. (6.1) and (6.2). Their directions coincide with the direction of the internal torque T on the cross section (Fig. 6.2). These stresses must be accompanied by equal shear stresses occurring on the axial planes of the shaft, since equal shear stresses always exist on mutually perpendicular planes.

Consider the state of stress as depicted in Fig. 6.5a. Note that τ is the only stress component acting on this element removed from the bar; the element is in the state of *pure shear*. Observe that $\tau = \tau_{xz} = \tau_{zx}$ denotes the shear stresses in the *tangential* and *axial directions*. So, the internal torque develops not only a **transverse shear stress** along radial lines or in the cross section but also an associated **axial shear stress** distribution along a longitudinal plane. The variation of these stresses on the mutually perpendicular planes is illustrated in Fig. 6.5b, where a portion of the shaft has been removed for the purposes of illustration. Consequently, if a material is weaker in shear axially than laterally (such as wood), the failure in a twisted bar occurs longitudinally along axial planes.

6.3 STRESSES ON INCLINED PLANES

Our treatment thus far has been limited to shearing stresses on planes at a point parallel or perpendicular to the axis of a shaft, defined by the torsion formula. This state of stress is depicted acting on a surface element on the left of the shaft in

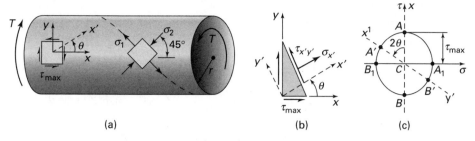

FIGURE 6.6. *(a) Stresses acting on the surface elements of a shaft in torsion; (b) free-body diagram of wedge cut from the shaft; (c) Mohr's circle for torsional loading.*

Fig. 6.6a. Alternatively, representation of the stresses at the same point may be given on an infinitesimal two-dimensional wedge whose outer normal (x') on its oblique face makes an angle θ with the axial axis (x), as shown in Fig. 6.6b.

We shall analyze the stresses $\sigma_{x'}$ and $\tau_{x'y'}$ that must act on the inclined plane to keep the isolated body, as discussed in Sections 1.7 and 1.9. The equilibrium conditions of forces in the x' and y' directions, Eqs. (1.18a and 1.18b) with $\sigma_x = \sigma_y = 0$ and $\tau_{xy} = -\tau_{max}$, give

$$\sigma_{x'} = -\tau_{max}\sin2\theta \qquad \textbf{(6.4a)}$$

$$\tau_{x'y'} = -\tau_{max}\cos2\theta \qquad \textbf{(6.4b)}$$

These are the *equations of transformation* for stress in a shaft under torsion.

Figure 6.6c, a *Mohr's circle for torsional loading* provides a convenient way of checking the preceding results. Observe that the center C of the circle is at the origin of the coordinates σ and τ. It is recalled that points A' and B' define the states of stress about any other set of x' and y' planes relative to the set through an angle θ. Similarly, the points $A(0, \tau_{max})$ and $B(0, \tau_{min})$ are located on the τ axis, while A_1 and B_1 define the principal planes and the principal stresses. The radius of the circle is equal to the magnitude of the maximum shearing stress.

The variation in the stresses $\sigma_{x'}$ and $\sigma_{x'y'}$ with the orientation of the inclined plane is demonstrated in Fig. 6.7. It is seen that $\tau_{max} = \tau$ and $\tau_{x'y'} = 0$ when $\theta = 45°$ and $\sigma_{min} = -\tau$, and $\tau_{x'y'} = 0$ when $\theta = 135°$. That is, the *maximum normal stress* equals $\sigma_1 = Tr/J$, and the minimum normal stress equals $\sigma_2 = -Tr/J$, acting in the directions shown in Fig. 6.6a. Also, no shearing stress greater than τ is developed.

It is clear that, for a brittle material such as cast iron, which is weaker in tension than in shear, failure occurs in tension along a helix as indicated by the dashed lines in Fig. 6.6a. Ordinary chalk behaves in this manner. On the contrary, shafts made of ductile materials weak in shearing strength (for example, mild steel) break along a line perpendicular to the axis. The preceding is consistent with the types of tensile failures discussed in Section 2.7.

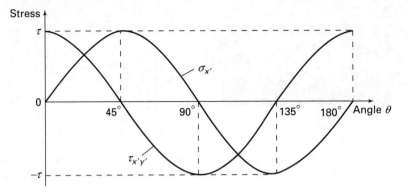

FIGURE 6.7. *Graph of normal stress $\sigma_{x'}$ and shear stress $\tau_{x'y'}$ versus angle θ of the inclined plane.*

6.4 GENERAL SOLUTION OF THE TORSION PROBLEM

Consider a prismatic bar of constant arbitrary cross section subjected to equal and opposite twisting moments applied at the ends, as in Fig. 6.8a. The origin of x, y, and z in the figure is located at the *center of twist* of the cross section, about which the cross section rotates during twisting. It is sometimes defined as the point at rest in every cross section of a bar in which one end is fixed and the other twisted by a couple. At this point, u and v, the x and y displacements, are thus zero. The location of the center of twist is a function of the shape of the cross section. Note that while the center of twist is referred to in the derivations of the basic relationships, it is not dealt with explicitly in the solution of torsion problems (see Prob. 6.19). The z passes through the centers of twist of all cross sections.

Geometry of Deformation

In general, the cross sections warp, as already noted. We now explore the problem of torsion with free warping, applying the *Saint-Venant semi-inverse method*

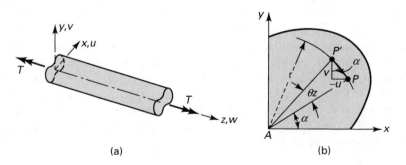

FIGURE 6.8. *Torsion member of arbitrary cross section.*

[Refs. 6.2 and 6.3]. As a fundamental assumption, the warping deformation is taken to be independent of axial location, that is, identical for any cross section:

$$w = f(x, y) \tag{a}$$

It is also assumed that the projection on the xy plane of any warped cross section rotates as a rigid body and that the *angle of twist per unit length of the bar*, θ, is *constant*.

We refer now to Fig. 6.8b, which shows the partial end view of the bar (and could represent any section). An arbitrary point on the cross section, point $P(x, y)$, located a distance r from center of twist A, has moved to $P'(x - u, y + v)$ as a result of torsion. Assuming that *no rotation* occurs at end $z = 0$ and that θ is *small*, the x and y displacements of P are, respectively,

$$\begin{aligned} u &= -(r\theta z)\sin\alpha = -y\theta z \\ v &= (r\theta z)\cos\alpha = x\theta z \end{aligned} \tag{b}$$

where the angular displacement of AP at a distance z from the left end is θz; then x, y, and z are the coordinates of point P, and α is the angle between AP and the x axis. Clearly, Eqs. (b) specify the rigid body rotation of any cross section through a small angle θz. By substituting Eqs. (a) and (b) into Eq. (2.4), we have

$$\varepsilon_x = \gamma_{xy} = \varepsilon_y = \varepsilon_z = 0$$

$$\gamma_{zx} = \frac{\partial w}{\partial x} - y\theta, \qquad \gamma_{zy} = \frac{\partial w}{\partial y} + x\theta \tag{c}$$

Equation (2.36) together with this expression leads to the following:

$$\sigma_x = \sigma_y = \sigma_z = \tau_{xy} = 0 \tag{d}$$

$$\tau_{zx} = G\left(\frac{\partial w}{\partial x} - y\theta\right)$$

$$\tau_{zy} = G\left(\frac{\partial w}{\partial y} + x\theta\right) \tag{e}$$

Equations of Equilibrium

By now substituting Eq. (d) into the equations of equilibrium (1.14), assuming negligible body forces, we obtain

$$\frac{\partial \tau_{zx}}{\partial z} = 0, \qquad \frac{\partial \tau_{zy}}{\partial z} = 0, \qquad \frac{\partial \tau_{zx}}{\partial x} + \frac{\partial \tau_{zy}}{\partial y} = 0 \tag{6.5}$$

Equations of Compatibility

Differentiating the first equation of (e) with respect to y and the second with respect to x and subtracting the second from the first, we obtain an equation of compatibility:

$$\frac{\partial \tau_{zx}}{\partial y} - \frac{\partial \tau_{zy}}{\partial x} = H \tag{6.6}$$

where

$$H = -2G\theta \tag{6.7}$$

The stress in a bar of arbitrary section may thus be determined by solving Eqs. (6.5) and (6.6) along with the given boundary conditions.

6.5 PRANDTL'S STRESS FUNCTION

As in the case of beams, the torsion problem formulated in the preceding is commonly solved by introducing a single stress function. If a function $\Phi(x, y)$, the *Prandtl stress function*, is assumed to exist, such that

$$\tau_{zx} = \frac{\partial \Phi}{\partial y}, \qquad \tau_{zy} = -\frac{\partial \Phi}{\partial x} \tag{6.8}$$

then the equations of equilibrium (6.5) are satisfied. The equation of compatibility (6.6) becomes, upon substitution of Eq. (6.8),

$$\frac{\partial^2 \Phi}{\partial x^2} + \frac{\partial^2 \Phi}{\partial y^2} = H \tag{6.9}$$

The stress function Φ must therefore satisfy *Poisson's equation* if the compatibility requirement is to be satisfied.

Boundary Conditions

We are now prepared to consider the boundary conditions, treating first the *load-free* lateral surface. Recall from Section 1.5 that τ_{xz} is a z-directed shearing stress acting on a plane whose normal is parallel to the x axis, that is, the yz plane. Similarly, τ_{zx} acts on the xy plane and is x-directed. By virtue of the symmetry of the stress tensor, we have $\tau_{xz} = \tau_{zx}$ and $\tau_{yz} = \tau_{zy}$. Therefore, the stresses given by Eq. (e) may be indicated on the xy plane near the boundary, as shown in Fig. 6.9. The boundary element is associated with arc length ds. Note that ds increases in the counterclockwise direction. When ds is zero, the element represents a point at the boundary. Then, referring to Fig. 6.9 together with Eq. (1.48), which relates the surface forces to the internal stress, and noting that the cosine of the angle between z and a unit normal $\boldsymbol{n}$ to the surface is zero [that is, $\cos(\boldsymbol{n}, z) = 0$], we have

$$\tau_{zx} l + \tau_{zy} m = 0 \tag{a}$$

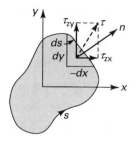

FIGURE 6.9. *Arbitrary cross section of a prismatic member in torsion.*

According to Eq. (a), the resultant shear stress τ must be tangent to the boundary (Fig. 6.9). From the figure, it is clear that

$$l = \cos(\mathbf{n}, x) = \frac{dy}{ds}, \qquad m = \cos(\mathbf{n}, y) = -\frac{dx}{ds} \qquad \textbf{(b)}$$

Note that as we proceed in the direction of increasing s, x decreases and y increases. This accounts for the algebraic sign in front of dx and dy in Fig. 6.9. Substitution of Eqs. (6.8) and (b) into Eq. (a) yields

$$\frac{\partial \Phi}{\partial y}\frac{dy}{ds} + \frac{\partial \Phi}{\partial x}\frac{dx}{ds} = \frac{d\Phi}{ds} = 0 \qquad (\text{on the boundary}) \qquad \textbf{(6.10)}$$

This expression states that the directional deviation along a boundary curve is zero. Thus, the function $\Phi(x, y)$ must be an *arbitrary* constant on the lateral surface of the prism. Examination of Eq. (6.8) indicates that the stresses remain the same regardless of additive constants; that is, if Φ + constant is substituted for Φ, the stresses will not change. For solid cross sections, we are therefore free to set Φ equal to zero at the boundary. In the case of multiply connected cross sections, such as hollow or tubular members, an arbitrary value may be assigned at the boundary of *only one* of the contours $s_0, s_1, \ldots, s_n$. For such members it is necessary to extend the mathematical formulation presented in this section. Solutions for thin-walled, multiply connected cross sections are treated in Section 6.8 by use of the membrane analogy.

Returning to the member of solid cross section, we complete discussion of the boundary conditions by considering the ends, at which the normals are parallel to the z axis and therefore $\cos(\mathbf{n}, z) = n = \pm 1, l = m = 0$. Equation (1.48) now gives, for $p_z = 0$,

$$p_x = \pm \tau_{zx}, \qquad p_y = \pm \tau_{zy} \qquad \textbf{(c)}$$

where the algebraic sign depends on the relationship between the outer normal and the positive z direction. For example, it is negative for the end face at the origin in Fig. 6.8a.

We now confirm that the summation of forces over the ends of the bar is zero:

$$\iint p_x \, dx \, dy = \iint \tau_{zx} \, dx \, dy = \iint \frac{\partial \Phi}{\partial y} \, dx \, dy$$

$$= \int dx \int_{y_1}^{y_2} \frac{\partial \Phi}{\partial y} \, dy = \int \Phi|_{y_1}^{y_2} \, dx = \int (\Phi_2 - \Phi_1) \, dx = 0$$

Here y_1 and y_2 represent the y coordinates of points located on the surface. Inasmuch as Φ = constant on the surface of the bar, the values of Φ corresponding to y_1 and y_2 must be equal to a constant, $\Phi_1 = \Phi_2$ = constant. Similarly it may be shown that

$$\iint \tau_{zy} \, dx \, dy = 0$$

The end forces, while adding to zero, must nevertheless provide the required twisting moment or externally applied torque about the z axis:

$$T = \iint (x\tau_{zy} - y\tau_{zx}) \, dx \, dy = -\iint x \frac{\partial \Phi}{\partial x} \, dx \, dy - \iint y \frac{\partial \Phi}{\partial y} \, dx \, dy$$

$$= -\int dy \int x \frac{\partial \Phi}{\partial x} \, dx - \int dx \int y \frac{\partial \Phi}{\partial y} \, dy$$

Integrating by parts,

$$T = -\int x\Phi|_{x_1}^{x_2} \, dy + \iint \Phi \, dx \, dy - \int y\Phi|_{y_1}^{y_2} \, dx + \iint \Phi \, dx \, dy$$

Since Φ = constant at the boundary and x_1, x_2, y_1, and y_2 denote points on the lateral surface, it follows that

$$T = 2 \iint \Phi \, dx \, dy \qquad \qquad \textbf{(6.11)}$$

Inasmuch as $\Phi(x, y)$ has a value at each point on the cross section, it is clear that Eq. (6.11) represents twice the volume beneath the Φ surface.

What has resulted from the foregoing development is a set of equations satisfying all the conditions of the prescribed torsion problem. Equilibrium is governed by Eqs. (6.8), compatibility by Eq. (6.9), and the boundary condition by Eq. (6.10). Torque is related to stress by Eq. (6.11). To ascertain the distribution of stress, it is necessary to determine a stress function that satisfies Eqs. (6.9) and (6.10), as is demonstrated in the following example.

EXAMPLE 6.3   Analysis of an Elliptical Torsion Bar

Consider a solid bar of elliptical cross section (Fig. 6.10a). Determine the maximum shearing stress and the angle of twist per unit length. Also, derive an expression for the warping $w(x, y)$.

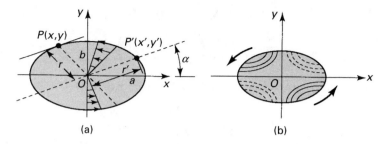

FIGURE 6.10. *Example 6.3. Elliptical cross section: (a) shear stress distribution; (b) out-of plane deformation or warping.*

Solution Equations (6.9) and (6.10) are satisfied by selecting the stress function

$$\Phi = k\left(\frac{x^2}{a^2} + \frac{y^2}{b^2} - 1\right)$$

where k is a constant. Substituting this into Eq. (6.9), we obtain

$$k = \frac{a^2 b^2}{2(a^2 + b^2)} H$$

Hence,

$$\Phi = \frac{a^2 b^2 H}{2(a^2 + b^2)}\left(\frac{x^2}{a^2} + \frac{y^2}{b^2} - 1\right) \tag{d}$$

and Eq. (6.11) yields

$$T = \frac{a^2 b^2 H}{a^2 + b^2}\left(\frac{1}{a^2}\iint x^2\, dx\, dy + \frac{1}{b^2}\iint y^2\, dx\, dy - \iint dx\, dy\right)$$

$$= \frac{a^2 b^2 H}{a^2 + b^2}\left(\frac{1}{a^2} I_y + \frac{1}{b^2} I_x - A\right)$$

where A is the cross-sectional area. Inserting expressions for I_x, I_y, and A (Table C.1) results in

$$T = \frac{a^2 b^2 H}{a^2 + b^2}\left(\frac{1}{a^2}\frac{\pi b a^3}{4} + \frac{1}{b^2}\frac{\pi a b^3}{4} - \pi a b\right) = \frac{-\pi a^3 b^3 H}{2(a^2 + b^2)} \tag{e}$$

from which

$$H = -\frac{2T(a^2 + b^2)}{\pi a^3 b^3} \tag{f}$$

The stress function is now expressed as

$$\Phi = -\frac{T}{\pi ab}\left(\frac{x^2}{a^2} + \frac{y^2}{b^2} - 1\right)$$

and the shearing stresses are found readily from Eq. (6.8):

$$\tau_{zx} = \frac{\partial \Phi}{\partial y} = -\frac{2Ty}{\pi ab^3} = -\frac{Ty}{2I_x}$$

$$\tau_{zy} = -\frac{\partial \Phi}{\partial x} = \frac{2Tx}{\pi a^3 b} = \frac{Tx}{2I_y} \tag{g}$$

The ratio of these stress components is proportional to y/x and thus constant along any radius of the ellipse:

$$\frac{\tau_{zx}}{\tau_{zy}} = -\frac{y}{x}\frac{a^2}{b^2} = -\frac{y}{x}\frac{I_y}{I_x}$$

The resultant shearing stress,

$$\tau_{z\alpha} = \left(\tau_{zx}^2 + \tau_{zy}^2\right)^{1/2} = \frac{2T}{\pi ab}\left(\frac{x^2}{a^4} + \frac{y^2}{b^4}\right)^{1/2} \tag{h}$$

has a direction parallel to a tangent drawn at the boundary at its point of intersection with the radius containing the point under consideration. Note that α represents an arbitrary angle (Fig. 6.10a).

To determine the location of the maximum resultant shear, which from Eq. (g) is somewhere on the boundary, consider a point $P'(x', y')$ located on a diameter conjugate to that containing $P(x, y)$ (Fig. 6.10a). Note that OP' is parallel to the tangent line at P. The coordinates of P and P' are related by

$$x = \frac{a}{b}y', \qquad y = \frac{b}{a}x'$$

When these expressions are substituted into Eq. (h), we have

$$\tau_{z\alpha} = \frac{2T}{\pi a^2 b^2}(x'^2 + y'^2)^{1/2} = \frac{2T}{\pi a^2 b^2}r' \tag{i}$$

Clearly, $\tau_{z\alpha}$ will have its maximum value corresponding to the largest value of the conjugate semidiameter r'. This occurs where $r' = a$ or $r = b$. The maximum resultant shearing stress thus occurs at $P(x, y)$, corresponding to the extremities of the minor axis as follows: $x = 0$, $y = \pm b$. From Eq. (i),

$$\tau_{max} = \frac{2T}{\pi ab^2} \tag{6.12a}$$

The angle of twist per unit length is obtained by substituting Eq. (f) into Eq. (6.7):

$$\theta = \frac{(a^2 + b^2)T}{\pi a^3 b^3 G} \tag{6.12b}$$

We note that the factor by which the twisting moment is divided to determine the twist per unit length is called the *torsional rigidity*, commonly denoted C. That is,

$$C = \frac{T}{\theta} \tag{6.13}$$

The torsional rigidity for an elliptical cross section, from Eq. (6.12b), is thus

$$C = \frac{\pi a^3 b^3}{a^2 + b^2} G = \frac{G}{4\pi^2} \frac{(A)^4}{J}$$

Here $A = \pi ab$ and $J = \pi ab(a^2 + b^2)/4$ are the area and polar moment of inertia of the cross section.

The components of displacement u and v are then found from Eq. (b) of Section 6.4. To obtain the warpage $w(x, y)$, consider Eq. (e) of Section 6.4 into which have been substituted the previously derived relations for τ_{zx}, τ_{zy}, and θ:

$$\tau_{zx} = -\frac{2Ty}{\pi ab^3} = G\left[\frac{\partial w}{\partial x} - \frac{y(a^2 + b^2)T}{\pi a^3 b^3 G}\right]$$

$$\tau_{zy} = \frac{2Tx}{\pi a^3 b} = G\left[\frac{\partial w}{\partial y} + \frac{x(a^2 + b^2)T}{\pi a^3 b^3 G}\right]$$

Integration of these equations leads to identical expressions for $w(x, y)$, except that the first also yields an arbitrary function of y, $f(y)$, and the second an arbitrary function of x, $f(x)$. Since $w(x, y)$ must give the same value for a given $P(x, y)$, we conclude that $f(x) = f(y) = 0$; what remains is

$$w(x, y) = \frac{T}{G} \frac{(b^2 - a^2)xy}{\pi a^3 b^3} \tag{6.14}$$

The contour lines, obtained by setting w = constant, are the hyperbolas shown in Fig. 6.10b. The solid lines indicate the portions of the section that become convex, and the dashed lines indicate the portions of the section that become concave when the bar is subjected to a torque in the direction shown.

Circular Cross Section

The results obtained in this example for an elliptical section may readily be reduced to the case of a circular section by setting a and b equal to the radius of a circle r (Fig. 6.6a). Equations (6.12a and b) thus become

$$\tau_{max} = \frac{2T}{\pi r^3} = \frac{Tr}{J}, \quad \phi = \theta L = \frac{TL}{JG}$$

where the polar moment of inertia, $J = \pi r^4/2$. Similarly, Eq. (6.14) gives $w = 0$, verifying assumption (1) of Section 6.2.

EXAMPLE 6.4 Equilateral Triangle Bar under Torsion

An equilateral cross-sectional solid bar is subjected to pure torsion (Fig. 6.11). What are the maximum shearing stress and the angle of twist per unit length?

Solution The equations of the boundaries are expressed as

$$x + \frac{h}{3} = 0 \qquad \text{on } BC$$

$$\frac{x}{\sqrt{3}} + y - \frac{2h}{3\sqrt{3}} = 0 \qquad \text{on } AB$$

$$\frac{x}{\sqrt{3}} - y - \frac{2h}{3\sqrt{3}} = 0 \qquad \text{on } AC$$

Therefore, the stress function that vanishes at the boundary may be written in the form:

$$\Phi = k\left(x + \sqrt{3}y - \frac{2h}{3}\right)\left(x - \sqrt{3}y - \frac{2h}{3}\right)\left(x + \frac{h}{3}\right) \qquad \text{(j)}$$

in which k is constant.

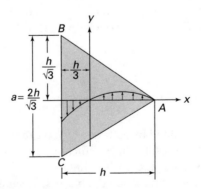

FIGURE 6.11. *Example 6.4. Equilateral triangle cross section.*

Proceeding as in Example 6.3, it can readily be shown (see Prob. 6.23) that the largest shearing stress at the middle of the sides of the triangle is equal to

$$\tau_{max} = \frac{15\sqrt{3}T}{2h^3} \tag{k}$$

At the corners of the triangle, the shearing stress is zero, as shown in Fig. 6.11. Angle of twist per unit length is given by

$$\theta = \frac{15\sqrt{3}T}{h^4 G} \tag{l}$$

The torsional rigidity is therefore $C = T/\theta = h^4 G/15\sqrt{3}$. Note that, for an equilateral triangle of sides a, we have $a = 2h/\sqrt{3}$ (Fig. 6.11).

EXAMPLE 6.5   Rectangular Bar Subjected to Torsion

A torque T acts on a bar having rectangular cross section of sides a and b (Fig. 6.12). Outline the derivations of the expressions for the maximum shearing stress τ_{max} and the angle of twist per unit length θ.

Solution   The indirect method employed in the preceding examples does not apply to the rectangular cross sections. Mathematical solution of the problem is lengthy. For situations that cannot be conveniently solved by applying the theory of elasticity, the governing equations are used in conjunction with the experimental methods, such as membrane analogy (Ref. 6.4). Finite element analysis is also very efficient for this purpose.

The shear stress distribution along three radial lines initiating from the center of a rectangular section of a bar in torsion are shown in Fig. 6.12, where the largest stresses are along the center line of each face. The difference in this stress distribution compared with that of a circular section is very clear. For the latter, the stress is a maximum at the most remote point, but for the former, the stress is zero at the most remote

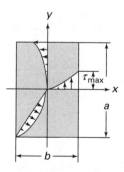

FIGURE 6.12.   *Example 6.5. Shear stress distribution in the rectangular cross section of a torsion bar.*

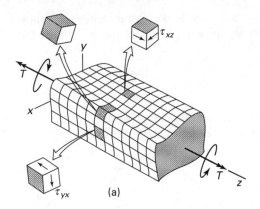

FIGURE 6.13. *Example 6.5. Deformation and stress in a rectangular bar segment under torsion. Note that the original plane cross sections have warped out of their own plane.*

point. The values of the maximum shearing stress τ_{max} in a rectangular cross section and the angle of twist per unit length θ are given in the next section (see Table 6.2).

Interestingly, a *corner element* of the cross section of a rectangular shaft under torsion does not distort at all, and hence the shear stresses are zero at the corners, as illustrated in Fig. 6.13. This is possible because outside surfaces are free of all stresses. The same considerations can be applied to the other points on the boundary. The shear stresses acting on three *outermost* cubic elements isolated from the bar are illustrated in the figure. Here stress-free surfaces are indicated as shaded. Observe that all shear stresses τ_{xy} and τ_{xz} in the plane of a cut near the boundaries act on them.

6.6 PRANDTL'S MEMBRANE ANALOGY

It is demonstrated next that the differential equation for the stress function, Eq. (6.9), is of the same form as the equation describing the deflection of a membrane or soap film subject to pressure. Hence, an analogy exists between the torsion and membrane problems, serving as the basis of a number of experimental techniques. Consider an edge-supported homogeneous membrane, given its boundary contour by a hole cut in a plate (Fig. 6.14a). The *shape* of the hole is the same as that of the twisted bar to be studied; the *sizes* need not be identical.

Equation of Equilibrium

The equation describing the z deflection of the membrane is derived from considerations of equilibrium applied to the isolated element *abcd*. Let the tensile forces per unit membrane length be denoted by S. From a small z deflection, the

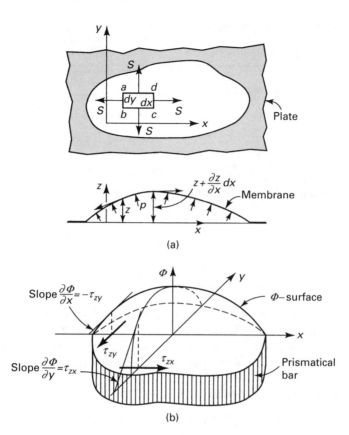

FIGURE 6.14. *Membrane analogy for torsion members of solid cross section.*

inclination of S acting on side ab may be expressed as $\beta \approx \partial z/\partial x$. Since z varies from point to point, the angle at which S is inclined on side dc is

$$\beta + \frac{\partial \beta}{\partial x} dx \approx \frac{\partial z}{\partial x} + \frac{\partial^2 z}{\partial x^2} dx$$

Similarly, on sides ad and bc, the angles of inclination for the tensile forces are $\partial z/\partial y$ and $\partial z/\partial y + (\partial^2 z/\partial y^2)\, dy$, respectively. In the development that follows, S is regarded as a constant, and the weight of the membrane is ignored. For a uniform lateral pressure p, the equation of vertical equilibrium is then

$$-(S\,dy)\frac{\partial z}{\partial x} + S\,dy\left(\frac{\partial z}{\partial x} + \frac{\partial^2 z}{\partial x^2} dx\right) - (S\,dx)\frac{\partial z}{\partial y}$$

$$+ (S\,dx)\left(\frac{\partial z}{\partial y} + \frac{\partial^2 z}{\partial y^2} dy\right) + p\,dx\,dy = 0$$

TABLE 6.1   *Analogy between Membrane and Torsion Problems*

Membrane Problem	Torsion Problem
z	Φ
$\dfrac{1}{S}$	G
p	2θ
$-\dfrac{\partial z}{\partial x},\ \dfrac{\partial z}{\partial y}$	$\tau_{zy},\ \tau_{zx}$
$2 \cdot$ (volume beneath membrane)	T

leading to

$$\frac{\partial^2 z}{\partial x^2} + \frac{\partial^2 z}{\partial y^2} = -\frac{p}{S} \tag{6.15}$$

This is again Poisson's equation. Upon comparison of Eq. (6.15) with Eqs. (6.9) and (6.8), the quantities shown in Table 6.1 are observed to be analogous. The membrane, subject to the conditions outlined, thus represents the Φ surface (Fig. 6.14b). In view of the derivation, the restriction with regard to smallness of slope must be borne in mind.

Shearing Stress and Angle of Twist

We outline next one method by which the foregoing theory can be reduced to a useful experiment. In two thin, stiff plates, bolted together, are cut two adjacent holes; one conforms to the outline of the irregular cross section and the other is circular. The plates are then separated and a thin sheet of rubber stretched across the holes (with approximately uniform and equal tension). The assembly is then bolted together. Subjecting one side of the membrane to a uniform pressure p causes a different distribution of deformation for each cross section, with the circular hole providing calibration data. The measured geometric quantities associated with the circular hole, together with the known solution, provide the needed proportionalities between pressure and angle of twist, slope and stress, volume and torque. These are then applied to the irregular cross section, for which the measured slopes and volume yield τ and T. The need for precise information concerning the membrane stress is thus obviated.

The membrane analogy provides more than a useful *experimental* technique. As is demonstrated in the next section, it also serves as the basis for obtaining approximate *analytical* solutions for bars of narrow cross section as well as for members of open thin-walled section.

For reference purposes, Table 6.2 presents the shearing stress and angle of twist for a number of commonly encountered shapes [Ref. 6.4]. Note that the values of coefficients α and β depend on the ratio of the length of the long side or depth a to the width b of the short side of a rectangular section. For thin sections, where a is much greater than b, their values approach 1/3. We observe that, in all cases, the maximum shearing stresses occur at a point on the edge of the cross section that is closest to the

TABLE 6.2 *Shear Stress and Angle of Twist of Various Members in Torsion*

Cross section	Maximum shearing stress	Angle of twist per unit length		
For circular bar: $a = b$	$\tau_A = \dfrac{2T}{\pi ab^2}$	$\theta = \dfrac{(a^2 + b^2)T}{\pi a^3 b^3 G}$		
Equilateral triangle	$\tau_A = \dfrac{20T}{a^3}$	$\theta = \dfrac{46.2T}{a^4 G}$		
	$\tau_A = \dfrac{T}{\alpha ab^2}$	$\theta = \dfrac{T}{\beta ab^3 G}$		
	a/b	β		α
	1.0	0.141		0.208
	1.5	0.196		0.231
	2.0	0.229		0.246
	2.5	0.249		0.256
	3.0	0.263		0.267
	4.0	0.281		0.282
	5.0	0.291		0.292
	10.0	0.312		0.312
	∞	0.333		0.333
	$\tau_A = \dfrac{T}{2abt_1}$ $\tau_B = \dfrac{T}{2abt}$	$\theta = \dfrac{(at + bt_1)T}{2tt_1 a^2 b^2 G}$		
For circular tube: $a = b$	$\tau_A = \dfrac{T}{2\pi abt}$	$\theta = \dfrac{\sqrt{2(a^2 + b^2)}T}{4\pi a^2 b^2 tG}$		
Hexagon	$\tau_A = \dfrac{5.7T}{a^3}$	$\theta = \dfrac{8.8T}{a^4 G}$		

center axis of the shaft. A *circular shaft* is the most efficient; it is subjected to both smaller maximum shear stress and a smaller angle of twist than the corresponding noncircular shaft of the same cross-sectional area and carrying the same torque.

6.6 Prandtl's Membrane Analogy **313**

EXAMPLE 6.6   Analysis of a Stepped Bar in Torsion

A rectangular bar of width b consists of two segments: one with depth $a_1 = 60$ mm and length $L_1 = 2.5$ m and the other with depth $a_2 = 45$ mm and length $L_2 = 1.5$ m (Fig. 6.15). The bar is to be designed using an allowable shearing stress $\tau_{all} = 50$ MPa and an allowable angle of twist per unit length $\theta_{all} = 1.5°$ per meter. Determine (a) the maximum permissible applied torque T_{max}, assuming $b = 30$ mm and $G = 80$ GPa, and (b) the corresponding angle of twist between the end sections, ϕ_{max}.

Solution   The values of the torsion parameters from Table 6.2 are

For segment $AC\left(\dfrac{a_1}{b} = 2\right)$: $\qquad \alpha_1 = 0.246 \qquad$ and $\qquad \beta_1 = 0.229$

For segment $CB\left(\dfrac{a_2}{b} = 1.5\right)$: $\qquad \alpha_2 = 0.231 \qquad$ and $\qquad \beta_2 = 0.196$

a. Segment BC governs because it is of smaller depth. The permissible torque T based on the allowable shearing stress is obtained from $\tau_{max} = T/\alpha ab^2$. Thus,

$$T = \alpha_2 a_2 b^2 \tau_{max} = 0.231(0.045)(0.03)^2(50 \times 10^6) = 468 \text{ N} \cdot \text{m}$$

The allowable torque T corresponding to the allowable angle of twist per unit length is determined from $\theta = T/\beta ab^3 G$:

$$T = \beta_2 a_2 b^3 G \theta_{all} = 0.196(0.045)(0.03)^3(80 \times 10^9)\left(\frac{1.5\pi}{180}\right) = 499 \text{ N} \cdot \text{m}$$

The maximum permissible torque, equal to the smaller of the two preceding values, is $T_{max} = 468$ N · m.

b. The angle of twist is equal to the sum of the angles of twist for the two segments:

$$\phi_{max} = \frac{T_{max}}{b^3 G}\left(\frac{L_1}{\beta_1 a_1} + \frac{L_2}{\beta_2 a_2}\right)$$

$$= \frac{468}{(0.03^3)(80 \times 10^9)}\left[\frac{2.5}{0.229(0.06)} + \frac{1.5}{0.196(0.045)}\right]$$

$$= 0.0763 \text{ rad} = 4.37°$$

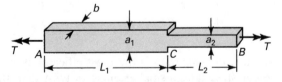

FIGURE 6.15.   *Example 6.6. A stepped rectangular torsion bar.*

Clearly, had the allowable torque been based on the angle of twist per unit length, we would have found that $\phi_{max} = \theta_{all}(L_1 + L_2) = 1.5(4) = 6°$.

6.7 TORSION OF NARROW RECTANGULAR CROSS SECTION

In applying the analogy to a bar of narrow rectangular cross section, it is usual to assume a constant cylindrical membrane shape over the entire dimension b (Fig. 6.16). Subject to this approximation, $\partial z/\partial y = 0$, and Eq. (6.15) reduces to $d^2z/dx^2 = -p/S$, which is twice integrated to yield the parabolic deflection

$$z = \frac{1}{2}\frac{P}{S}\left[\left(\frac{t}{2}\right)^2 - x^2\right] \tag{a}$$

To arrive at Eq. (a), the boundary conditions that $dz/dx = 0$ at $x = 0$ and $z = 0$ at $x = t/2$ have been employed. The volume bounded by the parabolic cylindrical membrane and the xy plane is given by $V = pbt^3/12S$. According to the analogy, p is replaced by 2θ and $1/S$ by G, and consequently $T = 2V = \frac{1}{3}bt^3G\theta$. The torsional rigidity for a thin rectangular section is therefore

$$C = \frac{T}{\theta} = \tfrac{1}{3}bt^3G = J_eG \tag{6.16}$$

Here J_e represents the *effective* polar moment of inertia of the section. The analogy also requires that

$$\tau_{zy} = -\frac{\partial z}{\partial x} = 2G\theta x \tag{b}$$

The angle of twist per unit length is, from Eq. (6.16),

$$\theta = \frac{3T}{bt^3G} \tag{6.17}$$

Maximum shear occurs at $\pm t/2$ is

$$\tau_{max} = G\theta t = \frac{3T}{bt^2} \tag{6.18}$$

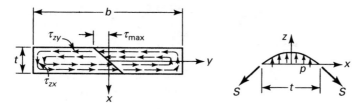

FIGURE 6.16.   *Membrane analogy for a torsional member of narrow rectangular cross section.*

or

$$T = \tfrac{1}{3}bt^2\tau_{max} \tag{6.19}$$

According to Eq. (b), the shearing stress is linear in x, as in Fig. 6.9, producing a twisting moment T about z given by

$$T = 2\left(\tfrac{1}{2}\tau_{max}\frac{t}{2}\right)\left(\frac{t}{3}\right)(b) = \tfrac{1}{6}bt^2\tau_{max}$$

This is exactly one-half the torque given by Eq. (6.19). The remaining applied torque is evidently resisted by the shearing stresses τ_{zx}, neglected in the original analysis in which the membrane is taken as cylindrical. The membrane slope at $y = \pm b/2$ is smaller than that at $x = \pm t/2$ or, equivalently, $(\tau_{zx})_{max} < (\tau_{zy})_{max}$. It is clear, therefore, that Eq. (6.18) represents the maximum shearing stress in the bar, of a magnitude unaffected by the original approximation. That the lower τ_{zx} stresses can provide a resisting torque equal to that of the τ_{zy} stresses is explained on the basis of the longer moment arm for the stresses near $y = \pm b/2$.

Thin-Walled Open Cross Sections

Equations (6.17) and (6.18) are also applicable to thin-walled open sections such as those shown in Fig. 6.17. Because the foregoing expressions neglect stress concentration, the points of interest should be reasonably distant from the corners of the section (Figs. 6.17b and c). The validity of the foregoing approach depends on the degree of similarity between the membrane shape of Fig. 6.16 and that of the geometry of the component section. Consider, for example, the I-section of Fig. 6.17c. Because the section is of varying thickness, the effective polar moment of inertia is written as

$$J_e = \sum \tfrac{1}{3}bt^3 \tag{6.20}$$

or

$$J_e = \tfrac{1}{3}b_1t_1^3 + \tfrac{2}{3}b_2t_2^3$$

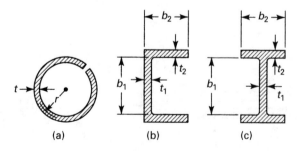

FIGURE 6.17. *Thin-walled open sections.*

We thus have

$$\theta = \frac{T}{J_e G} = \frac{3T}{G} \frac{1}{b_1 t_1^3 + 2b_2 t_2^3} \qquad \text{(6.21a)}$$

$$\tau_{\max} = G\theta t_i = \frac{3T t_i}{b_1 t_1^3 + 2b_2 t_2^3} \qquad \text{(6.21b)}$$

where t_i is the larger of t_1 and t_2. The effect of the stress concentrations at the corners will be examined in Section 6.9.

6.8 TORSION OF MULTIPLY CONNECTED THIN-WALLED SECTIONS

The membrane analogy may be applied to good advantage to analyze the torsion of thin tubular members, provided that some care is taken. Consider the deformation that would occur if a membrane subject to pressure were to span a hollow tube of arbitrary section (Fig. 6.18a). Since the membrane surface is to describe the stress function (and its slope, the stress at any point), arc *ab* cannot represent a meaningful stress function. This is simply because in the region *ab* the stress must be zero, because no material exists there. If the curved surface *ab* is now replaced by a plane representing constant Φ, the zero-stress requirement is satisfied. For bars containing multiply connected regions, each boundary is also a line of constant Φ of different value. The absolute value of Φ is meaningless, and therefore at one boundary, Φ may arbitrarily be equated to zero and the others adjusting accordingly.

Shearing Stress

Based on the foregoing considerations, the membrane analogy is extended to a thin tubular member (Fig. 6.18b), in which the fixed plate to which the membrane is attached has the same contour as the outer boundary of the tube. The membrane is also attached to a "weightless" horizontal plate having the same shape as

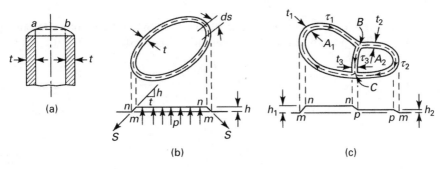

FIGURE 6.18. *Membrane analogy for tubular torsion members: (a) and (b) one-cell section; (c) two-cell section.*

the inner boundary *nn* of the tube, thus bridging the inner and outer contours over a distance *t*. The inner horizontal plate, made "weightless" by a counterbalance system, is permitted to seek its own vertical position but is guided so as not to experience sideward motion. Because we have assumed the tube to be thin walled, the membrane curvature may be disregarded; that is, lines *nn* may be considered straight. We are thus led to conclude that the slope is constant over a given thickness *t*, and consequently the shearing stress is likewise constant, given by

$$\tau = \frac{h}{t} \quad \text{or} \quad h = \tau t \tag{a}$$

where *h* is the membrane deflection and *t* the tube thickness. Note that the tube thickness may vary circumferentially.

The dashed line in Fig. 6.18b indicates the mean perimeter, which may be used to determine the volume bounded by the membrane. Letting *A* represent the *area enclosed by the mean perimeter*, the volume *mnnm* is simply *Ah*, and the analogy gives

$$T = 2Ah \quad \text{or} \quad h = \frac{T}{2A} \tag{b}$$

Combining Eqs. (a) and (b), we have

$$\tau = \frac{T}{2At} \tag{6.22}$$

The application of Eq. (6.22) is limited to thin-walled members displaying no abrupt variations in thickness and no reentrant corners in their cross sections.

Angle of Twist

To develop a relationship for the angle of twist from the membrane analogy, we again consider Fig. 6.18b, in which $h \ll t$ and consequently $\tan(h/t) \approx h/t = \tau$. Vertical equilibrium therefore yields

$$pA = \oint \left(\frac{h}{t} S\right) ds = \oint \tau S \, ds$$

Here *s* is the *length of the mean perimeter* of the tube. Since the membrane tension is constant, *h* is independent of *S*. The preceding is then written

$$\frac{P}{S} = \frac{1}{A} \oint \tau \, ds = \frac{h}{A} \oint \frac{ds}{t} = 2G\theta$$

where the last term follows from the analogy. The angle of twist per unit length is now found directly:

$$\theta = \frac{h}{2AG} \oint \frac{ds}{t} = \frac{1}{2AG} \oint \tau \, ds \tag{6.23}$$

Equations (6.22) and (6.23) are known as *Bredt's formulas*.

In Eqs. (a), (b), and (6.23), the quantity h possesses the dimensions of force per unit length, representing the resisting force per unit length along the tube perimeter. For this reason, h is referred to as the *shear flow*.

EXAMPLE 6.7   Rectangular Torsion Tube

An aluminum tube of rectangular cross section (Fig. 6.19a) is subjected to a torque of 56.5 kN·m along its longitudinal axis. Determine the shearing stresses and the angle of twist. Assume $G = 28$ GPa.

Solution   Referring to Fig. 6.19b, which shows the membrane surface *mnnm* (representing Φ), the applied torque is, according to Eq. (b),

$$T = 2Ah = 2(0.125h) = 56,500 \text{ N} \cdot \text{m}$$

from which $h = 226,000$ N/m. The shearing stresses are found from Eq. (a) as follows:

$$\tau_1 = \frac{h}{t_1} = \frac{226,000}{0.012} = 18.83 \text{ MPa}$$

$$\tau_2 = \frac{h}{t_2} = \frac{h}{t_4} = \frac{226,000}{0.006} = 37.67 \text{ MPa}$$

$$\tau_3 = \frac{h}{t_3} = \frac{226,000}{0.01} = 22.6 \text{ MPa}$$

Applying Eq. (6.23), the angle of twist per unit length is

$$\theta = \frac{h}{2AG} \oint \frac{ds}{t} = \frac{226,000}{2 \times 28 \times 10^9 \times 0.125} \left(\frac{0.25}{0.012} + 2\frac{0.5}{0.006} + \frac{0.25}{0.01} \right)$$
$$= 0.00686 \text{ rad/m}$$

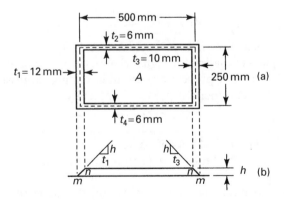

FIGURE 6.19.   *Example 6.7. (a) Cross section; (b) membrane.*

If multiply connected regions exist within a tubular member, as in Fig. 6.18c, the foregoing techniques are again appropriate. As before, the thicknesses are assumed small, so lines such as *mn, np*, and *pm* are regarded as straight. The stress function is then represented by the membrane surface *mnnppm*. As in the case of a simple hollow tube, lines *nn* and *pp* are straight by virtue of flat, weightless plates with contours corresponding to the inner openings. Referring to the figure, the shearing stresses are

$$\tau_1 = \frac{h_1}{t_1}, \qquad \tau_2 = \frac{h_2}{t_2} \tag{c}$$

$$\tau_3 = \frac{h_1 - h_2}{t_3} = \frac{t_1\tau_1 - t_2\tau_2}{t_3} \tag{d}$$

The stresses are produced by a torque equal to twice the volume beneath surface *mnnppm*,

$$T = 2A_1 h_1 + 2A_2 h_2 \tag{e}$$

or, upon substitution of Eqs. (c),

$$T = 2A_1 t_1 \tau_1 + 2A_2 t_2 \tau_2 \tag{f}$$

Assuming the thicknesses t_1, t_2, and t_3 constant, application of Eq. (6.23) yields

$$\tau_1 s_1 + \tau_3 s_3 = 2G\theta A_1 \tag{g}$$
$$\tau_2 s_2 - \tau_3 s_3 = 2G\theta A_2 \tag{h}$$

where s_1, s_2, and s_3 represent the paths of integration indicated by the dashed lines. Note the relationship among the algebraic sign, the assumed direction of stress, and the direction in which integration proceeds. There are thus four equations [(d), (f), (g), and (h)] containing four unknowns: τ_1, τ_2, τ_3, and θ.

If now Eq. (d) is written in the form

$$\tau_1 t_1 = \tau_2 t_2 + \tau_3 t_3 \tag{i}$$

it is observed that the shear flow $h = \tau t$ is constant and distributes itself in a manner analogous to a liquid circulating through a channel of shape identical with that of the tubular bar. This analogy proves very useful in writing expressions for shear flow in tubular sections of considerably greater complexity.

EXAMPLE 6.8   Three-Cell Torsion Tube

A multiply connected steel tube (Fig. 6.20) resists a torque of 12 kN·m. The wall thicknesses are $t_1 = t_2 = t_3 = 6$ mm and $t_4 = t_5 = 3$ mm. Determine the maximum shearing stresses and the angle of twist per unit length. Let $G = 80$ GPa. Dimensions are given in millimeters.

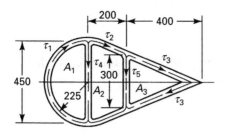

FIGURE 6.20. *Example 6.8. Three-cell section.*

Solution Assuming the shearing stresses directed as shown, consideration of shear flow yields

$$\tau_1 t_1 = \tau_2 t_2 + \tau_4 t_4, \qquad \tau_2 t_2 = \tau_5 t_5 + \tau_3 t_3 \qquad \textbf{(j)}$$

The torque associated with the shearing stresses must resist the
externally applied torque, and an expression similar to Eq. (f) is obtained:

$$T = 2A_1 t_1 \tau_1 + 2A_2 t_2 \tau_2 + 2A_3 t_3 \tau_3 = 12{,}000 \text{ N} \cdot \text{m} \qquad \textbf{(k)}$$

Three more equations are available through application of Eq. (6.23)
over areas A_1, A_2, and A_3:

$$\tau_1 s_1 + \tau_4 s_4 = 2G\theta A_1$$
$$-\tau_4 s_4 + 2\tau_2 s_2 + \tau_5 s_5 = 2G\theta A_2 \qquad \textbf{(l)}$$
$$-\tau_5 s_5 + 2\tau_3 s_3 = 2G\theta A_3$$

Here, $s_1 = 0.7069$ m, $s_2 = 0.2136$ m, $s_3 = 0.4272$ m, $s_4 = 0.45$ m,
$s_5 = 0.3$ m, $A_1 = 0.079522$ m², $A_2 = 0.075$ m², and $A_3 = 0.06$ m².
There are six equations in the five unknown stresses and the angle of
twist per unit length. Thus, simultaneous solution of Eqs. (j), (k), and (l)
leads to the following rounded values: $\tau_1 = 4.902$ MPa, $\tau_2 = 5.088$ MPa,
$\tau_3 = 3.809$ MPa, $\tau_4 = -0.373$ MPa, $\tau_5 = 2.558$ MPa, and $\theta = 0.0002591$
rad/m. The positive values obtained for τ_1, τ_2, τ_3, and τ_5 indicate that the
directions of these stresses have been correctly assumed in Fig. 6.20. The
negative sign of τ_4 means that the direction initially assumed was incor-
rect; that is, τ_4 is actually upward directed.

6.9 FLUID FLOW ANALOGY AND STRESS CONCENTRATION

Examination of Eq. 6.8 suggests a similarity between the stress function Φ and the
stream function ψ of fluid mechanics:

$$\tau_{zx} = \frac{\partial \Phi}{\partial y}, \qquad \tau_{zy} = -\frac{\partial \Phi}{\partial x}$$
$$v_x = \frac{\partial \psi}{\partial y}, \qquad v_y = -\frac{\partial \psi}{\partial x} \qquad \textbf{(6.24)}$$

In Eqs. (6.24), v_x and v_y represent the x and y components of the fluid velocity v.
Recall that, for an incompressible fluid, the equation of continuity may be written

$$\frac{\partial v_x}{\partial x} + \frac{\partial v_y}{\partial y} = 0$$

Continuity is thus satisfied when $\psi(x, y)$ is defined as in Eqs. (6.24). The vorticity $\omega = \frac{1}{2}(\nabla \times \mathbf{v})$ is for two-dimensional flow,

$$\omega = \frac{1}{2}\left(-\frac{\partial v_x}{\partial y} + \frac{\partial v_y}{\partial x} \right)$$

where $\nabla = (\partial/\partial x)\mathbf{i} + (\partial/\partial y)\mathbf{j}$. In terms of the stream function, we obtain

$$\frac{\partial^2 \psi}{\partial y^2} + \frac{\partial^2 \psi}{\partial x^2} = -2\omega \tag{6.25}$$

The expression is clearly analogous to Eq. (6.9) with -2ω replacing $-2G\theta$. The completeness of the analogy is assured if it can be demonstrated that ψ is constant along a streamline (and hence on a boundary), just as Φ is constant over a boundary. Since the equation of a streamline in two-dimensional flow is

$$\frac{dy}{dx} = \frac{v_y}{v_x} \qquad \text{or} \qquad v_x\, dy - v_y\, dx = 0$$

in terms of the stream function, we have

$$\frac{\partial \psi}{\partial y}dy + \frac{\partial \psi}{\partial x}dx = 0 \tag{6.26}$$

This is simply the total differential $d\psi$, and therefore ψ is constant along a streamline.

Based on the foregoing, experimental techniques have been developed in which the analogy between the motion of an ideal fluid of constant vorticity and the torsion of a bar is successfully exploited. The tube in which the fluid flows and the cross section of the twisted member are identical in these experiments and useful in visualizing stress patterns in torsion. Moreover, a vast body of literature exists that deals with flow patterns around bodies of various shapes, and the results presented are often directly applicable to the torsion problem.

The *fluid flow* or *hydrodynamic analogy* is especially valuable in dealing with *stress concentration in shear*, which we have heretofore neglected. In this regard, consider first the torsion of a circular bar containing a small circular hole (Fig. 6.21a). Figure 6.21b shows the analogous flow pattern produced by a solid

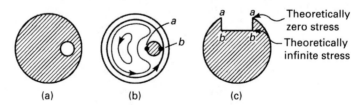

FIGURE 6.21.   *(a) Circular shaft with hole; (b) fluid-flow pattern around small cylindrical obstacle; (c) circular shaft with keyway.*

cylindrical obstacle placed in a circulating fluid. From hydrodynamic theory, it is found that the maximum velocity (at points a and b) is twice the value in the undisturbed stream at the respective radii. From this, it is concluded that a small hole has the effect of doubling the shearing stress normally found at a given radius.

Of great importance also is the shaft keyway shown in Fig. 6.21c. According to the hydrodynamic analogy, the points a ought to have zero stress, since they are stagnation points of the fluid stream. In this sense, the material in the immediate vicinity of points a is excess. On the other hand, the velocity at the points b is theoretically infinite and, by analogy, so is the stress. It is therefore not surprising that most torsional fatigue failures have their origins at these sharp corners, and the lesson is thus supplied that it is profitable to round such corners.

6.10 TORSION OF RESTRAINED THIN-WALLED MEMBERS OF OPEN CROSS SECTION

It is a basic premise of previous sections of this chapter that all cross sections of a bar subject to torques applied at the ends suffer free warpage. As a consequence, we must assume that the torque is produced by pure shearing stresses distributed over the ends as well as all other cross sections of the member. In this way, the stress distribution is obtained from Eq. (6.9) and satisfies the boundary conditions, Eq. (6.10).

If any section of the bar is held rigidly, it is clear that the rate of change of the angle of twist as well as the warpage will now vary in the longitudinal direction. The longitudinal fibers are therefore subject to tensile or compressive stresses. Equations (6.9) and (6.10) are, in this instance, applied with satisfactory results in regions away from the restrained section of the bar. While this restraint has negligible influence on the torsional resistance of members of solid section such as rectangles and ellipses, it is significant when dealing with open thin-walled sections such as channels of I-beams. Consider, for example, the case of a cantilever I-beam, shown in Fig. 6.22a. The applied torque causes each cross section to rotate about the axis of twist (z), thereby resulting in bending of the flanges. According to beam theory, the associated bending stresses in the flanges are zero at the juncture with the web. Consequently, the web does not depart from a state of simple torsion. In resisting the bending of the flanges or the warpage of a cross section, considerable torsional stiffness can, however, be imparted to the beam.

Torsional and Lateral Shears

Referring to Fig. 6.22a, the applied torque T is balanced in part by the action of torsional shearing stresses and in part by the resistance of the flanges to bending. At the representative section AB (Fig. 6.22b), consider the influence of torques T_1 and T_2. The former is attributable to pure torsional shearing stresses in the entire cross section, assumed to occur as though each cross section were free to warp. Torque T_1 is thus related to the angle of twist of section AB by the expression

$$T_1 = C \frac{d\phi}{dz} \tag{a}$$

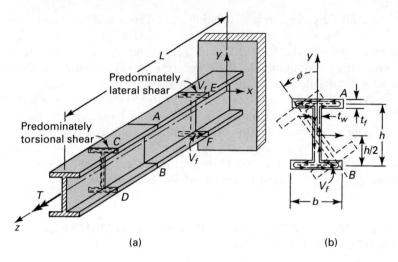

FIGURE 6.22. *I-section torsion member: (a) warping is prevented of the section at x = 0; (b) partly lateral shear and partly torsional shear at arbitrary cross section AB.*

in which C is the torsional rigidity of the beam. The right-hand rule should be applied to furnish the sign convention for both torque and angle of twist. A pair of lateral shearing forces *owing to bending of the flanges* acting through the moment arm h gives rise to torque T_2:

$$T_2 = V_f h \qquad \textbf{(b)}$$

An expression for V_f may be derived by considering the x displacement, u. Because the beam cross section is symmetrical and the deformation small, we have $u = (h/2)\phi$, and

$$\frac{du}{dz} = \frac{h}{2}\frac{d\phi}{dz} \qquad \textbf{(c)}$$

Thus, the bending moment M_f and shear V_f in the flange are

$$M_f = EI_f \frac{d^2u}{dz^2} = \frac{EI_f h}{2}\frac{d^2\phi}{dz^2} \qquad \textbf{(d)}$$

$$V_f = -EI_f \frac{d^3u}{dz^3} = -\frac{EI_f h}{2}\frac{d^3\phi}{dz^3} \qquad \textbf{(e)}$$

where I_f is the moment of inertia of one flange about the y axis. Now Eq. (b) becomes

$$T_2 = -\frac{EI_f h^2}{2}\frac{d^3\phi}{dz^3} \qquad \textbf{(f)}$$

The total torque is therefore

$$T = T_1 + T_2 = C\frac{d\phi}{dz} - \frac{EI_f h^2}{2}\frac{d^3\phi}{dz^3} \qquad \textbf{(6.27)}$$

Boundary Conditions

The conditions appropriate to the flange ends are

$$\left(\frac{d\phi}{dz}\right)_{z=0} = 0, \qquad \left(\frac{d^2\phi}{dz^2}\right)_{z=L} = 0$$

indicating that the slope and bending moment are zero at the fixed and free ends, respectively. The solution of Eq. (6.27) is, upon satisfying these conditions,

$$\frac{d\phi}{dz} = \frac{T}{C}\left[1 - \frac{\cosh\alpha(L-z)}{\cosh\alpha L}\right] \tag{g}$$

where

$$\alpha = \left(\frac{2C}{EI_f h^2}\right)^{1/2} \tag{6.28}$$

Long Beams

For a beam of *infinite length*, Eq. (g) reduces to

$$\frac{d\phi}{dz} = \frac{T}{C}(1 - e^{-\alpha z}) \tag{h}$$

By substituting Eq. (h) into Eqs. (a) and (f), the following expressions result:

$$\begin{aligned} T_1 &= T(1 - e^{-\alpha z}) \\ T_2 &= Te^{-\alpha z} \end{aligned} \tag{6.29}$$

From this, it is noted that at the fixed end $(z = 0)$ $T_1 = 0$ and $T_2 = T$. At this end, the applied torque is counterbalanced by the effect of shearing forces only, which from Eq. (b) are given by $V_f = T_2/h = T/h$. The torque distribution, Eq. (6.29), indicates that sections such as EF, close to the fixed end, contain predominantly lateral shearing forces (Fig. 6.22a). Sections such as CD, near the free end, contain mainly torsional shearing stresses (as Eq. 6.29 indicates for $z \to \infty$).

The flange bending moment, obtained from Eqs. (d) and (h), is a maximum at $z = 0$:

$$M_{f,\,\text{max}} = \frac{EI_f h\alpha}{2C}T \tag{i}$$

The maximum bending moment, occurring at the fixed end of the flange, is found by substituting the relations (6.28) into *(i)*:

$$M_{f,\,\text{max}} = \frac{T}{\alpha h} \tag{6.30}$$

An expression for the **angle of twist** is determined by integrating Eq. (h) and satisfying the condition $\phi = 0$ at $z = 0$:

$$\phi = \frac{T}{C}\left[z + \frac{1}{\alpha}(e^{-\alpha z} - 1)\right]$$ (j)

For relatively long beams, for which $e^{-\alpha z}$ may be neglected, the total angle of twist at the free end is, from Eq. (j),

$$\phi_{z=L} = \frac{T}{C}\left(L - \frac{1}{\alpha}\right)$$ (6.31)

In this equation the term $1/\alpha$ indicates the influence of flange bending on the angle of twist. Since for pure torsion the total angle of twist is given by $\phi = TL/C$, it is clear that end restraint increases the stiffness of the beam in torsion.

EXAMPLE 6.9 Analysis of I-Beam Under Torsion

A cantilever I-beam with the idealized cross section shown in Fig. 6.22 is subjected to a torque of 1.2 kN·m. Determine (a) the maximum longitudinal stress, and (b) the total angle of twist, ϕ. Take $G = 80$ GPa and $E = 200$ GPa. Let $t_f = 10$ mm, $t_w = 7$ mm, $b = 0.1$ m, $h = 0.2$ m, and $L = 2.4$ m.

Solution

a. The torsional rigidity of the beam is, from Eq. (6.21a),

$$C = \frac{T}{\theta} = (b_1 t_1^3 + 2b_2 t_2^3)\frac{G}{3} = (0.19 \times 0.007^3 + 2 \times 0.1 \times 0.01^3)\frac{G}{3}$$

$$= 8.839 \times 10^{-8}G$$

The flexural rigidity of one flange is

$$I_f E = \frac{0.01 \times 0.1^3}{12}E = 8.333 \times 10^{-7}E$$

Hence, from Eq. (6.28) we have

$$\frac{1}{\alpha} = h\sqrt{\frac{EI_f}{2}} = h\sqrt{\frac{2.5 \times 8.333 \times 10^{-7}}{8.839 \times 10^{-8} \times 2}} = 3.43h$$

From Eq. (6.30), the bending moment in the flange is found to be 3.43 times larger than the applied torque, T. Thus, the maximum longitudinal bending stress in the flange is

$$\sigma_{f,\max} = \frac{M_{f,\max}\, x}{I_f} = \frac{3.43T \times 0.05}{8.333 \times 10^{-7}} = 0.2058 \times 10^6 T = 247 \text{ MPa}$$

b. Since $e^{-\alpha L} = 0.03$, we can apply Eq. (6.31) to calculate the angle of twist at the free end:

$$\phi = \frac{T}{C}\left(L - \frac{1}{\alpha}\right) = \frac{1200}{8.839 \times 10^{-8} \times 80 \times 10^9}(2.4 - 3.43 \times 0.2)$$

$$= 0.2908 \text{ rad}$$

It is interesting to note that if the ends of the beam were both free, the total angle of twist would be $\phi = TL/C = 0.4073$ rad, and the beam would experience $\phi_{free}/\phi_{fixed} = 1.4$ times more twist under the same torque.

6.11 CURVED CIRCULAR BARS: HELICAL SPRINGS

The assumptions of Section 6.2 are also valid for a *curved, circular bar*, provided that the radius r of the bar is small in comparison with the radius of curvature R. When $r/R = \frac{1}{12}$, for example, the maximum stress computed on the basis of the torsion formula, $\tau = Tr/J$, is approximately 5% too low. On the other hand, if the diameter of the bar is large relative to the radius of curvature, the length differential of the longitudinal surface elements must be taken into consideration, and there is a stress concentration at the inner point of the bar. We are concerned here with the torsion of slender curved members for which $r/R \ll 1$.

Frequently, a curved bar is subjected to loads that at any cross section produce a twisting moment as well as a bending moment. Expressions for the strain energy in torsion and bending have already been developed (Secs. 2.14 and 5.12), and application of Castigliano's theorem (Sec. 10.4) leads readily to the displacements. Note that, in the design of springs manufactured from curved bars or wires, deflection is as important as strength. Springs are employed to apply forces or torques in a mechanism or to absorb the energy owing to suddenly applied loads.*

Consider the case of a cylindrical rod or bar bent into a quarter-circle of radius R, as shown in Fig. 6.23a. The rod is fixed at one end and loaded at the free end by a twisting moment T. The bending and twisting moments at any section are (Fig. 6.23b)

$$M_\theta = T \sin \theta, \qquad T_\theta = -T \cos \theta \qquad \text{(a)}$$

Substituting these quantities into Eqs. (2.63) and (2.61) together with $dx = ds = R\, d\theta$ yields

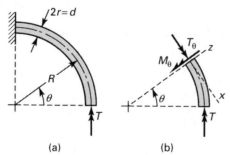

FIGURE 6.23.   *Twisting of a curved, circular bar.*

---

*For an extensive discussion of springs, see Refs. 6.5 and 6.6.

$$U = \frac{1}{2} \int \left(\frac{M_\theta^2}{EI} + \frac{T_\theta^2}{GJ} \right) ds \qquad \textbf{(b)}$$

or

$$U = \tfrac{1}{2}T^2R \int_0^{\pi/2} \left(\frac{\sin^2\theta}{EI} + \frac{\cos^2\theta}{GJ} \right) d\theta$$

where $J = \pi d^4/32 = 2I$. The strain energy in the entire rod is obtained by integrating:

$$U = \frac{8(2 + \nu)R}{d^4 E} T^2 \qquad \textbf{(6.32)}$$

Upon application of $\phi = \partial U/\partial T$, it is found that

$$\phi = \frac{16(2 + \nu)R}{d^4 E} T \qquad \textbf{(6.33)}$$

for the *angle of twist* at the free end.

A helical spring, produced by wrapping a wire around a cylinder in such a way that the wire forms a helix of uniformly spaced turns, as typifies a curved bar, is discussed in the following example.

EXAMPLE 6.10 Analysis of a Helical Spring

An *open-coiled helical spring* wound from wire of diameter d, with *pitch angle* α and n *number of coils* of radius R, is extended by an axial load P (Fig. 6.24). (a) Develop expressions for maximum stress and deflection. (b) Redo part (a) for the spring *closely coiled*. *Assumption:* Load is applied steadily at a rigid hook and loop at ends.

Solution An element of the spring located between two adjoining sections of the wire may be treated as a straight circular bar in torsion and bending. This is because a tangent to the coil at any point such as A is not perpendicular to the load. At cross section A, components $P \cos \alpha$ and $P \sin \alpha$ produce the following respective torque and moment:

$$T = PR \cos \alpha, \qquad M = PR \sin \alpha \qquad \textbf{(c)}$$

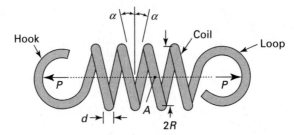

FIGURE 6.24. *Example 6.10. Helical spring under tension.*

a. The stresses, from Eq. (4.7), are given by

$$\sigma_{1,2} = \frac{16}{\pi d^3}(M \pm \sqrt{M^2 + T^2}), \qquad \tau_{max} = \frac{16}{\pi d^3}\sqrt{M^2 + T^2}$$

The *maximum normal stress* and the *maximum shear stress* are thus

$$\sigma_{max} = \frac{16PR}{\pi d^3}(1 + \sin\alpha) \qquad \text{(6.34a)}$$

and

$$\tau_{max} = \frac{16PR}{\pi d^3} \qquad \text{(6.34b)}$$

The deflection is computed by applying Castigliano's theorem together with Eqs. (b) and (c):

$$\delta = \int_0^L \left(\frac{PR^2 \sin^2\alpha}{EI} + \frac{PR^2 \cos^2\alpha}{JG}\right) ds$$

where the length of the coil $L = 2\pi Rn$. It follows that the relationship

$$\delta = 2\pi PR^3 n\left(\frac{\sin^2\alpha}{EI} + \frac{\cos^2\alpha}{JG}\right)$$

or

$$\delta = \frac{128PR^3 n}{d^4}\left(\frac{\sin^2\alpha}{E} + \frac{\cos^2\alpha}{2G}\right) \qquad \text{(6.35)}$$

defines the *axial end deflection* of an open-coil helical spring.

b. For a *closely coiled helical spring*, the angle of pitch α of the coil is very small. The deflection is now produced *entirely* by the torsional stresses induced in the coil. To derive expressions for the stress and deflection, let $\alpha = \sin\alpha = 0$ and $\cos\alpha = 1$ in Eqs. (6.34) and (6.35). In so doing, we obtain

$$\tau_{max} = \sigma_{max} = \frac{16PR}{\pi d^3} \qquad \text{(6.36)}$$

and

$$\delta = \frac{64PR^3 n}{Gd^4} \qquad \text{(6.37)}$$

Comment The foregoing results are applicable to *both tension and compression* helical springs, the wire diameters of which are small in relation to coil radius.

REFERENCES

6.1. UGURAL, A. C., *Mechanics of Materials*, Wiley, Hoboken, N. J., 2008, Sec. 6. 2.

6.2. TODHUNTER, I. and PEARSON, K., *History of the Theory of Elasticity and the Strength of Materials*, Dover, New York, 1960, Vols. I and II.

6.3. TIMOSHENKO, S. P., and GOODIER, J. N., *Theory of Elasticity*, 3rd ed., McGraw-Hill, New York, 1970.

6.4. YOUNG, W. C. and BUDYNAS, R. G., *Roark's Formulas for Stress and Strain*, 6th ed., McGraw-Hill, New York, 1989.

6.5. UGURAL, A. C., *Mechanical Design: An Integrated Approach*, McGraw-Hill, New York, 2004, Chap. 14.

6.6. WAHL, A. M., *Mechanical Springs*, McGraw-Hill, New York, 1963.

PROBLEMS

Sections 6.1 through 6.3

6.1. A hollow steel shaft of outer radius $c = 35$ mm is fixed at one end and subjected to a torque $T = 3$ kN $\cdot$ m at the other end. Calculate the required inner radius b, knowing that the average shearing stress is limited to 100 MPa.

6.2. A solid shaft of 40-mm diameter is to be replaced by a hollow circular tube of the same material, resisting the same maximum shear stress and the same torque. Determine the outer diameter D of the tube for the case in which its wall thickness is $t = D/25$.

6.3. A solid shaft of diameter d and a hollow shaft of outer diameter $D = 60$ mm and thickness $t = D/4$ are to transmit the same torque at the same maximum shear stress. What is the required diameter d the shaft?

6.4. Figure P6.4 shows four pulleys, attached to a solid stepped shaft, transmit the torques. Find the maximum shear stress for each shaft segment.

6.5. Resolve Prob. 6.4, for the case in which a hole of 16-mm diameter drilled axially through the shaft to form a tube.

6.6. As seen in Fig. P6.6, a stepped shaft ABC with built-in end at A is subjected to the torques $T_B = 3$ kN $\cdot$ m and $T_C = 1$ kN $\cdot$ m at sections B and C. Based on a stress concentration factor $K = 1.6$ at the step B, what is the maximum shearing stress in the shaft? *Given:* $d_1 = 50$ mm and $d_2 = 40$ mm.

6.7. A brass rod AB ($G_b = 42$ GPa) is bonded to an aluminum rod BC ($G_d = 28$ GPa), as illustrated in Fig. P6.6. Determine the angle of twist at C, for the case in which $T_B = 2T_C = 8$ kN $\cdot$ m, $d_1 = 2d_2 = 100$ mm, and $L_1 = 2L_2 = 0.7$ m.

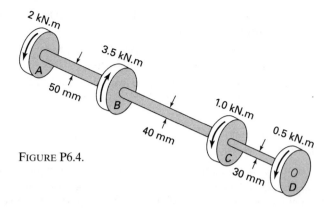

FIGURE P6.4.

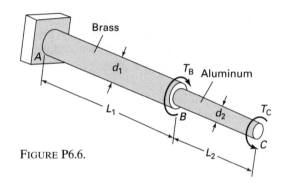

FIGURE P6.6.

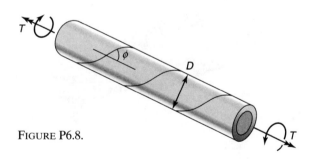

FIGURE P6.8.

6.8. A hollow shaft is made by rolling a metal plate of thickness t into a cylindrical form and welding the edges along the helical seams oriented at an angle of ϕ to the axis of the member (Fig. P6.8). Calculate the maximum torque that can be applied to the shaft. *Assumption:* The allowable tensile and shear stresses in the weld are 120 MPa and 50 MPa, respectively. *Given:* $D = 100$ mm, $t = 5$ mm, and $\phi = 50°$.

6.9. Resolve Prob. 6.8 for the case in which the helical seam is oriented at an angle of $\phi = 35°$ to the axis of the member (Fig. P6.8).

6.10. What is the required diameter d_1 for the segment AB of the shaft illustrated in Fig. P6.6 if the permissible shear stress is $\tau_{all} = 50$ MPa and the total angle of twist between A and C is limited to $\phi = 0.02$ rad? *Given:* $G = 39$ GPa, $T_B = 3$ kN·m, $T_C = 1$ kN·m, $L_1 = 2$ m, $L_2 = 1$ m, and $d_2 = 25$ mm.

6.11. Redo Prob. 6.10 for the case in which the torque applied at B is $T_B = 2$ kN·m and $T_C = 0.5$ kN·m.

6.12. A stepped shaft of diameters D and d is under a torque T, as shown in Fig. P6.12. The shaft has a fillet of radius r (see Fig. D.4). Determine (a) the maximum shear stress in the shaft for $r = 1.0$ mm; (b) the maximum shear stress in the shaft for $r = 5$ mm. *Given:* $D = 60$ mm, $d = 50$ mm, and $T = 2$ kN·m.

6.13. A stepped shaft having solid circular parts with diameters D and d is in pure torsion (Fig. P6.12). The two parts are joined with a fillet of radius r (see Fig. D.4). If the shaft is made of brass with allowable shear strength 80 MPa, determine the largest torque capacity of the shaft. *Given:* $D = 100$ mm, $d = 50$ mm, $r = 10$ mm.

6.14. Consider two bars, one having a circular section of radius b, the other an elliptic section with semiaxes a, b (Figure P6.14). Determine (a) for equal angles of twist, which bar experiences the larger shearing stress, and (b) for equal allowable shearing stresses, which bar resists a larger torque.

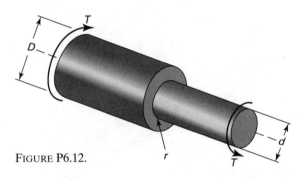

FIGURE P6.12.

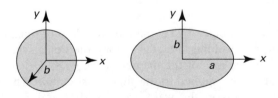

FIGURE P6.14.

6.15. A hollow $(r_i = b, r_0 = c)$ and a solid $(r_0 = a)$ cylindrical shaft are constructed of the same material. The shafts are of identical length and cross-sectional area and both are subjected to pure torsion. Determine the ratio of the largest torques that may be applied to the shafts for $c = 1.4b$ (a) if the allowable stress is τ_a and (b) if the allowable angle of twist is θ_a.

6.16. A solid circular shaft AB, held rigidly at both ends, has two different diameters (Fig. 6.4a). For a maximum permissible shearing stress $\tau_{all} = 150$ MPa, calculate the allowable torque T that may be applied at section C. Use $d_a = 20$ mm, $d_b = 15$ mm, and $a = 2b = 0.4$ m.

6.17. Redo Problem 6.16 for $\tau_{all} = 70$ MPa, $d_a = 25$ mm, $d_b = 15$ mm, and $a = 1.6b = 0.8$ m.

Sections 6.4 through 6.7

6.18. The stress function appropriate to a solid bar subjected to torques at its free ends is given by

$$\Phi = k(a^2 - x^2 + by^2)(a^2 + bx^2 - y^2)$$

where a and b are constants. Determine the value of k.

6.19. Show that Eqs. (6.6) through (6.11) are not altered by a shift of the origin of x, y, z from the center of twist to any point within the cross section defined by $x = a$ and $y = b$, where a and b are constants. [Hint: The displacements are now expressed as $u = -\theta z(y - b)$, $v = \theta z(x - a)$, and $w = w(x, y)$.]

6.20. Rederive Eq. (6.11) for the case in which the stress function $\Phi = c$ on the boundary, where c is a nonzero constant.

6.21. The thin circular ring of cross-sectional radius r, shown in Fig. P6.21, is subjected to a distributed torque per unit length, $T_\theta = T\cos^2\theta$. Determine the angle of twist at sections A and B in terms of T, a, and r. Assume that the radius a is large enough to permit the effect of curvature on the torsion formula to be neglected.

6.22. Consider two bars of the same material, one circular of radius c, the other of rectangular section with dimensions $a \times 2a$. Determine the radius c so

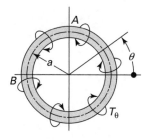

FIGURE P6.21.

that, for an applied torque, both the maximum shear stress and the angle of twist will not exceed the corresponding quantities in the rectangular bar.

6.23. The torsion solution for a cylinder of equilateral triangular section (Fig. 6.11) is derivable from the stress function, Eq. (j) of Section 6.5:

$$\Phi = k(x - \sqrt{3}y - \tfrac{2}{3}h)(x + \sqrt{3}y - \tfrac{2}{3}h)(x + \tfrac{1}{3}h)$$

Derive expressions for the maximum and minimum shearing stresses and the twisting angle.

6.24. The torsional rigidity of a circle, an ellipse, and an equilateral triangle (Fig. 6.11) are denoted by C_c, C_e, and C_t, respectively. If the cross-sectional areas of these sections are equal, demonstrate that the following relationships exist:

$$C_e = \frac{2ab}{a^2 + b^2} C_c, \qquad C_t = \frac{2\pi\sqrt{3}}{15} C_c$$

where a and b are the semiaxes of the ellipse in the x and y directions.

6.25. Two thin-walled circular tubes, one having a seamless section, the other (Fig. 6.17a) a split section, are subjected to the action of identical twisting moments. Both tubes have equal outer diameter d_o, inner diameter d_i, and thickness t. Determine the ratio of their angles of twist.

6.26. A steel bar of slender rectangular cross section (5 mm × 125 mm) is subjected to twisting moments of 80 N · m at the ends. Calculate the maximum shearing stress and the angle of twist per unit length. Take $G = 80$ GPa.

6.27. The torque T produces a rotation of 15° at free end of the steel bar shown in Fig. P6.27. Use $a = 24$ mm, $b = 16$ mm, $L = 400$ m, and $G = 80$ GPa. What is the maximum shearing stress in the bar?

6.28. Determine the largest permissible $b \times b$ square cross section of a steel shaft of length $L = 3$ m, for the shearing stress is not to exceed 120 MPa and the shaft is twisted through 25° (Fig. P6.27). Take $G = 75$ GPa.

6.29. A steel bar ($G = 200$ GPa) of cross section as shown in Fig. P6.29 is subjected to a torque of 500 N · m. Determine the maximum shearing stress and the angle of twist per unit length. The dimensions are $b_1 = 100$ mm, $b_2 = 125$ mm, $t_1 = 10$ mm, and $t_2 = 4$ mm.

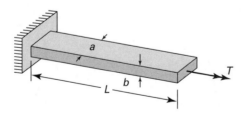

FIGURE P6.27.

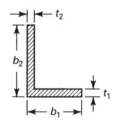

FIGURE P6.29.

6.30. Rework Example 6.4 for the case in which segments AC and CB of torsion member AB have as a cross section an equilateral triangular of sides $a_1 = 60$ mm and $a_2 = 45$ mm, respectively.

6.31. Derive an approximate expression for the twisting moment in terms of G, θ, b, and t_o for the thin triangular section shown in Fig. P6.31. Assume that at any y the expression for the stress function Φ corresponds to a parabolic membrane appropriate to the width at that y: $\Phi = G\theta[(t/2)^2 - x^2]$.

6.32. Consider the following sections: (a) a hollow tube of 50-mm outside diameter and 2.5-mm wall thickness, (b) an equal-leg angle, having the same perimeter and thickness as in (a), (c) a square box section with 50-mm sides and 2.5-mm wall thickness. Compare the torsional rigidities and the maximum shearing stresses for the same applied torque.

6.33. The cross section of a 3-m-long steel bar is an equilateral triangle with 50-mm sides. The bar is subjected to end twisting couples causing a maximum shearing stress equal to two-thirds of the elastic strength in shear ($\tau_{yp} = 420$ MPa). Using Table 6.2, determine the angle of twist between the ends. Let $G = 80$ GPa.

Sections 6.8 through 6.11

6.34. Show that when Eq. (6.2) is applied to a thin-walled tube, it reduces to Eq. (6.22).

6.35. A torque T is applied to a thin-walled tube of a cross section in the form of a regular hexagon of constant wall thickness t and mean side length a. Derive relationships for the shearing stress τ and the angle of twist θ per unit length.

6.36. Redo Example 6.7 with a 0.01-m-thick vertical wall at the middle of the section.

6.37. The cross section of a thin-walled aluminum tube is an equilateral triangular section of mean side length 50 mm and wall thickness 3.5 mm. If the tube is subjected to a torque of 40 N·m, what are the maximum shearing stress and angle of twist per unit length? Let $G = 28$ GPa.

6.38. A square thin-walled tube of mean dimensions $a \times a$ and a circular thin-walled tube of mean radius c, both of the same material, length, thickness t, and cross-sectional area, are subjected to the same torque. Determine the ratios of the shearing stresses and the angle of twist of the tubes.

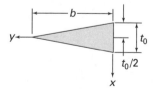

FIGURE P6.31.

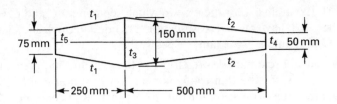

FIGURE P6.39.

6.39. A hollow, multicell aluminum tube (cross section shown in Fig. P6.39) resists a torque of 4 kN · m. The wall thicknesses are $t_1 = t_2 = t_4 = t_5 = 0.5$ mm and $t_3 = 0.75$ mm. Determine the maximum shearing stresses and the angle of twist per unit length. Let $G = 28$ GPa.

6.40. Consider two closely coiled helical springs, one made of steel, the other of copper, each 0.01 m in wire diameter, one fitting within the other. Each has an identical number of coils, $n = 20$, and ends constrained to deflect the same amount. The steel outer spring has a diameter of 0.124 m, and the copper inner spring, a diameter of 0.1 m. Determine (a) the total axial load the two springs can jointly sustain if the shear stresses in the steel and the copper are not to exceed 500 and 300 MPa, respectively, and (b) the ratio of spring constants. For steel and copper, use shear moduli of elasticity $G_s = 79$ GPa and $G_c = 41$ GPa, respectively.

CHAPTER 7

Numerical Methods

7.1 INTRODUCTION

This chapter is subdivided into two parts. The finite difference method is treated briefly first. Then, the most commonly employed numerical technique, the finite element method, is discussed. We shall apply both approaches to the solution of problems in elasticity and the mechanics of materials. The use of these *numerical methods* enables the engineer to expand his or her ability to solve practical design problems. The engineer may now treat real shapes as distinct from the somewhat limited variety of shapes amenable to simple analytic solution. Similarly, the engineer need no longer force a complex loading system to fit a more regular load configuration to conform to the dictates of a purely academic situation. Numerical analysis thus provides a tool with which the engineer may feel freer to undertake the solution of problems as they are found in practice.

Analytical solutions of the type discussed in earlier chapters have much to offer beyond the specific cases for which they have been derived. For example, they enable us to gain insight into the variation of stress and deformation with basic shape and property changes. In addition, they provide the basis for rough approximations in preliminary design even though there is only crude similarity between the analytical model and the actual case. In other situations, analytical methods provide a starting point or guide in numerical solutions.

Numerical analyses lead often to a system of linear algebraic equations. The most appropriate method of solution then depends on the nature and the number of such equations, as well as the type of computing equipment available. The techniques introduced in this chapter and applied in the chapters following have clear application to computation by means of electronic digital computer. Formulating and solving a problem (Appendix A), and tools used for computations, discussed in Section 7.16, are important. Observe that fundamentals of matrix algebra, a subset of the finite element method, will be extensively used.

Part A—Finite Difference Method

7.2 FINITE DIFFERENCES

The numerical solution of a differential equation is essentially a table in which values of the required function are listed next to corresponding values of the independent variable(s). In the case of an ordinary differential equation, the unknown function (y) is listed at specific *pivot* or *nodal* points spaced along the x axis. For a two-dimensional partial differential equation, the nodal points will be in the xy plane.

The basic *finite difference* expressions follow logically from the fundamental rules of calculus. Consider the definition of the first derivative with respect to x of a continuous function $y = f(x)$ (Fig. 7.1):

$$\left(\frac{dy}{dx}\right)_n = \lim_{\Delta x \to 0} \frac{y(x_n + \Delta x) - y(x_n)}{\Delta x} = \lim_{\Delta x \to 0} \frac{y_{n+1} - y_n}{\Delta x}$$

The subscript n denotes any point on the curve. If the increment in the independent variable does not become vanishingly small but instead assumes a finite $\Delta x = h$, the preceding expression represents an approximation to the derivative:

$$\left(\frac{dy}{dx}\right)_n \approx \frac{\Delta y_n}{h} = \frac{y_{n+1} - y_n}{h}$$

Here Δy_n is termed the first difference of y at point x_n:

$$\Delta y_n = y_{n+1} - y_n \approx h\left(\frac{dy}{dx}\right)_n \tag{7.1}$$

Because the relationship (Fig. 7.1) is expressed in terms of the numerical value of the function at the point in question (n) and a point ahead of it ($n + 1$), the difference is termed a *forward difference*. The *backward difference* at n, denoted ∇y_n, is given by

$$\nabla y_n = y_n - y_{n-1} \tag{7.2}$$

Central differences involve pivot points symmetrically located with respect to x_n and often result in more accurate approximations than forward or backward differences. The latter are especially useful where, because of geometrical limitations

FIGURE 7.1. *Finite difference approximation of f(x).*

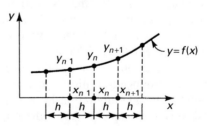

(as near boundaries), central differences cannot be employed. In terms of symmetrical pivot points, the derivative of y at x_n is

$$\left(\frac{dy}{dx}\right)_n \approx \frac{y(x_n + h) - y(x_n - h)}{2h} = \frac{1}{2h}(y_{n+1} - y_{n-1}) \qquad (7.3)$$

The *first central difference* δy is thus

$$\delta y = \tfrac{1}{2}(y_{n+1} - y_{n-1}) \approx h\left(\frac{dy}{dx}\right)_n \qquad (7.4)$$

A procedure similar to that just used will yield the higher-order derivatives.

The *second forward difference* at x_n is expressed in the form

$$\Delta^2 y_n = y_{n+2} - 2y_{n+1} + y_n \approx h^2\left(\frac{d^2 y}{dx^2}\right)_n \qquad (7.5)$$

The *second backward difference* is found in the same way:

$$\nabla^2 y_n = \nabla(\nabla y_n) = \nabla(y_n - y_{n-1}) = \nabla y_n - \nabla y_{n-1}$$
$$= (y - y_{n-1}) - (y_{n-1} - y_{n-2})$$

$$\nabla^2 y_n = y_n - 2y_{n-1} + y_{n-2} \approx h^2\left(\frac{d^2 y}{dx^2}\right)_n \qquad (7.6)$$

It is a simple matter to verify that the coefficients of the pivot values in the mth forward and backward differences are the same as the coefficients of the binomial expansion $(a - b)^m$. Using this scheme, higher-order forward and backward differences are easily written.

The *second central difference* at x_n is the difference of the first central differences. Therefore,

$$\delta^2 y_n = \Delta(\nabla y_n) = \Delta(y_n - y_{n-1}) = \Delta y_n - \Delta y_{n-1}$$
$$= (y_{n+1} - y_n) - (y_n - y_{n-1})$$
$$= y_{n+1} - 2y_n + y_{n-1} \approx h^2\left(\frac{d^2 y}{dx^2}\right)_n \qquad (7.7)$$

In a like manner, the *third* and *fourth central differences* are readily determined:

$$\delta^3 y_n = \delta(\delta^2 y_n) = \delta(y_{n+1} - 2y_n + y_{n-1}) = \delta y_{n+1} - 2\delta y_n + \delta y_{n-1}$$
$$= \tfrac{1}{2}(y_{n+2} - y_n) - (y_{n+1} - y_{n-1}) + \tfrac{1}{2}(y_n - y_{n-2})$$
$$= \tfrac{1}{2}(y_{n+2} - 2y_{n+1} + 2y_{n-1} - y_{n-2}) \approx h^3\left(\frac{d^3 y}{dx^3}\right)_n \qquad (7.8)$$

$$\delta^4 y_n = y_{n+2} - 4y_{n+1} + 6y_n - 4y_{n-1} + y_{n-2} \approx h^4\left(\frac{d^4 y}{dx^4}\right)_n \qquad (7.9)$$

Examination of Eqs. (7.7) and (7.9) reveals that for *even-order* derivatives, the coefficients of y_n, y_{n+1} are equal to the coefficients in the binomial expansion $(a - b)^m$.

Unless otherwise specified, we use the term *finite differences* to refer to *central differences*.

We now discuss a continuous function $w(x, y)$ of two variables. The *partial derivatives* may be approximated by the following procedures, similar to those discussed in the previous chapter. For purposes of illustration, consider a rectangular boundary as in Fig. 7.2. By taking $\Delta x = \Delta y = h$, a square *mesh* or *net* is formed by the horizontal and vertical lines. The intersection points of these lines are the nodal points. Equation (7.7) yields

$$\frac{\partial w}{\partial x} \approx \frac{1}{h}\delta_x w, \qquad \frac{\partial w}{\partial y} \approx \frac{1}{h}\delta_y w \tag{7.10}$$

$$\frac{\partial^2 w}{\partial x^2} \approx \frac{1}{h^2}\delta_x^2 w, \qquad \frac{\partial^2 w}{\partial y^2} \approx \frac{1}{h^2}\delta_y^2 w, \qquad \frac{\partial^2 w}{\partial x \partial y} \approx \frac{1}{h}\delta_x\left(\frac{\partial w}{\partial y}\right) \tag{7.11}$$

The subscripts x and y applied to the δ's indicate the coordinate direction appropriate to the difference being formed. The preceding expressions written for the point 0 are

$$\frac{\partial w}{\partial x} \approx \frac{1}{2h}[w(x + h, y) - w(x - h, y)] = \frac{1}{2h}(w_1 - w_3)$$

$$\frac{\partial w}{\partial y} \approx \frac{1}{2h}[w(x, y + h) - w(x, y - h)] = \frac{1}{2h}(w_2 - w_4) \tag{7.12}$$

and

$$\frac{\partial^2 w}{\partial x^2} \approx \frac{1}{h^2}[w(x + h, y) - 2w(x, y) + w(x - h, y)]$$

$$= \frac{1}{h^2}(w_1 - 2w_0 + w_3) \tag{7.13}$$

FIGURE 7.2. *Rectangular boundary divided into a square mesh.*

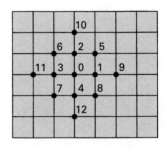

$$\frac{\partial^2 w}{\partial y^2} \approx \frac{1}{h^2}(w_2 - 2w_0 + w_4)$$

$$\frac{\partial^2 w}{\partial x\, \partial y} \approx \frac{1}{2h^2}(\delta_x w_2 - \delta_x w_4) = \frac{1}{4h^2}(w_5 - w_6 + w_7 - w_8) \tag{7.14}$$

Similarly, Eqs. (7.8) and (7.9) lead to expressions for approximating the third- and fourth-order partial derivatives.

7.3 FINITE DIFFERENCE EQUATIONS

We are now in a position to transform a differential equation into an algebraic equation. This is accomplished by substituting the appropriate finite difference expressions into the differential equation. At the same time, the boundary conditions must also be converted to finite difference form. The solution of a differential equation thus reduces to the simultaneous solution of a set of linear, algebraic equations, written for every nodal point within the boundary.

EXAMPLE 7.1 Torsion Bar with Square Cross Section
Analyze the torsion of a bar of square section using finite difference techniques.

Solution   The governing partial differential equation is (see Sec. 6.3)

$$\frac{\partial^2 \Phi}{\partial x^2} + \frac{\partial^2 \Phi}{\partial y^2} = -2G\theta \tag{7.15}$$

where Φ may be assigned the value of zero at the boundary. Referring to Fig. 7.2, the finite difference equation about the point 0, corresponding to Eq. (7.15), is

$$\Phi_1 + \Phi_2 + \Phi_3 + \Phi_4 - 4\Phi_0 = -2G\theta h^2 \tag{7.16}$$

A similar expression is written for every other nodal point within the section. The solution of the problem then requires the determination of those values of Φ that satisfy the system of algebraic equations.

The domain is now divided into a number of small squares, 16 for example. In labeling nodal points, it is important to take into account any conditions of symmetry that may exist. This has been done in Fig. 7.3. Note that $\Phi = 0$ has been substituted at the boundary. Equation (7.16) is now applied to nodal points b, c, and d, resulting in the following set of expressions:

$$4\Phi_d - 4\Phi_b = -2G\theta h^2$$
$$2\Phi_d - 4\Phi_c = -2G\theta h^2$$
$$2\Phi_c + \Phi_b - 4\Phi_d = -2G\theta h^2$$

FIGURE 7.3. *Example 7.1. A square cross section of a torsion bar.*

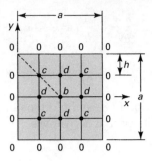

Simultaneous solution yields

$$\Phi_b = 2.25G\theta h^2, \qquad \Phi_c = 1.375G\theta h^2, \qquad \Phi_d = 1.75G\theta h^2 \qquad \textbf{(a)}$$

The results for points b and d are tabulated in the second column of Table 7.1.

To determine the partial derivatives of the stress function, we shall assume a smooth curve containing the values in Eq. (a) to represent the function Φ. *Newton's interpolation formula* [Ref. 7.1], used for fitting such a curve, is

$$\Phi(x) = \Phi_0 + \frac{x}{h}\Delta\Phi_0 + \frac{x(x-h)}{2!\,h^2}\Delta^2\Phi_0 + \frac{x(x-h)(x-2h)}{3!\,h^3}\Delta^3\Phi_0$$

$$+ \frac{x(x-h)(x-2h)(x-3h)}{4!\,h^4}\Delta^4\Phi_0 \qquad \textbf{(7.17)}$$

$$+ \cdots + \frac{x(x-h)\cdots(x-nh+h)}{n!\,h^n}\Delta^n\Phi_0$$

Here the $\Delta\Phi_0$'s are the forward differences calculated at $x = 0$ as follows:

$$\Delta\Phi_0 = \Phi_d - \Phi_0 = (1.75 - 0)G\theta h^2 = 1.75G\theta h^2$$
$$\Delta^2\Phi_0 = \Phi_b - 2\Phi_d + \Phi_0 = -1.25G\theta h^2$$
$$\Delta^3\Phi_0 = \Phi_d - 3\Phi_b + 3\Phi_d - \Phi_0 = 0.25G\theta h^2 \qquad \textbf{(b)}$$
$$\Delta^4\Phi_0 = \Phi_0 - 4\Phi_d + 6\Phi_b - 4\Phi_d + \Phi_0 = -0.5G\theta h^2$$

The differences are also calculated at $x = h$, $x = 2h$, and so on, and are listed in Table 7.1. Note that we can readily obtain the values given in Table 7.1

TABLE 7.1. *Values of the Forward Differences for Given Φ's*

x	$\Phi/G\theta h^2$	$\Delta/G\theta h^2$	$\Delta^2/G\theta h^2$	$\Delta^3/G\theta h^2$	$\Delta^4/G\theta h^2$
0	0	1.75	−1.25	0.25	−0.5
h	1.75	0.5	−1.0	−0.25	
2h	2.25	−0.5	−1.25		
3h	1.75	−1.75			
4h	0				

(for the given Φ's) by starting at node $x = 4h$: $0 - 1.75 = -1.75$, $1.75 - 2.25 = -0.5, -1.75 - (-0.5) = -1.25$, and so on.

The maximum shear stress, which occurs at $x = 0$, is obtained from $(\partial\Phi/\partial x)_0$. Thus, differentiating Eq. (7.17) with respect to x and then setting $x = 0$, the result is

$$\left(\frac{\partial\Phi}{\partial x}\right)_0 = \frac{1}{h}(\Delta\Phi_0 - \tfrac{1}{2}\Delta^2\Phi_0 + \tfrac{1}{3}\Delta^3\Phi_0 - \tfrac{1}{4}\Delta^4\Phi_0 + \cdots) \qquad \textbf{(c)}$$

Substituting the values in the first row of Table 7.1 into Eq. (c), we obtain

$$\left(\frac{\partial\Phi}{\partial x}\right)_0 = \tau_{max} = \tfrac{31}{12}G\theta h = 0.646G\theta a$$

The exact value, given in Table 6.2 as $\tau_{max} = 0.678G\theta a$, differs from this approximation by only 4.7%.

By means of a finer network, we expect to improve the result. For example, selecting $h = a/6$, six nodal equations are obtained. It can be shown that the maximum stress in this case, $0.661G\theta a$, is within 2.5% of the exact solution. On the basis of results for $h = a/4$ and $h = a/6$, a still better approximation can be found by applying extrapolation techniques.

7.4 CURVED BOUNDARIES

It has already been mentioned that one important strength of numerical analysis is its adaptability to irregular geometries. We now turn, therefore, from the straight and parallel boundaries of previous problems to situations involving curved or irregular boundaries. Examination of one segment of such a boundary (Fig. 7.4) reveals that the standard five-point operator, in which all arms are of equal length, is not appropriate because of the unequal lengths of arms bc, bd, be, and bf. When at least one arm is of nonstandard length, the pattern is referred to as an *irregular star*. One method for constructing irregular star operators is discussed next.

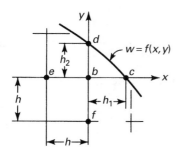

FIGURE 7.4. *Curved boundary and irregular star operator.*

Assume that, in the vicinity of point b, $w(x, y)$ can be approximated by the second-degree polynomial

$$w(x, y) = w_b + a_1 x + a_2 y + a_3 x^2 + a_4 y^2 + a_5 xy \qquad (7.18)$$

Referring to Fig. 7.4, this expression leads to approximations of the function w at points $c, d, e,$ and f:

$$
\begin{aligned}
w_c &= w_b + a_1 h_1 + a_3 h_1^2 \\
w_d &= w_b + a_2 h_2 + a_4 h_2^2 \\
w_e &= w_b - a_1 h + a_3 h^2 \\
w_f &= w_b - a_2 h + a_4 h^2
\end{aligned}
\qquad \textbf{(a)}
$$

At nodal point $b(x = y = 0)$, Eq. (7.18) yields

$$\left(\frac{\partial^2 w}{\partial x^2}\right)_b = 2a_3, \qquad \left(\frac{\partial^2 w}{\partial y^2}\right)_b = 2a_4 \qquad \textbf{(b)}$$

Combining Eqs. (a) and (b), we have

$$\left(\frac{\partial^2 w}{\partial x^2}\right)_b = 2\frac{h(w_c - w_b) + h_1(w_e - w_b)}{hh_1(h + h_1)}$$

$$\left(\frac{\partial^2 w}{\partial y^2}\right)_b = 2\frac{h(w_d - w_b) + h_2(w_f - w_b)}{hh_2(h + h_2)}$$

Introducing this into the Laplace operator, we obtain

$$h^2 \left(\frac{\partial^2 w}{\partial x^2} + \frac{\partial^2 w}{\partial y^2}\right)_b = \frac{2w_c}{\alpha_1(1 + \alpha_1)} + \frac{2w_d}{\alpha_2(1 + \alpha_2)}$$

$$+ \frac{2w_e}{1 + \alpha_1} + \frac{2w_f}{1 + \alpha_2} - \left(\frac{2}{\alpha_1} + \frac{2}{\alpha_2}\right)w_b \qquad (7.19)$$

In this expression, $\alpha_1 = h_1/h$ and $\alpha_2 = h_2/h$. It is clear that for irregular stars, $0 \leq \alpha_i \leq 1 (i = 1, 2)$.

The foregoing result may readily be reduced for one-dimensional problems with irregularly spaced nodal points. For example, in the case of a beam, Eq. (7.19), with reference to Fig. 7.4, simplifies to

$$h^2 \left(\frac{\partial^2 w}{\partial x^2}\right)_b = \frac{2w_c}{\alpha_1(1 + \alpha_1)} + \frac{2w_e}{1 + \alpha_1} - \frac{2w_b}{\alpha_1} \qquad (7.20)$$

where x represents the longitudinal direction. This expression, setting $\alpha = \alpha_1$, may be written

$$h^2 \frac{d^2 w}{dx^2} = \frac{2}{\alpha(\alpha + 1)} [\alpha w_{n-1} - (1 + \alpha)w_n + w_{n+1}] \qquad (7.21)$$

EXAMPLE 7.2 Elliptical Bar Under Torsion

Find the shearing stresses at the points A and B of the torsional member of the elliptical section shown in Fig. 7.5. Let $a = 15$ mm, $b = 10$ mm, and $h = 5$ mm.

Solution Because of symmetry, only a quarter of the section need be considered. From the equation of the ellipse with the given values of a, b, and h, it is found that $h_1 = 4.4$ mm, $h_2 = 2.45$ mm, $h_3 = 3$ mm. At points b, e, f, and g, the standard finite difference equation (7.18) applies, while at c and d, we use a modified equation found from Eq. (7.15) with reference to Eq. (7.19). We can therefore write six equations presented in the following matrix form:

$$\begin{bmatrix} 2 & 0 & 0 & 0 & 2 & -4 \\ 0 & 2 & 0 & 1 & -4 & 1 \\ 0 & 0 & 2 & -4 & 1 & 0 \\ -4 & 2 & 0 & 0 & 0 & 1 \\ 1 & -4.27 & 1 & 0 & 1.06 & 0 \\ 0 & 1.25 & -7.41 & 1.34 & 0 & 0 \end{bmatrix} \begin{Bmatrix} \Phi_b \\ \Phi_c \\ \Phi_d \\ \Phi_e \\ \Phi_f \\ \Phi_g \end{Bmatrix} = \begin{Bmatrix} -2h^2 G\theta \\ -2h^2 G\theta \\ -2h^2 G\theta \\ -2h^2 G\theta \\ -2h^2 G\theta \\ -2h^2 G\theta \end{Bmatrix}$$

These equations are solved to yield

$$\Phi_b = 2.075 G\theta h^2, \qquad \Phi_c = 1.767 G\theta h^2, \qquad \Phi_d = 0.843 G\theta h^2$$
$$\Phi_e = 1.536 G\theta h^2, \qquad \Phi_f = 2.459 G\theta h^2, \qquad \Phi_g = 2.767 G\theta h^2$$

The solution then proceeds as in Example 7.1. The following forward differences at point B are first evaluated:

$$\Delta \Phi_B = 2.075 G\theta h^2, \qquad \Delta^3 \Phi_B = -0.001 G\theta h^2$$
$$\Delta^2 \Phi_B = -1.383 G\theta h^2, \qquad \Delta^4 \Phi_B = 0.002 G\theta h^2$$

Similarly, for point A, we obtain

$$\Delta \Phi_A = 1.536 G\theta h^2, \qquad \Delta^4 \Phi_A = 0.001 G\theta h^2$$
$$\Delta^2 \Phi_A = -0.613 G\theta h^2, \qquad \Delta^5 \Phi_A = 0.001 G\theta h^2$$
$$\Delta^3 \Phi_A = -0.002 G\theta h^2, \qquad \Delta^6 \Phi_A = -0.002 G\theta h^2$$

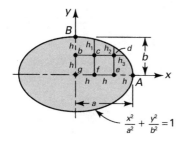

FIGURE 7.5. *Example 7.2. Elliptical cross section of a torsion member.*

Thus,

$$
\tau_B = \left(\frac{\partial \Phi}{\partial y}\right)_B = \left(2.075 + \frac{1.383}{2} - \frac{0.001}{3} + \frac{0.002}{4}\right) G\theta h
$$

$$
= 2.766 G\theta h = 1.383 G\theta b = 0.01383 G\theta
$$

$$
\tau_A = \left(\frac{\partial \Phi}{\partial x}\right)_A = \left(1.536 + \frac{0.613}{2} - \frac{0.002}{3} - \frac{0.001}{4} + \frac{0.001}{5} + \frac{0.002}{6}\right) G\theta h
$$

$$
= 1.842 G\theta h = 0.614 G\theta a = 0.00921 G\theta
$$

Comment Note that, according to the exact theory, the maximum stress occurs at $y = b$ and is equal to $1.384 G\theta b$ (see Example 6.3), indicating excellent agreement with τ_B.

7.5 BOUNDARY CONDITIONS

Our concern has thus far been limited to problems in which the boundaries have been assumed free of constraint. Many practical situations involve boundary conditions related to the deformation, force, or moment at one or more points. Application of numerical methods under these circumstances may become more complex.

To solve *beam deflection* problems, the boundary conditions as well as the differential equations must be transformed into central differences. Two types of homogeneous *boundary conditions*, obtained from $v = 0$, $dv/dx = 0$, and $v = 0$, $d^2v/dx^2 = 0$ at a support (n), are depicted in Fig. 7.6a and b, respectively. At a *free edge* (n), the finite difference boundary conditions are similarly written from $d^2v/dx^2 = 0$ and $d^3v/dx^3 = 0$ as follows

$$
v_{n+1} - 2v_n + v_{n-1} = 0
$$
$$
v_{n+2} - 2v_{n+1} + 2v_{n-1} - v_{n-2} = 0 \tag{a}
$$

Let us consider the deflection of a nonprismatic cantilever beam with depth varying arbitrarily and of constant width. We shall divide the beam into m segments of length $h = L/m$ and replace the variable loading by a load changing linearly between nodes (Fig. 7.7). The finite difference equations at a nodal point n may now be written as follows. Referring to Eq. (7.7), we find the difference equation corresponding to $EI(d^2v/dx^2) = M$ to be of the form

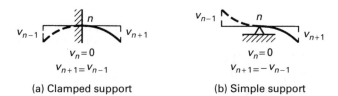

(a) Clamped support   (b) Simple support

FIGURE 7.6. *Boundary conditions in finite differences: (a) clamped or fixed support; (b) simple support.*

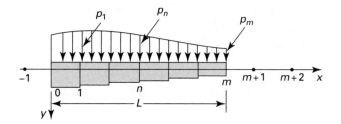

FIGURE 7.7. *Finite-difference representation of a nonprismatic cantilever beam with varying load.*

$$v_{n+1} - 2v_n + v_{n-1} = h^2 \left(\frac{M}{EI} \right)_n \qquad (7.22)$$

The quantities M_n and $(IE)_n$ represent the moment and the flexural rigidity, respectively, of the beam at n. In a like manner, the equation $EI(d^4v/dx^4) = p$ is expressed as

$$v_{n+2} - 4v_{n+1} + 6v_n - 4v_{n-1} + v_{n-2} = h^4 \left(\frac{p}{EI} \right)_n \qquad (7.23)$$

wherein p_n is the load intensity at n. Equations identical with the foregoing can be established at each remaining point in the beam. There will be m such expressions; the problem involves the solution of m unknowns, $v_1, \ldots, v_m$.

Applying Eq. (7.4), the *slope* at any point along the beam is

$$\theta_n = \left(\frac{dv}{dx} \right)_n = \frac{v_{n+1} - v_{n-1}}{2h} \qquad (7.24)$$

The following simple examples illustrate the method of solution.

EXAMPLE 7.3   Displacements of a Simple Beam
Use a finite difference approach to determine the deflection and the slope at the midspan of the beam shown in Fig. 7.8a.

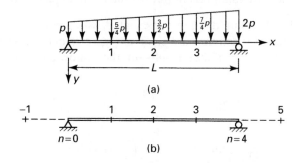

FIGURE 7.8. *Example 7.3. Simply supported beam with varying load.*

Solution For simplicity, take $h = L/4$ (Fig. 7.8b). The boundary conditions $v(0) = v(L) = 0$ and $v''(0) = v''(L) = 0$ are replaced by finite difference conditions by setting $v(0) = v_0$, $v(L) = v_4$ and applying Eq. (7.7):

$$v_0 = 0, \qquad v_1 = -v_{-1}, \qquad v_4 = 0, \qquad v_3 = -v_5 \qquad \textbf{(b)}$$

When Eq. (7.23) is used at points 1, 2, and 3, the following expressions are obtained:

$$v_3 - 4v_2 + 6v_1 - 4v_0 + v_{-1} = \frac{L^4}{256} \frac{5p}{4EI}$$

$$v_4 - 4v_3 + 6v_2 - 4v_1 + v_0 = \frac{L^4}{256} \frac{3p}{2EI} \qquad \textbf{(c)}$$

$$v_5 - 4v_4 + 6v_3 - 4v_2 + v_1 = \frac{L^4}{256} \frac{7p}{4EI}$$

Simultaneous solution of Eqs. (b) and (c) yields

$$v_1 = 0.0139 \frac{pL^4}{EI}$$

$$v_2 = 0.0198 \frac{pL^4}{EI}$$

$$v_3 = 0.0144 \frac{pL^4}{EI}$$

Then, from Eq. (7.24), we obtain

$$\theta_2 = \frac{v_3 - v_1}{2h} = 0.001 \frac{pL^3}{EI}$$

Note that, by successive integration of $EI d^4v/dx^4 = p(x)$, the result $v_2 = 0.0185 pL^4/EI$ is obtained. Thus, even a coarse segmentation leads to a satisfactory solution in this case.

EXAMPLE 7.4   Reactions of a Continuous Beam

Determine the redundant reaction R for the beam depicted in Fig. 7.9a.

Solution The bending diagrams associated with the applied loads $2P$ and the redundant reaction R are given in Figs. 7.9b and c, respectively.

For $h = L/2$, Eq. (7.22) results in the following expressions at points 1, 2, and 3:

$$v_0 - 2v_1 + v_2 = \frac{L^2}{4EI}\left(-PL + \frac{RL}{4}\right)$$

$$v_1 - 2v_2 + v_3 = \frac{L^2}{4EI}\left(-PL + \frac{RL}{2}\right) \qquad \textbf{(d)}$$

$$v_2 - 2v_3 + v_4 = \frac{L^2}{4EI}\left(-PL + \frac{RL}{4}\right)$$

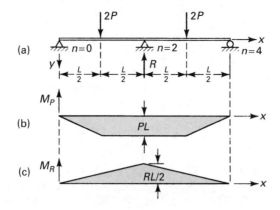

FIGURE 7.9. *Example 7.4. Statically indetermi-*
nate beam: (a) load diagram; (b)
and (c) moment diagrams.

The number of unknowns in this set of equations is reduced from five to three through application of the conditions of symmetry, $v_1 = v_3$, and the support conditions, $v_0 = v_2 = v_4 = 0$. Solution of Eq. (d) now yields $R = 8P/3$.

EXAMPLE 7.5 **Stepped Cantilever Beam**

Determine the deflection of the free end of the stepped cantilever beam loaded as depicted in Fig. 7.10a. Take $h = a/2$ (Fig. 7.10b).

Solution The boundary conditions $v(0) = 0$ and $v'(0) = 0$, referring to Fig. 7.6a, lead to

$$v_0 = 0 \qquad v_1 = v_{-1} \tag{e}$$

The moments are $M_0 = 3Pa$, $M_1 = 2Pa$, $M_2 = Pa$, and $M_3 = Pa/2$. Applying Eq. (7.22) at nodes 0, 1, 2, and 3, we obtain

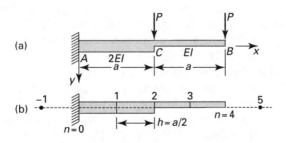

FIGURE 7.10. *Example 7.5. Nonprismatic cantilever*
beam under concentrated loads.

$$v_1 - 2v_0 + v_{-1} = \frac{3Pa}{2EI}h^2$$

$$v_2 - 2v_1 + v_0 = \frac{Pa}{EI}h^2$$

$$v_3 - 2v_2 + v_1 = \frac{Pa}{1.5EI}h^2 \qquad \text{(f)}$$

$$v_4 - 2v_3 + v_2 = \frac{Pa}{2EI}h^2$$

Observe that at point 2, the average flexural rigidity is used. Solving Eqs. (e) and (f) gives $v_1 = 3C/4$, $v_2 = 5C/2$, $v_3 = 59C/12$, and $v_4 = 47C/6$, in which $C = (Pa/EI)h^2$. Therefore, after setting $h = a/2$, we have

$$v_B = v_4 = \frac{47Pa^3}{24EI}$$

The foregoing deflection is about 2.2% larger than the exact value.

Part B—Finite Element Method

7.6 FUNDAMENTALS

Structural analysis involves the determination of the forces and deflections within a structure or its members. The earliest demands for structural analysis led to a host of so-called classical methods. The specialization of the classical methods was replaced by generalities of the modern *matrix methods*. The presentation of this chapter is limited to the most widely used of these techniques: the finite element *stiffness* or *displacement* method. Unless otherwise specified, we shall refer to it as the *finite element method* (FEM). The *finite element analysis* (FEA) is a numerical approach and well suited to digital computers. The method relies on the formulations of a simultaneous set of algebraic equations *relating forces to corresponding displacements* at discrete preselected points (called nodes) on the structure. These governing algebraic equations, also called the force–displacement relations, are expressed in matrix notations.

The powerful finite element method had its beginnings in the 1980s, and with the advent of high-speed, large-storage-capacity digital computers, it has gained great prominence throughout the industries in the solution of practical analysis and design problems of high complexity. The FEA offers many advantages. The structural geometry can be readily described, and combined load conditions can be easily handled. It offers the ability to treat discontinuities, to handle composite and anisotropic materials, to handle unlimited numbers and kinds of boundary conditions, to handle dynamic and thermal loadings, and to treat nonlinear structural problems. It also has the capacity for complete automation. The literature related to the FEA is extensive. See, for example, Refs. 7.2 through 7.20. Numerous commercial FEA software programs are available, as described in Section 7.16, including some directed at the learning process.

The *basic concept* of the finite element approach is that the real structure can be divided or *discretized* by a finite number of elements, connected not only at their nodes but along the interelement boundaries as well. Triangular, rectangular, tetrahedron, quadrilateral, or hexagonal *forms of elements* are often employed in the finite element method. The *types of elements* that are commonly used in structural idealization are the truss, beam, two-dimensional elements, shell and plate bending, and three-dimensional elements. Figure 7.11a depicts how a multistory hotel building is *modeled* using bar, beam, column, and plate elements, which can be employed for static, free vibration, earthquake response, and wind response analysis [Ref. 7.2]. A model of a nozzle in a thin-walled cylinder, created using triangular shell elements, is shown in Fig. 7.11b. Similarly, large structural systems, such as aircrafts or ships, are usually analyzed by dividing the structure into smaller units or substructures (e.g., wing of an airplane). When the stiffness of each unit has been determined, the analysis of the system follows the familiar procedure of matrix methods used in structural mechanics.

The network of elements and nodes that discretize the region is called a *mesh*. The density of a mesh increases as more elements are placed within a given region. *Mesh refinement* is when the mesh is modified in an analysis of a model to give improved solutions. Mesh density is increased in areas of high stress concentrations and when geometric transition zones are meshed smoothly (Fig. 7.11b). Usually, the FEA results converge toward the exact results as the mesh is continuously refined. To discuss adequately the subject of the FEA would require a far more lengthy

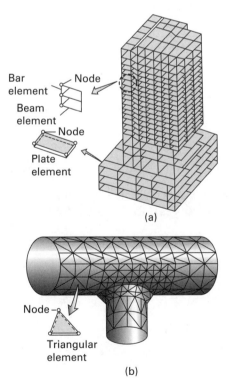

FIGURE 7.11. *Finite element models of two structures: (a) Multistory building; (b) pipe connection (Ref. 7.8).*

(a)

(b)

presentation than could be justified here. However, the subject is sufficiently important that engineers concerned with the analysis and design of members should have at least an understanding of the FEA. The fundamentals presented can clearly show the potential of the FEA as well as its complexities. It can be covered as an option, used as a "teaser" for a student's advance study of the topic, or used as a professional reference.

7.7 THE BAR ELEMENT

A *bar element*, also called a *truss bar element* or *axial element*, represents the simplest form of structural finite element. An element of this type of length L, modulus of elasticity E, and cross-sectional area A, is denoted by e (Fig. 7.12). The two ends or joints or *nodes* are numbered 1 and 2, respectively. A set of two equations in matrix form are required to relate the *nodal forces* ($\overline{F}_1$ and $\overline{F}_2$) to the *nodal displacements* ($\overline{u}_1$ and $\overline{u}_2$).

Equilibrium Method

The direct equilibrium approach is simple and physically clear, but it is practically well suited only for truss, beam, and frame elements. The equilibrium of the x-directed forces requires that $\overline{F}_1 = -\overline{F}_2$ (Fig. 7.12). In terms of the *spring rate* AE/L of the element, we have

$$\overline{F}_1 = \frac{AE}{L}(\overline{u}_1 - \overline{u}_2) \qquad\qquad \overline{F}_2 = \frac{AE}{L}(\overline{u}_2 - \overline{u}_1)$$

This may be written in matrix form,

$$\left\{\begin{matrix} \overline{F}_1 \\ \overline{F}_2 \end{matrix}\right\} = \frac{AE}{L}\begin{bmatrix} 1 & -1 \\ -1 & 1 \end{bmatrix}\left\{\begin{matrix} \overline{u}_1 \\ \overline{u}_2 \end{matrix}\right\} \qquad\qquad \textbf{(7.25a)}$$

or symbolically,

$$\{\overline{F}\} = [\overline{k}]_e\{\overline{u}\}_e \qquad\qquad \textbf{(7.25b)}$$

The quantity $[\overline{k}]_e$ is called the element *stiffness matrix* with dimensions of force per unit displacement. It relates the nodal displacements $\{\overline{u}\}_e$ to the nodal forces $\{\overline{F}\}_e$ on the element.

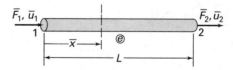

FIGURE 7.12.   *The one-dimensional axial element (e).*

Energy Method

The energy approach is more general, easier to apply, and more powerful than the direct method discussed in the foregoing, particularly for complex kinds of finite elements (see Ref. 7.2). Using this technique, it is necessary to first define a *displacement function* for the element (Fig. 7.12):

$$\bar{u} = a_1 + a_2 \bar{x} \tag{7.26}$$

where a_1 and a_2 are constants. Clearly, Eq. (7.26) represents a continuous linear displacement variation along the x axis of the element that corresponds to that of engineering formulation for a bar under axial loading. The axial displacements of joints 1 (at $x = 0$) and 2 (at $\bar{x} = L$), respectively, are

$$\bar{u}_1 = a_1 \qquad \bar{u}_2 = a_1 + a_2 L$$

Solving, we have $a_1 = \bar{u}_1$ and $a_2 = -(\bar{u}_1 - \bar{u}_2)/L$. Carrying these results into Eq. (7.26), we obtain

$$\bar{u} = \left(1 - \frac{\bar{x}}{L}\right)\bar{u}_1 + \frac{\bar{x}}{L}\bar{u}_2 \tag{7.27}$$

Using Eq. (7.27), we write

$$\varepsilon_x = \frac{d\bar{u}}{dx} = \frac{1}{L}(-\bar{u}_1 + \bar{u}_2) \tag{7.28}$$

Thus, the axial force in the element is

$$\bar{F} = (E\varepsilon_x)A = \frac{AE}{L}(-\bar{u}_1 + \bar{u}_2) \tag{7.29}$$

Substituting Eq. (7.29) into Eq. (2.58), the *strain energy in the element* may be expressed as follows

$$U = \int_0^L \frac{F^2 dx}{2AE} = \frac{AE}{2L}(\bar{u}_1^2 - 2\bar{u}_1 \bar{u}_2 + \bar{u}_2^2) \tag{7.30}$$

Then, *Castigliano's first theorem*, Eq. (10.22), gives

$$\bar{F}_1 = \frac{\partial U}{\partial \bar{u}_1} = \frac{AE}{L}(\bar{u}_1 - \bar{u}_2)$$

$$\bar{F}_2 = \frac{\partial U}{\partial \bar{u}_2} = \frac{AE}{L}(-\bar{u}_1 + \bar{u}_2)$$

The matrix form of these equations is the same as that given by Eqs. (7.25).

7.8   ARBITRARILY ORIENTED BAR ELEMENT

The global stiffness matrix for an element oriented arbitrarily in a two-dimensional plane is developed in this section. The *local coordinates* are chosen to conveniently represent the individual element. On the other hand, the reference or *global coordinates* are chosen to be convenient for the entire structure (Fig. 7.13). The local and global coordinates systems for an axial element are designated by $\overline{x}, \overline{y}$ and x, y, respectively.

Coordinate Transformation

A typical axial element e lying along the x axis, which is oriented at an angle θ, measured *counterclockwise* from the reference axis x, is shown in Fig. 7.14. In the local coordinate system, each joint has an axial force $\overline{F}_x$, a transverse force $\overline{F}_y$, an axial displacement $\overline{u}$, and a transverse displacement $\overline{v}$. Referring to the figure, Eq. (7.25a) may be expanded as follows:

$$\begin{Bmatrix} \overline{F}_{1x} \\ \overline{F}_{1y} \\ \overline{F}_{2x} \\ \overline{F}_{2y} \end{Bmatrix}_e = \frac{AE}{L} \begin{bmatrix} 1 & 0 & -1 & 0 \\ 0 & 0 & 0 & 0 \\ -1 & 0 & 1 & 0 \\ 0 & 0 & 0 & 0 \end{bmatrix} \begin{Bmatrix} \overline{u}_1 \\ \overline{v}_1 \\ \overline{u}_2 \\ \overline{v}_2 \end{Bmatrix}_e \qquad \textbf{(7.31a)}$$

or concisely,

$$\{F\} = [\overline{k}]_e \{\overline{\delta}\}_e \qquad \textbf{(7.31b)}$$

The quantities $[\overline{k}]_e$ and $\{\overline{\delta}\}_e$ are the stiffness and nodal displacement matrices, respectively, in the local coordinate system.

FIGURE 7.13.   *Global coordinates (x, y) for plane truss and local coordinates ($\overline{x}, \overline{y}$) for a bar element 1-2.*

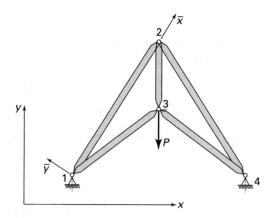

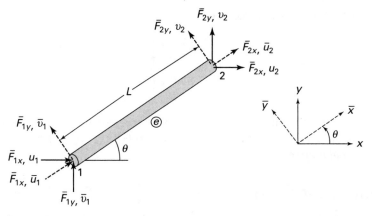

FIGURE 7.14.   *Local $(\bar{x}, \bar{y})$ and global (x, y) coordinates for a two-dimensional bar element (e).*

Force Transformation

It is seen from Fig. 7.14 that the two local and global forces at joint 1 may be related by the following expressions:

$$\bar{F}_{1x} = F_{1x} \cos\theta + F_{1y} \sin\theta$$
$$\bar{F}_{1y} = -F_{1x} \sin\theta + F_{1y} \cos\theta$$

Similar equations apply at joint 2. For brevity, let

$$c = \cos\theta \qquad \text{and} \qquad s = \sin\theta \qquad\qquad \textbf{(a)}$$

Hence, the local and global forces are related in the matrix form as

$$\begin{Bmatrix} \bar{F}_{1x} \\ \bar{F}_{1y} \\ \bar{F}_{2x} \\ \bar{F}_{2y} \end{Bmatrix}_e = \begin{bmatrix} c & s & 0 & 0 \\ -s & c & 0 & 0 \\ 0 & 0 & c & s \\ 0 & 0 & -s & c \end{bmatrix} \begin{Bmatrix} F_{1x} \\ F_{1y} \\ F_{2x} \\ F_{2y} \end{Bmatrix}_e \qquad\qquad \textbf{(7.32a)}$$

or symbolically,

$$\{\bar{F}\}_e = [T]\{F\}_e \qquad\qquad \textbf{(7.32b)}$$

In Eq. (7.32b), $[T]$ represents the *coordinate transformation matrix*:

$$[T] = \begin{bmatrix} c & s & 0 & 0 \\ -s & c & 0 & 0 \\ 0 & 0 & c & s \\ 0 & 0 & -s & c \end{bmatrix} \qquad\qquad \textbf{(7.33)}$$

The $\{F\}_e$ is the global nodal force matrix

$$\{F\}_e = \begin{Bmatrix} F_{1x} \\ F_{1y} \\ F_{2x} \\ F_{2y} \end{Bmatrix}_e \qquad (7.34)$$

We note that the coordinate transformation matrix satisfies the following *conditions of orthogonality*: $c^2 + s^2 = 1$, $(-s)^2 + c^2 = 1$, and $(-s)(c) + (c)(s) = 0$. Thus, the $[T]$ is an *orthogonal matrix*.

Displacement Transformation

Inasmuch as the displacement transforms in an identical way as forces, we can write

$$\begin{Bmatrix} \bar{u}_1 \\ \bar{v}_1 \\ \bar{u}_2 \\ \bar{v}_2 \end{Bmatrix}_e = [T] \begin{Bmatrix} u_1 \\ v_1 \\ u_2 \\ v_2 \end{Bmatrix}_e \qquad (7.35a)$$

The concise form of this is

$$\{\bar{\delta}\}_e = [T]\{\delta\}_e \qquad (7.35b)$$

where $\{\delta\}$ is the global nodal displacements. Carrying Eqs. (7.35b) and (7.32b) into (7.31b) results in

$$[T]\{F\}_e = [\bar{k}]_e[T]\{\delta\}_e$$

from which

$$\{F\}_e = [T]^{-1}[\bar{k}]_e[T]\{\delta\}_e$$

The inverse of the orthogonal matrix $[T]$ is the same as its transpose: $[T] = [T]^T$. Here the superscript T denotes the transpose. It is recalled that the transpose of a matrix is found by interchanging the rows and columns of the matrix.

Governing Equations

The global *force–displacement relations* or the governing equations for an element e are thus expressed in the form

$$\{F\}_e = [k]_e\{\delta\}_e \qquad (7.36)$$

where

$$[k]_e = [T]^T[\bar{k}]_e[T] \qquad (7.37)$$

The *global stiffness matrix* for the element, substituting Eq. (7.33) and $[k]$ from Eq. (7.31a) into Eq. (7.37), may be written as

$$[k]_e = \frac{AE}{L} \begin{bmatrix} c^2 & cs & -c^2 & -cs \\ cs & s^2 & -cs & -s^2 \\ -c^2 & -cs & c^2 & cs \\ -cs & -s^2 & cs & s^2 \end{bmatrix} = \frac{AE}{L} \begin{bmatrix} c^2 & cs & -c^2 & -cs \\ & s^2 & -cs & -s^2 \\ & & c^2 & cs \\ \text{Symmetric} & & & s^2 \end{bmatrix} \qquad \textbf{(7.38)}$$

The foregoing indicates that the element stiffness matrix depends on its dimensions, orientation, and material property.

7.9 AXIAL FORCE EQUATION

Reconsider the general case of a truss element 1–2 oriented arbitrarily in a two-dimensional plane, shown in Fig. 7.14. We multiply the third and fourth expressions in Eq. (7.36) to obtain the global forces at node 2 as

$$F_{2x} = \frac{AE}{L}[c^2(u_2 - u_1) + cs(v_2 - v_1)]$$

$$F_{2y} = \frac{AE}{L}[cs(u_2 - u_1) + s^2(v_2 - v_1)]$$

The axial tensile force in the element with nodes 1 and 2, designated by F_{12}, is then

$$F_{12} = F_{2x}c + F_{2y}s$$

in which $\cos\theta = c$ and $\sin\theta = s$. It is seen that this force is equivalent to the nodal force $\overline{F}_{1x}$ or $\overline{F}_{2x}$ associated with local coordinates $\overline{x}_1$ and $\overline{x}_2$.

Combining the preceding relationships leads to

$$F_{12} = \frac{AE}{L}(c^2 + s^2)[c(u_2 - u_1) + s(v_2 - v_1)]$$

$$= \frac{AE}{L}[c(u_2 - u_1) + s(v_2 - v_1)] \qquad \textbf{(a)}$$

or

$$F_{12} = (\text{spring rate})(\text{total axial elongation})$$

Here, the quantities $(u_2 - u_1)$ and $(v_2 - v_1)$ are, respectively, the horizontal and vertical components of the axial elongation. Equation (a) may be expressed in the matrix form

$$F_{12} = \frac{AE}{L}[c \quad s]\begin{Bmatrix} u_2 - u_1 \\ v_2 - v_1 \end{Bmatrix}$$

The *axial force* in the *bar element with nodes ij* is thus

$$F_{ij} = \left(\frac{AE}{L}\right)_{ij} [c \quad s]_{ij} \begin{Bmatrix} u_j - u_i \\ v_j - v_i \end{Bmatrix} \tag{7.39}$$

A positive (negative) value obtained for F_{ij} indicates that the element is in tension (compression). The axial stress, in element of cross-sectional area A, then is $\sigma_{ij} = Q_{ij}/A$.

EXAMPLE 7.6   Properties of a Truss Bar Element

The element 1–2 of the steel truss shown in Fig. 7.13 with a length L, cross-sectional area A, and modulus of elasticity E, is oriented at angle θ counterclockwise from the x axis. *Given:*

$$u_1 = 0.5 \text{ mm} \qquad v_1 = 0.625 \text{ mm} \qquad u_2 = -1.25 \text{ mm} \qquad v_2 = 0$$
$$\theta = 30° \qquad A = 600 \text{ mm}^2 \qquad L = 1.5 \text{ m} \qquad E = 200 \text{ GPa}$$

Calculate (a) the global stiffness matrix for the element; (b) the local displacements $\bar{u}_1, \bar{v}_1, \bar{u}_2$, and $\bar{v}_2$ of the element; (c) the axial stress in the element.

Solution   The free-body diagram of the element 1–2 is shown in Fig. 7.15. The spring rate of the element is

$$\frac{AE}{L} = \frac{1}{1.5}(600 \times 10^{-6})(200 \times 10^9) = 80 \times 10^6 \text{ N/m}$$

and

$$c = \cos 30° = \sqrt{3}/2 \qquad s = \sin 30° = 1/2$$

a. *Element Stiffness Matrix.*   Applying Eq. (7.38),

$$[k]_1 = 80(10^6) \begin{bmatrix} \frac{3}{4} & \frac{3}{4} & -\frac{3}{4} & -\frac{3}{4} \\ \frac{3}{4} & \frac{1}{4} & -\frac{3}{4} & -\frac{1}{4} \\ -\frac{3}{4} & -\frac{3}{4} & \frac{3}{4} & \frac{3}{4} \\ -\frac{3}{4} & -\frac{1}{4} & \frac{3}{4} & \frac{1}{4} \end{bmatrix}$$

or

$$[k]_1 = 80(10^6) \begin{bmatrix} 0.75 & 0.433 & -0.75 & -0.433 \\ 0.433 & 0.25 & -0.433 & -0.25 \\ -0.75 & -0.433 & 0.75 & 0.433 \\ -0.433 & -0.25 & 0.433 & 0.25 \end{bmatrix} \text{N/m}$$

FIGURE 7.15.   *Example 7.6. An axially loaded bar.*

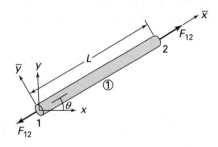

b. *Element Displacements.* Equations (7.35b), $\{\bar{\delta}\}_e = [T]\{\delta\}_e$, results in

$$\begin{Bmatrix} \bar{u}_1 \\ \bar{v}_1 \\ \bar{u}_2 \\ \bar{v}_2 \end{Bmatrix} = \begin{bmatrix} 0.833 & 0.5 & 0 & 0 \\ -0.5 & 0.833 & 0 & 0 \\ 0 & 0 & 0.833 & 0.5 \\ 0 & 0 & -0.5 & 0.833 \end{bmatrix} \begin{Bmatrix} 0.5 \\ 0.625 \\ -1.25 \\ 0 \end{Bmatrix} = \begin{Bmatrix} 0.73 \\ 0.27 \\ -1.04 \\ -0.63 \end{Bmatrix} \text{ mm}$$

c. *Axial Force.* Substituting the given numerical values into Eq. (7.39) with $i = 1$ and $j = 2$, we find

$$F_1 = F_{12} = 80(10^6)\begin{bmatrix} \dfrac{\sqrt{3}}{2} & \dfrac{1}{2} \end{bmatrix}\begin{Bmatrix} -1.25 & -0.5 \\ 0 & -0.625 \end{Bmatrix}(10^{-3}) = -146.2 \text{ kN}$$

It follows that *axial stress* in the element is equal to $\sigma_1 = F_1/A = -244$ MPa.

Comment The negative sign indicates a compression.

7.10 FORCE-DISPLACEMENT RELATIONS FOR A TRUSS

We now develop the finite element method by considering a plane truss. A truss is a structure composed of differently oriented straight bars connected at their joints (or nodes) by means of pins, such as shown Fig. 7.13. It is easier to understand the basic procedures of treatment by referring to this simple system. The general derivation of the FEM will be discussed in Section 7.13. To derive truss equations, the global element relations for the axial element given by Eq. (7.36) will be assembled. This gives the following *force-displacement relations* for the entire truss or the system equations:

$$\{F\} = [K]\{\delta\} \tag{7.40}$$

The global nodal matrix $\{F\}$ and the global stiffness matrix $[K]$ are expressed as

$$\{F\} = \sum_1^n \{F\}_e \tag{7.41}$$

and

$$[K] = \sum_1^n [k]_e \tag{7.42}$$

Here the quantity e designates an element and n is the number of elements comprising the truss. Observe that $[K]$ relates the global nodal force $\{F\}$ to the global displacement $\{\delta\}$ for the entire truss.

The Assembly Process

The element stiffness matrices $[k]_e$ in Eq. (7.38) must be properly added together or superimposed. To perform direct addition, a convenient method is to label the columns and rows of each element stiffness matrix in accordance to the displacement components related with it. The truss stiffness matrix $[K]$ is then found simply by summing terms from the individual element stiffness matrix into their corresponding locations in $[K]$. Alternatively, expand the $[k]_e$ for each element to the order of the truss stiffness matrix by *adding rows and columns of zeros*. We shall employ this process of assemblage of the element stiffness matrices. It is obvious that, for the problems involving a large number of elements, to implement the assembly of the $[K]$ requires a digital computer.

The use of the fundamental equations developed are illustrated in the solution of the following sample numerical problems.

EXAMPLE 7.7 Stepped Axially Loaded Bar

A bar consists of two prismatic parts, fixed at the left end, and supports a load P at the right end, as illustrated in Fig. 7.16a. Determine the nodal displacements and nodal forces in the bar.

Solution In this case, the *global* coordinates (x, y) *coincide* with *local* $(\bar{x}, \bar{y})$ coordinates. The bar is discretized into elements with *nodes* 1, 2, and 3 (Fig. 7.16b). The axial rigidities of the *elements* 1 and 2 are $2AE$ and AE, respectively. Therefore, $(AE/L)_1 = 2AE/L$ and $(AE/L)_2 = AE/L$.

Element Stiffness Matrices. Through the use of Eq. (7.25), the stiffness for the elements 1 and 2 is expressed as

$$[k]_1 = \frac{AE}{L}\begin{matrix}u_1 & u_2 \\ \begin{bmatrix} 2 & -2 \\ -2 & 2 \end{bmatrix} & \begin{matrix} u_1 \\ u_2 \end{matrix}\end{matrix} \qquad [k]_2 = \frac{AE}{L}\begin{matrix}u_2 & u_3 \\ \begin{bmatrix} 1 & -1 \\ -1 & 1 \end{bmatrix} & \begin{matrix} u_2 \\ u_3 \end{matrix}\end{matrix}$$

It is seen that the column and row of each stiffness matrix are labeled according to the nodal displacements associated with them. There are three displacement components (u_1, u_2, u_3), and hence the order of the

FIGURE 7.16. *Example 7.7. (a) A stepped bar under an axial load; (b) two-element model.*

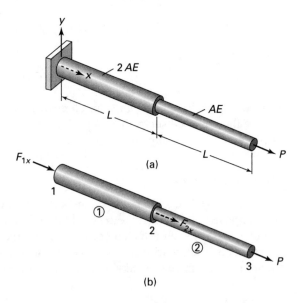

(a)

(b)

system matrix must be 3×3. In terms of the bar displacements, we write

$$[k]_1 = \frac{AE}{L} \begin{array}{c} u_1 u_2 u_3 \\ \begin{bmatrix} 2 & -2 & 0 \\ -2 & 2 & 0 \\ 0 & 0 & 0 \end{bmatrix} \begin{array}{c} u_1 \\ u_2 \\ u_3 \end{array} \end{array}$$

$$[k]_2 = \frac{AE}{L} \begin{array}{c} u_1 u_2 u_3 \\ \begin{bmatrix} 0 & 0 & 0 \\ 0 & 1 & -1 \\ 0 & -1 & 1 \end{bmatrix} \begin{array}{c} u_1 \\ u_2 \\ u_3 \end{array} \end{array}$$

In the foregoing matrices, the last row and first column of zeros are added, respectively (boxed in by the dashed lines).

Stiffness Matrix System. The superposition of the terms of each stiffness matrix leads to

$$[K] = \frac{AE}{L} \begin{array}{c} u_1 u_2 u_3 \\ \begin{bmatrix} 2 & -2 & 0 \\ -2 & 3 & -1 \\ 0 & -1 & 1 \end{bmatrix} \begin{array}{c} u_1 \\ u_2 \\ u_3 \end{array} \end{array}$$

Force–Displacement Relations. Equations (7.40) are expressed as

$$\begin{Bmatrix} F_{1x} \\ F_{2x} \\ F_{3x} \end{Bmatrix} = \frac{AE}{L} \begin{bmatrix} 2 & -2 & 0 \\ -2 & 3 & -1 \\ 0 & -1 & 1 \end{bmatrix} \begin{Bmatrix} u_1 \\ u_2 \\ u_3 \end{Bmatrix} \tag{a}$$

The displacement and force boundary conditions are $u_1 = 0$, $F_{2x} = 0$, and $F_{3x} = P$. Then, Eqs. (a), referring to Fig. 7.16b,

$$\begin{Bmatrix} F_{1x} \\ 0 \\ P \end{Bmatrix} = \frac{AE}{L} \begin{bmatrix} 2 & -2 & 0 \\ -2 & 3 & -1 \\ 0 & -1 & 1 \end{bmatrix} \begin{Bmatrix} 0 \\ u_2 \\ u_3 \end{Bmatrix}$$

Displacements. To determine u_2 and u_3, only the part of this equations is considered:

$$\begin{Bmatrix} 0 \\ P \end{Bmatrix} = \frac{AE}{L} \begin{bmatrix} 3 & -1 \\ -1 & 2 \end{bmatrix} \begin{Bmatrix} u_2 \\ u_3 \end{Bmatrix}$$

Solution of this equation equals

$$\begin{Bmatrix} u_2 \\ u_3 \end{Bmatrix} = \begin{Bmatrix} PL/2AE \\ 3PL/2AE \end{Bmatrix}$$

Comment With the displacements available, the axial force and stress in the element 1 (or 2) can readily be found as described in Example 7.6.

Nodal Forces. Equations (a) gives

$$\begin{Bmatrix} F_{1x} \\ F_{2x} \\ F_{3x} \end{Bmatrix} = \frac{AE}{L} \begin{bmatrix} 2 & -2 & 0 \\ -2 & 3 & 1 \\ 0 & -1 & 1 \end{bmatrix} \begin{Bmatrix} 0 \\ PL/2AE \\ 3PL/2AE \end{Bmatrix} = \begin{Bmatrix} -P \\ 0 \\ P \end{Bmatrix}$$

Comment The results indicate that the reaction $F_{1x} = -P$ is equal in magnitude but opposite in direction to the applied force at node 3, $F_{3x} = P$. Also, $F_{2x} = 0$ shows that no force is applied at node 2. Equilibrium of the bar assembly is thus satisfied.

EXAMPLE 7.8   Analysis of a Three-Bar Truss

A steel plane truss in which all members have the same axial rigidity AE supports a horizontal force P and a load W acting at joint 2, as shown in Fig. 7.17a. Find the nodal displacements, reactions, and stresses in each member. *Given:*

$$P = 24 \text{ kN} \quad W = 36 \text{ kN} \quad \sigma_{yp} = 250 \text{ MPa} \quad E = 200 \text{ GPa}$$

$$L_1 = L_2 = L = 2 \text{ m} \quad L_3 = 2\sqrt{2} \text{ m} \quad A = 400 \text{ mm}^2$$

Solution The reactions are marked and the nodes numbered arbitrarily for elements in Fig. 7.17a.

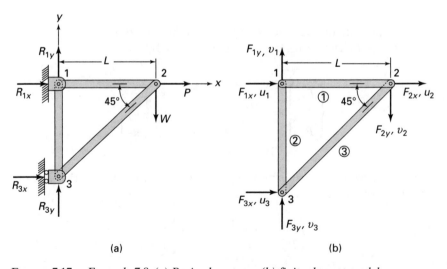

(a)   (b)

FIGURE 7.17.   *Example 7.8. (a) Basic plane truss; (b) finite element model.*

TABLE 7.2.   *Data for the Truss of Fig. 7.17*

Element	θ	c	s	c^2	cs	s^2	AE/L
1	0	1	0	1	0	0	$4(10^7)$
2	270°	0	−1	0	0	1	$4(10^7)$
3	225°	$-1\sqrt{2}$	$-1\sqrt{2}$	0.5	0.5	0.5	$2\sqrt{2}(10^7)$

Input Data.   At each node there are two displacements and two nodal force components (Fig. 7.17b). It is recalled that θ is measured *counterclockwise* from the positive x axis to each element (Table 7.2).

Element Stiffness Matrix.   Applying Eq. (7.38) and Table 7.2, we have for the bars 1, 2, and 3, respectively:

$$[k]_1 = 4(10^7) \begin{array}{cccc} u_1 & v_1 & u_2 & v_2 \end{array} \\ \begin{bmatrix} c^2 & cs & -c^2 & -cs \\ cs & s^2 & -cs & s^2 \\ -c^2 & -cs & c^2 & cs \\ cs & -s^2 & cs & s^2 \end{bmatrix} \begin{array}{c} u_1 \\ v_1 \\ u_2 \\ v_2 \end{array} = 4(10^7) \begin{bmatrix} 1 & 0 & -1 & 0 \\ 0 & 0 & 0 & 0 \\ -1 & 0 & 1 & 0 \\ 0 & 0 & 0 & 0 \end{bmatrix} \begin{array}{c} u_1 \\ v_1 \\ u_2 \\ v_2 \end{array}$$

$$[k]_2 = 4(10^7) \begin{array}{cccc} u_1 & v_1 & u_3 & v_3 \end{array} \\ \begin{bmatrix} 0 & 0 & 0 & 0 \\ 0 & 1 & 0 & -1 \\ 0 & 0 & 0 & 0 \\ 0 & -1 & 0 & 1 \end{bmatrix} \begin{array}{c} u_1 \\ v_1 \\ u_3 \\ v_3 \end{array}$$

$$[k]_3 = 2\sqrt{2}(10^7) \begin{array}{cccc} u_2 & v_2 & u_3 & v_3 \end{array} \\ \begin{bmatrix} 0.5 & 0.5 & -0.5 & -0.5 \\ -0.5 & 0.5 & -0.5 & -0.5 \\ -0.5 & -0.5 & 0.5 & 0.5 \\ -0.5 & -0.5 & 0.5 & 0.5 \end{bmatrix} \begin{array}{c} u_2 \\ v_2 \\ u_3 \\ v_3 \end{array}$$

In the preceding, the column and row of each stiffness matrix are labeled according to the nodal displacements associated with them. It is seen that displacements u_3 and v_3 are not involved in element 1; the u_2 and v_2 are not involved in element 2; the u_1 and v_1 are not involved in element 3. Thus, prior to adding $[k]_1$, $[k]_2$, and $[k]_3$ to obtain the system matrix, two rows and columns of zeros must be added to each of the element matrices to account for the absence of these displacements. It follows that, using a common factor 10^7, element stiffness matrices take the forms

$$[k]_1 = (10^7)\begin{array}{c}\begin{array}{cccccc} u_1 & v_1 & u_2 & v_2 & u_3 & v_3 \end{array}\\ \left[\begin{array}{cccccc} 4 & 0 & -4 & 0 & 0 & 0 \\ 0 & 0 & 0 & 0 & 0 & 0 \\ -4 & 0 & 4 & 0 & 0 & 0 \\ 0 & 0 & 0 & 0 & 0 & 0 \\ 0 & 0 & 0 & 0 & 0 & 0 \\ 0 & 0 & 0 & 0 & 0 & 0 \end{array}\right]\begin{array}{c} u_1 \\ v_1 \\ u_2 \\ v_2 \\ u_3 \\ v_3 \end{array}\end{array}$$

$$[k]_2 = (10^7)\begin{array}{c}\begin{array}{cccccc} u_1 & v_1 & u_2 & v_2 & u_3 & v_3 \end{array}\\ \left[\begin{array}{cccccc} 0 & 0 & 0 & 0 & 0 & 0 \\ 0 & 4 & 0 & 0 & 0 & -4 \\ 0 & 0 & 0 & 0 & 0 & 0 \\ 0 & 0 & 0 & 0 & 0 & 0 \\ 0 & 0 & 0 & 0 & 0 & 0 \\ 0 & -4 & 0 & 0 & 0 & 4 \end{array}\right]\begin{array}{c} u_1 \\ v_1 \\ u_2 \\ v_2 \\ u_3 \\ v_3 \end{array}\end{array}$$

$$[k]_3 = (10^7)\begin{array}{c}\begin{array}{cccccc} u_1 & v_1 & u_2 & v_2 & u_3 & v_3 \end{array}\\ \left[\begin{array}{cccccc} 0 & 0 & 0 & 0 & 0 & 0 \\ 0 & 0 & 0 & 0 & 0 & 0 \\ 0 & 0 & \sqrt{2} & -\sqrt{2} & -\sqrt{2} & \sqrt{2} \\ 0 & 0 & -\sqrt{2} & \sqrt{2} & \sqrt{2} & -\sqrt{2} \\ 0 & 0 & -\sqrt{2} & \sqrt{2} & \sqrt{2} & -\sqrt{2} \\ 0 & 0 & \sqrt{2} & -\sqrt{2} & -\sqrt{2} & \sqrt{2} \end{array}\right]\begin{array}{c} u_1 \\ v_1 \\ u_2 \\ v_2 \\ u_3 \\ v_3 \end{array}\end{array}$$

System Stiffness Matrix. There are a total of six components of displacement for the truss before boundary constraints are imposed. Hence, the *order of the truss stiffness matrix* must be 6 × 6. After adding the terms from each element stiffness matrix into their *corresponding locations* in [K], we readily determine the global stiffness matrix for the truss as

$$[K] = (10^7)\begin{array}{c}\begin{array}{cccccc} u_1 & v_1 & u_2 & v_2 & u_3 & v_3 \end{array}\\ \left[\begin{array}{cccccc} 4 & 0 & -4 & 0 & 0 & 0 \\ 0 & 4 & 0 & 0 & 0 & -4 \\ -4 & 0 & 4+\sqrt{2} & \sqrt{2} & -\sqrt{2} & \sqrt{2} \\ 0 & 0 & -\sqrt{2} & \sqrt{2} & \sqrt{2} & -\sqrt{2} \\ 0 & 0 & -\sqrt{2} & \sqrt{2} & \sqrt{2} & -\sqrt{2} \\ 0 & -4 & \sqrt{2} & -\sqrt{2} & -\sqrt{2} & 4+\sqrt{2} \end{array}\right]\begin{array}{c} u_1 \\ v_1 \\ u_2 \\ v_2 \\ u_3 \\ v_3 \end{array}\end{array} \qquad \textbf{(b)}$$

System Force–Displacement Relationship. With reference to Fig. 7.11, the boundary conditions are $u_1 = 0$, $v_1 = 0$, $u_3 = 0$. In addition, we have $F_{3y} = 0$. Then, Eq. (7.40) becomes

$$\begin{Bmatrix} R_{1x} \\ R_{1y} \\ P \\ W \\ R_{3x} \\ 0 \end{Bmatrix} = [K] \begin{Bmatrix} 0 \\ 0 \\ u_2 \\ v_2 \\ 0 \\ v_3 \end{Bmatrix} \qquad \textbf{(c)}$$

where $[K]$ is given by Eq. (b).

Displacements. To find u_2, v_2, and v_3, only part of Eq. (c) associated with these displacements is considered. In so doing, we have

$$\begin{Bmatrix} P \\ W \\ 0 \end{Bmatrix} = \begin{Bmatrix} 24{,}000 \\ -36{,}000 \\ 0 \end{Bmatrix} = (10^7) \begin{Bmatrix} (4 + \sqrt{2}) & \sqrt{2} & \sqrt{2} \\ \sqrt{2} & \sqrt{2} & \sqrt{2} \\ \sqrt{2} & \sqrt{2} & (4 + \sqrt{2}) \end{Bmatrix} \begin{Bmatrix} u_2 \\ v_2 \\ v_3 \end{Bmatrix} \qquad \textbf{(d)}$$

Inversion of the preceding gives

$$\begin{Bmatrix} u_2 \\ v_2 \\ v_3 \end{Bmatrix} = 2.5(10^{-8}) \begin{Bmatrix} 1 & -1 & 0 \\ -1 & 4.828 & 1 \\ 0 & 1 & 1 \end{Bmatrix} \begin{Bmatrix} 24 \\ -36 \\ 0 \end{Bmatrix} (10^3) = \begin{Bmatrix} 1.34 \\ -4.94 \\ -0.80 \end{Bmatrix} \text{mm}$$

Reactions. Carrying the foregoing values of u_2, v_2, and v_3 into Eq. (c) results in the reactional forces as

$$\begin{Bmatrix} R_{1x} \\ R_{1y} \\ R_{3x} \end{Bmatrix} = 10^7 \begin{Bmatrix} -4 & 0 & 0 \\ 0 & 0 & -4 \\ -\sqrt{2} & -\sqrt{2} & \sqrt{2} \end{Bmatrix} \begin{Bmatrix} 1.34 \\ -4.94 \\ -0.90 \end{Bmatrix} (10^{-3}) = \begin{Bmatrix} -60 \\ 36 \\ 36 \end{Bmatrix} \text{kN}$$

The results may be verified by applying the equilibrium equations to the free-body diagram of the entire truss, Fig. 7.17a.

Axial Forces in Elements. From Eqs. (7.39) and (d) and Table 7.2, we obtain

$$F_1 = F_{12} = \frac{AE}{L} \begin{bmatrix} 1 & 0 \end{bmatrix} \begin{Bmatrix} u_2 \\ v_2 \end{Bmatrix} = 4(10^7) \begin{bmatrix} 1 & 0 \end{bmatrix} \begin{Bmatrix} 1.34 \\ -4.94 \end{Bmatrix} (10^{-3}) = 60 \text{ kN}$$

$$F_1 = F_{13} = \frac{AE}{L}[0 \quad -1]\left\{\begin{matrix} 0 \\ v_3 \end{matrix}\right\} = 4(10^7)\,[0 \quad -1]\left\{\begin{matrix} 0 \\ -1.5 \end{matrix}\right\}(10^{-3}) = 36\text{ kN}$$

$$F_3 = F_{23} = \frac{AE}{\sqrt{2}L}[-1 \quad -1]\left\{\begin{matrix} -u_2 \\ v_3 - v_2 \end{matrix}\right\}$$

$$= 2(10^7)\,[-1 \quad -1]\left\{\begin{matrix} -1.34 \\ -0.90 + 4.94 \end{matrix}\right\}(10^{-3}) = -50.9\text{ kN}$$

Stresses in Elements. By dividing the preceding element forces by the cross-sectional area of each bar, we obtain

$$\sigma_1 = \frac{60(10^3)}{400(10^6)} = 150\text{ MPa} \qquad \sigma_2 = 150\left(\frac{36}{60}\right) = 90\text{ MPa}$$

$$\sigma_3 = 150\left(-\frac{50.9}{60}\right) = -127.2\text{ MPa}$$

The negative sign indicates a compressive stress.

Comment The results indicate that member axial stresses are well below the yield strength for the material considered. Observe that the FEA permits the calculation of displacements, forces, and stresses in the truss with unprecedented ease and precision. It is evident, however, that the FEA, even in the simplest cases, requires considerable algebra. For any significant problem, the electronic digital computer must be used.

7.11 BEAM ELEMENT

This section provides only a brief discussion of the development of a stiffness matrix for *beam elements*. Let us consider an initially straight beam element of uniform flexural rigidity EI and length L (Fig. 7.18a). The element has a transverse deflection v and a slope $\theta = dv/dx$ at each end or node. Corresponding to these displacements, a transverse shear force F and a bending moment M act at

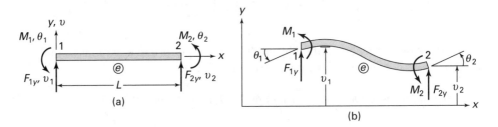

FIGURE 7.18. *The beam element with nodal forces and displacements: (a) before deformation; (b) after deformation.*

each node. The deflected configuration of the beam element is depicted in Fig. 7.18b.

The linearly elastic behavior of a beam element is governed by Eq. (5.32) as $d^4v/dx^4 = 0$. Observe that the right-hand side of this equation is zero because in the formulation of the stiffness matrix equations, we assume *no loading* between nodes. In the elements where there is a distributed load (see Example 7.9) or a concentrated load between the nodes, the *equivalent nodal load* components listed in Table D.5 of Appendix D are employed. Note that the equivalent nodal loads correspond to the (oppositely directed) reactions provided by a beam subject to the distributed or concentrated loading under fixed-fixed boundary conditions. For further details, see Ref. 7.6.

The solution of the governing equation is assumed, a cubic polynomial function of x, as follows

$$v = a_1 + a_2x + a_3x^2 + a_4x^3 \tag{a}$$

It is seen that the number of terms in the preceding expression is the same as the number of nodal displacements of the element. Equation (a) satisfies the basic beam equation, the conditions of displacement, and the continuity of interelement nodes. The coefficients a_1, a_2, a_3, and a_4 are obtained from the conditions at both nodes:

$$
\begin{aligned}
v = v_1 \quad \text{and} \quad \frac{dv_1}{dx} = \theta_1 \quad (\text{at } x = 0) \\
v = v_2 \quad \text{and} \quad \frac{dv_2}{dx} = \theta_2 \quad (\text{at } x = L)
\end{aligned}
\tag{b}
$$

Introducing Eq. (a) into (b) leads to

$$
\begin{Bmatrix} v_1 \\ \theta_1 \\ v_2 \\ \theta_2 \end{Bmatrix} =
\begin{bmatrix}
1 & 0 & 0 & 0 \\
1 & 1 & 0 & 0 \\
1 & L & L^2 & L^3 \\
1 & 1 & 2L & 3L^2
\end{bmatrix}
\begin{Bmatrix} a_1 \\ a_2 \\ a_3 \\ a_4 \end{Bmatrix}
$$

The inverse of these equations is

$$
\begin{Bmatrix} a_1 \\ a_2 \\ a_3 \\ a_4 \end{Bmatrix} = \frac{1}{L^3}
\begin{bmatrix}
L^3 & 0 & 0 & 0 \\
0 & L^3 & 0 & 0 \\
-3L & -2L^2 & 3L & -L^2 \\
2 & L & -2 & L
\end{bmatrix}
\begin{Bmatrix} v_1 \\ \theta_1 \\ v_2 \\ \theta_2 \end{Bmatrix}
$$

Substituting the foregoing expressions into Eq. (a) yields the displacement function in the form

$$
\begin{aligned}
v = v_1 + x\theta_1 - \frac{3x^2}{L}v_1 - \frac{2x^2}{L^2}\theta_1 + \frac{3x^2}{L^2}v_2 - \frac{x^2}{L}\theta_2 + \frac{2x^3}{L^3}v_1 \\
+ \frac{x^3}{L^2}\theta_1 - \frac{2x^3}{L^3}v_2 + \frac{x^3}{L^2}\theta_2
\end{aligned}
\tag{7.43}
$$

The bending moment M and shear force V in an elastic beam with cross section that is symmetrical about the plane of loading are related to the displacement function by Eqs. (5.32). Therefore,

$$F_{1y} = V_1 = EI \frac{d^3v}{dx^3} \quad \text{and} \quad M_1 = -EI \frac{d^2v}{dx^2} \quad (\text{at } x = 0)$$

$$F_{2y} = -V_2 = EI \frac{d^3v}{dx^3} \quad \text{and} \quad M_2 = EI \frac{d^2v}{dx^2} \quad (\text{at } x = L)$$

(7.44)

in which the minus signs in the second and third expressions are due to opposite to the sign conventions adopted for V and M in Figs. 5.7b and 7.18.

It can be verified [Ref. 7.2] that inserting Eq. (7.43) into Eqs. (7.44) gives the nodal *force* (moment)–*deflection* (slope) *relations* in the matrix form as

$$\begin{Bmatrix} F_{1y} \\ M_1 \\ F_{2y} \\ M_2 \end{Bmatrix} = \frac{EI}{L^3} \begin{bmatrix} 12 & 6L & -12 & 6L \\ 6L & 4L^2 & -6L & 2L^2 \\ -12 & -6L & 12 & -6L \\ 6L & 2L^2 & -6L & 4L^2 \end{bmatrix} \begin{Bmatrix} v_1 \\ \theta_1 \\ v_2 \\ \theta_2 \end{Bmatrix}$$

(7.45a)

or symbolically,

$$\{F\} = [k]_e \{\delta\}_e$$

(7.45b)

in which the matrix $[k]_e$ represents the force and moment components. Likewise, $\{\delta\}_e$ represents both deflections and slopes. The *element stiffness matrix* lying along a coordinate x is then

$$[k]_e = \frac{EI}{L^3} \begin{bmatrix} 12 & 6L & -12 & 6L \\ 6L & 4L^2 & -6L & 2L^2 \\ -12 & -6L & 12 & -6L \\ 6L & 2L^2 & -6L & 4L^2 \end{bmatrix} = \frac{EI}{L^3} \begin{bmatrix} 12 & 6L & -12 & 6L \\ & 4L^2 & -6L & 2L^2 \\ & & 12 & -6L \\ \text{Symmetric} & & & 4L^2 \end{bmatrix}$$

(7.46)

With development of the stiffness matrix, formulation and solution of problems involving beam elements proceeds like that of bar elements, as demonstrated in the examples to follow.

EXAMPLE 7.9 Displacements of a Uniformly Loaded Cantilever Beam

A cantilever beam of length L and flexural rigidity EI carries a uniformly distributed load of intensity p, as illustrated in Fig. 7.19a. Find the vertical deflection and rotation at the free end.

Solution Only one finite element to represent the entire beam is used. The distributed load is replaced by the equivalent forces and moments, as shown in Fig. 7.19b (see case 3 in Table D.4). The boundary conditions are given by

$$v_1 = 0 \qquad \theta_1 = 0$$

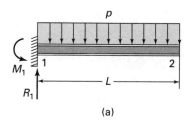
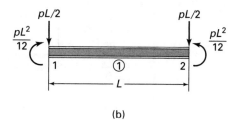

(a)   (b)

FIGURE 7.19.   *Example 7.9. (a) Beam with distributed load; (b) the equivalent nodal forces.*

It is noted that

$$F_{1y} = F_{2y} = -\frac{pL}{2} \qquad M_2 = -M_1 = \frac{pL^2}{12}$$

Then, the *force–displacement relations*, Eqs. (7.45a), simplifies to

$$\left\{\begin{array}{c} -\dfrac{pL}{2} \\[2mm] \dfrac{pL^2}{12} \end{array}\right\} = \frac{EI}{L^3}\begin{bmatrix} 12 & -6L \\ -6L & 4L \end{bmatrix}\left\{\begin{array}{c} v_2 \\ \theta_2 \end{array}\right\}$$

Solving for the *displacements*,

$$\left\{\begin{array}{c} v_2 \\ \theta_2 \end{array}\right\} = \frac{L}{6EI}\begin{bmatrix} 2L^2 & 3L \\ 3L & 6 \end{bmatrix}\left\{\begin{array}{c} -\dfrac{pL}{2} \\[2mm] \dfrac{pL^2}{12} \end{array}\right\}$$

Subsequent to the multiplication, we obtain

$$\left\{\begin{array}{c} v_2 \\ \theta_2 \end{array}\right\} = \left\{\begin{array}{c} -\dfrac{W L^4}{8EI} \\[3mm] -\dfrac{L^3}{6EI} \end{array}\right\}$$

Comment The minus sign indicates a downward deflection and a clockwise rotation at the right end, node 2.

EXAMPLE 7.10   Analysis of a Statically Indeterminate Stepped Beam

A propped cantilever beam of flexural rigidity EI and $2EI$ for the parts 1–2 and 2–3, respectively, supports a concentrated load P at point 2 (Fig. 7.20a). Calculate (a) the nodal displacements; (b) the nodal forces and moments.

Solution   The beam is discretized into elements 1 and 2 with nodes 1, 2, and 3, as illustrated in Fig. 7.20a. There are a total of six displacement components for the beam before the boundary conditions are applied. The order of the system stiffness matrix must therefore be 6 × 6.

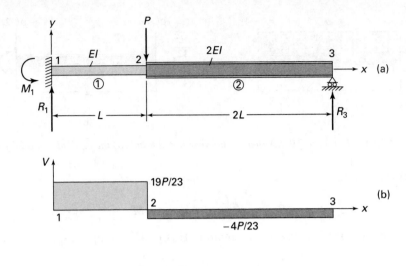

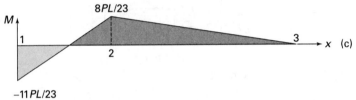

FIGURE 7.20. *Example 7.10. (a) Stepped beam with a load; (b) shear diagram; (c) bending moment diagram.*

Applying Eq. (7.46), the stiffness matrix for the *element 1*, with $(EI/L^3)_1 = EI/L^3$ and $L_1 = L$, may be written as

$$[k]_1 = \frac{EI}{L^3} \begin{array}{c} \begin{array}{cccccc} v_1 & \theta_1 & v_2 & \theta_2 & v_3 & \theta_3 \end{array} \\ \left[\begin{array}{cccc:cc} 12 & 6L & -12 & 6L & 0 & 0 \\ 6L & 4L^2 & -6L & 2L^2 & 0 & 0 \\ -12 & -6L & 12 & -6L & 0 & 0 \\ 6L & 2L^2 & -6L & 4L^2 & 0 & 0 \\ \hdashline 0 & 0 & 0 & 0 & 0 & 0 \\ 0 & 0 & 0 & 0 & 0 & 0 \end{array} \right] \begin{array}{c} v_1 \\ \theta_1 \\ v_2 \\ \theta_2 \\ v_3 \\ \theta_3 \end{array} \end{array} \quad \textbf{(c)}$$

Likewise, with $(EI/L^3)_2 = EI/4L^3$ and $L_2 = 2L$, for *element 2*, after rearrangement:

$$[k]_2 = \frac{EI}{L^3} \begin{array}{c} \begin{array}{cccccc} v_1 & \theta_1 & v_2 & \theta_2 & v_3 & \theta_3 \end{array} \\ \left[\begin{array}{cc:cccc} 0 & 0 & 0 & 0 & 0 & 0 \\ 0 & 0 & 0 & 0 & 0 & 0 \\ \hdashline 0 & 0 & 3 & 3L & -3 & 3L \\ 0 & 0 & 3L & 4L^2 & -3L & 2L^2 \\ 0 & 0 & -3 & -3L & 3 & -3L \\ 0 & 0 & 3L & -2L^2 & -3L & 4L^4 \end{array} \right] \begin{array}{c} v_1 \\ \theta_1 \\ v_2 \\ \theta_2 \\ v_3 \\ \theta_3 \end{array} \end{array}$$

Note that, in the preceding, the nodal displacements are shown to indicate the associativity of the rows and columns of the member stiffness matrices. Hence, rows and columns of zeros are added (boxed in by the dashed lines).

a. The *system stiffness matrix* of the beam can now be superimposed $[K] = [k]_1 + [k]_2$. The beam governing equations, with $F_{2y} = -P$ (Fig. 7.20b), are therefore

$$\begin{Bmatrix} R_1 \\ M_1 \\ -P \\ 0 \\ R_3 \\ 0 \end{Bmatrix} = \frac{EI}{L^3} \begin{bmatrix} 12 & 6L & -12 & 6L & 0 & 0 \\ 6L & 4L^2 & -6L & 2L^2 & 0 & 0 \\ -12 & -6L & 15 & -3L & 3 & 3L \\ 6L & 2L^2 & -3L & 8L^2 & -3L & 2L^2 \\ 0 & 0 & -3 & -3L & 3 & -3L \\ 0 & 0 & 3L & 2L^2 & -3L & 4L^2 \end{bmatrix} \begin{Bmatrix} v_1 \\ \theta_1 \\ v_2 \\ \theta_2 \\ v_3 \\ \theta_3 \end{Bmatrix}$$

(7.47)

The *boundary conditions* are $v_1 = 0$, $\theta_1 = 0$, and $v_3 = 0$. After multiplying these equations with the corresponding unknown displacements, we obtain

$$\begin{Bmatrix} -P \\ 0 \\ 0 \end{Bmatrix} = \frac{EI}{L^3} \begin{bmatrix} 15 & -3L & 3L \\ -3L & 8L^2 & 2L^2 \\ 3L & 2L^2 & 4L^2 \end{bmatrix} \begin{Bmatrix} v_2 \\ \theta_2 \\ \theta_3 \end{Bmatrix}$$ **(d)**

and

$$\begin{Bmatrix} R_1 \\ M_1 \\ R_3 \end{Bmatrix} = \frac{EI}{L^3} \begin{bmatrix} -12 & 6L & 0 \\ -6L & 2L^2 & 0 \\ -3 & -3L & -3L \end{bmatrix} \begin{Bmatrix} v_2 \\ \theta_2 \\ \theta_3 \end{Bmatrix}$$ **(e)**

Solving Eqs. (d), we find *deflection and slopes* as follows:

$$\begin{Bmatrix} v_2 \\ \theta_2 \\ \theta_3 \end{Bmatrix} = \frac{L^2}{276EI} \begin{bmatrix} 28L & 18 & -30 \\ 18 & 51/L & -39/L \\ -30 & -39/L & 111/L \end{bmatrix} \begin{Bmatrix} -P \\ 0 \\ 0 \end{Bmatrix} = \frac{PL^2}{276EI} \begin{Bmatrix} -28L \\ -18 \\ 30 \end{Bmatrix}$$

The minus sign means a downward deflection at node 2 and clockwise rotation of left end 1; the positive sign means a counterclockwise rotation at right end 3 (Fig. 7.20), as appreciated intuitively.

b. Substituting the displacements found into Eq. (e), after multiplying and simplifying, *nodal forces* and *moments* are

$$\begin{Bmatrix} R_1 \\ M_1 \\ R_3 \end{Bmatrix} = \frac{P}{276L} \begin{bmatrix} -12 & 6L & 0 \\ -6L & 2L^2 & 0 \\ -3 & -3L & -3L \end{bmatrix} \begin{Bmatrix} -28L \\ -18 \\ 30 \end{Bmatrix} = \frac{1}{23} \begin{Bmatrix} 19P \\ 11PL \\ 4P \end{Bmatrix}$$

Comment Usually, it is necessary to obtain the nodal forces and moments associated with *each element* to analyze the whole structure. For the case under consideration, it may readily be seen from a free-body diagram of element 2 that $M_2 = R_3 (2L) = 8PL/23$. So, we have the shear and moment diagrams for the beam, as shown in Figs. 7.20b and c, respectively.

7.12 PROPERTIES OF TWO-DIMENSIONAL ELEMENTS

Now we define a number of basic quantities relevant to an individual finite element of an isotropic elastic body. In the interest of simple presentation, in this section the relationships are written only for the two-dimensional case. The general formulation of the finite element method applicable to any structure is presented in the next section. Solutions of plane stress and plane strain problems are illustrated in detail in Section 7.14. The analyses of axisymmetric structures and thin plates employing the finite element are given in Chapters 8 and 13, respectively.

To begin with, the relatively thin, continuous body shown in Fig. 7.21a is replaced or discretized by an assembly of finite elements (triangles, for example) indicated by the dashed lines (Fig. 7.21b). These elements are connected not only at their corners or nodes but along the interelement boundaries as well. The basic unknowns are the nodal displacements.

Displacement Matrix

The nodal displacements are related to the internal displacements throughout the entire element by means of a *displacement function*. Consider the typical element *e* in Fig. 7.21b, shown isolated in Fig. 7.22a. Designating the nodes *i*, *j*, and *m*, the element nodal displacement matrix is

$$\{\delta\}_e = \{u_i, v_i, u_j, v_j, u_m, v_m\} \qquad \textbf{(7.48a)}$$

or, for convenience, expressed in terms of submatrices δ_u and δ_v,

$$\{\delta\}_e = \left\{\frac{\delta_u}{\delta_v}\right\} = \{u_i, u_j, u_m, v_i, v_j, v_m\} \qquad \textbf{(7.48b)}$$

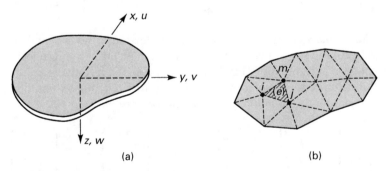

(a)   (b)

FIGURE 7.21. *Plane stress region (a) before and (b) after division into finite elements.*

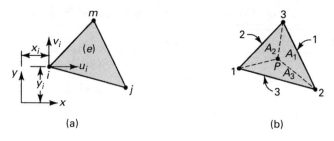

FIGURE 7.22.   *Triangular finite element.*

where the braces indicate a *column* matrix. The displacement function defining the displacement at any point within the element, $\{f\}_e$, is given by

$$\{f\}_e = \{u(x, y), v(x, y)\} \tag{7.49}$$

which may also be expressed as

$$\{f\}_e = [N]\{\delta\}_e \tag{7.50}$$

where the matrix $[N]$ is a function of position, to be obtained later for a specific element. It is desirable that a displacement function $\{f\}_e$ be selected such that the true displacement field will be represented as closely as possible. The approximation should result in a finite element solution that converges to the exact solution as the element size is progressively decreased.

Strain, Stress, and Elasticity Matrices

The strain, and hence the stress, are defined uniquely in terms of displacement functions (see Chap. 2). The strain matrix is of the form

$$\{\varepsilon\}_e = \{\varepsilon_x, \varepsilon_y, \gamma_{xy}\}$$

$$= \left\{ \frac{\partial u}{\partial x}, \frac{\partial v}{\partial y}, \frac{\partial u}{\partial y} + \frac{\partial v}{\partial x} \right\} \tag{7.51a}$$

or

$$\{\varepsilon\}_e = [B]\{\delta\}_e \tag{7.51b}$$

where $[B]$ is also yet to be defined.

Similarly, the state of stress throughout the element is, from Hooke's law,

$$\{\sigma\}_e = \frac{E}{1 - \nu^2} \begin{bmatrix} 1 & \nu & 0 \\ \nu & 1 & 0 \\ 0 & 0 & (1 - \nu)/2 \end{bmatrix} \begin{Bmatrix} \varepsilon_x \\ \varepsilon_y \\ \gamma_{xy} \end{Bmatrix} \tag{a}$$

Succinctly,

$$\{\sigma\}_e = [D]\{\varepsilon\}_e \tag{7.52}$$

TABLE 7.3.   *Elastic Constants for Two-Dimensional Problems*

Quantity	Plane strain	Plane stress
λ	$\dfrac{E}{1 - \nu^2}$	$\dfrac{E(1 - \nu)}{(1 + \nu)(1 - 2\nu)}$
D_{12}	ν	$\dfrac{\nu}{1 - \nu}$
D_{33}	$\dfrac{1 - \nu}{2}$	$\dfrac{1 - 2\nu}{2(1 - \nu)}$

where $[D]$, an *elasticity matrix*, contains material properties. If the element is subjected to thermal or *initial strain*, the stress matrix becomes

$$\{\sigma\}_e = [D](\{\varepsilon\} - \{\varepsilon_0\})_e \tag{7.53}$$

The thermal strain matrix, for the case of plane stress, is given by $\{\varepsilon_0\} = \{\alpha T, \alpha T, 0\}$ (Sec. 3.8). Comparing Eqs. (a) and (7.52), it is clear that

$$[D] = \frac{E}{1 - \nu^2} \begin{bmatrix} 1 & \nu & 0 \\ \nu & 1 & 0 \\ 0 & 0 & (1 - \nu)/2 \end{bmatrix} \tag{7.54}$$

In general, the elasticity matrix may be represented in the form

$$[D] = \lambda \begin{bmatrix} 1 & D_{12} & 0 \\ D_{12} & 1 & 0 \\ 0 & 0 & D_{33} \end{bmatrix} \tag{7.55}$$

It is recalled from Section 3.5 that two-dimensional problems are of two classes: plane stress and plane strain. The constants λ, D_{12}, and D_{33} for a two-dimensional problem are given in Table 7.3.

7.13   GENERAL FORMULATION OF THE FINITE ELEMENT METHOD

A convenient method for executing the finite element procedure relies on the minimization of the total potential energy of the system, expressed in terms of displacement functions. Consider again in this regard an elastic body (Fig. 7.21). The principle of potential energy, from Eq. (10.21), is expressed for the *entire body* as follows:

$$\begin{aligned} \Delta\Pi = {}& \sum_{1}^{n} \int_V (\sigma_x \, \Delta\varepsilon_x + \cdots + \sigma_z \, \Delta\varepsilon_z) \, dV \\ & - \sum_{1}^{n} \int_V (F_x \, \Delta u + F_y \, \Delta v + F_z \, \Delta w) \, dV \\ & - \sum_{1}^{n} \int_s (p_x \, \Delta u + p_y \, \Delta v + p_z \, \Delta w) \, ds = 0 \end{aligned} \tag{7.56}$$

Note that variation notation δ has been replaced by Δ to avoid confusing it with nodal displacement. In Eq. (7.56),

n = number of elements comprising the body

V = volume of a discrete element

s = portion of the boundary surface area over which forces are prescribed

F = body forces per unit volume

p = prescribed boundary force or surface traction per unit area

Through the use of Eq. (7.49), Eq. (7.56) may be expressed in the following matrix form:

$$\sum_{1}^{n} \int_{V} \left(\{\Delta\varepsilon\}_e^T \{\sigma\}_e - \{\Delta f\}_e^T \{F\}_e \right) dV - \sum_{1}^{n} \int_{s} \{\Delta f\}_e^T \{p\}_e \, ds = 0 \qquad \textbf{(a)}$$

where superscript T denotes the transpose of a matrix. Now, using Eqs. (7.50), (7.51), and (7.53), Eq. (a) becomes

$$\sum_{1}^{n} \{\Delta\delta\}_e^T ([k]_e\{\delta\}_e - \{Q\}_e) = 0 \qquad \textbf{(b)}$$

The element *stiffness* matrix $[k]_e$ and element *nodal force* matrix $\{Q\}_e$ (due to body force, initial strain, and surface traction) are

$$[k]_e = \int_{V} [B]^T [D][B] \, dV \qquad \textbf{(7.57)}$$

$$\{Q\}_e = \int_{V} [N]^T \{F\} \, dV + \int_{V} [B]^T [D]\{\varepsilon_0\} \, dV + \int_{s} [N]^T \{p\} \, ds \qquad \textbf{(7.58)}$$

It is clear that the variations in $\{\delta\}_e$ are independent and arbitrary, and from Eq. (b) we may therefore write

$$[k]_e\{\delta\}_e = \{Q\}_e \qquad \textbf{(7.59)}$$

We next derive the governing equations appropriate to the entire continuous body. The assembled form of Eq. (b) is

$$(\Delta\delta)^T ([K]\{\delta\} - \{Q\}) = 0 \qquad \textbf{(c)}$$

This expression must be satisfied for arbitrary variations of *all* nodal displacements $\{\Delta\delta\}$. This leads to the following equations of equilibrium for nodal forces for the entire structure, the *system* equations:

$$[K]\{\delta\} = \{Q\} \qquad \textbf{(7.60)}$$

where

$$[K] = \sum_{1}^{n} [k]_e, \qquad \{Q\} = \sum_{1}^{n} \{Q\}_e \qquad \textbf{(7.61)}$$

It is noted that structural matrix $[K]$ and the total or equivalent nodal force matrix $\{Q\}$ are found by proper superposition of all element stiffness and nodal force matrices, respectively, as discussed in Section 7.10 for trusses.

Outline of General Finite Element Analysis

We can now summarize the *general procedure* for solving a problem by application of the finite element method as follows:

1. Calculate $[k]_e$ from Eq. (7.57) in terms of the given element properties. Generate $[K] = \Sigma[k]_e$.
2. Calculate $\{Q\}_e$ from Eq. (7.58) in terms of the applied loading. Generate $\{Q\} = \Sigma\{Q\}_e$.
3. Calculate the nodal displacements from Eq. (7.60) by satisfying the boundary conditions: $\{\delta\} = [K]^{-1}\{Q\}$.
4. Calculate the element strain using Eq. (7.51): $\{\varepsilon\}_e = [B]\{\delta\}_e$.
5. Calculate the element stress using Eq. (7.53): $\{\sigma\}_e = [D](\{\varepsilon\} - \{\varepsilon_0\})_e$.

A simple block diagram of finite element analysis is shown in Fig. 7.23.

When the stress found is uniform throughout each element, this result is usually interpreted two ways: the stress obtained for an element is assigned to its centroid; if the material properties of the elements connected at a node are the same, the average of the stresses in the elements is assigned to the common node.

The foregoing outline will be better understood when applied to a triangular element in the next section. Formulation of the properties of a simple *one-dimensional* element, using relations developed in this section, is illustrated in the following example.

EXAMPLE 7.11   Deflection of a Bar under Combined Loading

A *bar element* of constant cross-sectional area A, length L, and modulus of elasticity E is subjected to a distributed load p_x per unit length and a uniform temperature change T (Fig. 7.24a). Determine (a) the stiffness matrix, (b) the total nodal force matrix, and (c) the deflection of the right end u_2 for fixed left end and $p_x = 0$ (Fig. 7.24b).

Solution

a. As before (see Sec. 7.7), we shall assume that the displacement u at any point within the element varies linearly with x:

$$f = u = a_1 + a_2 x \tag{d}$$

wherein a_1 and a_2 are constants. The axial displacements of nodes 1 and 2 are $u_1 = a_1$ and $u_2 = a_1 + a_2 L$ from which $a_2 = -(u_1 - u_2)/L$. Substituting this into Eq. (d),

$$u = \begin{bmatrix} 1 - \dfrac{x}{L} & \dfrac{x}{L} \end{bmatrix} \begin{Bmatrix} u_1 \\ u_2 \end{Bmatrix} = [N]\{\delta\} \tag{e}$$

Applying Eq. (7.51a), the strain in the element is

$$\varepsilon_x = \frac{u_2 - u_1}{L}$$

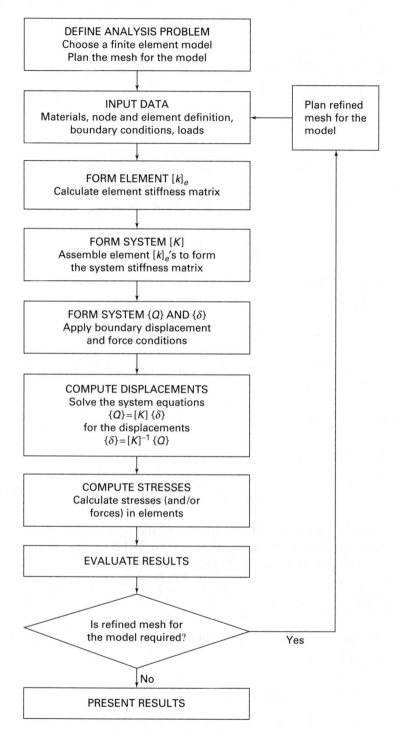

FIGURE 7.23.   *The finite element analysis block diagram [Ref. 7.8].*

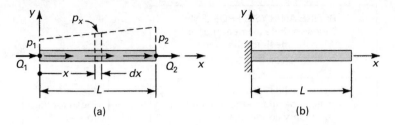

FIGURE 7.24. *Example 7.11. (a) Bar element subjected to an axial load p_x and uniform temperature change T; (b) fixed end bar.*

or in matrix form

$$\varepsilon_x = \frac{1}{L}[-1 \quad 1]\begin{Bmatrix} u_1 \\ u_2 \end{Bmatrix} = [B]\{\delta\} \qquad \textbf{(f)}$$

For a one-dimensional element, we have $D = E$, and the stress from Eq. (7.52) equals

$$\sigma_x = \frac{E}{L}[-1 \quad 1]\begin{Bmatrix} u_1 \\ u_2 \end{Bmatrix}$$

The element stiffness matrix is obtained upon introduction of $[B]$ from Eq. (f) into Eq. (7.57):

$$[k]_e = [B]^T E[B]AL = \begin{Bmatrix} -1 \\ 1 \end{Bmatrix}[-1 \quad 1]\frac{EA}{L} = \frac{EA}{L}\begin{bmatrix} 1 & -1 \\ -1 & 1 \end{bmatrix} \qquad \textbf{(7.62)}$$

b. Referring to Fig. 7.24a,

$$p_x = p_1 + \frac{p_2 - p_1}{L}x \qquad \textbf{(7.63)}$$

where p_1 and p_2 are the intensities of the load per unit length at nodes 1 and 2, respectively. Substitution of $[N]$ from Eq. (e) together with Eq. (7.63) into Eq. (7.58) yields

$$\{Q\}_e^p = \int_0^L [N]^T\{p_x\}dx = \int_0^L \begin{bmatrix} 1 - x/L \\ x/L \end{bmatrix}\left[p_1 + \frac{p_2 - p_1}{L}x \right]dx$$

The distributed load effects are obtained by integrating the preceding equations:

$$\{Q\}_e^p = \begin{Bmatrix} Q_1 \\ Q_2 \end{Bmatrix}_e^p = \frac{L}{6}\begin{Bmatrix} 2p_1 + p_2 \\ p_1 + 2p_2 \end{Bmatrix} \qquad \textbf{(g)}$$

The strain due to the temperature change is $\varepsilon_0 = \alpha T$, where α is the coefficient of thermal expansion. Inserting $[B]$ from Eq. (f) into Eq. (7.58),

$$\{Q\}_e^t = \int_0^L \int_A [B]^T [D] \{\varepsilon_0\} dA\, dx = \int_0^L \frac{1}{L} \begin{Bmatrix} -1 \\ 1 \end{Bmatrix} EA\alpha(T)\, dx$$

The thermal strain effects are then

$$\{Q\}_e^t = \begin{Bmatrix} Q_1 \\ Q_2 \end{Bmatrix}_e^t = EA\alpha(T) \begin{Bmatrix} -1 \\ 1 \end{Bmatrix} \tag{h}$$

The total element nodal matrix is obtained by adding Eqs. (g) and (h):

$$\{Q\}_e = \frac{L}{6} \begin{Bmatrix} 2p_1 + p_2 \\ p_1 + 2p_2 \end{Bmatrix} + EA\alpha(T) \begin{Bmatrix} -1 \\ 1 \end{Bmatrix} \tag{7.64}$$

c. The nodal force–displacement relations (7.60) now take the form

$$EA\alpha(T) \begin{Bmatrix} -1 \\ 1 \end{Bmatrix} = \frac{EA}{L} \begin{bmatrix} 1 & -1 \\ -1 & 1 \end{bmatrix} \begin{Bmatrix} 0 \\ u_2 \end{Bmatrix}$$

from which the elongation of the bar equals $u_2 = \alpha(T)L$, a predictable result.

7.14 TRIANGULAR FINITE ELEMENT

Because of the relative ease with which the region within an arbitrary boundary can be approximated, the triangle is used extensively in finite element assemblies. Before deriving the properties of the triangular element, we describe *area* or *triangular* coordinates, which are very useful for the simplification of the displacement functions.

In this section, we derive the basic *constant strain triangular (CST)* plane stress and strain element. Note that there are a variety of two-dimensional finite element types that lead to better solutions. Examples are *linear strain triangular (LST)* elements, triangular elements having additional side and interior nodes, rectangular elements with corner nodes, and rectangular elements having additional side nodes [Refs. 7.4, 7.6]. The LST element has six nodes: the usual corner nodes plus three additional nodes located at the midpoints of the sides. The procedures for the development of the LST element equations follow the identical steps as that of the CST element.

Consider the triangular finite element 1 2 3 (where $i = 1$, $j = 2$, and $m = 3$) shown in Fig. 7.22b, in which the *counterclockwise* numbering convention of nodes and sides is indicated. A point P located within the element, by connection with the corners of the element, forms three subareas denoted A_1, A_2, and A_3. The ratios of these areas to the total area A of the triangle locate P and represent the area coordinates:

$$L_1 = \frac{A_1}{A}, \qquad L_2 = \frac{A_2}{A}, \qquad L_3 = \frac{A_3}{A} \tag{7.65}$$

It follows from this that

$$L_1 + L_2 + L_3 = 1 \tag{7.66}$$

and consequently only two of the three coordinates are independent. A useful property of area coordinates is observed through reference to Fig. 7.22b and Eq. (7.65):

$$L_n = 0 \quad \text{on side } n, \qquad n = 1, 2, 3$$

and

$$
\begin{aligned}
L_i &= 1, && L_j = L_m = 0 \quad \text{at node } i \\
L_j &= 1, && L_i = L_m = 0 \quad \text{at node } j \\
L_m &= 1, && L_i = L_j = 0 \quad \text{at node } m
\end{aligned}
$$

The area A of the triangle may be expressed in terms of the coordinates of two sides, for example, 2 and 3:

$$A = \tfrac{1}{2}\{[(x_2 - x_1)\boldsymbol{i} + (y_2 - y_1)\boldsymbol{j}] \times [(x_3 - x_1)\boldsymbol{i} + (y_3 - y_1)\boldsymbol{j}]\} \cdot \boldsymbol{k}$$

or

$$2A = \det \begin{vmatrix} 1 & x_1 & y_1 \\ 1 & x_2 & y_2 \\ 1 & x_3 & y_3 \end{vmatrix} \tag{7.67a}$$

Here $\boldsymbol{i}, \boldsymbol{j}$, and $\boldsymbol{k}$ are the unit vectors in the x, y, and z directions, respectively. Two additional expressions are similarly found. In general, we have

$$2A = a_j b_i - a_i b_j = a_m b_j - a_j b_m = a_i b_m - a_m b_i \tag{7.67b}$$

where

$$
\begin{aligned}
a_i &= x_m - x_j, & b_i &= y_j - y_m \\
a_j &= x_i - x_m, & b_j &= y_m - y_i \\
a_m &= x_j - x_i, & b_m &= y_i - y_j
\end{aligned} \tag{7.68}
$$

Note that a_j, a_m, b_j, and b_m can be found from definitions of a_i and b_i with the permutation of the subscripts in the order $ijmijm$, and so on.

Similar equations are derivable for subareas A_1, A_2, and A_3. The resulting expressions, together with Eq. (7.65), lead to the following relationship between area and Cartesian coordinates:

$$
\begin{Bmatrix} L_1 \\ L_2 \\ L_3 \end{Bmatrix} = \frac{1}{2A} \begin{bmatrix} 2c_{23} & b_1 & a_1 \\ 2c_{31} & b_2 & a_2 \\ 2c_{12} & b_3 & a_3 \end{bmatrix} \begin{Bmatrix} 1 \\ x \\ y \end{Bmatrix} \tag{7.69}
$$

where

$$
\begin{aligned}
c_{23} &= x_2 y_3 - x_3 y_2 \\
c_{31} &= x_3 y_1 - x_1 y_3 \\
c_{12} &= x_1 y_2 - x_2 y_1
\end{aligned} \tag{7.70}
$$

Note again that, given any of these expressions for c_{ij}, the others may be obtained by permutation of the subscripts.

Now we explore the properties of an ordinary triangular element of a continuous body in a state of plane stress or plane strain (Fig. 7.22). The nodal displacements are

$$\{\delta\}_e = \{u_1, u_2, u_3, v_1, v_2, v_3\} \tag{a}$$

The displacement throughout the element is provided by

$$\{f\}_e = \begin{bmatrix} u_1 & u_2 & u_3 \\ v_1 & v_2 & v_3 \end{bmatrix} \begin{Bmatrix} L_1 \\ L_2 \\ L_3 \end{Bmatrix} \tag{7.71}$$

Matrices $[N]$ and $[B]$ of Eqs. (7.50) and (7.51b) are next evaluated, beginning with

$$\{f\}_e = [N]\{\delta\}_e \tag{b}$$

$$\{\varepsilon\}_e = [B]\{\delta\}_e \tag{c}$$

We observe that Eqs. (7.71) and (b) are equal, provided that

$$[N] = \begin{bmatrix} L_1 & L_2 & L_3 & 0 & 0 & 0 \\ 0 & 0 & 0 & L_1 & L_2 & L_3 \end{bmatrix} \tag{7.72}$$

The strain matrix is obtained by substituting Eqs. (7.71) and (7.69) into Eq. (7.51a):

$$\begin{Bmatrix} \varepsilon_x \\ \varepsilon_y \\ \gamma_{xy} \end{Bmatrix}_e = \frac{1}{2A} \begin{bmatrix} b_1 & b_2 & b_3 & 0 & 0 & 0 \\ 0 & 0 & 0 & a_1 & a_2 & a_3 \\ a_1 & a_2 & a_3 & b_1 & b_2 & b_3 \end{bmatrix} \begin{Bmatrix} u_1 \\ u_2 \\ u_3 \\ v_1 \\ v_2 \\ v_3 \end{Bmatrix} \tag{d}$$

Here A, a_i, and b_i are defined by Eqs. (7.67) and (7.68). The strain (stress) is observed to be constant throughout, and the element of Fig. 7.22 is thus referred to as a *constant strain triangle*. Comparing Eqs. (c) and (d), we have

$$[B] = \frac{1}{2A} \begin{bmatrix} b_1 & b_2 & b_3 & 0 & 0 & 0 \\ 0 & 0 & 0 & a_1 & a_2 & a_3 \\ a_1 & a_2 & a_3 & b_1 & b_2 & b_3 \end{bmatrix} \tag{7.73}$$

The stiffness of the element can now be obtained through the use of Eq. (7.57):

$$[k]_e = [B]^T[D][B]tA \tag{e}$$

Let us now define

$$[D^*] = \frac{t[D]}{4A} = \begin{bmatrix} D^*_{11} & D^*_{12} & D^*_{13} \\ & D^*_{22} & D^*_{23} \\ \text{Symm.} & & D^*_{33} \end{bmatrix} \tag{7.74}$$

where $[D]$ is given by Eq. (7.54) for plane stress. Assembling Eq. (e) together with Eqs. (7.73) and (7.74) and expanding, the stiffness matrix is expressed in the following partitioned form of order 6×6:

$$[k]_e = \begin{bmatrix} k_{uu,\,ln} & k_{uv,\,ln} \\ k^T_{uv,\,ln} & k_{vv,\,ln} \end{bmatrix}, \qquad l, n = 1, 2, 3 \tag{7.75}$$

where the submatrices are

$$k_{uu,\,ln} = D^*_{11}b_lb_n + D^*_{33}a_la_n + D^*_{13}(b_la_n + b_na_l)$$
$$k_{vv,\,ln} = D^*_{33}b_lb_n + D^*_{22}a_la_n + D^*_{23}(b_la_n + b_na_l) \tag{7.76}$$
$$k_{uv,\,ln} = D^*_{13}b_lb_n + D^*_{23}a_lb_n + D^*_{12}b_la_n + D^*_{33}b_na_l$$

Element Nodal Forces

Finally, we consider the determination of the element nodal force matrices. The nodal force owing to a constant *body force* per unit volume is, from Eqs. (7.58) and (7.72),

$$\{Q\}^b_e = \int_V [N]^T\{F\}\,dV = \int_V \begin{bmatrix} L_1 & 0 \\ L_2 & 0 \\ L_3 & 0 \\ 0 & L_1 \\ 0 & L_2 \\ 0 & L_3 \end{bmatrix} \begin{Bmatrix} F_x \\ F_y \end{Bmatrix} dV \tag{f}$$

For an element of constant thickness, this expression is readily integrated to yield*

$$\{Q\}_e = \tfrac{1}{3}At\{F_x, F_x, F_x, F_y, F_y, F_y\} \tag{7.77}$$

The nodal forces associated with the weight of an element are observed to be equally distributed at the nodes.

---

*For a general function $f = L^\alpha_1 L^\beta_2 L^\gamma_3$, defined in area coordinates, the integral of f over any triangular area A is given by

$$\int_A L^\alpha_1 L^\beta_2 L^\gamma_3\,dA = 2A\,\frac{\alpha!\,\beta!\,\gamma!}{(\alpha + \beta + \gamma)!}$$

where α, β, γ are constants.

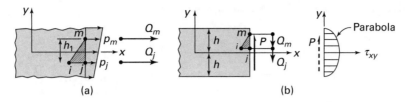

FIGURE 7.25. Nodal forces due to (a) linearly distributed load and (b) shear load.

The element nodal forces attributable to applied external loading may be determined either by evaluating the static resultants or by application of Eq. (7.58). Nodal force expressions for arbitrary nodes j and m are given next for a number of common cases (Prob. 7.40).

Linear load, $p(y)$ per unit area, Fig. 7.25a:

$$Q_j = \frac{h_1 t}{6}(2p_j + p_m), \qquad Q_m = \frac{h_1 t}{6}(2p_m + p_j) \qquad (7.78\text{a})$$

where t is the thickness of the element.

Uniform load is a special case of the preceding with $p_j = p_m = p$:

$$Q_j = Q_m = \tfrac{1}{2}ph_1 t \qquad (7.78\text{b})$$

End shear load, P, the resultant of a parabolic shear stress distribution defined by Eq. (3.24) (see Fig. 7.25b):

$$Q_m = \frac{3P}{4h}\left\{\tfrac{1}{2}(y_m - y_j) + \frac{1}{3h^2}\left[\frac{y_j}{4}(y_m^2 + y_j y_m + y_j^2) - \tfrac{3}{4}y_m^3\right]\right\}$$

$$Q_j = \frac{3P}{4h}\left\{\tfrac{1}{2}(y_m - y_j) - \frac{1}{3h^2}[(y_m - \tfrac{3}{4}y_j)(y_m^2 + y_j y_m + y_j^2) - \tfrac{3}{4}y_m^3]\right\} \quad (7.79)$$

Equation (7.75), together with those expressions given for the nodal forces, characterizes the constant strain element. These are substituted into Eq. (7.61) and subsequently into Eq. (7.60) in order to evaluate the nodal displacements by satisfying the boundary conditions.

The basic procedure employed in the finite element method is illustrated in the following simple problems.

EXAMPLE 7.12 Nodal Forces of a Plate Segment under Combined Loading

The element e shown in Fig. 7.26 represents a segment of a thin elastic plate having side 2–3 adjacent to its boundary. The plate is subjected to several loads as well as a uniform temperature rise of 50°C. Determine (a) the stiffness matrix and (b) the equivalent (or total) nodal force matrix for the element if a pressure of $p = 14$ MPa acts on side 2–3. Let $t = 0.3$ cm, $E = 200$ GPa, $\nu = 0.3$, specific weight $\gamma = 77$ kN/m³, and $\alpha = 12 \times 10^{-6}/°C$.

FIGURE 7.26.   *Example 7.12. A triangular plate.*

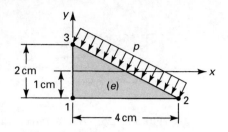

Solution   The origin of the coordinates is located at midlength of side
1–3, for convenience. However, it may be placed at any point in the x, y
plane. Applying Eq. (7.74), we have (in N/cm³)

$$[D^*] = \frac{t[D]}{4A} = \frac{10^6}{8} \begin{bmatrix} 3.33 & 0.99 & 0 \\ 0.99 & 3.3 & 0 \\ 0 & 0 & 1.16 \end{bmatrix} \qquad \textbf{(g)}$$

a. *Stiffness matrix:* The nodal points are located at

$$x_i = x_1 = 0, \qquad y_i = y_1 = -1$$
$$x_j = x_2 = 4, \qquad y_j = y_2 = -1 \qquad \textbf{(h)}$$
$$x_m = x_3 = 0, \qquad y_m = y_3 = 1$$

Using Eq. (7.68) and referring to Fig. 7.26, we obtain

$$a_i = a_1 = 0 - 4 = -4, \qquad b_i = b_1 = -1 - 1 = -2$$
$$a_j = a_2 = 0 - 0 = 0, \qquad b_j = b_2 = 1 + 1 = 2 \qquad \textbf{(i)}$$
$$a_m = a_3 = 4 - 0 = 4, \qquad b_m = b_3 = -1 + 1 = 0$$

Next, the first equation of (7.76), together with Eqs. (g) and (i), yields

$$k_{uu,\,11} = \frac{10^6}{8}[3.3(4) + 1.16(16)] = 3.97 \times 10^6$$

$$k_{uu,\,12} = k_{uu,\,21} = \frac{10^6}{8}[3.3(2)(-2) + 0] = -1.65 \times 10^6$$

$$k_{uu,\,13} = k_{uu,\,31} = \frac{10^6}{8}[0 + 1.16(4)(-4)] = -2.32 \times 10^6$$

$$k_{uu,\,22} = \frac{10^6}{8}[3.3(4) + 0] = 1.65 \times 10^6$$

$$k_{uu,\,23} = k_{uu,\,32} = 0$$

$$k_{uu,\,33} = \frac{10^6}{8}[0 + 1.16(16)] = 2.32 \times 10^6$$

The submatrix k_{uu} is thus

$$k_{uu} = 10^6 \begin{bmatrix} 3.97 & -1.65 & -2.32 \\ -1.65 & 1.65 & 0 \\ -2.32 & 0 & 2.32 \end{bmatrix}$$

Similarly, from the second and third equations of (7.76), we obtain the following matrices:

$$k_{uv} = 10^6 \begin{bmatrix} 2.15 & -1.16 & -0.99 \\ -0.99 & 0 & 0.99 \\ -1.16 & 1.16 & 0 \end{bmatrix}, \qquad k_{vv} = 10^6 \begin{bmatrix} 7.18 & -0.58 & -6.6 \\ -0.58 & 0.58 & 0 \\ -6.6 & 0 & 6.6 \end{bmatrix}$$

Assembling the preceding equations, the stiffness matrix of the element (in newtons per centimeter) is

$$[k]_e = 10^6 \left[\begin{array}{ccc:ccc} 3.97 & -1.65 & -2.32 & 2.15 & -1.16 & -0.99 \\ -1.65 & 1.65 & 0 & -0.99 & 0 & 0.99 \\ -2.32 & 0 & 2.32 & -1.16 & 1.16 & 0 \\ \hdashline 2.15 & -0.99 & -1.16 & 7.18 & -0.58 & -6.6 \\ -1.16 & 0 & 1.16 & -0.58 & 0.58 & 0 \\ -0.99 & 0.99 & 0 & -6.6 & 0 & 6.6 \end{array} \right] \qquad \text{(j)}$$

b. We next determine the nodal forces of the element owing to various loadings. The components of body force are $F_x = 0$ and $F_y = 0.077 \text{ N/cm}^3$.

Body force effects: Through the application of Eq. (7.77), it is found that

$$\{Q\}_e^b = \{0, \quad 0, \quad 0, \quad -0.0308, \quad -0.0308, \quad -0.0308\} \text{ N}$$

Surface traction effects: The total load, $st\{p\} = \sqrt{20}(0.3)$ $\{-2 \times 1400/\sqrt{20}, \ -4 \times 1400/\sqrt{20}\} = \{-840, -1680\}$, is equally divided between nodes 2 and 3. The nodal forces can therefore be expressed as

$$\{Q\}_e^p = \{0, \quad -420, \quad -420, \quad 0, \quad -840, \quad -840\} \text{ N}$$

Thermal strain effects: The initial strain associated with the 50°C temperature rise is $\varepsilon_0 = \alpha T = 0.0006$. From Eq. (7.59),

$$\{Q\}_e^t = [B]^T[D]\{\varepsilon_0\}(At)$$

Substituting matrix $[B]$, given by Eq. (7.73), into this equation, and the values of the other constants already determined, the nodal force is calculated as follows:

$$\{Q\}_e^t = \frac{1}{8} \begin{bmatrix} -2 & 0 & -4 \\ 2 & 0 & 0 \\ 0 & 0 & 4 \\ 0 & -4 & -2 \\ 0 & 0 & 2 \\ 0 & 4 & 0 \end{bmatrix} \frac{200 \times 10^5}{0.91} \begin{bmatrix} 1 & 0.3 & 0 \\ 0.3 & 1 & 0 \\ 0 & 0 & 0.35 \end{bmatrix} \begin{Bmatrix} 0.0006 \\ 0.0006 \\ 0 \end{Bmatrix} \qquad \text{(1.2)}$$

or

$$\{Q\}_e^t = \{-5142.85, 5142.85, 0, -10{,}285.70, 0, 10{,}285.70\} \text{ N}$$

Equivalent nodal force matrix: Summation of the nodal matrices due to the several effects yields the total element nodal force matrix:

$$\{Q\}_e = \begin{Bmatrix} Q_{x1} \\ Q_{x2} \\ Q_{x3} \\ Q_{y1} \\ Q_{y2} \\ Q_{y2} \end{Bmatrix} = \begin{Bmatrix} -5142.85 \\ 4742.85 \\ -400 \\ -10{,}285.73 \\ -840.03 \\ 9445.67 \end{Bmatrix} \text{ N}$$

If, in addition, there are any actual node forces, these must also be added to the value obtained.

7.15 CASE STUDIES IN PLANE STRESS

A good engineering case study includes all necessary data to analyze a problem in details and may come in many varieties [Refs. 7.8 and 7.9]. Obviously, in solid mechanics it deals with stress and deformation in load-carrying members or structures. In this section we briefly present four case studies limited to plane stress situations and CST finite elements. A cantilever beam supporting a concentrated load, a deep beam or plate in pure bending, a plate with a hole subjected to an axial loading, and a disk carrying concentrated diametral compression are members considered.

Recall from Chapter 3 that there are very few elasticity or "exact" solutions to two-dimensional problems, especially for any but the simplest shapes. As will become evident from the following discussion, the stress analyst and designer can reach a very accurate solution by employing proper techniques and modeling. Accuracy is often limited by the willingness to model all the significant features of the problem and pursue the FEA until convergence is reached.

CASE STUDY 7.1 Stresses in a Cantilever Beam under a Concentrated Load

A 0.3-cm-thick cantilever beam is subjected to a parabolically varying end shearing stress resulting in a total load of 5000 N (Fig. 7.27a). Divide the beam into two constant strain triangles and calculate the deflections. Let $E = 200$ GPa and $\nu = 0.3$.

Solution The discretized beam is shown in Fig. 7.27b.

Stiffness matrix: Inasmuch as the dimensions and material properties of element *a* are the same as those given in the previous example, $[k]_a$ is

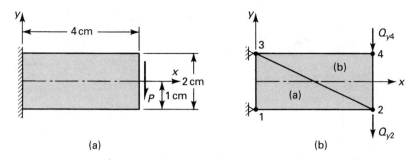

FIGURE 7.27. *Case Study 7.1. Cantilever beam (a) before and (b) after being discretized.*

defined by Eq. (j) of Example 7.12. For element b, assignment of $i = 2, j = 4$, and $m = 3$ [Eq. (7.68)] leads to

$$
\begin{aligned}
a_i &= a_2 = 0 - 4 = -4, & b_i &= b_2 = 1 - 1 = 0 \\
a_j &= a_4 = 4 - 0 = 4, & b_j &= b_4 = 1 + 1 = 2 \\
a_m &= a_3 = 4 - 4 = 0, & b_m &= b_3 = -1 - 1 = -2
\end{aligned}
$$

Substitution of these and Eq. (g) of Example 7.12 into Eqs. (7.76) yields

$$k_{uu,22} = \frac{10^6}{8}[3.3(0) + 1.16(16)] = 2.32 \times 10^6$$

$$k_{uu,44} = \frac{10^6}{8}[3.3(4) + 1.16(16)] = 3.97 \times 10^6$$

$$k_{uu,23} = k_{uu,32} = 0$$

$$k_{uu,24} = k_{uu,42} = \frac{10^6}{8}[3.3(0) + 1.16(-4)(4)] = -2.32 \times 10^6$$

$$k_{uu,43} = k_{uu,34} = \frac{10^6}{8}[3.3(2)(-2) + 1.16(0)] = -1.65 \times 10^6$$

$$k_{uu,33} = \frac{10^6}{8}[3.3(4) + 1.16(0)] = 1.65 \times 10^6$$

Thus,

$$
k_{uu} = 10^6 \begin{bmatrix} 2.32 & 0 & -2.32 \\ 0 & 1.65 & -1.65 \\ -2.32 & -1.65 & 3.97 \end{bmatrix}
$$

Similarly, we obtain

$$
k_{uv} = 10^6 \begin{bmatrix} 0 & 1.16 & -1.16 \\ 0.99 & 0 & -0.99 \\ -0.99 & -1.16 & 2.15 \end{bmatrix}
$$

$$k_{vv} = 10^6 \begin{bmatrix} 6.6 & 0 & -6.6 \\ 0 & 0.58 & -0.58 \\ -6.6 & -0.58 & 7.18 \end{bmatrix}$$

The stiffness matrix of element b is therefore

$$[k]_b = 10^6 \left[\begin{array}{ccc:ccc} 2.32 & 0 & -2.32 & 0 & 1.16 & -1.16 \\ 0 & 1.65 & -1.65 & 0.99 & 0 & -0.99 \\ -2.32 & -1.65 & 3.97 & -0.99 & -1.16 & 2.15 \\ \hdashline 0 & 0.99 & -0.99 & 6.6 & 0 & -6.6 \\ 1.16 & 0 & -1.16 & 0 & 0.58 & -0.58 \\ -1.16 & -0.99 & 2.15 & -6.6 & -0.58 & 7.18 \end{array} \right] \qquad \textbf{(a)}$$

The displacements u_4, v_4 and u_1, v_1 are not involved in elements a and b, respectively. Therefore, prior to the addition of $[k]_a$ and $[k]_b$ to form the system stiffness matrix, rows and columns of zeros must be added to each element matrix to account for the absence of these displacements. In doing so, Eqs. (j) of Example 7.12 and (a) become

$$[k]_a = 10^6 \left[\begin{array}{cccc:cccc} 3.97 & -1.65 & -2.32 & 0 & 2.15 & -1.16 & -0.99 & 0 \\ -1.65 & 1.65 & 0 & 0 & -0.99 & 0 & 0.99 & 0 \\ -2.32 & 0 & 2.32 & 0 & -1.16 & 1.16 & 0 & 0 \\ 0 & 0 & 0 & 0 & 0 & 0 & 0 & 0 \\ \hdashline 2.15 & -0.99 & -1.16 & 0 & 7.18 & -0.58 & -6.6 & 0 \\ -1.16 & 0 & 1.16 & 0 & -0.58 & 0.58 & 0 & 0 \\ -0.99 & 0.99 & 0 & 0 & -6.6 & 0 & 6.6 & 0 \\ 0 & 0 & 0 & 0 & 0 & 0 & 0 & 0 \end{array} \right]$$

$$\textbf{(b)}$$

and

$$[k]_b = 10^6 \left[\begin{array}{cccc:cccc} 0 & 0 & 0 & 0 & 0 & 0 & 0 & 0 \\ 0 & 2.32 & 0 & -2.32 & 0 & 0 & 1.16 & -1.16 \\ 0 & 0 & 1.65 & -1.65 & 0 & 0.99 & 0 & -0.99 \\ 0 & -2.32 & -1.65 & 3.97 & 0 & -0.99 & -1.16 & 2.15 \\ \hdashline 0 & 0 & 0 & 0 & 0 & 0 & 0 & 0 \\ 0 & 0 & 0.99 & -0.99 & 0 & 6.6 & 0 & -6.6 \\ 0 & 1.16 & 0 & -1.16 & 0 & 0 & 0.58 & -0.58 \\ 0 & -1.16 & -0.99 & 2.15 & 0 & -6.6 & -0.58 & 7.18 \end{array} \right]$$

$$\textbf{(c)}$$

Then, addition of Eqs. (b) and (c) yields the system matrix (in newtons per centimeter):

$$[K] = 10^6 \begin{bmatrix} 3.97 & -1.65 & -2.32 & 0 & 2.15 & -1.16 & -0.99 & 0 \\ -1.65 & 3.97 & 0 & -2.32 & -0.99 & 0 & 2.15 & -1.16 \\ -2.32 & 0 & 3.97 & -1.65 & -1.16 & 2.15 & 0 & -0.99 \\ 0 & -2.32 & -1.65 & 3.97 & 0 & -0.99 & -1.16 & 2.15 \\ 2.15 & -0.99 & -1.16 & 0 & 7.18 & -0.58 & -6.6 & 0 \\ -1.16 & 0 & 2.15 & -0.99 & -0.58 & 7.18 & 0 & -6.6 \\ -0.99 & 2.15 & 0 & -1.16 & -6.6 & 0 & 7.18 & -0.58 \\ 0 & -1.16 & -0.99 & 2.15 & 0 & -6.6 & -0.58 & 7.18 \end{bmatrix}$$

(d)

Nodal forces: Referring to Fig. 7.27b and applying Eq. (7.79), we obtain

$$Q_{y4} = \frac{3(-5000)}{4(1)}\{\tfrac{1}{2}(1 + 1) + \tfrac{1}{3}[-\tfrac{1}{4}(1 - 1 + 1) - \tfrac{3}{4}]\} = -2500 \text{ N}$$

$$Q_{y2} = \frac{3(-5000)}{4(1)}\{\tfrac{1}{2}(1 + 1) - \tfrac{1}{3}[1 + \tfrac{3}{4}(1 - 1 + 1) - \tfrac{3}{4}]\} = -2500 \text{ N}$$

Because no other external force exists, the system nodal force matrix is

$$\{Q\} = \{0, 0, 0, 0, 0 -2500, 0 -2500\}$$

Nodal displacements: The boundary conditions are

$$u_1 = u_3 = v_1 = v_3 = 0$$

The force–displacement relationship of the system is therefore

$$\{0, 0, 0, 0, 0, -2500, 0, -2500\} = [K]\{0, u_2, 0, u_4, 0, v_2, 0, v_4\} \quad \text{(e)}$$

Equation (e) is readily reduced to the form

$$\begin{Bmatrix} 0 \\ 0 \\ -2500 \\ -2500 \end{Bmatrix} = 10^6 \begin{bmatrix} 3.97 & -2.32 & 0 & -1.16 \\ -2.32 & 3.97 & -0.99 & 2.15 \\ 0 & -0.99 & 7.18 & -6.6 \\ -1.16 & 2.15 & -6.6 & 7.18 \end{bmatrix} \begin{Bmatrix} u_2 \\ u_4 \\ v_2 \\ v_4 \end{Bmatrix} \quad \text{(f)}$$

From this we obtain

$$\begin{Bmatrix} u_2 \\ u_4 \\ v_2 \\ v_4 \end{Bmatrix} = 10^{-6} \begin{bmatrix} 0.429 & 0.180 & 0.252 & 0.247 \\ 0.180 & 0.483 & -0.256 & -0.351 \\ 0.252 & -0.256 & 1.366 & 1.373 \\ 0.247 & -0.351 & 1.373 & 1.546 \end{bmatrix} \begin{Bmatrix} 0 \\ 0 \\ -2500 \\ -2500 \end{Bmatrix}$$

$$= \begin{Bmatrix} -0.0012 \\ 0.0015 \\ -0.0068 \\ -0.0073 \end{Bmatrix} \text{cm} \qquad \text{(g)}$$

The strains $\{\varepsilon\}_a$ may now be found upon introduction of Eqs. (i) of Example 7.12 and (g) into Eq. (d) of Section 7.14 as

$$\begin{Bmatrix} \varepsilon_x \\ \varepsilon_y \\ \gamma_{xy} \end{Bmatrix} = \frac{1}{8} \begin{bmatrix} -2 & 2 & 0 & 0 & 0 & 0 \\ 0 & 0 & 0 & -4 & 0 & 4 \\ -4 & 0 & 4 & -2 & 2 & 0 \end{bmatrix} \begin{Bmatrix} 0 \\ -0.0012 \\ 0 \\ 0 \\ -0.0068 \\ 0 \end{Bmatrix} = \begin{Bmatrix} -3 \\ 0 \\ -17 \end{Bmatrix} 10^{-4}$$

Finally, the stress is determined by multiplying $[D]$ by $\{\varepsilon\}_a$:

$$\begin{Bmatrix} \sigma_x \\ \sigma_y \\ \tau_{xy} \end{Bmatrix}_a = \frac{200 \times 10^9}{0.91} \begin{bmatrix} 1 & 0.3 & 0 \\ 0.3 & 1 & 0 \\ 0 & 0 & 0.35 \end{bmatrix} \begin{Bmatrix} -3 \\ 0 \\ -17 \end{Bmatrix} 10^{-4} = \begin{Bmatrix} -66.67 \\ -20.00 \\ -132.22 \end{Bmatrix} \text{MPa}$$

Element b is treated similarly.

Note that the model employed in the foregoing solution is quite crude. The effect of element size on solution accuracy is illustrated in the following case study.

CASE STUDY 7.2   Analysis of a Deep Beam by the Theory of Elasticity and FEM

By means of (a) exact and (b) finite element approaches, investigate the stresses and displacements in a thin beam subjected to end moments applied about the centroidal axis (Fig. 7.28a). Let $L = 76.2$ mm, $h = 50.8$ mm, thickness $t = 25.4$ mm, $p = 6895$ kPa, $E = 207$ GPa, and $\nu = 0.15$. Neglect the weight of the member.

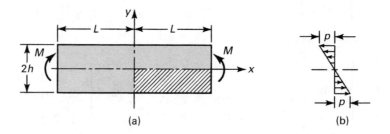

FIGURE 7.28. *Case Study 7.2. (a) Beam in pure bending; (b) moment is replaced by a statically equivalent load.*

Solution

a. *Exact solution:* Replacing the end moments with the statically equivalent load per unit area $p = Mh/I$ (Fig. 7.28b), the stress distribution from Eq. (5.5) is

$$\sigma_x = -\frac{y}{h}p, \qquad \sigma_y = \tau_{xy} = 0 \tag{h}$$

From Hooke's law and Eq. (2.4), we have

$$\frac{\partial u}{\partial x} = -\frac{yp}{Eh}, \qquad \frac{\partial v}{\partial y} = \frac{\nu y p}{Eh}, \qquad \frac{\partial u}{\partial y} + \frac{\partial v}{\partial x} = 0$$

By now following a procedure similar to that of Section 5.4, satisfying the conditions $u(0,0) = v(0,0) = 0$ and $u(L,0) = 0$, we obtain

$$u = -\frac{p}{Eh}xy, \qquad v = -\frac{p}{2Eh}(x^2 + \nu y^2) \tag{i}$$

Substituting the data into Eqs. (h) and (i), the results are

$$\sigma_x = -\frac{1}{0.0508}yp, \qquad \sigma_{x,\max} = 6895 \text{ kPa}$$

$$u(0.0762, -0.0254) = 25.4 \times 10^{-6}\text{m}, \qquad v(0.0762, 0) = 1.905 \times 10^{-6}\text{m} \tag{j}$$

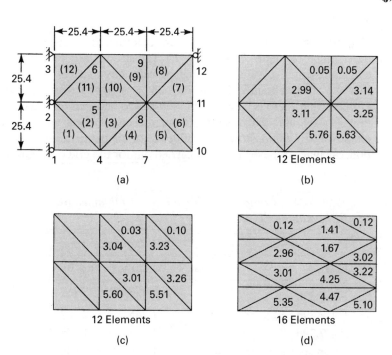

FIGURE 7.29. *Case Study 7.2. Influence of element size and orientation on stress (megapascals) in beam shown in Fig. 7.28.*

b. *Finite element solution:* Considerations of symmetry and antisymmetry indicate that only any one-quarter (shown as shaded portion in the figure) of the beam need be analyzed.

Displacement boundary conditions: Figure 7.29a shows the quarter-plate discretized to contain 12 triangular elements. The origin of coordinates is located at node 3. As no axial deformation occurs along the x and y axes, nodes 1, 2, 6, 9, and 12 are restrained against u deformation; node 3 is restrained against both u and v deformation. The boundary conditions are thus

$$u_1 = u_2 = u_3 = u_6 = u_9 = u_{12} = 0, \qquad v_3 = 0$$

Nodal forces: For the loading system of Fig. 7.28b, Eq. (7.78a) applies. Upon substitution of numerical values,

$$Q_{x10} = \frac{0.0254 \times 0.0254}{6}(2 \times 6895 + 3447.5) = 1853.5 \text{ N}$$

$$Q_{x11} = \frac{0.0254 \times 0.0254}{6}(2 \times 3447.5 + 6895)$$

$$+ \frac{0.0254 \times 0.0254}{6}(2 \times 3447.5 + 0) = 2224 \text{ N}$$

$$Q_{x12} = \frac{0.0254 \times 0.0254}{6}(0 + 3447.5) = 370.5 \text{ N}$$

The remaining Q's are zero.

Results: The nodal displacements are determined following a procedure similar to that of Case Study 7.1. The stresses are then evaluated and representative values (in megapascals) given in Fig. 7.29b. Note that the stress obtained for an element is assigned to the centroid. Observe that there is considerable difference between the exact solution, Eq. (j), and that resulting from the coarse mesh arrangement employed. To demonstrate the influence of element size and orientation, calculations have also been carried out for the grid configurations

TABLE 7.4. *Comparison of Deflections Obtained by FEM and the Elasticity Theory*

		Deflection (m)	
Case	Number of nodes	$v_{12}(10^6)$	$u_{10}(10^6)$
Fig. 7.28b	12	1.547	2.133
Fig. 7.29c	12	1.745	2.062
Fig. 7.29d	15	1.572	1.976
Exact solution	—	1.905	2.540

shown in Figs. 7.29c and d. For purposes of comparison, the deflections corresponding to Figs. 7.29b through d are presented in Table 7.4.

Comment   The effect of element orientation is shown in Figs. 7.29b and c for an equal number of elements and node locations. Figure 7.29d reveals that elements characterized by large differences between their side lengths, *weak elements*, lead to unfavorable results even though the number of nodes is larger than those of Figs. 7.29b and c. The employment of equilateral or nearly equilateral *well-formed elements* of finer mesh leads to solutions approaching the exact values.

Note that, owing to the approximate nature of the finite element method, nonzero values are found for σ_y and τ_{xy}. These are not listed in the figures. As the mesh becomes finer, these stresses do essentially vanish.

It is clear that we cannot reduce element size to extremely small values, as this would tend to increase to significant magnitudes the computer error incurred. An "exact" solution is thus unattainable, and we seek instead an *acceptable* solution. The goal is then the establishment of a finite element that ensures convergence to the exact solution in the absence of round-off error. The literature contains many comparisons between the various basic elements. The efficiency of a finite element solution can, in certain situations, be enhanced through the use of a "mix" of elements. For example, a denser mesh within a region of severely changing or localized stress may save much time and effort.

CASE STUDY 7.3   Stress Concentration in a Plate with a Hole in Uniaxial Tension

A thin aluminum alloy 6061-T6 plate containing a small circular hole of radius a is under uniform tensile stress σ_o at its edges (Fig. 7.30a). Through the use of the finite element analysis, outline the determination of the stress concentration factor K. Compare the result with that obtained by the theory of elasticity in Section 3.12. *Given:* $L = 600$ mm, $a = 50$ mm, $h = 500$ mm, $\sigma_o = 42$ MPa, $\gamma = 0.3$, and from Table D.1: $E = 70$ GPa.

Solution   Due to symmetry, only any one-quarter of the plate need be analyzed, as illustrated in Fig. 7.30b. We note that for this case, the quarter plate is discretized into 202 CST elements [Ref. 7.2]. The roller boundary conditions are also shown in the figure. The values of the edge stress σ_x, calculated by the FEM and the theory of elasticity are plotted in Fig. 7.30c for comparison [Ref. 7.9]. Observe that the agreement is reasonably good. The stress concentration factor for σ_x is equal to $K \approx 3$ $\sigma_o/\sigma_o = 3$.

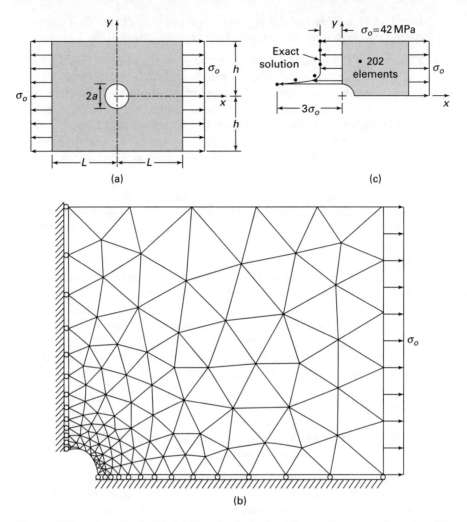

FIGURE 7.30. *Case Study 7.3. (a) Circular hole in a plate under uniaxial tension; (b) one-quadrant plate model; (c) uniaxial stress (σ_x) distribution.*

7.16 COMPUTATIONAL TOOLS

Various computational tools can be employed to carry out analysis calculations with success. A quality scientific calculator may be the best tool for solving most of the problems in this text. General-purpose analysis tools such as spreadsheets and equation solvers are very useful for certain computational tasks. Mathematical software packages of these types include MATLAB, TK Solver, and MathCAD. The tools offer the advantage of allowing the user to document and save the detailed completed work

The computer-aided drafting or design (CAD) software packages can produce re-alistic three-dimensional representations of a member. Most CAD software provides an interface to one or more FEA programs. They permit direct transfer of the member's geometry to an FEA package for analysis of stress and vibration as well as fluid and thermal analysis. Computer programs (such as NASTRAN, ANSYS, ABAQUS, GT-STRUDL) are used widely for performing the numerical computations required in the analysis and design of structural and mechanical systems.

With the proper use of *computer aided engineering* (CAE) software, problems can be solved more quickly and more accurately. Clearly, the results are subject to the accuracy of the various assumptions that must necessarily be made in the analysis and design. The foregoing computer-based software may be used as a tool to assist students with lengthy homework assignments. But it is important that basics be thoroughly understood, and analysts must make checks on computer solutions.

REFERENCES

7.1. SOKOLNIKOFF, I. S. and REDHEFFER, R. M., *Mathematics of Physics and Modern Engineering*, 2nd ed., McGraw-Hill, New York, 1966, p. 665.

7.2. YANG, T. Y., *Finite Element Structural Analysis*, Prentice Hall, Upper Saddle River, N.J., 1986.

7.3. WEAVER, W. JR. and JOHNSTON, P. R., *Finite Element for Structural Analysis*, Prentice Hall, Upper Saddle River, N.J., 1984.

7.4. GALLAGHER, R. H., *Finite Element Analysis: Fundamentals*, Prentice Hall, Englewood Cliffs, N.J., 1975.

7.5. MARTIN, H. C. and CAREY, G. F., *Introduction to Finite Element Analysis*, McGraw-Hill, New York, 1973.

7.6. LOGAN, D. L., *A First Course in the Finite Element Method*, PWS-Kent, Boston, Mass., 1986.

7.7. KNIGHT, E., *The Finite Element Method in Mechanical Design*, PWS-Kent, Boston, Mass., 1993.

7.8. UGURAL, A. C., *Mechanics of Materials*, Wiley, Hoboken, N.J., 2008.

7.9. UGURAL, A. C., *Mechanical Design: An Integrated Approach*, McGraw-Hill, New York, 2004.

7.10. BORESI, A. P. and SCHMIDT, R. J., *Advanced Mechanics of Materials*, 6th ed., Wiley, New York, 2003.

7.11. UGURAL, A. C., *Stresses in Beams, Plates and Shells*, 3rd ed., CRC Press, Taylor & Francis, Boca Raton, Fla. 2010.

7.12. ZIENKIEWICZ, O. C. and TAYLOR, R. I., *The Finite Element Method*, 4th ed., Vol. 2 (Solid and Fluid Mechanics, Dynamics and Nonlinearity), McGraw-Hill, London, 1991.

7.13. COOK, R. D. and MALKUS, D. S., *Concepts and Applications of Finite Element Analysis*, 3rd ed., Wiley, Hoboken, N.J., 1989.

7.14. SEGERLIND, L. J., *Applied Finite Element Analysis*, 2nd ed., Wiley, Hoboken, N.J., 1984.

7.15. BATHE, K. I., *Finite Element Procedures in Engineering Analysis*, Prentice Hall, Upper Saddle River, N.J., 1996.

7.16. SEGERLIND, L. J., *Applied Finite Element Analysis*, 2nd ed., Wiley, New York, 1984.

7.17. BAKER, A. J. and PEPPER, D. W., *Finite Elements*, McGraw-Hill, New York, 1991.

7.18. BERNADOU, M., *Finite Element Methods for Thin Shell Problems*, Wiley, Chichester, U.K., 1996.

7.19. DUNHAM, R. S. and NICKELL, R. E., "Finite element analysis of axisymmetric solids with arbitrary loadings," Report AD 655 253, National Technical Information Service, Springfield, Va., June 1967.

7.20. UTKU, S., Explicit expressions for triangular torus element stiffness matrix, AIAA Journal, 6/6, 1174–1176, June 1968.

PROBLEMS

Sections 7.1 through 7.4

7.1. Referring to Fig. 7.2, demonstrate that the biharmonic equation

$$\nabla^4 w = \frac{\partial^4 w}{\partial x^4} + 2\,\frac{\partial^4 w}{\partial x^2\,\partial y^2} + \frac{\partial^4 w}{\partial y^4} = 0$$

takes the following finite difference form:

$$h^4 \nabla^4 w = 20w_0 - 8(w_1 + w_2 + w_3 + w_4)$$
$$+ 2(w_5 + w_6 + w_7 + w_8) + w_9 + w_{10} + w_{11} + w_{12} \qquad \textbf{(P7.1)}$$

7.2. Consider a torsional bar having rectangular cross section of width $4a$ and depth $2a$. Divide the cross section into equal nets with $h = a/2$. Assume that the origin of coordinates is located at the centroid. Find the shear stresses at points $x = \pm 2a$ and $y = \pm a$. Use the direct finite difference approach. Note that the exact value of stress at $y = \pm a$ is, from Table 6.2, $\tau_{max} = 1.860G\theta a$.

7.3. For the torsional member of cross section shown in Fig. P7.3, find the shear stresses at point B. Take $h = 5$ mm and $h_1 = h_2 = 3.5$ mm.

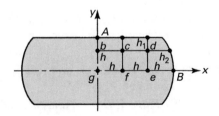

FIGURE P7.3.

7.4. Redo Prob. 7.3 to find the shear stress at point A. Let $h = 4.25$ mm; then $h_1 = h$ and $h_2 = 2.25$ mm.

7.5. Calculate the maximum shear stress in a torsional member of rectangular cross section of sides a and b ($a = 1.5b$). Employ the finite difference method, taking $h = a/4$. Compare the results with that given in Table 6.2.

Section 7.5

7.6. A force P is applied at the free end of a stepped cantilever beam of length L (Fig. P7.6). Determine the deflection of the free end using the finite difference method, taking $n = 3$. Compare the result with the exact solution $v(L) = 3PL^3/16EI$.

7.7. A stepped simple beam is loaded as shown in Fig. P7.7. Apply the finite difference approach, with $h = L/4$, to determine (a) the slope at point C; (b) the deflection at point C.

7.8. A stepped simple beam carries a uniform loading of intensity p, as shown in Fig. P7.8. Use the finite difference method to calculate the deflection at point C. Let $h = a/2$.

7.9. Cantilever beam AB carries a distributed load that varies linearly as shown in Fig. P7.9. Determine the deflection at the free end by applying the finite difference method. Use $n = 4$. Compare the result with the "exact" solution $w(L) = 11p_oL^4/120EI$.

7.10. Applying Eq. (7.22), determine the deflection at points 1 through 5 for beam and loading shown in Fig. P7.10.

7.11. Employ the finite difference method to obtain the maximum deflection and the slope of the simply supported beam loaded as shown in Fig. P7.11. Let $h = L/4$.

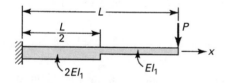

FIGURE P7.6.

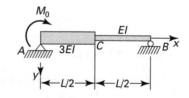

FIGURE P7.7.

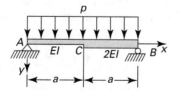

FIGURE P7.8.

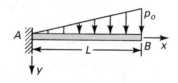

FIGURE P7.9.

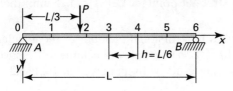

FIGURE P7.10.

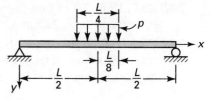

FIGURE P7.11.

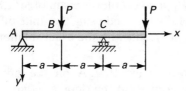

FIGURE P7.12.

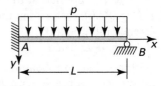

FIGURE P7.14.

7.12. Determine the deflection at a point B and the slope at point A of the overhanging beam loaded as shown in Fig. P7.12. Use the finite difference approach, with $n = 6$. Compare the deflection with its "exact" value $v_B = Pa^3/12EI$.

7.13. Redo Prob. 7.6 with the beam subjected to a uniform load p per unit length and $P = 0$. The exact solution is $v(L) = 3pL^4/32EI$.

7.14. A beam is supported and loaded as depicted in Fig. P7.14. Use the finite difference approach, with $h = L/4$, to compute the maximum deflection and slope.

7.15. A fixed-ended beam supports a concentrated load P at its midspan as shown in Fig. P7.15. Apply the finite difference method to determine the reactions. Let $h = L/4$.

7.16. Use the finite difference method to calculate the maximum deflection and the slope of a fixed-ended beam of length L carrying a uniform load of intensity p (Fig. P7.16). Let $h = L/4$.

Sections 7.6 through 7.16

7.17. The bar element 4–1 of length L and the cross-sectional area A is oriented at an angle α clockwise from the x axis (Fig. P7.17). Calculate (a) the global stiffness matrix of the bar; (b) the axial force in the bar; (c) the local displacements at the ends of the bar. *Given:* $A = 1350$ mm^2, $L = 1.7$ m, $\alpha = 60°$, $E = 96$ GPa, $u_4 = -1.1$ mm, $v_4 = -1.2$ mm, $u_1 = 2$ mm, and $v_1 = 1.5$ mm.

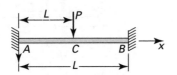

FIGURE P7.15.

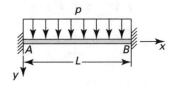

FIGURE P7.16.

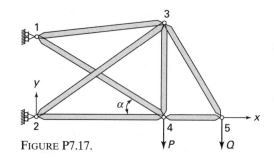

FIGURE P7.17.

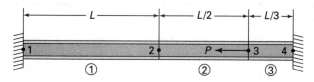

FIGURE P7.18.

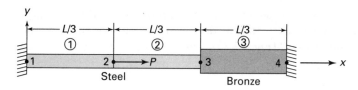

FIGURE P7.19.

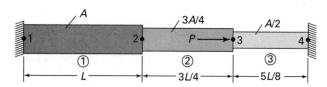

FIGURE P7.20.

7.18. The axially loaded bar 1–4 of constant axial rigidity AE is held between two rigid supports and under a concentrated load P at node 3, as illustrated in Fig. P7.18. Find (a) the system stiffness matrix; (b) the displacements at nodes 2 and 3; (c) the nodal forces and reactions at the supports.

7.19. The axially loaded composite bar 1–4 is held between two rigid supports and subjected to a concentrated load P at node 2, as depicted in Fig. P7.19. The steel bar 1–3 has cross-sectional area A and modulus of elasticity E. The brass bar 3–4 is with cross-sectional area $2A$ and elastic modulus $E/2$. Determine (a) the system stiffness matrix; (b) the displacements of nodes 2 and 3; (c) the nodal forces and reactions at the supports.

7.20. A stepped bar 1–4 is held between rigid supports and carries a concentrated load P at node 3, as illustrated in Fig. P7.20. Find (a) the system stiffness matrix; (b) the displacements of nodes 2 and 3; (c) the nodal forces and reactions at the supports.

7.21. A planar truss containing five members with axial rigidity AE is supported at joints 1 and 4, as shown in Fig. P7.21. What is the global stiffness matrix for each element?

7.22. The plane truss is loaded and supported as illustrated in Fig. P7.22. Each element has an axial rigidity AE. Find (a) the global stiffness matrix for each element; (b) the system stiffness matrix; (c) the system force–displacement relations.

7.23. A two-bar planar truss is supported by a spring of stiffness k at joint 1, as depicted in Fig. P7.23. Each element has an axial rigidity AE. Calculate (a) the stiffness matrix for bars 1 and 2, and spring 3; (b) the system stiffness matrix; (c) the force–displacement equations. *Given:* $L_2 = L$, $(AE)_2 = AE$, $L_1 = \frac{3}{4}L$, $(AE)_1 = \frac{3}{4}AE$.

7.24. A vertical concentrated load $P = 6$ kN is applied at joint 2 of the two-bar plane truss supported as shown in Fig. P7.24. Take $AE = 20$ MN for each member. Find (a) the global stiffness matrix of each bar; (b) the system stiffness matrix; (c) the nodal displacements; (d) the support reactions; (e) the axial forces in each bar.

7.25. In a two-bar plane truss, its support at joint 1 settles vertically by an amount of $u = 15$ mm downward when loaded by a horizontal concentrated load P (Fig. P7.25). Calculate (a) the global stiffness matrix of each

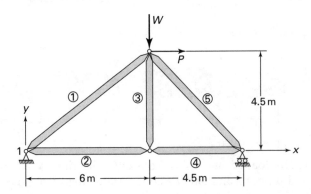

FIGURE P7.21.

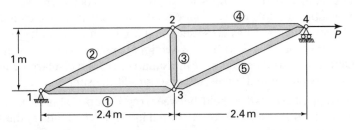

FIGURE P7.22.

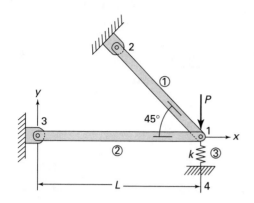

FIGURE P7.23.

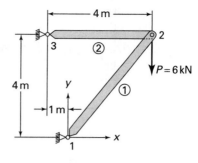

FIGURE P7.24.

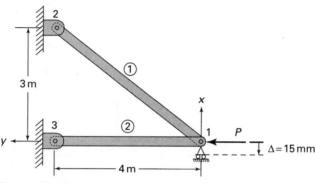

FIGURE P7.25.

element; (b) the system stiffness matrix; (c) the nodal displacements; (d) the support reactions. *Given:* $E = 105$ GPa, $A = 10 \times 10^{-4}$ m^2, $P = 10$ kN.

7.26. A cantilever of constant flexural rigidity EI carries a concentrated load P at its free end, as shown in Fig. P7.26. Find (a) the deflection v_1 and angle of rotation θ_1 at the free end; (b) the reactions R_2 and M_2 at the fixed end.

7.27. A simple beam 1–3 of length L and flexural rigidity EI is supports a uniformly distributed load of intensity p, as illustrated in Fig. P7.27. Determine the deflection of the beam at midpoint 2 by replacing the applied load with the equivalent nodal loads (see Table D.5).

7.28. A beam supported by a pin, a spring of stiffness k, and a roller at points 1, 2, and 3, respectively, is under a concentrated load P at point 2 (Fig. P7.28). Calculate (a) the nodal displacements; (b) the nodal forces and spring force. *Given:* $L = 4$ m, $P = 12$ kN, $EI = 14$ MN $\cdot$ m^2, and $k = 200$ kN/m.

7.29. A cantilever beam 1–2 of length L and constant flexural rigidity EI is subjected to a concentrated load P at the midspan, as shown in Fig. P7.29. Find the vertical deflection v_2 and angle of rotation θ_2 at the free end by replacing the applied load with the equivalent nodal loads acting at each end of the beam (see Table D.5).

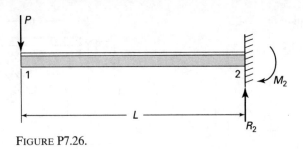

FIGURE P7.26.

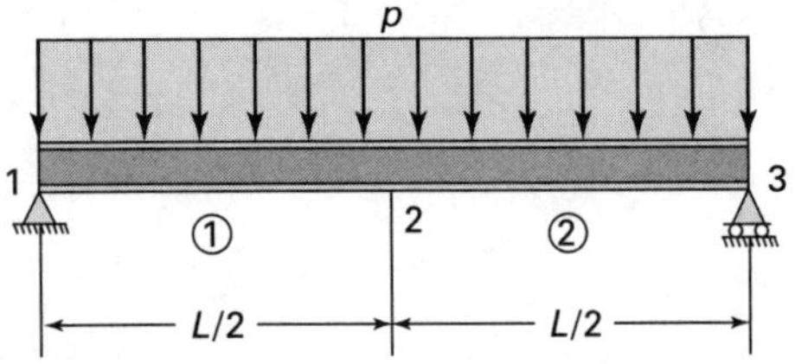

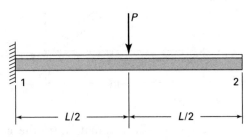

FIGURE P7.27.

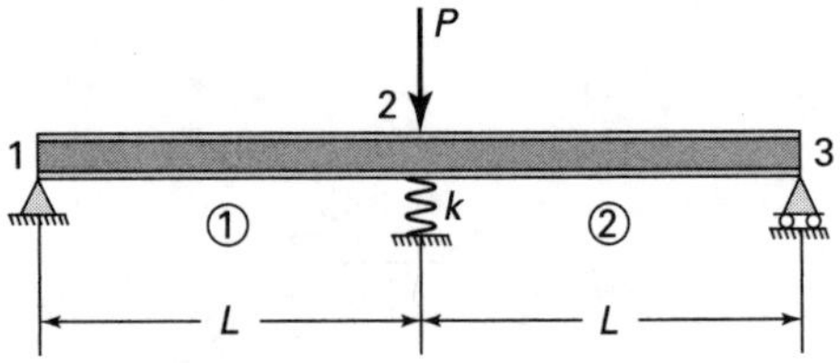

FIGURE P7.28.

FIGURE P7.29.

7.30. A propped cantilever beam with flexural rigidities EI and $EI/2$ for the parts 1–2 and 2–3, respectively, carries concentrated load P and moment $3PL$ at point 2 (Fig. P7.30). Calculate the displacements v_2, θ_2, and θ_3. *Given:* $L = 1.2$ m, $P = 30$ kN, $E = 207$ GPa, and $I = 15 \times 10^6$ mm^4.

7.31. A continuous beam of constant flexural rigidity EI is loaded and supported as seen in Fig. P7.31. Determine (a) the stiffness matrix for each element; (b) the system stiffness matrix and the force-displacement relations.

7.32. A propped cantilever beam with an overhang is subjected to a concentrated load P, as illustrated in Fig. P7.32. The beam has a constant flexural rigidity EI. Determine (a) the stiffness matrix for each element; (b) the system stiffness matrix; (c) the nodal displacements; (d) the forces and moments at the ends of each member; (e) the shear and moment diagrams.

7.33. A plane truss consisting of five members having the same axial rigidity AE is supported at joints 1 and 4, as seen in Fig. P7.33. Find the global stiffness matrix for each element.

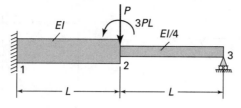

FIGURE P7.30.

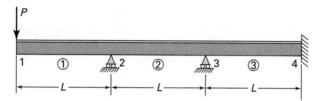

FIGURE P7.31.

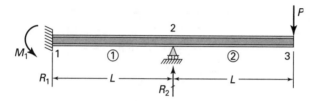

FIGURE P7.32.

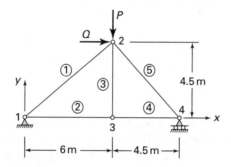

FIGURE P7.33.

7.34. The planar truss, with the axial rigidity AE the same for each element, is loaded and supported as illustrated in Fig. P7.34. Determine (a) the global stiffness matrix for each element; (b) the system matrix and the system force–displacement equations.

7.35. A vertical load $P = 20$ kN acts at joint 2 of the two-bar (1–2 and 2–3) truss shown in Fig. P7.35. Find (a) the global stiffness matrix for each member; (b) the system stiffness matrix; (c) the nodal displacements; (d) reactions; (e) the axial forces in each member, and show the results on a sketch of each member. *Assumption:* The axial rigidity $AE = 60$ MN is the same for each bar.

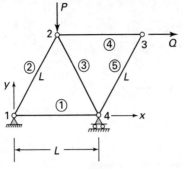

FIGURE P7.34.

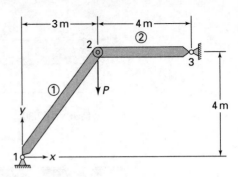

FIGURE P7.35.

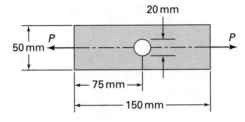

FIGURE P7.36.

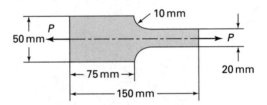

FIGURE P7.37.

7.36. A plate with a hole is under an axial tension loading P (Fig. P7.36). Dimensions are in millimeters. *Given*: $P = 5$ kN and plate thickness $t = 12$ mm. (a) Analyze the stresses using a computer program with the CST (or LST) elements. (b) Compare the stress concentration factor K obtained in part (a) with that found from Fig. D.8.

7.37. Resolve Prob. 7.36 for the plate shown in Fig. P7.37.

7.38. A simple beam is under a uniform loading of intensity p (Fig. P7.38). Let $L = 10h$, $t = 1$, and $v = 0.3$. Refine meshes to calculate the stress and deflection within 5% accuracy, by using a computer program with the CST (or LST) elements. *Given*: Exact solution [Ref. 7.7] is of the form:

$$\sigma_{x,max} = \frac{3pL^2}{4th^2}, \qquad v_{max} = -\frac{5pL^4}{16Eth^3} - \frac{3pL^2(1 + v)}{5Eht} \qquad \textbf{(P7.38)}$$

where t represents the thickness.

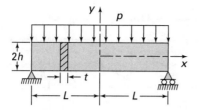

FIGURE P7.38.

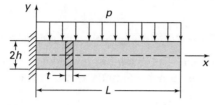

FIGURE P7.39.

7.39. Resolve Prob. 7.38 for the case in which a cantilever beam is under a uniform loading of intensity p (Fig. P7.39). *Given:* Exact solution [Ref. 7.7] is given by

$$\sigma_{x,\,\text{max}} = \frac{3pL^2}{4th^2}, \qquad v_{\text{max}} = -\frac{3pL^4}{16Eth^3} - \frac{3pL^2(1+\nu)}{5Eht} \qquad \textbf{(P7.39)}$$

in which t is the thickness.

7.40. Verify Eqs. (7.78) and (7.79) by determining the static resultant of the applied loading. [Hint: For Eq. (7.79), apply the principle of virtual work.]

$$Q_j \delta v_j + Q_m \delta v_m - \int_{y_j}^{y_m} \tau_{xy} t \, \delta v \, dy = 0$$

with

$$v = v_j + \frac{v_m - v_j}{y_m - y_j}(y - y_j)$$

to obtain

$$Q_j + Q_m = \int_{y_j}^{y_m} \tau_{xy} t \, dy, \qquad Q_m = \frac{1}{y_m - y_j}\left[\int_{y_j}^{y_m} \tau_{xy} t (y - y_j) \, dy\right]$$

7.41. Redo Case Study 7.1 for the beam subjected to a uniform additional load throughout its span, $p = -7$ MPa, and a temperature rise of 50°C. Let $\gamma = 77$ kN/m^3 and $\alpha = 12 \times 10^{-6}$/°C.

7.42. A $\frac{2}{7}$-cm-thick cantilever beam is subjected to a parabolically varying end shear stress resulting in a load of P N and a linearly distributed load p N/cm (Fig. P7.42). Dividing the beam into two triangles as shown, calculate the

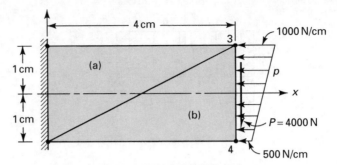

FIGURE P7.42.

stresses in the member. The beam is made of a transversely isotropic material, in which a rotational symmetry of properties exists within the xz plane:

$$E_1 = 210 \text{ GPa}, \qquad \nu_2 = 0.1$$
$$E_2 = 70 \text{ GPa}, \qquad G_2 = 28 \text{ GPa}$$

Here E_1 is associated with the behavior in the xz plane, and E_2, G_2, and ν_2 with the direction perpendicular to the xz plane. Now the elasticity matrix, Eq. (7.54), becomes

$$[D] = \frac{E_2}{1 - n\nu_2^2} \begin{bmatrix} n & n\nu_2 & 0 \\ n\nu_2 & 1 & 0 \\ 0 & 0 & m(1 - n\nu_2^2) \end{bmatrix}$$

where $n = E_1/E_2$ and $m = G_2/E_2$.

7.43. Redo Case Study 7.1 if the discretized beam consists of triangular elements 1 4 3 and 1 2 4 (Fig. 7.27b). Assume the remaining data to be unchanged.

CHAPTER 8

Axisymmetrically Loaded Members

8.1 INTRODUCTION

In the class of axisymmerically loaded members, the fundamental problem may be defined in terms of the radial coordinate. There are numerous practical situations in which the distribution of stress manifests symmetry about an axis. Examples include pressure vessels, compound cylinders, clad reactor elements, chemical reaction vessels, heat exchanger tubes, solid or hollow spherical structures, turbine disks, and components of many other machines used in aerospace to household. This chapter concerns mainly "exact" stress distribution in various axisymmetrically loaded machine and structural components. The methods of mechanics of materials, the theory of elasticity, and finite elements are used. The displacements, strains, and stresses at locations far removed from the ends due to pressure, thermal, and dynamic loadings are discussed. Applications to compound press or shrink-fit cylinders, disk flywheels, and design of hydraulic cylinders are included. Before doing these, however, we begin with an analysis of stress developed in thick-walled vessels.

Basic Relations

Consider a large thin plate having a small circular hole subjected to uniform pressure, as shown in Fig. 8.1. Note that axial loading is absent, and therefore $\sigma_z = 0$. The stresses are clearly symmetrical about the z axis, and the deformations likewise display θ independence. The symmetry argument also dictates that the shearing stresses $\tau_{r\theta}$ must be zero. Assuming z independence for this thin plate, the polar equations of equilibrium (3.31), reduce to

$$\frac{d\sigma_r}{dr} + \frac{\sigma_r - \sigma_\theta}{r} + F_r = 0 \tag{8.1}$$

FIGURE 8.1. *Large thin plate with a small circular hole.*

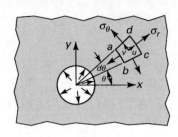

Here σ_θ and σ_r denote the tangential (circumferential) and radial stresses acting normal to the sides of the element, and F_r represents the radial body force per unit volume, for example, the inertia force associated with rotation. In the absence of body forces, Eq. (8.1) reduces to

$$\frac{d\sigma_r}{dr} + \frac{\sigma_r - \sigma_\theta}{r} = 0 \qquad (8.2)$$

Consider now the radial and tangential displacements u and v, respectively. There can be *no tangential displacement* in the symmetrical field; that is, $v = 0$. A point represented by the shaded element *abcd* in the figure will thus move radially as a consequence of loading, but not tangentially. On the basis of displacements indicated, the strains given by Eqs. (3.33) become

$$\varepsilon_r = \frac{du}{dr}, \qquad \varepsilon_\theta = \frac{u}{r}, \qquad \gamma_{r\theta} = 0 \qquad (8.3)$$

Substituting $u = r\varepsilon_\theta$ into the first expression in Eq. (8.3), a simple compatibility equation is obtained:

$$\frac{du}{dr} - \varepsilon_r = \frac{d}{dr}(r\varepsilon_\theta) - \varepsilon_r = 0$$

or

$$r\frac{d\varepsilon_\theta}{dr} + \varepsilon_\theta - \varepsilon_r = 0 \qquad (8.4)$$

The equation of equilibrium [Eq. (8.1) or (8.2)], the strain–displacement or compatibility relations [Eqs. (8.3) or (8.4)], and Hooke's law are sufficient to obtain a unique solution to any axisymmetrical problem with specified boundary conditions.

8.2 THICK-WALLED CYLINDERS

The circular cylinder, of special importance in engineering, is usually divided into *thin-walled* and *thick-walled* classifications. A thin-walled cylinder is defined as one in which the tangential stress may, within certain prescribed limits, be regarded as constant with thickness. The following familiar expression applies to the case of a thin-walled cylinder subject to internal pressure:

$$\sigma_\theta = \frac{pr}{t}$$

Here p is the internal pressure, r the mean radius (see Sec. 13.13), and t the thickness. If the wall thickness exceeds the inner radius by more than approximately 10%, the cylinder is generally classified as thick walled, and the variation of stress with radius can no longer be disregarded (see Prob. 8.1).

In the case of a thick-walled cylinder subject to uniform internal or external pressure, the deformation is symmetrical about the axial (z) axis. Therefore, the equilibrium and strain–displacement equations, Eqs. (8.2) and (8.3), apply to any point on a ring of unit length cut from the cylinder (Fig. 8.2). Assuming that the ends of the cylinder are *open* and *unconstrained*, $\sigma_z = 0$, as is subsequently demonstrated. Thus, the cylinder is in a condition of *plane stress* and, according to Hooke's law (3.34), the strains are given by

$$\frac{du}{dr} = \frac{1}{E}(\sigma_r - \nu\sigma_\theta)$$

$$\frac{u}{r} = \frac{1}{E}(\sigma_\theta - \nu\sigma_r)$$

(8.5)

From these, σ_r and σ_θ are as follows:

$$\sigma_r = \frac{E}{1 - \nu^2}(\varepsilon_r + \nu\varepsilon_\theta) = \frac{E}{1 - \nu^2}\left(\frac{du}{dr} + \nu\frac{u}{r}\right)$$

$$\sigma_\theta = \frac{E}{1 - \nu^2}(\varepsilon_\theta + \nu\varepsilon_r) = \frac{E}{1 - \nu^2}\left(\frac{u}{r} + \nu\frac{du}{dr}\right)$$

(8.6)

Substituting this into Eq. (8.2) results in the following *equidimensional equation* in radial displacement:

$$\frac{d^2u}{dr^2} + \frac{1}{r}\frac{du}{dr} - \frac{u}{r^2} = 0$$

(8.7)

having a solution

$$u = c_1 r + \frac{c_2}{r}$$

(a)

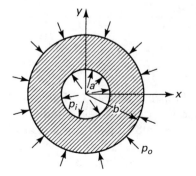

FIGURE 8.2.   *Thick-walled cylinder.*

The radial and tangential stresses may now be written in terms of the constants of integration c_1 and c_2 by combining Eqs. (a) and (8.6):

$$\sigma_r = \frac{E}{1 - \nu^2}\left[c_1(1 + \nu) - c_2\left(\frac{1 - \nu}{r^2}\right)\right] \tag{b}$$

$$\sigma_\theta = \frac{E}{1 - \nu^2}\left[c_1(1 + \nu) + c_2\left(\frac{1 - \nu}{r^2}\right)\right] \tag{c}$$

The constants are determined from consideration of the conditions pertaining to the inner and outer surfaces.

Observe that the sum of the radial and tangential stresses is constant, regardless of radial position: $\sigma_r + \sigma_\theta = 2Ec_1/(1 - \nu)$. Hence, the longitudinal strain is constant:

$$\varepsilon_z = -\frac{\nu}{E}(\sigma_r + \sigma_\theta) = \text{constant}$$

We conclude, therefore, that *plane sections remain plane* subsequent to loading. Then $\sigma_z = E\varepsilon_z = \text{constant} = c$. But if the ends of the cylinder are open and free,

$$\int_a^b \sigma_z \cdot 2\pi r\, dr = \pi c(b^2 - a^2) = 0$$

or $c = \sigma_z = 0$, as already assumed previously.

For a cylinder subjected to internal and external pressures p_i and p_o, respectively, the boundary conditions are

$$(\sigma_r)_{r=a} = -p_i, \qquad (\sigma_r)_{r=b} = -p_o \tag{d}$$

where the negative sign connotes compressive stress. The constants are evaluated by substitution of Eqs. (d) into (b):

$$c_1 = \frac{1 - \nu}{E}\frac{a^2p_i - b^2p_o}{b^2 - a^2}, \qquad c_2 = \frac{1 + \nu}{E}\frac{a^2b^2(p_i - p_o)}{b^2 - a^2} \tag{e}$$

leading finally to

$$\sigma_r = \frac{a^2p_i - b^2p_o}{b^2 - a^2} - \frac{(p_i - p_o)a^2b^2}{(b^2 - a^2)r^2}$$

$$\sigma_\theta = \frac{a^2p_i - b^2p_o}{b^2 - a^2} + \frac{(p_i - p_o)a^2b^2}{(b^2 - a^2)r^2} \tag{8.8}$$

$$u = \frac{1 - \nu}{E}\frac{(a^2p_i - b^2p_o)r}{b^2 - a^2} + \frac{1 + \nu}{E}\frac{(p_i - p_o)a^2b^2}{(b^2 - a^2)r}$$

These expressions were first derived by French engineer G. Lamé in 1833, for whom they are named. The maximum numerical value of σ_r is found at $r = a$ to be p_i, provided that p_i exceeds p_o. If $p_o > p_i$, the maximum σ_r occurs at $r = b$ and equals p_o. On the other hand, the maximum σ_θ occurs at either the inner or outer edge according to the pressure ratio, as discussed in Section 8.3.

Recall that the maximum shearing stress at any point equals one-half the algebraic difference between the maximum and minimum principal stresses. At any point in the cylinder, we may therefore state that

$$\tau_{\max} = \tfrac{1}{2}(\sigma_\theta - \sigma_r) = \frac{(p_i - p_o)a^2b^2}{(b^2 - a^2)r^2} \tag{8.9}$$

The largest value of $\tau_{\max}$ is found at $r = a$, the inner surface. The effect of reducing p_o is clearly to increase $\tau_{\max}$. Consequently, the greatest $\tau_{\max}$ corresponds to $r = a$ and $p_o = 0$:

$$\tau_{\max} = \frac{p_i b^2}{b^2 - a^2} \tag{8.10}$$

Because σ_r and σ_θ are principal stresses, $\tau_{\max}$ occurs on planes making an angle of $45°$ with the plane on which σ_r and σ_θ act, as depicted in Fig. 8.3. This is quickly confirmed by a Mohr's circle construction. The pressure p_{yp} that *initiates yielding* at the inner surface is obtained by setting $\tau_{\max} = \sigma_{yp}/2$ in Eq. (8.10):

$$p_{yp} = \frac{(b^2 - a^2)\sigma_{yp}}{2b^2} \tag{8.11}$$

Here σ_{yp} is the yield stress in uniaxial tension.

Special Cases

Internal Pressure Only. If only internal pressure acts, Eqs. (8.8) reduce to

$$\sigma_r = \frac{a^2 p_i}{b^2 - a^2}\left(1 - \frac{b^2}{r^2}\right) \tag{8.12}$$

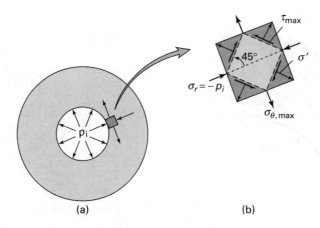

(a)   (b)

FIGURE 8.3.   *(a) Thick-walled cylinder under p_i; (b) an element at the inner edge in which τ_{max} occurs.*

$$\sigma_\theta = \frac{a^2 p_i}{b^2 - a^2}\left(1 + \frac{b^2}{r^2}\right)$$ (8.13)

$$u = \frac{a^2 p_i r}{E(b^2 - a^2)}\left[(1 - \nu) + (1 + \nu)\frac{b^2}{r^2}\right]$$ (8.14)

Since $b^2/r^2 \geq 1, \sigma_r$ is negative (compressive) for all r except $r = b$, in which case $\sigma_r = 0$, the maximum radial stress occurs at $r = a$. As for σ_θ, it is positive (tensile) for all radii and also has a maximum at $r = a$.

To illustrate the variation of stress and radial displacement for the case of zero external pressure, dimensionless stress and displacement are plotted against dimensionless radius in Fig. 8.4a for $b/a = 4$.

External Pressure Only. In this case, $p_i = 0$, and Eq. (8.8) becomes

$$\sigma_r = -\frac{p_o b^2}{b^2 - a^2}\left(1 - \frac{a^2}{r^2}\right)$$ (8.15)

$$\sigma_\theta = -\frac{p_o b^2}{b^2 - a^2}\left(1 + \frac{a^2}{r^2}\right)$$ (8.16)

$$u = -\frac{b^2 p_o r}{E(b^2 - a^2)}\left[(1 - \nu) + (1 + \nu)\frac{a^2}{r^2}\right]$$ (8.17)

The maximum radial stress occurs at $r = b$ and is compressive for all r. The maximum σ_θ is found at $r = a$ and is likewise compressive.

For a cylinder with $b/a = 4$ and subjected to external pressure only, the stress and displacement variations over the wall thickness are shown in Fig. 8.4b.

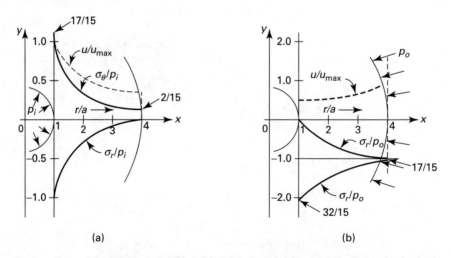

FIGURE 8.4. *Distribution of stress and displacement in a thick-walled cylinder with $b/a = 4$: (a) under internal pressure; (b) under external pressure.*

EXAMPLE 8.1   Analysis of a Thick-Walled Cylinder

A thick-walled cylinder with 0.3-m and 0.4-m internal and external diameters is fabricated of a material whose ultimate strength is 250 MPa. Let $v = 0.3$. Determine (a) for $p_o = 0$, the maximum internal pressure to which the cylinder may be subjected without exceeding the ultimate strength, (b) for $p_i = 0$, the maximum external pressure to which the cylinder can be subjected without exceeding the ultimate strength, and (c) the radial displacement of a point on the inner surface for case (a).

Solution

a. From Eq. (8.13), with $r = a$,

$$\sigma_{\theta,\,max} = p_i \frac{b^2 + a^2}{b^2 - a^2} \qquad (8.18)$$

or

$$p_i = \sigma_{\theta,\,max} \frac{b^2 - a^2}{b^2 + a^2} = (250 \times 10^6) \frac{0.2^2 - 0.15^2}{0.2^2 + 0.15^2} = 70 \text{ MPa}$$

b. From Eq. (8.16), with $r = a$,

$$\sigma_{\theta,\,max} = -2p_o \frac{b^2}{b^2 - a^2} \qquad (8.19)$$

Then

$$p_o = -\sigma_{\theta,\,max} \frac{b^2 - a^2}{2b^2} = -(-250 \times 10^6) \frac{0.2^2 - 0.15^2}{(2)0.2^5} = 54.7 \text{ MPa}$$

c. Using Eq. (8.14), we obtain

$$(u)_{r=a} = \frac{0.15^3 \times 70 \times 10^6}{E(0.2^2 - 0.15^2)} \left[0.7 + 1.3 \frac{0.2^2}{0.15^2} \right] = 4.065 \times 10^7/E \text{ m}$$

Closed-Ended Cylinder

In the case of a closed-ended cylinder subjected to internal and external pressures, longitudinal or z-directed stresses exist in addition to the radial and tangential stresses. For a transverse section some distance from the ends, this stress may be assumed uniformly distributed over the wall thickness. The magnitude of σ_z is then determined by equating the net force acting on an end attributable to pressure loading to the internal z-directed force in the cylinder wall:

$$p_i \pi a^2 - p_o \pi b^2 = (\pi b^2 - \pi a^2) \sigma_z$$

The resulting expression for *longitudinal stress*, applicable only away from the ends, is

$$\sigma_z = \frac{p_i a^2 - p_o b^2}{b^2 - a^2} \qquad (8.20)$$

Clearly, here it is again assumed that the ends of the cylinder are not constrained: $\varepsilon_z \neq 0$ (see Prob. 8.13).

8.3 MAXIMUM TANGENTIAL STRESS

An examination of Figs. 8.4a and b shows that if either internal pressure or external pressure acts alone, the maximum tangential stress occurs at the innermost fibers, $r = a$. This conclusion is not always valid, however, if both internal and external pressures act simultaneously. There are situations, explored next, in which the maximum tangential stress occurs at $r = b$ [Ref. 8.1].

Consider a thick-walled cylinder, as in Fig. 8.2, subject to p_i and p_o. Denote the ratio b/a by R, p_o/p_i by P, and the ratio of tangential stress at the inner and outer surfaces by S. The tangential stress, given by Eqs. (8.8), is written

$$\sigma_\theta = p_i \frac{1 - PR^2}{R^2 - 1} + p_i b^2 \frac{1 - P}{(R^2 - 1)r^2} \tag{a}$$

Hence,

$$S = \frac{\sigma_{\theta i}}{\sigma_{\theta o}} = \frac{1 + \dfrac{R^2(1 - P)}{1 - PR^2}}{1 + \dfrac{1 - P}{1 - PR^2}} \tag{8.21}$$

The variation of the tangential stress σ_θ over the wall thickness is shown in Fig. 8.5 for several values of S and P. Note that for pressure ratios P, indicated by dashed lines, the maximum magnitude of the circumferential stress occurs at the *outer surface* of the cylinder.

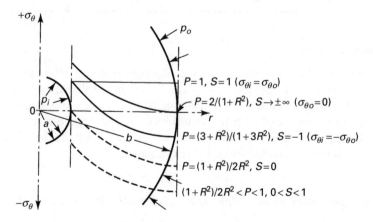

FIGURE 8.5. *Maximum tangential stress distribution in a thick-walled cylinder under internal and external pressures.*

8.4 APPLICATION OF FAILURE THEORIES

Unless we are content to grossly overdesign, it is necessary to predict, as best possible, the most probable failure mechanism. Thus, while examination of Fig. 8.5 indicates that failure is likely to originate at the innermost or outermost fibers of the cylinder, it cannot predict at what pressure or stresses failure will occur. To do this, consideration must be given the stresses determined from Lamé's equations, the material strength, and an appropriate theory of failure consistent with the nature of the material.

EXAMPLE 8.2   Thick-Walled Cylinder Pressure Requirement

A steel cylinder is subjected to an internal pressure four times greater than the external pressure. The tensile elastic strength of the steel is $\sigma_{yp} = 340$ MPa, and the shearing elastic strength $\tau_{yp} = \sigma_{yp}/2 = 170$ MPa. Calculate the allowable internal pressure according to the various yielding theories of failure. The dimensions are $a = 0.1$ m and $b = 0.15$ m. Let $\nu = 0.3$.

Solution   The maximum stresses occur at the innermost fibers. From Eqs. (8.8), for $r = a$ and $p_i = 4p_o$, we have

$$\sigma_\theta = \frac{p_i(a^2 + b^2) - 2p_o b^2}{b^2 - a^2} = 1.7p_i, \qquad \sigma_r = -p_i \qquad \textbf{(a)}$$

The value of internal pressure at which yielding begins is predicted according to the various theories of failure, as follows:

a. *Maximum shearing stress theory:*

$$\frac{\sigma_\theta - \sigma_r}{2} = 1.35p_i = 170 \times 10^6, \qquad p_i = 125.9 \text{ MPa}$$

b. *Energy of distortion theory* [Eq. (4.5a)]:

$$\sigma_{yp} = (\sigma_\theta^2 + \sigma_r^2 - \sigma_\theta\sigma_r)^{1/2} = p_i[(1.7)^2 + (-1)^2 - (-1.7)]^{1/2}$$
$$340 \times 10^6 = 2.364p_i, \qquad p_i = 143.8 \text{ MPa}$$

c. *Octahedral shearing stress theory:* By use of Eqs. (4.6) and (1.38), we have

$$\frac{\sqrt{2}}{3}\sigma_{yp} = \tfrac{1}{3}\left[(\sigma_\theta - \sigma_r)^2 + (\sigma_r)^2 + (-\sigma_\theta)^2\right]^{1/2}$$

$$\frac{\sqrt{2}}{3}(340 \times 10^6) = \frac{p_i}{3}[(2.7)^2 + (-1)^2 + (-1.7)^2]^{1/2}, \qquad p_i = 143.8 \text{ MPa}$$

Comment   The results found in (b) and (c) are identical as expected (Sec. 4.8). The onset of inelastic action is governed by the maximum shearing stress: the allowable value of internal pressure is limited to 125.9 MPa, modified by an appropriate factor of safety.

8.5 COMPOUND CYLINDERS: PRESS OR SHRINK FITS

If properly designed, a system of multiple cylinders resists relatively large pressures more efficiently, that is, requires less material, than a single cylinder. To assure the integrity of the compound cylinder, one of several methods of prestressing is employed. For example, the inner radius of the outer member or jacket may be made smaller than the outer radius of the inner cylinder. The cylinders are assembled after the outer cylinder is heated, contact being effected upon cooling. The magnitude of the resulting *contact pressure p*, or *interface pressure*, between members may be calculated by use of the equations of Section 8.2. Examples of compound cylinders, carrying very high pressures, are seen in compressors, extrusion presses, and the like.

Referring to Fig. 8.6, assume the external radius of the inner cylinder to be larger, in its unstressed state, than the internal radius of the jacket, by an amount δ. The quantity δ is called the *shrinking allowance*, or also known as the *radial interference*. Subsequent to assembly, the contact pressure, acting equally on both members, causes the sum of the increase in the inner radius of the jacket and decrease in the outer radius of the inner member to exactly equal δ. By using Eqs. (8.14) and (8.17), we obtain

$$\delta = \delta_o + \delta_i \frac{bp}{E_o}\left(\frac{b^2 + c^2}{c^2 - b^2} + \nu_o\right) + \frac{bp}{E_i}\left(\frac{a^2 + b^2}{b^2 - a^2} - \nu_i\right) \tag{8.22}$$

Here E_o, ν_o and E_i, ν_i represent the material properties of the outer and inner cylinders, respectively. When both cylinders are made of the *same materials*, we have $E_o = E_i = E$ and $\nu_o = \nu_i = \nu$. In such a case, from Eq. (8.22),

$$p = \frac{E\delta}{b}\frac{(b^2 - a^2)(c^2 - b^2)}{2b^2(c^2 - a^2)} \tag{8.23}$$

The stresses in the jacket are then determined from Eqs. (8.12) and (8.13) by treating the contact pressure as p_i. Similarly, by regarding the contact pressure as p_o, the stresses in the inner cylinder are calculated from Eqs. (8.15) and (8.16).

EXAMPLE 8.3   Stresses in a Compound Cylinder under Internal Pressure
A compound cylinder with $a = 150$ mm, $b = 200$ mm, $c = 250$ mm, $E = 200$ GPa, and $\delta = 0.1$ mm is subjected to an internal pressure of 140 MPa. Determine the distribution of tangential stress throughout the composite wall.

FIGURE 8.6.   *Compound cylinder.*

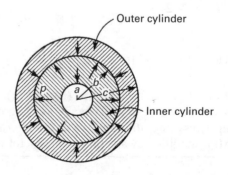

Outer cylinder

Inner cylinder

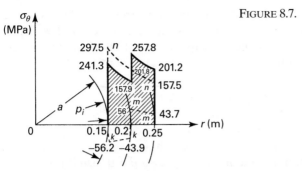

FIGURE 8.7. *Example 8.3. Tangential stress distribution produced by a combination of shrink-fit and internal pressure in a thick-walled compound cylinder. Dashed lines nn represent stresses due to p_i alone.*

Solution In the absence of applied internal pressure, the contact pressure is, from Eq. (8.23),

$$p = \frac{200 \times 10^9 \times 0.0001}{0.2} \frac{(0.2^2 - 0.15^2)(0.25^2 - 0.2^2)}{2(0.2^2)(0.25^2 - 0.15^2)} = 12.3 \text{ MPa}$$

The tangential stresses in the outer cylinder associated with this pressure are found by using Eq. (8.13)

$$(\sigma_\theta)_{r=0.2} = p \frac{b^2 + c^2}{c^2 - b^2} = 12.3 \times 10^6 \frac{0.2^2 + 0.25^2}{0.25^2 - 0.2^2} = 56.0 \text{ MPa}$$

$$(\sigma_\theta)_{r=0.25} = \frac{2pb^2}{c^2 - b^2} = \frac{2(12.3 \times 10^6)(0.2^2)}{0.25^2 - 0.2^2} = 43.7 \text{ MPa}$$

The stresses in the inner cylinder are, from Eq. (8.16),

$$(\sigma_\theta)_{r=0.15} = -\frac{2pb^2}{b^2 - a^2} = -\frac{2(12.3 \times 10^6)(0.2^2)}{0.2^2 - 0.15^2} = -56.2 \text{ MPa}$$

$$(\sigma_\theta)_{r=0.2} = -p \frac{b^2 + a^2}{b^2 - a^2} = -12.3 \times 10^6 \frac{0.2^2 + 0.15^2}{0.2^2 - 0.15^2} = -43.9 \text{ MPa}$$

These stresses are plotted in Fig. 8.7, indicated by the dashed lines *kk* and *mm*. The stresses owing to internal pressure alone, through the use of Eq. (8.13) with $b = c$, are found to be $(\sigma_\theta)_{r=0.15} = 297.5$ MPa, $(\sigma_\theta)_{r=0.2} = 201.8$ MPa, $(\sigma_\theta)_{r=0.25} = 157.5$ MPa, and are shown as the dashed line *nn*. The stress resultant is obtained by superposition of the two distributions, represented by the solid line. The use of a compound prestressed cylinder has thus reduced the maximum stress from 297.5 to 257.8 MPa. Based on the maximum principal stress theory of elastic failure, significant weight savings can apparently be effected through such configurations.

It is interesting to note that additional jackets prove not as effective, in that regard, as the first one. Multilayered shrink-fit cylinders, each of small wall thickness, are, however, considerably stronger than a single jacket of the same total thickness. These assemblies can, in fact, be designed so that prestressing owing to shrinking combines with stresses due to loading to produce a nearly uniform distribution of stress throughout.* The closer this uniform stress is to the allowable stress for the given material, the more efficiently is the material utilized. A single cylinder cannot be uniformly stressed and consequently must be stressed considerably below its allowable value, contributing to inefficient use of material.

EXAMPLE 8.4 Design of a Duplex Hydraulic Conduit

Figure 8.8 shows a duplex *hydraulic conduit* composed of a *thin* steel cylindrical liner in a concrete pipe. For the steel and concrete, the properties are E_s and E_c, ν_c, respectively. The conduit is subjected to an internal pressure p_i. Determine the interface pressure p transmitted to the concrete shell.

Solution The radial displacement at the bore $(r = a)$ of the concrete pipe is, from Eq. (8.14),

$$u = \frac{pa}{E_c}\left(\frac{a^2 + b^2}{b^2 - a^2} + \nu_c\right) \tag{a}$$

The steel sleeve experiences an internal pressure p_i and an external pressure p. Thus,

$$\sigma_\theta = \frac{p_i - p}{t}\left(a - \frac{t}{2}\right) = \frac{p_i - p}{2}\left(\frac{2a}{t} - 1\right) \tag{b}$$

On the other hand, Hooke's law together with Eq. (8.2) yields at $r = a$

$$\sigma_\theta = E_s\,\varepsilon_\theta = E_s\frac{u}{a} \tag{c}$$

Finally, evaluation of u from Eqs. (b) and (c) and substitution into Eq. (a) result in the interface pressure in the form:

$$p = \frac{p_i}{1 + \left(\dfrac{E_s}{E_c}\right)\left(\dfrac{2t}{2a - t}\right)\left(\dfrac{a^2 + b^2}{b^2 - a^2} + \nu_c\right)} \tag{8.24}$$

Comment Note that, for practical purposes, we can use $E_s/E_c = 15$, $\nu_c = 0.2$, and $2t/(2a - t) = t/a$. Equation (8.24) then becomes

FIGURE 8.8. *Example 8.4. Thick-walled concrete pipe with a steel cylindrical liner.*

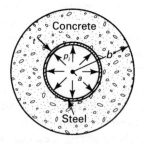

$$p = \frac{p_i}{1 + 15\left(\dfrac{t}{a}\right)\left(\dfrac{R^2 + 1}{R^2 - 1} + 0.2\right)} \tag{8.25}$$

where $R = b/a$. From the foregoing, observe that, as the thickness t of the liner increases, the pressure p transmitted to the concrete decreases. However, for any given t/a ratio, the p increases as the diameter ratio (R) of the concrete increases.

8.6 ROTATING DISKS OF CONSTANT THICKNESS

The equation of equilibrium, Eq. (8.1), can be used to treat the case of a rotating disk, provided that the centrifugal "inertia force" is included as a body force. Again, stresses induced by rotation are distributed symmetrically about the axis of rotation and assumed independent of disk thickness. Thus, application of Eq. (8.1), with the body force per unit volume F_r equated to the centrifugal force $\rho\omega^2 r$, yields

$$\frac{d\sigma_r}{dr} + \frac{\sigma_r - \sigma_\theta}{r} + \rho\omega^2 r = 0 \tag{8.26}$$

where ρ is the mass *density* and ω the constant *angular speed* of the disk in *radians per second*. Note that the gravitational body force ρg has been neglected. Substituting Eq. (8.6) into Eq. (8.26), we have

$$\frac{d^2 u}{dr^2} + \frac{1}{r}\frac{du}{dr} - \frac{u}{r^2} = -(1 - \nu^2)\frac{\rho\omega^2 r}{E} \tag{a}$$

requiring a homogeneous and particular solution. The former is given by Eq. (a) of Section 8.2.

It is easily demonstrated that the particular solution is

$$u_p = -(1 - \nu^2)\frac{\rho\omega^2 r^3}{8E}$$

The complete solution is therefore

$$u = -\frac{\rho\omega^2 r^3(1 - \nu^2)}{8E} + c_1 r + \frac{c_2}{r} \tag{8.27a}$$

which, upon substitution into Eq. (8.6), provides the following expressions for radial and tangential stress:

$$\sigma_r = \frac{E}{1 - \nu^2}$$
$$\times \left[\frac{-(3 + \nu)(1 - \nu^2)\rho\omega^2 r^2}{8E} + (1 + \nu)c_1 - (1 - \nu)\frac{c_2}{r^2}\right] \tag{8.27b}$$

$$\sigma_\theta = \frac{E}{1 - \nu^2}$$

$$\times \left[\frac{-(1 + 3\nu)(1 - \nu^2)\rho\omega^2 r^2}{8E} + (1 + \nu)c_1 + (1 - \nu)\frac{c_2}{r^2} \right] \quad \textbf{(8.27c)}$$

The constants of integration may now be evaluated on the basis of the boundary conditions.

Annular Disk

In the case of an annular disk with zero pressure at the inner $(r = a)$ and outer $(r = b)$ boundaries (Fig. 8.9) the distribution of stress is due entirely to rotational effects. The boundary conditions are

$$(\sigma_r)_{r=a} = 0, \qquad (\sigma_r)_{r=b} = 0 \qquad \textbf{(b)}$$

These conditions, combined with Eq. (8.27b), yield two equations in the two unknown constants,

$$0 = -\rho\omega^2 \frac{a^2}{E} \frac{(1 - \nu^2)(3 + \nu)}{8} + (1 + \nu)c_1 - (1 - \nu)\frac{c_2}{a^2}$$

$$0 = -\rho\omega^2 \frac{b^2}{E} \frac{(1 - \nu^2)(3 + \nu)}{8} + (1 + \nu)c_1 - (1 - \nu)\frac{c_2}{b^2} \qquad \textbf{(c)}$$

from which

$$c_1 = \rho\omega^2 \frac{(a^2 + b^2)}{E} \frac{(1 - \nu)(3 + \nu)}{8}$$

$$c_2 = \rho\omega^2 \left(\frac{a^2 b^2}{E} \right) \frac{(1 + \nu)(3 + \nu)}{8} \qquad \textbf{(d)}$$

The stresses and displacement are therefore

$$\sigma_r = \frac{3 + \nu}{8} \left(a^2 + b^2 - r^2 - \frac{a^2 b^2}{r^2} \right) \rho\omega^2$$

$$\sigma_\theta = \frac{3 + \nu}{8} \left(a^2 + b^2 - \frac{1 + 3\nu}{3 + \nu} r^2 + \frac{a^2 b^2}{r^2} \right) \rho\omega^2 \qquad \textbf{(8.28 a–c)}$$

$$u = \frac{(3 + \nu)(1 - \nu)}{8E} \left(a^2 + b^2 - \frac{1 + \nu}{3 + \nu} r^2 + \frac{1 + \nu}{1 - \nu} \frac{a^2 b^2}{r^2} \right) \rho\omega^2 r$$

FIGURE 8.9. *Annular rotating disk of constant thickness.*

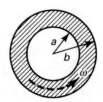

Applying the condition $d\sigma_r/dr = 0$ to the first of these equations, it is readily verified that the maximum radial stress occurs at $r = \sqrt{ab}$. Then, substituting the foregoing into Eq. (8.28a), the *maximum radial stress* is given by

$$\sigma_{r,\,max} = \frac{3 + \nu}{8}(b - a)^2 \rho\omega^2 \tag{8.29}$$

Figure 8.10a is a dimensionless representation of stress and displacement as a function of radius for an annular disk described by $b/a = 4$.

Solid Disk

In this case, $a = 0$, and the boundary conditions are

$$(\sigma_r)_{r=b} = 0, \qquad (u)_{r=0} = 0 \tag{e}$$

To satisfy the condition on the displacement, it is clear from Eq. (8.27a) that c_2 must be zero. The remaining constant is now evaluated from the first expression of Eq. (d):

$$c_1 = \rho\omega^2 \frac{b^2}{E} \frac{(1 - \nu)(3 + \nu)}{8}$$

Combining these constants with Eqs. (8.27), the following results are obtained:

$$\sigma_r = \frac{3 + \nu}{8}(b^2 - r^2)\,\rho\omega^2$$

$$\sigma_\theta = \frac{3 + \nu}{8}\left(b^2 - \frac{1 + 3\nu}{3 + \nu}r^2\right)\rho\omega^2 \tag{8.30 a–c}$$

$$u = \frac{1 - \nu}{8E}\left[(3 + \nu)b^2 - (1 + \nu)r^2\right]\rho\omega^2 r$$

The stress and displacement of a solid rotating disk are displayed in a dimensionless representation (Fig. 8.10b) as functions of radial location.

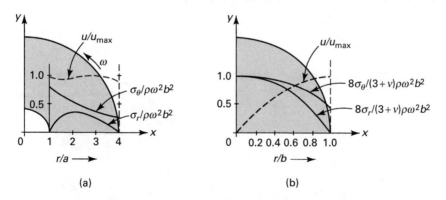

FIGURE 8.10. *Distribution of stress and displacement in a rotating disk of constant thickness: (a) annular disk; (b) solid disk.*

The constant thickness disks discussed in this section are generally employed when stresses or speeds are low and, as is clearly shown in Fig. 8.10b, do not make optimum use of material. Other types of rotating disks, offering many advantages over flat disks, are discussed in the sections to follow.

Clearly, the sharp rise of the tangential stress near $r = a$ is observed in Fig. 8.10a. It can be shown that, by setting $r = a$ in Eq. (8.29b) and letting a approach zero, the resulting σ_θ is twice that given by Eq. (8.30b). The foregoing conclusion applies to the radial stress as well. Thus, a small hole *doubles the stress* over the case of no hole.

8.7 DESIGN OF DISK FLYWHEELS

A *flywheel* is usually employed to smooth out changes in the speed of a shaft caused by torque fluctuations (Fig. 8.11). Flywheels are thus found in small and large machinery, such as compressors, internal combustion engines, punch presses, and rock crushers. At high speeds, considerable stresses may be induced in these components. Since failure of a rotating disk is particularly hazardous, analysis of the stress effects is important. Designing of energy-storing flywheels for hybrid-electric cars is an active area of contemporary research. *Disk flywheels*, that is, rotating annular disks of constant thickness, are made of high-strength steel plate.

An interference fit produces stress concentration in the *shaft* and in the (hub of) the disk, owing to the abrupt change from uncompressed to compressed material. Various design modifications are usually made in the faces of the disk close to the shaft diameter to decrease the stress concentrations at each sharp corner. Often, for a press or shrink fit, a *stress concentration factor K* is used. The value of K, depending on the contact pressure, the design of the disk, and the maximum bending stress in the shaft, rarely exceeds 2 [Ref. 8.3]. It should be noted that an approximation of the torque capacity of the assembly may be made on the basis of a *coefficient of friction* of about $f = 0.15$ between shaft and disk. The American Gear and Manufacturing Association (AGMA) standard suggests a value of $0.15 < f < 0.20$ for shrink or press fits having a smooth finish on both surfaces.

FIGURE 8.11. *A flywheel shrunk onto a shaft.*

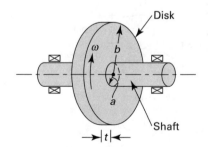

The Method of Superposition

Combined radial stress, tangential stress, and displacement of an annular *disk* due to internal pressure p between the disk and the shaft and angular speed ω may readily be obtained through the use of superposition of the results obtained previously. Therefore, we have

$$\sigma_r = (\sigma_r)_p + \frac{3+\nu}{8}\left(a^2 + b^2 - \frac{a^2 b^2}{r^2} - r^2\right)\rho\omega^2 \tag{8.31a}$$

$$\sigma_\theta = (\sigma_\theta)_p + \frac{3+\nu}{8}\left(a^2 + b^2 - \frac{1+3\nu}{3+\nu}r^2 + \frac{a^2 b^2}{r^2}\right)\rho\omega^2 \tag{8.31b}$$

$$u = (u)_p + \frac{(3+\nu)(1-\nu)}{8E}\left(a^2 + b^2 - \frac{1+\nu}{3+\nu}r^2 + \frac{1+\nu}{1-\nu}\frac{a^2 b^2}{r^2}\right)\rho\omega^2 r \tag{8.31c}$$

In the foregoing, the quantities $(\sigma_r)_p$, $(\sigma_\theta)_p$, and $(u)_p$ are given by Eqs. (8.8). The *tangential stress* σ_θ often *controls the design*. It is a maximum at the inner boundary $(r = a)$. Hence,

$$\sigma_{\theta,\,\text{max}} = p\frac{a^2 + b^2}{b^2 - a^2} + \frac{\rho\omega^2}{4}[(1-\nu)a^2 + (3+\nu)b^2] \tag{8.32}$$

As noted previously, owing to *rotation* alone, maximum radial stress occurs at $r = \sqrt{ab}$ and is given by Eq. (8.29). On the other hand, due to the internal *pressure* only, the largest radial stress is at the inner boundary and equals $\sigma_{r,\,\text{max}} = -p$.

It is customary to *neglect* the inertial stress and displacement of a shaft. Hence, for a *shaft*, we approximately have

$$\sigma_r = \sigma_\theta = -p$$
$$u = -\frac{1-\nu}{E_s}pr \tag{8.33}$$

We note, however, that the contact pressure p depends on angular speed ω. For a prescribed contact pressure p at angular speed ω, the required initial radial interference δ may be found from Eqs. (8.31c) and (8.33) for the displacement u. In so doing, with $r = a$, we obtain

$$\delta = \frac{ap}{E_d}\left(\frac{a^2 + b^2}{b^2 - a^2} + \nu\right) + \frac{ap}{E_s}(1-\nu) + \frac{a\rho\omega^2}{4E_d}[(1-\nu)a^2 + (3+\nu)b^2] \tag{8.34}$$

Here the quantities E_d and E_s represent the moduli of elasticity of the disk and shaft, respectively. Clearly, the preceding equation is valid as long as a positive contact pressure is maintained.

EXAMPLE 8.5   Torque and Power Capacity of a Flywheel

A flat steel disk with $t = 20$ mm and $a = 50$ mm is shrunk onto a shaft, causing a contact pressure p (Fig. 8.11). The coefficient of static friction is $f = 0.15$. Determine the torque T carried by the fit and power P

transmitted to the disk when the assembly is rotating at $n = 2400$ rpm and $p = 25$ Mpa.

Solution The *force* (axial or tangential) required for the assembly may be written in the form

$$F = 2\pi a f p t \qquad \text{(8.35a)}$$

The *torque capacity* of the fit is therefore

$$T = Fa = 2\pi a^2 f p t \qquad \text{(8.35b)}$$

Substituting the given numerical values,

$$T = 2\pi(50^2)(0.15)(25)(0.02) = 1.178 \text{ kN} \cdot \text{m}$$

Power transmitted is expressed by

$$P = T\omega \qquad \text{(8.35c)}$$

Here the angular velocity of the disk $\omega = 24000(2\pi)/60 = 251.3$ rad/s. We thus have $P = (1.178)(251.3) = 296$ kW.

EXAMPLE 8.6 Rotating Shrink-Fit Performance Analysis

A flat 0.5-m outer diameter, 0.1-m inner diameter, and 0.075-m-thick steel disk is shrunk onto a steel shaft (Fig. 8.12). If the assembly is to run at *speeds* up to $n = 6900$ rpm, determine (a) the shrinking allowance, (b) the maximum stress when not rotating, and (c) the maximum stress when rotating. The material properties are $\rho = 7.8$ kN $\cdot$ s^2/m^4, $E = 200$ GPa, and $\nu = 0.3$.

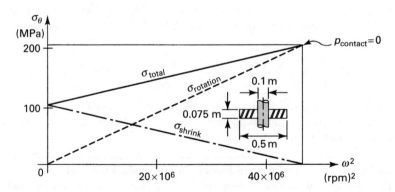

FIGURE 8.12. *Example 8.6. Tangential stresses produced by the combination of a shrink fit and rotation in a disk of constant thickness.*

Solution

a. The radial displacements of the disk (u_d) and shaft (u_s) are, from Eqs. (8.28) and (8.30),

$$u_d = 0.05 \times \frac{3.3 \times 0.7}{8E} \left(0.0025 + 0.0625 - \frac{1.3}{3.3} \times 0.0025 + \frac{1.3}{0.7} \times 0.0625 \right) \rho \omega^2$$

$$= 0.0026 \rho \frac{\omega^2}{E}$$

$$u_s = 0.05 \times \frac{0.7}{8E} (3.3 \times 0.0025 - 1.3 \times 0.0025) \rho \omega^2$$

$$= 2.1875 \times 10^{-5} \rho \frac{\omega^2}{E}$$

We observe that u_s may be neglected, as it is less than 1% of u_d of the disk at the common radius. The exact allowance is

$$\delta = u_d - u_s = \frac{(0.0026 - 2.1875 \times 10^{-5}) \times 7.8 \times 10^3 (6900 \times 2\pi/60)^2}{200 \times 10^9}$$

$$= 5.25 \times 10^{-5} \text{ m}$$

b. Applying Eq. (8.23), we have

$$p = \frac{E\delta}{2b} \frac{c^2 - b^2}{c^2} = \frac{200 \times 10^9 \times 5.25 \times 10^{-5} (0.0625 - 0.0025)}{2 \times 0.05 \times 0.0625} = 100.8 \text{ MPa}$$

Therefore, from Eq. (8.18),

$$\sigma_{\theta, \max} = p \frac{c^2 + b^2}{c^2 - b^2} = 100.8 \times 10^6 \frac{0.0625 + 0.0025}{0.0625 - 0.0025} = 109.2 \text{ MPa}$$

c. From Eq. (8.28), for $r = 0.05$,

$$\sigma_{\theta, \max} = \frac{3.3}{8} \left(0.0025 + 0.0625 - \frac{1.9}{3.3} \times 0.0025 + 0.0625 \right) \rho \omega^2$$

$$= 0.052 \rho \omega^2 = 211.78 \text{ MPa}$$

A plot of the variation of stress in the rotating disk is shown in Fig. 8.12.

EXAMPLE 8.7   Maximum Speed of a Flywheel Assembly

A flywheel of 380-mm diameter is to be shrunk onto a 60-mm diameter shaft (Fig. 8.11). Both members are made of steel with $\rho = 7.8 \text{ kN} \cdot \text{s}^2/\text{m}^4$, $E = 200 \text{ GPa}$, and $\nu = 0.3$. At a maximum speed of $n = 6000$ rpm, a contact pressure of $p = 20$ MPa is to be maintained. Find (a) the required radial interference; (b) the maximum tangential stress in the assembly; (c) the speed at which the fit loosens, that is, contact pressure becomes zero.

a. Through the use of Eq. (8.34), we obtain

$$\delta = \frac{30(10^{-3})p}{200(10^9)}\left(\frac{30^2 + 190^2}{190^2 - 30^2} + 1\right) + \frac{30(7.8)\omega^2}{4(200 \times 10^9)}\left[0.7(0.03)^2 + 3.3(0.19)^2\right] \quad \textbf{(a)}$$

$$= (0.308p + 35.03\omega^2)10^{-12}$$

Substituting the given numerical values $p = 20$ MPa and $\omega = 6000(2\pi/60) = 628.32$ rad/s, Eq. (a) results in $\delta = 0.02$ mm.

b. Applying Eq. (8.32), we obtain

$$\sigma_{\theta,\,max} = 20\frac{30^2 + 190^2}{190^2 - 30^2} + \frac{7800(628.32)}{4}\left[0.7(0.03)^2 + 3.3(0.19)^2\right]$$

$$= 21.02 + 92.23 = 113.2 \text{ MPa}$$

c. Carrying $\delta = 0.02 \times 10^{-3}$ m and $p = 0$ into Eq. (a) leads to

$$\omega = \left[\frac{0.02 \times 10^9}{32.282}\right]^{1/2} = 806.5 \text{ rad/s}$$

We thus have $n = 806.5\dfrac{60}{2\pi} = 7702$ rpm. Note that, at this speed, the shrink fit becomes completely ineffective.

8.8 ROTATING DISKS OF VARIABLE THICKNESS

In Section 8.6, the maximum stress in a flat rotating disk was observed to occur at the innermost fibers. This explains the general shape of many disks: thick near the hub, tapering down in thickness toward the periphery, as in a steam turbine. This not only has the effect of reducing weight but also results in lower rotational inertia.

The approach employed in the analysis of flat disks can be extended to variable thickness disks. Let the profile of a radial section be represented by the *general hyperbola* (Fig. 8.13),

$$t = t_1 r^{-s} \tag{8.36}$$

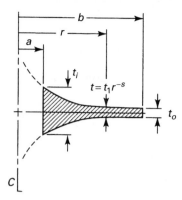

FIGURE 8.13. *Disk of hyperbolic profile.*

where t_1 represents a constant and s a positive number. The shape of the curve depends on the value selected for s; for example, for $s = 1$, the profile is that of an *equilateral hyperbola*. The constant t_1 is simply the thickness at radius equal to unity. If the thickness at $r = a$ is t_i and that at $r = b$ is t_o, as shown in the figure, the hyperbolic curve is fitted by forming the ratio

$$\frac{t_i}{t_o} = \frac{t_1 a^{-s}}{t_1 b^{-s}}$$

and solving for s. Clearly, Eq. (8.36) does not apply to solid disks, as all values of s except zero yield infinite thickness at $r = 0$.

In a turbine application, the actual configuration may have a thickened outer rim to which blades are affixed and a hub for attachment to a shaft. The hyperbolic relationship cannot describe such a situation exactly, but sometimes serves as an adequate approximation. If greater accuracy is required, the hub and outer ring may be approximated as flat disks, with the elements of the assembly related by the appropriate boundary conditions.

The differential equation of equilibrium (Eq. 8.26) must now include $t(r)$ and takes the form

$$\frac{d}{dr}(tr\sigma_r) - t\sigma_\theta + t\rho\omega^2 r^2 = 0 \qquad \textbf{(8.37)}$$

Equation (8.37) is satisfied by a stress function of the form

$$\Phi = tr\sigma_r, \qquad \frac{d\Phi}{dr} + t\rho\omega^2 r^2 = t\sigma_\theta \qquad \textbf{(a)}$$

Then the compatibility equation (8.4), using Eqs. (a) and Hooke's law, becomes

$$r^2 \frac{d^2\Phi}{dr^2} + \left(1 - \frac{r}{t}\frac{dt}{dr}\right) r \frac{d\Phi}{dr} + \left(\nu \frac{r}{t}\frac{dt}{dr} - 1\right)\Phi = -(3 + \nu)\rho\omega^2 tr^3 \qquad \textbf{(b)}$$

Introducing Eq. (8.36), we have

$$r^2 \frac{d^2\Phi}{dr^2} + (1 + s)r \frac{d\Phi}{dr} - (1 + \nu s)\,\Phi = -(3 - \nu)\,\rho\omega^2 t_1 r^{3-s} \tag{c}$$

This is an *equidimensional equation*, which the transformation $r = e^\alpha$ reduces to a linear differential equation with constant coefficients:

$$\frac{d^2\Phi}{d\alpha^2} + s\frac{d\Phi}{d\alpha} - (1 + \nu s)\,\Phi = -(3 + \nu)\,t_1\rho\,\omega^2\,e^{(3-s)\,\alpha} \tag{d}$$

The auxiliary equation corresponding to Eq. (d) is given by

$$m^2 + sm - (1 + \nu s) = 0$$

and has the roots

$$m_{1,2} = -\frac{s}{2} \pm \left[\left(\frac{s}{2}\right)^2 + (1 + \nu s)\right]^{1/2} \tag{8.38}$$

The general solution of Eq. (d) is then

$$\Phi = c_1 r^{m_1} + c_2 r^{m_2} - \frac{3+\nu}{8 - (3+\nu)s}\,t_1\,\rho\omega^2 r^{3-s} \tag{e}$$

The stress components for a disk of variable thickness are therefore, from Eqs. (a),

$$\sigma_r = \frac{c_1}{t_1} r^{m_1+s-1} + \frac{c_2}{t_1} r^{m_2+s-1} - \frac{3+\nu}{8 - (3+\nu)s}\,\rho\omega^2 r^2$$

$$\sigma_\theta = \frac{c_1}{t_1} m_1 r^{m_1+s-1} + \frac{c_2}{t_1} m_2 r^{m_2+s-1} - \frac{1+3\nu}{8 - (3+\nu)s}\,\rho\omega^2 r^2 \tag{8.39}$$

Note that, for a flat disk, $t = $ constant; consequently, $s = 0$ in Eq. (8.36) and $m = 1$ in Eq. (8.38). Thus, Eqs. (8.39) reduce to Eqs. (8.30), as expected. The constants c_1 and c_2 are determined from the boundary conditions

$$(\sigma_r)_{r=a} = (\sigma_r)_{r=b} = 0 \tag{f}$$

The evaluation of the constants is illustrated in the next example.

EXAMPLE 8.8   Rotating Hyperbolic Disk

The cross section of the disk in the assembly given in Example 8.6 is hyperbolic with $t_i = 0.075$ m and $t_o = 0.015$ m; $a = 0.05$ m, $b = 0.25$ m, and $\delta = 0.05$ mm. The rotational speed is 6900 rpm. Determine (a) the maximum stress owing to rotation and (b) the maximum radial displacement at the bore of the disk.

Solution

a. The value of the positive number s is obtained by the use of Eq. (8.36):

$$\frac{t_i}{t_o} = \frac{t_1 a^{-s}}{t_1 b^{-s}} = \left(\frac{b}{a}\right)^s$$

Substituting $t_i/t_o = 5$ and $b/a = 5$, we obtain $s = 1$. The profile will thus be given by $t = t_i/r$. From Eq. (8.38), we have

$$m_{1,2} = -\tfrac{1}{2} \pm [(\tfrac{1}{2})^2 + (1 + 0.3 \times 1)]^{1/2}, \qquad m_1 = 0.745, \qquad m_2 = -1.745$$

Hence the radial stresses, using Eqs. (8.39) and (f) for $r = 0.05$ and 0.25, are

$$(\sigma_r)_{r=0.05} = 0 = \frac{c_1}{t_1} 0.05^{0.745} + \frac{c_2}{t_1} 0.05^{-1.745} - 0.00176\rho\omega^2$$

$$(\sigma_r)_{r=0.25} = 0 = \frac{c_1}{t_1} 0.25^{0.745} + \frac{c_2}{t_1} 0.25^{-1.745} - 0.0439\,\rho\omega^2$$

from which

$$\frac{c_1}{t_1} = 0.12529\rho\omega^2, \qquad \frac{c_2}{t_1} = -6.272 \times 10^{-5}\rho\omega^2$$

The stress components in the disk, substituting these values into Eqs. (8.39), are therefore

$$\sigma_r = (0.12529r^{0.745} - 6.272 \times 10^{-5}r^{-1.745} - 0.70r^2)\rho\omega^2$$
$$\sigma_\theta = (0.09334r^{0.745} + 1.095 \times 10^{-4}r^{-1.745} - 0.40r^2)\rho\omega^2 \qquad \text{(g)}$$

The maximum stress occurs at the bore of the disk and from Eqs. (g) is equal to $(\sigma_\theta)_{r=0.05} = 0.0294\,\rho\omega^2$. Note that it was $0.052\,\rho\omega^2$ in Example 8.6. For the same speed, we conclude that the maximum stress is reduced considerably by tapering the disk.

b. The radial displacement is obtained from the second equation of (8.5), which together with Eq. (g) gives $u_{r=0.05} = (r\sigma_\theta/E)_{r=0.05} = 0.00147\rho\omega^2/E$. Again, this is advantageous relative to the value of $0.0026\,\rho\omega^2/E$ found in Example 8.6.

8.9 ROTATING DISKS OF UNIFORM STRESS

If every element of a rotating disk is stressed to a prescribed allowable value, presumed constant throughout, the disk material will clearly be used in the most efficient manner. For a given material, such a design is of minimum mass, offering the distinct advantage of reduced inertia loading as well as lower weight. What is sought, then, is a thickness variation $t(r)$ such that $\sigma_r = \sigma_\theta = \sigma = $ constant everywhere in the body. Under such a condition of stress, the strains, according to Hooke's law, are $\varepsilon_r = \varepsilon_\theta = \varepsilon = $ constant, and the compatibility equation (8.4) is satisfied.

The equation of equilibrium (8.37) may, under the conditions outlined, be written

$$\sigma \frac{d}{dr}(rt) - t\sigma + \rho\omega^2 tr^2 = 0$$

FIGURE 8.14. *Profile of disk of uniform strength.*

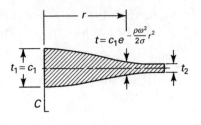

or

$$\frac{dt}{dr} + \frac{\rho\omega^2}{\sigma} rt = 0 \qquad\qquad (8.40)$$

which is easily integrated to yield

$$t = c_1 e^{-(\rho\omega^2/2\sigma)r^2} \qquad\qquad (8.41)$$

This variation assures that $\sigma_\theta = \sigma_r = \sigma =$ constant throughout the disk. To obtain the value of the constant in Eq. (8.41), the boundary condition $t = t_1$ at $r = 0$ is applied, resulting in $c_1 = t_1$ (Fig. 8.14).

EXAMPLE 8.9   Rotating Disk of Uniform Strength

A steel disk of the same outer radius, $b = 0.25$ m, and rotational speed, 6900 rpm, as the disk of Example 8.6 is to be designed for uniform stress. The thicknesses are $t_1 = 0.075$ m at the center and $t_2 = 0.015$ m at the periphery. Determine the stress and disk profile.

Solution   From Eq. (8.41),

$$\frac{t_2}{t_1} = \frac{c_1 e^{-(\rho\omega^2/2\sigma)b^2}}{c_1} = e^{-\rho(\omega^2/2\sigma)b^2} = \frac{1}{5}$$

or

$$\ln\left(\frac{1}{5}\right) = -\rho\frac{\omega^2}{2\sigma}b^2, \qquad \sigma = \rho\frac{\omega^2 b^2}{3.218} = 0.0194\rho\omega^2 = 79.08 \text{ MPa}$$

Thus,

$$t = c_1 e^{-\rho(\omega^2/2\sigma)r^2} = 0.075 e^{-25.752r^2}$$

Recall that the maximum stress in the hyperbolic disk of Example 8.8 was $0.0294\,\rho\omega^2$. The uniformly stressed disk is thus about 34% stronger than a hyperbolic disk with a small hole at its center.

In actual practice, fabrication and design constraints make it impractical to produce a section of exactly constant stress in a solid disk. On the other hand, in an annular disk, if the boundary condition is applied such that the radial stress is zero at the inner radius, constancy of stress dictates that σ_θ and σ_r be zero everywhere. This is clearly not a useful result for the situation as described. For these reasons, the hyperbolic variation in thickness is often used.

8.10 THERMAL STRESSES IN THIN DISKS

In this section, our concern is with the stresses associated with a radial temperature field $T(r)$ that is independent of the axial dimension. The practical applications are numerous and include annular fins and turbine disks. Because the temperature field is symmetrical with respect to a central axis, it is valid to assume that the stresses and displacements are distributed in the same way as those of Section 8.1, and therefore the equations of that section apply here as well.

In this case of plane stress, the applicable equations of stress and strain are obtained from Eq. (3.37) with reference to Eq. (3.26):

$$\sigma_r = \frac{E}{1 - v^2}[\varepsilon_r + v\varepsilon_\theta - (1 + v)\alpha T]$$

$$\sigma_\theta = \frac{E}{1 - v^2}[\varepsilon_\theta + v\varepsilon_r - (1 + v)\alpha T] \tag{8.42}$$

The equation of equilibrium, Eq. (8.2), is now

$$r\frac{d}{dr}(\varepsilon_r + v\varepsilon_\theta) + (1 - v)(\varepsilon_r - \varepsilon_\theta) = (1 + v)\alpha r\frac{dT}{dr} \tag{a}$$

Introduction of Eq. (8.3) into expression (a) yields the following differential equation in radial displacement:

$$\frac{d^2u}{dr^2} + \frac{1}{r}\frac{du}{dr} - \frac{u}{r^2} = (1 + v)\alpha\frac{dT}{dr} \tag{b}$$

This is rewritten

$$\frac{d}{dr}\left[\frac{1}{r}\frac{d(ru)}{dr}\right] = (1 + v)\alpha\frac{dT}{dr} \tag{c}$$

to render it easily integrable. The solution is

$$u = \frac{(1 + v)\alpha}{r}\int_a^r Tr\,dr + c_1 r + \frac{c_2}{r} \tag{8.43}$$

where a, the inner radius of an annular disk, is taken as zero for a solid disk, and v and α have been treated as constants.

Annular Disk

The radial and tangential stresses in the annular disk of inner radius a and an outer radius b may be found by substituting Eq. (8.43) into Eq. (8.3) and the results into Eq. (8.42):

$$\sigma_r = -\frac{\alpha E}{r^2}\int_a^r Tr\,dr + \frac{E}{1 - v^2}\left[c_1(1 + v) - \frac{c_2(1 - v)}{r^2}\right] \tag{d}$$

$$\sigma_\theta = \frac{\alpha E}{r^2}\int_a^r Tr\,dr - \alpha ET + \frac{E}{1 - v^2}\left[c_1(1 + v) + \frac{c_2(1 - v)}{r^2}\right] \tag{e}$$

The constants c_1 and c_s are determined on the basis of the boundary conditions $(\sigma_r)_{r=a} = (\sigma_r)_{r=b} = 0$. Equation (d) thus gives

$$c_1 = \frac{(1-\nu)\alpha}{b^2 - a^2} \int_a^b Tr\, dr, \qquad c_2 = \frac{(1+\nu)a^2\alpha}{b^2 - a^2} \int_a^b Tr\, dr$$

The stresses are therefore

$$\sigma_r = \alpha E \left[-\frac{1}{r^2} \int_a^r Tr\, dr + \frac{r^2 - a^2}{r^2(b^2 - a^2)} \int_a^b Tr\, dr \right]$$

$$\sigma_\theta = \alpha E \left[-T + \frac{1}{r^2} \int_a^r Tr\, dr + \frac{r^2 + a^2}{r^2(b^2 - a^2)} \int_a^b Tr\, dr \right] \tag{8.44}$$

Solid Disk

In the case of a solid disk of radius b, the displacement must vanish at $r = 0$ in order to preserve the continuity of material. The value of c_2 in Eq. (8.43) must therefore be zero. To evaluate c_1, the boundary condition $(\sigma_r)_{r=b} = 0$ is employed, and Eq. (d) now gives

$$c_1 = \frac{(1-\nu)\alpha}{b^2} \int_0^b Tr\, dr$$

Substituting c_1 and c_2 into Eqs. (d) and (e), the stresses in a solid disk are found to be

$$\sigma_r = \alpha E \left[\frac{1}{b^2} \int_0^b Tr\, dr - \frac{1}{r^2} \int_0^r Tr\, dr \right]$$

$$\sigma_\theta = \alpha E \left[-T + \frac{1}{b^2} \int_0^b Tr\, dr + \frac{1}{r^2} \int_0^r Tr\, dr \right] \tag{8.45}$$

Given a temperature distribution $T(r)$, the stresses in a solid or annular disk can thus be determined from Eqs. (8.44) or (8.45). Note that $T(r)$ need not be limited to those functions that can be analytically integrated. A numerical integration can easily be carried out for σ_r, σ_θ to provide results of acceptable accuracy.

8.11 THERMAL STRESS IN LONG CIRCULAR CYLINDERS

Consider a long cylinder with *ends assumed restrained* so that $w = 0$. This is another example of plane strain, for which $\varepsilon_z = 0$. The stress–strain relations are, from Hooke's law,

$$\varepsilon_r = \frac{1}{E} [\sigma_r - \nu(\sigma_\theta + \sigma_z)] + \alpha T$$

$$\varepsilon_\theta = \frac{1}{E} [\sigma_\theta - \nu(\sigma_r + \sigma_z)] + \alpha T \tag{8.46}$$

$$\varepsilon_z = \frac{1}{E} [\sigma_z - \nu(\sigma_r + \sigma_\theta)] + \alpha T$$

For $\varepsilon_z = 0$, the final expression yields

$$\sigma_z = \nu(\sigma_r + \sigma_\theta) - \alpha ET \tag{a}$$

Substitution of Eq. (a) into the first two of Eqs. (8.46) leads to the following forms in which z stress does not appear:

$$\varepsilon_r = \frac{1 + \nu}{E}[(1 - \nu)\sigma_r - \nu\sigma_\theta + \alpha ET]$$

$$\varepsilon_\theta = \frac{1 + \nu}{E}[(1 - \nu)\sigma_\theta - \nu\sigma_r + \alpha ET] \tag{8.47}$$

Inasmuch as Eqs. (8.2) and (8.3) are valid for the case under discussion, the solutions for u, σ_r, and σ_θ proceed as in Section 8.10, resulting in

$$u = \frac{(1 + \nu)\alpha}{(1 - \nu)r}\int_a^r Tr\,dr + c_1 r + \frac{c_2}{r} \tag{b}$$

$$\sigma_r = \frac{E}{1 + \nu}\left[-\frac{(1 + \nu)\alpha}{(1 - \nu)r^2}\int_a^r Tr\,dr + \frac{c_1}{1 - 2\nu} - \frac{c_2}{r^2}\right] \tag{c}$$

$$\sigma_\theta = \sigma_r + r\frac{d\sigma_r}{dr} \tag{d}$$

Finally, from Eq. (a), we obtain

$$\sigma_z = -\frac{\alpha ET}{1 - \nu} + \frac{2\nu E c_1}{(1 + \nu)(1 - 2\nu)} \tag{e}$$

Solid Cylinder

For the radial displacement of a solid cylinder of radius b to vanish at $r = 0$, the constant c_2 in Eq. (b) must clearly be zero. Applying the boundary condition $(\sigma_r)_{r=b} = 0$, Eq. (c) may be solved for the remaining constant of integration,

$$c_1 = \frac{(1 + \nu)(1 - 2\nu)\alpha}{(1 - \nu)b^2}\int_0^b Tr\,dr \tag{f}$$

and the stress distributions determined from Eqs. (c), (d), and (e):

$$\sigma_r = \frac{\alpha E}{1 - \nu}\left[\frac{1}{b^2}\int_0^b Tr\,dr - \frac{1}{r^2}\int_0^r Tr\,dr\right]$$

$$\sigma_\theta = \frac{\alpha E}{1 - \nu}\left[-T + \frac{1}{b^2}\int_0^b Tr\,dr + \frac{1}{r^2}\int_0^r Tr\,dr\right] \tag{8.48}$$

$$\sigma_z = \frac{\alpha E}{1 - \nu}\left[\frac{2\nu}{b^2}\int_0^b Tr\,dr - T\right] \tag{8.49a}$$

To derive an expression for the radial displacement, $c_2 = 0$ and Eq. (f) are introduced into Eq. (b).

The longitudinal stress given by Eq. (8.49a) is valid only for the case of a fixed-ended cylinder. In the event the *ends are free,* a uniform axial stress $\sigma_z = s_0$ may be superimposed to cause the force resultant at each end to vanish:

$$s_0 \pi b^2 - \int_0^b \sigma_z(2\pi r)\, dr = 0$$

This expression together with Eq. (e) yields

$$s_0 = \frac{2\alpha E}{b^2(1 - \nu)} \int_0^b Tr\, dr - \frac{2\nu E c_1}{(1 + \nu)(1 - 2\nu)} \tag{g}$$

The longitudinal stress for a free-ended cylinder is now obtained by adding s_0 to the stress given by Eq. (e):

$$\sigma_z = \frac{\alpha E}{1 - \nu}\left(\frac{2}{b^2}\int_0^b Tr\, dr - T\right) \tag{8.49b}$$

Stress components σ_r and σ_θ remain as before. The axial displacement is obtained by adding $u_0 = -\nu s_0 r/E$, a displacement due to uniform axial stress, s_0, to the right side of Eq. (b).

Cylinder with Central Circular Hole

When the inner $(r = a)$ and outer $(r = b)$ surfaces of a hollow cylinder are free of applied load, the boundary conditions $(\sigma_r)_{r=a} = (\sigma_r)_{r=b} = 0$ apply. Introducing these into Eq. (c), the constants of integration are

$$c_1 = \frac{(1 + \nu)(1 - 2\nu)\alpha}{(1 - \nu)(b^2 - a^2)} \int_a^b Tr\, dr - \nu\varepsilon_z$$

$$\tag{h}$$

$$c_2 = \frac{(1 + \nu)\, a^2\alpha}{(1 - \nu)(b^2 - a^2)} \int_a^b Tr\, dr$$

Equations (c), (d), and (e) thus provide

$$\sigma_r = \frac{E}{1 - \nu}\left[-\frac{\alpha}{r^2}\int_a^r Tr\, dr + \frac{(r^2 - a^2)\,\alpha}{r^2(b^2 - a^2)}\int_a^b Tr\, dr\right]$$

$$\sigma_\theta = \frac{E}{1 - \nu}\left[-T + \frac{\alpha}{r^2}\int_a^r Tr\, dr + \frac{(r^2 + a^2)\,\alpha}{r^2(b^2 - a^2)}\int_a^b Tr\, dr\right] \tag{8.50}$$

$$\sigma_z = \frac{\alpha E}{1 - \nu}\left[\frac{2\nu}{b^2 - a^2}\int_a^b Tr\, dr - T\right] \tag{8.51a}$$

If the *ends are free,* proceeding as in the case of a solid cylinder, the longitudinal stress is described by

$$\sigma_z = \frac{\alpha E}{1 - \nu} \left[\frac{2}{b^2 - a^2} \int_a^b Tr\,dr - T \right] \tag{8.51b}$$

Implementation of the foregoing analyses depends on a knowledge of the radial distribution of temperature $T(r)$.

EXAMPLE 8.10 Thermal Stresses in a Cylinder

Determine the stress distribution in a hollow free-ended cylinder, subject to constant temperatures T_a and T_b at the inner and outer surfaces, respectively.

Solution The radial steady-state heat flow through an arbitrary internal cylindrical surface is given by Fourier's law of conduction:

$$Q = -2\pi r K \frac{dT}{dr} \tag{i}$$

Here Q is the heat flow per unit axial length and K the thermal conductivity. Assuming Q and K to be constant,

$$r \frac{dT}{dr} = -\frac{Q}{2\pi K} = \text{constant} = c$$

This is easily integrated upon separation:

$$T = \int c_1 \frac{dr}{r} + c_2 = c_1 \ln r + c_2 \tag{j}$$

Here c_1 and c_2, the constants of integration, are determined by applying the temperature boundary conditions $(T)_{r=a} = T_a$ and $(T)_{r=b} = T_b$. By so doing, Eq. (j) may be written in the form

$$T = \frac{T_a - T_b}{\ln(b/a)} \ln \frac{b}{r} \tag{8.52}$$

When this is substituted into Eqs. (8.50) and (8.51a), the following stresses are obtained:

$$\sigma_r = \frac{\alpha E(T_a - T_b)}{2(1 - \nu)\ln(b/a)} \left[-\ln \frac{b}{r} - \frac{a^2(r^2 - b^2)}{r^2(b^2 - a^2)} \ln \frac{b}{a} \right]$$

$$\sigma_\theta = \frac{\alpha E(T_a - T_b)}{2(1 - \nu)\ln(b/a)} \left[1 - \ln \frac{b}{r} - \frac{a^2(r^2 + b^2)}{r^2(b^2 - a^2)} \ln \frac{b}{a} \right] \tag{8.53}$$

$$\sigma_z = \frac{\alpha E(T_a - T_b)}{2(1 - \nu)\ln(b/a)} \left[1 - 2\ln \frac{b}{r} - \frac{2a^2}{b^2 - a^2} \ln \frac{b}{a} \right]$$

We note that, for $\nu = 0$, Eqs. (8.53) provide a solution for a thin hollow disk. In the event the heat flow is outward, that is, $T_a > T_b$, examination of Eqs. (8.53) indicates that the stresses σ_θ and σ_z are compressive (negative) on the inner surface and tensile (positive) on the outer surface. They exhibit their maximum values at the inner and outer surfaces. On the other hand, the radial stress is compressive at all points and becomes zero at the inner and outer surfaces of the cylinder.

In practice, a pressure loading is often superimposed on the thermal stresses, as in the case of chemical reaction vessels. An approach is to follow similarly to that already discussed, with boundary conditions modified to reflect the pressure; for example, $(\sigma_r)_{r=a} = -p_i$, $(\sigma_r)_{r=b} = 0$. In this instance, the internal pressure results in a circumferential tensile stress (Fig. 8.4a), causing a partial cancellation of compressive stress owing to temperature.

Note that, when a rotating disk is subjected to inertia, loading combined with an axisymmetrical distribution of temperature $T(r)$, the stresses and displacements may be determined by superposition of the two cases.

8.12 FINITE ELEMENT SOLUTION

In the previous sections, the cases of axisymmetry discussed were ones in which along the axis of revolution (z) there was uniformity of structural geometry and loading. In this section, the finite element approach of Chapter 7 is applied for computation of displacement, strain, and stress in a general axisymmetric structural system, formed as a solid of revolution having material properties, support conditions, and loading, all of which are symmetrical about the z axis, but that may vary along this axis. A simple illustration of this situation is a sphere uniformly loaded by gravity forces.

"Elements" of the body of revolution (rings or, more generally, tori) are used to discretize the axisymmetric structure. We shall here employ an element of triangular cross section, as shown in Fig. 8.15. Note that a node is now in fact a circle; for example, node i is the circle with r_i as radius. Thus, the elemental volume dV appearing in the expressions of Section 7.13 is the volume of the ring element $(2\pi r\, dr\, dz)$.

The element clearly lies in three-dimensional space. Any randomly selected vertical cross section of the element, however, is a *plane* triangle. As already discussed in Section 8.1, no tangential displacement can exist in the symmetrical system; that is, $v = 0$. Inasmuch as only the radial displacement u and the axial displacement w in a plane are involved (rz plane), the expressions for displacement established for plane stress and plane strain may readily be extended to the axisymmetric analysis [Ref. 8.4].

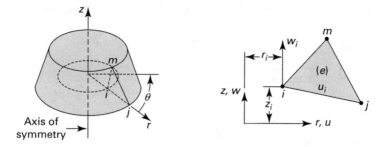

FIGURE 8.15. *Triangular cross section of axisymmetric solid finite element.*

8.13 AXISYMMETRIC ELEMENT

The theoretical development follows essentially the procedure given in Chapter 7, with the exception that, in the present case, cylindrical coordinates are employed (r, θ, z), as shown in Fig. 8.15.

Strain, Stress, and Elasticity Matrices

The strain matrix, from Eqs. (2.4), (3.33a), and (3.33b), may be defined as follows:

$$\{\varepsilon\}_e = \{\varepsilon_r, \varepsilon_z, \varepsilon_\theta, \gamma_{rz}\} = \left\{\frac{\partial u}{\partial r}, \frac{\partial w}{\partial z}, \frac{u}{r}, \frac{\partial u}{\partial z} + \frac{\partial w}{\partial r}\right\} \tag{8.54}$$

The initial strain owing to a temperature change is expressed in the form

$$\{\varepsilon_0\}_e = \{\alpha T, \alpha T, \alpha T, 0\}$$

It is observed from Eq. (8.54) that the tangential strain ε_θ becomes infinite for a zero value of r. Thus, if the structural geometry is continuous at the z axis, as in the case of a solid sphere, r is generally assigned a small value (for example, 0.1 mm) for the node located at this axis.

It can be demonstrated that the state of stress throughout the element $\{\sigma\}_e$ is expressible as follows:

$$\{\sigma_r, \sigma_z, \sigma_\theta, \tau_{rz}\}_e = \frac{E}{(1 + \nu)(1 - 2\nu)}$$

$$\times \begin{bmatrix} 1 - \nu & \nu & \nu & 0 \\ \nu & 1 - \nu & \nu & 0 \\ \nu & \nu & 1 - \nu & 0 \\ 0 & 0 & 0 & (1 - 2\nu)/2 \end{bmatrix} \begin{Bmatrix} \varepsilon_r - \alpha T \\ \varepsilon_z - \alpha T \\ \varepsilon_\theta - \alpha T \\ \gamma_{rz} \end{Bmatrix} \tag{8.55}$$

A comparison of Eqs. (8.55) and (7.58) yields the elasticity matrix

$$[D] = \frac{E}{(1+\nu)(1-2\nu)} \begin{bmatrix} 1-\nu & \nu & \nu & 0 \\ & 1-\nu & \nu & 0 \\ & & 1-\nu & 0 \\ \text{Symm.} & & & (1-2\nu)/2 \end{bmatrix} \qquad (8.56)$$

Displacement Function

The nodal displacements of the element are written in terms of submatrices δ_u and δ_v:

$$\{\delta\}_e = \{u_i, u_j, u_m, w_i, w_j, w_m\} \qquad \textbf{(a)}$$

The displacement function $\{f\}_e$, which describes the behavior of the element, is given by

$$\{f\}_e = \begin{Bmatrix} u \\ w \end{Bmatrix} = \begin{Bmatrix} \alpha_1 + \alpha_2 r + \alpha_3 z \\ \alpha_4 + \alpha_5 r + \alpha_6 z \end{Bmatrix} \qquad (8.57a)$$

or

$$\{f\}_e = \begin{bmatrix} 1 & r & z & 0 & 0 & 0 \\ 0 & 0 & 0 & 1 & r & z \end{bmatrix} \{\alpha_1, \alpha_2, \alpha_3, \alpha_4, \alpha_5, \alpha_6\} \qquad (8.57b)$$

Here the α's are the constants, which can be evaluated as follows. First, we express the nodal displacements $\{\delta\}_e$:

$$u_i = \alpha_1 + \alpha_2 r_i + \alpha_3 z_i$$
$$u_j = \alpha_1 + \alpha_2 r_j + \alpha_3 z_j$$
$$u_m = \alpha_1 + \alpha_2 r_m + \alpha_3 z_m$$
$$w_i = \alpha_4 + \alpha_5 r_i + \alpha_6 z_i$$
$$w_j = \alpha_4 + \alpha_5 r_j + \alpha_6 z_j$$
$$w_m = \alpha_4 + \alpha_5 r_m + \alpha_6 z_m$$

Then, by the inversion of these linear equations,

$$\begin{Bmatrix} \alpha_1 \\ \alpha_2 \\ \alpha_3 \end{Bmatrix} = \frac{1}{2A} \begin{bmatrix} a_i & a_j & a_m \\ b_i & b_j & b_m \\ c_i & c_j & c_m \end{bmatrix} \begin{Bmatrix} u_i \\ u_j \\ u_m \end{Bmatrix}$$

$$\begin{Bmatrix} \alpha_4 \\ \alpha_5 \\ \alpha_6 \end{Bmatrix} = \frac{1}{2A} \begin{bmatrix} a_i & a_j & a_m \\ b_i & b_j & b_m \\ c_i & c_j & c_m \end{bmatrix} \begin{Bmatrix} w_i \\ w_j \\ w_m \end{Bmatrix} \qquad (8.58)$$

where A is defined by Eq. (7.67) and

$$
\begin{aligned}
a_i &= r_j z_m - z_j r_m, & b_i &= z_j - z_m, & c_i &= r_m - r_j \\
a_j &= r_m z_i - z_m r_i, & b_j &= z_m - z_i, & c_j &= r_i - r_m \\
a_m &= r_i z_j - z_i r_j, & b_m &= z_i - z_j, & c_m &= r_j - r_i
\end{aligned}
\tag{8.59}
$$

Finally, upon substitution of Eqs. (8.58) into (8.57), the displacement function is represented in the following convenient form:

$$
\{f\}_e = \begin{bmatrix} N_i & N_j & N_m & 0 & 0 & 0 \\ 0 & 0 & 0 & N_i & N_j & N_m \end{bmatrix} \cdot \{u_i, u_j, u_m, w_i, w_j, w_m\}
\tag{8.60}
$$

or

$$
\{f\}_e = [N]\{\delta\}_e
$$

with

$$
\begin{aligned}
N_i &= \frac{1}{2A}(a_i + b_i r + c_i z) \\
N_j &= \frac{1}{2A}(a_j + b_j r + c_j z) \\
N_m &= \frac{1}{2A}(a_m + b_m r + c_m z)
\end{aligned}
\tag{8.61}
$$

The element strain matrix is found by introducing Eq. (8.60) together with (8.61) into (8.54):

$$
\{\varepsilon_r, \varepsilon_z, \varepsilon_\theta, \gamma_{rz}\}_e = [B]\{u_i, u_j, u_m, w_i, w_j, w_m\}
\tag{8.62}
$$

where

$$
[B] = \frac{1}{2A} \begin{bmatrix} b_i & b_j & b_m & 0 & 0 & 0 \\ 0 & 0 & 0 & c_i & c_j & c_m \\ d_i & d_j & d_m & 0 & 0 & 0 \\ c_i & c_j & c_m & b_i & b_j & b_m \end{bmatrix}
\tag{8.63}
$$

with

$$
d_n = \frac{a_n}{r} + b_n + \frac{c_n z}{r}, \qquad n = i, j, m
$$

It is observed that the matrix $[B]$ includes the coordinates r and z. Thus, *the strains are not constant*, as is the case with plane stress and plane strain.

The Stiffness Matrix

The element stiffness matrix, from Eq. (7.61), is given by

$$
[k]_e = \int_V [B]^T [D][B] \, dV
\tag{8.64a}
$$

and must be integrated along the circumferential or ring boundary. This may thus be rewritten

$$[k]_e = 2\pi \int r[B]^T[D][B]dr\,dz \tag{8.64b}$$

where the matrices $[D]$ and $[B]$ are defined by Eqs. (8.56) and (8.63), respectively. It is observed that integration is not easily performed as in the case of plane stress problems, because $[B]$ is also a function of r and z. Although tedious, the integration can be carried out explicitly. Alternatively, approximate numerical approaches may be used. In a simple approximate procedure, $[B]$ is evaluated for a centroidal point of the element. To accomplish this, we substitute fixed centroidal coordinates of the element

$$\bar{r} = \tfrac{1}{3}(r_i + r_j + r_m), \qquad \bar{z} = \tfrac{1}{3}(z_i + z_j + z_m) \tag{8.65}$$

into Eq. (8.63) in place of r and z to obtain $[\bar{B}]$. Then, by letting $dV = 2\pi\bar{r}A$, from Eqs. (8.64), the element stiffness is found to be

$$[k]_e = 2\pi\bar{r}A[B]^T[D][\bar{B}] \tag{8.66}$$

This simple procedure leads to results of acceptable accuracy.

External Nodal Forces

In the axisymmetric case, "concentrated" or "nodal" forces are actually loads axisymmetrically located around the body. Let q_r and q_z represent the radial and axial components of force per unit length, respectively, of the circumferential boundary of a node or a radius r. The total nodal force in the radial direction is

$$Q_r = 2\pi r q_r \tag{8.67a}$$

Similarly, the total nodal force in the axial direction is

$$Q_z = 2\pi r q_z \tag{8.67b}$$

Other external load components can be treated analogously. When the approximation leading to Eq. (8.66) is used, we can, from Eq. (7.58), readily obtain expressions for nodal forces owing to the initial strains, body forces, and any surface tractions (see Prob. 8.49).

In summary, the solution of an axisymmetric problem can be obtained, having generated the total stiffness matrix $[K]$, from Eq. (8.64) or (8.66), and the load matrices $\{Q\}$. Then Eq. (7.60) provides the numerical values of nodal displacements $\{\delta\} = [K]^{-1}\{Q\}$. The expression (8.62) together with Eq. (8.63) yields values of the element strains. Finally, Eq. (8.55), upon substitution of Eq. (8.62), is used to determine the element stresses.

CASE STUDY 8.1 **Thick-Walled High-Pressure Steel Cylinder**

Figure 8.16a illustrates a thick-walled pipe having the inner radius a and outer radius b under external pressure p_o. Plot the distribution of stress, as found by the FEA and equations obtained by the theory of elasticity in Section 8.2, across the wall of the cylinder. *Given:* $b = 2a$ and $\nu = 0.3$.

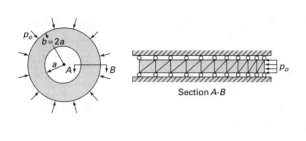

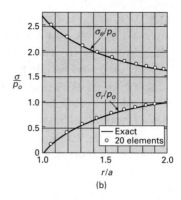

(a)

(b)

FIGURE 8.16. *Case Study 8.1. (a) Thick-walled cylinder subjected to external pressure and modeling of a slice at section A–B; (b) distribution of tangential and radial stresses in the cylinder for b = 2a.*

Solution   Only a thin segment or slice of the cylinder need be analyzed. We shall use a total of 20 triangular elements, gradually decreasing in size toward the region under pressure, as shown at a section *A–B* in Fig. 8.16a. The boundary conditions for the mesh system may be depicted by rollers at the nodes along the top and bottom faces of the slice, so that at these nodes the displacements *w* are to be constrained to vanish.

Employing a general-purpose finite element program [Refs. 8.5 through 8.9], the tangential stresses σ_θ and radial stresses σ_r are calculated. The nondimensional results are sketched in Fig. 8.16b. In addition, the exact results given by Eqs. (8.15) and (8.16) are shown. Observe the excellent agreement between the solutions found by the two approaches.

REFERENCES

8.1. RANOV, T. and PARK, F. R., On the numerical value of the tangential stress on thick-walled cylinders, J. Appl. Mech., March 1953.

8.2. BECKER, S. J. and MOLLICK, L., The theory of the ideal design of a compound vessel, Trans. ASME, J. Engineering Industry, May 1960, 136.

8.3. UGURAL, A. C., *Mechanical Design: An Integrated Approach*, McGraw-Hill, New York, 2004, Chap. 16.

8.4. YANG, T. Y., *Finite Element Structural Analysis*, Prentice Hall, Englewood Cliffs, N.J., 1986, Chap. 10.

8.5. ZIENKIEWICZ, O. C. and TAYLOR, R. I., *The Finite Element Method*, 4th ed., Vol. 2 (Solid and Fluid Mechanics, Dynamics and Nonlinearity), McGraw-Hill, London, 1991.

8.6. COOK, R. D., *Concepts and Applications of Finite Element Analysis*, 2nd ed., Wiley, New York, 1980.

8.7. BATHE, K. I., *Finite Element Procedures in Engineering Analysis*, Prentice Hall, Upper Saddle River, N.J., 1996.

8.8. DUNHAM, R. S. and NICKELL, R. E., "Finite element analysis of axisymmetric solids with arbitrary loadings," Report AD 655 253, National Technical Information Service, Springfield, VA, June 1967.

8.9. UTKU, S., Explicit expressions for triangular torus element stiffness matrix, AIAA Journal, 6/6, 1174–1176, June 1968.

PROBLEMS

Sections 8.1 through 8.5

8.1. A cylinder of internal radius a and external radius $b = 1.10a$ is subjected to (a) internal pressure p_i only and (b) external pressure p_o only. Determine for each case the ratio of maximum to minimum tangential stress.

8.2. A thick-walled cylinder with closed ends is subjected to internal pressure p_i only. Knowing that $a = 0.6$ m, $b = 1$ m, $\sigma_{all} = 140$ MPa, and $\tau_{all} = 80$ MPa, determine the allowable value of p_i.

8.3. A cylinder of inner radius a and outer radius na, where n is an integer, has been designed to resist a specific internal pressure, but reboring becomes necessary. (a) Find the new inner radius r_x required so that the maximum tangential stress does not exceed the previous value by more than $\Delta\sigma_\theta$, while the internal pressure is the same as before. (b) If $a = 25$ mm and $n = 2$ and after reboring the tangential stress is increased by 10%, determine the new diameter.

8.4. A steel tank having an internal diameter of 1.2 m is subjected to an internal pressure of 7 MPa. The tensile and compressive elastic strengths of the material are 280 MPa. Assuming a factor of safety of 2, determine the wall thickness.

8.5. Two thick-walled, closed-ended cylinders of the same dimensions are subjected to internal and external pressure, respectively. The outer diameter of each is twice the inner diameter. What is the ratio of the pressures for the following cases? (a) The maximum tangential stress has the same absolute value in each cylinder. (b) The maximum tangential strain has the same absolute value in each cylinder. Take $\nu = \frac{1}{3}$.

8.6. Determine the radial displacement of a point on the inner surface of the tank described in Prob. 8.4. Assume that outer diameter $2b = 1.2616$ m, $E = 200$ GPa, and $\nu = 0.3$.

8.7. Derive expressions for the maximum circumferential strains of a thick-walled cylinder subject to internal pressure p_i only, for two cases: (a) an open-ended cylinder and (b) a closed-ended cylinder. Then, assuming that the allowable strain is limited to $1000\ \mu$, $p_i = 60$ MPa, and $b = 2$ m, determine the required wall thickness. Use $E = 200$ GPa and $\nu = \frac{1}{3}$.

8.8. An aluminum cylinder ($E = 72$ GPa, $\nu = 0.3$) with closed and free ends has a 500-mm external diameter and a 100-mm internal diameter. Determine maximum tangential, radial, shearing, and longitudinal stresses and the change in the internal diameter at a section away from the ends for an internal pressure of $p_i = 60$ MPa.

8.9. Rework Prob. 8.8 with $p_i = 0$ and $p_o = 60$ MPa.

8.10. A steel cylinder of 0.3-m radius is shrunk over a solid steel shaft of 0.1-m radius. The shrinking allowance is 0.001 m/m. Determine the external pressure p_o on the outside of the cylinder required to reduce to zero the circumferential tension at the inside of the cylinder. Use $E_s = 200$ GPa.

8.11. A steel cylinder is subjected to an internal pressure only. (a) Obtain the ratio of the wall thickness to the inner diameter if the internal pressure is three-quarters of the maximum allowable tangential stress. (b) Determine the increase in inner diameter of such a cylinder, 0.15 m in internal diameter, for an internal pressure of 6.3 MPa. Take $E = 210$ GPa and $\nu = \frac{1}{3}$.

8.12. Verify the results shown in Fig. 8.5 using Eqs. (a) and (8.21) of Section 8.3.

8.13. A thick-walled cylinder is subjected to internal pressure p_i and external pressure p_o. Find (a) the longitudinal stress σ_z if the longitudinal strain is zero and (b) the longitudinal strain if σ_z is zero.

8.14. A cylinder, subjected to internal pressure only, is constructed of aluminum having a tensile strength σ_{yp}. The internal radius of the cylinder is a, and the outer radius is $2a$. Based on the maximum energy of distortion and maximum shear stress theories of failure, predict the limiting values of internal pressure.

8.15. A cylinder, subjected to internal pressure p_i only, is made of cast iron having ultimate strengths in tension and compression of $\sigma_u = 350$ MPa and $\sigma_u' = 630$ MPa, respectively. The inner and outer radii are a and $3a$. Determine the allowable value of p_i using (a) the maximum principal stress theory and (b) the Coulomb–Mohr theory.

8.16. A flywheel of 0.5-m outer diameter and 0.1-m inner diameter is pressed onto a solid shaft. The maximum tangential stress induced in the flywheel is 35 MPa. The length of the flywheel parallel to the shaft axis is 0.05 m. Assuming a coefficient of static friction of 0.2 at the common surface, find the maximum torque that may be transmitted by the flywheel without slippage.

8.17. A solid steel shaft of 0.1-m diameter is pressed onto a steel cylinder, inducing a contact pressure p_1 and a maximum tangential stress $2p_1$ in the cylinder. If an axial tensile load of $P_L = 45$ kN is applied to the shaft, what change in contact pressure occurs? Let $\nu = \frac{1}{3}$.

8.18. A brass cylinder of outer radius c and inner radius b is to be press-fitted over a steel cylinder of outer radius $b + \delta$ and inner radius a (Fig. 8.6). Calculate the maximum stresses in both materials for $\delta = 0.02$ mm, $a = 40$ mm, $b = 80$ mm, and $c = 140$ mm. Let $E_b = 110$ GPa, $\nu_b = 0.32$, $E_s = 200$ GPa, and $\nu_s = 0.28$.

8.19. An aluminum alloy cylinder ($E_a = 72$ GPa, $\nu_a = 0.33$) of outer and inner diameters of 300 and 200 mm is to be press-fitted over a solid steel shaft ($E_s = 200$ GPa, $\nu_s = 0.29$) of diameter 200.5 mm. Calculate (a) the interface pressure p and (b) the change in the outer diameter of the aluminum cylinder.

8.20. A thin circular disk of inner radius a and outer radius b is shrunk onto a rigid plug of radius $a + \delta_o$ (Fig. P8.20). Determine (a) the interface pressure; (b) the radial and tangential stresses.

8.21. When a steel sleeve of external diameter $3b$ is shrunk onto a solid steel shaft of diameter $2b$, the internal diameter of the sleeve is increased by an amount δ_0. What reduction occurs in the diameter of the shaft? Let $\nu = \frac{1}{3}$.

8.22. A cylinder of inner diameter b is shrunk onto a solid shaft. Find (a) the initial difference in diameters if the contact pressure is p and the maximum tangential stress is $2p$ in the cylinder and (b) the axial compressive load that should be applied to the shaft to increase the contact pressure from p to p_1. Let $\nu = \frac{1}{3}$.

8.23. A brass solid cylinder is a firm fit within a steel tube of inner diameter $2b$ and outer diameter $4b$ at a temperature $T_1°$C. If now the temperature of both elements is increased to $T_2°$C, find the maximum tangential stresses in the cylinder and in the tube. Take $\alpha_s = 11.7 \times 10^{-6}/°$C, $\alpha_b = 19.5 \times 10^{-6}/°$C, and $\nu = \frac{1}{3}$, and neglect longitudinal friction forces at the interface.

8.24. A thick-walled, closed-ended cylinder of inner radius a and outer radius b is subjected to an internal pressure p_i only (Fig. P8.24). The cylinder is made of a material with permissible tensile strength σ_{all} and shear strength τ_{all}. Determine the allowable value of p_i. *Given:* $a = 0.8$ m, $b = 1.2$ m, $\sigma_{\text{all}} = 80$ MPa, and $\tau_{\text{all}} = 50$ MPa.

8.25. A thick-walled cylindrical tank of inner radius a and outer radius b is made of ASTM A-48 cast iron (see Table D.1) having a modulus of elasticity E and Poisson's ratio of v. Calculate the maximum radial displacement of the tank, if it is under an internal pressure of p_i, as illustrated in Fig. P8.24. *Given:* $a = 0.5$ m, $b = 0.8$ m, $p_i = 60$ MPa, $E = 70$ GPa and $v = 0.3$.

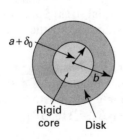

FIGURE P8.20.

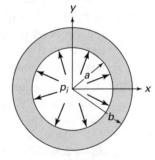

FIGURE P8.24.

8.26. A steel cylinder (σ_{yp} = 350 MPa) having inner radius a and outer radius $3a$ is subjected to an internal pressure p_i (Fig. P8.24). Find the limiting values of the p_i, using (a) the maximum shear stress theory; (b) the maximum energy of distortion theory.

8.27. A cylinder with inner radius a and outer radius $2a$ is subjected to an internal pressure p_i (Fig. P8.24). Determine the allowable value of p_i, applying (a) the maximum principal stress theory; (b) the Coulomb–Mohr theory. *Assumptions*: The cylinder is made of aluminum of σ_u = 320 MPa and σ'_u = 620 MPa.

8.28. A bronze bushing 60 mm in outer diameter and 40 mm in inner diameter is to be pressed into a hollow steel cylinder of 120-mm outer diameter. Calculate the tangential stresses for steel and bronze at the boundary between the two parts. *Given*: E_b = 105 GPa, E_s = 210 GPa, v = 0.3. *Assumption:* The radial interference is equal to δ = 0.05 mm.

8.29. A cast-iron disk is to be shrunk on a 100-mm-diameter steel shaft. Find (a) the contact pressure; (b) the minimum allowable outside diameter of the disk. *Assumption:* The tangential stress in the disk is limited to 80 MPa. *Given*: The radial interference is δ = 0.06 mm, E_c = 100 GPa, E_s = 200 GPa, and v = 0.3.

8.30. A cast-iron cylinder of outer radius 140 mm is to be shrink-fitted over a 50-mm-radius steel shaft. Find the maximum tangential and radial stresses in both parts. *Given*: E_c = 120 GPa, v_c = 0.2, E_s = 210 G Pa, v_s = 0.3. *Assumption*: The radial interference is δ = 0.04 mm.

8.31. In the case of which a steel disk (v = 0.3) of external diameter $4b$ is shrunk onto a steel shaft of diameter of b, internal diameter of the disk is increased by an amount λ. Find the reduction in the diameter of the shaft.

8.32. A gear of inner and outer radii 0.1 and 0.15 m, respectively, is shrunk onto a hollow shaft of inner radius 0.05 m. The maximum tangential stress induced in the gear wheel is 0.21 MPa. The length of the gear wheel parallel to shaft axis is 0.1 m. Assuming a coefficient of static friction of f = 0.2 at the common surface, what maximum torque may be transmitted by the gear without slip?

Sections 8.6 through 8.13

8.33. A solid steel shaft of radius b is pressed into a steel disk of outer radius $2b$ and the length of hub engagement t = $3b$ (Figure 8.11). Calculate the value of the radial interference in terms of b. *Given*: The shearing stress in the shaft caused by the torque that the joint is to carry equals 120 MPa; E = 200 GPa, and f = 0.2.

8.34. A cast-iron gear with 100-mm effective diameter and t = 40 mm hub engagement length is to transmit a maximum torque of 120 N · m at low speeds (Fig. 8.11). Determine (a) the required radial interference on a 20-mm diameter steel shaft; (b) the maximum stress in the gear due to a press fit. *Given*: E_c = 100 GPa, E_s = 200 GPa, v = 0.3, and f = 0.16.

8.35. Show that for an annular rotating disk, the ratio of the maximum tangential stress to the maximum radial stress is given by

$$\frac{\sigma_{\theta,\,max}}{\sigma_{r,\,max}} = \frac{2}{(b-a)^2}\left[b^2 + \frac{1-\nu}{3+\nu}a^2\right]$$ **(P8.35)**

8.36. Determine the allowable speed ω_{all} in rpm of a flat solid disk employing the maximum energy of distortion criterion. The disk is constructed of an aluminum alloy with $\sigma_{yp} = 260$ MPa, $\nu = \frac{1}{3}$, $\rho = 2.7$ kN·s²/m⁴, and $b = 125$ mm.

8.37. A rotating flat disk has 60-mm inner diameter and 200-mm outer diameter. If the maximum shearing stress is not to exceed 90 MPa, calculate the allowable speed in rpm. Let $\nu = 1/3$ and $\rho = 7.8$ kN·s²/m⁴.

8.38. A flat disk of outer radius $b = 125$ mm and inner radius $a = 25$ mm is shrink-fitted onto a shaft of radius 25.05 mm. Both members are made of steel with $E = 200$ GPa, $\nu = 0.3$, and $\rho = 7.8$ kN·s²/m⁴. Determine (a) the speed in rpm at which the interface pressure becomes zero and (b) the maximum stress at this speed.

8.39. A flat annular steel disk ($\rho = 7.8$ kN·s²/m⁴, $\nu = 0.3$) of $4c$ outer diameter and c inner diameter rotates at 5000 rpm. If the maximum radial stress in the disk is not to exceed 50 MPa, determine (a) the radial wall thickness and (b) the corresponding maximum tangential stress.

8.40. A flat annular steel disk of 0.8-m outer diameter and 0.15-m inner diameter is to be shrunk around a solid steel shaft. The shrinking allowance is 1 part per 1000. For $\nu = 0.3$, $E = 210$ GPa, and $\rho = 7.8$ kN·s²/m⁴, determine, neglecting shaft expansion, (a) the maximum stress in the system at standstill and (b) the rpm at which the shrink fit will loosen as a result of rotation.

8.41. Show that in a solid disk of diameter $2b$, rotating with a tangential velocity v, the maximum stress is $\sigma_{max} = \frac{5}{12}\rho v^2$. Take $\nu = \frac{1}{3}$.

8.42. A steel disk of 500-mm outer diameter and 50-mm inner diameter is shrunk onto a solid shaft of 50.04-mm diameter. Calculate the speed (in rpm) at which the disk will become loose on the shaft. Take $\nu = 0.3$, $E = 210$ GPa, and $\rho = 7.8$ kN·s²/m⁴.

8.43. Consider a steel rotating disk of hyperbolic cross section (Fig. 8.13) with $a = 0.125$ m, $b = 0.625$ m, $t_i = 0.125$ m, and $t_o = 0.0625$ m. Determine the maximum tangential force that can occur at the outer surface in newtons per meter of circumference if the maximum stress at the bore is not to exceed 140 MPa. Assume that outer and inner edges are free of pressure.

8.44. A steel turbine disk with $b = 0.5$ m, $a = 0.0625$ m, and $t_o = 0.05$ m rotates at 5000 rpm carrying blades weighing a total of 540 N. The center of gravity of each blade lies on a circle of 0.575-m radius. Assuming zero pressure at the bore, determine (a) the maximum stress for a disk of constant thickness and (b) the maximum stress for a disk of hyperbolic cross section. The thickness at the hub and tip are $t_i = 0.4$ m and $t_o = 0.05$ m, respectively. (c) For a thickness at the axis $t_i = 0.02425$ m, determine the thickness at the outer edge, t_o, for a disk under uniform stress, 84 MPa. Take $\rho = 7.8$ kN$\cdot$s^2/m^4 and $g = 9.81$ m/s^2.

8.45. A solid, thin flat disk is restrained against displacement at its outer edge and heated uniformly to temperature T. Determine the radial and tangential stresses in the disk.

8.46. Show that for a *hollow disk* or *cylinder*, when subjected to a temperature distribution given by $T = (T_a - T_b) \ln (b/r)/ \ln (b/a)$, the maximum radial stress occurs at

$$r = ab\left(\frac{2}{b^2 - a^2} \ln \frac{b}{a}\right)^{1/2} \qquad \text{(P8.46)}$$

8.47. Calculate the maximum thermal stress in a gray cast-iron cylinder for which the inner temperature is $T_a = -8°C$ and the outer temperature is zero. Let $a = 10$ mm, $b = 15$ mm, $E = 90$ GPa, $\alpha = 10.4 \times 10^{-6}$ per °C, and $\nu = 0.3$.

8.48. Verify that the distribution of stress in a solid disk in which the temperature varies linearly with the radial dimension, $T(r) = T_0(b - r)/b$, is given by

$$\sigma_r = \tfrac{1}{3}T_0\left(\frac{r}{b} - 1\right)\alpha E \qquad \text{(P8.48)}$$

Here T_0 represents the temperature rise at $r = 0$.

8.49. Redo Example 7.12 with the element shown in Fig. 7.26 representing a segment adjacent to the boundary of a sphere subjected to external pressure $p = 14$ MPa.

8.50. A cylinder of hydraulic device having inner radius a and outer radius $4a$ is subjected to an internal pressure p_i. Using a finite element program with CST (or LST) elements, determine the distribution of the tangential and radial stresses. Compare the results with the exact solution shown in Fig. 8.4a.

8.51. A cylinder of a hydraulic device with inner radius a and outer radius $4a$ is under an external pressure p_o. Employing a finite element computer program with CST (or LST) elements, calculate the distribution of tangential and radial stresses.

8.52. Redo Prob. 8.51 if the cylinder is under an internal pressure p_i. Compare the results with the exact solution shown in Fig. 8.4.

8.53. Resolve Prob. 8.51 for the case in which the cylinder is subjected to internal pressure p_i and external pressure $p_o = 0.5p_i$.

CHAPTER 9

Beams on Elastic Foundations

9.1 INTRODUCTION

In the problems involving beams previously considered, support was provided at a number of discrete locations, and the beam was usually assumed to suffer no deflection at these points of support. We now explore the case of a prismatic beam supported continuously along its length by a foundation, itself assumed to experience elastic deformation. We shall take the reaction forces of the elastic foundation to be *linearly* proportional to the beam deflection at any point. This simple analytical model of a continuous elastic foundation is often referred to as the *Winkler model*.

The foregoing assumption not only leads to equations amenable to solution but also represents an idealization closely approximating many real situations. Examples include a railroad track, where the elastic support consists of the cross ties, the ballast, and the subgrade; concrete footings on an earth foundation; long steel pipes resting on earth or on a series of elastic springs; ship hulls; or a bridgedeck or floor structure consisting of a network of closely spaced bars.

9.2 GENERAL THEORY

Let us consider a beam on elastic foundation subject to a variable loading, as depicted in Fig. 9.1. The force q per unit length, resisting the displacement of the beam, is equal to $-kv$. Here v is the beam deflection, positive downward as in the figure. The quantity k represents a constant, usually referred to as the *modulus of the foundation*, possessing the dimensions of force per unit length of beam per unit of deflection (for example, newtons per square meter or pascals).

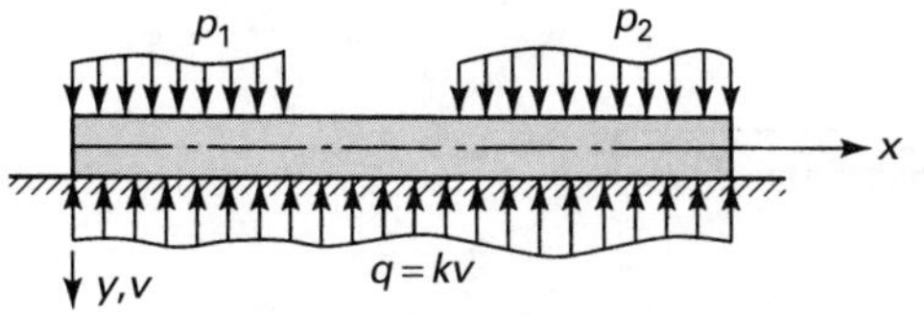

FIGURE 9.1. *Beam on elastic (Winkler) foundation.*

The analysis of a beam whose length is very much greater than its depth and width serves as the basis of the treatment of all beams on elastic foundations. Referring again to Fig. 9.1, which shows a beam of constant section supported by an elastic foundation, the x axis passes through the centroid, and the y axis is a principal axis of the cross section. The deflection v, subject to reaction q and applied load per unit length p, for a condition of small slope, must satisfy the beam equation:

$$EI \frac{d^4v}{dx^4} + kv = p \qquad (9.1)$$

For those parts of the beam on which no distributed load acts, $p = 0$, and Eq. (9.1) takes the form

$$EI \frac{d^4v}{dx^4} + kv = 0 \qquad (9.2)$$

It will suffice to consider the general solution of Eq. (9.2) only, requiring the addition of a particular integral to satisfy Eq. (9.1) as well. Selecting $v = e^{ax}$ as a trial solution, it is found that Eq. (9.2) is satisfied if

$$\left(a^4 + \frac{k}{EI} \right) v = 0$$

requiring that

$$a = \pm \beta (1 \pm i)$$

where

$$\beta = \left(\frac{k}{4EI} \right)^{1/4} \qquad (9.3)$$

The general solution of Eq. (9.2) may now be written as

$$v = e^{\beta x}[A \cos \beta x + B \sin \beta x] + e^{-\beta x}[C \cos \beta x + D \sin \beta x] \qquad (9.4)$$

where A, B, C, and D are the constants of integration.

In the developments that follow, the case of a single load acting on an infinitely long beam is treated first. The solution of problems involving a variety of loading combinations will then rely on the principle of superposition.

9.3 INFINITE BEAMS

Consider an infinitely long beam resting on a continuous elastic foundation, loaded by a concentrated force P (Fig. 9.2). The variation of the reaction kv is unknown, and the equations of static equilibrium are not sufficient for its determination. The

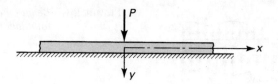

FIGURE 9.2. *Infinite beam on an elastic foundation and loaded at the origin.*

problem is therefore *statically indeterminate* and requires additional formulation, which is available from the equation of the deflection curve of the beam. Owing to beam symmetry, only that portion to the right of the load P need be considered. The two boundary conditions for this segment are deduced from the fact that as $x \to \infty$, the deflection and all derivatives of v with respect to x must vanish. On this basis, it is clear that the constants A and B in Eq. (9.4) must equal zero. What remains is

$$v = e^{-\beta x}(C \cos \beta x + D \sin \beta x) \tag{9.5}$$

The conditions applicable a very small distance to the right of P are

$$v'(0) = 0, \qquad V = -EIv'''(0) = -\frac{P}{2} \tag{a}$$

where the minus sign is consistent with the general convention adopted in Section 1.3. Substitution of Eq. (a) into Eq. (9.5) yields

$$C = D = \frac{P}{8\beta^3 EI} = \frac{P\beta}{2k}$$

Introduction of the expressions for the constants into Eq. (9.5) provides the following equation, applicable to an infinite beam subject to a concentrated force P at midlength:

$$v = \frac{P\beta}{2k}e^{-\beta x}(\cos \beta x + \sin \beta x) \tag{9.6a}$$

or

$$v = \frac{P\beta}{2k} e^{-\beta x}\left[\sqrt{2} \sin\left(\beta x + \frac{\pi}{4}\right)\right] \tag{9.6b}$$

Equation (9.6b) indicates clearly that the characteristic of the deflection is an exponential decay of a sine wave of wavelength

$$\lambda = \frac{2\pi}{\beta} = 2\pi\left(\frac{4EI}{k}\right)^{1/4}$$

To simplify the equations for deflection, rotation, moment, and shear, the following notations and relations are introduced:

$$f_1(\beta x) = e^{-\beta x}(\cos \beta x + \sin \beta x)$$

$$f_2(\beta x) = e^{-\beta x} \sin \beta x = -\frac{1}{2\beta} f'_1$$

$$f_3(\beta x) = e^{-\beta x}(\cos \beta x - \sin \beta x) = \frac{1}{\beta} f'_2 = -\frac{1}{2\beta^2} f''_1$$

$$f_4(\beta x) = e^{-\beta x} \cos \beta x = -\frac{1}{2\beta} f'_3 = -\frac{1}{2\beta^2} f''_2 = \frac{1}{4\beta^3} f'''_1 \qquad (9.7)$$

$$f_1(\beta_x) = -\frac{1}{\beta} f'_4$$

Table 9.1 lists numerical values of the foregoing functions for various values of the argument βx. The solution of Eq. (9.5) for specific problems is facilitated by this table and the graphs of the functions of βx [Ref. 9.1]. Equation (9.6) and its derivatives, together with Eq. (9.7), yield the following expressions for deflection, slope, moment, and shearing force:

$$v = \frac{P\beta}{2k} f_1$$

$$\theta = v' = -\frac{P\beta^2}{k} f_2$$

$$M = EIv'' = -\frac{P}{4\beta} f_3 \qquad (9.8)$$

$$V = -EIv''' = -\frac{P}{2} f_4$$

These expressions are valid for $x \geq 0$.

EXAMPLE 9.1   Long Beam with a Partial Uniform Load

A very long rectangular beam of width 0.1 m and depth 0.15 m (Fig. 9.3) is subject to a uniform loading over 4 m of its length of $p = 175$ kN/m. The beam is supported on an elastic foundation having a modulus $k = 14$ MPa. Derive an expression for the deflection at an arbitrary point Q within length L. Calculate the maximum deflection and the maximum force per unit length between beam and foundation. Use $E = 200$ GPa.

Solution   The deflection Δv at point Q due to the load $P_x = p \, dx$ is, from Eq. (9.8);

$$\Delta v = \frac{p \, dx}{2k} \beta e^{-\beta x}(\cos \beta x + \sin \beta x)$$

TABLE 9.1. *Selected Values of the Functions Defined by Eqs. (9.7)*

βx	$f_1(\beta x)$	$f_2(\beta x)$	$f_3(\beta x)$	$f_4(\beta x)$
0	1	0	1	1
0.02	0.9996	0.0196	0.9604	0.9800
0.04	0.9984	0.0384	0.9216	0.9600
0.05	0.9976	0.0476	0.9025	0.9501
0.10	0.9907	0.0903	0.8100	0.9003
0.20	0.9651	0.1627	0.6398	0.8024
0.30	0.9267	0.2189	0.4888	0.7077
0.40	0.8784	0.2610	0.3564	0.6174
0.50	0.8231	0.2908	0.2415	0.5323
0.60	0.7628	0.3099	0.1431	0.4530
0.70	0.6997	0.3199	0.0599	0.3798
$\pi/4$	0.6448	0.3224	0	0.3224
0.80	0.6354	0.3223	−0.0093	0.3131
0.90	0.5712	0.3185	−0.0657	0.2527
1.00	0.5083	0.3096	−0.1108	0.1988
1.10	0.4476	0.2967	−0.1457	0.1510
1.20	0.3899	0.2807	−0.1716	0.1091
1.30	0.3355	0.2626	−0.1897	0.0729
1.40	0.2849	0.2430	−0.2011	0.0419
1.50	0.2384	0.2226	−0.2068	0.0158
$\pi/2$	0.2079	0.2079	−0.2079	0
1.60	0.1959	0.2018	−0.2077	−0.0059
1.70	0.1576	0.1812	−0.2047	−0.0235
1.80	0.1234	0.1610	−0.1985	−0.0376
1.90	0.0932	0.1415	−0.1899	−0.0484
2.00	0.0667	0.1231	−0.1794	−0.0563
2.20	0.0244	0.0896	−0.1548	−0.0652
$3\pi/4$	0	0.0670	−0.1340	−0.0670
2.40	−0.0056	0.0613	−0.1282	−0.0669
2.60	−0.0254	0.0383	−0.1019	−0.0636
2.80	−0.0369	0.0204	−0.0777	−0.0573
3.00	−0.0423	0.0070	−0.0563	−0.0493
π	−0.0432	0	−0.0432	−0.0432
3.20	−0.0431	−0.0024	−0.0383	−0.0407
3.40	−0.0408	−0.0085	−0.0237	−0.0323
3.60	−0.0366	−0.0121	−0.0124	−0.0245
3.80	−0.0314	−0.0137	−0.0040	−0.0177
$5\pi/4$	−0.0279	−0.0139	0	−0.0139
4.00	−0.0258	−0.0139	0.0019	−0.0120
4.50	−0.0132	−0.0108	0.0085	−0.0023
$3\pi/2$	−0.0090	−0.0090	0.0090	0
5.00	−0.0046	−0.0065	0.0084	0.0019
$7\pi/4$	0	−0.0029	0.0058	0.0029
6.00	0.0017	−0.0007	0.0031	0.0024
2π	0.0019	0	0.0019	0.0019
6.50	0.0018	0.0003	0.0012	0.0018
7.00	0.0013	0.0006	0.0001	0.0007
$9\pi/4$	0.0012	0.0006	0	0.0006
7.50	0.0007	0.0005	−0.0003	0.0002
$5\pi/2$	0.0004	0.0004	−0.0004	0

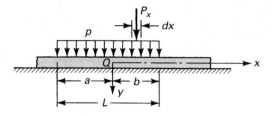

FIGURE 9.3. *Example 9.1. Uniformly distributed load segment on an infinite beam on an Elastic foundation.*

The deflection at point Q resulting from the entire distributed load is then

$$v_Q = \int_0^a \frac{p\,dx}{2k}\,\beta e^{-\beta x}(\cos\beta x + \sin\beta x) + \int_0^b \frac{p\,dx}{2k}\,\beta e^{-\beta x}(\cos\beta x + \sin\beta x)$$

$$= \frac{p}{2k}(2 - e^{-\beta a}\cos\beta a - e^{-\beta b}\cos\beta b)$$

or

$$v_Q = \frac{p}{2k}[2 - f_4(\beta a) - f_4(\beta b)] \tag{b}$$

Although the algebraic sign of the distance a in Eq. (b) is negative, in accordance with the placement of the origin in Fig. 9.3, we shall treat it as a positive number because Eq. (9.8) gives the deflection for positive x only. This is justified on the basis that the beam deflection under a concentrated load is the same at equal distances from the load, whether these distances are positive or negative. By the use of Eq. (9.3),

$$\beta = \left(\frac{k}{4EI}\right)^{1/4} = \left(\frac{14 \times 10^6}{4 \times 200 \times 10^9 \times 0.1 \times 0.15^3/12}\right)^{1/4} = 0.888 \text{ m}^{-1}$$

From this value of β, $\beta L = (0.888)(4) = 3.552 = \beta(a + b)$. We are interested in the *maximum* deflection and therefore locate the origin at point Q, the center of the distributed loading. Now a and b represent equal lengths, so $\beta a = \beta b = 1.776$, and Eq. (b) gives

$$v_{max} = \frac{175}{2(14{,}000)}[2 - (-0.0345) - (-0.0345)] = 0.0129 \text{ m}$$

The maximum force per unit of length between beam and foundation is then $kv_{max} = 14 \times 10^6(0.0129) = 180.6 \text{ kN/m}$.

EXAMPLE 9.2   Long Beam with a Moment
A very long beam is supported on an elastic foundation and is subjected to a concentrated moment M_o (Fig. 9.4). Determine the equations describing the deflection, slope, moment, and shear.

FIGURE 9.4. *Example 9.2. Infinite beam resting on an elastic foundation subjected to M.*

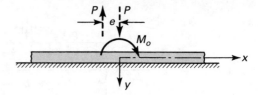

Solution Observe that the couple $P \cdot e$ is equivalent to M_o for the case in which e approaches zero (indicated by the dashed lines in the figure). Applying Eq. (9.8), we have therefore

$$v = \frac{P\beta}{2k}\{f_1(\beta x) - f_1[\beta(x+e)]\} = -\frac{M_o\beta}{2k}\frac{f_1[\beta(x+e)] - f_1(\beta x)}{e}$$

$$= -\frac{M_o\beta}{2k}\lim_{e\to 0}\frac{f_1[\beta(x+e)] - f_1(\beta x)}{e} = -\frac{M_o\beta}{2k}\frac{df_1(\beta x)}{dx} = \frac{M_o\beta^2}{k}f_2(\beta x)$$

Successive differentiation yields

$$v = \frac{M_o}{k}\beta^2 f_2(\beta x)$$

$$\theta = \frac{M_o}{k}\beta^3 f_3(\beta x)$$

$$M = EIv'' = -\frac{M_o}{2}f_4(\beta x) \qquad \textbf{(9.9)}$$

$$V = -EIv''' = -\frac{M_o\beta}{2}f_1(\beta x)$$

which are the deflection, slope, moment, and shear, respectively.

9.4 SEMI-INFINITE BEAMS

The theory of Section 9.3 is now applied to a semi-infinite beam, having one end at the origin and the other extending indefinitely in a positive x direction, as in Fig. 9.5. At $x = 0$, the beam is subjected to a concentrated load P and a moment M_A. The constants C and D of Eq. (9.5) can be ascertained by applying the following conditions at the left end of the beam:

$$EIv'' = M_A, \qquad EIv''' = -V = P$$

The results are

$$C = \frac{P + \beta M_A}{2\beta^3 EI}, \qquad D = -\frac{M_A}{2\beta^2 EI}$$

FIGURE 9.5. *Semi-infinite end-loaded beam on elastic foundation.*

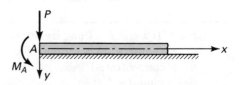

The deflection is now found by substituting C and D into Eq. (9.5) as

$$v = \frac{e^{-\beta x}}{2\beta^3 EI}[P\cos\beta x + \beta M_A(\cos\beta x - \sin\beta x)] \qquad \textbf{(9.10)}$$

At $x = 0$,

$$\delta = v(0) = \frac{2\beta}{k}(P + \beta M_A) \qquad \textbf{(9.11)}$$

Finally, successively differentiating Eq. (9.10) yields expressions for slope, moment, and shear:

$$v = \frac{2\beta}{k}[Pf_4(\beta x) + \beta M_A f_3(\beta x)]$$

$$\theta = -\frac{2\beta^2}{k}[Pf_1(\beta x) + 2\beta M_A f_4(\beta x)] \qquad \textbf{(9.12)}$$

$$M = \frac{P}{\beta}f_2(\beta x) + M_A f_1(\beta x)$$

$$V = -Pf_3(\beta x) + 2\beta M_A f_2(\beta x)$$

Application of these equations together with the principle of superposition permits the solution of more complex problems, as is illustrated next.

EXAMPLE 9.3   Semi-Infinite Beam with a Concentrated Load Near Its End

Determine the equation of the deflection curve of a semi-infinite beam on an elastic foundation loaded by concentrated force P a distance c from the free end (Fig. 9.6a).

Solution   The problem may be restated as the sum of the cases shown in Figs. 9.6b and c. Applying Eqs. (9.8) and the conditions of symmetry, the reactions appropriate to the infinite beam of Fig. 9.6b are

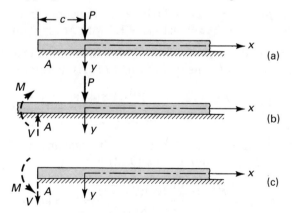

FIGURE 9.6.   *Example 9.3. Semi-infinite beam on an elastic foundation under load P.*

$$M = -\frac{P}{4\beta} f_3(\beta c)$$

$$V = -\frac{P}{2} f_4(\beta c)$$

(a)

Superposition of the deflections of Fig. 9.6b and c [see Eqs. (9.8) and (9.12)] results in

$$v = v_{\text{inf}} + v_{\text{semi-inf}} = \frac{P\beta}{2k} f_1(\beta x) + \frac{2\beta}{k} \{-Vf_4[\beta(x+c)] + \beta M f_3[\beta(x+c)]\}$$

Introducing Eqs. (a) into this, the following expression for deflection, applicable for positive x, is obtained:

$$v = \frac{P\beta}{2k} f_1(\beta x) + \frac{P\beta}{k} \{f_4(\beta c)f_4[\beta(x+c)] - \tfrac{1}{2}f_3(\beta c)f_3[\beta(x+c)]\}$$

This is clearly applicable for negative x as well, provided that x is replaced by $|x|$.

EXAMPLE 9.4 Semi-Infinite Beam Loaded at Its End
A 2-m-long steel bar ($E = 210$ GPa) of 75-mm by 75-mm square cross section rests with a side on a rubber foundation ($k = 24$ MPa). If a concentrated load $P = 20$ kN is applied at the left end of the beam (Fig. 9.5), determine (a) the maximum deflection and (b) the maximum bending stress.

Solution Applying Eq. (9.3), we have

$$\beta = \left(\frac{k}{4EI}\right)^{1/4} = \left[\frac{24(10^6)}{4(210 \times 10^9)(0.075)^4/12}\right]^{1/4} = 1.814 \text{ m}^{-1}$$

Inasmuch as $pL = 1.814(2) = 3.638 > 3$, the beam can be considered to be a long beam (see Sec. 9.4); Eqs. (9.12) with $M_A = 0$ thus apply.

a. The maximum deflection occurs at the left end for which $f_4(\beta x)$ is a maximum or $\beta x = 0$. The first of Eqs. (9.12) is therefore

$$v_{\text{max}} = \frac{2P\beta}{k} = \frac{2(20 \times 10^3)(1.814)}{24(10^6)} = 3.02 \text{ mm}$$

b. Referring to Table 9.1, $f_2(\beta x)$ has its maximum of 0.3224 at $\beta x = \pi/4$. Using the third of Eqs. (9.12),

$$M_{\text{max}} = \frac{Pf_2}{\beta} = \frac{20(10^3)(0.3224)}{1.814} = 3.55 \text{ kN} \cdot \text{m}$$

The maximum stress in the beam is obtained from the flexure formula:

$$\sigma_{max} = \frac{M_{max} c}{I} = \frac{3.55(10^3)(0.0375)}{(0.075)^4/12} = 50.49 \text{ MPa}$$

The location of this stress is at $x = \pi/4\beta = 433$ mm from the left end.

9.5 FINITE BEAMS

The bending of a finite beam on elastic foundation may also be treated by application of the general solution, Eq. (9.4). In this instance, four constants of integration must be evaluated. To accomplish this, two boundary conditions at each end may be applied, usually resulting in rather lengthy formulations. Results have been obtained applying this approach and are tabulated for numerous cases.*

An alternative approach to the solution of problems of finite beams with simply supported and clamped ends employs equations derived for infinite and semi-infinite beams, together with the principle of superposition. The use of this method was demonstrated in connection with semi-inverse beams in Example 9.3. Energy methods can also be employed in the analysis of beams whose ends are subjected to any type of support conditions [Ref. 9.4]. Solution by trigonometric series results in formulas that are particularly simple, as will be seen in Example 10.12. Design tables for short beams with free ends on elastic foundation have been provided by Iyengar and Ramu [Ref. 9.5].

Let us consider a finite beam on an elastic foundation, centrally loaded by a concentrated force P (Fig. 9.7) and compare the deflections occurring at the center and end of the beam. Note that the beam deflection is symmetrical with respect to C.

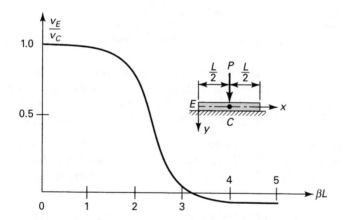

FIGURE 9.7. *Comparison of the center and end deflections of a finite beam on an elastic foundation subjected to a concentrated center load.*

---

*For a detailed presentation of a number of practical problems, see Refs. 9.1 through 9.3.

The appropriate boundary conditions are, for $x \geq 0$, $v'(L/2) = 0$, $EIv'''(L/2) = P/2$, $EIv''(0) = 0$, and $EIv'''(0) = 0$. Substituting these into the proper derivatives of Eq. (9.4) leads to four equations with unknown constants A, B, C, and D. After routine but somewhat lengthy algebraic manipulation, the following expressions are determined:

$$v_C = \frac{P\beta}{2k} \frac{2 + \cos \beta L + \cosh \beta L}{\sin \beta L + \sinh \beta L} \tag{9.13}$$

$$v_E = \frac{2P\beta}{k} \frac{\cos (\beta L/2)\cosh(\beta L/2)}{\sin \beta L + \sinh \beta L} \tag{9.14}$$

From Eqs. (9.13) and (9.14), we have

$$\frac{v_E}{v_C} = \frac{4 \cos (\beta L/2) \cosh(\beta L/2)}{2 + \cos \beta L + \cosh \beta L} \tag{9.15}$$

This result is sketched in Fig. 9.7.

9.6 CLASSIFICATION OF BEAMS

The plot of Eq. (9.15) permits the establishment of a stiffness criterion for beams resting on elastic foundation. It also serves well to readily discern a rationale for the classification of beams on elastic foundation. Referring to Fig. 9.7, we conclude that

1. *Short beams, $\beta L < 1$:* Inasmuch as the end deflection is essentially equal to that at the center, the deflection of the foundation can be determined to good accuracy by regarding the beam as infinitely rigid.
2. *Intermediate beams, $1 < \beta L < 3$:* In this region, the influence of the central force at the ends of the beam is substantial, and the beam must be treated as finite in length.
3. *Long beams, $\beta L > 3$:* It is clear from the figure that the ends are not affected appreciably by the central loading. Therefore, if we are concerned with one end of the beam, the other end or the middle may be regarded as being an infinite distance away; that is, the beam may be treated as infinite in length.

The foregoing groups do not relate only to the special case of loading shown in Fig. 9.7 but are quite general. Should greater accuracy be required, the upper limit of group 1 may be placed at $\beta L = 0.6$ and the lower limit of group 3 at $\beta L = 5$.

9.7 BEAMS SUPPORTED BY EQUALLY SPACED ELASTIC ELEMENTS

In the event that a long beam is supported by individual elastic elements, as in Fig. 9.8a, the problem is simplified if the separate supports are replaced by an equivalent continuous elastic foundation. To accomplish this, it is assumed that the

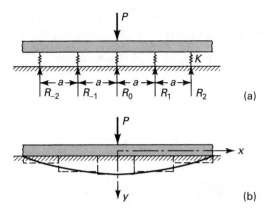

FIGURE 9.8. *Infinite beam supported by equally spaced elastic springs. Loading diagram: (a) with concentrated reactions; (b) with average continuous reaction force distribution.*

distance a between each support and the next is small, and that the concentrated reactions $R_i = Kv_i$ are replaced by equivalent uniform or stepped distributed forces shown by the dashed lines of Fig. 9.8b. Here K represents a spring constant (for example, newtons per meter).

For practical calculations, the usual limitation is $a \leq \pi/4\beta$. The average continuous reaction force distribution is shown by the solid line in the figure. The intensity of the latter distribution is ascertained as follows:

$$\frac{R}{a} = \frac{K}{a}v = q$$

or

$$q = kv \tag{a}$$

where the foundation modulus of the *equivalent* continuous elastic support is

$$k = \frac{K}{a} \tag{9.16}$$

The solution for the case of a beam on individual elastic support is then obtained through the use of Eq. 9.2, in which the value of k is that given by Eq. (9.16).

EXAMPLE 9.5 Finite-Length Beam with a Concentrated Load Supported by Springs

A series of springs, spaced so that $a = 1.5$ m, supports a long thin-walled steel tube having $E = 206.8$ GPa. A weight of 6.7 kN acts down at midlength of the tube. The average diameter of the tube is 0.1 m, and the moment of inertia of its section is 6×10^{-6} m⁴. Take the spring constant of each support to be $K = 10 \, \text{kN/m}$. Find the maximum moment and the maximum deflection, assuming negligible tube weight.

Solution Applying Eqs. (9.3) and (9.16), we obtain

$$k = \frac{K}{a} = \frac{10,000}{1.5} = 6667 \, \text{Pa}$$

$$\beta = \left(\frac{6667}{4 \times 206.8 \times 10^9 \times 6 \times 10^{-6}} \right)^{1/4} = 0.1913 \, \text{m}^{-1}$$

From Eq. (9.8),

$$\sigma_{max} = \frac{Mc}{I} = \frac{Pc}{4\beta I} = \frac{6700 \times 0.05}{4 \times 0.1913 \times 6 \times 10^{-6}} = 72.97 \, \text{MPa}$$

$$\upsilon_{max} = \frac{P\beta}{2k} = \frac{6700 \times 0.1913}{2 \times 6667} = 96.1 \, \text{mm}$$

9.8 SIMPLIFIED SOLUTIONS FOR RELATIVELY STIFF BEAMS

Examination of the analyses of the previous sections and of Fig. 9.7 leads us to conclude that the distribution of force acting on the beam by the foundation is, in general, a nonlinear function of the beam length coordinate. This distribution approaches linearity as the beam length decreases or as the beam becomes stiffer. Reasonably good results can be expected, therefore, by assuming a linearized elastic foundation pressure for stiff beams. The foundation pressure is then predicated on beam displacement in the manner of a rigid body [Ref. 9.6], and the reaction is, as a consequence, statically determinate.

To illustrate the approach, consider once more the beam of Fig. 9.7, this time with a linearized foundation pressure (Fig. 9.9). Because of loading symmetry, the foundation pressure is, in this case, not only linear but constant as well. We shall compare the results thus obtained with those found earlier.

The *exact theory* states that points E and C deflect in accordance with Eqs. (9.13) and (9.14). The relative deflection of these points is simply

$$\upsilon = \upsilon_C - \upsilon_E \tag{a}$$

FIGURE 9.9. *Relatively stiff finite beam on elastic foundation and loaded at the center.*

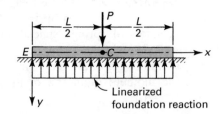

For the simplified load configuration shown in Fig. 9.9, the relative beam deflection may be determined by considering the elementary solution for a beam subjected to a uniformly distributed loading and a concentrated force. For this case, we label the relative deflection v_1 as follows:

$$v_1 = \frac{PL^3}{48EI} - \frac{5(P/L)L^4}{384EI} = \frac{PL^3}{128EI} \tag{b}$$

The ratio of the relative deflections obtained by the exact and approximate analyses now serves to indicate the validity of the approximations. Consider

$$\frac{v}{v_1} = \frac{32}{(\beta L)^3} \frac{\frac{1}{2}\cosh \beta L + \frac{1}{2}\cos \beta L + 1 - 2\cosh \frac{1}{2}\beta L \cos \frac{1}{2}\beta L}{\sinh \beta L + \sin \beta L} \tag{c}$$

where v and v_1 are given by Eqs. (a) and (b). The trigonometric and hyperbolic functions may be expanded as follows:

$$\sin \beta L = \beta L - \frac{(\beta L)^3}{3!} + \frac{(\beta L)^5}{5!} - \frac{(\beta L)^7}{7!} + \cdots$$

$$\cos \beta L = 1 - \frac{(\beta L)^2}{2!} + \frac{(\beta L)^4}{4!} - \frac{(\beta L)^6}{6!} + \cdots$$

$$\sinh \beta L = \beta L + \frac{(\beta L)^3}{3!} + \frac{(\beta L)^5}{5!} + \cdots \tag{d}$$

$$\cosh \beta L = 1 + \frac{(\beta L)^2}{2!} + \frac{(\beta L)^4}{4!} + \cdots$$

Introducing these into Eq. (c), we obtain

$$\frac{v}{v_1} = 1 - \frac{23}{120}\frac{(\beta L)^4}{4!} + \frac{51}{20}\frac{(\beta L)^3}{8!} + \cdots \tag{e}$$

Substituting various values of βL into Eq. (e) discloses that, for $\beta L < 1.0$, v/v_1 differs from unity by no more than 1%, and the linearization is seen to yield good results. It can be shown that for values of $\beta L < 1$, the ratio of the moment (or slope) obtained by the linearized analysis to that obtained from the exact analysis differs from unity by less than 1%.

Analysis of a finite beam, centrally loaded by a concentrated moment, also reveals results similar to those given here. We conclude, therefore, that when βL is small (< 1.0) no significant error is introduced by assuming a linear distribution of foundation pressure.

9.9 SOLUTION BY FINITE DIFFERENCES

Because of the considerable time and effort required in the analytical solution of practical problems involving beams on elastic foundations, approximate methods employing numerical analysis are frequently applied. A solution utilizing the method of finite differences is illustrated in the next example.

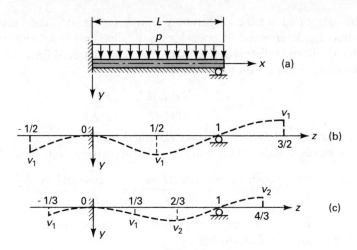

FIGURE 9.10.   *Example 9.6. (a) Uniformly loaded beam on an elastic foundation; (b) deflection curve for m = 2; (c) deflection curve for m = 3.*

EXAMPLE 9.6   Finite Beam under Uniform Loading

Determine the deflection of the built-in beam on an elastic foundation shown in Fig. 9.10a. The beam is subjected to a uniformly distributed loading p and is simply supported at $x = L$.

Solution   The deflection is governed by Eq. (9.1), for which the applicable boundary conditions are

$$v(0) = v(L) = 0, \qquad v'(0) = 0, \qquad v''(L) = 0 \qquad \textbf{(a)}$$

The solution will be obtained by replacing Eqs. (9.1) and (a) by a system of finite difference equations. It is convenient to first transform Eq. (9.1) into dimensionless form through the introduction of the following quantities:

$$z = \frac{x}{L}, \qquad \frac{d}{dx} = \frac{1}{L}\frac{d}{dz}, \quad \text{at} \quad x = 0,\ z = 0;\ x = L,\ z = 1$$

The deflection equation is therefore

$$EI\frac{d^4v}{dz^4} + kL^4v = pL^4 \qquad \textbf{(b)}$$

We next divide the interval of $z(0, 1)$ into n equal parts of length $h = 1/m$, where m represents an integer. Multiplying Eq. (b) by $h^4 = 1/m^4$, we have

$$h^4\frac{d^4v}{dz^4} + \frac{kL^4}{m^4EI}v = \frac{pL^4}{m^4EI} \qquad \textbf{(c)}$$

Employing Eq. (7.23), Eq. (c) assumes the following finite difference form:

$$v_{n-2} - 4v_{n-1} + 6v_n - 4v_{n+1} + v_{n+2} + \frac{kL^4}{m^4 EI} v_n = \frac{pL^4}{m^4 EI} \qquad \text{(d)}$$

Upon setting $C = kL^4/EI$, this becomes

$$v_{n-2} - 4v_{n-1} + \left[\frac{C}{m^4} + 6\right] v_n - 4v_{n+1} + v_{n+2} = \frac{pL^4}{m^4 EI} \qquad \text{(e)}$$

The boundary conditions, Eqs. (a), are transformed into difference conditions by employing Eq. (7.4):

$$v_0 = 0, \qquad v_{-1} = v_1, \qquad v_n = 0, \qquad v_{n-1} = -v_{n+1} \qquad \text{(f)}$$

Equations (e) and (f) represent the set required for a solution, with the degree of accuracy increased as the magnitude of m is increased. Any desired accuracy can thus be attained.

For purposes of illustration, let $k = 2.1$ MPa, $E = 200$ GPa, $I = 3.5 \times 10^{-4}$ m^4, $L = 3.8$ m, and $p = 540$ kN/m. Determine the deflections for $m = 2$, $m = 3$, and $m = 4$. Equation (e) thus becomes

$$v_{n-2} - 4v_{n-1} + 6\left(\frac{m^4 + 1}{m^4}\right) v_n - 4v_{n+1} + v_{n+2} = \frac{1.6}{m^4} \qquad \text{(g)}$$

For $m = 2$, the deflection curve, satisfying Eq. (f), is sketched in Fig. 9.10b. At $z = \frac{1}{2}$, we have $v_n = v_1$. Equation (g) then yields

$$v_1 - 4(0) + 6\frac{2^4 + 1}{2^4} v_1 - 4(0) - v_1 = \frac{1.6}{2^4}$$

from which $v_1 = 16$ mm.

For $m = 3$, the deflection curve satisfying Eq. (f) is now as in Fig. 9.10c. Hence, Eq. (g) at $z = \frac{1}{3}$ (by setting $v_n = v_1$) and at $z = \frac{2}{3}$ (by setting $v_n = v_2$) leads to

$$v_1 + 6\frac{3^4 + 1}{3^4} v_1 - 4v_2 = \frac{1.6}{3^4}$$

$$-4v_1 + 6\frac{3^4 + 1}{3^4} v_2 - v_2 = \frac{1.6}{3^4}$$

from which $v_1 = 9$ mm and $v_2 = 11$ mm.

For $m = 4$, a similar procedure yields $v_1 = 5.3$ mm, $v_2 = 9.7$ mm, and $v_3 = 7.5$ mm.

9.10 APPLICATIONS

The theory of beams on elastic foundation is applicable to many problems of practical importance, of which one is discussed next. The concept of a beam on elastic foundation may also be employed to approximate stress and deflection in axisymmetrically loaded cylindrical shells, to be discussed in Chapter 13. The governing equations of the two problems are of the same form. This means that solutions of one problem become solutions of the other problem through a simple change of constants [Ref. 9.7].

Grid Configurations of Beams

The ability of a floor to sustain extreme loads without undue deflection, as in a machine shop, is significantly enhanced by combining the floor beams in a particular array or grid configuration. Such a design is illustrated in the ensuing problem.

EXAMPLE 9.7 Machine Room Floor

A single concentrated load P acts at the center of a machine room floor composed of 79 transverse beams (spaced $a = 0.3$ m apart) and one longitudinal beam, as shown in Fig. 9.11. If all beams have the same modulus of rigidity EI, determine the deflection and the distribution of load over the various transverse beams supporting the longitudinal beam. Assume that the transverse and longitudinal beams are attached so that they deform together.

Solution The spring constant K of an individual elastic support such as beam AA is

$$K = \frac{R_C}{v_C} = \frac{R_C}{R_C L_t^3/48EI} = \frac{48EI}{L_t^3}$$

FIGURE 9.11. *Example 9.7. Long beam supported by several identical and equally spaced cross-beams. All beams are simply supported at their ends.*

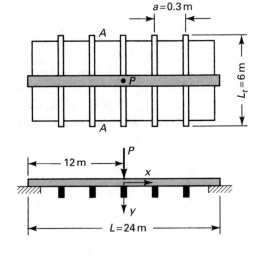

where v_C is the central deflection of a simply supported beam of length L_t carrying a center load R_C. From Eq. (9.16), the modulus k of the equivalent continuous elastic foundation is found to be

$$k = \frac{K}{a} = \frac{48EI}{aL_t^3}$$

Thus,

$$\beta = \sqrt[4]{\frac{k}{4EI}} = \sqrt[4]{\frac{12}{aL_t^3}} = \frac{3.936}{L_t}$$

and

$$\beta L = \frac{3.936}{L_t}(4L_t) = 15.744, \qquad \beta a = \frac{3.936}{L_t}\frac{L_t}{20} = 0.1968$$

In accordance with the criteria discussed in Sections 9.4 and 9.5, the longitudinal beam may be classified as a long beam resting on a continuous elastic support of modulus k. Consequently, from Eqs. (9.8), the deflection at midspan is

$$v_P = \frac{P\beta}{2k}(1) = \frac{3.936P}{2L_t}\frac{aL_t^3}{48EI} = \frac{1.968PL_t^2a}{48EI}$$

The deflection of a transverse beam depends on its distance x from the center of the longitudinal beam, as shown in the following tabulation:

x	0	2a	4a	6a	8a	10a	12a	16a	20a	24a
$f_1(\beta x)$	1	0.881	0.643	0.401	0.207	0.084	−0.0002	−0.043	−0.028	−0.009
v	v_P	$0.881v_P$	$0.643v_P$	$0.401v_P$	$0.207v_P$	$0.084v_P$	$-0.0002v_P$	$-0.043v_P$	$-0.028v_P$	$-0.009v_P$

We are now in a position to calculate the load R_{cc} supported by the central transverse beam. Since the midspan deflection v_M of the central transverse beam is equal to v_P, we have

$$v_M = \frac{R_{CC}L_t^3}{48EI} = \frac{1.968PL_t^2a}{48EI}$$

and

$$R_{CC} = 1.968P\frac{a}{L_t} = 0.0984P$$

The remaining transverse beam loads are now readily calculated on the basis of the deflections in the previous tabulation, recalling that the loads are linearly proportional to the deflections.

We observe that, beyond beam 11, it is possible for a transverse beam to be pulled up as a result of the central loading. This is indicated by the negative value of the deflection. The longitudinal beam thus serves to decrease transverse beam deflection only if it is sufficiently rigid.

REFERENCES

9.1. FLUGGE, W., ed., *Handbook of Engineering Mechanics*, McGraw-Hill, New York, 1968, Chap. 13.

9.2. HETENYI, M., *Beams on Elastic Foundations*, McGraw-Hill, New York, 1960.

9.3. TING, B. Y., Finite beams on elastic foundation with restraints, J. Structural Division, Proc. of the American Society of Civil Engineers, Vol. 108, No. ST 3. March 1982, pp. 611–621.

9.4. FLUGGE, W., ed., *Handbook of Engineering Mechanics*, McGraw-Hill, New York, 1968, Sec. 31.5.

9.5. IYENGAR, K. T., SUNDARA RAJA, and RAMU, S. A., *Design Tables for Beams on Elastic Foundation and Related Problems*. Applied Science Publications, London, 1979.

9.6. SHAFFER, B. W., Some simplified solutions for relatively simple stiff beams on elastic foundations, Trans. ASME J. Engineering Industry, February 1963, pp. 1–5.

9.7. UGURAL, A. C., *Stresses in Beams, Plates, and Shells*, 3rd ed., CRC Press, Taylor and Francis Group, Boca Raton, Fla., 2010, Chap. 14.

PROBLEMS

Sections 9.1 through 9.4

9.1. A very long S127 $\times$ 15 steel I-beam, 0.127-m deep, resting on a foundation for which $k = 1.4$ MPa, is subjected to a concentrated load at midlength. The flange is 0.0762-m wide, and the cross-sectional moment of inertia is 5.04×10^{-6} m^4. What is the maximum load that can be applied to the beam without causing the elastic limit to be exceeded? Assume that $E = 200$ GPa and $\sigma_{yp} = 210$ MPa.

9.2. A long steel beam ($E = 200$ GPa) of depth $2.5b$ and width b is to rest on an elastic foundation ($k = 20$ MPa) and support a 40-kN load at its center (Fig. 9.2). Design the beam (compute b) if the bending stress is not to exceed 250 MPa.

9.3. A long beam on an elastic foundation is subjected to a sinusoidal loading $p = p_1 \sin(2\pi x/L)$, where p_1 and L are the peak intensity and wave-

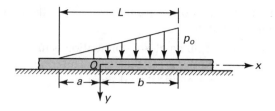

FIGURE P9.11.

length of loading, respectively. Determine the equation of the elastic deflection curve in terms of k and β.

9.4. If point Q is taken to the right of the loaded portion of the beam shown in Fig. 9.3, what is the deflection at this point?

9.5. A single train wheel exerts a load of 135 kN on a rail assumed to be supported by an elastic foundation. For a modulus of foundation $k = 16.8$ MPa, determine the maximum deflection and maximum bending stress in the rail. The respective values of the section modulus and modulus of rigidity are $S = 3.9 \times 10^{-4}$ m^3 and $EI = 8.437$ MN·m^2.

9.6. Resolve Prob. 9.1 based on a safety factor of $n = 2.5$ with respect to yielding of the beam and using the modulus of foundation of $k = 12$ MPa.

9.7. An infinite 6061-T6 aluminum alloy beam of $b \times b$ square cross section resting on an elastic foundation of k carries a concentrated load P at its center (Fig. 9.2). Using a factor of safety of n with respect to yielding of the beam, compute the allowable value of b. Given: $E = 70$ GPa, $\sigma_{yp} = 260$ MPa (Table D.1), $k = 7$ MPa, $P = 50$ kN, and $n = 1.8$.

9.8. Redo Prob. 9.5 for the case in which a single train wheel exerts a concentrated load of $P = 250$ kN on a rail resting on an elastic foundation having $k = 15$ MPa.

9.9. A long rail is subjected to a concentrated load at its center (Fig. 9.2). Determine the effect on maximum deflection and maximum stress of overestimating the modulus of foundation k by (a) 25% and (b) 40%.

9.10. Calculate the maximum resultant bending moment and deflection in the rail of Prob. 9.5 if two wheel loads spaced 1.66 m apart act on the rail. The remaining conditions of the problem are unchanged.

9.11. Determine the deflection at any point Q under the triangular loading acting on an infinite beam on an elastic foundation (Fig. P9.11).

9.12. What are the reactions acting on a semi-infinite beam built in at the left end and subjected to a uniformly distributed loading p? Use the method of superposition. [Hint: At a large distance from the left end, the deflection is p/k.]

9.13. A semi-infinite beam on elastic foundation is subjected to a moment M_A at its end (Fig. 9.5 with $P = 0$). Determine (a) the ratio of the maximum upward and maximum downward deflections; (b) the ratio of the maximum and minimum moments.

9.14. A semi-infinite beam on an elastic foundation is hinged at the left end and subjected to a moment M_L at that end. Determine the equation of the deflection curve, slope, moment, and shear force.

Sections 9.5 through 9.10

9.15. A machine base consists partly of a 5.4-m-long S127 × 15 steel I-beam supported by coil springs spaced $a = 0.625$ m apart. The constant for each spring is $K = 180$ kN/m. The moment of inertia of the I-section is 5.4×10^{-6} m⁴, the depth is 0.127 m, and the flange width is 0.0762 m. Assuming that a concentrated force of 6.75 kN transmitted from the machine acts at midspan, determine the maximum deflection, maximum bending moment, and maximum stress in the beam.

9.16. A steel beam of 0.75-m length and 0.05-m square cross section is supported on three coil springs spaced $a = 0.375$ m apart. For each spring, $K = 18$ kN/m. Determine (a) the deflection of the beam if a load $P = 540$ N is applied at midspan and (b) the deflection at the ends of the beam if a load $P = 540$ N acts 0.25 m from the left end.

9.17. A finite beam with $EI = 8.4$ MN · m² rests on an elastic foundation for which $k = 14$ MPa. The length L of the beam is 0.6 m. If the beam is subjected to a concentrated load $P = 4.5$ kN at its midpoint, determine the maximum deflection.

9.18. A finite beam is subjected to a concentrated force $P = 9$ kN at its mid-length and a uniform loading $p = 7.5$ kN/m. Determine the maximum deflection and slope if $L = 0.15$ m, $EI = 8.4$ MN · m², and $k = 14$ MPa.

9.19. A finite beam of length $L = 0.8$ m and $EI = 10$ MN m resting on an elastic foundation ($k = 8$ MPa) is under a concentrated load $P = 15$ kN at its midlength. What is the maximum deflection?

9.20. A finite cast-iron beam of width b, depth h, and length L, resting on an elastic foundation of modulus k, is subjected to a concentrated load P at midlength and a uniform load of intensity p. Calculate the maximum deflection and the slope. *Given:* $E = 70$ GPa (Table D.1), $b = 100$ mm, $h = 180$ mm, $L = 400$ mm, $P = 8$ kN, $p = 7$ kN/m, and $k = 20$ MPa.

9.21. Redo Example 9.6 for the case in which both ends of the beam are simply supported.

9.22. Assume that all the data of Example 9.7 are unchanged except that a uniformly distributed load p replaces the concentrated force on the longitudinal beam. Compute the load R_{CC} supported by the central transverse beam.

CHAPTER 10

Applications of Energy Methods

10.1 INTRODUCTION

As an alternative to the methods based on differential equations as outlined in Section 3.1, the analysis of stress and deformation can be accomplished through the use of energy methods. The latter are predicated on the fact that the equations governing a given stress or strain configuration are derivable from consideration of the minimization of energy associated with deformation, stress, or deformation and stress. *Applications of energy methods* are effective in situations involving a variety of shapes and variable cross sections and in complex problems involving elastic stability and multielement structures. In particular, strain energy methods offer concise and relatively simple approaches for computation of the displacements of slender structural and machine elements subjected to combined loading.

We shall deal with two principal energy methods.* The first is concerned with the finite deformation experienced by an element under load (Secs. 10.2 to 10.7). The second relies on a hypothetical or *virtual* variation in stress or deformation and represents one of the variational methods (Secs. 10.8 to 10.11). Energy principles are in widespread use in determining solutions to elasticity problems and obtaining deflections of structures and machines. In this chapter, energy methods are used to determine elastic displacements of statically determinate structures as well as to find the redundant reactions and deflections of statically indeterminate systems. With the exception of Section 10.6, our consideration is limited to linearly elastic material behavior and small displacements.

---

*For a more rigorous mathematical treatment and an extensive exposition of energy methods, see Refs. 10.1 through 10.8.

10.2 WORK DONE IN DEFORMATION

Consider a set of forces (applied forces and reactions) P_k ($k \approx 1, 2, \ldots, m$), acting on an elastic body (Fig. 10.1), for example, a beam, truss, or frame. Let the displacement in the direction of P_k of the point at which the force P_k is applied be designated δ_k. It is clear that δ_k is attributable to the action of the entire force set, and not to P_k alone. Suppose that all the forces are applied statically, and let the final values of load and displacement be designated P_k and δ_k. Based on the linear relationship of load and deflection, the work W done by the external force system in deforming the body is given by $\frac{1}{2}\Sigma P_k \delta_k$.

If no energy is dissipated during loading (which is certainly true of a conservative system), we may equate the work done on the body to the *strain energy U* gained by the body. Thus

$$U = W = \frac{1}{2} \sum_{k=1}^{m} P_k \delta_k \qquad (10.1)$$

While the force set P_k ($k = 1, 2, \ldots, m$) includes applied forces and reactions, it is noted that the support displacements are zero, and therefore the support reactions do no work and do not contribute to the preceding summation. Equation (10.1) states simply that the work done by the forces acting on the body manifests itself as elastic strain energy.

To further explore the foregoing concept, consider the body as a combination of small cubic elements. Owing to surface loading, the faces of an element are displaced, and stresses acting on these faces do work equal to the strain energy stored in the element. Consider two adjacent elements within the body. The work done by the stresses acting on two contiguous internal faces is equal but of opposite algebraic sign. We conclude therefore, that the work done on all adjacent faces of the elements will cancel. All that remains is the work done by the stresses acting on the faces that lie on the surface of the body. As the internal stresses balance the external forces at the boundary, the work, whether expressed in terms of external forces (W) or internal stresses (U), is the same.

FIGURE 10.1. *Displacements of an elastic body acted on by several forces.*

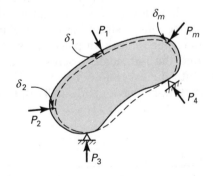

10.3 RECIPROCITY THEOREM

Consider now two sets of applied forces and reactions: P'_k ($k = 1, 2, \ldots, m$), set 1; P''_j ($j = 1, 2, \ldots, n$), set 2. If *only* the first set is applied, the strain energy is, from Eq. (10.1),

$$U_1 = \frac{1}{2} \sum_{k=1}^{m} P'_k \delta'_k \qquad \textbf{(a)}$$

where δ'_k are the displacements corresponding to the set P'_k. Application of *only* set 2 results in the strain energy

$$U_2 = \frac{1}{2} \sum_{j=1}^{n} P''_j \delta''_j \qquad \textbf{(b)}$$

in which δ''_j corresponds to the set P''_j.

Suppose that the first force system P'_k is applied, *followed* by the second force system P''_j. The total strain energy is

$$U = U_1 + U_2 + U_{1,2} \qquad \textbf{(c)}$$

where $U_{1,2}$ is the strain energy attributable to the work done by the first force system as a result of deformations associated with the application of the second force system. Because the forces comprising the first set are unaffected by the action of the second set, we may write

$$U_{1,2} = \sum_{k=1}^{m} P'_k \delta''_k \qquad \textbf{(d)}$$

Here δ''_k represents the displacements caused by the forces of the second set at the points of application of P'_k, the first set. If now the forces are applied in reverse order, we have

$$U = U_2 + U_1 + U_{2,1} \qquad \textbf{(e)}$$

where

$$U_{2,1} = \sum_{j=1}^{n} P''_j \delta'_j \qquad \textbf{(f)}$$

Here δ'_j represents the displacements caused by the forces of set 1 at the points of application of the forces P''_j, set 2.

The loading processes described must, according to the principle of superposition, cause identical stresses within the body. The strain energy must therefore be independent of the order of loading, and it is concluded from Eqs. (c) and (e) that $U_{1,2} = U_{2,1}$. We thus have

$$\sum_{k=1}^{m} P'_k \delta''_k = \sum_{j=1}^{n} P''_j \delta'_j \qquad \textbf{(10.2)}$$

Expression (10.2) is the *reciprocity* or *reciprocal theorem* credited to E. Betti and Lord Rayleigh: The work done by one set of forces owing to displacements due to a second set is equal to the work done by the second system of forces owing to displacements due to the first.

The utility of the reciprocal theorem lies principally in its application to the derivation of various approaches rather than as a method in itself.

10.4 CASTIGLIANO'S THEOREM

First formulated in 1879, Castigliano's theorem is in widespread use because of the ease with which it is applied to a variety of problems involving the deformation of structural elements, especially those classed as statically indeterminate. There are two theorems credited to Castigliano. In this section we discuss the one restricted to structures composed of *linearly elastic* materials, that is, those obeying Hooke's law. For these materials, the strain energy is equal to the complementary energy: $U = U^*$. In Section 10.9, another form of Castigliano's theorem is introduced that is appropriate to structures that behave *nonlinearly* as well as linearly. Both theorems are valid for the cases where any change in structure geometry owing to loading is so small that the action of the loads is not affected.

Refer again to Fig. 10.1, which shows an elastic body subjected to applied forces and reactions, $P_k(k = 1, 2, \ldots, m)$. This set of forces will be designated set 1. Now let one force of set 1, P_i, experience an infinitesimal increment ΔP_i. We designate as set 2 the increment ΔP_i. According to the reciprocity theorem, Eq. (10.2), we may write

$$\sum_{k=1}^{m} P_k \, \Delta \, \delta_k = \Delta P_i \delta_i \qquad \text{(a)}$$

where $\Delta \delta_k$ is the displacement in the direction and, at the point of application, of P_k attributable to the forces of set 2, and δ_i is the displacement in the direction and, at the point of application, of P_i due to the forces of set 1.

The incremental strain energy $\Delta U = \Delta U_2 + \Delta U_{1,2}$ associated with the application of ΔP_i is, from Eqs. (b) and (d) of Section 10.3, $\Delta U = \frac{1}{2} \Delta P_i \, \Delta \, \delta_i + \Sigma \, P_k \, \Delta \, \delta_k$. Substituting Eq. (a) into this, we have $\Delta U = \frac{1}{2} \Delta P_i \Delta \delta_i + \Delta P_i \delta_i$. Now divide this expression by ΔP_i and take the limit as the force ΔP_i approaches zero. In the limit, the displacement $\Delta \delta_i$ produced by ΔP_i vanishes, leaving

$$\frac{\partial U}{\partial P_i} = \delta_i \qquad \text{(10.3)}$$

This is known as *Castigliano's second theorem*: For a linear structure, the partial derivative of the strain energy with respect to an applied force is equal to the component of displacement at the point of application of the force that is in the direction of the force.

It can similarly be demonstrated that

$$\frac{\partial U}{\partial C_i} = \theta_i \qquad (10.4)$$

where C_i and θ_i are, respectively, the couple (bending or twisting) moment and the associated rotation (slope or angle of twist) at a point.

In applying Castigliano's theorem, the strain energy must be expressed as a function of the load. For example, the expression for the strain energy in a straight or curved slender bar (Sec. 5.13) subjected to a number of common loads (axial force N, bending moment M, shearing force V, and torque T) is, from Eqs. (2.59), (2.63), (5.64), and (2.62),

$$U = \int \frac{N^2\,dx}{2AE} + \int \frac{M^2\,dx}{2EI} + \int \frac{\alpha V^2\,dx}{2AG} + \int \frac{T^2\,dx}{2JG} \qquad (10.5)$$

in which the integrations are carried out over the length of the bar. Recall that the term given by the last integral is valid only for a circular cross-sectional area. The displacement at any point in the bar may then readily be found by applying Castigliano's theorem. Inasmuch as the force P is not a function of x, we can perform the differentiation of U with respect to P under the integral. In so doing, the displacement is obtained in the following convenient form:

$$\delta_i = \frac{1}{AE}\int N\frac{\partial N}{\partial P_i}\,dx + \frac{1}{EI}\int M\frac{\partial M}{\partial P_i}\,dx + \frac{1}{AG}\int \alpha V\frac{\partial V}{\partial P_i}\,dx + \frac{1}{JG}\int T\frac{\partial T}{\partial P}\,dx \qquad (10.6)$$

Similarly, an expression may be written for the angle of rotation:

$$\theta_i = \frac{1}{AE}\int N\frac{\partial N}{\partial C_i}\,dx + \frac{1}{EI}\int M\frac{\partial M}{\partial C_i}\,dx + \frac{1}{AG}\int \alpha V\frac{\partial V}{\partial C_i}\,dx + \frac{1}{JG}\int T\frac{\partial T}{\partial C_i}\,dx \qquad (10.7)$$

For a slender beam, as observed in Section 5.4, the contribution of the shear force V to the displacement is negligible.

Referring to Section 2.13, in the case of a *plane truss* consisting of n members of length L_j, axial rigidity A_jE_j, and internal axial force N_j, the strain energy can be found from Eq. (2.59) as follows:

$$U = \sum_{j=1}^{n} \frac{N_j^2 L_j}{2A_j E_j} \qquad (10.8)$$

The displacement δ_i of the point of application of load P is therefore

$$\delta_i = \frac{\partial U}{\partial P_i} = \sum_{j=1}^{n} \frac{N_j L_j}{A_j E_j}\frac{\partial N_j}{\partial P_i} \qquad (10.9)$$

Here the axial force N_j in each member may readily be determined using the *method of joints* or *method of sections*. The former consists of analyzing the truss joint by joint to find the forces in members by applying the equations of equilibrium at each joint. The latter is often employed for problems where the forces in only a few members are to be obtained.

FIGURE 10.2. *A cantilever beam with a uniform load.*

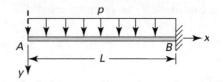

A final point to be noted that, when it is necessary to determine the deflection at a point at which no load acts, the problem is treated as follows. A fictitious load Q (or C) is introduced at the point in question in the direction of the desired displacement δ (or θ). The displacement is then found by applying Castigliano's theorem, setting $Q = 0$ (or $C = 0$) in the final result. Consider, for example, the cantilever beam AB supporting a uniformly distributed load of intensity p (Fig. 10.2). To determine the deflection at A, we apply a fictitious load Q, as shown by the dashed line in the figure. Expressions for the moment and its derivative with respect to Q are, respectively,

$$M = Qx + \frac{1}{2}px^2 \quad \text{and} \quad \frac{\partial M}{\partial Q} = x$$

Substituting these into Eq. (10.6) and making $Q = 0$ results in

$$\delta_A = \frac{1}{EI} \int M \frac{\partial M}{\partial Q} dx$$

$$= \frac{1}{EI} \int_0^L \left(\frac{1}{2}px^2\right)(x)\, dx = \frac{pL^4}{8EI}$$

Since the fictitious load was directed downward, the positive sign means that the deflection δ_A is also downward.

EXAMPLE 10.1   Slope of a Beam with an Overhang
Determine the slope of the elastic curve at the left support of the uniformly loaded beam shown in Fig. 10.3.

Solution   As a slope is sought, a fictitious couple moment C is introduced at point A. Applying the equations of statics, the reactions are found to be

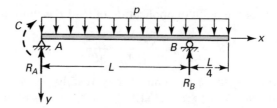

FIGURE 10.3.   *Example 10.1. Uniformly loaded beam with an overhang.*

$$R_A = \tfrac{15}{32}pL - \frac{C}{L}, \qquad R_B = \tfrac{25}{32}pL + \frac{C}{L}$$

The following expressions for the moments are thus available:

$$M_1 = -\left(\tfrac{15}{32}pL - \frac{C}{L}\right)x + \frac{px^2}{2} - C, \qquad 0 \le x \le L$$

$$M_2 = -\left(\tfrac{15}{32}pL - \frac{C}{L}\right)x + \frac{px^2}{2} - C - \left(\tfrac{25}{32}pL + \frac{C}{L}\right)(x - L), \qquad L \le x \le \tfrac{5}{4}L$$

The slope at A is now found from Eq. (10.7):

$$\theta_A = \frac{1}{EI}\int_0^L \left(-\tfrac{15}{32}pLx + \frac{Cx}{L} + \frac{px^2}{2} - C\right)\left(\frac{x}{L} - 1\right)dx + \frac{1}{EI}\int_L^{5L/4} M_2(0)\,dx$$

Setting $C = 0$ and integrating, we obtain $\theta_A = 7pL^3/192\,EI$.

EXAMPLE 10.2   Deflection of a Pin-Connected Structure

A load P is applied at B to two bars of equal length L but different cross-sectional areas and modulii of elasticity (Fig. 10.4a). Determine the horizontal displacement δ_B of point B.

Solution   A fictitious load Q is applied at B. The forces in the bars are determined by considering the equilibrium of the free-body diagram of pin B (Fig. 10.4b):

$$(N_{BC} + N_{BD})\sin\theta = P$$
$$(N_{BC} - N_{BD})\cos\theta = Q$$

or

$$N_{BC} = \frac{1}{2}\left(\frac{P}{\sin\theta} + \frac{Q}{\cos\theta}\right) \tag{b}$$

$$N_{BD} = \frac{1}{2}\left(\frac{P}{\sin\theta} - \frac{Q}{\cos\theta}\right)$$

FIGURE 10.4.   *Example 10.2. Two-bar structure carries a load P.*

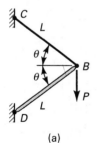

(a)

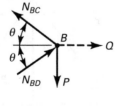

(b)

Differentiating these expressions with respect to Q,

$$\frac{\partial N_{BC}}{\partial Q} = \frac{1}{2\cos\theta}, \qquad \frac{\partial N_{BD}}{\partial Q} = -\frac{1}{2\cos\theta}$$

Applying Castigliano's theorem, Eq. (10.6),

$$\delta_B = \left(\frac{N_{BC}}{A_{BC}E_{BC}}\frac{\partial N_{BC}}{\partial Q} + \frac{N_{BD}}{A_{BD}E_{BD}}\frac{\partial N_{BD}}{\partial Q}\right)L$$

Introducing Eqs. (b) and setting $Q = 0$ yields the horizontal displacement of the pin under the given load P:

$$\delta_B = \frac{PL}{4\cos\theta\sin\theta}\left(\frac{1}{A_{BC}E_{BC}} - \frac{1}{A_{BD}E_{BD}}\right) \tag{c}$$

A check is provided in that, in the case of two *identical* rods, the preceding expression gives $\delta_B = 0$, a predictable result.

EXAMPLE 10.3   Three-Bar Truss

The simple pin-connected truss shown in Fig. 10.5 supports a force P. If all members are of equal rigidity AE, what is the deflection of point D?

Solution Applying the method of joints at points A and C and taking symmetry into account, we obtain $N_1 = N_2 = 5P/8$, $N_4 = N_5 = 3P/8$, and $N_3 = P$. Castigliano's theorem, Eq. (10.9), may be written

$$\delta_i = \frac{1}{AE}\sum_{j=1}^{n}N_j\left(\frac{\partial N_j}{\partial P_i}\right)L_j \tag{d}$$

where $n = 5$. Substituting the values of axial forces in terms of applied load into Eq. (d) leads to

$$\delta_D = \frac{1}{AE}[2(\tfrac{5}{8}P)(\tfrac{5}{8})5 + P(1)4 + 2(\tfrac{3}{8}P)(\tfrac{3}{8})3]$$

from which $\delta_D = 35p/4AE$.

FIGURE 10.5.   *Example 10.3. A truss supports a load P.*

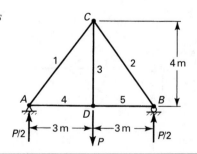

EXAMPLE 10.4   Thick-Walled Half Ring

A load P of 5 kN is applied to a steel curved bar as depicted in Fig. 10.6a. Determine the vertical deflection of the free end by considering the effects of the internal normal and shear forces in addition to the bending moment. Let $E = 200$ GPa and $G = 80$ GPa.

Solution   A free-body diagram of a portion of the bar subtended by angle θ is shown in Fig. 10.6b, where the internal forces (N and V) and moment (M) are positive as indicated. Referring to the figure,

$$M = PR(1 - \cos\theta), \quad V = P\sin\theta, \quad N = P\cos\theta \quad \textbf{(10.10)}$$

Thus,

$$\frac{\partial M}{\partial P} = R(1 - \cos\theta), \quad \frac{\partial V}{\partial P} = \sin\theta, \quad \frac{\partial N}{\partial P} = \cos\theta$$

The form factor for shear for the rectangular section is $\alpha = \frac{6}{5}$ (Table 5.1). Substituting the preceding expressions into Eq. (10.6) with $dx = R\,d\theta$, we have

$$\delta_v = \frac{PR^3}{EI}\int_0^\pi (1 - \cos\theta)^2\,d\theta + \frac{6PR}{5AG}\int_0^\pi \sin^2\theta\,d\theta + \frac{PR}{AE}\int_0^\pi \cos^2\theta\,d\theta$$

Integration of the foregoing results in

$$\delta_v = \frac{3\pi PR^3}{2EI} + \frac{3\pi PR}{5AG} + \frac{\pi PR}{2AE} \quad \textbf{(10.11)}$$

Geometric properties of the section of the bar are

$$I = \frac{0.01(0.02)^3}{12} = 66.7 \times 10^{-10}\,\text{m}^4,$$

$$A = 0.01(0.02) = 2 \times 10^{-4}\,\text{m}^2, \quad R = 0.05\,\text{m}$$

Insertion of the data into Eq. (10.11) gives

$$\delta_v = (2.21 + 0.03 + 0.01) \times 10^{-3} = 2.25\,\text{mm}$$

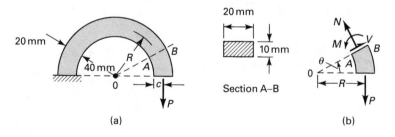

FIGURE 10.6.   *Example 10.4. (a) A thick-walled curved bar is fixed at one end and supports a load P at its free end; (b) segment of the bar.*

Comment Note that if the effects of the normal and shear forces are omitted

$$\delta_v = 2.21 \text{ mm}$$

with a resultant error in deflection of approximately 1.8%. For this curved bar, in which $R/c = 5$, the contribution of V and N to the displacement can thus be neglected. It is common practice *to omit the first and the third terms* in Eqs. (10.6) and (10.7) when $R/c > 4$ (Sec. 5.13).

EXAMPLE 10.5 Analysis of a Piping System

A piping system *expansion loop* is fabricated of pipe of constant size and subjected to a temperature differential ΔT (Fig. 10.7). The overall length of the loop and the coefficient of thermal expansion of the tubing material are L and α, respectively.

Determine, for each end of the loop, the restraining bending moment M and force N induced by the temperature change.

Solution In labeling the end points for each segment, the symmetry about a vertical axis through point A is taken into account, as shown in the figure. Expressions for the moments, associated with segments DC, CB, and BA, are, respectively,

$$\begin{aligned} M_1 &= Nr(1 - \cos\theta) - M \\ M_2 &= N[r + (r + R)(-\cos\phi) + R\cos\theta] - M \qquad \text{(e)} \\ M_3 &= N[r + R + (r + R)(-\cos\phi)] - M \end{aligned}$$

Upon application of Eqs. (10.3) and (10.4), the end deflection and end slope are found to be

$$\delta_{GN} = \frac{2}{EI} \left[\int_0^\phi M_1 \frac{\partial M_1}{\partial N} \cdot r\, d\theta + \int_0^\phi M_2 \frac{\partial M_2}{\partial N} \cdot R\, d\theta + M_3 \frac{\partial M_3}{\partial N} \cdot \frac{b}{2} \right] \qquad \text{(f)}$$

$$0 = \frac{2}{EI} \left[\int_0^\phi M_1 \frac{\partial M_1}{\partial M} \cdot r\, d\theta + \int_0^\phi M_2 \frac{\partial M_2}{\partial M} \cdot R\, d\theta + M_3 \frac{\partial M_3}{\partial M} \cdot \frac{b}{2} + M \frac{\partial M}{\partial M} \cdot a \right] \qquad \text{(g)}$$

FIGURE 10.7. *Example 10.5. Expansion loop is subjected to a temperature change.*

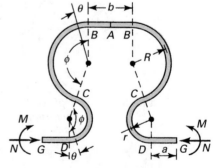

Substitution of Eqs. (e) into Eq. (f) results in

$$\delta_{GN} = \frac{2N}{EI} \left\{ r^3(\tfrac{3}{2}\phi - 2\sin\phi + \tfrac{1}{4}\sin 2\phi) + rR[r\phi - 2\phi\cos\phi(r + R) + 2R\sin\phi] \right.$$

$$\left. + (r + R)^2 \left[\frac{b}{2}(1 - \cos\phi) + R\phi\cos^2\phi \right] + R^2(\tfrac{1}{2}R\phi - \tfrac{3}{4}R\sin\phi - r\sin^2\phi) \right\}$$

$$- \frac{2M}{EI} \left\{ r^2(\phi - \sin\phi) + R^2[\sin\phi - \phi\cos\phi] \right.$$

$$\left. + rR\phi(-\cos\phi) + \frac{b}{2}(r + R)(1 - \cos\phi) \right\} \qquad \textbf{(h)}$$

Similarly, Eqs. (g) and (e) lead to

$$M = \frac{1}{a + (b/2) + N(r + R)} \left\{ N \left[(r^2 + rR)\phi + \sin\phi(R^2 - r^2) \right. \right.$$

$$\left. \left. + \frac{b}{2}(r + R) - \cos\phi(r + R)\left(\frac{b}{2} + R\phi\right) \right] \right\} \qquad \textbf{(10.12a)}$$

The deflections at G owing to the temperature variation and end restraints must be equal; that is,

$$\delta_{GN} = \delta_{GT} = \alpha(\Delta T)L \qquad \textbf{(10.12b)}$$

Expressions (h) and (10.12) are then solved to yield the unknown reactions N and M in terms of the given properties and loop dimensions.

10.5   UNIT- OR DUMMY-LOAD METHOD

Recall that the deformation at a point in an elastic body subjected to external loading P_i, expressed in terms of the moment produced by the force system, is, according to Castigliano's theorem,

$$\delta_i = \frac{\partial U}{\partial P_i} = \int \frac{1}{EI} M \frac{\partial M}{\partial P_i} dx \qquad \textbf{(a)}$$

For small deformations of linearly elastic materials, the moment is linearly proportional to the external loads, and consequently we are justified in writing $M = mP_i$, with m independent of P_i. It follows that $\partial M/\partial P_i = m$, the change in the bending moment per unit change in P_i, that is, the moment caused by a unit load.

The foregoing considerations lead to the *unit-load* or *dummy-load* approach, which finds extensive application in structural analysis. From Eq. (a),

$$\delta_i = \int \frac{Mm}{EI} dx \tag{10.13}$$

In a similar manner, the following expression is obtained for the change of slope:

$$\theta_i = \int \frac{Mm'}{EI} dx \tag{10.14}$$

Here $m' = \partial M/\partial C_i$ represents the change in the bending moment per unit change in C_i, that is, the change in bending moment caused by a unit-couple moment.

Analogous derivations can be made for the effects of axial, shear, and torsional deformations by replacing m in Eq. (10.13) with $n = \partial N/\partial P_i$, $v = \partial V/\partial P_i$, and $t = \partial T/\partial P_i$, respectively. It can also readily be demonstrated that, *in the case of truss*, Eq. (10.9) has the form

$$\delta_i = \sum_{j=1}^{n} \frac{n_j N_j L_j}{A_j E_j} \tag{10.15}$$

wherein AE is the axial rigidity. The quantity $n_j = \partial N_j/\partial P_i$ represents the change in the axial forces N_j owing to a unit value of load P_i.

EXAMPLE 10.6   Deflection of a Simple Beam with Two Loads

Derive an expression for the deflection of point C of the simply supported beam shown in Fig. 10.8a.

Solution   Figure 10.8b shows the dummy load of 1 N and the reactions it produces. Note that the unit load is applied at C because it is the deflection of C that is required. Referring to the figure, the following moment distributions are obtained:

$$M_1 = -Px, \qquad m_1 = -\tfrac{2}{3}x, \qquad 0 \le x \le \frac{L}{3}$$

$$M_2 = -Px + P\left(x - \frac{L}{3}\right) = -\frac{PL}{3}$$

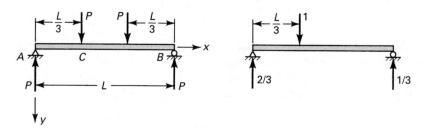

FIGURE 10.8.   *Example 10.6. (a) Actual loading; (b) unit loading.*

$$m_2 = -\tfrac{2}{3}x + 1\left(x - \frac{L}{3}\right) = -\frac{L-x}{3}, \qquad \frac{L}{3} \le x \le \frac{2L}{3}$$

$$M_3 = -Px + P\left(x - \frac{L}{3}\right) + P\left(x - \frac{2L}{3}\right) = -P(L-x)$$

$$m_3 = -\tfrac{1}{3}(L-x), \qquad \frac{2L}{3} \le x \le L$$

Here the M's refer to Fig. 10.8a and the m's to Fig. 10.8b. The vertical deflection at C is then, from Eq. (10.14),

$$\delta_C = \frac{1}{EI}\left[\int_0^{L/3} Px\left(\frac{2x}{3}\right)dx + \int_{L/3}^{2L/3} \frac{PL}{3}\left(\frac{L-x}{3}\right)dx\right.$$

$$\left. + \int_{2L/3}^{L} P(L-x)\left(\frac{L-x}{3}\right)dx\right]$$

The solution, after integration, is found to be $\delta_C = 5PL^3/162EI$.

10.6 CROTTI–ENGESSER THEOREM

Consider a set of forces acting on a structure that behaves *nonlinearly*. Let the displacement of the point at which the force P_i is applied, in the direction of P_i, be designated δ_i. This displacement is to be determined. The problem is the same as that stated in Section 10.4, but now it will be expressed in terms of P_i and the complementary energy U^* of the structure, the latter being given by Eq. (2.49).

In deriving the theorem, a procedure is employed similar to that given in Section 10.4. Thus, U is replaced by U^* in Castigliano's second theorem, Eq. (10.3), to obtain

$$\frac{\partial U^*}{\partial P_i} = \delta_i \qquad\qquad (10.16)$$

This equation is known as the *Crotti–Engesser theorem:* The partial derivative of the complementary energy with respect to an applied force is equal to the component of the displacement at the point of application of the force that is in the direction of the force. Obviously, here the complementary energy must be expressed in terms of the loads.

EXAMPLE 10.7 Nonlinearly Elastic Structure

A simple truss, constructed of pin-connected members 1 and 2, is subjected to a vertical force P at joint B (Fig. 10.9a). The bars are made of a nonlinearly elastic material displaying the stress–strain relation $\sigma = K\varepsilon^{1/2}$ equally in tension and compression (Fig. 10.9b). Here the strength coefficient K is a constant. The cross-sectional area of each member is A. Determine the vertical deflection of joint B.

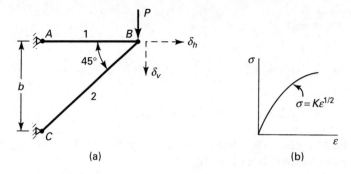

FIGURE 10.9. *Example 10.7. Two-bar structure with material nonlinearity.*

Solution The volume of member 1 is Ab and that of member 2 is $Ab\sqrt{2}$. The total complementary energy of the structure is therefore

$$U^* = (U_{o1}^* + U_{o2}^*\sqrt{2})Ab \qquad \textbf{(a)}$$

The complementary energy densities are, from Eq. (2.49),

$$U_{o1}^* = \int_0^{\sigma_1} \frac{\sigma^2}{K^2}\,d\sigma = \frac{\sigma_1^3}{3K^2}$$
$$U_{o2}^* = \int_0^{\sigma_2} \frac{\sigma^2}{K^2}\,d\sigma = \frac{\sigma_2^3}{3K^2} \qquad \textbf{(b)}$$

where σ_1 and σ_2 are the stresses in bars 1 and 2. Upon introduction of Eqs. (b) into (a), we have

$$U^* = \frac{Ab}{3K^2}(\sigma_1^3 + \sqrt{2}\,\sigma_2^3) \qquad \textbf{(c)}$$

From static equilibrium, the axial forces in 1 and 2 are found to be P and $P\sqrt{2}$, respectively. Thus, $\sigma_1 = P/A$ (tension) and $\sigma_2 = \sqrt{2}\,P/A$ (compression), which when introduced into Eq. (c) yields

$$U^* = 5P^3b/3A^2K^2 \qquad \textbf{(d)}$$

Applying Eq. (10.16), the vertical deflection of B is found to be

$$\delta_v = \frac{5P^2b}{A^2K^2}$$

Another approach to the solution of this problem is given in Example 10.11.

10.7 STATICALLY INDETERMINATE SYSTEMS

To supplement the discussion of statically indeterminate systems given in Section 5.11, energy methods are now applied to obtain the unknown, redundant forces (moments) in such systems. Consider, for example, the beam system of Fig. 10.3 rendered statically indeterminate by the addition of an extra or redundant support at the right end (not shown in the figure). The strain energy is, as before, written as a function of all external forces, including both applied loads and reactions. Castigliano's theorem may then be applied to derive an expression for the deflection at point B, which is clearly zero:

$$\frac{\partial U}{\partial R_B} = \delta_B = 0 \tag{a}$$

Expression (a) and the two equations of statics available for this force system provide the three equations required for the determination of the three unknown reactions.

Extending this reasoning to the case of a statically indeterminate beam with n redundant reactions, we write

$$\frac{\partial U}{\partial R_n} = \delta_n = 0 \tag{b}$$

The equations of statics together with the equations of the type given by Eq. (b) constitute a set sufficient for solution of all the reactions. This basic concept is fundamental to the analysis of structures of considerable complexity.

EXAMPLE 10.8   Spring-Propped Cantilever Beam
The built-in beam shown in Fig. 10.10 is supported at one end by a spring of constant k. Determine the redundant reaction.

Solution   The expressions for the moments are

$$M_1 = -R_D x, \qquad 0 \le x \le \frac{L}{2}$$

$$M_2 = -R_D x + P\left(x - \frac{L}{2}\right), \qquad \frac{L}{2} \le x \le L$$

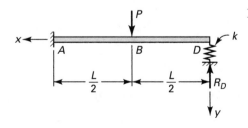

FIGURE 10.10.   *Example 10.8. A propped cantilever beam with load P.*

Applying Castigliano's theorem to obtain the deflection at point D, $\delta_D = \partial U/\partial R_D$, we have

$$\delta_D = \frac{1}{EI}\int_0^{L/2} R_D x (x\,dx) + \frac{1}{EI}\int_{L/2}^{L}\left[R_D x - P\left(x - \frac{L}{2}\right)\right]x\,dx = -\frac{R_D}{k}$$

from which $R_D = \frac{5}{16}P/(1 + 3EI/kL^3)$. Equilibrium of vertical forces yields

$$R_A = P - \frac{5P}{16(1 + 3EI/kL^3)} \qquad \textbf{(10.17)}$$

Note that were the right end rigidly supported, δ_D would be equated to zero.

EXAMPLE 10.9   Rectangular Frame

A rectangular frame of constant EI is loaded as shown in Fig. 10.11a. Assuming the strain energy to be attributable to bending alone, determine the increase in distance between the points of application of the load.

Solution   The situation described is statically indeterminate. For reasons of symmetry, we need analyze only one quadrant. Because the slope B is zero before and after application of the load, the segment may be treated as fixed at B (Fig. 10.11b). The moment distributions are

$$M_1 = -M_A, \qquad 0 \le x \le a$$
$$M_2 = -M_A + \tfrac{1}{2}Px, \qquad 0 \le x \le b$$

Since the slope is zero at A, we have

$$\theta_A = \frac{1}{EI}\int M\frac{\partial M}{\partial M_A}dx = \frac{1}{EI}\left[\int_0^{a} M_A\,dx + \int_0^{b}\left(M_A - \frac{1}{2}Px\right)dx\right] = 0$$

from which $M_A = Pb^2/4(a + b)$. By applying Castigliano's theorem, the relative displacement between the points of application of load is then found to be $\delta = Pb^3(4a + b)/12EI(a + b)$.

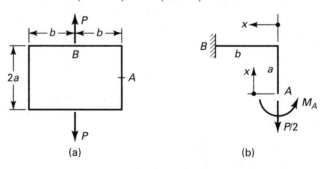

(a)   (b)

FIGURE 10.11.   *Example 10.9. (a) A frame; (b) free-body diagram of part AB.*

EXAMPLE 10.10   Member Forces in a Three-Bar Truss

A symmetric plane structure constructed of three bars of equal axial rigidity AE is subjected to a load P at joint D, as illustrated in Fig. 10.12a. Using Castigliano's theorem, find the force in each bar.

Solution   The structure is statically indeterminate to the first degree. The reaction R at B, which is equal to the force in the bar BD, is selected as redundant. Due to symmetry, the forces AD and CD are each equal to N. The condition of equilibrium of forces at joint D (Fig. 10.12b), gives

$$N = \frac{5}{8}(P - R) \qquad \text{(c)}$$

Substituting the foregoing relation and the given numerical values into Eq. (10.34), we find the deflection at joint B in the form

$$\delta_B = 2\,\frac{NL}{AE}\frac{\partial N}{\partial R} + \frac{R(0.8L)}{AE}\frac{\partial R}{\partial R}$$

$$= -\frac{25}{32AE}(P - R)L + \frac{0.8RL}{AE} \qquad \text{(d)}$$

Since the support B does not move, we set $\delta_B = 0$ in Eq. (d). In so doing, the reaction at B is determined as follows $R = 0.494P$. The *forces in the members*, from Eq. (c), are thus

$$N_{AD} = N_{CD} = \frac{5}{8}(1 - 0.494)P = 0.316P$$

and

$$N_{BD} = R = 0.494P$$

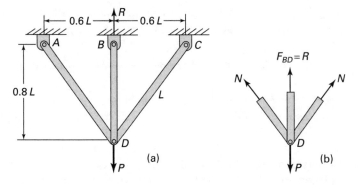

FIGURE 10.12.   *Example 10.10. (a) A plane structure; (b) free-body diagram of joint D.*

10.8 PRINCIPLE OF VIRTUAL WORK

In this section a second type of energy approach is explored, based on a hypothetical variation of deformation. This method, as is demonstrated, lends itself to the expeditious solution of a variety of important problems.

Consider a body in equilibrium under a system of forces. Accompanying a small displacement within the body, we expect a change in the original force system. Suppose now that an *arbitrary* incremental displacement occurs, termed a *virtual displacement*. This displacement need not actually take place and need not be infinitesimal. If we assume the displacement to be infinitesimal, as is usually done, it is reasonable to regard the system of forces as unchanged. In connection with virtual work, we shall use the symbol δ to denote a virtual infinitesimally small quantity.

Recall from *particle* mechanics that for a point mass, *unconstrained* and thereby free to experience arbitrary virtual displacements δu, δv, and δw, the *virtual work* accompanying these displacements is $\Sigma F_x\,\delta u$, $\Sigma F_y\,\delta v$, and $\Sigma F_z\,\delta w$, where ΣF_x, ΣF_y, and ΣF_z are the force resultants. If the particle is in equilibrium, it follows that the *virtual work must vanish*, since $\Sigma F_x = \Sigma F_y = \Sigma F_z = 0$. This is the principle of virtual work.

For an elastic body, it is necessary to impose a number of restrictions on the arbitrary virtual displacements. To begin with, these displacements must be continuous and their derivatives must exist. In this way, material continuity is assured. Because certain displacements on the boundary may be dictated by the circumstances of a given situation (boundary conditions), the virtual displacements at such points on the boundary must be zero. A virtual displacement results in no alteration in the magnitude or direction of external and internal forces. The imposition of a virtual displacement field on an elastic body does, however, result in the imposition of an increment in the strain field.

To determine the virtual strains, replace the displacements u, v, and w by virtual displacements δu, δv, and δw in the definition of the actual strains, Eq. (2.4):

$$\delta\varepsilon_x = \frac{\partial}{\partial x}\delta u, \ldots, \qquad \delta\gamma_{xy} = \frac{\partial}{\partial y}\delta u + \frac{\partial}{\partial x}\delta v, \ldots$$

The strain energy δU acquired by a body of volume V as a result of virtual straining is, by application of Eq. (2.47) together with the second of Eqs. (2.44),

$$\delta U = \int_V (\sigma_x\,\delta\varepsilon_x + \sigma_y\,\delta\varepsilon_y + \sigma_z\,\delta\varepsilon_z + \tau_{xy}\,\delta\gamma_{xy} + \tau_{yz}\,\delta\gamma_{yz} + \tau_{xz}\,\delta\gamma_{xz})\,dV \quad \textbf{(10.18)}$$

Note the absence in this equation of any term involving a variation in stress. This is attributable to the assumption that the *stress remains constant* during application of virtual displacement.

The variation in strain energy may be viewed as the work done against the mutual actions between the infinitesimal elements composing the body, owing to the virtual displacements (Sec. 10.2). The virtual work done in an elastic body by these mutual actions is therefore $-\delta U$.

Consider next the virtual work done by *external* forces. Again suppose that the body experiences virtual displacements δu, δv, and δw. The virtual work done by a body force **F** per unit volume and a surface force **p** per unit area is

$$\delta W = \int_V (F_x\,\delta u + F_y\,\delta v + F_z\,\delta w)\,dV$$
$$+ \int_A (p_x\,\delta u + p_y\,\delta v + p_z\,\delta w)\,dA \tag{10.19}$$

where A is the boundary surface. We have already stated that the total work done during the virtual displacement is zero: $\delta W - \delta U = 0$. The *principle of virtual work* for an elastic body is therefore expressed as follows:

$$\delta W = \delta U \tag{10.20}$$

10.9 PRINCIPLE OF MINIMUM POTENTIAL ENERGY

As the virtual displacements result in no geometric alteration of the body and as the external forces are regarded as constants, Eq. (10.20) may be rewritten

$$\delta\Pi = \delta\left[U - \int_A (p_x u + p_y v + p_z w)\,dA - \int_V (F_x u + F_y v + F_z w)\,dV \right] = 0 \tag{10.21a}$$

or, briefly,

$$\delta\Pi = \delta(U - W) = 0 \tag{10.21b}$$

Here it is noted that δ has been removed from under the integral sign. The term $\Pi = U - W$ is called the *potential energy*, and Eq. (10.21) represents a condition of *stationary* potential energy of the system.

It can be demonstrated that, for *stable equilibrium*, the potential energy is a minimum. Only for displacements that satisfy the boundary conditions and the equilibrium conditions will Π assume a minimum value. This is called the *principle of minimum potential energy*.

Consider now the case in which the loading system consists only of forces applied at points on the surface of the body, denoting each point force by P_i and the displacement in the direction of this force by δ_i (corresponding to the equilibrium state). From Eq. (10.21), we have

$$\delta(U - P_i\delta_i) = 0 \qquad \text{or} \qquad \delta U = P_i\delta(\delta_i)$$

The principle of minimum potential energy thus leads to

$$\frac{\partial U}{\partial \delta_i} = P_i \tag{10.22}$$

The preceding means that the partial derivative of the stain energy with respect to a displacement δ_i equals the force acting in the direction of δ_i at the point of application of P_i. Equation (10.22) is known as *Castigliano's first theorem*. This theorem, as with the Crotti–Engesser theorem, may be applied to any structure, linear or nonlinear.

EXAMPLE 10.11 Nonlinearly Elastic Basic Truss

Determine the vertical displacement δ_v and the horizontal displacement δ_h of the joint B of the truss described in Example 10.7.

Solution First introduce the unknown vertical and horizontal displacements at the joint shown by dashed lines in Fig. 10.9a. Under the influence of δ_v, member 1 does not deform, while member 2 is contracted by $\delta_v/2b$ per unit length. Under the influence of δ_h, member 1 elongates by δ_h/b, and member 2 by $\delta_h/2b$ per unit length. The strains produced in members 1 and 2 under the effect of both displacements are then calculated from

$$\varepsilon_1 = \frac{\delta_h}{b}, \qquad \varepsilon_2 = \frac{\delta_v - \delta_h}{2b} \qquad \textbf{(a)}$$

where ε_1 is an elongation and ε_2 is a shortening. Members 1 and 2 have volumes Ab and $Ab\sqrt{2}$, respectively. Next, the total strain energy of the truss, from Eq. (a) of Section 2.13, is determined as follows:

$$U = Ab \int_0^{\varepsilon_1} \sigma \, d\varepsilon + Ab\sqrt{2} \int_0^{\varepsilon_2} \sigma \, d\varepsilon$$

Upon substituting $\sigma = K\varepsilon^{1/2}$ and Eqs. (a), and integrating, this becomes

$$U = \frac{AK}{3\sqrt{b}} [2\delta_h^{3/2} + (\delta_v - \delta_h)^{3/2}]$$

Now we apply Castigliano's theorem in the horizontal and vertical directions at B, respectively:

$$\frac{\partial U}{\partial \delta_h} = \frac{AK}{2\sqrt{b}} \left[2\delta_h^{1/2} - (\delta_v - \delta_h)^{1/2} \right] = 0$$

$$\frac{\partial U}{\partial \delta_v} = \frac{AK}{2\sqrt{b}} (\delta_v - \delta_h)^{1/2} = P$$

Simplifying and solving these expressions simultaneously, the joint displacements are found to be

$$\delta_h = \frac{P^2 b}{A^2 K^2}, \qquad \delta_v = \frac{5P^2 b}{A^2 K^2} \qquad \textbf{(b)}$$

The stress–strain law, together with Eqs. (a) and (b), yields the stresses in the members if required:

$$\sigma_1 = K\varepsilon_1^{1/2} = \frac{P}{A}, \qquad \sigma_2 = K\varepsilon_2^{1/2} = \frac{\sqrt{2}P}{A}$$

The axial forces are therefore

$$N_1 = \sigma_1 A = P, \qquad N_2 = \sigma_2 A = \sqrt{2}P$$

Here N_1 is tensile and N_2 is compressive. It is noted that, for the statically determinate problem under consideration, the axial forces could readily be determined from static equilibrium. The solution procedure given here applies similarly to statically indeterminate structures as well as to linearly elastic structures.

10.10 DEFLECTIONS BY TRIGONOMETRIC SERIES

Certain problems in the analysis of structural deformation, mechanical vibration, heat transfer, and the like, are amenable to solution by means of trigonometric series. This approach offers an important advantage: a single expression may apply to the entire length of the member. The method is now illustrated using the case of a simply supported beam subjected to a moment at point A (Fig. 10.13a). The solution by trigonometric series can also be employed in the analysis of beams having any other type of end condition and beams under combined loading [Refs. 10.5 and 10.9], as demonstrated in Examples 10.12 and 10.13.

The deflection curve can be represented by a *Fourier* sine series:

$$v = a_1\sin\frac{\pi x}{L} + a_2\sin\frac{2\pi x}{L} + \cdots = \sum_{n=1}^{\infty} a_n\sin\frac{n\pi x}{L} \tag{10.23}$$

The end conditions of the beam ($v = 0$, $v'' = 0$ at $x = 0$, $x = L$) are observed to be satisfied by each term of this infinite series. The first and second terms of the

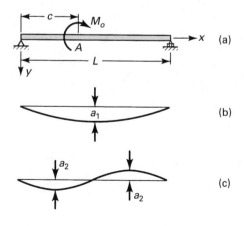

FIGURE 10.13. *(a) Simple beam subjected to a moment M_o at an arbitrary distance c from the left support; (b) and (c) deflection curve represented by the first and second terms of a Fourier series, respectively.*

series are represented by the curves in Fig. 10.13b and c, respectively. As a physical interpretation of Eq. (10.23), consider the true deflection curve of the beam to be the superposition of sinusoidal curves of n different configurations. The coefficients a_n of the series are the maximum coordinates of the sine curves, and the n's indicate the number of half-waves in the sine curves. It is demonstrable that, when the coefficients a_n are determined properly, the series given by Eq. (10.23) can be used to represent any deflection curve [Ref. 10.10]. By increasing the number of terms in the series, the accuracy can be improved.

To evaluate the coefficients, the principle of virtual work will be applied. The strain energy of the system, from Eqs. (5.62) and (10.23), is written

$$U = \frac{EI}{2} \int_0^L \left(\frac{d^2v}{dx^2} \right)^2 dx = \frac{EI}{2} \int_0^L \left[\sum_{n=1}^{\infty} a_n \left(\frac{n\pi}{L} \right)^2 \sin \frac{n\pi x}{L} \right]^2 dx \qquad \text{(a)}$$

Expanding the term in brackets,

$$\left[\sum_{n=1}^{\infty} a_n \left(\frac{n\pi}{L} \right)^2 \sin \frac{n\pi x}{L} \right]^2 = \sum_{m=1}^{\infty} \sum_{n=1}^{\infty} a_m a_n \left(\frac{m\pi}{L} \right)^2 \left(\frac{n\pi}{L} \right)^2 \sin \frac{m\pi x}{L} \sin \frac{n\pi x}{L}$$

Since for the orthogonal functions $\sin(m\pi x/L)$ and $\sin(n\pi x/L)$ it can be shown by direct integration that

$$\int_0^l \sin \frac{m\pi x}{L} \sin \frac{n\pi x}{L} dx = \begin{cases} 0, & m \neq n \\ L/2, & m = n \end{cases} \qquad \text{(10.24)}$$

Eq. (a) gives then

$$\delta U = \frac{\pi^4 EI}{2L^3} \sum_{n=1}^{\infty} n^4 a_n \delta a_n \qquad \text{(10.25)}$$

The virtual work done by a moment M_o acting through a virtual rotation at A increases the strain energy of the beam by δU:

$$M_o \left(\delta \frac{\partial v}{\partial x} \right)_A = \delta U \qquad \text{(b)}$$

Therefore, from Eqs. (10.25) and (b), we have

$$M_o \sum_{n=1}^{\infty} \frac{n\pi}{L} \cos \frac{n\pi c}{L} \delta a_n = \frac{\pi^4 EI}{2L^3} \sum_{n=1}^{\infty} n^4 a_n \delta a_n$$

which leads to

$$a_n = \frac{2M_o L^2}{\pi^3 EI} \frac{1}{n^3} \cos \frac{n\pi c}{L}$$

Upon substitution of this for a_n in the series given by Eq. (10.23), the equation for the deflection curve is obtained in the form

$$v = \frac{2M_oL^2}{\pi^3 EI} \sum_{n=1}^{\infty} \frac{1}{n^3} \cos\frac{n\pi c}{L} \sin\frac{n\pi x}{L} \tag{10.26}$$

Through the use of this infinite series, the deflection for any given value of x can be calculated.

EXAMPLE 10.12   Cantilever Beam under an End Load

Derive an expression for the deflection of a cantilever beam of length L subjected to a concentrated force P at its free end (Fig. 10.14).

Solution   The origin of the coordinates is located at the fixed end. Let us represent the deflection by the infinite series

$$v = \sum_{n=1,3,5,\ldots}^{\infty} a_n \left(1 - \cos\frac{n\pi x}{2L}\right) \tag{10.27}$$

It is clear that Eq. (10.27) satisfies the conditions related to the slope and deflection at $x = 0$: $v = 0$, $dv/dx = 0$. The strain energy of the system is

$$U = \frac{EI}{2} \int_0^L \left(\frac{d^2v}{dx^2}\right)^2 dx = \frac{EI}{2} \int_0^L \left[\sum_{n=1,3,5,\ldots}^{\infty} a_n \left(\frac{n\pi}{2L}\right)^2 \cos\frac{n\pi x}{2L}\right]^2 dx$$

Squaring the bracketed term and noting that the orthogonality relationship yields

$$\int_0^L \cos\frac{m\pi x}{2L} \cos\frac{n\pi x}{2L}\, dx = \begin{cases} 0, & m \neq n \\ L/2, & m = n \end{cases} \tag{10.28}$$

we obtain

$$\delta U = \frac{\pi^4 EI}{32L^3} \sum_{n=1,3,5,\ldots}^{\infty} n^4 a_n \delta a_n \tag{c}$$

Application of the principle of virtual work gives $P\,\delta v_E = \delta U$. Thus,

$$P \sum_{n=1,3,5,\ldots}^{\infty} \left(1 - \cos\frac{n\pi L}{2L}\right)\delta a_n = \frac{EI\pi^4}{32L^3} \sum_{n=1,3,5,\ldots}^{\infty} n^4 a_n \delta a_n$$

from which $a_n = 32PL^3/n^4\pi^4 EI$. The beam deflection is obtained by substituting the value of a_n obtained into Eq. (10.27). At $x = L$, disregarding terms beyond the first three, we obtain $v_{\max} = PL^3/3.001EI$. The exact solution due to bending is $PL^3/3EI$ (Sec. 5.4).

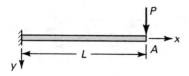

FIGURE 10.14.   *Example 10.12. Cantilever beam with a load P at its end.*

EXAMPLE 10.13   Beam on Elastic Foundation with a Concentrated Load

Determine the equation of the deflection curve of a beam supported at its ends and lying on an elastic foundation of modulus k, subjected to concentrated load P and axial tensile load N (Fig. 10.15).

Solution   Assume, as the deflection curve, a Fourier sine series, given by Eq. (10.23), which satisfies the end conditions. Denote the arc length of beam segment by ds (Fig. 10.15). To determine the work done by this load, we require the displacement $du = ds - dx$ experienced by the beam in the direction of axial load N:

$$u = \int_0^L (\sqrt{dx^2 + dv^2} - dx) = \int_0^L \left\{ \left[1 + \left(\frac{dv}{dx} \right)^2 \right]^{1/2} - 1 \right\} dx$$

Noting that for $(dv/dx)^2 \ll 1$,

$$\left[1 + \left(\frac{dv}{dx} \right)^2 \right]^{1/2} = \left[1 + \frac{1}{2} \left(\frac{dv}{dx} \right)^2 + \cdots \right]$$

we obtain

$$u \approx \frac{1}{2} \int_0^L \left(\frac{dv}{dx} \right)^2 dx \tag{10.29}$$

The total energy will be composed of four principal sources: beam bending, tension, transverse loading, and foundation displacement. Strain energy in bending in the beam, from Eq. (10.25), is

$$\delta U_1 = \frac{\pi^4 EI}{2L^3} \sum_{n=1}^{\infty} n^4 a_n \delta a_n \tag{c}$$

Strain energy owing to the deformation of the elastic foundation is determined as follows:

$$\delta U_2 = \frac{kL}{2} \sum_{n=1}^{\infty} a_n \delta a_n^2 \tag{10.30}$$

Work done against the axial load due to shortening u of the span, using Eq. (10.29), is

$$W_1 = N(-u) = -\frac{N}{2} \int_0^L \left(\frac{dv}{dx} \right)^2 \delta W_1 = \frac{\pi^2 N}{2L} \sum_{n=1}^{\infty} n^2 a_n \delta a_n \tag{10.31}$$

FIGURE 10.15.   *Example 10.13. A simply supported beam resting on an elastic foundation.*

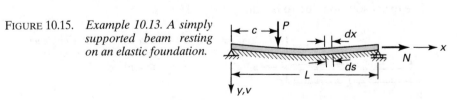

Work done by the load P is

$$\delta W_2 = P \cdot (\delta v)_{x=c} = P \sum_{n=1}^{\infty} \delta a_n \sin \frac{n\pi c}{L} \tag{10.32}$$

Application of the principle of virtual work, Eq. (10.20), leads to the following expression:

$$-\frac{\pi^2 N}{2L} n^2 a_n \delta a_n + P \sin \frac{n\pi c}{L} \delta a_n = \frac{n^4 EI}{2L^3} n^4 a_n \delta a_n + \frac{kL}{4} a_n \delta a_n^2$$

After determining a_n from the preceding and substituting in the series Eq. (10.23), we obtain the deflection curve of the beam:

$$v = 2PL^3 \sum_{n=1}^{\infty} \frac{\sin(n\pi c/L) \sin(n\pi x/L)}{n^4 \pi^4 EI + n^2 L^2 N + kL^4} \tag{10.33}$$

When the axial load is compressive, contrary to this example, we need to reverse the sign of N in Eq. (10.33). Clearly, the same relationship applies if there is no axial load present ($N = 0$) or when the foundation is absent ($k = 0$) and the beam is merely supported at its ends.

Consider, for instance, the case in which load P is *applied at the center* of the span of a simply supported beam. To calculate the deflection under the load, the values $x = c = L/2$, $N = 0$, and $k = 0$ are substituted into Eq. (10.33):

$$v_{\text{max}} = \frac{2PL^3}{\pi^4 EI} \left(1 + \frac{1}{3^4} + \frac{1}{5^4} + \cdots \right)$$

The series is rapidly converging, and the first few terms provide the deflection to a high degree of accuracy. Using only the first term of the series, we have

$$v_{\text{max}} = \frac{2PL^3}{\pi^4 EI} = \frac{PL^3}{48.7EI}$$

Comment Comparing this with the exact solution, we obtain 48.7 instead of 48 in the denominator of the expression. The error in using only the first term of the series is thus approximately 1.5%; the accuracy is sufficient for many practical purposes.

10.11 RAYLEIGH–RITZ METHOD

The *Rayleigh–Ritz method* offers a convenient procedure for obtaining solutions by the principle of minimum potential energy. This method was originated by Lord Rayleigh, and a generalization was contributed by W. Ritz. In this section, application of the method to the determination of the beam displacements is discussed. In Chapters 11 and 13, respectively, the procedure will also be applied to problems involving buckling of columns and bending of plates.

The essentials of the Rayleigh–Ritz method may be described as follows. First assume an expression for the *deflection curve*, also called the *displacement function* of the beam, in the form of a series containing unknown parameters $a_n(n = 1, 2, \ldots)$. This series function is such that it satisfies the *geometric* boundary conditions. These describe any end constraints pertaining to deflections and slopes. Another kind of condition, a *static* boundary condition, which equates the internal forces (and moments) at the edges of the member to prescribed external forces (and moments), need not be fulfilled. Next, using the assumed solution, determine the potential energy Π in terms of a_n. This indicates that the a_n's govern the variation of the potential energy. As the potential energy must be a minimum at equilibrium, the Rayleigh–Ritz method is stated as follows:

$$\frac{\partial \Pi}{\partial a_1} = 0, \ldots, \frac{\partial \Pi}{\partial a_n} = 0 \qquad \textbf{(10.34)}$$

This condition represents a set of algebraic equations that are solved to yield the parameters a_n. Substituting these values into the assumed function, we obtain the solution for a given problem. In general, only a finite number of parameters can be employed, and the solution found is thus only approximate. The accuracy of approximation depends on how closely the assumed deflection shape matches the exact shape. With some experience, the analyst will be able to select a satisfactory displacement function.

The method is illustrated in the solution of the following sample problem.

EXAMPLE 10.14 Analysis of a Simple Beam under Uniform Load
A simply supported beam of length L is subjected to uniform loading p per unit length. Determine the deflection $v(x)$ by employing (a) a power series and (b) a Fourier series.

Solution Let the origin of the coordinates be placed at the left support (Fig. 10.16).

a. Assume a solution of polynomial form:

$$v = a_1 x(L - x) + a_2 x^2(L - x)^2 + \cdots \qquad \textbf{(a)}$$

Note that this choice enables the deflection to vanish at either boundary. Consider now only the first term of the series:

$$v = a_1 x(L - x) \qquad \textbf{(b)}$$

FIGURE 10.16. *Example 10.14. A simple beam carries a load of intensity p.*

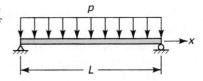

The corresponding potential energy, $\Pi = U - W$, is

$$\Pi = \int_0^L \left[\frac{EI}{2} \left(\frac{d^2 v}{dx^2} \right)^2 - pv \right] dx = \int_0^L \left[\frac{EI}{2} (-2a_1)^2 - a_1 px(L - x) \right] dx$$

From the minimizing condition, Eq. (10.34), we obtain $a_1 = pL^2/24EI$. The approximate displacement is therefore

$$v = \frac{pL^4}{24EI} \left(\frac{x}{L} - \frac{x^2}{L^2} \right)$$

which at midspan becomes $v_{max} = pL^4/96EI$. This result may be compared with the exact solution due to bending, $v_{max} = pL^4/76.8EI$, indicating an error in maximum deflection of roughly 17%. An improved approximation is obtained when two terms of the series given by Eq. (a) are retained. The same procedure now yields $a_1 = pL^2/24EI$ and $a_2 = p/24EI$, so that

$$v = \frac{pL^2}{24EI} [x(L - x)] + \frac{p}{24EI} [x^2(L - x)^2] \qquad \textbf{(10.35)}$$

At midspan, expression (10.35) provides the exact solution. The foregoing is laborious and not considered practical when compared with the approach given next.

b. Now suppose a solution of the form

$$v = \sum_{n=1}^{\infty} a_n \sin \frac{n\pi x}{L} \qquad \textbf{(c)}$$

The boundary conditions are satisfied inasmuch as v and v'' both vanish at either end of the beam. We now substitute v and its derivatives into $\Pi = U - W$. Employing Eq. (10.24), we obtain, after integration,

$$\Pi = \frac{EI}{2} \left(\frac{L}{2} \right) \sum_{n=1}^{\infty} a_n^2 \left(\frac{n\pi}{L} \right)^4 + p \left(\frac{L}{\pi} \right) \sum_{n=1}^{\infty} \frac{1}{n} a_n \cos \frac{n\pi x}{L} \Big|_0^L$$

Observe that if n is even, the second term vanishes. Thus,

$$\Pi = \frac{\pi^4 EI}{4L^3} \sum_{n=1}^{\infty} a_n^2 n^4 - \frac{2pL}{\pi} \sum_{n=1,3,5,\ldots}^{\infty} \frac{a_n}{n}$$

and Eq. (10.34) yields $a_n = 4pL^4/EI(n\pi)^5$, $n = 1, 3, 5, \ldots$. The deflection at midspan is, from Eq. (c),

$$v_{max} = \frac{4pL^4}{EI\pi^5} \left(1 - \frac{1}{3^5} + \frac{1}{5^5} - \cdots \right) \qquad \textbf{(10.36)}$$

Dropping all but the first term, $v_{max} = pL^4/76.5EI$. The exact solution is obtained when all terms in the series (c) are retained. Evaluation of *all* terms in the series may not always be possible, however.

Comment It should be noted that the results obtained in this example, based on only one or two terms of the series, are remarkably accurate. So few terms will not, in general, result in such accuracy when applying the Rayleigh–Ritz method.

REFERENCES

10.1. LANGHAAR, H. L., *Energy Methods in Applied Mechanics,* Krieger, Melbourne, Fla., 1989.

10.2. SOKOLNIKOFF, I. S., *Mathematical Theory of Elasticity,* 2nd ed., Krieger, Melbourne, Fla., 1986, Chap. 7.

10.3. ODEN, J. T., and RIPPERGER, E. A., *Mechanics of Elastic Structures*, 2nd ed., McGraw-Hill, New York, 1951.

10.4. UGURAL, A. C., *Stresses in Beams, Plates, and Shells*, 3rd ed., CRC Press, Taylor and Francis Group, Boca Raton, Fla., 2010.

10.5. FLUGGE, W., ed., *Handbook of Engineering Mechanics,* McGraw-Hill, New York, 1968, Sec. 31.5.

10.6. FAUPPEL, J. H., and FISHER, F. E., *Engineering Design*, 2nd ed., Wiley, Hoboken, N.J., 1986.

10.7. UGURAL, A. C., *Mechanics of Materials*, Wiley, Hoboken, N.J., 2008.

10.8. BURR, A. H., and CHEATHAM, J. B., *Mechanical Design*, 2nd ed., Prentice Hall, Upper Saddle River, N.J., 1995.

10.9. TIMOSHENKO, S. P., and GERE, J. M., *Theory of Elastic Stability*, 3rd ed., McGraw-Hill, New York, 1970, Sec. 1.11.

10.10. SOKOLNIKOFF, I. S., and REDHEFFER, R. M., *Mathematics of Physics and Modern Engineering*, 2nd ed., McGraw-Hill, New York, 1966, Chap. l.

PROBLEMS

Sections 10.1 through 10.7

10.1. A cantilever beam of constant AE and EI is loaded as shown in Fig. P10.1. Determine the vertical and horizontal deflections and the angular rotation of the free end, considering the effects of normal force and bending moment. Employ Castigliano's theorem.

10.2. The truss shown in Fig. P10.2 supports concentrated forces of $P_1 = P_2 = P_3 = 45$ kN. Assuming all members are of the same cross section and material, find the vertical deflection of point B in terms of AE. Take $L = 3$ m. Use Castigliano's theorem.

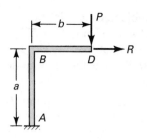

FIGURE P10.1.

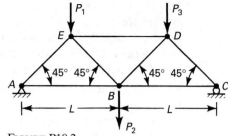

FIGURE P10.2.

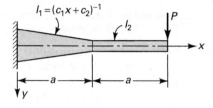

FIGURE P10.3.

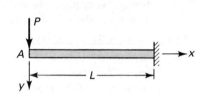

FIGURE P10.4.

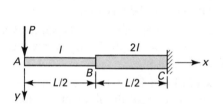

FIGURE P10.6.

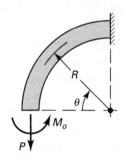

FIGURE P10.7.

10.3. The moments of inertia of the tapered and constant area segments of the cantilever beam shown in Fig. P10.3 are given by $I_1 = (c_1 x + c_2)^{-1}$ and I_2, respectively. Determine the deflection of the beam under a load P. Use Castigliano's theorem.

10.4. Determine the vertical deflection and slope at point A of the cantilever loaded as shown in Fig. P10.4. Use Castigliano's theorem.

10.5. Redo Prob. 10.4 for the case in which an additional uniformly distributed load of intensity p is applied to the beam.

10.6. Rework Prob. 10.4 for the stepped cantilever shown in Fig. P10.6.

10.7. A square slender bar in the form of a quarter ring of radius R is fixed at one end (Fig. P10.7). At the free end, force P and moment M_o are applied, both in the plane of the bar. Using Castigliano's theorem, determine (a) the horizontal and the vertical displacements of the free end and (b) the rotation of the free end.

10.8. The curved steel bar shown in Fig. 10.6 is subjected to a rightward horizontal force F of 4 kN at its free end and $P = 0$. Using $E = 200\,\text{GPa}$ and $G = 80\,\text{GPa}$, compute the horizontal deflection of the free end taking into account the effects of V and N, as well as M. What is the error in deflection if the contributions of V and N are omitted?

10.9. It is required to determine the horizontal deflection of point D of the frame shown in Fig. P10.9, subject to downward load F, applied at the top. The moment of inertia of segment BC is twice that of the remaining sections. Use Castigliano's theorem.

10.10. A steel spring of constant flexural rigidity is described by Fig. P10.10. If a force P is applied, determine the increase in the distance between the ends. Use Castigliano's theorem.

10.11. A cylindrical circular rod in the form of a quarter-ring of radius R is fixed at one end (Fig. P10.11). At the free end, a concentrated force P is applied in a diametral plane perpendicular to the plane of the ring. What is the deflection of the free end? Use the unit load method. [Hint: At any section, $M_\theta = PR \sin \theta$ and $T_\theta = PR(1 - \cos \theta)$.]

10.12. Redo Prob. 10.11 if the curved bar is a *split circular ring*, as shown in Fig. P10.12.

10.13. For the cantilever beam loaded as shown in Fig. P10.13, using Castigliano's theorem, find the vertical deflection δ_A of the free end.

10.14. A five-member plane truss in which all members have the same axial rigidity AE is loaded as shown in Fig. P10.14. Apply Castigliano's theorem to obtain the vertical displacement of point D.

10.15. For the beam and loading shown in Fig. P10.15, use Castigliano's theorem to determine (a) the deflection at point C; (b) the slope at point C.

10.16. and 10.17. A beam is loaded and supported as illustrated in Figs. P10.16 and P10.17. Apply Castigliano's theorem to find the deflection at point C.

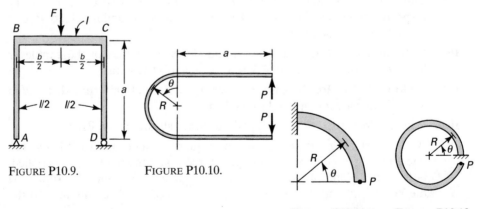

FIGURE P10.9. FIGURE P10.10.

FIGURE P10.11. FIGURE P10.12.

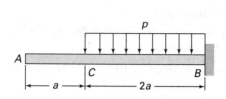

FIGURE P10.13.

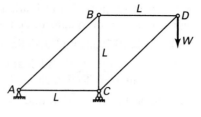

FIGURE P10.14.

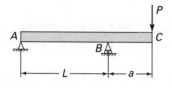

FIGURE P10.15.

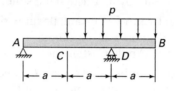

FIGURE P10.16.

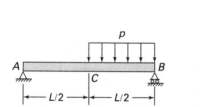

FIGURE P10.17.

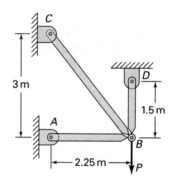

FIGURE P10.18.

10.18. A load P is carried at joint B of a structure consisting of three bars of equal axial rigidity AE, as shown in Fig. P10.18. Apply Castigliano's theorem to determine the force in each bar.

10.19. Applying the unit-load method, determine the support reactions R_A and M_A for the beam loaded and supported as shown in Fig. P10.19.

10.20. A propped cantilever beam AB is supported at one end by a spring of constant stiffness k and subjected to a uniform load of intensity p, as shown in Fig. P10.20. Use the unit-load method to find the reaction at A.

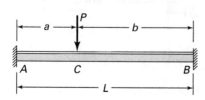

FIGURE P10.19.

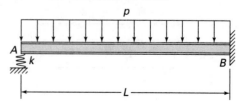

FIGURE P10.20.

10.21. A beam is supported and loaded as illustrated in Figs. P10.21. Use Castigliano's theorem to determine the reactions.

10.22. Redo Prob. 10.19, using Castigliano's theorem.

10.23. A beam is supported and loaded as shown in Fig. P10.23. Apply Castigliano's theorem to determine the reactions.

10.24. Determine the deflection and slope at midspan of the beam described in Example 10.8.

10.25. A circular shaft is fixed at both ends and subjected to a torque T applied at point C, as shown in Fig. P10.25. Determine the reactions at supports A and B, employing Castigliano's theorem.

10.26. A steel rod of constant flexural rigidity is described by Fig. P10.26. For force P applied at the simply supported end, derive a formula for roller reaction Q. Apply Castigliano's theorem.

10.27. A cantilever beam of length L subject to a linearly varying loading per unit length, having the value zero at the free end and p_o at the fixed end, is supported on a roller at its free end (Fig. P10.27). Find the reactions using Castigliano's theorem.

10.28. Using Castigliano's theorem, find the slope of the deflection curve at midlength C of a beam due to applied couple moment M_o (Fig. P10.28).

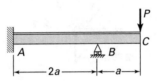

FIGURE P10.21.

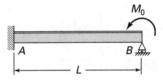

FIGURE P10.23.

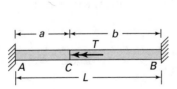

FIGURE P10.25.

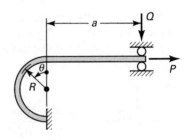

FIGURE P10.26.

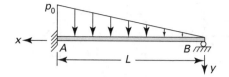

FIGURE P10.27.

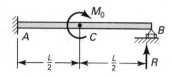

FIGURE P10.28.

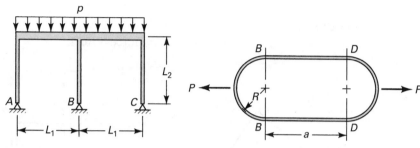

FIGURE P10.29.   FIGURE P10.30.

10.29. The symmetrical frame shown in Fig. P10.29 supports a uniform loading of p per unit length. Assume that each horizontal and vertical member has the modulus of rigidity E_1I_1 and E_2I_2, respectively. Determine the resultant reaction R_A at the left support, employing Castigliano's theorem.

10.30. Forces P are applied to a *compound loop* or *link* of constant flexural rigidity EI (Fig. P10.30). Assuming that the dimension perpendicular to the plane of the page is small in comparison with radius R and taking into account only the strain energy due to bending, determine the maximum moment.

10.31. A planar truss shown in Fig. P10.31 carries two vertical loads P acting at points B and E. Use Castigliano's theorem to obtain (a) the horizontal displacement of joint E; (b) the rotation of member DE.

10.32. Calculate the vertical displacement of joint E of the truss depicted in Fig. P10.32. Each member is made of a nonlinearly elastic material having the stress–strain relation $\sigma = K\varepsilon^{1/3}$ and the cross-sectional area A. Apply the Crotti–Engesser theorem.

10.33. A basic truss supports a load P, as shown in Fig. P10.33. Determine the horizontal displacement of joint C, applying the unit-load method.

10.34. A frame of constant flexural rigidity EI carries a concentrated load P at point E (Fig. P10.34). Determine (a) the reaction R at support A, using Castigliano's theorem; (b) the horizontal displacement δ_h at support A, using the unit-load method.

10.35. A bent bar ABC with fixed and roller supported ends is subjected to a bending moment M_o, as shown in Fig. P10.35. Obtain the reaction R at the roller. Apply the unit-load method.

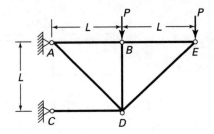

FIGURE P10.31.

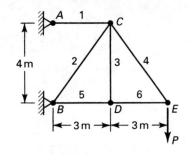

FIGURE P10.32.

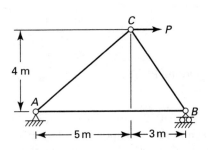

FIGURE P10.33.

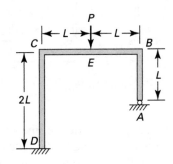

FIGURE P10.34.

10.36. A planar curved frame having rectangular cross section and mean radius R is fixed at one end and supports a load P at the free end (Fig. P10.36). Employ the unit-load method to determine the vertical component of the deflection at point B, taking into account the effects of normal force, shear, and bending.

10.37. A large ring is loaded as shown in Fig. P10.37. Taking into account only the strain energy associated with bending, determine the bending moment and the force within the ring at the point of application of P. Employ the unit-load method.

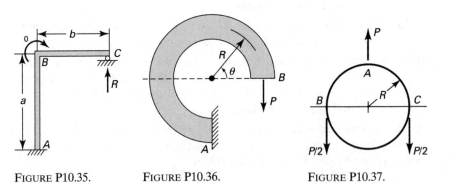

FIGURE P10.35.   FIGURE P10.36.   FIGURE P10.37.

Sections 10.8 through 10.11

10.38. Redo Problem 10.25, employing Castigliano's first theorem. [Hint: Strain energy is expressed by

$$U = \frac{1}{2}\int GJ\left(\frac{d\theta}{dx}\right)^2 dx \qquad \text{(P10.38)}$$

where θ is the angle of twist and GJ represents the torsional rigidity.]

10.39. Apply Castigliano's first theorem to compute the force P required to cause a vertical displacement of 5 mm in the hinge-connected structure of Fig. P10.39. Let $\alpha = 45°$, $L_o = 3$ m, and $E = 200$ GPa. The area of each member is 6.25×10^{-4} m².

10.40. A hinge-ended beam of length L rests on an elastic foundation and is subjected to a uniformly distributed load of intensity p. Derive the equation of the deflection curve by applying the principle of virtual work.

10.41. Determine the deflection of the free end of the cantilever beam loaded as shown in Fig. 10.14. Assume that the deflection shape of the beam takes the form

$$v = \frac{a_1 x^2}{2L^3}(3L - x) \qquad \text{(P10.41)}$$

where a_1 is an unknown coefficient. Use the principle of virtual work.

10.42. A cantilever beam carries a uniform load of intensity p (Fig. P10.42). Take the displacement v in the form

$$v = a\left(1 - \cos\frac{\pi x}{2L}\right) \qquad \text{(P10.42)}$$

where a is an unknown constant. Apply the Rayleigh–Ritz method to find the deflection at the free end.

10.43. Determine the equation of the deflection curve of the cantilever beam loaded as shown in Fig. 10.14. Use as the deflection shape of the loaded beam

$$v = a_1 x^2 + a_2 x^3 \qquad \text{(P10.43)}$$

where a_1 and a_2 are constants. Apply the Rayleigh–Ritz method.

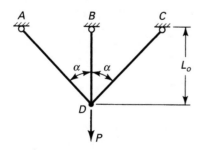

FIGURE P10.39.

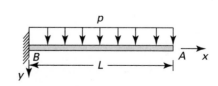

FIGURE P10.42.

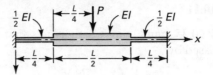

FIGURE P10.45.

10.44. A simply supported beam carries a load P at a distance c away from its left end. Obtain the beam deflection at the point where P is applied. Use the Rayleigh–Ritz method. Assume a deflection curve of the form $v = ax(L - x)$, where a is to be determined.

10.45. Determine the midspan deflection for the fixed-ended symmetrical beam of stepped section shown in Fig. P10.45. Take $v = a_1x^3 + a_2x^2 + a_3x + a_4$. Employ the Rayleigh–Ritz method.

10.46. Redo Prob. 10.45 for the case in which the beam is *uniform* and of flexural rigidity EI. Use

$$v = \sum_{n=1}^{\infty} a_n\left(1 - \cos\frac{n\pi x}{L}\right) \qquad \textbf{(P10.46)}$$

where the a_n's are unknown coefficients.

CHAPTER 11

Stability of Columns

11.1 INTRODUCTION

We have up to now dealt primarily with the prediction of stress and deformation in structural elements subject to various load configurations. Failure criteria have been based on a number of theories relying on the attainment of a particular stress, strain, or energy level within the body. This chapter demonstrates that the beginnings of structural failure may occur prior to the onset of any seriously high levels of stress. We thus explore failure owing to *buckling* or *elastic instability* and seek to determine those conditions of load and geometry that lead to a compromise of structural integrity.

We are concerned only with beams and slender members subject to axial compression. The problem is essentially one of ascertaining those configurations of the system that lead to sustainable patterns of deformation. The principal difference between the theories of linear elasticity and linear stability is that, in the former, equilibrium is based on the undeformed geometry, whereas in the latter the deformed geometry must be considered. Our treatment primarily concerns ideal columns. We define an *ideal column* as a perfectly constructed, perfectly straight column that is built of material that follows Hooke's law and that is under centric loading. Practical formulas and graphs are presented to facilitate improved understanding of the behavior of both elastic and inelastic actual columns.

11.2 CRITICAL LOAD

To demonstrate the concepts of stability and critical load, consider the *idealized structure*: a rigid, weightless bar AB, shown in Fig. 11.1a. This member is pinned at B and acted on by a force P. In the *absence* of restoring influences such as the

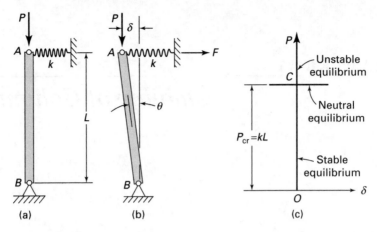

FIGURE 11.1. *Buckling behavior of a rigid bar: (a) vertical configuration; (b) displacement configuration; (c) equilibrium diagram.*

spring shown, any small lateral disturbance (causing a displacement δ) will result in rotation of the bar, as illustrated in Fig. 11.1b, with no possibility of return to the original configuration. *Without* the spring, therefore, the situation depicted in the figure is one of *unstable equilibrium.* With a spring present, different possibilities arise.

Equilibrium Method

A small *momentary* disturbance δ can now be sustained by the system (provided that P is also small) because the disturbing moment $P\delta$ is smaller than the restoring moment $FL = k\delta L$ (where k represents the linear spring constant, force per unit of deformation). For a small enough value of P, the moment $k\delta L$ will thus be sufficient to return the bar to $\delta = 0$. Since the system reacts to a small disturbance by creating a counterbalancing effect acting to diminish the disturbance, the configuration is in *stable equilibrium.*

If now the load is increased to the point where

$$P\delta = kL\,\delta \qquad \textbf{(a)}$$

it is clear that any small disturbance δ will be neither diminished nor amplified. The system is now in *neutral equilibrium* at any small value of δ. Expression (a) defines the *critical load*:

$$P_{cr} = kL \qquad \textbf{(b)}$$

If $P > P_{cr}$, the net moment acting will be such as to increase δ, tending to further increase the disturbing moment $P\delta$, and so on. For $P > P_{cr}$, the system is in *unstable equilibrium* because *any* lateral disturbance will be amplified, as in the springless case discussed earlier.

The equilibrium regimes are shown in Fig. 11.1c. Note that C, termed the *bifurcation point*, marks the two branches of the equilibrium solution. One is the vertical branch ($P \leq P_{cr}$, $\delta = 0$), and the other is the horizontal ($P = P_{cr}$, $\delta > 0$).

Energy Method

Stability may also be interpreted in terms of energy concepts, however. Referring again to Fig. 11.1b, the work done by P as it acts through a distance $L(1 - \cos \theta)$ is

$$\Delta W = PL(1 - \cos \theta)$$

$$= PL\left(1 - 1 + \frac{\theta^2}{2} - \cdots\right) \approx \frac{PL\,\theta^2}{2}$$

The elastic energy acquired as a result of the corresponding spring elongation $L\theta$ is

$$\Delta U = \tfrac{1}{2}k(L\theta)^2$$

If $\Delta U > \Delta W$, the configuration is *stable*; the work done is insufficient to result in a displacement that grows subsequent to a lateral disturbance. However, if $\Delta W > \Delta U$, the system is *unstable* because now the work is large enough to cause the displacement to grow following a disturbance. The boundary between stable and unstable configurations corresponds to $\Delta W = \Delta U$,

$$P_{\mathrm{cr}} = kL$$

as before.

The buckling analysis of compression members usually follows in essentially the same manner. Either the static equilibrium or the energy approach may be used for determination of the critical load. The choice depends on the particulars of the situation under analysis. Although the static equilibrium method leads to exact solutions, the results offered by the energy approach (sometimes approximate) are often preferable because of the physical insights that may be more readily gained.

11.3 BUCKLING OF PINNED-END COLUMNS

Consideration is now given to a relatively slender straight bar subject to axial compression. This member, a *column*, is similar to the element shown in Fig. 11.1a, in that it too can experience unstable behavior. In the case of a column, the restoring force or moment is provided by elastic forces established within the member rather than by springs external to it.

Refer to Fig. 11.2a, in which is shown a straight, homogeneous, pin-ended column. For such a column, the end moments are zero. This is regarded as the fundamental or most basic case. It is reasonable to suppose that the column can be held in a deformed configuration by a load P while remaining in the elastic range (Fig. 11.2b). Note that the requisite axial motion is permitted by the movable end support. In Fig. 11.2c, the postulated deflection is shown, having been caused by collinear forces P acting at the centroid of the cross section. The bending moment at any section, $M = -Pv$, when inserted into the equation for the elastic behavior of a beam, $EIv'' = M$, yields

$$EI\,\frac{d^2v}{dx^2} + Pv = 0 \tag{11.1}$$

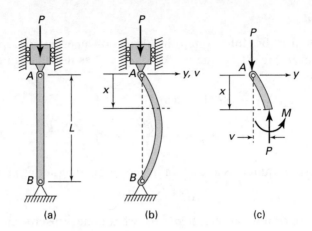

FIGURE 11.2. *Pinned-end column: (a) initial form; (b) buckled form; (c) free-body diagram of column segment.*

The solution of this differential equation is

$$v = c_1 \sin \sqrt{\frac{P}{EI}}\, x + c_2 \cos \sqrt{\frac{P}{EI}}\, x \qquad \textbf{(a)}$$

where the constants of integration, c_1 and c_2, are determined from the end conditions: $v(0) = v(L) = 0$. From $v(0) = 0$, we find that $c_2 = 0$. Substituting the second condition into Eq. (a), we obtain

$$c_1 \sin \sqrt{\frac{P}{EI}}\, L = 0 \qquad \textbf{(b)}$$

It must be concluded that either $c_1 = 0$, in which case $v = 0$ for all x and the column remains straight regardless of load, or $\sin \sqrt{P/EI}\, L = 0$. The case of $c_1 = 0$ corresponds to a condition of no buckling and yields a trivial solution [the energy approach (Sec. 11.10) sheds further light on this case]. The latter is the acceptable alternative because it is consistent with column deflection. It is satisfied if

$$\sqrt{\frac{P}{EI}}\, L = n\pi, \qquad n = 1, 2, \dots \qquad \textbf{(c)}$$

The value of P ascertained from Eq. (c), that is, the load for which the column may be maintained in a deflected shape, is the *critical load*,

$$(P_{cr})_n = \frac{n^2 \pi^2 EI}{L^2} \qquad \textbf{(11.2)}$$

where L represents the original length of the column. Assuming that column deflection is in no way restricted to a particular plane, the deflection may be expected to occur about an axis through the centroid for which the second moment of area is a minimum. The lowest *critical load* or *Euler buckling load* of the pinended column is of greatest interest; for $n = 1$,

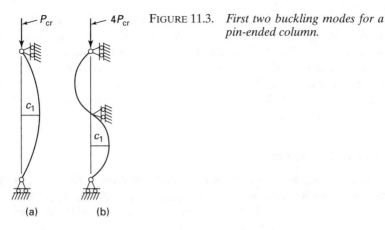

FIGURE 11.3. *First two buckling modes for a pin-ended column.*

$$P_{cr} = \frac{\pi^2 EI}{L^2} \tag{11.3}$$

The preceding, after L. Euler (1707–1783), represents the maximum theoretical load that an *idealized* column can expect to resist. Recall that a centrically loaded column in which the deflections are small, the construction is perfect, and the material follows Hooke's law is called an *ideal column*.

The deflection is found by combining Eqs. (a) and (c) and inserting the values of c_1 and c_2:

$$v = c_1 \sin\frac{n\pi x}{L} \tag{11.4}$$

in which c_1 represents the amplitude of the elastic curve. It is noted that a slender pin-ended elastic column has more than one critical load, as indicated in Eq. (11.2). When the column is restrained from buckling into a single lobe by a lateral support, it may buckle at a load higher than P_{cr}. A column with pinned ends can buckle in the shape of a single sine lobe ($n = 1$) at the critical load P_{cr}. But if it is prevented from bending in the form of one lobe by one lateral support, the load can increase until it buckles into two lobes ($n = 2$). Applying Eq. (11.2), for $n = 2$, the critical load is $(P_{cr})_2 = 4(\pi^2 EI/L) = 4P_{cr}$. Figure 11.3 illustrates the buckling modes for $n = 1$ and 2. The higher buckling modes ($n = 3, \dots$) are sketched similarly. Observe that, as n increases, the number of nodal points and critical buckling load also increase. However, usually the lowest buckling mode is significant.

We conclude this section by recalling that the boundary conditions employed in the solution of the differential equation led to an infinite set of discrete values of load, $(P_{cr})_n$. These solutions, typical of many engineering problems, are termed *eigenvalues*, and the corresponding deflections v are the *eigenfunctions*.

11.4 DEFLECTION RESPONSE OF COLUMNS

The conclusions of the preceding section are predicated on the linearized beam theory with which the analysis began. Recall from Section 5.2 that, in Eq. (11.1), the term d^2v/dx^2 is actually an approximation to the curvature. Inasmuch as c_1 in

Eq. (11.4) is undefined (and independent of P_{cr}), the critical load and deflection are independent, and P_{cr} will sustain any small lateral deflection, a condition represented by the horizontal line AB in Fig. 11.4. It is clear that, in this figure, we show only the right-hand half of the diagram; however, the two halves are symmetric about the vertical axis. Note that the effects of temperature and time on buckling are not considered in the following discussion. Essentially, these effects may be very significant and often act to lower the critical load.

Effects of Large Deflections

Were the exact curvature used, the differential equation derived would apply to large deformations within the elastic range, and the result would be less restricted. For this case, it is found that P depends on the magnitude of the deflection or c_1. Exact or large deflection analysis also reveals values of P exceeding P_{cr}, as shown with the curve AC in Fig. 11.4. That is, after an elastic column begins to buckle, a larger and larger load is required to cause an increase in the deflections. However, the bending stresses accompanying large deflection could carry the material into inelastic regime, thus leading to diminishing buckling loads [Ref. 11.1]. So, the load-deflection curve AC drops (as indicated by the dashed lines), and the column fails either by excessive yielding or fracture. Thus, in most applications, P_{cr} is usually regarded as the maximum load sustainable by a column.

Effects of Imperfections

The column that poses deviations from ideal conditions assumed in the preceding is called an *imperfect column*. Such a column is not constructed perfectly; for example, it could have an imperfection in the form of a misalignment of load or a small initial curvature or crookedness (see Secs. 11.8 and 11.9). An imperfection produces deflections from the start of loading, as depicted by curve OD in Fig. 11.4. For the

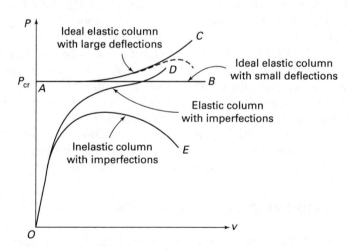

FIGURE 11.4. *Relation between load and deflection for columns.*

case in which deflections are small, curve OD approaches line AB; as the deflections become large, it approaches curve AC. The larger the imperfections, the further curve OD moves to the right. For the case in which the column is constructed with great accuracy, curve OD approaches more closely to the straight lines OA and AB. Observe from these lines and curves AC and OD that P_{cr} represents the maximum load-carrying capacity of an elastic column for practical purposes, because large deflections generally are not permitted to occur in structures.

Effects of Inelastic Behavior

We now consider the situation where the stresses exceed the proportional limit and the column material no longer follows Hooke's law. Obviously, the load-deflection curve is unchanged up to the level of load at which the proportional limit is reached. Then the curve for inelastic behavior (curve OE) deviates from the elastic curve, continues upward, reaches a maximum, and turns downward, as seen in Fig. 11.4. The specific forms of these curves depend on the material properties and column dimensions; however, the usual nature of the behavior is exemplified by the curves depicted. Empirical methods are often used in conjunction with analysis to develop an efficient design criteria.

Very *slender* columns alone remain elastic up to the critical load P_{cr}. If the load is slightly eccentric, a slender column will undergo lateral deflection as soon as load is applied. For *intermediate* columns, instability and inelastic collapse take place at relatively small lateral deflections. *Short* columns behave inelastically and follow a curve such as OD. Observe that the maximum load P that can be carried by an inelastic column may be considerably less than the critical load P_{cr}. A final point to note is that the descending part of curve OE leads to a plastic collapse or fracture, because it takes smaller and smaller loads to maintain larger and larger deflections. On the contrary, the curves for elastic columns are quite stable, since they continue upward as the deflections increase. Thus, it takes larger and larger loads to produce an increase in deflection.

11.5 COLUMNS WITH DIFFERENT END CONDITIONS

It is evident from the foregoing derivation that P_{cr} depends on the end conditions of the column. For other than the pin-ended, fundamental case discussed, we need only substitute the appropriate conditions into Eq. (a) of Section 11.3 and proceed as before.

Consider an alternative approach, beginning with the following revised form of the Euler buckling formula for a pin-ended column and applicable to a variety of end conditions:

$$P_{cr} = \frac{\pi^2 EI}{L_e^2} \tag{11.5}$$

Here L_e denotes the *effective* column length, which for the *pin-ended* column is the column length L. The effective length, shown in Fig. 11.5 for several end conditions, is determined by noting the length of a segment corresponding to a pin-ended column. In so doing we seek the distance between points of inflection on the elastic curve or the distance between hinges, if any exist.

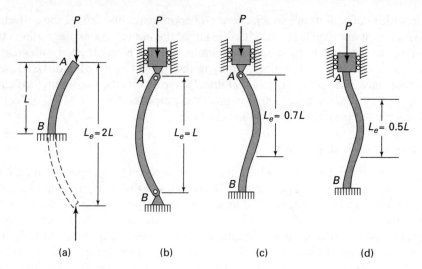

FIGURE 11.5. *Effective lengths of columns with different restraints: (a) fixed-free; (b) pinned-pinned; (c) fixed-pinned; (d) fixed-fixed.*

Regardless of end condition, it is observed that the critical load depends not on material strength but on the *flexural rigidity, EI.* Buckling resistance can thus be enhanced by deploying material so as to maximize the moment of inertia, but not to the point where the section thickness is so small as to result in local buckling or wrinkling. Setting $I = Ar^2$, the preceding equation may be written as

$$P_{cr} = \frac{\pi^2 EA}{(L_e/r)^2} \tag{11.6}$$

where A is the cross-sectional area and r is the *radius of gyration.* The quantity L_e/r, called the *effective slenderness ratio*, is an important parameter in the classification of compression members.

EXAMPLE 11.1 Load-Carrying Capacity of a Wood Column

A pinned-end wood bar of width b by depth h rectangular cross section (Fig. 11.6) and length L is subjected to an axial compressive load. Determine (a) the slenderness ratio; (b) the allowable load, using a factor safety of n. Data: $b = 60$ mm, $h = 120$ mm, $L = 1.8$ m, $n = 1.4$, $E = 12$ GPa, and $\sigma_u = 55$ MPa (by Table D.1).

Solution The properties of the cross-sectional area are $A = bh$, $I_x = hb^3/12$, $I_y = hb^3/12$, and $r = \sqrt{I/A}$.

a. *Slenderness ratio:* The smallest value of r is found when the centroidal axis is parallel to the longer side of the rectangle. Therefore,

$$r_y = \sqrt{\frac{I_y}{A}} = \frac{b}{\sqrt{12}} \tag{a}$$

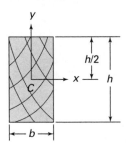

FIGURE 11.6. *Example 11.1. Cross section of wooden column.*

Substituting the given numerical value, we obtain

$$r_y = \frac{60}{\sqrt{12}} = 17.32 \text{ mm}$$

Since $L = L_e$, it follows that

$$\frac{L}{r_y} = \frac{1800}{17.32} = 103.9$$

b. *Permissible load:* Applying Eq. (11.6), the Euler buckling load with $r_y = r$ and $A = 60 \times 120 = 7.2 \times 10^3 \text{ mm}^2$ is

$$P_{cr} = \frac{\pi^2 E A}{(L/r)^2} = \frac{\pi^2(12 \times 10^9)(7.2 \times 10^{-3})}{(103.9)^2} = 79 \text{ kN}$$

Comment The largest load the column can support equals, then, $P = 79/1.4 = 56.4 \text{ kN}$. Observe that, based on material strength, the allowable load equals

$$P_{all} = \frac{\sigma_u A}{n} = \frac{(55)(7.2 \times 10^3)}{(103.9)^2} = 283 \text{ kN}$$

This, compared with 56.4 kN, shows the importance of buckling analysis in predicting the safe working load.

11.6 CRITICAL STRESS: CLASSIFICATION OF COLUMNS

The behavior of an ideal column is often represented on a plot of average compressive stress P/A versus slenderness ratio L_e/r (Fig. 11.7). Such a representation offers a clear rationale for the classification of compression bars. Tests of columns verify each portion of the curve with reasonable accuracy. The range of L_e/r is a function of the material under consideration. We note that the regions AB, BC, and CD represent the strength limit in compression, inelastic buckling limit, and elastic buckling limit, respectively.

Long Columns

For a member of sufficiently large slenderness ratio or *long column*, buckling occurs elastically at a stress that does *not* exceed the proportional limit of the material.

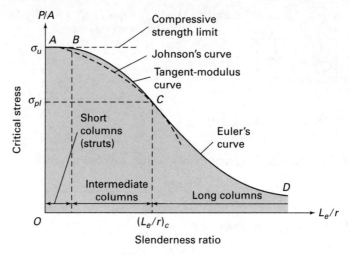

FIGURE 11.7. *Critical stress versus slenderness ratio.*

Consequently, the Euler's load of Eq. (11.6) is appropriate to this case. The *critical stress* is therefore

$$\sigma_{cr} = \frac{P_{cr}}{A} = \frac{\pi^2 E}{(L_e/r)^2} \tag{11.7}$$

The associated portion *CD* of the curve (Fig. 11.7) is labeled as *Euler's curve.* The *critical value* of slenderness ratio that fixes the lower limit of this curve, found by equating σ_{cr} to the proportional limit (σ_{pl}) of the specific material, is given by

$$\left(\frac{L_e}{r}\right)_c = \pi \sqrt{\frac{E}{\sigma_{pl}}} \tag{a}$$

In the case of a *structural steel* with modulus of elasticity $E = 210$ GPa and yield strength $\sigma_{yp} \approx \sigma_{pl} = 250$ MPa (Table D.1), for instance, the foregoing gives $(L_e/r)_c = 91$.

It is seen from Fig. 11.7 that very slender columns buckle at low levels of stress; they are much less stable than short columns. Use of a higher strength material does not improve this situation. Interestingly, Eq. (11.7) shows that the critical stress is increased by using a material of higher modulus elasticity E or by increasing the radius of gyration r.

Short Columns

Compression members having low slenderness ratios (for example, steel rods with $L/r < 30$) exhibit essentially no instability and are referred to as *short columns* or *struts*. For these bars, failure occurs by crushing, without buckling, at stresses exceeding the proportional limit of the material. The maximum stress is thus

$$\sigma_{max} = \frac{P}{A} \tag{11.8}$$

This represents the *strength limit* of a short column, represented by horizontal line AB in Fig. 11.7. Clearly, it is equal to the ultimate stress in compression.

Intermediate Columns: Inelastic Buckling

Most structural columns lie in a region between short and long classifications represented by the part BC in Fig. 11.7. Such *intermediate-length columns** do not fail by direct compression or by elastic instability. Hence, Eqs. (11.7) and (11.8) do not apply, and a separate analysis is required. The failure of an intermediate column occurs by *inelastic buckling* at stress levels exceeding the proportional limit. Over the years, numerous empirical formulas have been developed for the intermediate slenderness ratios. Presented here are two practical methods for finding critical stress in inelastic buckling.

Tangent Modulus Theory   Consider the concentric compression of an intermediate column, and imagine the loading to occur in small increments until such time as the *buckling load* P_t is achieved. As we would expect, the column does not remain perfectly straight but displays slight curvature as the increments in load are imposed. It is fundamental to the tangent modulus theory to assume that accompanying the increasing loads and curvature is a *continuous increase, or no decrease*, in the longitudinal stress and strain in *every* fiber of the column. That this should happen is not at all obvious, for it is reasonable to suppose that fibers on the convex side of the member might *elongate*, thereby reducing the stress. Accepting the former assumption, the stress distribution is as shown in Fig 11.8a. The increment of stress, $\Delta\sigma$, is attributable to bending effects; σ_{cr} is the value of stress associated with the attainment of the critical load P_t. The distribution of strain will display a pattern similar to that shown in Fig. 11.8b.

For small deformations Δv, the increments of stress and strain will likewise be small, and as in the case of elastic bending, sections originally plane are assumed to *remain plane* subsequent to bending. The change of stress $\Delta\sigma$ is thus assumed proportional to the increment of strain, $\Delta\varepsilon$; that is, $\Delta\sigma = E_t\,\Delta\varepsilon$. The constant of proportionality E_t is the *slope* of the stress–strain diagram beyond the proportional limit, termed the *tangent modulus* (Fig. 11.8b). Note that within the linearly elastic range, $E_t = E$. The stress–strain relationship beyond the proportional limit during the change in strain from ε to $\varepsilon + \Delta\varepsilon$ is thus assumed linear, as in the case of elastic buckling. The critical or *Engesser stress* may, on the basis of the foregoing rationale, be expressed by means of a modification of Eq. (11.6) in which E_t replaces E. Hence the critical stress may be expressed by the generalized Euler buckling formula, or the *tangent modulus formula*:

$$\sigma_{cr} = \frac{P_t}{A} = \frac{\pi^2 E_t}{(L_e/r)^2} \tag{11.9}$$

---

*The range of L_e/r depends on the material under consideration. In the case of structural steel, for example, long columns are those for which about $L_e/r > 100$; for intermediate columns, $30 < L_e/r < 100$, and for short struts, $L_e/r < 30$.

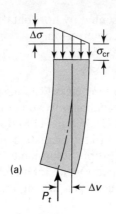

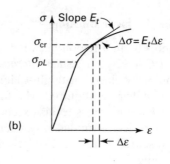

(a)

(b)

FIGURE 11.8.   (a) Stress distribution for tangent-modulus load;
(b) stress–strain diagram.

This is the so-called *Engesser formula*, proposed by F. Engesser in 1889. The buckling load P_t predicted by Eq. (11.9) is in good agreement with test results and therefore recommended for design purposes [Ref. 11.2].

When the critical stress is known and the slenderness ratio required, the application of Eq. (11.9) is straightforward. The value of E_t corresponding to σ_{cr} is read from the stress–strain curve obtained from a simple compression test, following which L_e/r is calculated using Eq. (11.9). If, however, L_e/r is known and σ_{cr} is to be ascertained, a trial-and-error approach is necessary (see Prob. 11.20).

Johnson's Buckling Criterion   Many formulas are based on the use of linear or parabolic relationship between the slenderness ratio and the critical stress for intermediate columns. The most widely employed modification of the parabola, proposed by J. B. Johnson around 1900, is the *Johnson formula*:

$$\sigma_{cr} = \sigma_{yp} - \frac{1}{E}\left(\frac{\sigma_{yp}}{2\pi}\frac{L_e}{r}\right)^2 \tag{11.10a}$$

In terms of the critical load, we write

$$P_{cr} = \sigma_{cr}A = \sigma_{yp}A\left[1 - \frac{\sigma_{yp}(L_e/r)^2}{4\pi^2 E}\right] \tag{11.10b}$$

where A is the cross-sectional area of the column. The Johnson's equation has been found to agree reasonably well with experimental results. But the dimensionless form of tangent modulus curves have very distinct advantages when structures of new materials are analyzed [Refs. 11.3 and 11.4].

Specific Johnson formula for regular cross-sectional shapes may be developed by substituting pertinent values of radius of gyration r and area A into Eq. (11.10). It can be readily shown that (see Prob. 11.8), in case of a solid *circular section* of diameter d:

$$d = 2\left(\frac{P_{cr}}{\pi\sigma_{yp}} + \frac{\sigma_{yp}^2 L_e^2}{\pi^2 E}\right)^{1/2} \tag{11.11}$$

Likewise, for a *rectangular section* of height h and depth b:

$$b = \frac{P_{cr}}{h\sigma_{yp}\left(1 - \dfrac{3L_e^2\sigma_{yp}}{\pi^2 E h^2}\right)} \tag{11.12}$$

in which, it is assumed that $h \leq b$.

We note that Eqs. (11.7) through (11.10) determine the *ultimate stresses*, not the working stresses. It is thus necessary to divide the right side of each formula by an appropriate factor of safety, usually 2 to 3, depending on the material, in order to obtain the allowable values. Some typical relationships for allowable stress are introduced in the next section.

11.7 ALLOWABLE STRESS

The foregoing discussion and analysis have related to ideal, homogeneous, concentrically loaded columns. Inasmuch as such columns are not likely candidates for application in structures, actual design requires the use of *empirical formulas* based on a strong background of test and experience. Care must be exercised in applying such special-purpose formulas. The designer should be prepared to respond to the following questions:

To what material does the formula apply?

Does the formula already contain a factor of safety, or is the factor of safety separately applied?

For what range of slenderness ratios is the formula valid?

Included among the many special-purpose relationships developed are the following, recommended by the American Institute of Steel Construction (AISC) and valid for a structural steel column:*

$$\sigma_{all} = \frac{1 - (L_e/r)^2/2C_c^2}{n}\sigma_{yp}, \qquad \frac{L_e}{r} < C_c$$

$$\sigma_{all} = \frac{\pi^2 E}{1.92(L_e/r)^2} \qquad C_c \leq \frac{L_e}{r} \leq 200 \tag{11.13}$$

Here σ_{all}, σ_{yp}, C_c, and n denote, respectively, the allowable and yield stresses, a material constant, and the factor of safety. The values of C_c and n are given by

$$C_c = \sqrt{\frac{2\pi^2 E}{\sigma_{yp}}}, \qquad n = \frac{5}{3} + \frac{3(L_e/r)}{8C_c} - \frac{(L_e/r)^3}{8C_c^3} \tag{11.14}$$

This relationship provides a smaller n for a short strut than for a column of higher L_e/r, recognizing that the former fails by yielding and the latter by buckling. The use of a variable factor of safety provides a consistent buckling formula for various

---

*The specifications of the AISC are given in Ref. 11.5. Similar formulas are available for aluminum and timber columns; see, for example, Ref. 11.6.

ranges of L_e/r. The second equation of (11.13) includes a constant factor of safety and gives the value of allowable stress in pascals. Both formulas apply to principal load-carrying (main) members.

EXAMPLE 11.2   Buckling of the Boom of a Crane

The boom of a crane, shown in Fig. 11.9, is constructed of steel, $E = 210$ GPa; the yield point stress is 250 MPa. The cross section is rectangular with a depth of 100 mm and a thickness of 50 mm. Determine the buckling load of the column.

Solution   The moments of inertia of the section are $I_z = 0.05(0.1)^3/12 = 4.17 \times 10^{-6}$ m^4 and $I_y = 0.1(0.05)^3/12 = 1.04 \times 10^{-6}$ m^4. The least radius of gyration is thus $r = \sqrt{I_y/A} = 14$ mm, and the slenderness ratio is $L/r = 194$. The Euler formula is applicable in this range. From statics, the axial force in terms of W is $P = W/\tan 15° = 3.732W$. Applying the formula for a hinged-end column, Eq. (11.5), for buckling in the yx plane, we have

$$P_{cr} = \frac{\pi^2 E I_z}{L^2} = \frac{9.86 \times 210 \times 10^9 \times 4.17 \times 10^{-6}}{(2.75)^2} = 1141.739 = 3.732W$$

or

$$W = 305.9 \text{ kN}$$

To calculate the load required for buckling in the xz plane, we must take note of the fact that the line of action of the compressive force passes through the joint and thus causes no moment about the y axis at the fixed end. Therefore, Eq. (11.5) may again be applied:

$$P_{cr} = \frac{\pi^2 E I_y}{L^2} = \frac{9.86 \times 210 \times 10^9 \times 1.04 \times 10^{-6}}{(2.75)^2} = 284.75 = 3.732W$$

or

$$W = 76.3 \text{ kN}$$

The member will thus fail by lateral buckling when the load W exceeds 76.3 kN. Note that the critical stress $P_{cr}/A = 76.3/0.005 = 15.26$ MPa. This, compared with the yield strength of 250 MPa, indicates the importance of buckling analysis in predicting the safe working load.

FIGURE 11.9.   *Example 11.2. A boom and cable assembly carrying a load W.*

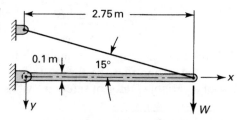

11.8   IMPERFECTIONS IN COLUMNS

As might be expected, the load-carrying capacity and deformation under load of a column are significantly affected by even a small initial curvature (see Fig. 11.4). In the preceding section, we described how to account for the effects of imperfections in construction of the column. To ascertain the extent of this influence, consider a pin-ended column for which the unloaded shape is described by

$$v_o = a_o \sin \frac{\pi x}{L} \qquad \text{(a)}$$

as shown by the dashed lines in Fig. 11.10. Here a_o is the maximum initial deflection, or *crookedness* of the column. An additional deflection v_1 will accompany a subsequently applied load P, so the total deflection is

$$v = v_o + v_1 \qquad \text{(b)}$$

The differential equation of the column is thus

$$EI \frac{d^2 v_1}{dx^2} = -P(v_o + v_1) = -Pv \qquad \text{(c)}$$

which together with Eq. (a) becomes

$$\frac{d^2 v_1}{dx^2} + \frac{P}{EI} v_1 = -\frac{P}{EI} a_o \sin \frac{\pi x}{L}$$

When the trial particular solution

$$v_{1P} = B \sin \frac{\pi x}{L}$$

is substituted into Eq. (c), it is found that

$$B = \frac{Pa_o}{(EI\pi^2/L^2) - P} = \frac{a_o}{(P_{cr}/P) - 1} \qquad \text{(d)}$$

The general solution of Eq. (c) is therefore

$$v_1 = c_1 \sin \sqrt{\frac{P}{EI}} x + c_2 \cos \sqrt{\frac{P}{EI}} x + B \sin \frac{\pi x}{L} \qquad \text{(e)}$$

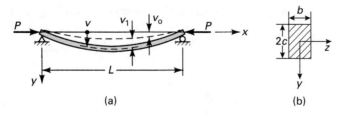

FIGURE 11.10.   *(a) Initially curved column with pinned ends; (b) the cross section.*

The constants c_1 and c_2 are evaluated upon consideration of the end conditions $v_1(0) = v_1(L) = 0$. The result of substituting these conditions is $c_1 = c_2 = 0$; the column deflection is thus

$$v = v_o + v_1 = a_o \sin\frac{\pi x}{L} + B \sin\frac{\pi x}{L} = \frac{a_o}{1 - (P/P_{cr})} \sin\frac{\pi x}{L} \qquad \textbf{(11.15)}$$

As for the critical stress, we begin with the expression applicable to combined axial loading and bending: $\sigma_x = (P/A) \pm (My/I)$, where A and I are the cross-sectional area and the moment of inertia. On substitution of Eqs. (c) and (11.15) into this expression, the maximum compressive stress at midspan is found to be

$$\sigma_{max} = \frac{P}{A}\left[1 + \frac{a_o A}{S}\frac{1}{1 - (P/P_{cr})}\right] \qquad \textbf{(11.16)}$$

Here S is the section modulus I/c, where c represents the distance measured in the y direction from the centroid of the cross section to the extreme fibers. In Eq. (11.16), σ_{max} is limited to the proportional or yield stress of the column material. Thus, setting $\sigma_{max} = \sigma_{yp}$ and $P = P_L$, we rewrite Eq. (11.16) as follows:

$$\sigma_{yp} = \frac{P_L}{A}\left[1 + \frac{a_o A}{S}\frac{1}{1 - (P_L/P_{cr})}\right] \qquad \textbf{(11.17)}$$

where P_L is the *limit load* that results in impending yielding and subsequent failure. Given σ_{yp}, a_o, E, and the column dimensions, Eq. (11.17) may be solved exactly by solving a quadratic or by trial and error for P_L. The allowable load P_{all} can then be found by dividing P_L by an appropriate factor of safety, n. Interestingly, in Eqs. (11.16) and (11.17), the term $a_o A/S = a_o c/r^2$ is known as the *imperfection ratio*, where r is the radius of gyration. Another approach to account for the effect of imperfections due to eccentricities in the line of application of the load is discussed in the next section.

11.9   ECCENTRICALLY LOADED COLUMNS: SECANT FORMULA

In contrast with the cases considered up to this point, we now analyze columns that are loaded eccentrically, that is, those in which the load is not applied at the centroid of the cross section. This situation is clearly of great practical importance, since we ordinarily have no assurance that loads are truly concentric. As seen in Fig. 11.11a, the bending moment at any section is $-P(v + e)$, where e is the eccentricity, defined in the figure. It follows that the beam equation is given by

$$EI\frac{d^2v}{dx^2} + P(v + e) = 0 \qquad \textbf{(a)}$$

or

$$EI\frac{d^2v}{dx^2} + Pv = -Pe$$

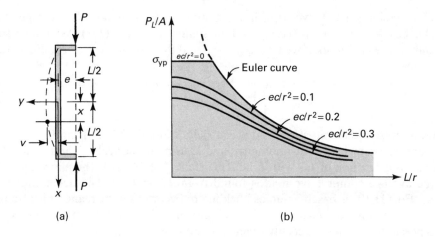

FIGURE 11.11.   *(a) Eccentrically loaded column; (b) graph of secant formula.*

The general solution is

$$v = c_1 \sin \sqrt{\frac{P}{EI}}\, x + c_2 \cos \sqrt{\frac{P}{EI}}\, x - e \qquad \textbf{(b)}$$

To determine the constants c_1 and c_2, the end conditions $v(L/2) = v(-L/2) = 0$ are applied, with the result

$$c_1 = 0, \qquad c_2 = \frac{e}{\cos\left[\sqrt{P/EI}\,(L/2)\right]}$$

Substituting these values into Eq. (b) provides an expression for the column deflection:

$$v = \frac{e}{\cos \sqrt{PL^2/(4EI)}} \cos \sqrt{\frac{P}{EI}}\, x - e \qquad \textbf{(c)}$$

In terms of the critical load $P_{cr} = \pi^2 EI/L^2$, the midspan deflection is

$$v_{\max} = e\left[\sec\left(\frac{\pi}{2}\sqrt{\frac{P}{P_{cr}}}\right) - 1\right] \qquad \textbf{(11.18)}$$

As P approaches P_{cr}, the maximum deflection is thus observed to approach infinity.

The maximum compressive stress, $(P/A) + (Mc/I)$, occurs at $x = 0$ on the concave side of the column; it is given by the so-called *secant formula*:

$$\sigma_{\max} = \frac{P}{A}\left[1 + \frac{ec}{r^2}\sec\left(\frac{\pi}{2}\sqrt{\frac{P}{P_{cr}}}\right)\right] \qquad \textbf{(11.19)}$$

where r represents the radius of gyration and c is the distance from the centroid of the cross section to the extreme fibers, both in the direction of eccentricity. When the eccentricity is zero, this formula no longer applies. Then, we have an ideal column with centric loading, and Eq. (11.7) is to be used.

11.9   Eccentrically Loaded Columns: Secant Formula

Expression (11.19) gives σ_{max} in the column as a function of the average stress (P/A), the *eccentricity ratio* (ec/r^2), and the slenderness ratio (L/r). As in the case of initially curved columns, if we let $\sigma_{max} = \sigma_{yp}$ and the limit load $P = P_L$, Eq. (11.19) becomes

$$\sigma_{yp} = \frac{P_L}{A}\left[1 + \frac{ec}{r^2}\sec\left(\frac{\pi}{2}\sqrt{\frac{P_L}{P_{cr}}}\right)\right] \tag{11.20}$$

For any prescribed yield stress and eccentricity ratio, Eq. (11.20) can be solved by trial and error, or by a root-finding technique using numerical methods, and P_L/A plotted as a function of L/r with E held constant (Fig. 11.11b). The allowable value of the average compressive load is found from $P_{all} = P_L/n$. The development on which Eq. (11.19) is based assumes buckling to occur in the xy plane. It is also necessary to investigate buckling in the xz plane, for which Eq. (11.19) does not apply. This possibility relates especially to narrow columns.

The behavior of a beam subjected to simultaneous axial and lateral loading, the *beam column*, is analogous to that of the bar shown in Fig. 11.1a, with an additional force acting transverse to the bar. For problems of this type, energy methods are usually more efficient than the equilibrium approach previously employed.

11.10 ENERGY METHODS APPLIED TO BUCKLING

Energy techniques usually offer considerable ease of solution compared with equilibrium approaches to the analysis of elastic stability and the determination of critical loads. Recall from Chapter 10 that energy methods are especially useful in treating members of variable cross section, where the variation can be expressed as a function of the beam axial coordinate.

To illustrate the method, let us apply the principle of virtual work in analyzing the stability of a straight, pin-ended column. Locate the origin of coordinates at the stationary end. Recall from Section 10.8 that the principle of virtual work may be stated as

$$\delta W = \delta U \tag{a}$$

where W and U are the virtual work and strain energy, respectively. Consider the configuration of the column in the first buckling mode, denoting the arc length of a column segment by ds. The displacement, $\delta u = ds - dx$, experienced by the column in the direction of applied load P is given by Eq. (10.29). Since the load remains constant, the work done is therefore

$$\delta W = \frac{1}{2}P\int_0^L \left(\frac{dv}{dx}\right)^2 dx \tag{11.21}$$

Next, the strain energy must be evaluated. There are components of strain energy associated with column bending, compression, and shear. We shall neglect the last. From Eq. (5.62), the bending component is

$$U_1 = \int_0^L \frac{M^2}{2EI}\,dx = \int_0^L \frac{EI}{2}\left(\frac{d^2v}{dx^2}\right)^2 dx \qquad \textbf{(11.22)}$$

The energy due to a uniform compressive loading P is, according to Eq. (2.59),

$$U_2 = \frac{P^2L}{2AE} \qquad \textbf{(11.23)}$$

Inasmuch as U_2 is constant, it plays no role in the analysis. The change in the strain energy as the column proceeds from its original to its buckled configuration is therefore

$$\delta U = \int_0^L \frac{EI}{2}\left(\frac{d^2v}{dx^2}\right)^2 dx \qquad \textbf{(b)}$$

since the initial strain energy is zero. Substituting Eqs. (11.21) and (b) into Eq. (a), we have

$$\frac{1}{2}\int_0^L P\left(\frac{dv}{dx}\right)^2 dx = \frac{1}{2}\int_0^L EI\left(\frac{d^2v}{dx^2}\right)^2 dx \qquad \textbf{(11.24a)}$$

from which

$$P_{cr} = \frac{\displaystyle\int_0^L EI(v'')^2\,dx}{\displaystyle\int_0^L (v')^2\,dx} \qquad \textbf{(11.24b)}$$

This result applies to a column with any end condition. The end conditions specific to this problem will be satisfied by a solution

$$v = a_1 \sin\frac{n\pi x}{L}$$

where a_1 is a constant. After substituting this assumed deflection into Eq. (11.24b) and integrating, we obtain

$$(P_{cr})_n = \left(\frac{n\pi}{L}\right)^2 EI$$

The minimum critical load and the deflection to which this corresponds are

$$P_{cr} = \frac{\pi^2 EI}{L^2}, \qquad v = a_1 \sin\frac{\pi x}{L} \qquad \textbf{(c)}$$

It is apparent from Eq. (11.24a) that for $P > P_{cr}$, the work done by P exceeds the strain energy stored in the column. The assertion can therefore be made that a straight column is unstable when $P > P_{cr}$. This point, with regard to stability, corresponds to $c_1 = 0$ in Eq. (b) of Section 11.3; it could not be obtained as readily from the equilibrium approach. In the event that $P = P_{cr}$, the column exists in neutral equilibrium. For $P < P_{cr}$, a straight column is in stable equilibrium.

EXAMPLE 11.3   Deflection of a Beam Column

A simply supported beam is subjected to a moment M_o at point A and axial loading P, as shown in Fig. 11.12. Determine the equation of the elastic curve.

Solution   The displacement of the right end, which occurs during the deformation of the beam from its initially straight configuration to the equilibrium curve, is given by Eq. (b). The total work done is evaluated by adding to Eq. (11.21) the work due to the moment. In Section 10.9, we already solved this problem for $P = 0$ by using the following Fourier series for displacement:

$$v = \sum_{n=1}^{\infty} a_n \sin \frac{n\pi x}{L} \qquad \text{(d)}$$

Proceeding in the same manner, Eq. (b) of Section 10.9, representing $\delta U = \delta W$, now takes the form

$$\frac{\pi^4 EI}{4L^3} \sum_{n=1}^{\infty} n^4 \delta(a_n^2) = \frac{\pi^2 P}{4L} \sum_{n=1}^{\infty} n^2 \delta(a_n^2) + \frac{\pi M_o}{L} \sum_{n=1}^{\infty} n \cos \frac{n\pi c}{L} \delta(a_n)$$

From this expression,

$$a_n = \frac{2M_o}{\pi} \sum_{n=1}^{\infty} \frac{1}{n} \frac{\cos(n\pi c/L)}{(n^2\pi^2 EI/L^2) - P} \qquad \text{(e)}$$

For purposes of simplification, let b denote the ratio of the axial force to its critical value:

$$b = \frac{PL^2}{\pi^2 EI} \qquad \text{(11.25)}$$

Then, by substituting Eqs. (11.25) and (e) into Eq. (d), the following expression for deflection results:

$$v = \frac{2M_o L^2}{\pi^3 EI} \sum_{n=1,3,5,\ldots}^{\infty} \frac{\cos(n\pi c/L)}{n(n^2 - b)} \sin \frac{n\pi x}{L}, \qquad 0 \le x \le L \quad \text{(11.26)}$$

Note that, when P approaches its critical value in Eq. (11.25), $b \to 1$. The first term in Eq. (11.26) is then

$$v = \frac{2M_o L^2}{\pi^3 EI} \frac{1}{1-b} \cos \frac{\pi c}{L} \sin \frac{\pi x}{L} \qquad \text{(11.27)}$$

indicating that the deflection becomes infinite, as expected.

Comparison of Eq. (11.27) with the solution found in Section 10.10 (corresponding to $P = 0$ and $n = 1$) indicates that the axial force P serves

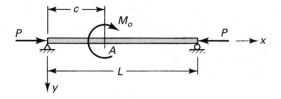

FIGURE 11.12. *Example 11.3. A beam column supports loads M_o and P.*

to increase the deflection produced by the lateral load (moment M_o) by a factor of $1/(1 - b)$.

In general, if we have a beam subjected to several moments or lateral loads in addition to an axial load P, the deflections owing to the lateral moments or forces are found for the $P = 0$ case. This usually involves superposition. The resulting deflection is then multiplied by the factor $1/(1 - b)$ to account for the deflection effect due to P. This procedure is valid for any lateral load configuration composed of moments, concentrated forces, and distributed forces.

EXAMPLE 11.4   Critical Load of a Fixed-Free Column

Apply the Rayleigh–Ritz method to determine the buckling load of a straight, uniform cantilever column carrying a vertical load (Fig. 11.13).

Solution The analysis begins with an assumed parabolic deflection curve,

$$v = \frac{ax^2}{L^2} \tag{f}$$

where a represents the deflection of the free end and L the column length. (The parabola is actually a very poor approximation to the true curve, since it describes a beam of constant curvature, whereas the curvature of the actual beam is zero at the top and a maximum at the bottom.) The assumed deflection satisfies the geometric boundary conditions

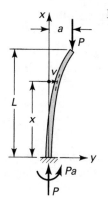

FIGURE 11.13.   *Example 11.4. A column fixed at the base and free at the top.*

pertaining to deflection and slope: $v(0) = 0$, $v'(0) = 0$. In accordance with the Rayleigh–Ritz procedure (see Sec. 10.11), it may therefore be used as a trial solution. The static boundary conditions, such as $v''(0) \neq 0$ or $M \neq 0$, need not be satisfied.

The work done by the load P and the strain energy gained are given by Eqs. (11.21) and (11.22). The potential energy function Π is thus given by

$$\Pi = U - W = \int_0^L \frac{EI}{2} \left(\frac{d^2v}{dx^2} \right)^2 dx - \int_0^L \frac{P}{2} \left(\frac{dv}{dx} \right)^2 dx \tag{g}$$

Substituting Eq. (f) into this expression and integrating, we have

$$\Pi = \frac{2EIa^2}{L^3} - \frac{2Pa^2}{3L} \tag{h}$$

Applying Eq. (10.34), $\partial\Pi/\partial a = 0$, we find that

$$P_{cr} = 3 \frac{EI}{L^2} \tag{i}$$

Let us rework this problem by replacing the strain–energy expression due to bending,

$$U_1 = \int_0^L \frac{EI}{2} \left(\frac{d^2v}{dx^2} \right)^2 dx \tag{j}$$

with one containing the moment deduced from Fig. 11.13, $M = P(a - v)$:

$$U_1 = \int_0^L \frac{M^2 \, dx}{2EI} = \int_0^L \frac{P^2(a - v)^2}{2EI} \, dx \tag{k}$$

Equation (h) becomes

$$\Pi = \int_0^L \frac{P^2(a - v)^2}{2EI} \, dx - \int_0^L \frac{P}{2} \left(\frac{dy}{dx} \right)^2 dx$$

Substituting Eq. (f) into this expression and integrating, we obtain

$$\Pi = \frac{8}{30} \frac{P^2 a^2 L}{EI} - \frac{2Pa^2}{3L}$$

Now, $\partial\Pi/\partial a = 0$ yields

$$P_{cr} = 2.50 \frac{EI}{L^2} \tag{l}$$

Comparison with the exact solution, $2.4674EI/L^2$, reveals errors for the solutions (i) and (l) of about 22% and 1.3%, respectively. The latter result is satisfactory, although it is predicated on an assumed deflection curve differing considerably in shape from the true curve.

It is apparent from the foregoing example and a knowledge of the exact solution that one solution is quite a bit more accurate than the other. It can be shown that the expressions (j) and (k) will be identical only when the true deflection is initially assumed. Otherwise, Eq. (k) will give better accuracy. This is because when we choose Eq. (k), the accuracy of the solution depends on the closeness of the assumed deflection to the actual deflection; with Eq. (j), the accuracy depends instead on the rate of change of slope, d^2v/dx^2.

An additional point of interest relates to the consequences, in terms of the critical load, of selecting a deflection that departs from the actual curve. When other than the true deflection is used, not every beam element is in equilibrium. The approximate beam curve can be maintained only through the introduction of additional constraints that tend to render the beam more rigid. The critical loads thus obtained will be higher than those derived from exact analysis. It may be concluded that *energy methods always yield buckling loads higher than the exact values if the assumed beam deflection differs from the true curve.*

More efficient application of energy techniques may be realized by selecting a series approximation for the deflection, as in Example 11.3. Inasmuch as a series involves a number of parameters, as for instance in Eq. (d), the approximation can be varied by appropriate manipulation of these parameters, that is, by changing the number of terms in the series.

EXAMPLE 11.5   Critical Load of a Tapered Column

A pin-ended, tapered bar of constant thickness is subjected to axial compression (Fig. 11.14). Determine the critical load. The variation of the moment of inertia is given by

$$I(x) = I_1\left(1 + \frac{3x}{L}\right), \qquad 0 \le x \le \frac{L}{2}$$

$$= I_1\left(4 - \frac{3x}{L}\right), \qquad \frac{L}{2} \le x \le L$$

where I_1 is the constant moment of inertia at $x = 0$ and $x = L$.

Solution   As before, we begin by representing the deflection by

$$v = a_1 \sin\frac{\pi x}{L}$$

Taking symmetry into account, the variation of strain energy and the work done [Eqs. (11.21) and (11.22)] are expressed by

$$\delta U = \delta \int_0^L \frac{EI}{2}\left(\frac{d^2v}{dx^2}\right)^2 dx = 2\delta \int_0^{L/2} \frac{EI_1}{2}\left(1 + \frac{3x}{L}\right) a_1^2 \frac{\pi^4}{L^4} \sin^2\frac{\pi x}{L}\, dx$$

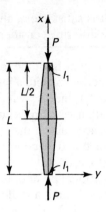

FIGURE 11.14. *Example 11.5. Tapered column with pinned ends.*

$$= \frac{\pi^4 EI_1}{4L^3}\,\delta(a_1^2) + \frac{4.215\pi^4 EI_1}{16L^3}\,\delta(a_1^2) = \frac{8.215\pi^4 EI_1}{16L^3}\,\delta(a_1^2)$$

$$\delta W = \delta \int_0^L \frac{P}{2}\left(\frac{dv}{dx}\right)^2 dx = 2\delta \int_0^{L/2} \frac{P}{2}\, a_1^2 \frac{\pi^2}{L^2}\cos^2\frac{\pi x}{L}\,dx = \frac{\pi^2 P}{4L}\,\delta(a_1^2)$$

From the principle of virtual work, $\delta W = \delta U$, we have

$$\frac{8.215\pi^4 EI_1}{16L^3} = \frac{\pi^2 P}{4L}$$

or

$$P_{cr} = 20.25\,\frac{EI_1}{L^2}$$

In Example 11.8, this problem is solved by numerical analysis, revealing that the preceding solution overestimates the buckling load.

EXAMPLE 11.6   **Critical Load of a Rail on a Track**
Consider the beam of Fig. 11.15, representing a rail on a track experiencing compression owing to a rise in temperature. Determine the critical load.

Solution   We choose the deflection shape in the form of a series given by Eq. (10.23). The solution proceeds as in Example 10.13. Now the total energy consists of three sources: beam bending and compression and foundation displacement. The strain energy in bending, from Eq. (10.25), is

$$U_1 = \frac{\pi^4 EI}{4L^3}\sum_{n=1}^{\infty} n^4 a_n^2$$

The strain energy due to the deformation of the elastic foundation is given by Eq. (10.30):

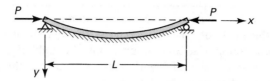

FIGURE 11.15. *Example 11.6. Beam with hinged ends, embedded in an elastic foundation of modulus k, subjected to compressive end forces.*

$$U_2 = \frac{kL}{4} \sum_{n=1}^{\infty} a_n^2$$

The work done by the forces P in shortening the span, using Eq. (10.31), is

$$W = \frac{\pi^2 P}{4L} \sum_{n=1}^{\infty} n^2 a_n^2$$

Applying the principle of virtual work, $\delta W = \delta(U_1 + U_2)$, we obtain

$$\frac{\pi^2 P}{4L} \sum_{n=1}^{\infty} n^2 \delta a_1^2 = \frac{\pi^4 EI}{4L} \sum_{n=1}^{\infty} n^4 \delta a_1^2 + \frac{kL}{4} \sum_{n=1}^{\infty} \delta a_1^2$$

The foregoing results in the critical load:

$$P_{cr} = \frac{\pi^2 EI}{L^2} \sum \left(n^2 + \frac{kL^4}{n^2 \pi^2 EI} \right) \tag{11.28}$$

Note that this expression yields exact results. The lowest critical load may occur with $n = 1, 2, 3, \ldots$, depending on the properties of the beam and the foundation [Ref. 11.1].

Interestingly, taking $n = 1$, Eq. (11.28) becomes

$$P_{cr} = \frac{\pi^2 EI}{L^2} + \frac{kL^2}{\pi^2} \tag{m}$$

which is substantially greater than the Euler load. The Euler buckling load is the first term; it is augmented by kL^2/π^2 due to the foundation.

11.11 SOLUTION BY FINITE DIFFERENCES

The equilibrium analysis of buckling often leads to differential equations that resist solution. Even energy methods are limited in that they require that the moment of inertia be expressible as a function of the column length coordinate. Therefore, to enable the analyst to cope with the numerous and varied columns of practical interest, numerical techniques must be relied on. The examples that follow apply the method of finite differences to the differential equation of a column.

EXAMPLE 11.7   Analysis of a Simple Column
Determine the buckling load of a pin-ended column of length L and constant cross section. Use four subdivisions of equal length. Denote the nodal points by 0, 1, 2, 3, and 4, with 0 and 4 located at the ends. Locate the origin of coordinates at the stationary end.

Solution   The governing differential equation (11.1) may be put into the form

$$\frac{d^2v}{dx^2} + \lambda^2 v = 0$$

by substituting $\lambda^2 = P/EI$. The boundary conditions are $v(0) = v(L) = 0$. The corresponding finite difference equation is, according to Eq. (7.9),

$$v_{m+1} + (\lambda^2 h^2 - 2)v_m + v_{m-1} = 0 \qquad \textbf{(11.29)}$$

valid at every node along the length. Here the integer m denotes the nodal points, and h the segment length. Applying Eq. (11.29) at points 1, 2, and 3, we have

$$(\lambda^2 h^2 - 2)v_1 + v_2 = 0$$
$$v_1 + (\lambda^2 h^2 - 2)v_2 + v_3 = 0 \qquad \textbf{(a)}$$
$$v_2 + (\lambda^2 h^2 - 2)v_3 = 0$$

or, in convenient matrix form,

$$\begin{bmatrix} \lambda^2 h^2 - 2 & 1 & 0 \\ 1 & \lambda^2 h^2 - 2 & 1 \\ 0 & 1 & \lambda^2 h^2 - 2 \end{bmatrix} \begin{Bmatrix} v_1 \\ v_2 \\ v_3 \end{Bmatrix} = 0 \qquad \textbf{(b)}$$

This set of simultaneous equations has a nontrivial solution for v_1, v_2, and v_3 only if the determinant of the coefficients vanishes:

$$\begin{vmatrix} \lambda^2 h^2 - 2 & 1 & 0 \\ 1 & \lambda^2 h^2 - 2 & 1 \\ 0 & 1 & \lambda^2 h^2 - 2 \end{vmatrix} = 0 \qquad \textbf{(c)}$$

The solution of the buckling problem is thus reduced to determination of the roots (the λ's of the characteristic equation) resulting from the expansion of the determinant.

To expedite the solution, we take into account the symmetry of deflection. For the lowest critical load, the buckling configuration is given by the first mode, shown in Fig. 11.3a. Thus, $v_1 = v_3$, and the Eqs. (a) become

$$(\lambda^2 h^2 - 2)v_1 + v_2 = 0$$
$$2v_1 + (\lambda^2 h^2 - 2)v_2 = 0$$

Setting equal to zero the characteristic determinant of this set, we find that $(\lambda^2 h^2 - 2)^2 - 2 = 0$, which has the solution $\lambda^2 h^2 = 2 \pm \sqrt{2}$. Selecting $2 - 2\sqrt{2}$ to obtain a minimum critical value and letting $h = L/4$, we obtain $\lambda^2 = (2 - \sqrt{2})16/L^2$. Thus,

$$P_{cr} = 9.373 \frac{EI}{L^2}$$

This result differs from the exact solution by approximately 5%. By increasing the number of segments, the accuracy may be improved.

If required, the critical load corresponding to second-mode buckling, indicated in Fig. 11.3b, may be determined by recognizing that for this case $v_0 = v_2 = v_4 = 0$ and $v_1 = -v_3$. We then proceed from Eqs. (a) as before. For buckling of higher than second mode, a similar procedure is followed, in which the number of segments is increased and the appropriate conditions of symmetry satisfied.

EXAMPLE 11.8   Critical Load of a Tapered Column

What load will cause buckling of the tapered pin-ended column shown in Fig. 11.14?

Solution   The finite difference equation is given by Eq. (11.29):

$$v_{m+1} + \left(\frac{Ph^2}{EI(x)} - 2 \right) v_m + v_{m-1} = 0 \tag{d}$$

The foregoing becomes, after substitution of $I(x)$,

$$v_{m+1} + \left[\frac{\lambda^2 h^2}{1 + (3x/L)} - 2 \right] v_m + v_{m-1} = 0, \qquad 0 \le x \le \frac{L}{2} \tag{e}$$

where $\lambda^2 = P/EI_1$. Note that the coefficient of v_m is a variable, dependent on x. This introduces no additional difficulties, however.

Dividing the beam into two segments, we have $h = L/2$ (Fig. 11.16a). Applying Eq. (e) at $x = L/2$,

$$\left[\frac{\lambda^2 L^2}{4(1 + 3/2)} - 2 \right] v_1 = 0$$

The nontrivial solution corresponds to $v_1 \ne 0$; then $(\lambda^2 L^2/10) - 2 = 0$ or, by letting $\lambda^2 = P/EI$,

$$P_{cr} = \frac{20EI_1}{L^2} \tag{f}$$

Similarly, for three segments $h = L/3$ (Fig. 11.16b). From symmetry, we have $v_1 = v_2$, and $v_0 = v_3 = 0$. Thus, Eq. (e) applied at $x = L/3$ yields

$$v_1 + \left[\frac{\lambda^2 L^2}{9(1 + 1)} - 2 \right] v_1 = 0, \quad \text{or} \quad \left(\frac{\lambda^2 L^2}{18} - 1 \right) v_1 = 0$$

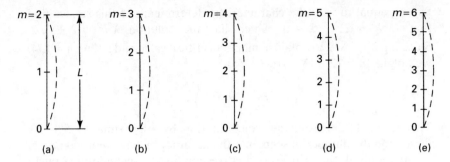

FIGURE 11.16. *Example 11.8. Division of the column of Fig. 11.14 into a number of segments.*

The nontrivial solution is

$$P_{cr} = \frac{18EI_1}{L^2} \qquad \text{(g)}$$

For $h = L/4$, referring to Fig. 11.16c, Eq. (e) leads to

$$\left(\frac{\lambda^2 L^2}{28} - 2\right)v_1 + v_2 = 0, \qquad 2v_1 + \left(\frac{\lambda^2 L^2}{40} - 2\right)v_2 = 0$$

For a nontrivial solution, the characteristic determinant is zero:

$$\begin{vmatrix} \lambda^2 L^2/28 - 2 & 1 \\ 2 & \lambda^2 L^2/40 - 2 \end{vmatrix} = 0$$

Expansion yields

$$\lambda^4 L^4 - 136\lambda^2 L^2 + 2240 = 0$$

for which

$$P_{cr} = \frac{19.55EI_1}{L^2} \qquad \text{(h)}$$

Similar procedures considering the symmetry shown in Figs. 11.16d and e lead to the following results:
For $h = L/5$,

$$P_{cr} = \frac{18.96EI_1}{L^2} \qquad \text{(i)}$$

For $h = L/6$,

$$P_{cr} = \frac{19.22EI_1}{L^2} \qquad \text{(j)}$$

Results (f) through (j) indicate that, for columns of variable moment of inertia, increasing the number of segments does not necessarily lead to improved P_{cr}. An energy approach to this problem (Example 11.5) gives the result $P_{cr} = 20.25EI_1/L^2$. Because this value is higher than those obtained here, we conclude that the column does not deflect into the half sine curve assumed in Example 11.5.

EXAMPLE 11.9   Critical Load of a Fixed-Pinned Column

Figure 11.17 shows a column of constant moment of inertia I and of length L, fixed at the left end and simply supported at the right end, subjected to an axial compressive load P. Determine the critical value of P using $m = 3$.

Solution   The characteristic value problem is defined by

$$\frac{d^4v}{dx^4} + \frac{P}{EI}\frac{d^2v}{dx^2} = 0 \qquad \text{(k)}$$

$$v(0) = v'(0) = v(L) = v''(L) = 0$$

where the first equation is found from Eq. (P11.34) by setting $p = 0$; the second expression represents the end conditions related to deflection, slope, and moment. Equations (k), referring to Section 7.2 and letting $\lambda^2 = P/EI$, may be written in the finite difference form as follows:

$$v_{m+2} + (\lambda^2 h^2 - 4)v_{m+1} + (6 - 2\lambda^2 h^2)v_m + (\lambda^2 h^2 - 4)v_{m-1} + v_{m-2} = 0 \qquad \text{(l)}$$

and

$$v_0 = 0, \qquad v_{-1} = v_1, \qquad v_m = 0, \qquad v_{m+1} = -v_{m-1} \qquad \text{(m)}$$

The quantities v_{-1} and v_{m+1} represent the deflections at the nodal points of the column prolonged by h beyond the supports. By dividing the column into three subintervals, the pattern of the deflection curve and the conditions (m) are represented in the figure by dashed lines. Now, Eq. (l) is applied at nodes 1 and 2 to yield, respectively,

$$v_1 + (6 - 2\lambda^2 h^2)\,v_1 + (\lambda^2 h^2 - 4)\,v_2 = 0$$

$$(\lambda^2 h^2 - 4)\,v_1 + (6 - 2\lambda^2 h^2)\,v_2 - v_2 = 0$$

We have a nonzero solution if the determinant of the coefficients of these equations vanishes:

$$\begin{vmatrix} 7 - 2\lambda^2 h^2 & \lambda^2 h^2 - 4 \\ \lambda^2 h^2 - 4 & 5 - 2\lambda^2 h^2 \end{vmatrix} = 3\lambda^4 h^4 - 16\lambda^2 h^2 + 19 = 0$$

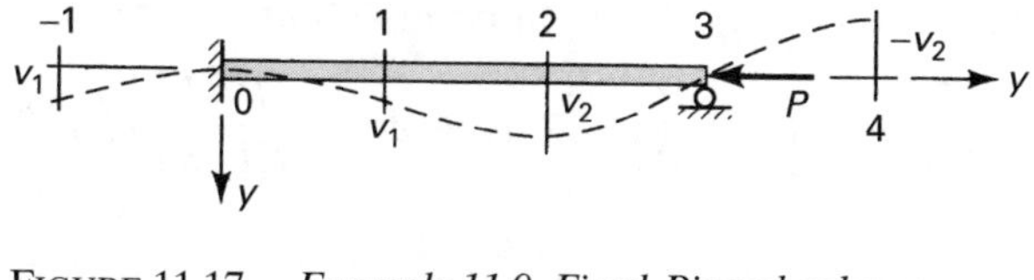

FIGURE 11.17.   *Example 11.9. Fixed-Pinned column.*

From this and setting $h = L/3$, we obtain $\lambda^2 = 16.063/L$. Thus,

$$P_{cr} = 16.063 \frac{EI}{L^2} \tag{n}$$

The exact solution is $20.187EI/L^2$. By increasing the number of segments and by employing an extrapolation technique, the results may be improved.

11.12 FINITE DIFFERENCE SOLUTION FOR UNEVENLY SPACED NODES

It is often advantageous, primarily because of geometrical considerations, to divide a structural element so as to produce uneven spacing between nodal points. In some problems, uneven spacing provides more than a saving in time and effort. As seen in Section 7.4, some situations cannot be solved without resort to this approach.

Consider the problem of the buckling of a straight pin-ended column governed by

$$\frac{d^2v}{dx^2} + \frac{P}{EI} v = 0, \qquad v(0) = v(L) = 0$$

Upon substitution from Eq. (7.21), the following corresponding finite difference equation is obtained:

$$\frac{2}{\alpha(\alpha + 1)} [\alpha v_{m-1} - (1 + \alpha)v_m + v_{m+1}] + \frac{Ph^2}{EI} v_m = 0 \tag{11.30}$$

where

$$h = x_m - x_{m-1}, \qquad \alpha = \frac{x_{m+1} - x_m}{x_m - x_{m-1}} \tag{11.31}$$

Equation (11.30), valid throughout the length of the column, is illustrated in the example to follow.

EXAMPLE 11.10 Critical Load of a Stepped Column
Determine the buckling load of a stepped pin-ended column (Fig. 11.18a). The variation of the moment of inertia is indicated in the figure.

Solution The nodal points are shown in Fig. 11.18b and are numbered in a manner consistent with the symmetry of the beam. Note that the nodes are unevenly spaced. From Eq. (11.31), we have $\alpha_1 = 2$ and $\alpha_2 = 1$. Application of Eq. (11.30) at points 1 and 2 leads to

$$\frac{2}{2(2 + 1)} [2(0) - (1 + 2)v_1 + v_2] + \frac{P}{EI_1} \frac{L^2}{36} v_1 = 0$$

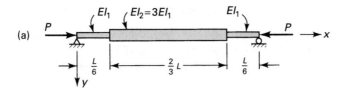

(a)

(b)

FIGURE 11.18. *Example 11.10. Stepped column with pinned ends.*

$$\frac{2}{1(1 + 1)} [v_1 - (1 + 1)v_2 + v_1] + \frac{P}{3EI_1} \frac{4L^2}{36} v_2 = 0$$

or

$$\left(\frac{PL^2}{36EI_1} - 1\right)v_1 + \tfrac{1}{3}v_2 = 0$$

$$2v_1 + \left(\frac{PL^2}{27EI_1} - 2\right)v_2 = 0$$

For a nontrivial solution, it is required that

$$\begin{vmatrix} (PL^2/36EI_1) - 1 & 1/3 \\ 2 & (PL^2/27EI_1) - 2 \end{vmatrix} = 0$$

Solving, we find that the root corresponding to minimum P is

$$P_{cr} = 18 \frac{EI}{L^2}$$

Employing additional nodal points may result in greater accuracy. This procedure lends itself to columns of arbitrarily varying section and various end conditions.

We conclude our discussion by noting that column buckling represents but one case of structural instability. Other examples include the lateral buckling of a narrow beam; the buckling of a flat plate compressed in its plane; the buckling of a circular ring subject to radial compression; the buckling of a cylinder loaded by torsion, compression, bending, or internal pressure; and the snap buckling of arches. Buckling analyses for these cases are often not performed as readily as in the examples presented in this chapter. The solutions more often involve considerable difficulty and subtlety.*

*For a complete discussion of this subject, see Refs. 11.7 through 11.10.

REFERENCES

11.1. TIMOSHENKO, S. P., and GERE, J. M., *Theory of Elastic Stability*, 3rd ed., McGraw-Hill, New York, 1970, Secs. 2.7 and 2.10.

11.2. SHANLEY, R. F., *Mechanics of Materials*, McGraw-Hill, New York, 1967, Sec. 11.3.

11.3. PEERY, D. J., and AZAR, J. J., *Aircraft Structures*, 2nd ed., McGraw-Hill, New York, 1993, Chap.12.

11.4. UGURAL, A. C., *Mechanical Design: An Integrated Approach*, McGraw-Hill, New York, 2004.

11.5. *Manual of Steel Construction*, latest ed., American Institute of Steel Construction, Inc., Chicago.

11.6. UGURAL, A. C., *Mechanics of Materials*, Wiley, Hoboken, N. J., 2008, Sec. 11.7.

11.7. TIMOSHENKO, S. P., and GERE, J. M., *Theory of Elastic Stability*, 3rd ed., McGraw-Hill, New York, 1970, Chaps. 2 through 6.

11.8. BORESI, A. P., and SCHMIDTH, R. J., *Advanced Mechanics of Materials*, 6th ed., Wiley, Hoboken, N.J., 2003.

11.9. UGURAL, A. C., *Stresses in Beams, Plates, and Shells*, 3rd ed., CRC Press, Taylor and Francis Group, Boca Raton, Fla., 2010.

11.10. BRUSH, D. O., and ALMROTH, B. O., *Buckling of Bars, Plates, and Shells*, McGraw-Hill, New York, 1975.

PROBLEMS

Sections 11.1 through 11.7

11.1. In the assembly shown in Figure P11.1, a 50-mm outer-diameter and 40-mm inner-diameter steel pipe ($E = 200$ GPa) that is 1-m long acts as a spreader member. For a factor of safety $n = 2.5$, what is the value of F that will cause buckling of the pipe?

11.2. Figure P11.2 shows the cross sections of two aluminum alloy 2114-T6 bars that are used as compression members, each with effective length of L_e. Find (a) the wall thickness the hollow square bar so that the bars have the same cross-sectional area; (b) the critical load of each bar. *Given:* $L_e = 3$ m and $E = 72$ GPa (from Table D.1).

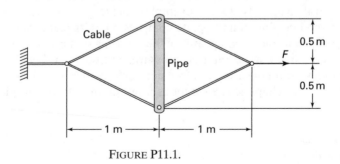

FIGURE P11.1.

11.3. Based on a factor of safety $n = 1.8$, determine the maximum load F that can be applied to the truss shown in Figs. P11.3. *Given:* Each column is of 50-mm-diameter aluminum bar having $E = 70$ GPa.

11.4. Figure P11.4 shows a column AB of length L and flexural rigidity EI with a support at its base that does not permit rotation but allows horizontal displacement and is pinned at its top. What is the critical load P_{cr}?

11.5. Figure P11.5 shows the tube of uniform thickness $t = 25$ mm cross section of a fixed-ended rectangular steel ($E = 200$ GPa) of a 9-m-long column. Determine the critical stress in the column.

11.6. Resolve Prob. 11.5 for the case in which the column is pinned at one end and fixed at the other.

11.7. Brace BD of the structure illustrated in Fig. P11.7 is made of a steel rod ($E = 210$ GPa and $\sigma_{yp} = 250$ MPa) with a square cross section (50 mm on a side). Calculate the factor of safety n against failure by buckling.

11.8. Verify the specific Johnson's formulas, given by Eqs. (11.11) and (11.12), for the intermediate sizes of columns with solid circular and rectangular cross sections, respectively.

11.9. A steel column with pinned ends supports an axial load $P = 90$ kN. Calculate the longest allowable column length. *Given:* Cross-sectional area $A = 3 \times 10^3$ mm^2 and radius of gyration $r = 25$ mm; the material properties are $E = 200$ GPa and $\sigma_{yp} = 250$ MPa.

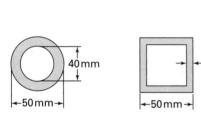

FIGURE P11.2.

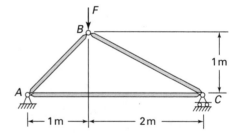

FIGURE P11.3.

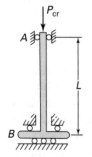

FIGURE P11.4.

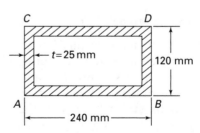

FIGURE P11.5.

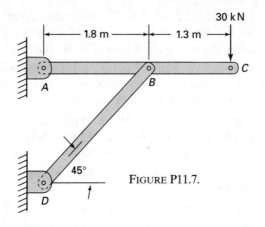

30 kN

FIGURE P11.7.

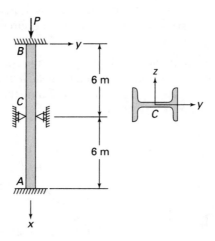

FIGURE P11.10.

11.10. Compute the allowable axial load P for a wide-flange steel column with fixed-fixed ends and a length of 12 m, braced at midpoint C (Fig. P11.10). *Given:* Material properties $E = 210$ GPa, $\sigma_{yp} = 280$ MPa; cross-sectional area $A = 27 \times 10^3$ mm^2 and radius of gyration $r = 100$ mm.

Sections 11.1 through 11.7

11.11. A column of length $3L$ is approximated by three bars of equal length connected by a torsional spring of appropriate stiffness k at each joint. The column is supported by a torsional spring of stiffness k at one end and is free at the other end. Derive an expression for determining the critical load of the system. Generalize the problem to the case of n connected bars.

11.12. A 3-m-long fixed-ended column $(L_e = 1.5\,\text{m})$ is made of a solid bronze rod $(E = 110\,\text{GPa})$ of diameter $D = 30$ mm. To reduce the weight of the column by 25%, the solid rod is replaced by the hollow rod of cross section shown in Fig. P11.12. Compute (a) the percentage of reduction in the critical load and (b) the critical load for the hollow rod.

11.13. A 2-m-long pin-ended column of square cross section is to be constructed of timber for which $E = 11$ GPa and $\sigma_{all} = 15$ MPa for compression parallel to the grain. Using a factor of safety of 2 in computing Euler's buckling load, determine the size of the cross section if the column is to safely support (a) a 100-kN load and (b) a 200-kN load.

11.14. A horizontal rigid bar AB is supported by a pin-ended column CD and carries a load F (Fig. P11.14). The column is made of steel bar having 50- by 50-mm square cross section, 3 m length, and $E = 200$ GPa. What is the allowable value of F based a factor of safety of $n = 2.2$ with respect to buckling of the column?

11.15. A uniform steel column, with fixed- and hinge-connected ends, is subjected to a vertical load $P = 450$ kN. The cross section of the column is 0.05 by 0.075 m and the length is 3.6 m. Taking $\sigma_{yp} = 280$ MPa and $E = 210$ GPa, calculate (a) the critical load and critical Euler stress, assuming a factor of safety of 2, and (b) the allowable stress according to the AISC formula, Eq. (11.13).

11.16. Figure P11.16 shows a square frame. Determine the critical value of the compressive forces P. All members are of equal length L and of equal modulus of rigidity EI. Assume that symmetrical buckling, indicated by the dashed lines in the figure, occurs.

11.17. A rigid block of weight W is to be supported by three identical steel bars. The bars are fixed at each end (Fig. P11.17). Assume that sidesway is not prevented and that, when an additional downward force of $2W$ is applied at

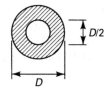

FIGURE P11.12.

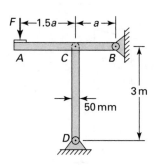

FIGURE P11.14.

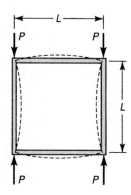

FIGURE P11.16.

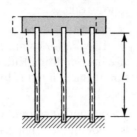

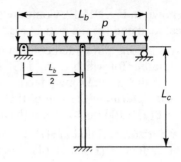

FIGURE P11.17.   FIGURE P11.18.

the middle of the block, buckling will take place as indicated by the dashed lines in the figure. Find the effective lengths of the columns by solving the differential equation for deflection of the column axis.

11.18. A simply supported beam of flexural rigidity EI_b is propped up at its center by a column of flexural rigidity EI_c (Fig. P11.18). Determine the midspan deflection of the beam if it is subjected to a uniform load p per unit length.

11.19. Two in-line identical cantilevers of cross-sectional area A, rigidity EI, and coefficient of thermal expansion α are separated by a small gap δ. What temperature rise will cause the beams to (a) just touch and (b) buckle elastically?

11.20. A W203 × 25 column fixed at both ends has a minimum radius of gyration $r = 29.4$ mm, cross-sectional area $A = 3230$ mm^2, and length 1.94 m. It is made of a material whose compression stress–strain diagram is given in Fig. P11.20 by dashed lines. Find the critical load. The stress–strain diagram may be approximated by a series of tangentlike segments, the accuracy improving as the number of segments increases. For simplicity, use four segments, as indicated in the figure. The modulus of elasticity and various tangent moduli (the slopes) are labeled.

11.21. The pin-jointed structure shown in Fig. P11.21 is constructed of two 0.025-m-diameter tubes having the following properties: $A = 5.4 \times 10^{-5}$ m^2, $I = 3.91 \times 10^{-9}$ m^4, $E = 210$ GPa. The stress–strain curve for the tube

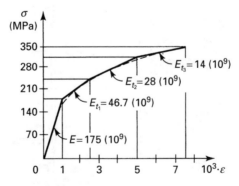

FIGURE P11.20.

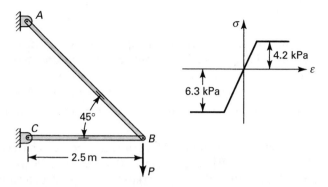

FIGURE P11.21.

material can be accurately approximated by three straight lines as shown. If the load P is increased until the structure fails, which tube fails first? Describe the nature of the failure and determine the critical load.

11.22. Two 0.075- by 0.075-m equal leg angles, positioned with the legs 0.025 m apart back to back, as shown in Fig. P11.22, are used as a column. The angles are made of structural steel with $\sigma_{yp} = 203$ MPa and $E = 210$ GPa. The area properties of an angle are thickness $t = 0.0125$ m, $A = 1.719 \times 10^{-3}$ m^2, $I_c = 8.6 \times 10^{-7}$ m^4, $I/c = S = 1.719 \times 10^{-5}$ m^3, $r_c = 0.0225$ m, and $\bar{z} = \bar{y} = 0.02325$ m. Assume that the columns are connected by lacing bars that cause them to act as a unit. Determine the critical stress of the column by using the AISC formula, Eq. (11.13), for effective column lengths (a) 2.1 m and (b) 4.2 m.

11.23. Redo Prob. 11.22(b) using the Euler formula.

11.24. A 1.2-m-long, 0.025- by 0.05-m rectangular column with rounded ends fits perfectly between a rigid ceiling and a rigid floor. Compute the change in temperature that will cause the column to buckle. Let $\alpha = 10 \times 10^{-6}/°$C, $E = 140$ GPa, and $\sigma_{yp} = 280$ MPa.

11.25. A pin-ended W150 × 24 rolled-steel column of cross section shown in Fig. P11.25 ($A = 3.06 \times 10^3$ mm^2, $r_x = 66$ mm, and $r_y = 24.6$ mm) carries an axial load of 125 kN. What is the longest allowable column length according to the AISC formula? Use $E = 200$ GPa and $\sigma_{yp} = 250$ Pa.

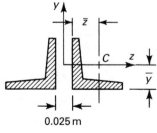

FIGURE P11.22.

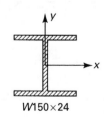

FIGURE P11.25.

Sections 11.8 and 11.9

11.26. A pinned-end rod AB of diameter d supports an eccentrically applied load of P, as depicted in Fig. P11.26. Assuming that the maximum deflection at the midlength is v_{max}, determine (a) the eccentricity e; (b) the maximum stress in the rod. *Given:* $d = 40$ mm, $P = 80$ kN, $v_{max} = 1.0$ mm, and $E = 200$ GPa.

11.27. Redo Prob. 11.26 for the case in which column AB is fixed at base B and free at top A.

11.28. A cast-iron hollow-box column of length L is fixed at its base and free at its top, as depicted in Fig. P11.28. Assuming that an eccentric load P acts at the middle of side AB (that is, $e = 50$ mm) of the free end, determine the maximum stress σ_{max} in the column. *Given:* $P = 250$ kN, $L = 1.8$ m, $E = 70$ GPa, and $e = 50$ mm.

11.29 Resolve Prob. 11.28, knowing that the load acts at the middle of side AC (that is, $e = 100$ mm).

11.30. A steel bar $(E = 210$ GPa$)$ of $b = 50$ mm by $h = 25$ mm rectangular cross section (Fig. P11.30) and length $L = 1.5$ m is eccentrically compressed by axial loads $P = 10$ kN. The forces are applied at the middle of

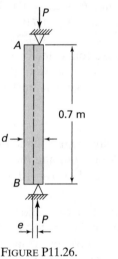

FIGURE P11.26.

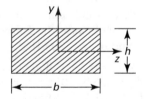

FIGURE P11.30.

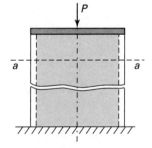

FIGURE P11.28.

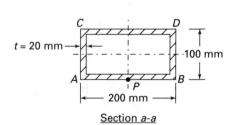

Section a-a

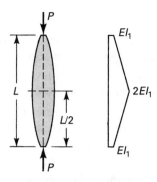

FIGURE P11.36.

the long edge of the cross section. What are the maximum deflection v_{max} and maximum bending moment M_{max}?

11.31. A 0.05-m square, horizontal steel bar, 9-m long, is simply supported at each end. The only force acting is the weight of the bar. (a) Find the maximum stress and deflection. (b) Assume that an axial compressive load of 4.5 kN is also applied at each end through the centroid of cross-sectional area. Determine the stress and deflection under this combined loading. For steel, the specific weight is 77 kN/m³, $E = 210$ GPa, and $v = 0.3$.

11.32. The properties of a W203 × 46 steel link are $A = 5880$ mm², $I_z \approx 45.66 \times 10^6$ mm⁴, $I_y \approx 15.4 \times 10^6$ mm⁴, depth = 203.2 mm, width of flange = 203.2 mm, and $E = 210$ GPa. What maximum end load P can be applied at both ends, given an eccentricity of 0.05 m along axis yy? A stress of 210 MPa is not to be exceeded. Assume that the effective column length of the link is 4.5 m.

11.33. A hinge-ended bar of length L and stiffness EI has an initial curvature expressed by $v_0 = a_1 \sin(\pi x/L) + 5a_1 \sin(2\pi x/L)$. If this bar is subjected to an increasing axial load P, what value of the load P, expressed in terms of $L, E,$ and I, will result in zero deflection at $x = 3L/4$?

11.34. Employing the equilibrium approach, derive the following differential equation for a simply supported beam column subjected to an arbitrary distributed transverse loading $p(x)$ and axial force P:

$$\frac{d^4v}{dx^4} + \frac{P}{EI}\frac{d^2v}{dx^2} = \frac{p}{EI} \tag{P11.34}$$

Demonstrate that the homogeneous solution for this equation is

$$v = c_1 \sin\sqrt{\frac{P}{EI}}\,x + c_2 \cos\sqrt{\frac{P}{EI}}\,x + c_3 x + c_4$$

where the four constants of integration will require, for evaluation, four boundary conditions.

Section 11.10

11.35. Assuming $v = a_0[1 - (2x/L)^2]$, determine the buckling load of a pin-ended column. Employ the Rayleigh–Ritz method, placing the origin at midspan.

11.36. The cross section of a pin-ended column varies as in Fig. P11.36. Determine the critical load using an energy approach.

11.37. A cantilever column has a moment of inertia varying as $I = I_1(1 - x/2L)$, where I_1 is the constant moment of inertia at the fixed end $(x = 0)$. Find the buckling load by choosing $v = v_1(x/L)^2$. Here v_1 is the deflection of the free end.

11.38. Derive an expression for the deflection of the uniform pin-ended beam-column of length L, subjected to a uniform transverse load p and axial compressive force P. Use an energy approach.

11.39. A simply supported beam column of length L is subjected to compression forces P at both ends and lateral loads F and $2F$ at quarter-length and midlength, respectively. Employ the Rayleigh–Ritz method to determine the beam deflection.

11.40. Determine the critical compressive load P that can be carried by a cantilever at its free end $(x = L)$. Use the Rayleigh–Ritz method and let $v = x^2(L - x)(a + bx)$, where a and b are constants.

Sections 11.11 and 11.12

11.41. A stepped cantilever beam with a hinged end, subjected to the axial compressive load P, is shown in Fig. P11.41. Determine the critical value of P, applying the method of finite differences. Let $m = 3$ and $L_1 = L_2 = L/2$.

11.42. A uniform cantilever column is subjected to axial compression at the free end $(x = L)$. Determine the critical load. Employ the finite differences, using $m = 2$. [Hint: The boundary conditions are $v(0) = v'(0) = v''(L) = v'''(L) = 0$.]

11.43. The cross section of a pin-ended column varies as in Fig. P11.36. Determine the critical load using the method of finite differences. Let $m = 4$.

11.44. Find the critical value of the load P in Fig. P11.41 if both ends of the beam are simply supported. Let $L_1 = L/4$ and $L_2 = 3L/4$. Employ the method of finite differences by taking the nodes at $x = 0$, $x = L/4$, $x = L/2$, and $x = L$.

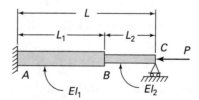

FIGURE P11.41.

CHAPTER 12

Plastic Behavior of Materials

12.1 INTRODUCTION

Thus far, we have considered loadings that cause the material of a member to behave elastically. We are now concern with the behavior of machine and structural components when stresses exceed the proportional limit. In such cases it is necessary to make use of the stress–strain diagrams obtained from an actual test of the material, or an *idealized stress–strain diagram*. For purposes of analysis, a single mathematical expression will often be employed for the entire stress–strain diagram. Deformations and stresses in members made of elastic–plastic and rigid-plastic materials having various forms will be determined under single and combined loadings. Applications include collapse load of structures, limit design, membrane analogy, rotating disks, and pressure vessels.

The *plasticity* describes the inelastic behavior of a material that retains permanent yielding on complete unloading. The subject of plasticity is perhaps best introduced by recalling the principal characteristics of elastic behavior. First, a material subjected to stressing within the elastic regime will return to its original state upon removal of those external influences causing application of load or displacement. Second, the deformation corresponding to a given stress depends solely on that stress and not on the history of strain or load. In plastic behavior, opposite characteristics are observed. The permanent distortion that takes place in the plastic range of a material can assume considerable proportions. This distortion depends not only on the final state of stress but on the stress states existing from the start of the loading process as well.

The equations of equilibrium (1.14), the conditions of compatibility (2.12), and the strain–displacement relationships (2.4) are all valid in plastic theory. New relationships must, however, be derived to connect stress and strain. The various yield criteria (discussed in Chap. 4), which strictly speaking are not required in solving a problem in elasticity, play a direct and important role in plasticity. This chapter can provide only an introduction to what is an active area of contemporary design and research in the mechanics of solids.* The basics presented can, however, indicate the potential of the field, as well as its complexities.

12.2 PLASTIC DEFORMATION

We shall here deal with the permanent alteration in the shape of a polycrystalline solid subject to external loading. The crystals are assumed to be randomly oriented. As has been demonstrated in Section 2.15, the stresses acting on an elemental cube can be resolved into those associated with change of volume (dilatation) and those causing distortion or change of shape. The distortional stresses are usually referred to as the *deviator stresses*.

The dilatational stresses, such as hydrostatic pressures, can clearly decrease the volume while they are applied. The volume change is recoverable, however, upon removal of external load. This is because the material cannot be compelled to assume, in the absence of external loading, interatomic distances different from the initial values. When the dilatational stresses are removed, therefore, the atoms revert to their original position. Under the conditions described, no plastic behavior is noted, and the volume is essentially unchanged upon removal of load.

In contrast with the situation described, during change of shape, the atoms within a crystal of a polycrystalline solid slide over one another. This slip action, referred to as *dislocation*, is a complex phenomenon. Dislocation can occur only by the shearing of atomic layers, and consequently it is primarily the shear component of the deviator stresses that controls plastic deformation. Experimental evidence supports the assumption that associated with plastic deformation essentially *no volume change* occurs; that is, the material is incompressible (Sec. 2.10):

$$\varepsilon_x + \varepsilon_y + \varepsilon_z = \varepsilon_1 + \varepsilon_2 + \varepsilon_3 = 0 \tag{12.1}$$

Therefore, from Eq. (2.39), *Poisson's ratio $v = \frac{1}{2}$ for a plastic material.*

Slip begins at an imperfection in the lattice, for example, along a plane separating two regions, one having one more atom per row than the other. Because slip does not occur simultaneously along every atomic plane, the deformation appears discontinuous on the microscopic level of the crystal grains. The overall effect, however, is plastic shear along certain slip planes, and the behavior described is approximately that of the ideal plastic solid. As the deformation continues, a locking of the dislocations takes place, resulting in *strain hardening* or *cold working*.

In performing engineering analyses of stresses in the plastic range, we do not usually need to consider dislocation theory, and the explanation offered previously, while overly simple, will suffice. What is of great importance to the analyst, however, is the experimentally determined curves of stress and strain.

12.3 IDEALIZED STRESS–STRAIN DIAGRAMS

The bulk of present-day analysis in plasticity is predicated on materials displaying idealized stress–strain curves as in Fig. 12.1a and c. Such materials are referred to as rigid, *perfectly plastic* and elastic-perfectly plastic, respectively. Examples include

---

*For a detailed discussion of problems in plasticity, see Refs. 12.1 through 12.3.

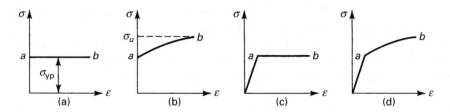

FIGURE 12.1. *Idealized stress–strain diagrams: (a) rigid, perfectly plastic material; (b) rigid-plastic material; (c) elastic, perfectly plastic material; and (d) elastic, rigid-plastic material.*

mild steel, clay, and nylon, which exhibit negligible elastic strains in comparison with large plastic deformations at practically constant stress. A more realistic portrayal including strain hardening is given in Fig. 12.1b for what is called a *rigid-plastic* solid. In the curves, *a* and *b* designate the tensile yield and ultimate stresses, σ_{yp} and σ_u, respectively. The curves of Fig. 12.1c and d do not ignore the elastic strain, which must be included in a more general stress–strain depiction. The latter figures thus represent idealized *elastic–plastic* diagrams for the perfectly plastic and plastic materials, respectively. The material in Figs. 12.c and d is also called a *strain-hardening* or *work-hardening material*. For a linearly strain-hardening material, the regions *ab* of the diagrams become a *sloped straight line*.

True Stress–True Strain Relationships

Consider the idealized true stress–strain diagram given in Fig. 12.2. The fracture stress and fracture strain are denoted by σ_f and ε_f, respectively. The fundamental *design* utility of the plot ends at the ultimate stress σ_u. Note that a comparison of a true and nominal $\sigma - \varepsilon$ plot is shown in Fig. 2.12. In addition to the preceding

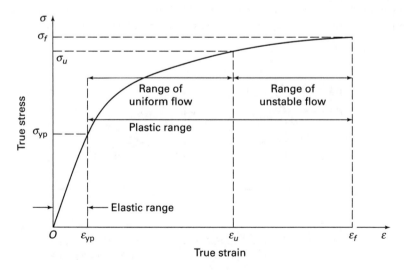

FIGURE 12.2. *Idealized true stress–strain diagram.*

idealized cases, various theoretical analyses have been advanced to predict plastic behavior. Equations relating stress and strain beyond the proportional limit range from the rather empirical to those leading to complex mathematical approaches applicable to materials of specific type and structure. For purposes of an accurate analysis, a single mathematical expression is frequently employed for the entire stress–strain curve. An equation representing the range of $\sigma - \varepsilon$ diagrams, especially useful for aluminum and magnesium, has been developed by Ramberg and Osgood [Ref. 12.4]. We confine our discussion to that of a perfectly plastic material displaying a horizontal straight-line relationship and to the parabolic relationship described next.

For many materials, the *entire* true stress–true strain curve may be represented by the parabolic form

$$\sigma = K\varepsilon^n \tag{12.2}$$

where n and K are the *strain-hardening index* and the *strength coefficient*, respectively. The definitions of true stress and true strain are given in Section 2.7. The curves of Eq. (12.2) are shown in Fig. 12.3a. We observe from the figure that the slope $d\sigma/d\varepsilon$ grows without limit as ε approaches zero for $n \neq 1$. Thus, Eq. (12.2) should not be used for small strains when $n \neq 1$. The stress–strain diagram of a perfectly plastic material is represented by this equation when $n = 0$ (and hence $K = \sigma_{yp}$). Clearly, for elastic materials ($n = 1$ and hence $K = E$), Eq. (12.2) represents Hooke's law, wherein E is the modulus of elasticity.

For a particular material, with true stress–true strain data available, K and n are readily evaluated inasmuch as Eq. (12.2) plots as a straight line on *logarithmic coordinates*. We can thus rewrite Eq. (12.2) in the form

$$\log \sigma = \log K + n \log \varepsilon \tag{a}$$

Here n is the *slope* of the line and K the true stress associated with the true strain at 1.0 on the *log–log plot* (Fig. 12.3b). The strain-hardening coefficient n for commercially used materials falls between 0.2 and 0.5. Fortunately, with the use of computers, much refined modeling of the stress–strain relations for real material is possible.

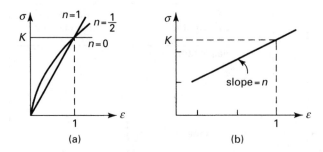

FIGURE 12.3. (a) Graphical representation of $\sigma = K\varepsilon^n$; (b) true stress versus true strain on log–log coordinates.

12.4 INSTABILITY IN SIMPLE TENSION

We now describe an instability phenomenon in uniaxial tension of practical importance in predicting the maximum allowable plastic stress in a rigid-plastic material. At the ultimate stress σ_u in a tensile test (Fig. 12.2), an *unstable flow* results from the effects of strain hardening and the decreasing cross-sectional area of the specimen. These tend to weaken the material. When the rate of the former effect is less than the latter, an *instability* occurs. This point corresponds to the *maximum tensile load* and is defined by

$$dP = 0 \qquad \text{(a)}$$

Since axial load P is a function of both the true stress and the area $(P = \sigma A)$, Eq. (a) is rewritten

$$\sigma \, dA + A \, d\sigma = 0 \qquad \text{(b)}$$

The condition of incompressibility, $A_o L_o = AL$, also yields

$$L \, dA + A \, dL = 0 \qquad \text{(c)}$$

as the original volume $A_o L_o$ is constant. Expressions (b) and (c) result in

$$\frac{d\sigma}{\sigma} = \frac{dL}{L} = d\varepsilon \qquad \text{(d)}$$

From Eqs. (2.25) and (d), we thus obtain the relationships

$$\frac{d\sigma}{d\varepsilon} = \sigma \quad \text{or} \quad \frac{d\sigma}{d\varepsilon_o} = \frac{\sigma}{1 + \varepsilon_o} = \sigma_o \qquad \textbf{(12.3)}$$

for the *instability of a tensile member*. Here the subscript o denotes the engineering strain and stress (Sec. 2.7).

Introduction of Eq. (12.2) into Eq. (12.3) results in

$$\sigma = K\varepsilon^n = \frac{d}{d\varepsilon}(K\varepsilon^n) = nK\varepsilon^{n-1}$$

or

$$\varepsilon = n \qquad \textbf{(12.4)}$$

That is, at the instant of instability of flow *in tension*, the *true strain ε has the same numerical value as the strain-hardening index*. The state of true stress and the true strain under uniaxial tension are therefore

$$\sigma_1 = Kn^n, \quad \sigma_2 = \sigma_3 = 0$$

$$\varepsilon_1 = n, \quad \varepsilon_2 = \varepsilon_3 = -\frac{n}{2} \qquad \text{(e)}$$

The problem of instability under *simple compression* or *plastic buckling* is discussed in Section 11.6. The instability condition for cases involving *biaxial tension* is derived in Sections 12.12 and 12.13.

FIGURE 12.4.   *Example 12.1. Pin-connected
structure in axial tension.*

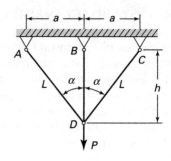

EXAMPLE 12.1   Three-Bar Structure

Determine the maximum allowable plastic stress and strain in the pin-jointed structure sustaining a vertical load P, shown in Fig. 12.4. Assume that $\alpha = 45°$ and that each element is constructed of an aluminum alloy with the following properties:

$$\sigma_{yp} = 350 \text{ MPa}, \quad K = 840 \text{ MPa}, \quad n = 0.2$$

$$A_{AD} = A_{CD} = 10 \times 10^{-5} \text{ m}^2, \quad A_{BD} = 15 \times 10^{-5} \text{ m}^2, \quad h = 3 \text{ m}$$

Solution   The structure is elastically statically indeterminate, and the solution may readily be obtained on applying Castigliano's theorem (Sec. 10.7). Plastic yielding begins upon loading:

$$P = \sigma_{yp} A_{BD} + 2\sigma_{yp} A_{AD} \cos \alpha$$
$$= 350 \times 10^6 [15 + 2 \times 10 \cos 45°] 10^{-5} = 101,990 \text{ N}$$

On applying Eqs. (e), the maximum allowable stress

$$\sigma = Kn^n = 840 \times 10^6 (0.2)^{0.2} = 608.9 \text{ MPa}$$

occurs at the following axial and transverse strains:

$$\varepsilon_1 = n = 0.2, \quad \varepsilon_2 = \varepsilon_3 = -0.1$$

We have $L = h/\cos \alpha$. The total elongations for instability of the bars are thus $\delta_{BD} = 3(0.2) = 0.6$ m and $\delta_{AB} = \delta_{CD} = (3/\cos 45°)(0.2) = 0.85$ m.

EXAMPLE 12.2   Tube in Axial Tension

A tube of original mean diameter d_o and thickness t_o is subjected to axial tensile loading. Assume a true stress–engineering strain relation of the form $\sigma = K_1 \varepsilon_o^n$ and derive expressions for the thickness and diameter at the instant of instability. Let $n = 0.3$.

Solution   Differentiating the given expression for stress,

$$\frac{d\sigma}{d\varepsilon_o} = nK_1 \varepsilon_o^{n-1} = \frac{n\sigma}{\varepsilon_o}$$

This result and Eq. (12.3) yield the engineering axial strain at instability:

$$\varepsilon_o = \frac{n}{1-n} \qquad \text{(f)}$$

The transverse strains are $-\varepsilon_o/2$, and hence the decrease of wall thickness equals $nt_o/2(1-n)$. The thickness at instability is thus

$$t = t_o - \frac{nt_o}{2(1-n)} = \frac{2-3n}{2(1-n)}t_o \qquad \text{(g)}$$

Similarly, the diameter at instability is

$$d = d_o - \frac{nd_o}{2(1-n)} = \frac{2-3n}{2(1-n)}d_o \qquad \text{(h)}$$

From Eqs. (g) and (h), with $n = 0.3$, we have $t = 0.79t_o$ and $d = 0.79d_o$. Thus, for the tube under axial tension, the diameter and thickness decrease approximately 21% at the instant of instability.

12.5 PLASTIC AXIAL DEFORMATION AND RESIDUAL STRESS

As pointed out earlier, if the stress in any part of the member exceeds the yield strength of the material, *plastic deformations* occur. The stress that remains in a structural member upon removal of external loads is referred to as the *residual stress*. The presence of residual stress may be very harmful or, if properly controlled, may result in substantial benefit.

Plastic behavior of ductile materials may be conveniently represented by considering an idealized *elastoplastic* material shown in Fig. 12.5, where σ_{yp} and ε_{yp} designate the yield strength and yield strain, respectively. The Y corresponds to the onset of yield in the material. The part OB is the strain corresponding to the plastic deformation that results from the loading and unloading of the specimen along line AB parallel to the initial portion OY of the loading curve.

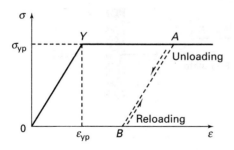

FIGURE 12.5. *Tensile stress–strain diagram for elastoplastic material.*

The distribution of residual stresses can be found by superposition of the stresses owing to loading and the *reverse*, or rebound, *stresses* due to unloading. (The strains corresponding to the latter are the reverse, or rebound, strains.) The reverse stress pattern is assumed to be *fully elastic* and consequently can be obtained applying Hooke's law. That means the linear relationship between σ and ε remains valid, as illustrated by line AB in the figure. We note that this superposition approach is not valid if the residual stresses thereby found exceed the yield strength.

The nonuniform deformations that may be caused in a material by plastic bending and plastic torsion are considered in Sections 12.7 and 12.9. This section is concerned with a restrained or *statically indeterminate* structure that is axially loaded beyond the elastic range. For such a case, some members of the structure experience different plastic deformation, and these members retain stress following the release of load.

The magnitude of the axial load at the onset of yielding, or *yield load* P_{yp}, in a statically determinate ductile bar of cross-sectional area A is $\sigma_{yp} A$. This also is equal to the *plastic, limit*, or *ultimate load* P_u of the bar. For a statically indeterminate structure, however, after one member yields, additional load is applied until the remaining members also reach their yield limits. At this time, *unrestricted* or *uncontained plastic flow* occurs, and the limit load P_u is reached. Therefore, the ultimate load is the load at which yielding begins in *all* materials.

Examples 12.3, 12.7, and 12.12 illustrate how plastic deformations and residual stresses are produced and how the limit load is determined in axially loaded members, beams, and torsional members, respectively.

EXAMPLE 12.3   Residual Stresses in an Assembly

Figure 12.6a shows a steel bar of 750-mm^2 cross-sectional area placed between two aluminum bars, each of 500-mm^2 cross-sectional area. The ends of the bars are attached to a rigid support on one side and a rigid thick plate on the other. *Given:* $E_s = 210$ GPa, $(\sigma_s)_{yp} = 240$ MPa, $E_a = 70$ GPa, and $(\sigma_a)_{yp} = 320$ MPa. *Assumption:* The material is elastic–plastic.

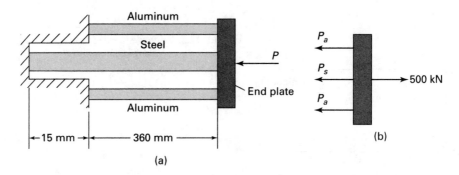

(a)

(b)

FIGURE 12.6.   *Example 12.3. (a) Plastic analysis of a statically indeterminate three-bar structure; (b) free-body diagram of end plate.*

Chapter 12   Plastic Behavior of Materials

Calculate the residual stress, for the case in which applied load P is increased from zero to P_u and removed.

Solution *Material Behavior.* At ultimate load P_u both materials yield. Either material yielding by itself will not result in failure because the other material is still in the elastic range. We therefore have

$$P_a = 500(320) = 160 \text{ kN}, \quad P_s = 750(240) = 180 \text{ kN}$$

Hence,

$$P_u = 2(160) + 180 = 500 \text{ kN}$$

Applying an equal and opposite load of this amount, equivalent to a release load (Fig. 12.6b), causes each bar to rebound elastically.

Geometry of Deformation. Condition of geometric fit, $\delta_a = \delta_s$ gives

$$\frac{P_a'(360)}{500(700)} = \frac{P_s'(375)}{750(210)} \quad \text{or} \quad P_s' = 4.32 P_a' \tag{a}$$

Condition of Equilibrium. From the free-body diagram of Fig. 12.6b,

$$2P_a' + P_s' = 500 \text{ kN} \tag{b}$$

Solving Eqs. (a) and (b) we obtain

$$P_a' = 79.1 \text{ kN} \qquad P_s' = 341.7 \text{ kN}$$

Superposition of the initial forces at ultimate load P_u and the elastic rebound forces owing to release of P_u results in:

$$(P_a)_{\text{res}} = 79.1 - 160 = -80.9 \text{ kN}$$

$$(P_s)_{\text{res}} = 341.7 - 180 = 161.7 \text{ kN}$$

The associated residual stresses are thus

$$(\sigma_a)_{\text{res}} = -\frac{80.9 \times 10^3}{500} = -162 \text{ MPa}$$

$$(\sigma_s)_{\text{res}} = -\frac{161.7 \times 10^3}{750} = 216 \text{ MPa}$$

Comment We note that after this prestressing process, the assembly remains elastic as long as the value of $P_u = 500$ kN is not exceeded.

12.6 PLASTIC DEFLECTION OF BEAMS

In this section we treat the inelastic deflection of a beam, employing the mechanics of materials approach. Consider a beam of rectangular section, as in Fig. 12.7a, wherein the bending moment M produces a radius of curvature r. The longitudinal

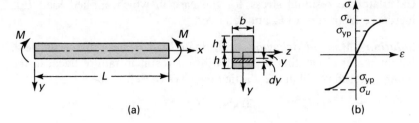

FIGURE 12.7.   (a) *Inelastic bending of a rectangular beam in pure bending;*
(b) *stress–strain diagram for an elastic-rigid plastic material.*

strain of any fiber located a distance y from the neutral surface, from Eq. (5.9), is given by

$$\varepsilon = \frac{y}{r} \tag{12.5}$$

Assume the beam material to possess equal properties in tension and compression. Then, the longitudinal tensile and compressive forces cancel, and the equilibrium of axial forces is satisfied. The following describes the equilibrium of moments about the z axis (Fig. 12.7a):

$$\int_A y\sigma \, dA = b \int_{-h}^{h} y\sigma \, dy = M \tag{a}$$

For any specific distribution of stress, as, for example, that shown in Fig. 12.7b, Eq. (a) provides M and then the deflection, as is demonstrated next.

Consider the true stress–true strain relationship of the form $\sigma = K\varepsilon^n$. Introducing this together with Eq. (12.5) into Eq. (a), we obtain

$$M = b \int_{-h}^{h} \frac{1}{r^n} K y^{n+1} dy = \frac{1}{r^n} K I_n \tag{b}$$

where

$$I_n = b \int_{-h}^{h} y^{n+1} dy \tag{12.6}$$

From Eqs. (12.5), $\sigma = K\varepsilon^n$, and (b) the following is derived:

$$\frac{\sigma}{y^n} = \frac{K}{r^n} = \frac{M}{I_n} \tag{c}$$

In addition, on the basis of the elementary beam theory, we have, from Eq. (5.7),

$$\frac{1}{r} = \frac{d^2 v}{dx^2} \tag{d}$$

Upon substituting Eq. (c) into Eq. (d), we obtain the following equation for a rigid plastic beam:

$$\frac{d^2v}{dx^2} = \left(\frac{M}{KI_n}\right)^{1/n} \tag{12.7}$$

It is noted that when $n = 1$ (and hence $K = E$), this expression, as expected, reduces to that of an elastic beam [Eq. (5.10)].

EXAMPLE 12.4   Rigid-Plastic Simple Beam

Determine the deflection of a rigid-plastic simply supported beam subjected to a downward concentrated force P at its midlength. The beam has a rectangular cross section of depth $2h$ and width b (Fig. 12.8).

Solution   The bending moment for segment AC is given by

$$M = -\tfrac{1}{2}Px, \qquad 0 \le x \le \frac{L}{2} \tag{e}$$

where the minus sign is due to the sign convention of Section 5.2.

Substituting Eq. (e) into Eq. (12.7) and integrating, we have

$$\frac{dv}{dx} = -\frac{\lambda x^{(1/n)+1}}{(1/n)+1} + c_1$$

$$v = -\frac{\lambda x^{(1/n)+2}}{[(1/n)+1][(1/n)+2]} + c_1 x + c_2 \tag{f}$$

where

$$\lambda = \left(\frac{P}{2KI_n}\right)^{1/n} \tag{g}$$

The constants of integration c_1 and c_2 depend on the boundary conditions $v(0) = dv/dx(L/2) = 0$:

$$c_2 = 0, \quad c_1 = \frac{\lambda(L/2)^{(1/n)+1}}{(1/n)+1}$$

Upon introduction of c_1 and c_2 into Eq. (f), the beam deflection is found to be

$$v = -\frac{\lambda}{(1/n)+1}\left[\frac{x^{(1/n)+2}}{(1/n)+2} - \left(\frac{L}{2}\right)^{(1/n)+1} x\right], \quad 0 \le x \le \frac{L}{2} \tag{12.8}$$

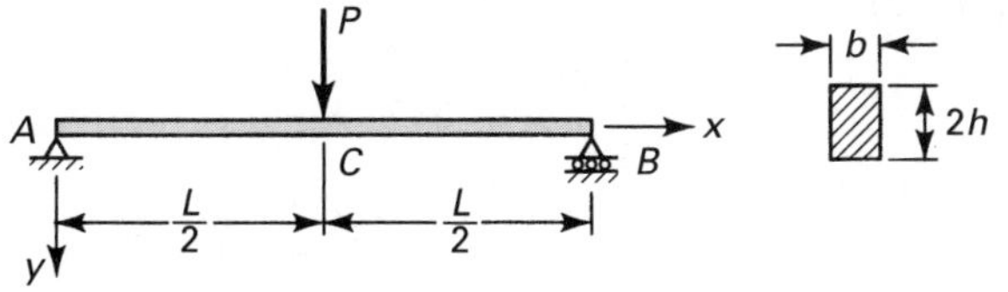

FIGURE 12.8.   *Examples 12.4 and 12.8.*

Interestingly, in the case of an elastic beam, this becomes

$$v = -\frac{P}{4EI_1}\left(\frac{x^3}{3} - \frac{L^2}{4}x\right), \quad 0 \le x \le \frac{L}{2}$$

For $x = L/2$, the familiar result is

$$v = v_{\max} = \frac{PL^3}{48EI_1}$$

The foregoing procedure is applicable to the determination of the deflection of beams subject to a variety of end conditions and load configurations. It is clear, however, that owing to the nonlinearity of the stress law, $\sigma = K\varepsilon^n$, the principle of superposition cannot validly be applied.

12.7 ANALYSIS OF PERFECTLY PLASTIC BEAMS

By neglecting strain hardening, that is, by assuming a perfectly plastic material, considerable simplification can be realized. We shall, in this section, focus our attention on the analysis of a perfectly plastic straight beam of rectangular section subject to *pure bending* (Fig. 12.7a).

The bending moment at which plastic deformation impends, M_{yp}, may be found directly from the flexure relationship:

$$M_{\mathrm{yp}} = \frac{\sigma_{\mathrm{yp}}I}{h} = \tfrac{2}{3}bh^2\sigma_{\mathrm{yp}} \tag{12.9a}$$

or

$$M_{\mathrm{yp}} = S\sigma_{\mathrm{yp}} \tag{12.9b}$$

Here σ_{yp} represents the stress at which yielding begins (and at which deformation continues in a perfectly plastic material). The quantity S is the *elastic modulus* of the cross section. Clearly, for the beam considered, $S = I/c = [b(2h)^3/12]/h = \tfrac{2}{3}bh^2$. The stress distribution corresponding to M_{yp}, assuming identical material properties in tension and compression, is shown in Fig. 12.9a. As the bending moment is increased, the region of the beam that has yielded progresses in toward the neutral surface (Fig. 12.9b). The distance from the neutral surface to the point at which yielding begins is denoted by the symbol e, as shown.

It is clear, upon examining Fig. 12.9b, that the normal stress varies in accordance with the relations

$$\sigma_x = \frac{\sigma_{\mathrm{yp}}y}{e}, \quad -e \le y \le e \tag{a}$$

and

$$\sigma_x = \sigma_{\mathrm{yp}}, \qquad e \le y \le h$$
$$\sigma_x = -\sigma_{\mathrm{yp}}, \quad -e \ge y \ge -h \tag{b}$$

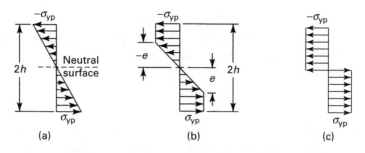

FIGURE 12.9. *Stress distribution in a rectangular beam with increase in bending moment: (a) elastic; (b) partially plastic; (c) fully plastic.*

It is useful to determine the manner in which the bending moment M relates to the distance e. To do this, we begin with a statement of the x equilibrium of forces:

$$\int_{-h}^{-e} -\sigma_{yp}b\,dy + \int_{-e}^{e} \sigma_x b\,dy + \int_{e}^{h} \sigma_{yp}b\,dy = 0$$

Cancelling the first and third integrals and combining the remaining integral with Eq. (a), we have

$$\frac{\sigma_{yp}}{e}\int_{-e}^{e} yb\,dy = 0$$

This expression indicates that the *neutral and centroidal axes of the cross section coincide*, as in the case of an entirely elastic distribution of stress. Note that in the case of a *nonsymmetrical* cross section, the neutral axis is generally in a location different from that of the centroidal axis [Ref. 12.5].

Next, the equilibrium of moments about the neutral axis provides the following relation:

$$\int_{-h}^{-e} -\sigma_{yp}yb\,dy + \int_{-e}^{e} \sigma_x yb\,dy + \int_{e}^{h} \sigma_{yp}yb\,dy = M$$

Substituting σ_x from Eq. (a) into this equation gives the *elastic–plastic moment*, after integration,

$$M = b\sigma_{yp}\left(h^2 - \frac{e^2}{3}\right) \tag{12.10a}$$

Alternatively, using Eq. (12.9a), we have

$$M = \tfrac{3}{2}M_{yp}\left[1 - \frac{1}{3}\left(\frac{e}{h}\right)^2\right] \tag{12.10b}$$

The general stress distribution is thus defined in terms of the applied moment, inasmuch as e is connected to σ_x by Eq. (a). For the case in which $e = h$, Eq. (12.10) reduces to Eq. (12.9) and $M = M_{yp}$. For $e = 0$, which applies to a totally plastic beam, Eq. (12.10a) becomes

$$M_u = bh^2\sigma_{yp} \tag{12.11}$$

where M_u is the *ultimate moment*. It is also referred to as the *plastic moment*.

Through application of the foregoing analysis, similar relationships can be derived for *other* cross-sectional shapes. In general, for any cross section, the plastic or ultimate resisting moment for a beam is

$$M_u = \sigma_{yp}Z \tag{12.12}$$

where Z is the *plastic section modulus*. Clearly, for the rectangular beam analyzed here, $Z = bh^2$. The *Steel Construction Manual* (Sec. 11.7) lists plastic section moduli for many common geometries. Interestingly, the ratio of the ultimate moment to yield moment for a beam depends on the geometric form of the cross section, called the *shape factor*:

$$f = \frac{M_u}{M_{yp}} = \frac{Z}{S} \tag{12.13}$$

For instance, in the case of a perfectly plastic beam with ($b \times 2h$) with rectangular cross section, we have $S = I/h = \frac{2}{3}bh^2$ and $Z = bh^2$. Substitution of these area properties into the preceding equation results in $f = 3/2$. Observe that the plastic modulus represents the sum of the *first moments of the areas* (defined in Sec. C.1) of the cross section above and below the neutral axis z (Fig. 12.7a): $Z = bh(h/2 + h/2) = bh^2$.

As noted earlier, when a beam is bent inelastically, some plastic deformation is produced, and the beam does not return to its initial configuration after load is released. There will be some *residual stresses* in the beam (see Example 12.7). The foregoing stresses are found by using the principle of superposition in a way similar to that described in Section 12.5 for axial loading. The unloading phase may be analyzed by assuming the beam to be fully plastic, using the flexure formula, Eq. (5.4).

Plastic Hinge

We now consider the *moment–curvature relation* of an elastoplastic rectangular beam in pure bending (Fig. 12.7a). As described in Section 5.2, the applied bending moment M produces a curvature which is the reciprocal of the radius of curvature r. At the elastic–plastic boundary of the beam (Fig. 12.9b), the yield strain is $\varepsilon_x = \varepsilon_{yp}$. Then, Eq. (12.5) leads to

$$\frac{1}{r} = -\frac{\varepsilon_{yp}}{e} \quad \text{and} \quad \frac{1}{r_{yp}} = -\frac{\varepsilon_{yp}}{h} \tag{c}$$

The quantities $1/r$ and $1/r_{yp}$ correspond to the curvatures of the beam *after* and *at the onset* of yielding, respectively. Through the use of Eqs. (c),

$$\frac{1/r_{yp}}{1/r} = \frac{r}{r_{yp}} = \frac{e}{h} \tag{12.14}$$

Carrying Eqs. (12.14) and (12.9) into Eq. (12.10b) give the required moment–curvature relationship in the form:

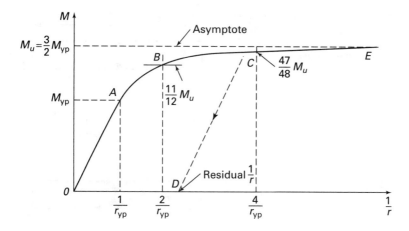

FIGURE 12.10. *Moment–curvature relationship for a rectangular beam.*

$$M = \frac{3}{2} M_{yp} \left[1 - \frac{1}{3} \left(\frac{r}{r_{yp}} \right)^2 \right] = M_u \left[1 - \frac{1}{3} \left(\frac{r}{r_{yp}} \right)^2 \right] \qquad \textbf{(12.15)}$$

We note that Eq. (12.15) is valid only if the bending moment becomes greater than M_{yp}. When $M < M_{yp}$, Eq. (5.9a) applies. The variation of M with curvature is illustrated in Fig. 12.10. Observe that M rapidly approaches the asymptotic value $\frac{3}{2} M_{yp}$, which is the plastic moment M_u. If $1/r = 2/r_{yp}$ or $e = h/2$, eleven-twelfths of M_u has already been reached. Obviously, positions A, B, and E of the curve correspond to the stress distributions depicted in Fig. 12.9. Removing the load at C, for example, elastic rebound occurs along the line CD, and point D is the residual curvature.

Figure 12.10 depicts the three stages of loading. There is an *initial range* of linear elastic response. This is followed by a curved line representing the region in which the member is partially plastic and partially elastic. This is the region of *contained plastic flow*. Finally, the member continues to yield with no increase of applied bending moment. Note the rapid ascent of each curve toward its *asymptote* as the section approaches the fully plastic condition. At this stage, *unrestricted plastic flow* occurs and the corresponding moment is the ultimate or *plastic hinge moment* M_u. Therefore, the cross section will abruptly continue to rotate; the beam is said to have developed a *plastic hinge*. The rationale for the term *hinge* becomes apparent upon describing the behavior of a beam under a concentrated loading, discussed next.

Consider a simply supported beam of rectangular cross section, subjected to a load P at its midspan (Fig. 12.11a). The corresponding bending moment diagram is shown in Fig. 12.11b. Clearly, when $M_{yp} < |PL/4| < M_u$, a region of plastic deformation occurs, as indicated in the figure by the shaded areas. The depth of penetration of these zones can be found from $h - e$, where e is determined using Eq. (12.10a), because M at midspan is known. The length of the middle portion of the beam where plastic deformation occurs can be readily determined with reference to the figure. The magnitude of bending moment at the edge of the plastic zone is $M_{yp} = (P/2)(L - L_p)/2$, from which

FIGURE 12.11.
Bending of a rectangular beam: (a) plastic region; (b) moment diagram; (c) plastic hinge.

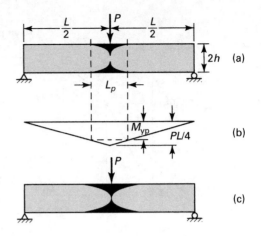

$$L_p = L - \frac{4M_{yp}}{P} \tag{d}$$

With the increase of P, $M_{max} \to M_u$, and the plastic region extends farther inward. When the magnitude of the maximum moment $PL/4$ is equal to M_u, the cross section at the midspan becomes fully plastic (Fig. 12.11c). Then, as in the case of pure bending, the curvature at the center of the beam grows without limit, and the beam fails. The beam halves on either side of the midspan experience rotation in the manner of a rigid body about the neutral axis, as about a *plastic hinge*, under the influence of the constant ultimate moment M_u. For a plastic hinge, $P = 4M_u/L$ is substituted into Eq. (d), leading to $L_p = L(1 - M_{yp}/M_u)$.

The capacity of a beam to resist collapse is revealed by comparing Eqs. (12.9a) and (12.11). Note that the M_u is 1.5 times as large as M_{yp}. Elastic design is thus conservative. Considerations such as this lead to concepts of *limit design* in structures, discussed in the next section.

EXAMPLE 12.5 Elastic–Plastic Analysis of a Link: Interaction Curves

A link of rectangular cross section is subjected to a load N (Fig. 12.12a). Derive general relationships involving N and M that govern, first, the

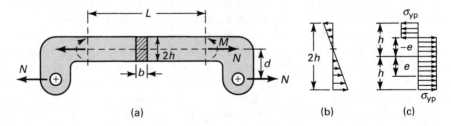

FIGURE 12.12. *Example 12.5. (a) A link of rectangular cross section carries load P at its ends; (b) elastic stress distribution; (c) fully plastic stress distribution.*

case of initial yielding and, then, fully plastic deformation for the straight part of the link of length L.

Solution   Suppose N and M are such that the state of stress is as shown in Fig. 12.12b at any straight beam section. The maximum stress in the beam is then, by superposition of the axial and bending stresses,

$$\sigma_{yp} = \frac{N_1}{2hb} + \frac{3}{2}\frac{M_1}{bh^2} \tag{e}$$

The upper limits on N ($M = 0$) and M ($N = 0$), corresponding to the condition of yielding, are

$$N_{yp} = 2hb\sigma_{yp}, \quad M_{yp} = \frac{I}{h}\sigma_{yp} = \tfrac{2}{3}bh^2\sigma_{yp} \tag{f}$$

Substituting $2hb$ and I/h from Eq. (f) into Eq. (e) and rearranging terms, we have

$$\frac{N_1}{N_{yp}} + \frac{M_1}{M_{yp}} = 1 \tag{12.16}$$

If N_1 is zero, then M_1 must achieve its maximum value M_{yp} for yielding to impend. Similarly, for $M_1 = 0$, it is necessary for N_1 to equal N_{yp} to initiate yielding. Between these extremes, Eq. (12.16) provides the infinity of combinations of N_1 and M_1 that will result in σ_{yp}.

For the *fully plastic* case (Fig. 12.12c), we shall denote the state of loading by N_2 and M_2. It is apparent that the stresses acting within the range $-e < y < e$ contribute pure axial load only. The stresses within the range $e < y < h$ and $-e > y > -h$ form a couple, however. For the total load system described, we may write

$$N_2 = 2eb\sigma_{yp}, \quad e = \frac{N_2}{2b\sigma_{yp}} \tag{g}$$

$$M_2 = (h - e)b\sigma_{yp} \cdot 2\left(e + \frac{h - e}{2}\right) = b(h^2 - e^2)\sigma_{yp} \tag{h}$$

Introducing Eqs. (12.11) and (g) into Eq. (h), we have

$$M_2 = M_u - \frac{N_2^2}{4b\sigma_{yp}}$$

Finally, dividing by M_u and noting that $M_u = \tfrac{3}{2}M_{yp} = bh^2\sigma_{yp}$ and $N_{yp} = 2bh\sigma_{yp}$, we obtain

$$\frac{2}{3}\frac{M_2}{M_{yp}} + \left(\frac{N_2}{N_{yp}}\right)^2 = 1 \tag{12.17}$$

Figure 12.13 is a plot of Eqs. (12.16) and (12.17). By employing these *interaction curves*, any combination of limiting values of bending moment and axial force is easily arrived at.

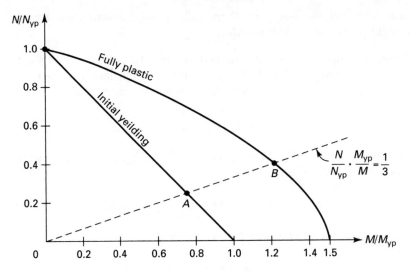

FIGURE 12.13. *Example 12.5 (continued). Interaction curves for N and M for a rectangular cross-sectional member.*

Let, for instance, $d = h$, $h = 2b = 24$ mm (Fig. 12.12a), and $\sigma_{yp} = 280$ MPa. Then the value of $N/M = \frac{1}{24}$ and, from Eq. (e), $M_{yp}/N_{yp} = h/3 = 8$. The radial line representing $(N/M)(M_{yp}/N_{yp}) = \frac{1}{3}$ is indicated by the dashed line in the figure. This line intersects the interaction curves at $A(0.75, 0.25)$ and $B(1.24, 0.41)$. Thus, yielding impends for $N_1 = 0.25N_{yp} = 0.25(2bh \cdot \sigma_{yp}) = 40.32$ kN, and for fully plastic deformation, $N_2 = 0.41N_{yp} = 66.125$ kN.

Note that the distance d is assumed constant and the values of N found are conservative. If the link deflection were taken into account, d would be smaller and the calculations would yield larger N.

EXAMPLE 12.6   Shape Factor of an I-Beam

An I-beam (Fig. 12.14a) is subjected to pure bending resulting from end couples. Determine the moment causing initial yielding and that results in complete plastic deformation.

Solution   The moment corresponding to σ_{yp} is, from Eq. (12.9a),

$$M_{yp} = \frac{I}{h}\sigma_{yp} = \frac{1}{h}(\tfrac{2}{3}bh^3 - \tfrac{2}{3}b_1h_1^3)\sigma_{yp}$$

Refer now to the completely plastic stress distribution of Fig. 12.14b. The moments of force owing to σ_{yp}, taken about the neutral axis, provide

$$M_u = (bh^2 - b_1h_1^2)\sigma_{yp}$$

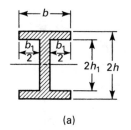

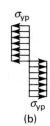

FIGURE 12.14. *Example 12.6. (a) Cross section; (b) fully plastic stress distribution.*

(a)   (b)

Combining the preceding equations, we have

$$\frac{M_u}{M_{yp}} = \frac{3}{2} \frac{1 - (b_1 h_1^2/bh^2)}{1 - (b_1 h_1^3/bh^3)}$$ **(i)**

From this expression, it is seen that $M_u/M_{yp} < \frac{3}{2}$, while it is $\frac{3}{2}$ for a beam of rectangular section ($h_1 = 0$). We conclude, therefore, that if a rectangular beam and an I-beam are designed plastically, the former will be more resistant to complete plastic failure.

EXAMPLE 12.7   Residual Stresses in a Rectangular Beam

Figure 12.15 shows an elastoplastic beam of rectangular cross section 40 mm by 100 mm carrying a bending moment of M. Determine (a) the thickness of the elastic core; (b) the residual stresses following removal of the bending moment. *Given:* $b = 40$ mm, $h = 50$ mm, $M = 21$ kN $\cdot$ m, $\sigma_{yp} = 240$ MPa, and $E = 200$ GPa.

Solution

a. Through the use of Eq. (12.9a), we have

$$M_{yp} = \frac{2}{3} bh^2 \sigma_{yp} = \frac{2}{3} (0.04)(0.05)^3 (240 \times 10) = 16 \text{ kN·m}$$

Then Eq. (12.10b) leads to

$$21 = \frac{3}{2} (16) \left[1 - \frac{1}{3} \left(\frac{e}{0.05} \right)^2 \right]$$

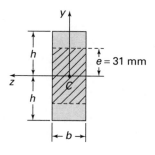

FIGURE 12.15. *Example 12.7. Finding the thickness (2e) of the elastic core.*

$e = 31$ mm

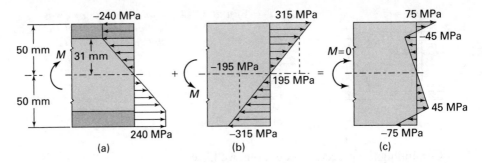

FIGURE 12.16. *Example 12.7 (continued). Stress distribution in a rectangular beam: (a) elastic–plastic; (b) elastic rebound; (c) residual.*

Solving,

$$e = 31 \times 10^{-3}\,\text{m} = 31\,\text{mm} \quad \text{and} \quad 2e = 62\,\text{mm}$$

Elastic core depicted as shaded in Fig. 12.15.

b. The stress distribution corresponding to moment $M = 21\,\text{kN} \cdot \text{m}$ is illustrated in Fig. 12.16a. The release of moment M produces elastic stresses, and the flexure formula applies (Fig. 12.16b). Equation (5.5) is therefore

$$\sigma'_{\max} = \frac{Mc}{I} = \frac{21 \times 10^3\,(0.05)}{0.04(0.1)^2/12} = 315\,\text{MPa}$$

By the superimposition of the two stress distributions, we can find the residual stresses (Fig. 12.16c). Observe that both tensile and compressive residual stresses remain in the beam.

EXAMPLE 12.8   Deflection of a Perfectly Plastic Simple Beam

Determine the maximum deflection due to an applied force P acting on the perfectly plastic simply supported rectangular beam (Fig. 12.8).

Solution   The center deflection in the *elastic range* is given by

$$v_{\max} = \frac{PL}{48EI} \tag{j}$$

At the *start of yielding*,

$$P = P_{yp}, \quad M_{\max} = M_{yp} = \frac{P_{yp}L}{4} \tag{k}$$

Expression (j), together with Eqs. (k) and (12.9a), leads to

$$v_{\max} = \frac{M_{yp}L^2}{12EI} = \frac{1}{12}\left(\frac{L^2}{Eh}\right)\sigma_{yp} \tag{l}$$

In a like manner, we obtain

$$v_{\max} = \frac{M_u L^2}{12EI} = \frac{1}{8}\left(\frac{L^2}{Eh}\right)\sigma_{\text{yp}} \qquad \textbf{(m)}$$

for the center deflection at the instant of *plastic collapse*.

12.8 COLLAPSE LOAD OF STRUCTURES: LIMIT DESIGN

On the basis of the simple examples in the previous section, it may be deduced that structures may withstand loads in excess of those that lead to initial yielding. We recognize that, while such loads need not cause structural collapse, they will result in some amount of permanent deformation. If no permanent deformation is to be permitted, the load configuration must be such that the stress does not attain the yield point anywhere in the structure. This is, of course, the basis of elastic design.

When a limited amount of permanent deformation may be tolerated in a structure, the design can be predicated on higher loads than correspond to initiation of yielding. On the basis of the ultimate or plastic load determination, safe dimensions can be determined in what is termed *limit design*. Clearly, such design requires higher than usual factors of safety. Examples of ultimate load determination are presented next.

Consider first a built-in beam subjected to a concentrated load at midspan (Fig. 12.17a). The general bending moment variation is sketched in Fig. 12.17b. As the load is progressively increased, we may anticipate plastic hinges at points 1, 2, and 3, because these are the points at which maximum bending moments are found. The configuration indicating the assumed location of the plastic hinges (Fig. 12.17c) is the *mechanism of collapse*. At every hinge, the hinge moment must clearly be the same.

The equilibrium and the energy approaches are available for determination of the ultimate loading. Electing the latter, we refer to Fig. 12.17c and note that the change in energy associated with rotation at points 1 and 2 is $M_u \cdot \delta\theta$, while at point 3, it is $M_u(2\delta\theta)$. The work done by the concentrated force is $P \cdot \delta v$. According to the principle of virtual work, we may write

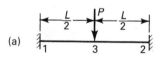

(a)

(b)

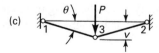

(c)

FIGURE 12.17. *(a) Beam with built-in ends; (b) elastic bending moment diagram; (c) mechanism of collapse with plastic hinges at 1 through 3.*

$$P_u(\delta v) = M_u(\delta\theta) + M_u(\delta\theta) + M_u(2\delta\theta) = 4M_u(\delta\theta)$$

where P_u represents the *ultimate load*. Because the deformations are limited to small values, it may be stated that $v \approx \frac{1}{2}L\theta$ and $\delta v = \frac{1}{2}L\delta\theta$. Substituting in the preceding expression for δv, it is found that

$$P_u = \frac{8M_u}{L} \qquad \textbf{(a)}$$

where M_u is calculated for a given beam using Eq. (12.12). It is interesting that, by introduction of the plastic hinges, the originally statically indeterminate beam is rendered determinate. The determination of P_u is thus simpler than that of P_{yp}, on which elastic analysis is based. An advantage of limit design may also be found in noting that a small rotation at either end of the beam or a slight lowering of a support will not influence the value of P_u. Moderate departures from the ideal case, such as these, will, however, have a pronounced effect on the value of P_{yp} in a statically indeterminate system.

While the positioning of the plastic hinges in the preceding problem is limited to the single possibility shown in Fig. 12.17c, more than one possibility will exist for situations in which several forces act. Correspondingly, a number of collapse mechanisms may exist, and it is incumbent on the designer to select from among them the one associated with the lowest load.

EXAMPLE 12.9 Collapse Analysis of a Continuous Beam

Determine the collapse load of the continuous beam shown in Fig. 12.18a.

Solution The four possibilities of collapse are indicated in Figs. 12.18b through d. We first consider the mechanism of Fig. 12.18b. In this system, motion occurs because of rotations at hinges 1, 2, and 3. The remainder of the beam remains rigid. Applying the principle of virtual work, noting that the moment at point 1 is zero, we have

$$P(\delta v) = M_u(2\,\delta\theta) + M_u(\delta\theta) = 3M_u(\delta\theta)$$

Because

$$v \approx \tfrac{1}{2}L\theta, \qquad \delta v = \tfrac{1}{2}L\,\delta\theta$$

this equation yields $P_u = 6M_u/L$.

For the collapse mode of Fig. 12.18c,

$$2P(\delta v) = M_u(\tfrac{1}{2}\delta\theta) + M_u(\tfrac{3}{2}\delta\theta) = 2M_u(\delta\theta)$$

and thus $P_u = 2M_u/L$.

The collapse mechanisms indicated by the solid and dashed lines of Fig. 12.18d are unacceptable because they imply a zero bending moment at section 3. We conclude that collapse will occur as in Fig. 12.17c, when $P \to 2M_u/L$.

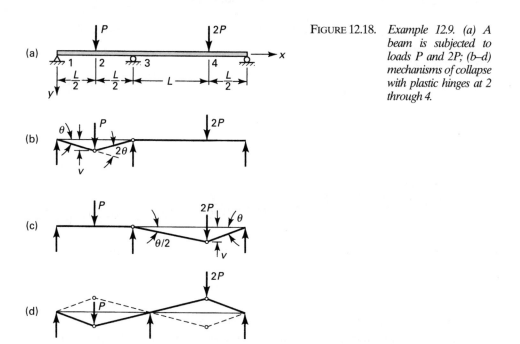

FIGURE 12.18. *Example 12.9. (a) A beam is subjected to loads P and 2P; (b–d) mechanisms of collapse with plastic hinges at 2 through 4.*

EXAMPLE 12.10   **Collapse Load of a Continuous Beam**

Determine the collapse load of the beam shown in Fig. 12.19a.

Solution   There are a number of collapse possibilities, of which one is indicated in Fig. 12.19b. Let us suppose that there exists a hinge at point 2, a distance e from the left support. Then examination of the geometry

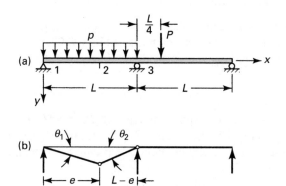

FIGURE 12.19. *Example 12.10. (a) Variously loaded beam; (b) mechanism of collapse with plastic hinges at 2 and 3.*

leads to $\theta_1 = \theta_2(L - e)/e$ or $\theta_1 + \theta_2 = L\theta_2/e$. Applying the principle of virtual work,

$$\delta \int_0^e (\theta_1 x)p \, dx + \delta \int_e^L [\theta_1 e - \theta_2(x - e)]p \, dx = M_u\left(\frac{L}{e}\delta\theta_2\right) + M_u\delta\theta_2$$

or

$$\frac{L(L - e)}{2}p = \left(\frac{L}{e} + 1\right)M_u$$

from which

$$p = \frac{2(e + L)M_u}{e(L - e)L} \tag{b}$$

The minimization condition for p in Eq. (b), $dp/de = 0$, results in

$$e = L\left(\frac{\sqrt{28}}{2} - 1\right) \tag{c}$$

Thus, Eqs. (b) and (c) provide a possible collapse configuration. The remaining possibilities are similar to those discussed in the previous example and should be checked to ascertain the minimum collapse load.

Determination of the collapse load of frames involves much the same analysis. For complex frames, however, the approaches used in the foregoing examples would lead to extremely cumbersome calculations. For these, special-purpose methods are available to provide approximate solutions [Ref. 12.6].

EXAMPLE 12.11 Collapse Analysis of a Frame

Apply the method of virtual work to determine the collapse load of the structure shown in Fig. 12.20a. Assume that the rigidity of member BC is 1.2 times greater than that of the vertical members AB and CD.

Solution Of the several collapse modes, we consider only the two given in Figs. 12.20b and c. On the basis of Fig. 12.20b, plastic hinges will be formed at the ends of the vertical members. Thus, from the principle of virtual work,

$$P(\tfrac{1}{2}\delta u) = 4M_u(\delta\theta)$$

Substituting $u = L\theta$, this expression leads to $P_u = 8M_u/L$.

Referring to Fig. 12.20c, we have $M_{uE} = 1.2M_u$, where M_u is the collapse moment of the vertical elements. Applying the principle of virtual work,

$$P(\tfrac{1}{2}\delta u) + 4P(\delta v) = 4M_u(\delta\theta) + 2(1.2M_u\delta\theta)$$

Noting that $u = L\theta$ and $v = \tfrac{1}{2}L\theta$, this equation provides the following expression for the collapse load: $P_u = 2.56M_u/L$.

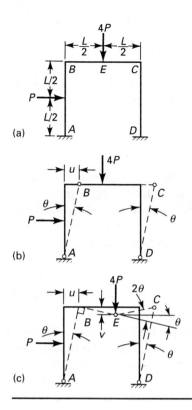

FIGURE 12.20. *Example 12.11. (a) A frame with concentrated loads; (b and c) mechanism of collapse with plastic hinges at A, B, C, and D.*

12.9 ELASTIC–PLASTIC TORSION OF CIRCULAR SHAFTS

We now consider the torsion of circular bars of ductile materials, which are idealized as elastoplastic, stressed into the plastic range. In this case, the first two basic assumptions associated with small deformations of circular bars in torsion (see Sec. 6.2) are still valid. This means that the circular cross sections remain plane and their radii remain straight. Consequently, strains vary linearly from the shaft axis. The shearing

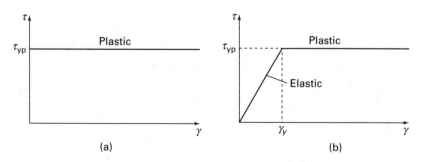

FIGURE 12.21. *Idealized shear stress–shear strain diagrams for (a) perfectly plastic materials; (b) elastoplastic materials.*

stress–strain curve of plastic materials is shown in Fig. 12.21. Referring to this diagram, we can proceed as discussed before and determine the stress distribution across a section of the shaft for any given value of the torque T.

The basic relationships given in Section 6.2 are applicable as long as the shear strain in the bar does not exceed the yield strain γ_{yp}. It is recalled that the condition of torque equilibrium for the entire shaft (Fig. 6.2) requires

$$T = \int_A \rho \tau \, dA = 2\pi \int_A \tau \rho^2 d\rho \tag{a}$$

Here ρ, τ are any arbitrary distance and shearing stress from the center O, respectively, and A the entire area of a cross section of the shaft. Increasing in the applied torque, yielding impends on the boundary and moves progressively toward the interior. The cross-sectional stress distribution will be as shown in Fig. 12.22.

At the start of yielding (Fig. 12.22a), the torque T_{yp}, through the use of Eq. (6.1), may be written in the form:

$$T_{yp} = \frac{\pi c^3}{2} \tau_{yp} = \frac{J}{c} \tau_{yp} \tag{12.18}$$

The quantity $J = \pi c^4/2$ is the polar moment of inertia for a solid shaft with radius $r = c$. Equation (12.18) is called the *maximum elastic torque*, or *yield torque*. It represents the largest torque for which the deformation remains fully elastic.

If the twist is increased further, an inelastic or *plastic portion* develops in the bar around an elastic core of radius ρ_0 (Fig. 12.22b). Using Eq. (a), we obtain that the torque resisted by the elastic core equals

$$T_1 = \frac{\pi \rho_0^3}{2} \tau_{yp} \tag{b}$$

The outer portion is subjected to constant yield stress τ_{yp} and resists the torque,

$$T_2 = 2\pi \int_{\rho_0}^c \tau_{yp} \rho^2 d\rho = \frac{2\pi}{3} (c^3 - \rho_0^3) \tau_{yp} \tag{c}$$

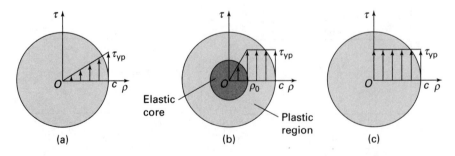

FIGURE 12.22.   *Stress distribution in a shaft as torque is increased: (a) onset of yield; (b) partially plastic; and (c) fully plastic.*

The *elastic–plastic* or *total torque* T, the sum of T_1 and T_2, may now be expressed as follows

$$T = \frac{\pi c^3}{6}\left(4 - \frac{\rho_0^3}{c^3}\right)\tau_{yp} = \frac{4}{3}T_{yp}\left(1 - \frac{1}{4}\frac{\rho_0^3}{c^3}\right) \tag{12.19}$$

When twisting becomes very large, the region of yielding will approach the middle of the shaft and will approach zero (Fig. 12.22c). The corresponding torque T_u is the plastic, or ultimate, shaft torque, and its value from the foregoing equation is

$$T_u = \frac{2}{3}\pi c^3 \tau_{yp} = \frac{4}{3}T_{yp} \tag{12.20}$$

It is thus seen that only one-third of the torque-carrying capacity remains after τ_{yp} is reached at the outermost fibers of a shaft.

The radius of elastic core (Fig. 12.22b) is found, referring to Fig. 6.2, by setting $\gamma = \gamma_{yp}$ and $\rho = \rho_0$. It follows that

$$\rho_0\frac{L\gamma_{yp}}{\phi} \tag{12.21a}$$

in which L is the length of the shaft. The angle of twist at the onset of yielding ϕ_{yp} (when $\rho_0 = c$) is therefore

$$c = \frac{L\gamma_{yp}}{\phi_y} \tag{12.21b}$$

Equations (12.21) lead to the relation,

$$\frac{\rho_0}{c} = \frac{\phi_{yp}}{\phi} \tag{12.22}$$

Using Eq. (12.19), the ultimate torque may then be expressed in the form:

$$T_u = \frac{4}{3}T_{yp}\left(1 - \frac{1}{4}\frac{\phi_{yp}^3}{\phi^3}\right) \tag{12.23}$$

This is valid for $\phi > \phi_{yp}$. When $\phi < \phi_{yp}$, linear relation (6.3) applies.

A sketch of Eqs. (6.3) and (12.23) is illustrated in Fig. 12.23. Observe that after yielding torque T_{yp} is reached, T and ϕ are related nonlinearly. As T approaches T_u,

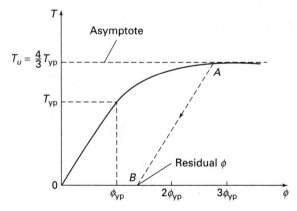

FIGURE 12.23. *Torque–angle of twist relationship for a circular shaft.*

the angle of twist grows without limit. A final point to be noted, however, is that the value of T_u is approached very rapidly (for instance, $T_u = 1.32T_{yp}$ when $\phi = 3\phi_{yp}$).

When a shaft is strained beyond the elastic limit (point A in Fig. 12.23) and the applied torque is then removed, rebound is assumed to follow Hooke's law. Thus, once a portion of a shaft has yielded, *residual stresses* and *residual rotations* (ϕ_B) will develop. This process and the application of the preceding relationships are demonstrated in Example 12.12. Statically indeterminate, inelastic torsion problems are dealt with similarly to those of axial load, as was discussed in Section 12.5.

EXAMPLE 12.12   Residual Stress in a Shaft

Figure 12.24 shows a solid circular steel shaft of diameter d and length L carrying a torque T. Determine (a) the radius of the elastic core; (b) the angle of twist of the shaft; (c) the residual stresses and the residual rotation when the shaft is unloaded. *Assumption:* The steel is taken to be an elastoplastic material. *Given:* $d = 60$ mm, $L = 1.4$ m, $T = 7.75$ kN·m, $\tau_{yp} = 145$ MPa, and $G = 80$ GPa.

Solution   We have $c = 30$ mm and $J = \pi(0.03)^4/2 = 1272 \times 10^{-9}$ m⁴.

a. *Radius of Elastic Core.* The yield torque, applying Eq. (12.18), equals

$$T_{yp} = \frac{J\tau_{yp}}{c} = \frac{1272 \times 10^{-9}(145 \times 10^6)}{0.03} = 6.15 \text{ kN·m}$$

Equation (12.19), substituting the values of T and τ_{yp}, gives

$$\left(\frac{\rho_0}{c}\right)^3 = 4 - \frac{3T}{T_{yp}} = 4 - \frac{3(7.75 \times 10^3)}{6.15 \times 10^3} = 0.22$$

Solving, $\rho_0 = 0.604\,(30) = 18.1$ mm. The elastic–plastic stress distribution in the loaded shaft is illustrated in Fig. 12.25a.

b. *Yield Twist Angle.* Through the use of Eq. (6.3), the angle of twist at the onset of yielding,

$$\phi_{yp} = \frac{T_{yp}L}{GJ} = \frac{6.15 \times 10^3(1.4)}{1272 \times 10^{-9}(80 \times 10^9)} = 0.0846 \text{ rad}$$

FIGURE 12.24.   *Example 12.13. Torsion of a circular bar of elastoplastic material.*

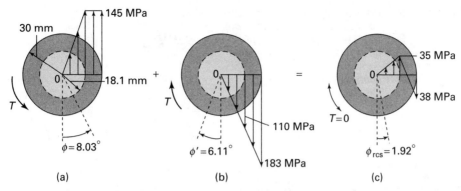

FIGURE 12.25.   *Example 5.13 (continued). (a) Partial plastic stresses; (b) elastic rebound stresses; (c) residual stresses.*

Introducing the value found for ϕ_{yp} into Eq. (12.22), we have

$$\phi = \frac{C\phi_{yp}}{\rho_0} = \frac{30(0.0846)}{18.1} = 0.1402 \text{ rad} = 8.03°$$

c. *Residual Stresses and Rotation.* The removal of the torque produces elastic stresses as depicted in Fig. 12.25b, and the torsion formula, Eq. (6.1), leads to reversed stress as

$$\tau'_{max} = \frac{Tc}{J} = \frac{7.75 \times 10^3(30 \times 10^{-3})}{1272(10^{-9})} = 183 \text{ MPa}$$

Superposition of the two distributions of stress results in the residual stresses (Fig. 12.25c).

Permanent Twist. The elastic rebound rotation, using Eq. (6.3), equals

$$\phi' = \frac{TL}{GJ} = \frac{7.75 \times 10^3(1.4)}{1272 \times 10^{-9}(80 \times 10^9)} = 0.1066 \text{ rad} = 6.11°$$

The preceding results indicate that residual rotation of the shaft is

$$\phi_{res} = 8.03° - 6.11° = 1.92°$$

Comment   We see that even though the reversed stresses τ'_{max} exceed the yield strength τ_{yp}, the assumption of linear distribution of these stresses is valid, inasmuch as they do not exceed $2\tau_{yp}$.

12.10   PLASTIC TORSION: MEMBRANE ANALOGY

Recall from Chapter 6 that the maximum shearing stress in a slender bar of arbitrary section subject to pure torsion is always found on the boundary. As the applied torque is increased, we expect yielding to occur on the boundary and to

FIGURE 12.26. *(a) Partially yielded rectangular section; (b) membrane–roof analogy applied to elastic–plastic torsion of a rectangular bar; (c) sand hill analogy applied to plastic torsion of a circular bar.*

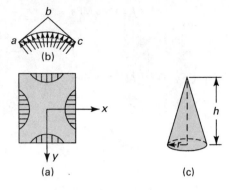

move progressively toward the interior, as sketched in Fig. 12.26a for a bar of rectangular section. We now determine the ultimate torque T_u that can be carried. This torque corresponds to the totally plastic state of the bar, as was the case of the beams previously discussed. Our analysis treats only perfectly plastic materials.

The stress distribution within the elastic region of the bar is governed by Eq. (6.9),

$$\frac{\partial^2 \Phi}{\partial x^2} + \frac{\partial^2 \Phi}{\partial y^2} = -2G\theta \tag{12.24}$$

where Φ represents the stress function ($\Phi = 0$ at the boundary) and θ is the angle of twist. The shearing stresses, in terms of Φ, are

$$\tau_{zx} = \frac{\partial \Phi}{\partial y}, \quad \tau_{zy} = -\frac{\partial \Phi}{\partial x} \tag{a}$$

Inasmuch as the bar is in a state of pure shear, the stress field in the plastic region is, according to the *Mises yield criterion*, expressed by

$$\left(\frac{\partial \Phi}{\partial x}\right)^2 + \left(\frac{\partial \Phi}{\partial y}\right)^2 = \tau_{yp}^2 \tag{12.25}$$

where τ_{yp} is the yield stress in shear. This expression indicates that the slope of the Φ surface remains constant throughout the plastic region and is equal to τ_{yp}.

Membrane–Roof Analogy

Bearing in mind the condition imposed on Φ by Eq. (12.25), the membrane analogy (Sec. 6.6) may be extended from the purely elastic to the elastic–plastic case. As shown in Fig. 12.26b, a roof *abc* of constant slope is erected with the membrane as its base. Figure 12.26c shows such a roof for a circular section. As the pressure acting beneath the membrane increases, more and more contact is made between the membrane and the roof. In the fully plastic state, the membrane is in total contact with the roof, membrane and roof being of identical slope. Whether the membrane makes partial or complete contact with the roof clearly depends on the pressure. The membrane–roof analogy thus permits solution of elastic–plastic torsion problems.

TABLE 12.1. *Torque Capacity for Various Common Sections*

Cross section	Radius or sides	Torque T_u for full plasticity
Circular	r	$\frac{2}{3}\pi r^3 \tau_{yp}$
Equilateral triangle	a	$\frac{1}{12}a^3 \tau_{yp}$
Rectangle	a, b	$\frac{1}{6}a^2(3b - a)\tau_{yp}$
	$(b > a)$	
Square	a	$\frac{1}{3}a^3 \tau_{yp}$
Thick-walled tube	b: outer	$\frac{2}{3}\pi(b^3 - a^3)\tau_{yp}$
	a: inner	

Sand Hill Analogy

For the case of a totally yielded bar, the membrane–roof analogy leads quite naturally to the *sand hill analogy*. We need not construct a roof at all, using this method. Instead, sand is heaped on a plate whose outline is cut into the shape of the cross section of the torsion member. The torque is, according to the membrane analogy, proportional to twice the volume of the sand figure so formed. The ultimate torque corresponding to the fully plastic state is thus found.

Referring to Fig. 12.26c, let us apply the sand hill analogy to determine the ultimate torque for a circular bar of radius r. The volume of the corresponding cone is $V = \frac{1}{3}\pi r^2 h$, where h is the height of the sand hill. The slope h/r represents the yield point stress τ_{yp}. The ultimate torque is therefore

$$T_u = \frac{2}{3}\pi r^3 \tau_{yp} \tag{12.26}$$

Note that the maximum elastic torque is $T_{yp} = (\pi r^3/2)\tau_{yp}$. We may thus form the ratio

$$\frac{T_u}{T_{yp}} = \frac{4}{3} \tag{12.27}$$

Other solid sections may be treated similarly [Ref. 12.7]. Table 12.1 lists the ultimate torques for bars of various cross-sectional geometry.

The procedure may also be applied to members having a symmetrically located hole. In this situation, the plate representing the cross section must contain the same hole as the actual cross section.

12.11 ELASTIC–PLASTIC STRESSES IN ROTATING DISKS

This section treats the stresses in a flat disk fabricated of a *perfectly plastic material*, rotating at constant angular velocity. The maximum elastic stresses for this geometry are, from Eqs. (8.30) and (8.28) as follows:

For the *solid disk* at $r = 0$,

$$\sigma_\theta = \sigma_r = \frac{\rho\omega^2(3 + \nu)b^2}{8} \tag{a}$$

For the *annular disk* at $r = a$,

$$\sigma_\theta = \frac{3 + \nu}{4} \rho\omega^2 \left(b^2 + \frac{1 - \nu}{3 + \nu} a^2 \right) \qquad \textbf{(b)}$$

Here a and b represent the inner and outer radii, respectively, ρ the mass density, and ω the angular speed. The following discussion relates to initial, partial, and complete yielding of an annular disk. Analysis of the solid disk is treated in a very similar manner.

Initial Yielding

According to the Tresca yield condition, yielding impends when the maximum stress is equal to the yield stress. Denoting the critical speed as ω_0 and using $\nu = \frac{1}{3}$, we have, from Eq. (b),

$$\omega_0 = \left(\frac{6}{5b^2 + a^2} \frac{\sigma_{yp}}{\rho} \right)^{1/2} \qquad \textbf{(12.28)}$$

Partial Yielding

For angular speeds in excess of ω_0 but lower than speeds resulting in total plasticity, the disk contains both an elastic and a plastic region, as shown in Fig. 12.27a. In the plastic range, the equation of radial equilibrium, Eq. (8.26), with σ_{yp} replacing the maximum stress σ_θ, becomes

$$r\frac{d\sigma_r}{dr} + \sigma_r - \sigma_{yp} + \rho\omega^2 r^2 = 0 \qquad \textbf{(12.29)}$$

or

$$\frac{d}{dr}(r\sigma_r) - \sigma_{yp} + \rho\omega^2 r^2 = 0$$

The solution is given by

$$r\sigma_r - \sigma_{yp} r + \frac{\rho\omega^2 r^3}{3} + c_1 = 0 \qquad \textbf{(c)}$$

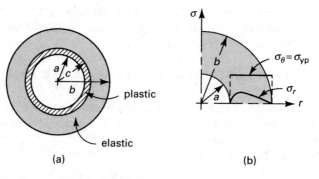

(a)　　　　　　　　　　　　　(b)

FIGURE 12.27.　*(a) Partially yielded rotating annular disk;*
(b) stress distribution in complete yielding.

By satisfying the boundary condition $\sigma_r = 0$ at $r = a$, Eq. (c) provides an expression for the constant c_1, which when introduced here results in

$$r\sigma_r - \sigma_{yp}(r - a) + \frac{\rho\omega^2}{3}(r^3 - a^3) = 0 \tag{d}$$

The stress within the plastic region is now determined by letting $r = c$ in Eq. (d):

$$\sigma_c = \frac{\rho\omega^2}{3}\frac{a^3 - c^3}{c} + \frac{c - a}{c}\sigma_{yp} \tag{12.30}$$

Referring to the elastic region, the distribution of stress is determined from Eq. (8.27) with $\sigma_r = \sigma_c$ at $r = c$, and $\sigma_r = 0$ at $r = b$. Applying these conditions, we obtain

$$c_1 = -\frac{c^2(1 - \nu)\sigma_c}{E(b^2 - c^2)} + \frac{\rho\omega^2(1 - \nu)(3 + \nu)}{8E}(b^2 + c^2)$$

$$c_2 = -\frac{b^2c^2(1 + \nu)\sigma_c}{E(b^2 - c^2)} + \frac{\rho\omega^2(1 + \nu)(3 + \nu)b^2c^2}{8E} \tag{e}$$

The stresses in the outer region are then obtained by substituting Eqs. (e) into Eq. (8.27):

$$\sigma_r = \frac{c^2}{b^2 - c^2}\left(-1 + \frac{b^2}{r^2}\right)\sigma_c + \frac{\rho\omega^2}{8}(3 + \nu)\left(b^2 + c^2 - \frac{b^2c^2}{r^2} - r^2\right)$$

$$\sigma_\theta = -\frac{c^2}{b^2 - c^2}\left(1 + \frac{b^2}{r^2}\right)\sigma_c + \frac{\rho\omega^2}{8}(3 + \nu) \tag{12.31}$$

$$\times \left(b^2 + c^2 - \frac{1 + 3\nu}{3 + \nu}r^2 + \frac{b^2c^2}{r^2}\right)$$

To determine the value of ω that causes yielding up to radius c, we need only substitute σ_θ for σ_{yp} in Eq. (12.31) and introduce σ_c as given by Eq. (12.30).

Complete Yielding

We turn finally to a determination of the speed ω_1 at which the disk becomes fully plastic. First, Eq. (c) is rewritten

$$\sigma_r - \sigma_{yp} + \frac{\rho\omega^2 r^2}{3} + \frac{c_1}{r} = 0 \tag{f}$$

Applying the boundary conditions, $\sigma_r = 0$ at $r = a$ and $r = b$ in Eq. (f), we have

$$c_1 = a\sigma_{yp} - \frac{\rho^2\omega^2 a^3}{3} \tag{g}$$

and the critical speed ($\omega = \omega_1$) is given by

$$\omega_1 = \left(\frac{3\sigma_{yp}}{\rho} - \frac{b - a}{b^3 - a^3}\right)^{1/2} \tag{12.32}$$

Substitution of Eqs. (g) and (12.32) into Eq. (f) provides the radial stress in a fully plastic disk:

$$\sigma_r = \left(1 - \frac{a}{r} - \frac{1 - a^2/b^2}{1 - a^3/b^3} - \frac{r^2}{b^2}\frac{1 - a/b}{1 - a^3/b^3}\right)\sigma_{yp} \qquad \textbf{(12.33)}$$

The distributions of radial and tangential stress are plotted in Fig. 12.27b.

12.12   PLASTIC STRESS–STRAIN RELATIONS

Consider an element subject to true stresses σ_1, σ_2, and σ_3 with corresponding true straining. The true strain, which is plastic, is denoted ε_1, ε_2, and ε_3. A simple way to derive expressions relating true stress and strain is to replace the elastic constants E and v by E_s and $\frac{1}{2}$, respectively, in Eqs. (2.34). In so doing, we obtain equations of the *total strain theory* or the *deformational theory*, also known as *Hencky's plastic stress–strain relations*:

$$\varepsilon_1 = \frac{1}{E_s}[\sigma_1 - \tfrac{1}{2}(\sigma_2 + \sigma_3)]$$

$$\varepsilon_2 = \frac{1}{E_s}[\sigma_2 - \tfrac{1}{2}(\sigma_1 + \sigma_3)] \qquad \textbf{(12.34a)}$$

$$\varepsilon_3 = \frac{1}{E_s}[\sigma_3 - \tfrac{1}{2}(\sigma_1 + \sigma_2)]$$

The foregoing may be restated as

$$\frac{\varepsilon_1}{\sigma_1 - \tfrac{1}{2}(\sigma_2 + \sigma_3)} = \frac{\varepsilon_2}{\sigma_2 - \tfrac{1}{2}(\sigma_1 + \sigma_3)} = \frac{\varepsilon_3}{\sigma_3 - \tfrac{1}{2}(\sigma_1 + \sigma_2)} = \frac{1}{E_s} \quad \textbf{(12.34b)}$$

Here E_s, a function of the state of plastic stress, is termed the *modulus of plasticity* or *secant modulus*. It is defined by Fig. 12.28

$$E_s = \frac{\sigma_e}{\varepsilon_e} \qquad \textbf{(12.35)}$$

in which the quantities σ_e and ε_e are the *effective stress* and the *effective strain*, respectively.

FIGURE 12.28.   *True stress–true strain diagram for a rigid-plastic material.*

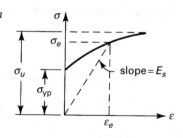

Although other yield theories may be employed to determine σ_e, the maximum energy of distortion or Mises theory (Sec. 4.7) is most suitable. According to the Mises theory, the following relationship connects the uniaxial yield stress to the general state of stress at a point:

$$\sigma_e = \frac{1}{\sqrt{2}}[(\sigma_1 - \sigma_2)^2 + (\sigma_2 - \sigma_3)^2 + (\sigma_3 - \sigma_1)^2]^{1/2} \tag{12.36}$$

where the effective stress σ_e is also referred to as the *von Mises stress*. It is assumed that expression (12.36) applies not only to yielding or the beginning of inelastic action ($\sigma_e = \sigma_{yp}$) but to any stage of plastic behavior. That is, σ_e has the value σ_{yp} at yielding, and as inelastic deformation progresses, σ_e increases in accordance with the right side of Eq. (12.36). Equation (12.36) then represents the logical *extension* of the yield condition to describe plastic deformation after the yield stress is exceeded.

Collecting terms of Eqs. (12.34), we have

$$[\tfrac{2}{3}(\varepsilon_1^2 + \varepsilon_2^2 + \varepsilon_3^2)]^{1/2} = \frac{1}{E_s\sqrt{2}}[(\sigma_1 - \sigma_2)^2 + (\sigma_2 - \sigma_3)^2 + (\sigma_3 - \sigma_1)^2]^{1/2} \tag{a}$$

The foregoing, together with Eqs. (12.35) and (12.36), leads to the definition

$$\varepsilon_e = [\tfrac{2}{3}(\varepsilon_1^2 + \varepsilon_2^2 + \varepsilon_3^2)]^{1/2} \tag{12.37a}$$

or (on the basis of $\varepsilon_1 + \varepsilon_2 + \varepsilon_3 = 0$), in different form,

$$\varepsilon_e = \frac{\sqrt{2}}{3}[(\varepsilon_1 - \varepsilon_2)^2 - (\varepsilon_2 - \varepsilon_3)^2 + (\varepsilon_3 - \varepsilon_1)^2]^{1/2} \tag{12.37b}$$

relating the effective plastic strain and the true strain components. Note that, for simple tension, $\sigma_2 = \sigma_3 = 0$, $\varepsilon_2 = \varepsilon_3 = -\varepsilon_1/2$, and Eqs. (12.36) and (12.37) result in

$$\sigma_e = \sigma_1, \qquad \varepsilon_e = \varepsilon_1 \tag{b}$$

Therefore, for a given σ_e, ε_e can be read directly from true stress–strain diagram (Fig. 12.28).

Hencky's equations as they appear in Eqs. (12.34) have little utility. To give these expressions generality and convert them to a more convenient form, it is useful to employ the empirical relationship (12.2)

$$\sigma_e = K(\varepsilon_e)^n$$

from which

$$\frac{\varepsilon_e}{\sigma_e} = \frac{(\sigma_e)^{1/n-1}}{K^{1/n}} \tag{c}$$

The *true stress–strain relations*, upon substitution of Eqs. (12.36) and (c) into (12.34), then assume the following more useful form:

$$\varepsilon_1 = \left(\frac{\sigma_1}{K}\right)^{1/n} [\alpha^2 - \beta^2 - \alpha\beta - \alpha - \beta + 1]^{(1-n)/2n} \left(1 - \frac{\alpha}{2} - \frac{\beta}{2}\right) \qquad \textbf{(12.38a)}$$

$$\varepsilon_2 = \left(\frac{\sigma_1}{K}\right)^{1/n} [\alpha^2 - \beta^2 - \alpha\beta - \alpha - \beta + 1]^{(1-n)/2n} \left(\alpha - \frac{\beta}{2} - \frac{1}{2}\right) \qquad \textbf{(12.38b)}$$

$$\varepsilon_3 = \left(\frac{\sigma_1}{K}\right)^{1/n} [\alpha^2 - \beta^2 - \alpha\beta - \alpha - \beta + 1]^{(1-n)/2n} \left(\beta - \frac{\alpha}{2} - 1\right) \qquad \textbf{(12.38c)}$$

where $\alpha = \sigma_2/\sigma_1$ and $\beta = \sigma_3/\sigma_1$.

In the case of an elastic material ($K = E$ and $n = 1$), it is observed that Eqs. (12.38) reduce to the familiar generalized Hooke's law.

EXAMPLE 12.13 Analysis of Cylindrical Tube by Hencky's Relations

A thin-walled cylindrical tube of initial radius r_o is subjected to internal pressure p. Assume that the values of r_o, p, and the material properties (K and n) are given. Apply Hencky's relations to determine (a) the maximum allowable stress and (b) the initial thickness t_o for the cylinder to become unstable at internal pressure p.

Solution   The current radius, thickness, and the length are denoted by r, t, and L, respectively. In the plastic range, the hoop, axial, and radial stresses are

$$\sigma_\theta = \sigma_1 = \frac{pr}{t}, \qquad \sigma_z = \sigma_2 = \frac{pr}{2t}, \qquad \sigma_r = \sigma_3 = 0 \qquad \textbf{(d)}$$

We thus have $\alpha = \sigma_2/\sigma_1$ and $\beta = 0$ in Eqs. (12.38).

Corresponding to these stresses, the components of true strain are, from Eq. (2.24),

$$\varepsilon_\theta = \varepsilon_1 = \ln\frac{r}{r_o}$$

$$\varepsilon_z = \varepsilon_2 = \ln\frac{L}{L_o} \qquad \textbf{(e)}$$

$$\varepsilon_r = \varepsilon_3 = \ln\frac{t}{t_o}$$

Based on the constancy of volume, Eq. (12.1), we then have

$$\varepsilon_3 = -(\varepsilon_1 + \varepsilon_2) = \ln\frac{t}{t_o}$$

or

$$t = t_o e^{-\varepsilon_1 - \varepsilon_2} = t_o \ln^{-1} \varepsilon_3 \qquad \textbf{(f)}$$

The first of Eqs. (e) gives

$$r = r_o e^{\varepsilon_1} = r_o \ln^{-1} \varepsilon_1 \tag{g}$$

The tangential stress, the first of Eqs. (d), is therefore

$$\sigma_1 = pr_o e^{\varepsilon_1} \frac{1}{t_o e^{-\varepsilon_1 - \varepsilon_2}}$$

from which

$$p = \sigma_1 \frac{t_o}{r_o} e^{-\varepsilon_1(2 + \varepsilon_2/\varepsilon_1)} \tag{h}$$

Simultaneous solution of Eqs. (12.38) leads readily to

$$\frac{\varepsilon_1}{\varepsilon_2} = \frac{2 - \alpha}{2\alpha - 1} \tag{i}$$

Equation (h) then appears as

$$p = \sigma_1 \frac{t_o}{r_o} e^{-3\varepsilon_1/(2-\alpha)} \tag{j}$$

For material instability,

$$dp = \frac{\partial p}{\partial \sigma_1} d\sigma_1 + \frac{\partial p}{\partial \varepsilon_1} d\varepsilon_1 = 0$$

which, upon substitution of $\partial p/\partial \sigma_1$ and $\partial p/\partial \varepsilon_1$ derived from Eq. (j), becomes

$$dp = \frac{t_o}{r_o} e^{-3\varepsilon_1/(2-\alpha)} d\sigma_1 + \sigma_1 \frac{t_o}{r_o} e^{-3\varepsilon_1/(2-\alpha)} \left(-\frac{3}{2 - \alpha} \right) d\varepsilon_1 = 0$$

or

$$\frac{d\sigma_1}{d\varepsilon_1} = \sigma_1 \frac{3}{2 - \alpha} \tag{k}$$

In Eqs. (12.38) it is observed that σ_1 depends on α and ε_1^n. That is,

$$\sigma_1 = f(\alpha)\varepsilon_1^n \tag{l}$$

Differentiating, we have

$$\frac{d\sigma_1}{d\varepsilon_1} = nf(\alpha)\varepsilon_1^{n-1} \tag{m}$$

Expressions (k), (l), and (m) lead to the *instability condition*:

$$\varepsilon_1 = \frac{2 - \alpha}{3} n \tag{12.39}$$

a. Equating expressions (12.39) and (12.38a), we obtain

$$\frac{2 - \alpha}{3} n = \left(\frac{\sigma_1}{K} \right)^{1/n} (\alpha^2 - \alpha + 1)^{(1-n)/2n} \frac{2 - \alpha}{2}$$

and the true tangential stress is thus

$$\sigma_1 = K\left(\frac{2n}{3}\right)^n \left(\frac{1}{\alpha^2 - \alpha + 1}\right)^{(1-n)/2} \tag{12.40}$$

b. On the other hand, Eqs. (d), (f), (g), and (12.38) yield

$$\sigma_1 = \frac{pr_o}{t_o} \frac{\ln^{-1}(\sigma_1/K)^{1/n}(\alpha^2 - \alpha + 1)^{(1-n)/2n}[(2-\alpha)/2]}{\ln^{-1}(\sigma_1/K)^{1/n}(\alpha^2 - \alpha + 1)^{(1-n)/2n}(-1-\alpha)}$$

From this expression, the required original thickness is found to be

$$t_o = \frac{pr_o}{\sigma_1} \ln^{-1} \frac{\alpha - 2}{2(1 + \alpha)} \tag{12.41}$$

wherein σ_1 is given by Eq. (12.40).

In the case under consideration, $\alpha = 1/2$ and Eqs. (12.39), (12.40), and (12.41) thus become

$$\varepsilon_1 = \frac{n}{2}$$

$$\sigma_1 = K\left(\frac{2n}{3}\right)^n \left(\frac{4}{3}\right)^{(1-n)/2} = \frac{2K}{\sqrt{3}}\left(\frac{n}{\sqrt{3}}\right)^n \tag{12.42}$$

$$t_0 = \frac{0.606\,pr_o}{(2K/\sqrt{3})(n/\sqrt{3})^n}$$

For a thin-walled *spherical shell* under internal pressure, the two principal stresses are equal and hence $\alpha = 1$. Equations (12.39), (12.40), and (12.41) then reduce to

$$\varepsilon_1 = \frac{n}{3}, \qquad \sigma_1 = K\left(\frac{2n}{3}\right)^n, \qquad t_o = \frac{0.717\,pr_o}{K(2n/3)^n} \tag{12.43}$$

Based on the relations derived in this example, the effective stress and the effective strain are determined readily. Table 12.2 furnishes the maximum true and effective stresses and strains in a thin-walled cylinder and a thin-walled sphere. For purposes of comparison, the table also lists the results (Sec. 12.4) pertaining to simple tension. We observe that at instability, the maximum true strains in a sphere and cylinder are much lower than the corresponding longitudinal strain in uniaxial tension.

It is significant that, for loading situations in which the components of stress do not increase continuously, Hencky's equations provide results that are somewhat in error, and the incremental theory (Sec. 12.13) must be used. Under these circumstances, Eqs. (12.34) or (12.38) cannot describe the *complete plastic behavior* of the

TABLE 12.2. *Stress and Strain in Pressurized Plastic Tubes*

Member	Tensile bar	Cylindrical tube	Spherical tube
ε_1	n	$\dfrac{n}{2}$	$\dfrac{n}{3}$
ε_2	$-\dfrac{n}{2}$	0	$\dfrac{n}{3}$
ε_3	$-\dfrac{n}{2}$	$-\dfrac{n}{2}$	$-\dfrac{2n}{3}$
σ_1/K	$(n)^n$	$\dfrac{2}{\sqrt{3}}\left(\dfrac{n}{\sqrt{3}}\right)^n$	$\left(\dfrac{2n}{3}\right)^n$
σ_2/K	0	$\dfrac{1}{\sqrt{3}}\left(\dfrac{n}{\sqrt{3}}\right)^n$	$\left(\dfrac{2n}{3}\right)^n$

material. The latter is made clearer by considering the following. Suppose that subsequent to a given plastic deformation the material is unloaded, either partially or completely, and then reloaded to a new state of stress that does not result in yielding. We expect no change to occur in the plastic strains; but Hencky's equations indicate different values of the plastic strain, because the stress components have changed. The latter cannot be valid because during the unloading–loading process, the plastic strains have, in reality, not been affected.

12.13 PLASTIC STRESS–STRAIN INCREMENT RELATIONS

We have already discussed the limitations of the deformational theory in connection with a situation in which the loading does not continuously increase. The *incremental theory* offers another approach, treating not the total strain associated with a state of stress but rather the increment of strain.

Suppose now that the true stresses at a point experience very small changes in magnitude $d\sigma_1$, $d\sigma_2$, and $d\sigma_3$. As a consequence of these increments, the effective stress σ_e will be altered by $d\sigma_e$ and the effective strain ε_e by $d\varepsilon_e$. The plastic strains thus suffer increments $d\varepsilon_1$, $d\varepsilon_2$, and $d\varepsilon_3$.

The following modification of Hencky's equations, due to Lévy and Mises, describe the foregoing and give good results in metals:

$$d\varepsilon_1 = \frac{d\varepsilon_e}{\sigma_e}[\sigma_1 - \tfrac{1}{2}(\sigma_2 + \sigma_3)]$$

$$d\varepsilon_2 = \frac{d\varepsilon_e}{\sigma_e}[\sigma_2 - \tfrac{1}{2}(\sigma_1 + \sigma_3)] \qquad \textbf{(12.44a)}$$

$$d\varepsilon_3 = \frac{d\varepsilon_e}{\sigma_e}[\sigma_3 - \tfrac{1}{2}(\sigma_1 + \sigma_2)]$$

An alternative form is

$$\frac{d\varepsilon_1}{\sigma_1 - \frac{1}{2}(\sigma_2 + \sigma_3)} = \frac{d\varepsilon_2}{\sigma_2 - \frac{1}{2}(\sigma_1 + \sigma_3)} = \frac{d\varepsilon_3}{\sigma_3 - \frac{1}{2}(\sigma_1 + \sigma_2)} = \frac{d\varepsilon_e}{\sigma_e} \quad \textbf{(12.44b)}$$

The plastic strain is, as before, to occur at constant volume; that is,

$$d\varepsilon_1 + d\varepsilon_2 + d\varepsilon_3 = 0$$

The *effective strain increment*, referring to Eq. (12.37b), may be written

$$d\varepsilon_e = \frac{\sqrt{2}}{3}[(d\varepsilon_1 - d\varepsilon_2)^2 + (d\varepsilon_2 - d\varepsilon_3)^2 + (d\varepsilon_3 - d\varepsilon_1)^2]^{1/2} \quad \textbf{(12.45)}$$

The effective stress σ_e is given by Eq. (12.36). Alternatively, to ascertain $d\varepsilon_e$ from a uniaxial true stress–strain curve such as Fig. 12.28, it is necessary to know the increase of equivalent stress $d\sigma_e$. Given σ_e and $d\varepsilon_e$, it then follows that at any point in the loading process, application of Eqs. (12.44) provides the increment of strain as a unique function of the state of stress and the increment of the stress. According to the Lévy–Mises theory, therefore, the deformation suffered by an element varies in accordance with the specific loading path taken.

In a particular case of straining of sheet metal under *biaxial tension*, the Lévy–Mises equations (12.44) become

$$\frac{d\varepsilon_1}{2 - \alpha} = \frac{d\varepsilon_2}{2\alpha - 1} = -\frac{d\varepsilon_3}{1 + \alpha} \quad \textbf{(12.46)}$$

where $\alpha = \sigma_2/\sigma_1$ and σ_3 taken as zero. The effective stress and strain increment, Eqs. (12.36) and (12.45), is now written

$$\sigma_e = \sigma_1(1 - \alpha + \alpha^2)^{1/2} \quad \textbf{(12.47)}$$

$$d\varepsilon_e = d\varepsilon_1\left(\frac{2}{2 - \alpha}\right)(1 - \alpha + \alpha^2)^{1/2} \quad \textbf{(12.48)}$$

Combining Eqs. (12.46) and (12.48) and integrating yields

$$\frac{\varepsilon_e}{2(1 - \alpha + \alpha^2)^{1/2}} = \frac{\varepsilon_1}{2 - \alpha} = \frac{\varepsilon_2}{2\alpha - 1} = -\frac{\varepsilon_3}{1 + \alpha} \quad \textbf{(12.49)}$$

To generalize this result, it is useful to employ Eq. (12.2), $\sigma_e = K(\varepsilon_e)^n$, to include strain-hardening characteristics. Differentiating this expression, we have

$$\frac{d\sigma_e}{d\varepsilon_e} = \frac{n\sigma_e}{\varepsilon_e} \quad \textbf{(12.50)}$$

Note that, for simple tension, $n = \varepsilon_e = \varepsilon$ and Eq. (12.50) reduces to Eq. (12.3).

The utility of the foregoing development is illustrated in Examples 12.14 and 12.15.

In closing, we note that the total (elastic–plastic) strains are determined by adding the elastic strains to the plastic strains. The elastic–plastic strain relations, together with the equations of equilibrium and compatibility and appropriate boundary conditions, completely describe a given situation. The general form of the Lévy–Mises relationships, including the elastic incremental components of strain, are referred to as the *Prandtl* and *Reuss* equations.

EXAMPLE 12.14 Analysis of Tube by Lévy–Mises Equations

Redo Example 12.13 employing the Lévy–Mises stress–strain increment relations.

Solution For the thin-walled cylinder under internal pressure, the plastic stresses are

$$\sigma_\theta = \sigma_1 = \frac{pr}{t}, \quad \sigma_z = \sigma_2 = \frac{pr}{2t}, \quad \sigma_r = \sigma_3 = 0 \tag{a}$$

where r and t are the current radius and the thickness. At instability,

$$dp = \frac{\partial p}{\partial \sigma_i} d\sigma_i + \frac{\partial p}{\partial r} dr + \frac{\partial p}{\partial t} dt = 0, \quad i = 1, 2 \tag{b}$$

Because $\alpha = \frac{1}{2}$, introduction of Eqs. (a) into Eq. (b) provides

$$\frac{d\sigma_1}{\sigma_1} = \frac{d\sigma_2}{\sigma_2} = \frac{dr}{r} - \frac{dt}{t} = d\varepsilon_2 - d\varepsilon_3 \tag{12.51}$$

Clearly, dr/r is the hoop strain increment $d\varepsilon_2$, and dt/t is the incremental thickness strain or radial strain increment $d\varepsilon_3$. Equation (12.51) is the condition of instability for the cylinder material.

Upon application of Eqs. (12.47) and (12.48), the effective stress and the effective strains are found to be

$$\sigma_e = \frac{\sqrt{3}}{2}\sigma_1$$

$$d\varepsilon_e = \frac{2}{\sqrt{3}}d\varepsilon_1 = -\frac{2}{\sqrt{3}}d\varepsilon_3, \quad d\varepsilon_2 = 0 \tag{c}$$

It is observed that axial strain does not occur and the situation is one of *plane strain*. The first of Eqs. (c) leads to $d\sigma_e = (\sqrt{3}/2)d\sigma_1$, and condition (12.51) gives

$$d\sigma_e = \frac{\sqrt{3}}{2}\sigma_1(d\varepsilon_1 - d\varepsilon_3) = \sqrt{3}\sigma_e d\varepsilon_e$$

from which

$$\frac{d\sigma_e}{d\varepsilon_e} = \sqrt{3}\sigma_e$$

A comparison of this result with Eq. (12.50) shows that

$$\varepsilon_e = \frac{n}{2\sqrt{3}} \tag{12.52}$$

The true stresses and true strains are then obtained from Eqs. (c) and $\sigma_e = K(\varepsilon_e)^n$ and the results found to be identical with that obtained using Hencky's relations (Table 12.2).

For a *spherical shell* subjected to internal pressure $\sigma_1 = \sigma_2 = pr/2t$, $\alpha = 1$ and $d\varepsilon_1 = d\varepsilon_2 = -d\varepsilon_3/2$. At stability $dp = 0$, and we now have

$$\frac{d\sigma_1}{\sigma_1} = \frac{dr}{r} - \frac{dt}{t} = d\varepsilon_1 - d\varepsilon_3 \qquad \textbf{(d)}$$

Equations (12.47) and (12.48) result in

$$\sigma_e = \sigma_1$$
$$d\varepsilon_e = 2d\varepsilon_1 = 2d\varepsilon_2 = -d\varepsilon_3 \qquad \textbf{(e)}$$

Equations (d) and (e) are combined to yield

$$\frac{d\sigma_e}{d\varepsilon_e} = \tfrac{2}{3}\sigma_e \qquad \textbf{(f)}$$

From Eqs. (f) and (12.50), it is concluded that

$$\varepsilon_e = \tfrac{2}{3}n \qquad \textbf{(12.53)}$$

True stress and true strain are easily obtained, and are the same as the values determined by a different method (Table 12.2).

12.14 STRESSES IN PERFECTLY PLASTIC THICK-WALLED CYLINDERS

The case of a thick-walled cylinder under internal pressure alone was considered in Section 8.2. Equation (8.11) was derived for the onset of yielding at the inner surface of the cylinder owing to maximum shear. This was followed by a discussion of the strengthening of a cylinder by shrinking a jacket on it (Sec. 8.5). The same goal can be achieved by applying sufficient pressure to cause some or all of the material to deform plastically and then releasing the pressure. This is briefly described next [Ref. 12.8].

A continuing increase in internal pressure will result in yielding at the inner surface. As the pressure increases, the plastic zone will spread toward the outer surface, and an elastic–plastic state will prevail in the cylinder with a limiting radius c beyond which the cross section remains elastic. As the pressure increases further, the radius c also increases until, eventually, the entire cross section becomes fully plastic at the ultimate pressure. When the pressure is reduced, the material unloads elastically. Thus, elastic and plastic stresses are superimposed to produce a *residual stress* pattern (see Example 12.15). The generation of such stresses by plastic action is called *autofrettage*. Upon reloading, the pressure needed to produce renewed yielding is *greater* than the pressure that produced initial yielding; the cylinder is thus strengthened by autofrettage.

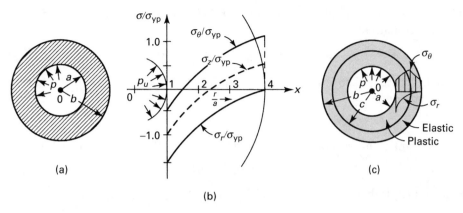

FIGURE 12.29. (a) Thick-walled cylinder of perfectly plastic material under internal pressure; (b) fully plastic stress distribution in the cylinder at the ultimate pressure for $b = 4a$; (c) partially yielded cylinder.

This section concerns the *fully plastic* and *elastic–plastic actions* in a thick-walled cylinder fabricated of a perfectly plastic material of yield strength σ_{yp} under internal pressure p as shown in Fig. 12.29a.

Complete Yielding

The fully plastic or ultimate pressure as well as the stress distribution corresponding to this pressure is determined by application of the Lévy–Mises relations with $\sigma_e = \sigma_{yp}$. In polar coordinates, the axial strain increment is

$$d\varepsilon_z = \frac{d\varepsilon_e}{\sigma_{yp}}[\sigma_z - \tfrac{1}{2}(\sigma_r + \sigma_\theta)] \tag{12.54}$$

If the ends of the cylinder are restrained so that the axial displacement $w = 0$, the problem may be regarded as a case of *plane strain*, for which $\varepsilon_z = 0$. It follows that $d\varepsilon_z = 0$ and Eq. (12.54) gives

$$\sigma_z = \tfrac{1}{2}(\sigma_r + \sigma_\theta) \tag{a}$$

The equation of equilibrium is, from Eq. (8.2),

$$\frac{d\sigma_r}{dr} + \frac{\sigma_r - \sigma_\theta}{r} = 0 \tag{b}$$

subject to the following boundary conditions:

$$(\sigma_r)_{r=a} = -p, \qquad (\sigma_r)_{r=b} = 0 \tag{c}$$

Based on the *maximum energy of distortion theory of failure*, Eqs. (4.4b) or (12.36), setting $\sigma_1 = \sigma_\theta$, $\sigma_3 = \sigma_r$, and $\sigma_2 = \tfrac{1}{2}(\sigma_r + \sigma_\theta)$ results in $\sigma_{yp}^2 = (\tfrac{3}{4})(\sigma_\theta - \sigma_r)^2$. From this, we obtain the yield condition:

$$\sigma_\theta - \sigma_r = k\sigma_{yp}, \qquad \text{where} \quad k = \frac{2}{\sqrt{3}} \tag{12.55}$$

Alternatively, according to the *maximum shearing stress theory of failure* (Sec. 4.6), the yield condition is

$$\sigma_\theta - \sigma_r = k\sigma_{yp}, \qquad \text{where} \quad k = 1 \qquad (12.56)$$

Introducing Eq. (12.55) or (12.56) into Eq. (b), we obtain $d\sigma_r/dr = k\sigma_{yp}/r$, which has the solution

$$\sigma_r = k\sigma_{yp} \ln r + c_1 \qquad (d)$$

The constant of integration is determined by applying the second of Eqs. (c):

$$c_1 = -k\sigma_{yp} \ln b$$

Equation (d) is thus

$$\sigma_r = -k\sigma_{yp} \ln \frac{b}{r} \qquad (e)$$

The first of conditions (c) now leads to the *ultimate pressure*:

$$p_u = k\sigma_{yp} \ln \frac{b}{a} \qquad (12.57)$$

An expression for σ_θ can now be obtained by substituting Eq. (e) into Eqs. (12.55) or (12.56). Consequently, Eq. (a) provides σ_z. The *complete plastic stress distribution* for a specified σ_{yp} is thus found to be

$$\sigma_r = -k\sigma_{yp} \ln \frac{b}{r}$$

$$\sigma_\theta = k\sigma_{yp}\left(1 - \ln \frac{b}{r}\right) \qquad (12.58)$$

$$\sigma_z = k\sigma_{yp}\left(\frac{1}{2} - \ln \frac{b}{r}\right)$$

The stresses given by Eqs. (12.58) are plotted in Fig. 12.29, whereas the distribution of elastic tangential and radial stresses is shown in Fig. 8.4a.

EXAMPLE 12.15   Residual Stresses in a Pressurized Cylinder

A perfectly plastic closed-ended cylinder ($\sigma_{yp} = 400$ MPa) with 50- and 100-mm internal and external diameters is subjected to an internal pressure p (Fig. 12.29a). On the basis of the maximum distortion energy theory of failure, determine (a) the complete plastic stresses at the inner surface and (b) the residual tangential and axial stresses at the inner surface when the cylinder is unloaded from the ultimate pressure p_u.

Solution   The magnitude of the ultimate pressure is, using Eq. (12.57),

$$p_u = \frac{2}{\sqrt{3}}(400) \ln \frac{50}{25} = 320.2 \text{ MPa}$$

a. From Eqs. (12.58), with $r = a$,

$$\sigma_r = -p_u = -320.2 \text{ MPa}$$

$$\sigma_\theta = \frac{2}{\sqrt{3}}(400)\left(1 - \ln\frac{50}{25}\right) = 141.7 \text{ MPa}$$

$$\sigma_z = \frac{2}{\sqrt{3}}(400)\left(\frac{1}{2} - \ln\frac{50}{25}\right) = -89.2 \text{ MPa}$$

Note, as a check, that $\sigma_z = (\sigma_r + \sigma_\theta)/2$ yields the same result.

b. Unloading is assumed to be *linearly elastic*. Thus, we have the following elastic stresses at $r = a$, using Eqs. (8.13) and (8.20):

$$\sigma_\theta = p_u\frac{b^2 + a^2}{b^2 - a^2} = (320.2)\frac{50^2 + 25^2}{50^2 - 25^2} = 533.7 \text{ MPa}$$

$$\sigma_z = p_u\frac{a^2}{b^2 - a^2} = (320.2)\frac{25^2}{50^2 - 25^2} = 106.7 \text{ MPa}$$

The residual stresses at the inner surface are therefore

$$(\sigma_\theta)_{\text{res}} = 141.7 - 533.7 = -392 \text{ MPa}$$
$$(\sigma_z)_{\text{res}} = -89.2 - 106.7 = -195.9 \text{ MPa}$$

The stresses at any other location may be obtained in a like manner.

Partial Yielding

For the *elastic segment* for which $c \leq r \leq b$ (Fig. 12.29c), the tangential and radial stresses are determined using Eqs. (8.12) and (8.13) with $a = c$. In so doing, we obtain

$$\sigma_r = \frac{p_c c^2}{b^2 - c^2}\left(1 - \frac{b^2}{r^2}\right) \tag{12.59a}$$

$$\sigma_\theta = \frac{p_c c^2}{b^2 - c^2}\left(1 + \frac{b^2}{r^2}\right) \tag{12.59b}$$

Here p_c is the magnitude of the (compressive) radial stress at the elastic–plastic boundary $r = c$ where yielding impends. Accordingly, by substituting these expressions into the yield condition $\sigma_\theta - \sigma_r = k\sigma_{yp}$, we have

$$p_c = k\sigma_{yp}\frac{b^2 - c^2}{2b^2} \tag{12.60}$$

This pressure represents the boundary condition for a fully plastic segment with inner radius a and outer radius c. That is, constant c_1 in Eq. (d) is obtained by applying $(\sigma_r)_{r=c} = -p_c$:

$$c_1 = -k\sigma_{yp}\ln c - k\sigma_{yp}\frac{b^2 - c^2}{2b^2}$$

Substituting this value of c_1 into Eq. (d), the radial stress in the *plastic zone* becomes

$$\sigma_r = k\sigma_{yp}\left(\ln\frac{r}{c} - \frac{b^2 - c^2}{2b^2}\right) \qquad (12.61a)$$

Then, using the yield condition, the tangential stress in the *plastic zone* is obtained in the following form:

$$\sigma_\theta = k\sigma_{yp}\left(1 + \ln\frac{r}{c} - \frac{b^2 - c^2}{2b^2}\right) \qquad (12.61b)$$

Equations (12.59) and (12.61), for a given elastic–plastic boundary radius c, provide the relationships necessary for calculation of the elastic–plastic stress distribution in the cylinder wall (Fig. 12.29c).

REFERENCES

12.1. FORD, H., *Advanced Mechanics of Materials*, 2nd ed., Ellis Horwood, Chichester, England, 1977, Part 4.

12.2. FAUPEL, J. H., and FISHER, F. E., *Engineering Design*, 2nd ed., Wiley, Hoboken, N.J. 1981, Chap. 6.

12.3. MENDELSSON, A., *Plasticity: Theory and Applications*, Macmillan, New York, 1968 (reprinted, R. E. Krieger, Melbourne, Fla., 1983).

12.4. RAMBERG, W., and OSGOOD, W. R., *Description of Stress–Strain Curves by Three Parameters*, National Advisory Committee of Aeronautics, TN 902, 1943, Washington, D.C.

12.5. UGURAL, A. C., *Mechanics of Materials*, Wiley, Hoboken, N.J. 2008, Secs. 4.3 and 7.18.

12.6. HODGE, P C., *Plastic Design Analysis of Structures*, McGraw-Hill, New York, 1963.

12.7. NADAI, A., *Theory of Flow and Fracture of Solids*, McGraw-Hill, New York, 1950, Chap. 35.

12.8. HOFFMAN, O., and SACHS, G., *Theory of Plasticity for Engineers*, McGraw-Hill, New York, 1953.

PROBLEMS

Sections 12.1 through 12.6

12.1. A solid circular cylinder of 100-mm diameter is subjected to a bending moment $M = 3.375$ kN·m, an axial tensile force $P = 90$ kN, and a twisting end couple $T = 4.5$ kN·m. Determine the stress deviator tensor. [Hint: Refer to Sec. 2.15.]

12.2. In the pin-connected structure shown in Fig. 12.4, the true stress–engineering strain curves of the members are expressed by $\sigma_{AD} = \sigma_{DC} = K_1\varepsilon_o^{n_1}$ and $\sigma_{BD} = K_2\varepsilon_o^{n_2}$. Verify that, for all three bars to reach tensile instability simultaneously, they should be set initially at an angle described by

$$\cos \alpha = \frac{L_{BD}}{L_{AD}} = \frac{1 - n_2}{1 - n_1}\left[\frac{n_1(2 - n_1)}{n_2(2 - n_2)}\right]^{1/2} \qquad \textbf{(P12.2)}$$

Calculate the value of this initial angle for $n_1 = 0.2$ and $n_2 = 0.3$.

12.3. Determine the deflection of a uniformly loaded rigid-plastic cantilever beam of length L. Locate the origin of coordinates at the fixed end, and denote the loading by p.

12.4. Redo Prob. 12.3 for $p = 0$ and a concentrated load P applied at the free end.

12.5. Consider a beam of rectangular section, subjected to end moments as shown in Fig. 12.7a. Assuming that the relationship for tensile and compressive stress for the material is approximated by $\sigma = K\varepsilon^{1/4}$, determine the maximum stress.

12.6. A simply supported rigid-plastic beam is described in Fig. P12.6. Compute the maximum deflection. Reduce the result to the case of a linearly elastic material. $EIv_{\max} = \left(\frac{1}{8}\right)PaL^2 - \left(\frac{1}{6}\right)Pa^3$. Let $P = 8$ kN, $E = 200$ GPa, $L = 1.2$ m, and $a = 0.45$ m. Cross-sectional dimensions shown are in millimeters.

12.7. Figure P12.7 shows a stepped steel bar ABC, axially loaded until it elongates 10 mm, and then unloaded. Determine (a) the largest value of P; (b) the plastic axial deformation of segments AB and BC. Given: $d_1 = 60$ mm, $d_2 = 50$ mm, $E = 210$ GPa, and $\sigma_{yp} = 240$ MPa.

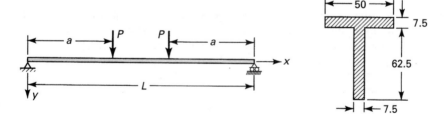

FIGURE P12.6.

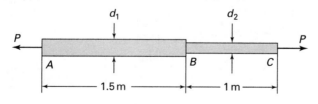

FIGURE P12.7.

12.8. Resolve Prob. 12.7, knowing that $d_1 = 30$ mm, $d_2 = 20$ mm, and $\sigma_{yp} = 280$ MPa.

12.9. The assembly of three steel bars shown in Fig. 12.4 supports a vertical load P. It can be verified that [see Ref. 12.5], the forces in the bars $AD = CD$ and BD are, respectively,

$$N_{AD} = N_{CD} = \frac{P\cos^2\alpha}{1 + 2\cos^3\alpha} \qquad N_{BD} = \frac{P}{1 + 2\cos^3\alpha} \qquad \textbf{(P12.9)}$$

Using these equations, calculate the value of ultimate load P_u. *Given:* Each member is made of mild steel with $\sigma_{yp} = 250$ MPa and has the same cross-sectional area $A = 400$ mm^2, and $\alpha = 40°$.

12.10. When a load $P = 400$ kN is applied and then removed, calculate the residual stress in each bar of the assembly described in Example 12.3.

12.11. Figure P12.11 depicts a cylindrical rod of cross-sectional area A inserted into a tube of the same length L and of cross-sectional area A_t; the left ends of the members are attached to rigid support and the right ends to a rigid plate. When an axial load P is applied as shown, determine the maximum deflection and draw the load-deflection diagram of the rod–tube assembly. *Given:* $L = 1.2$ m, $A_r = 45$ mm^2, $A_t = 60$ mm^2, $E_r = 200$ GPa, $E_t = 100$ GPa, $(\sigma_r)_{yp} = 250$ MPa, and $(\sigma_t)_{yp} = 310$ MPa. *Assumption:* The rod and tube are both made of elastoplastic materials.

Sections 12.7 and 12.8

12.12. A ductile bar $(\sigma_{yp} = 350$ MPa$)$ of square cross section with sides $a = 12$ mm is subjected to bending moments M about the z axis at its ends (Fig. P12.12). Determine the magnitude of M at which (a) yielding impends; (b) the plastic zones at top and bottom of the bar are 2-mm thick.

12.13. A beam of rectangular cross section (width a, depth h) is subjected to bending moments M at its ends. The beam is constructed of a material displaying the stress–strain relationship shown in Fig. P12.13. What value of M can be carried by the beam?

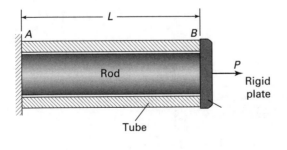

FIGURE P12.11.

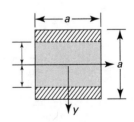

FIGURE P12.12.

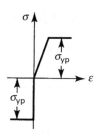

FIGURE P12.13.

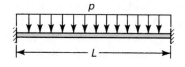

FIGURE P12.14.

12.14. A perfectly plastic beam is supported as shown in Fig. P12.14. Determine the maximum deflection at the start of yielding.

12.15. Figure P12.15 shows the cross section of a rectangular beam made of mild steel with $\sigma_{yp} = 240$ MPa. For bending about the z axis, find (a) the yield moment; (b) the moment producing a $e = 20$-mm-thick plastic zone at the top and bottom of the beam. *Given:* $b = 60$ mm and $h = 40$ mm.

12.16. A steel rectangular beam, the cross section shown in Fig. P12.15, is subjected to a moment about the z axis 1.3 times greater than M. Calculate (a) the distance from the neutral axis to the point at which elastic core ends, e; (b) the residual stress pattern following release of loading.

12.17. A singly symmetric aluminum beam has the cross section shown in Fig. P12.17. (Dimensions are in millimeters.) Determine the ultimate moment M_u. *Assumption:* The aluminum is to be elastoplastic with a yield stress of $\sigma_{yp} = 260$ MPa.

12.18 and 12.19. Find the shape factor f for an elastoplastic beam of the cross sections shown in Figs. P12.18 and P12.19 (Refer to Table C.1).

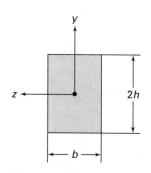

FIGURE P12.15.

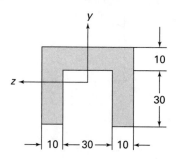

FIGURE P12.17.

FIGURE P12.18.

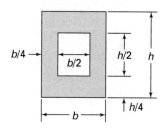

FIGURE P12.19.

12.20. A rectangular beam with $b = 70$ mm and $h = 120$ mm (Fig. P12.15) is subjected to an ultimate moment M_u. Knowing that $\sigma_{yp} = 250$ MPa for this beam of ductile material, determine the residual stresses at the upper and lower faces if the loading has been removed.

12.21. Consider a uniform bar of solid circular cross section with radius r, subjected to axial tension and bending moments at both ends. Derive general relationships involving N and M that govern, first, the case of initial yielding and, then, fully plastic deformation. Sketch the interaction curves.

12.22. Figure P12.22 shows a hook made of steel with $\sigma_{yp} = 280$ MPa, equal in tension and compression. What load P results in complete plastic deformation in section A–B? Neglect the effect of curvature on the stress distribution.

12.23. A propped cantilever beam made of ductile material is loaded as shown in Fig. P12.23. What are the values of the collapse load p_u and the distance x?

12.24. Obtain the interaction curves for the beam cross section shown in Fig. 12.14a. The beam is subjected to a bending moment M and an axial load N at both ends. Take $b = 2h$, $b_1 = 1.8h$, and $h_1 = 0.7h$.

12.25. Obtain the collapse load of the structure shown in Fig. P12.25. Assume that plastic hinges form at 1, 3, and 4.

12.26. What is the collapse load of the beam shown in Fig. P12.26? Assume two possible modes of collapse such that plastic hinges form at 2, 3, and 4.

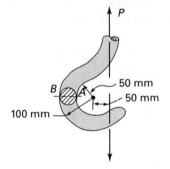

FIGURE P12.22.

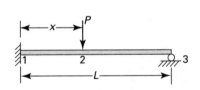

FIGURE P12.23.

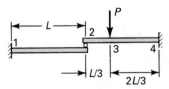

FIGURE P12.25.

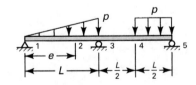

FIGURE P12.26.

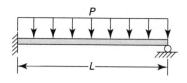

FIGURE P12.27.

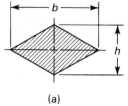

(a)   (b)

FIGURE P12.28.

12.27. A propped cantilever beam AB, made of ductile material, supports a uniform load of intensity p (Fig. P12.27). What is the ultimate limit load p_u?

12.28. Figure P12.28 shows two beam cross sections. Determine M_u/M_{yp} for each case.

Sections 12.9 through 12.14

12.29. Figure 12.24 shows a circular elastoplastic shaft with yield strength in shear τ_{yp}, shear modulus of elasticity G, diameter d, and length L. The shaft is twisted until the maximum shearing strain equals 6000μ. Determine (a) the magnitude of the corresponding angle of twist ϕ; (b) the value of the applied torque T. *Given:* $L = 0.5$ m, $d = 60$ mm, $G = 70$ GPa, and $\tau_{yp} = 180$ MPa.

12.30. A circular shaft of diameter d and length L is subjected to a torque of T, as shown in Fig. 12.24. The shaft is made of 6061-T6 aluminum alloy (see Table D.1), which is assumed to be elastoplastic. Find (a) the radius of the elastic core ρ_0; (b) the angle of twist ϕ. *Given:* $d = 25$ mm, $L = 1.2$ m, and $T = 4.5$ kN·m.

12.31. The fixed-ended shaft illustrated in Fig. P12.31 is made of an elastoplastic material for which the shear modulus of elasticity is G and the yield stress in shear is τ_{yp}. Find the magnitude of the applied torque T. *Assumption:* The angle of twist at step C is $\phi_{yp} = 0.25$ rad. *Given:* $a = 2$ m, $b = 1.5$ m, $d_1 = 80$ mm, $d_2 = 50$ mm, $G = 80$ MPa, and $\tau_{yp} = 240$ MPa.

12.32. Determine the elastic–plastic stresses in a rotating solid disk.

12.33. For a rectangular bar of sides a and b, determine the ultimate torque corresponding to the fully plastic state (Table 12.1). Use the sand hill analogy.

12.34. For an equilateral triangular bar of sides $2a$, determine (a) the ultimate torque corresponding to the fully plastic state (Table 12.1) (use the sand hill analogy) and (b) the maximum elastic torque by referring to Table 6.2. (c) Compare the results found in (a) and (b).

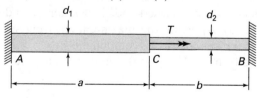

FIGURE P12.31.

12.35. An annular disk of equilateral hyperbolic profile with outer and inner radii b and a (Fig. 8.13) is shrunk onto a solid shaft so that the interfacial pressure is p_i. Demonstrate that, according to the Tresca yield criterion, when the disk becomes fully plastic,

$$p_i = \frac{a(b-r)}{rb}\sigma_{yp} \tag{P12.35}$$

Here σ_{yp} is the yield point stress and r represents any arbitrary radius.

12.36. Consider a thin-walled cylinder of original radius r_o, subjected to internal pressure p. Determine the value of the required original thickness at instability employing Hencky's relations. Use the following:

$$K = 900 \text{ MPa}, \qquad n = 0.2$$
$$r_o = 0.5 \text{ m}, \qquad p = 14 \text{ MPa}$$

12.37. Redo Example 12.13 for the cylinder under uniform axial tensile load P and $p = 0$.

12.38. A thin-walled tube of original thickness t_o and outer radius R_o just fits over a rigid rod of radius r_o. Employ the Lévy–Mises relations to verify that the axial load the tube can sustain before instability occurs is represented by

$$P = 2\pi r_o t_o e^{(-2\pi K/\sqrt{3})}(2n/\sqrt{3})^n \tag{P12.38}$$

Assume the tube–rod interface to be frictionless. Use $\sigma = K\varepsilon^n$ as the true stress–true strain relationship of the tube material in simple tension.

12.39. A thick-walled cylinder ($\sigma_{yp} = 250$ MPa) of inner radius $a = 50$ mm is subjected to an internal pressure of $p_i = 60$ MPa. Determine the outer radius b such as to provide (a) a factor of safety $n = 2.5$ against yielding; (b) a factor of safety $n = 3$ against ultimate collapse based on maximum shearing stress theory of failure.

12.40. A thick-walled cylinder has an inner radius a and outer radius $b = 2a$. What is the internal pressure at which the elastic–plastic boundary is at $r = 1.4a$, based on maximum energy of distortion criterion? Let $\sigma_{yp} = 260$ MPa.

12.41. Consider a perfectly plastic pipe ($\sigma_{yp} = 420$ MPa) having an outer radius of 60 mm and inner radius of 50 mm. Determine the maximum internal pressures at the onset of yielding and for complete yielding on the basis of a factor of safety $n = 3$ and the following theories of failure: (a) maximum shearing stress and (b) maximum energy of distortion.

12.42. A perfectly plastic, closed-ended cylinder is under internal pressure p (Fig. 12.29a). Applying the maximum shearing stress criterion of failure, calculate the residual stress components at $r = 0.25$ m when the cylinder is unloaded from p_u. Use $a = 0.2$ m, $b = 0.3$ m, and $\sigma_{yp} = 400$ MPa.

12.43. A thick-walled compound cylinder having $a = 20$ mm, $b = 30$ mm, and $c = 50$ mm is subjected to internal pressure. Material yield strengths are 280 MPa and 400 MPa for inner and outer cylinders, respectively. Determine the fully plastic pressure on the basis of the maximum shearing stress criterion of failure.

12.44. A perfectly plastic cylinder for which $b/a = 3$ is subjected to internal pressure causing yielding of the material to the mid-depth $c = (a + b)/2$. In terms of k and σ_{yp}, determine (a) the pressure between the elastic and plastic zones, (b) the radial stress at $r = a$, and (c) the tangential stresses at $r = b$, $r = c$, and $r = a$. Note: Relationships derived (Sec. 12.14) depend on the ratios of the radii rather than on their magnitudes. Accordingly, convenient numbers such as $a = 1$, $c = 2$, and $b = 3$ may be employed.

CHAPTER 13

Plates and Shells

13.1 INTRODUCTION

This chapter is subdivided into two parts. In Part A, we develop the governing equations and methods of solution of deflection for rectangular and circular plates. Applications of the energy and finite element methods for computation of deflection and stress in plates are also included. Membrane stresses in shells are taken up in Part B. Load-carrying mechanism of a shell, which differs from that of other elements, is demonstrated first. This is followed with a discussion of stress distribution in spherical, conical, and cylindrical shells. Thermal stresses in compound cylindrical shells are also considered.

Part A—Bending of Thin Plates

13.2 BASIC ASSUMPTIONS

Plates and shells are initially flat and curved structural elements, respectively, with thicknesses small compared with the remaining dimensions. We first consider plates, for which it is usual to divide the thickness t into equal halves by a plane parallel to the faces. This plane is termed the *midsurface* of the plate. The plate thickness is measured in a direction normal to the midsurface at each point under consideration. Plates of technical significance are often defined as *thin* when the ratio of the thickness to the smaller span length is less than 1/20. We here treat the small deflection theory of homogeneous, uniform thin plates, leaving for the numerical approach of Section 13.10 a discussion of plates of nonuniform thickness and irregular shape.

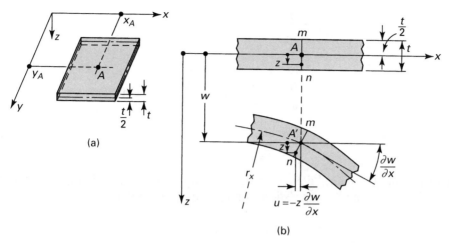

FIGURE 13.1. *Deformation of a plate in bending.*

Consider now a plate prior to deformation, shown in Fig. 13.1a, in which the xy plane coincides with the midsurface and hence the z deflection is zero. When, owing to external loading, deformation occurs, the midsurface at any point x_A, y_A suffers a deflection w. Referring to the coordinate system shown, the fundamental assumptions of the *small deflection* theory of bending for *isotropic, homogeneous, thin* plates may be summarized as follows:

1. The deflection of the midsurface is small in comparison with the thickness of the plate. The slope of the deflected surface is much less than unity.
2. Straight lines initially normal to the midsurface remain straight and normal to that surface subsequent to bending. This is equivalent to stating that the vertical shear strains γ_{xz} and γ_{yz} are negligible. The deflection of the plate is thus associated principally with bending strains, with the implication that the normal strain ε_z owing to vertical loading may also be neglected.
3. No midsurface straining or in-plane straining, stretching, or contracting occurs as a result of bending.
4. The component of stress normal to the midsurface, σ_z, is negligible.

These presuppositions are analogous to those associated with simple bending theory of beams.

13.3 STRAIN–CURVATURE RELATIONS

We here develop fundamental relationships between the strains and curvatures of the midsurface of thin plates. On the basis of assumption 2 stated in the preceding section, the strain–displacement relations of Eq. (2.4) reduce to

$$\varepsilon_x = \frac{\partial u}{\partial x}, \qquad \varepsilon_z = \frac{\partial w}{\partial z} = 0$$

$$\varepsilon_y = \frac{\partial v}{\partial y}, \qquad \gamma_{xz} = \frac{\partial w}{\partial x} + \frac{\partial u}{\partial z} = 0 \tag{a}$$

$$\gamma_{xy} = \frac{\partial u}{\partial y} + \frac{\partial v}{\partial x}, \qquad \gamma_{yz} = \frac{\partial w}{\partial y} + \frac{\partial v}{\partial z} = 0$$

Integration of $\varepsilon_z = \partial w/\partial z = 0$ yields

$$w = f_1(x, y) \tag{13.1}$$

indicating that the lateral deflection does not vary throughout the plate thickness. Similarly, integrating the expressions for γ_{xz} and γ_{yz}, we obtain

$$u = -z\frac{\partial w}{\partial x} + f_2(x, y), \quad v = -z\frac{\partial w}{\partial y} + f_3(x, y) \tag{b}$$

It is clear that $f_2(x, y)$ and $f_3(x, y)$ represent, respectively, the values of u and v corresponding to $z = 0$ (the midsurface). Because assumption 3 precludes such in-plane straining, we conclude that $f_2 = f_3 = 0$, and therefore

$$u = -z\frac{\partial w}{\partial x}, \quad v = -z\frac{\partial w}{\partial y} \tag{13.2}$$

where $\partial w/\partial x$ and $\partial w/\partial y$ are the slopes of the midsurface. The expression for u is represented in Fig. 13.1b at section mn passing through arbitrary point $A(x_A, y_A)$. A similar interpretation applies for v in the zy plane. It is observed that Eqs. (13.2) are consistent with assumption 2. Combining the first three equations of (a) with Eq. (13.2), we have

$$\varepsilon_x = -z\frac{\partial^2 w}{\partial x^2}, \quad \varepsilon_y = -z\frac{\partial^2 w}{\partial y^2}, \quad \gamma_{xy} = -2z\frac{\partial^2 w}{\partial x\, \partial y} \tag{13.3a}$$

which provide the strains at any point.

Because in small deflection theory the square of a slope may be regarded as negligible, the partial derivatives of Eqs. (13.3a) represent the *curvatures* of the plate (see Eq. 5.7). Therefore, the curvatures at the midsurface in planes parallel to the xz (Fig. 13.1b), yz, and xy planes are, respectively,

$$\frac{1}{r_x} = \frac{\partial}{\partial x}\left(\frac{\partial w}{\partial x}\right) = \frac{\partial^2 w}{\partial x^2}$$

$$\frac{1}{r_y} = \frac{\partial}{\partial y}\left(\frac{\partial w}{\partial y}\right) = \frac{\partial^2 w}{\partial y^2}$$

$$\frac{1}{r_{xy}} = \frac{1}{r_{yx}} = \frac{\partial}{\partial x}\left(\frac{\partial w}{\partial y}\right) \tag{13.4}$$

$$= \frac{\partial}{\partial y}\left(\frac{\partial w}{\partial x}\right) = \frac{\partial^2 w}{\partial x\, \partial y}$$

The foregoing are simply the *rates* at which the slopes vary over the plate.

In terms of the radii of curvature, the strain–deflection relations (13.3a) may be written

$$\varepsilon_x = -z\,\frac{1}{r_x}, \quad \varepsilon_y = -z\,\frac{1}{r_y}, \quad \gamma_{xy} = -2z\,\frac{1}{r_{xy}} \tag{13.3b}$$

Examining these equations, we are led to conclude that a circle of curvature can be constructed similarly to Mohr's circle of strain. The curvatures therefore transform in the same manner as the strains. It can be verified by employing Mohr's circle that $(1/r_x) + (1/r_y) = \nabla^2 w$. The sum of the curvatures in perpendicular directions, called the *average curvature*, is invariant with respect to rotation of the coordinate axis. This assertion is valid at any location on the midsurface.

13.4 STRESS, CURVATURE, AND MOMENT RELATIONS

The stress components $\sigma_x, \sigma_y,$ and $\tau_{xy} = \tau_{yx}$ are related to the strains by Hooke's law, which for a thin plate becomes

$$\sigma_x = \frac{E}{1 - \nu^2}(\varepsilon_x + \nu\varepsilon_y) = -\frac{Ez}{1 - \nu^2}\left(\frac{\partial^2 w}{\partial x^2} + \nu\frac{\partial^2 w}{\partial y^2}\right)$$

$$\sigma_y = \frac{E}{1 - \nu^2}(\varepsilon_y + \nu\varepsilon_x) = -\frac{Ez}{1 - \nu^2}\left(\frac{\partial^2 w}{\partial y^2} + \nu\frac{\partial^2 w}{\partial x^2}\right) \tag{13.5}$$

$$\tau_{xy} = \frac{E}{2(1 + \nu)}\gamma_{xy} = -\frac{Ez}{1 + \nu}\frac{\partial^2 w}{\partial x\,\partial y}$$

These expressions demonstrate clearly that the stresses vanish at the midsurface and vary linearly over the thickness of the plate.

The stresses distributed over the side surfaces of the plate, while producing no net force, do result in bending and twisting moments. These moment resultants per unit length (for example, newtons times meters divided by meters, or simply newtons) are denoted M_x, M_y, and M_{xy}. Referring to Fig. 13.2a,

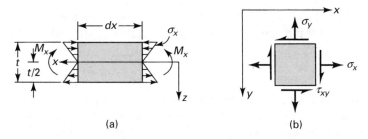

FIGURE 13.2.   (a) Plate segment in pure bending; (b) positive stresses on an element in the bottom half of a plate.

$$\int_{-t/2}^{t/2} z\sigma_x \, dy \, dz = dy \int_{-t/2}^{t/2} z\sigma_x \, dz = M_x \, dy$$

Expressions involving M_y and $M_{xy} = M_{yx}$ are similarly derived. The bending and twisting moments per unit length are thus

$$M_x = \int_{-t/2}^{t/2} z\sigma_x \, dz$$

$$M_y = \int_{-t/2}^{t/2} z\sigma_y \, dz \qquad (13.6)$$

$$M_{xy} = \int_{-t/2}^{t/2} z\tau_{xy} \, dz$$

Introducing into Eq. (13.6) the stresses given by Eqs. (13.5), and taking into account the fact that $w = w(x, y)$, we obtain

$$M_x = -D\left(\frac{\partial^2 w}{\partial x^2} + \nu \frac{\partial^2 w}{\partial y^2}\right)$$

$$M_y = -D\left(\frac{\partial^2 w}{\partial y^2} + \nu \frac{\partial^2 w}{\partial x^2}\right) \qquad (13.7)$$

$$M_{xy} = -D(1 - \nu)\frac{\partial^2 w}{\partial x \, \partial y}$$

where

$$D = \frac{Et^3}{12(1 - \nu^2)} \qquad (13.8)$$

is the *flexural rigidity* of the plate. Note that, if a plate element of unit width were free to expand sidewise under the given loading, anticlastic curvature would not be prevented; the flexural rigidity would be $Et^3/12$. The remainder of the plate does not permit this action, however. Because of this, a plate manifests *greater stiffness* than a narrow beam by a factor $1/(1 - \nu^2)$ or about 10%. Under the sign convention, a positive moment is one that results in positive stresses in the positive (bottom) half of the plate (Sec. 1.5), as shown in Fig. 13.2b.

Substitution of $z = t/2$ into Eq. (13.5), together with the use of Eq. (13.7), provides expressions for the maximum stress (which occurs on the surface of the plate):

$$\sigma_{x,\,max} = \frac{6M_x}{t^2}, \qquad \sigma_{y,\,max} = \frac{6M_y}{t^2}, \qquad \tau_{xy,\,max} = \frac{6M_{xy}}{t^2} \qquad (13.9)$$

Employing a new set of coordinates in which x', y' replaces x, y and $z' = z$, we first transform $\sigma_x, \sigma_y, \tau_{xy}$ into $\sigma_{x'}, \sigma_{y'}, \tau_{x'y'}$ through the use of Eq. (1.18). These are then substituted into Eq. (13.6) to obtain the corresponding $M_{x'}, M_{y'}, M_{x'y'}$. Examination of Eqs. (13.6) and (13.9) indicates a direct correspondence between the moments and stresses. It is concluded, therefore, that the equation for transforming the stresses should be identical with that used for the moments. Mohr's circle may thus be applied to moments as well as to stresses.

13.5 GOVERNING EQUATIONS OF PLATE DEFLECTION

Consider now a plate element $dx\,dy$ subject to a uniformly distributed lateral load per unit area p (Fig. 13.3). In addition to the moments M_x, M_y, and M_{xy} previously discussed, we now find vertical shearing forces Q_x and Q_y (force per unit length) acting on the sides of the element. These forces are related directly to the vertical shearing stresses:

$$Q_x = \int_{-t/2}^{t/2} \tau_{xz}\,dz, \qquad Q_y = \int_{-t/2}^{t/2} \tau_{yz}\,dz \qquad (13.10)$$

The sign convention associated with Q_x and Q_y is identical with that for the shearing stresses τ_{xz} and τ_{yz}: A positive shearing force acts on a positive face in the positive z direction (or on a negative face in the negative z direction). The bending moment sign convention is as previously given. On this basis, all forces and moments shown in Fig. 13.3 are positive.

It is appropriate to emphasize that while the simple theory of thin plates neglects the effect on bending of σ_z, $\gamma_{xz} = \tau_{xz}/G$, and $\gamma_{yz} = \tau_{yz}/G$ (as discussed in Sec. 13.2), the vertical forces Q_x and Q_y resulting from τ_{xz} and τ_{yz} are not negligible. In fact, they are of the same order of magnitude as the lateral loading and moments.

It is our next task to obtain the equation of equilibrium for an element and eventually to reduce the system of equations to a single expression involving the deflection w. Referring to Fig. 13.3, we note that body forces are assumed negligible relative to the surface loading and that no horizontal shear and normal forces act on the sides of the element. The equilibrium of z-directed forces is governed by

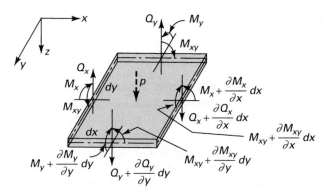

FIGURE 13.3.   *Positive moments and shear forces (per unit length) and distributed lateral load (per unit area) on a plate element.*

$$\frac{\partial Q_x}{\partial x}\,dx\,dy + \frac{\partial Q_y}{\partial y}\,dy\,dx + p\,dx\,dy = 0$$

or

$$\frac{\partial Q_x}{\partial x} + \frac{\partial Q_y}{\partial y} + p = 0 \qquad \text{(a)}$$

For the equilibrium of moments about the x axis,

$$\frac{\partial M_{xy}}{\partial x}\,dx\,dy + \frac{\partial M_y}{\partial y}\,dx\,dy - Q_y\,dx\,dy = 0$$

from which

$$\frac{\partial M_{xy}}{\partial x} + \frac{\partial M_y}{\partial y} - Q_y = 0 \qquad \text{(b)}$$

Higher-order terms, such as the moment of p and the moment owing to the change in Q_y, have been neglected. The equilibrium of moments about the y axis yields an expression similar to Eq. (b):

$$\frac{\partial M_{xy}}{\partial y} + \frac{\partial M_x}{\partial x} - Q_x = 0 \qquad \text{(c)}$$

Equations (b) and (c), when combined with Eq. (13.7), lead to

$$Q_x = -D\frac{\partial}{\partial x}\left(\frac{\partial^2 w}{\partial x^2} + \frac{\partial^2 w}{\partial y^2}\right), \quad Q_y = -D\frac{\partial}{\partial y}\left(\frac{\partial^2 w}{\partial x^2} + \frac{\partial^2 w}{\partial y^2}\right) \qquad \text{(13.11)}$$

Finally, substituting Eq. (13.11) into Eq. (a) results in the governing equation of plate theory (Lagrange, 1811):

$$\frac{\partial^4 w}{\partial x^4} + 2\frac{\partial^4 w}{\partial x^2 \partial y^2} + \frac{\partial^4 w}{\partial y^4} = \frac{p}{D} \qquad \text{(13.12a)}$$

or, in concise form,

$$\nabla^4 w = \frac{p}{D} \qquad \text{(13.12b)}$$

The bending of plates subject to a lateral loading p per unit area thus reduces to a single differential equation. Determination of $w(x, y)$ relies on the integration of Eq. (13.12) with the constants of integration dependent on the identification of appropriate boundary conditions. The shearing stresses τ_{xz} and τ_{yz} can be readily determined by applying Eqs. (13.11) and (13.10) once $w(x, y)$ is known. These stresses display a parabolic variation over the thickness of the plate. The maximum shearing stress, as in the case of a beam of rectangular section, occurs at $z = 0$:

$$\tau_{xz,\,\text{max}} = \frac{3}{2}\frac{Q_x}{t}, \qquad \tau_{yz,\,\text{max}} = \frac{3}{2}\frac{Q_y}{t} \qquad \text{(13.13)}$$

The key to evaluating all the stresses, employing Eqs. (13.5) or (13.9) and (13.3), is thus the solution of Eq. (13.12) for $w(x, y)$. As already indicated, τ_{xz} and τ_{yz} are regarded as small compared with the remaining plane stresses.

13.6 BOUNDARY CONDITIONS

Solution of the plate equation requires that two boundary conditions be satisfied at each edge. These may relate to deflection and slope, or to forces and moments, or to some combination. The principal feature distinguishing the boundary conditions applied to plates from those applied to beams relates to the existence along the plate edge of twisting moment resultants. These moments, as demonstrated next, may be replaced by an equivalent vertical force, which when added to the vertical shearing force produces an *effective* vertical force.

Consider two successive elements of lengths dy on edge $x = a$ of the plate shown in Fig. 13.4a. On the right element, a twisting moment $M_{xy} \, dy$ acts, while the left element is subject to a moment $[M_{xy} + (\partial M_{xy}/\partial y) \, dy]dy$. In Fig. 13.4b, we observe that these twisting moments have been replaced by equivalent force couples that produce only *local* differences in the distribution of stress on the edge $x = a$. The stress distribution elsewhere in the plate is unaffected by this substitution. Acting at the left edge of the right element is an upward directed force M_{xy}. Adjacent to this force is a downward directed force $M_{xy} + (\partial M_{xy}/\partial y) \, dy$ acting at the right edge of the left element. The difference between these forces (expressed per unit length), $\partial M_{xy}/\partial y$, may be combined with the transverse shearing force Q_x to produce an effective transverse edge force per unit length, V_x, known as *Kirchhoff's force* (Fig. 13.4c):

$$V_x = Q_x + \frac{\partial M_{xy}}{\partial y}$$

Substitution of Eqs. (13.7) and (13.11) into this leads to

$$V_x = -D\left[\frac{\partial^3 w}{\partial x^3} + (2 - \nu)\frac{\partial^3 w}{\partial x \partial y^2}\right] \qquad \textbf{(13.14)}$$

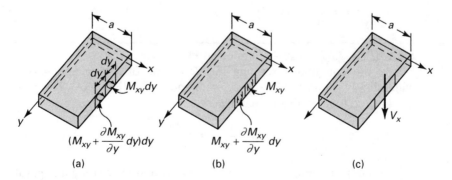

FIGURE 13.4. *Edge effect of twisting moment.*

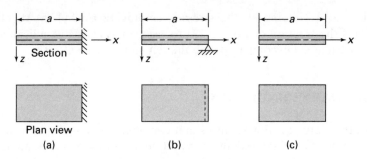

FIGURE 13.5. *Various boundary conditions: (a) fixed edge; (b) simple support; (c) free edge.*

We are now in a position to formulate a variety of commonly encountered situations. Consider first the conditions that apply along the *clamped* edge $x = a$ of the rectangular plate with edges parallel to the x and y axes (Fig. 13.5a). As both the deflection and slope are zero,

$$w = 0, \qquad \frac{\partial w}{\partial x} = 0, \qquad \text{for } x = a \qquad \textbf{(13.15)}$$

For the *simply supported* edge (Fig. 13.5b), the deflection and bending moment are both zero:

$$w = 0, \qquad M_x = -D\left(\frac{\partial^2 w}{\partial x^2} + \nu \frac{\partial^2 w}{\partial y^2}\right) = 0, \qquad \text{for } x = a \qquad \textbf{(13.16a)}$$

Because the first of these equations implies that along edge $x = a$, $\partial w/\partial y = 0$ and $\partial^2 w/\partial y^2 = 0$, the conditions expressed by Eq. (13.16a) may be restated in the following equivalent form:

$$w = 0, \qquad \frac{\partial^2 w}{\partial x^2} = 0, \qquad \text{for } x = a \qquad \textbf{(13.16b)}$$

For the case of the *free edge* (Fig. 13.5c), the moment and vertical edge force are zero:

$$\frac{\partial^2 w}{\partial x^2} + \nu \frac{\partial^2 w}{\partial y^2} = 0, \qquad \frac{\partial^3 w}{\partial x^3} + (2 - \nu)\frac{\partial^3 w}{\partial x \partial y^2} = 0, \qquad \text{for } x = a \qquad \textbf{(13.17)}$$

EXAMPLE 13.1   Simply Supported Plate Strip

Derive the equation describing the deflection of a long and narrow plate, simply supported at edges $y = 0$ and $y = b$ (Fig. 13.6). The plate is subjected to nonuniform loading

$$p(y) = p_o \sin\left(\frac{\pi y}{b}\right) \qquad \textbf{(a)}$$

so that it deforms into a cylindrical surface with its generating line parallel to the x axis. The constant p_o thus represents the load intensity along the line passing through $y = b/2$, parallel to x.

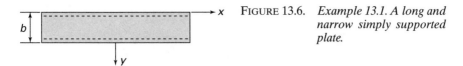

FIGURE 13.6. *Example 13.1. A long and narrow simply supported plate.*

Solution Because for this situation $\partial w/\partial x = 0$ and $\partial^2 w/\partial x \partial y = 0$, Eq. (13.7) reduces to

$$M_x = -\nu D \frac{d^2 w}{dy^2}, \quad M_y = -D \frac{d^2 w}{dy^2} \qquad \textbf{(b)}$$

while Eq. (13.12) becomes

$$\frac{d^4 w}{dy^4} = \frac{p}{D} \qquad \textbf{(c)}$$

The latter expression is the same as the *wide beam equation*, and we conclude that the solution proceeds as in the case of a beam. A wide beam shall, in the context of this chapter, mean a rectangular plate supported on one edge or on two opposite edges in such a way that these edges are free to approach one another as deflection occurs.

Substituting Eq. (a) into Eq. (c), integrating, and satisfying the boundary conditions at $y = 0$ and $y = b$, we obtain

$$w = \left(\frac{b}{\pi}\right)^4 \frac{p_o}{D} \sin\left(\frac{\pi y}{b}\right) \qquad \textbf{(d)}$$

The stresses are now readily determined through application of Eqs. (13.5) or (13.9) and Eq. (13.13).

13.7 SIMPLY SUPPORTED RECTANGULAR PLATES

In general, the solution of the plate problem for a geometry as in Fig. 13.7a, with simple supports along all edges, may be obtained by the application of Fourier series for load and deflection:*

$$p(x, y) = \sum_{m=1}^{\infty} \sum_{n=1}^{\infty} p_{mn} \sin \frac{m\pi x}{a} \sin \frac{n\pi y}{b} \qquad \textbf{(13.18a)}$$

$$w(x, y) = \sum_{m=1}^{\infty} \sum_{n=1}^{\infty} a_{mn} \sin \frac{m\pi x}{a} \sin \frac{n\pi y}{b} \qquad \textbf{(13.18b)}$$

Here p_{mn} and a_{mn} represent coefficients to be determined. The problem at hand is described by

$$\frac{\partial^4 w}{\partial x^4} + 2 \frac{\partial^4 w}{\partial x^2 \partial y^2} + \frac{\partial^4 w}{\partial y^4} = \frac{p(x, y)}{D} \qquad \textbf{(a)}$$

---

*This approach was introduced by Navier in 1820. For details, see Refs. 13.1 and 13.2.

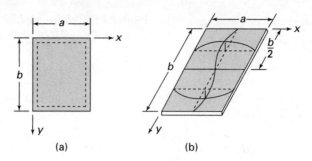

FIGURE 13.7. *Simply supported rectangular plate: (a)
location of coordinate system for Navier's
method; (b) deflection of the simply supported plate into half-sine curves of m = 1
and n = 2.*

and

$$w = 0, \qquad \frac{\partial^2 w}{\partial x^2} = 0, \qquad \text{for } x = 0, a$$

$$w = 0, \qquad \frac{\partial^2 w}{\partial y^2} = 0, \qquad \text{for } y = 0, b$$

(b)

The boundary conditions given are satisfied by Eq. (13.18b), and the coefficients a_{mn} must be such as to satisfy Eq. (a). The solution corresponding to the loading $p(x, y)$ thus requires a determination of p_{mn} and a_{mn}. Let us consider, as an *interpretation* of Eq. (13.18b), the true deflection surface of the plate to be the superposition of sinusoidal curves m and n different configurations in the x and y directions. The coefficients a_{mn} of the series are the maximum central coordinates of the sine series, and the m's and the n's mean the number of the half-sine curves in the x and y directions, respectively. For instance, the terms a $a_{12} \sin(\pi x/a) \sin(2\pi y/b)$ of the series is shown in Fig. 13.7b. By increasing the number of terms in the series, the accuracy can be improved.

We proceed by dealing first with a general load configuration, subsequently treating specific loadings. To determine the coefficients p_{mn}, each side of Eq. (13.18a) is multiplied by

$$\sin \frac{m'\pi x}{a} \sin \frac{n'\pi y}{b} dx\, dy$$

Integrating between the limits $0, a$ and $0, b$ yields

$$\int_0^b \int_0^a p(x, y) \sin \frac{m'\pi x}{a} \sin \frac{n'\pi y}{b} dx\, dy$$

$$= \sum_{m=1}^{\infty} \sum_{n=1}^{\infty} p_{mn} \int_0^b \int_0^a \sin \frac{m\pi x}{a} \sin \frac{m'\pi x}{a} \sin \frac{n\pi y}{b} \sin \frac{n'\pi y}{b} dx\, dy$$

Applying the orthogonality relation (10.24) and integrating the right side of the preceding equation, we obtain

$$p_{mn} = \frac{4}{ab} \int_0^b \int_0^a p(x, y) \sin\frac{m\pi x}{a} \sin\frac{n\pi y}{b} \, dx \, dy \qquad (13.19)$$

Evaluation of a_{mn} in Eq. (13.18b) requires substitution of Eqs. (13.18a) and (13.18b) into Eq. (a), with the result

$$\sum_{m=1}^{\infty} \sum_{n=1}^{\infty} \left\{ a_{mn}\left[\left(\frac{m\pi}{a}\right)^4 + 2\left(\frac{m\pi}{a}\right)^2 \left(\frac{n\pi}{b}\right)^2 + \left(\frac{n\pi}{b}\right)^4 \right] - \frac{p_{mn}}{D} \right\}$$

$$\times \sin\frac{m\pi x}{a} \sin\frac{n\pi y}{b} = 0$$

This expression must apply for all x and y; we conclude therefore that

$$a_{mn}\pi^4 \left(\frac{m^2}{a^2} + \frac{n^2}{b^2}\right)^2 - \frac{p_{mn}}{D} = 0 \qquad (c)$$

Solving for a_{mn} and substituting into Eq. (13.18b), the equation of the deflection surface of a thin plate is

$$w = \frac{1}{\pi^4 D} \sum_{m=1}^{\infty} \sum_{n=1}^{\infty} \frac{p_{mn}}{[(m/a)^2 + (n/b)^2]^2} \sin\frac{m\pi x}{a} \sin\frac{n\pi y}{b} \qquad (13.20)$$

in which p_{mn} is given by Eq. (13.19).

EXAMPLE 13.2 Analysis of Uniformly Loaded Rectangular Plate

(a) Determine the deflections and moments in a simply supported rectangular plate of thickness t (Fig. 13.7a). The plate is subjected to a uniformly distributed load p_o. (b) Setting $a = b$, obtain the deflections, moments, and stresses in the plate.

Solution

a. For this case, $p(x, y) = p_o$, and Eq. (13.19) is thus

$$p_{mn} = \frac{4p_o}{ab} \int_0^b \int_0^a \sin\frac{m\pi x}{a} \sin\frac{n\pi y}{b} \, dx \, dy = \frac{16p_o}{\pi^2 mn}$$

It is seen that, because $p_{mn} = 0$ for even values of m and n, they can be taken as odd integers. Substituting p_{mn} into Eq. (13.20) results in

$$w = \frac{16p_o}{\pi^6 D} \sum_m^{\infty} \sum_n^{\infty} \frac{\sin(m\pi x/a) \sin(n\pi y/b)}{mn[(m/a)^2 + (n/b)^2]^2}, \qquad m, n = 1, 3, 5, \ldots \quad (13.21)$$

We know that on physical grounds, the uniformly loaded plate must deflect into a symmetrical shape. Such a configuration results when m and n are odd. The maximum deflection occurs at $x = a/2$, $y = b/2$. From Eq. (13.21), we thus have

$$w_{max} = \frac{16p_o}{\pi^6 D} \sum_m^\infty \sum_n^\infty \frac{(-1)^{(m+n)/2-1}}{mn[(m/a)^2 + (n/b)^2]^2}, \quad m, n = 1, 3, 5, \ldots \quad \textbf{(13.22)}$$

By substituting Eq. (13.21) into Eq. (13.7), the bending moments M_x, M_y are obtained:

$$M_x = \frac{16p_o}{\pi^4} \sum_m^\infty \sum_n^\infty \frac{(m/a)^2 + \nu(n/b)^2}{mn[(m/a)^2 + (n/b)^2]^2} \sin\frac{m\pi x}{a} \sin\frac{n\pi y}{b}$$

$$M_y = \frac{16p_o}{\pi^4} \sum_m^\infty \sum_n^\infty \frac{\nu(m/a)^2 + (n/b)^2}{mn[(m/a)^2 + (n/b)^2]^2} \sin\frac{m\pi x}{a} \sin\frac{n\pi y}{b}$$

$$\textbf{(13.23)}$$

b. For the case of a square plate (setting $a = b$), substituting $\nu = 0.3$, the first term of Eq. (13.22) gives

$$w_{max} = 0.0454 p_o \frac{a^4}{Et^3}$$

The rapid convergence of Eq. (13.22) is demonstrated by noting that retaining the first four terms gives the results $w_{max} = 0.0443 p_o(a^4/Et^3)$.

The bending moments occurring at the center of the plate are found from Eq. (13.23). Retaining only the first term, the result is

$$M_{x,max} = M_{y,max} = 0.0534 p_o a^2$$

while the first four terms yield

$$M_{x,max} = M_{y,max} = 0.0472 p_o a^2$$

Observe from a comparison of these values that the series given by Eq. (13.23) does not converge as rapidly as that of Eq. (13.22).

13.8 AXISYMMETRICALLY LOADED CIRCULAR PLATES

The deflection w of a circular plate will manifest dependence on radial position r only if the applied load and conditions of end restraint are independent of the angle θ. In other words, symmetry in plate deflection follows from symmetry in applied load. For this case, only radial and tangential moments M_r and M_θ per unit length and force Q_r per unit length act on the circular plate element shown in Fig. 13.8. To derive the fundamental equations of a circular plate, we need only transform the appropriate formulations of previous sections from Cartesian to polar coordinates.

Through the application of the coordinate transformation relationships of Section 3.8, the bending moments and vertical shear forces are found from Eqs. (13.7) and (13.11) to be $Q_\theta = 0$, $M_{r\theta} = 0$, and

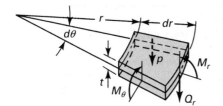

FIGURE 13.8. *Axisymmetrically loaded circular plate element. (Only shear force and moments on the positive faces are shown.)*

$$M_r = -D\left(\frac{d^2w}{dr^2} + \frac{\nu}{r}\frac{dw}{dr}\right) \tag{13.24a}$$

$$M_\theta = -D\left(\frac{1}{r}\frac{dw}{dr} + \nu\frac{d^2w}{dr^2}\right) \tag{13.24b}$$

$$Q_r = -D\frac{d}{dr}\left(\frac{d^2w}{dr^2} + \frac{1}{r}\frac{dw}{dr}\right) \tag{13.24c}$$

The differential equation describing the surface deflection is obtained from Eq. (13.12) in a similar fashion:

$$\nabla^4 w = \left(\frac{d^2}{dr^2} + \frac{1}{r}\frac{d}{dr}\right)\left(\frac{d^2w}{dr^2} + \frac{1}{r}\frac{dw}{dr}\right) = \frac{p}{D} \tag{13.25a}$$

where p, as before, represents the load acting per unit of surface area, and D is the plate rigidity. By introducing the identity

$$\frac{d^2w}{dr^2} + \frac{1}{r}\frac{dw}{dr} = \frac{1}{r}\frac{d}{dr}\left(r\frac{dw}{dr}\right) \tag{a}$$

Eq. (13.25a) assumes the form

$$\frac{1}{r}\frac{d}{dr}\left\{r\frac{d}{dr}\left[\frac{1}{r}\frac{d}{dr}\left(r\frac{dw}{dr}\right)\right]\right\} = \frac{p}{D} \tag{13.25b}$$

For applied loads varying with radius, $p(r)$, representation (13.25b) is preferred.

The *boundary conditions* at the edge of the plate of radius a may readily be written by referring to Eqs. (13.15) to (13.17) and (13.24):

$$\text{Clamped edge:}\quad w = 0,\quad \frac{\partial w}{\partial r} = 0 \tag{13.26a}$$

$$\text{Simply supported edge:}\quad w = 0,\quad M_r = 0 \tag{13.26b}$$

$$\text{Free edge:}\quad M_r = 0,\quad Q_r = 0 \tag{13.26c}$$

Equation (13.25), together with the boundary conditions, is sufficient to solve the axisymmetrically loaded circular plate problem.

EXAMPLE 13.3 Circular Plate with Fixed-Edge

Determine the stress and deflection for a built-in circular plate of radius a subjected to uniformly distributed loading p_o (Fig. 13.9).

Solution The origin of coordinates is located at the center of the plate. The displacement w is obtained by successive integration of Eq. (13.25b):

$$w = \int \frac{1}{r} \int r \int \frac{1}{r} \int \frac{rp}{D} \, dr \, dr \, dr \, dr$$

or

$$\frac{D}{p_o} w = \int \frac{1}{r} \int r \int \left[\frac{1}{r} \left(\frac{r^2}{2} + c_1 \right) \right] dr \, dr \, dr$$

$$= \frac{r^4}{64} + \frac{c_1 r^2}{4} (\ln r - 1) + \frac{c_2 r^2}{4} + c_3 \ln r + c_4 \quad \textbf{(b)}$$

where the c's are constants of integration.

The boundary conditions are

$$w = 0, \qquad \frac{dw}{dr} = 0 \qquad \text{for } r = a \qquad \textbf{(c)}$$

The terms involving logarithms in Eq. (b) lead to an infinite displacement at the center of the plate ($r = 0$) for all values of c_1 and c_3 except zero; therefore, $c_1 = c_3 = 0$. Satisfying the boundary conditions (c), we find that $c_2 = -a^2/8$ and $c_4 = a^4/64$. The deflection is then

$$w = \frac{p_o}{64D} (a^2 - r^2)^2 \qquad \textbf{(13.27)}$$

The maximum deflection occurs at the center of the plate:

$$w_{max} = \frac{p_o a^4}{64D}$$

Substituting the deflection given by Eq. (13.27) into Eq. (13.24), we have

$$M_r = \frac{p_o}{16} [(1 + \nu)a^2 - (3 + \nu)r^2]$$

$$M_\theta = \frac{p_o}{16} [(1 + \nu)a^2 - (1 + 3\nu)r^2] \qquad \textbf{(13.28)}$$

The extreme values of the moments are found at the center and edge. At the center,

$$M_r = M_\theta = (1 + \nu) \frac{p_o a^2}{16} \quad \text{for } r = 0$$

At the edges, Eq. (13.28) yields

$$M_r = -\frac{p_o a^2}{8}, \quad M_\theta = -\frac{\nu p_o a^2}{8} \quad \text{for } r = a$$

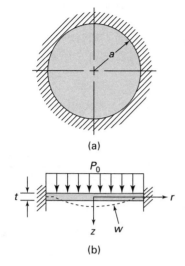

(a)

FIGURE 13.9.   *Example 13.3. Uniformly loaded clamped circular plate: (a) top view; (b) front view.*

P_0

t

r

z w

(b)

Examining these results, it is clear that the maximum stress occurs at the edge:

$$\sigma_{r,\max} = \frac{6M_r}{t^2} = -\frac{3}{4}\frac{p_o a^2}{t^2} \tag{13.29}$$

A similar procedure may be applied to symmetrically loaded circular plates subject to different end conditions.

13.9   DEFLECTIONS OF RECTANGULAR PLATES BY THE STRAIN-ENERGY METHOD

Based on the assumptions of Section 13.2, for thin plates, stress components σ_z, τ_{xz}, and τ_{yz} can be neglected. Therefore, the *strain energy in pure bending of plates*, from Eq. (2.52), has the form

$$U = \iiint_V \left[\frac{1}{2E}\left(\sigma_x^2 + \sigma_y^2 - 2\nu\sigma_x\,\sigma_y\right) + \frac{1}{2G}\tau_{xy}^2 \right] \tag{a}$$

Integration extends over the entire plate volume. For a plate of *uniform thickness*, this expression may be written in terms of deflection w, applying Eqs. (13.5) and (13.8), as follows:

$$U = \frac{D}{2} \iint_A \left[\left(\frac{\partial^2 w}{\partial x^2}\right)^2 + \left(\frac{\partial^2 w}{\partial y^2}\right)^2 + 2\nu\frac{\partial^2 w}{\partial x^2}\frac{\partial^2 w}{\partial y^2} \right.$$
$$\left. + 2(1-\nu)\left(\frac{\partial^2 w}{\partial x \partial y}\right)^2 \right] dx\,dy \tag{13.30}$$

Alternatively,

$$U = \frac{D}{2} \iint_A \left\{ \left(\frac{\partial^2 w}{\partial x^2} + \frac{\partial^2 w}{\partial y^2} \right)^2 - 2(1 - \nu) \left[\frac{\partial^2 w}{\partial x^2} \frac{\partial^2 w}{\partial y^2} - \left(\frac{\partial^2 w}{\partial x \partial y} \right)^2 \right] \right\} dx\, dy \quad \textbf{(13.31)}$$

where A is the area of the plate surface.

If the edges of the plate are fixed or simply supported, the second term on the right of Eq. (13.31) becomes zero [Ref. 13.1]. Hence, the strain energy simplifies to

$$U = \frac{D}{2} \iint_A \left(\frac{\partial^2 w}{\partial x^2} + \frac{\partial^2 w}{\partial y^2} \right)^2 dx\, dy \quad \textbf{(13.32)}$$

The work done by the lateral load p is

$$W = \iint_A wp\, dx\, dy \quad \textbf{(13.33)}$$

The potential energy of the system is then

$$\Pi = \iint_A \left[\frac{D}{2} \left(\frac{\partial^2 w}{\partial x^2} + \frac{\partial^2 w}{\partial y^2} \right)^2 - wp \right] dx\, dy \quad \textbf{(13.34)}$$

Application of a strain-energy method in calculating deflections of a rectangular plate is illustrated next.

EXAMPLE 13.4   Deflection of a Rectangular Plate

Rework Example 13.2 using the Rayleigh–Ritz method.

Solution   From the discussion of Section 13.7, the deflection of the simply supported plate (Fig. 13.7a) can always be represented in the form of a double trigonometric series given by Eq. (13.18b). Introducing this series for w and setting $p = p_o$, Eq. (13.34) becomes

$$\Pi = \int_0^a \int_0^b \sum_m^\infty \sum_n^\infty \left\{ \frac{D}{2} \left[a_{mn} \left(\frac{m^2 \pi^2}{a^2} + \frac{n^2 \pi^2}{b^2} \right) \sin \frac{m\pi x}{a} \sin \frac{n\pi y}{b} \right]^2 \right.$$

$$\left. - p_o a_{mn} \sin \frac{m\pi x}{a} \sin \frac{n\pi y}{b} \right\} dx\, dy \quad \textbf{(b)}$$

Observing the orthogonality relation [Eq. (10.24)], we conclude that, in calculating the first term of the preceding integral, we need to consider only the squares of the terms of the infinite series in the parentheses (Prob. 13.19). Hence, integrating Eq. (b), we have

$$\Pi = \sum_m^\infty \sum_n^\infty \left[\frac{\pi^4 abD}{8} a_{mn}^2 \left(\frac{m^2}{a^2} + \frac{n^2}{b^2} \right)^2 - \frac{4 p_o ab}{\pi^2 mn} a_{mn} \right], \quad m, n = 1, 3, \ldots \quad \textbf{(c)}$$

From the minimizing conditions $\partial \Pi / \partial a_{mn} = 0$, it follows that

$$a_{mn} = \frac{16p_o}{\pi^6 mnD[(m/a)^2 + (n/b)^2]^2}, \qquad m, n = 1, 3, \ldots \qquad \textbf{(d)}$$

Substitution of these coefficients into Eq. (13.18b) leads to the same expression as given by Eq. (13.21).

13.10 FINITE ELEMENT SOLUTION

In this section we present the finite element method (Chapter 7) for computation of deflection, strain, and stress in a thin plate. For details, see, for example, Refs. 13.3 through 13.6. The plate, in general, may have any irregular geometry and loading. The derivations are based on the assumptions of small deformation theory, described in Section 13.2. A triangular plate element ijm coinciding with the xy plane will be employed as the finite element model (Fig. 13.10). Each nodal displacement of the element possesses three components: a displacement in the z direction, w; a rotation about the x axis, θ_x; and a rotation about the y axis, θ_y. Positive directions of the rotations are determined by the right-hand rule, as illustrated in the figure. It is clear that rotations θ_x and θ_y represent the slopes of w: $\partial w / \partial y$ and $\partial w / \partial x$, respectively.

Strain, Stress, and Elasticity Matrices

Referring to Eqs. (13.3), we define, for the finite element analysis, a generalized "strain"–displacement matrix as follows:

$$\{\varepsilon\}_e = \left\{ -\frac{\partial^2 w}{\partial x^2}, \quad -\frac{\partial^2 w}{\partial y^2}, \quad -2\frac{\partial^2 w}{\partial x \, \partial y} \right\} \qquad \textbf{(13.35)}$$

The moment generalized "strain" relationship, from Eq. (13.7), is given in matrix form as

$$\{M\}_e = [D]\{\varepsilon\}_e \qquad \textbf{(13.36)}$$

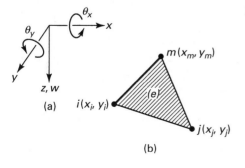

(a)

(b)

FIGURE 13.10. *(a) Positive directions of displacements; (b) triangular finite plate element.*

where

$$[D] = \frac{Et^3}{12(1 - \nu^2)} \begin{bmatrix} 1 & \nu & 0 \\ \nu & 1 & 0 \\ 0 & 0 & (1 - \nu)/2 \end{bmatrix} \qquad (13.37)$$

The stresses $\{\sigma\}_e$ and the moments $\{M\}_e$ are related by Eq. (13.9).

Displacement Function

The nodal displacement can be defined, for node i, as follows:

$$\{\delta_i\} = \{w_i, \theta_{xi}, \theta_{yi}\} = \left\{ w_i, \left(\frac{\partial w}{\partial y}\right)_i, \left(\frac{\partial w}{\partial x}\right)_i \right\} \qquad (a)$$

The element nodal displacements are expressed in terms of submatrices θ_i, θ_j, and θ_m:

$$\{\delta\}_e = \begin{Bmatrix} \delta_i \\ \hline \delta_j \\ \hline \delta_m \end{Bmatrix} = \{w_i, \quad \theta_{xi}, \quad \theta_{yi}, \quad w_j, \quad \theta_{xj}, \quad \theta_{yj}, \quad w_m, \quad \theta_{xm}, \quad \theta_{ym}\} \qquad (b)$$

The displacement function $\{f\} = w$ is assumed to be a modified third-order polynomial of the form

$$w = a_1 + a_2 x + a_3 y + a_4 x^2 + a_5 xy + a_6 y^2$$
$$+ a_7 x^3 + a_8(x^2 y + xy^2) + a_9 y^3 \qquad (c)$$

Note that the number of terms is the same as the number of nodal displacements of the element. This function satisfies displacement compatibility at interfaces between elements but does not satisfy the compatibility of slopes at these boundaries. Solutions based on Eq. (c) do, however, yield results of acceptable accuracy.

The constants a_1 through a_9 can be evaluated by writing the nine equations involving the values of w and θ at the nodes:

$$\begin{Bmatrix} w_i \\ \theta_{xi} \\ \theta_{yi} \\ w_j \\ \theta_{xj} \\ \theta_{yj} \\ w_m \\ \theta_{xm} \\ \theta_{ym} \end{Bmatrix}_e = \begin{bmatrix} 1 & x_i & y_i & x_i^2 & x_i y_i & y_i^2 & x_i^3 & x_i^2 y_i + x_i y_i^2 & y_i^3 \\ 0 & 0 & 1 & 0 & x_i & 2y_i & 0 & x_1^2 + 2x_i y_i & 3y_i^2 \\ 0 & 1 & 0 & 2x_i & y_i & 0 & 3x_i^3 & 2x_i y_i + y_i^2 & 0 \\ 1 & x_j & y_j & x_j^2 & x_j y_j & y_i^2 & x_j^3 & x_j^2 y_j + x_j y_j^2 & y_j^3 \\ 0 & 0 & 1 & 0 & x_j & 2y_j & 0 & x_j^2 + 2x_j y_i & 3y_j^2 \\ 0 & 1 & 0 & 2x_j & y_j & 0 & 3x_j^2 & 2x_j y_j + y_j^2 & 0 \\ 1 & x_m & y_m & x_m^2 & x_m y_m & y_m^2 & x_m^3 & x_m^2 y_m + x_m y_m^2 & y_m^3 \\ 0 & 0 & 1 & 0 & x_m & 2y_m & 0 & x_m^2 + 2x_m y_m & 3y_m^2 \\ 0 & 1 & 0 & 2x_m & y_m & 0 & 3x_m^2 & 2x_m y_m + y_m^2 & 0 \end{bmatrix} \begin{Bmatrix} a_1 \\ a_2 \\ a_3 \\ a_4 \\ a_5 \\ a_6 \\ a_7 \\ a_8 \\ a_9 \end{Bmatrix}$$

$$(13.38a)$$

or

$$\{\theta\}_e = [C]\{a\} \qquad (13.38b)$$

Inverting,

$$\{a\} = [C]^{-1}\{\theta\}_e \qquad (13.39)$$

It is observed in Eq. (13.38) that the matrix $[C]$ depends on the coordinate dimensions of the nodal points.

Note that the displacement function may now be written in the usual form of Eq. (7.50) as

$$\{f\}_e = w = [N]\{\theta\}_e = [P][C]^{-1}\{\theta\}_e \qquad (13.40)$$

in which

$$[P] = [1, x, y, x^2, xy, y^2, x^3, (x^2y + xy^2), y^3] \qquad (13.41)$$

Introducing Eq. (c) into Eq. (13.35), we have

$$\{\varepsilon\}_e = \begin{bmatrix} 0 & 0 & 0 & -2 & 0 & 0 & -6x & -2y & 0 \\ 0 & 0 & 0 & 0 & 0 & -2 & 0 & -2x & -6y \\ 0 & 0 & 0 & 0 & -2 & 0 & 0 & -4(x+y) & 0 \end{bmatrix}$$
$$\times \{a_1, a_2, \ldots, a_9\} \qquad (13.42a)$$

or

$$\{\varepsilon\}_e = [H]\{a\} \qquad (13.42b)$$

Upon substituting the values of the constants $\{a\}$ from Eq. (13.39) into (13.42b), we can obtain the generalized "strain"–displacement matrix in the following common form:

$$\{\varepsilon\}_e = [B]\{\theta\}_e = [H][C]^{-1}\{\theta\}_e$$

Thus,

$$[B] = [H][C]^{-1} \qquad (13.43)$$

Stiffness Matrix

The element stiffness matrix given by Eq. (7.34), treating the thickness t as a constant within the element and introducing $[B]$ from Eq. (13.43), becomes

$$[k]_e = [[C]^{-1}]^T \left(\int \int [H]^T[D][H] dx \, dy \right)[C]^{-1} \qquad (13.44)$$

where the matrices $[H]$, $[D]$, and $[C]^{-1}$ are defined by Eqs. (13.42), (13.37), and (13.38), respectively. After expansion of the expression under the integral sign, the integrations can be carried out to obtain the element stiffness matrix.

External Nodal Forces

As in two-dimensional and axisymmetrical problems, the nodal forces due to the distributed *surface* loading may also be obtained through the use of Eq. (7.58) or by physical intuition.

The standard finite element method procedure described in Section 7.13 may now be followed to obtain the unknown displacement, strain, and stress in any element of the plate.

Part B—Membrane Stresses in Thin Shells

13.11 THEORIES AND BEHAVIOR OF SHELLS

Structural elements resembling curved plates are referred to as shells. Included among the more familiar examples of shells are soap bubbles, incandescent lamps, aircraft fuselages, pressure vessels, and a variety of metal, glass, and plastic containers. As was the case for plates, we limit our treatment to isotropic, homogeneous, elastic shells having a constant thickness that is small relative to the remaining dimensions. The surface bisecting the shell thickness is referred to as the *midsurface*. To specify the geometry of a shell, we need only know the configuration of the midsurface and the thickness of the shell at each point. According to the criterion often applied to define a thin shell (for purposes of technical calculations), the ratio of thickness t to radius of curvature r should be equal to or less than $\frac{1}{20}$.

The stress analysis of shells normally embraces two distinct theories. The *membrane theory* is limited to moment-free membranes, which often applies to a rather large proportion of the entire shell. The *bending theory* or *general theory* includes the influences of bending and thus enables us to treat discontinuities in the field of stress occurring in a limited region in the vicinity of a load application or a structural discontinuity. This method generally involves a membrane solution, corrected in those areas in which discontinuity effects are pronounced. The principal objective is thus not the improvement of the membrane solution but the analysis of stresses associated with edge loading, which cannot be accomplished by the membrane theory alone [Ref. 13.1].

The following assumptions are generally made in the *small deflection analysis* of thin shells:

1. The ratio of the shell thickness to the radius of curvature of the midsurface is small compared with unity.
2. Displacements are very small compared with the shell thickness.
3. Straight sections of an element, which are perpendicular to the midsurface, remain perpendicular and straight to the *deformed* midsurface subsequent to bending. The implication of this assumption is that the strains γ_{xz} and γ_{yz} are negligible. Normal strain, ε_z, due to transverse loading may also be omitted.
4. The z-directed stress σ_z is negligible.

13.12 SIMPLE MEMBRANE ACTION

As testimony to the fact that the load-carrying mechanism of a shell differs from that of other elements, we have only to note the extraordinary capacity of an eggshell to withstand normal forces, despite its thinness and fragility. This contrasts markedly with a similar material in a plate configuration subjected to lateral loading.

To understand the phenomenon, consider a portion of a spherical shell of radius r and thickness t, subjected to a uniform pressure p (Fig. 13.11). Denoting by N the normal force per unit length required to maintain the shell in a state of equilibrium, static equilibrium of vertical forces is expressed by

$$2\pi r_0 N \sin \phi = p\pi r_0^2$$

or

$$N = \frac{pr_0}{2 \sin \phi} = \frac{pr}{2}$$

This result is valid anywhere in the shell, because N is observed not to vary with ϕ. Note that, in contrast with the case of plates, it is the midsurface that sustains the applied load.

Once again referring to the simple shell shown in Fig. 13.11, we demonstrate that the bending stresses play an insignificant role in the load-carrying mechanism. On the basis of the symmetry of the shell and the loading, the stresses (equal at any point) are given by

$$\sigma_n = -\frac{N}{t} = -\frac{pr}{2t} \tag{a}$$

Here σ_n represents the compressive, in-plane stress. The stress normal to the midsurface is negligible, and thus the in-plane strain involves only σ_n:

$$\varepsilon_n = \frac{\sigma_n}{E} - \nu\frac{\sigma_n}{E} = -(1 - \nu)\frac{pr}{2tE} \tag{b}$$

The reduced circumference associated with this strain is

$$2\pi r' = 2\pi(r + r\varepsilon_n)$$

so

$$r' = r(1 + \varepsilon_n)$$

The change in curvature is therefore

$$\Delta\left(\frac{1}{r}\right) = \frac{1}{r'} - \frac{1}{r} = \frac{1}{r}\left(\frac{1}{1 + \varepsilon_n} - 1\right)$$

$$= -\frac{\varepsilon_n}{r}\left(\frac{1}{1 + \varepsilon_n}\right) = -\frac{\varepsilon_n}{r}(1 - \varepsilon_n + \varepsilon_n^2 - \cdots)$$

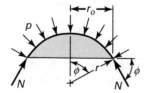

FIGURE 13.11. *Truncated spherical shell under uniform pressure.*

Dropping higher-order terms because of their negligible magnitude and substituting Eq. (b), this expression becomes

$$\Delta\left(\frac{1}{r}\right) \approx -\frac{\varepsilon_n}{r} = \frac{(1-\nu)p}{2tE} \tag{c}$$

The bending moment in the shell is determined from the plate equations. Noting that $1/r_x$ and $1/r_y$ in Eq. (13.7) refer to the change in plate curvatures between the undeformed and deformed conditions, we see that for the spherical shell under consideration, $\Delta(1/r) = 1/r_x = 1/r_y$. Therefore, Eq. (13.7) yields

$$M_b = -D\left(\frac{1}{r_x} + \frac{\nu}{r_y}\right) = -D(1-\nu^2)\frac{p}{2tE} = -\frac{pt^2}{24} \tag{d}$$

and the maximum corresponding stress is

$$\sigma_b = \frac{6M_b}{t^2} = -\frac{p}{4} \tag{e}$$

Comparing σ_b and σ_n [Eqs. (e) and (a)], we have

$$\frac{\sigma_n}{\sigma_b} = \frac{2r}{t} \tag{13.45}$$

demonstrating that the in-plane or direct stress is very much larger than the bending stress, inasmuch as $t/2r \ll 1$. It may be concluded, therefore, that the *applied load is resisted primarily by the in-plane stressing* of the shell.

Although the preceding discussion relates to the simplest shell configuration, the conclusions drawn with respect to the fundamental mechanism apply to any shape and loading at locations away from the boundaries or points of concentrated load application. If there are asymmetries in load or geometry, shearing stresses will exist in addition to the normal and bending stresses.

In the following sections we discuss the membrane theory of two common structures: the shell of revolution and cylindrical shells.

13.13   SYMMETRICALLY LOADED SHELLS OF REVOLUTION

A surface of revolution, such as in Fig. 13.12a, is formed by the rotation of a *meridian curve (eo')* about the OO axis. As shown, a point on the shell is located by coordinates θ, ϕ, r_0. This figure indicates that the elemental surface *abcd* is defined by two meridian and two parallel circles. The planes containing the principal radii of curvature at any point on the surface of the shell are the meridian plane and a plane perpendicular to it at the point in question. The meridian plane will thus contain r_θ, which is related to side *ab*. The other principal radius of curvature, r_ϕ, is found in the perpendicular plane and is therefore related to side *bd*. Thus, length $ab = (r_\theta \sin\theta)d\theta = r_0\, d\theta$, and length $bd = r_\phi\, d\phi$.

The condition of symmetry prescribes that no shearing forces act on the element and that the normal forces N_θ and N_ϕ per unit length display no variation with θ

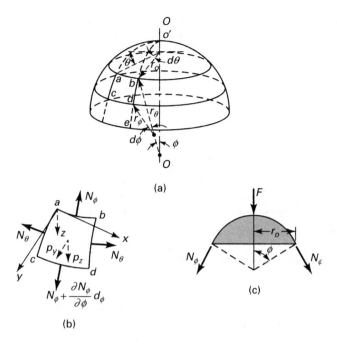

(a)

(b)

(c)

FIGURE 13.12. *Diagrams for analysis of symmetrically loaded shells of revolution: (a) geometry of the shell; (b) membrane forces (per unit length) and distributed loading (per unit area) acting on a shell element; (c) meridian forces and resultant of the loading acting on a truncated shell.*

(Fig. 13.12b). These *membrane* forces are also referred to as the *tangential* or *hoop* and *meridian forces*, respectively. The name arises from the fact that forces of this kind exist in true membranes, such as soap films or thin sheets of rubber. The externally applied load is represented by the perpendicular components p_y and p_z. We turn now to a derivation of the equations governing the force equilibrium of the element.

Equations of Equilibrium

To describe equilibrium in the z direction, it is necessary to consider the z components of the external loading as well as of the forces acting on each edge of the element. The external z-directed load acting on an element of area $(r_0 \, d\theta)(r_\phi \, d\phi)$ is

$$p_z r_0 r_\phi \, d\theta \, d\phi \tag{a}$$

The force acting on the top edge of the element is $N_\phi r_0 \, d\theta$. Neglecting terms of higher order, the force acting on the bottom edge is also $N_\phi r_0 \, d\theta$. The z component at each edge is then $N_\phi r_0 \, d\theta \sin(d\phi/2)$, which may be approximated by $N_\phi r_0 \, d\theta \, d\phi/2$, leading to a resultant for both edges of

$$N_\phi r_0 \, d\theta \, d\phi \tag{b}$$

The force on each side of the element is $N_\theta r_\phi\, d\phi$. The radial resultant for both such forces is $(N_\theta r_\phi\, d\phi)d\theta$, having a z-directed component

$$N_\theta r_\phi\, d\phi\, d\theta \sin \phi \qquad\qquad \textbf{(c)}$$

Adding the z forces, equating to zero, and canceling $d\theta\, d\phi$, we obtain

$$N_\phi r_0 + N_\theta r_\phi \sin \phi + p_z r_0 r_\phi = 0$$

Dividing by $r_0 r_\phi$ and replacing r_0 by $r_\theta \sin \phi$, this expression is converted to the following form:

$$\frac{N_\phi}{r_\phi} + \frac{N_\theta}{r_\theta} = -p_z \qquad\qquad \textbf{(13.46a)}$$

Similarly, an equation for the y equilibrium of the element of the shell may also be derived. But instead of solving the z and y equilibrium equations simultaneously, it is more convenient to obtain N_ϕ from the equilibrium of the portion of the shell corresponding to the angle ϕ (Fig. 13.12c):

$$2\pi r_0 N_\phi \sin \phi + F = 0 \qquad\qquad \textbf{(13.46b)}$$

Then calculate N_θ from Eq. (13.46a). Here F represents the resultant of all external loading acting on the portion of the shell.

Conditions of Compatibility

For the axisymmetrical shells considered, owing to their freedom of motion in the z direction, strains are produced such as to assure consistency with the field of stress. These strains are compatible with one another. It is clear that, when a shell is subjected to concentrated surface loadings or is constrained at its boundaries, membrane theory cannot satisfy the conditions on deformation everywhere. In such cases, however, the departure from membrane behavior is limited to a narrow zone in the vicinity of the boundary or the loading. Membrane theory remains valid for the major portion of the shell, but the complete solution can be obtained only through application of bending theory.

13.14 SOME COMMON CASES OF SHELLS OF REVOLUTION

The membrane stresses in any particular axisymmetrically loaded shell in the form of a surface of revolution may be obtained readily from Eqs. (13.46). Treated next are three common structural members. It is interesting to note that, for the circular cylinder and the cone, the meridian is a straight line.

Spherical Shell (Fig. 13.13a)

We need only set the *mean radius* $r = r_\theta = r_\phi$ and hence $r_0 = r \sin \phi$. Equations (13.46) then become

$$N_\phi + N_\theta = -p_z r$$

$$N_\phi = -\frac{F}{2\pi r \sin^2\phi} \qquad\qquad \textbf{(13.47)}$$

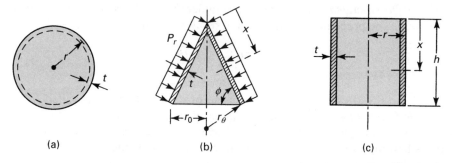

(a)   (b)   (c)

FIGURE 13.13. *Typical shells of revolution: (a) spherical shells; (b) conical shell under uniform pressure; (c) cylindrical shell.*

Example 13.6 illustrates an application of this equation.

The simplest case is that of a spherical shell subjected to internal pressure p. We now have $p = -p_z$, $\phi = 90°$, and $F = -\pi r^2 p$. As is evident from the symmetry of a spherical shell, $N_\phi = N_\theta = N$. These quantities, introduced into Eqs. (13.47), lead to the membrane stress in a *spherical pressure vessel*:

$$\sigma = \frac{N}{t} = \frac{pr}{2t} \tag{13.48}$$

where t is the thickness of the shell. Clearly, this equation represents uniform stresses in all directions of a pressurized sphere.

Conical Shell (Fig. 13.13b)

We need only set $r_\phi = \infty$ in Eq. (13.46a). This, together with Eq. (13.46b), provides the following pair of equations for determining the membrane forces under uniformly distributed load $p_z = p_r$:

$$N_\theta = p_r r_\theta = -\frac{p_r r_\theta}{\sin \phi}$$

$$N_\phi = N_x = -\frac{F}{2\pi r_0 \sin \phi} \tag{13.49}$$

Here x represents the direction of the generator. The hoop and meridian stresses in a conical shell are found by dividing N_θ and N_x by t, wherein t is the thickness of the shell.

Circular Cylindrical Shell (Fig. 13.13c)

To determine the membrane forces in a circular cylindrical shell, we begin with the cone equations, setting $\phi = \pi/2$, $p_z = p_r$, and *mean radius* $r = r_0 =$ constant. Equations (13.49) are therefore

$$N_x = -\frac{F}{2\pi r}, \qquad N_\theta = -p_r r \tag{13.50}$$

Here x represents the axial direction of the cylinder.

13.14 Some Common Cases of Shells of Revolution

For a *closed-ended cylindrical vessel* under internal pressure, $p = -p_r$ and $F = -\pi r^2 p$. The preceding equations then result in the membrane stresses:

$$\sigma_\theta = \frac{pr}{t}, \quad \sigma_x = \frac{pr}{2t} \tag{13.51}$$

Here t is the thickness of the shell. Because of its direction, σ_θ is called the *tangential, circumferential*, or *hoop stress*; similarly, σ_x is the *axial* or the *longitudinal stress*. Note that for thin-walled cylindrical pressure vessels, $\sigma_\theta = 2\sigma_x$. From Hooke's law, the *radial extension* θ of the cylinder, under the action of tangential and axial stresses σ_θ and σ_x, is

$$\theta = \frac{r}{E}(\sigma_\theta + \nu\sigma_x) = \frac{pr}{2Et}(2 - \nu) \tag{13.52}$$

where E and ν represent modulus of elasticity and Poisson's ratio.

Observe that tangential stresses in the sphere are half the magnitude of the tangential stresses in the cylinder. Thus, a sphere is an *optimum shape* for an internally pressurized closed vessel.

EXAMPLE 13.5   Compressed Air Tank

A steel cylindrical vessel with hemispherical ends or so-called heads, supported by two cradles (Fig. 13.14), contains air at a pressure of p. Calculate (a) the stresses in the tank if each portion has the same mean radius r and the thickness t; (b) radial extension of the cylinder. *Given:* $r = 0.5$ m, $t = 10$ mm, $p = 1.5$ MPa, $E = 200$ GPa, and $\nu = 0.3$. *Assumptions:* One of the cradles is designed so that it does not exert any axial force on the vessel; the cradles act as simple supports. The weight of the tank may be disregarded.

Solution

a. The axial stress in the cylinder and the tangential stresses in the spherical heads are the same.

Referring to Eqs. (13.51), we write

$$\sigma_a = \sigma = \frac{pr}{2t} = \frac{1.5(10^6)(0.5)}{2(0.01)} = 37.5 \text{ MPa}$$

FIGURE 13.14.   *Example 13.5. Cylindrical vessel with spherical ends.*

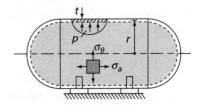

The tangential stress in the cylinder is therefore

$$\sigma_\theta = \frac{pr}{t} = 2\sigma_a = 75 \text{ MPa}$$

b. Through the use of Eq. (13.52), we have

$$\theta = \frac{pr^2}{2Et}(1 - \nu) = \frac{1.5(10^6)(0.5)^2(1.7)}{2(200 \times 10^9)(0.01)} = 0.16 \text{ mm}$$

Comment The largest radial stress, occurring on the surface of the tank, $\sigma_r = p$, is negligibly small compared with the tangential and axial stresses. The state of stress in the wall of a thin-walled vessel is thus taken to be *biaxial*.

EXAMPLE 13.6 Hemispherical Dome

Derive expressions for the stress resultants in a hemispherical dome of radius a and thickness t, loaded only by its own weight, p per unit area (Fig. 13.15).

Solution The weight of that portion of the dome intercepted by ϕ is

$$F = \int_0^\phi p(2\pi a \sin \phi \, a \cdot d\phi) = 2\pi a^2 p(1 - \cos \phi)$$

In addition,

$$p_z = p \cos \phi$$

Substituting into Eqs. (13.46) for p_z and F, we obtain

$$N_\phi = \frac{ap(1 - \cos \phi)}{\sin^2\phi} = -\frac{ap}{1 + \cos \phi}$$

$$N_\theta = -ap\left(a \cos \phi - \frac{1}{1 + \cos \phi}\right) \qquad \textbf{(13.53)}$$

where the negative signs indicate compression. It is clear that N_ϕ is always compressive. The sign of N_θ, on the other hand, depends on ϕ. From the second expression, when $N_\theta = 0$, $\phi = 51°50'$. For ϕ smaller than this value, N_θ is compressive. For $\phi > 51°50'$, N_θ is tensile.

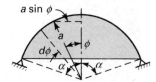

FIGURE 13.15. *Example 13.6. Simply supported hemispherical dome carries its own weight.*

13.15 THERMAL STRESSES IN COMPOUND CYLINDERS

This section provides the basis for design of composite or compound multishell cylinders, constructed of a number of concentric, thin-walled shells. In the development that follows, each component shell is assumed homogeneous and isotropic. Each may have a different thickness and different material properties and be subjected to different uniform or variable temperature differentials [Refs. 13.7 and 13.8].

In the case of a free-edged multishell cylinder undergoing a *uniform temperature change* ΔT, the free motion of any shell having different material properties is restricted by components adjacent to it. Only a tangential or hoop stress σ_θ and corresponding strain ε_θ are then produced in the cylinder walls. Through the use of Eqs. (3.26a), with $\sigma_y = \sigma_\theta$, $\varepsilon_y = \varepsilon_\theta$, and $T = \Delta T$, we write

$$\sigma_\theta = E[\varepsilon_\theta - \alpha(\Delta T)] \qquad (13.54)$$

where α is the coefficient of thermal expansion. In a cylinder subjected to a *temperature gradient*, both axial and hoop stresses occur, and $\sigma_x = \sigma_\theta = \sigma$. Hence, Eq. (3.26b) leads to

$$\sigma = \frac{E}{1 - \nu^2}\left[\varepsilon_\theta - \alpha(\Delta T)\right] \qquad (13.55)$$

Let us consider a compound cylinder consisting of three components, each under different temperature gradients (the ΔTs), as depicted in Fig. 13.16. Applying Eq. (13.55), the stresses may be expressed in the following form:

$$\begin{Bmatrix} \sigma_{a1} \\ \sigma_{a2} \end{Bmatrix} = \frac{E_a}{1 - \nu_a}\begin{Bmatrix} [\varepsilon_\theta - \alpha_a(\Delta T_1)] \\ [\varepsilon_\theta - \alpha_a(\Delta T_2)] \end{Bmatrix}$$

$$\begin{Bmatrix} \sigma_{b2} \\ \sigma_{b3} \end{Bmatrix} = \frac{E_b}{1 - \nu_b}\begin{Bmatrix} [\varepsilon_\theta - \alpha_b(\Delta T_2)] \\ [\varepsilon_\theta - \alpha_b(\Delta T_3)] \end{Bmatrix} \qquad (13.56a\text{-}f)$$

$$\begin{Bmatrix} \sigma_{c3} \\ \sigma_{c4} \end{Bmatrix} = \frac{E_c}{1 - \nu_c}\begin{Bmatrix} [\varepsilon_\theta - \alpha_c(\Delta T_3)] \\ [\varepsilon_\theta - \alpha_c(\Delta T_4)] \end{Bmatrix}$$

Here the subscripts $a, b,$ and c refer to individual shells.

The *axial forces* corresponding to each layer are next determined from the stresses given by Eqs. (13.56) as

$$N_a = \frac{1}{2}(\sigma_{a1} + \sigma_{a2})A_a$$

FIGURE 13.16. *Compound cylinder under a uniform temperature change.*

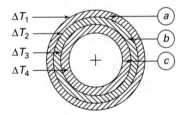

$$N_b = \frac{1}{2}(\sigma_{b2} + \sigma_{b3})A_b \tag{a}$$

$$N_c = \frac{1}{2}(\sigma_{c3} + \sigma_{b4})A_c$$

in which the A's are the cross-sectional area of each component. The condition that the sum of the axial forces be equal to zero is satisfied when

$$N_a + N_b + N_c = 0 \tag{b}$$

Equations (13.56) together with Eqs. (a) and (b), for the case in which $\nu_a = \nu_b = \nu_c$, lead to the following expression for the tangential strain:

$$\varepsilon_\theta = \frac{A_a E_a \alpha_a (\Delta T)_a + A_b E_b \alpha_b (\Delta T)_b + A_c E_c \alpha_c (\Delta T)_c}{A_a E_a + A_b E_b + A_c E_c} \tag{13.57}$$

Here the quantities $(\Delta T)_a$, $(\Delta T)_b$, and $(\Delta T)_c$ represent the *average* of the temperature differentials at the boundaries of elements a, b, and c, respectively, for instance, $(\Delta T)_a = \frac{1}{2}(\Delta T_1 + \Delta T_2)$. The stresses at the inner and outer surfaces of each shell may be obtained readily on carrying Eq. (13.57) into Eqs. (13.56).

A final point to be noted is that near the ends there will often be some bending of the composite cylinder, and the *total thermal stresses* will be obtained by *superimposing* on Eqs. (13.56) such stresses as may be necessary to satisfy the boundary conditions [Ref. 13.1].

EXAMPLE 13.7   Brass-Steel Cylinder

The cylindrical part of a jet nozzle is made by just slipping a steel tube over a brass tube with each shell uniformly heated and the temperature raised by ΔT (Figure 13.17). Determine the hoop stress that develops in each component on cooling. *Assumptions:* Coefficient of thermal expansion of the brass is larger than that of steel (see Table D.1). Each tube is a thin-walled shell with a mean radius of r.

Solution For the condition described, the composite cylinder is contracting. Equation (13.54) is therefore

$$\sigma_\theta = -E[\varepsilon_\theta + \alpha(\Delta T)] \tag{13.58}$$

from which

$$\varepsilon_\theta = -\frac{1}{E}\sigma_\theta - \alpha(\Delta T) \tag{13.59}$$

We have $A_b = 2\pi r t_b$ and $A_s = 2\pi r t_s$, in which t_b and t_s represent the wall thickness of the brass and steel tubes, respectively. Hoop strain, referring to Eqs. (13.57) and (13.59), is written as

$$\varepsilon_\theta = -\frac{(\Delta T)[t_s E_s \alpha_s + t_b E_b \alpha_b]}{t_s E_s + t_b E_b} \tag{13.60}$$

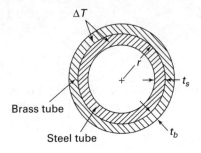

FIGURE 13.17. *Example 13.7. Two-layer compound cylinder.*

Finally, hoop stresses, through the use of Eqs. (13.58) and (13.60) and after simplification, are determined as follows:

$$(\sigma_\theta)_b = -E_b[\varepsilon_\theta + \alpha_b(\Delta T)] = -\frac{E_b(\Delta T)(\alpha_b - \alpha_s)}{1 + (t_b E_b/t_s E_s)} \qquad \textbf{(13.61)}$$

$$(\sigma_\theta)_s = -E_s[\varepsilon_\theta + \alpha_s(\Delta T)] = -\frac{E_s(\Delta T)(\alpha_s - \alpha_b)}{1 + (t_b E_b/t_s E_s)} \qquad \textbf{(13.62)}$$

Comment Observe that the stresses in the brass and steel tubes are tensile and compressive, respectively, since $\alpha_b > \alpha_s$, and ΔT is a negative quantity because of cooling.

13.16 CYLINDRICAL SHELLS OF GENERAL SHAPE

A cylindrical shell generated as a straight line, the *generator*, moves parallel to itself along a closed path. Figure 13.18 shows an element isolated from a *cylindrical shell of arbitrary cross section*. The element is located by coordinates x (axial) and θ on the cylindrical surface.

The forces acting on the sides of the element are depicted in the figure. The x and θ components of the externally applied forces per unit area are denoted p_x and p_θ and are shown to act in the directions of increasing x and θ (or y). In addition, a

FIGURE 13.18. *Membrane forces (per unit length) and distributed axial, radial, and tangential loads (per unit area) on a cylindrical shell element.*

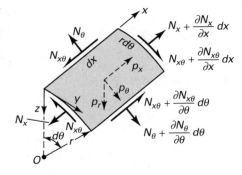

radial (or normal) component of the external loading p_r acts in the positive z direction. The following expressions describe the requirements for equilibrium in the x, θ, and r directions:

$$\frac{\partial N_x}{\partial x}dx(r\,d\theta) + \frac{\partial N_{\theta x}}{\partial \theta}d\theta(dx) + p_x(dx)r\,d\theta = 0$$

$$\frac{\partial N_\theta}{\partial \theta}d\theta(dx) + \frac{\partial N_{x\theta}}{\partial x}dx(r\,d\theta) + p_\theta(dx)r\,d\theta = 0$$

$$N_\theta dx(d\theta) + p_r(dx)r\,d\theta = 0$$

Canceling the differential quantities, we obtain the equations of a cylindrical shell:

$$N_\theta = -p_r r$$

$$\frac{\partial N_{x\theta}}{\partial x} + \frac{1}{r}\frac{\partial N_\theta}{\partial \theta} = -p_\theta \qquad\qquad (13.63)$$

$$\frac{\partial N_x}{\partial x} + \frac{1}{r}\frac{\partial N_{x\theta}}{\partial \theta} = -p_x$$

Given the external loading, N_θ is readily determined from the first equation of (13.63). Following this, by integrating the second and the third equations, $N_{x\theta}$ and N_x are found:

$$N_\theta = -p_r r$$

$$N_{x\theta} = -\int\left(p_\theta + \frac{1}{r}\frac{\partial N_\theta}{\partial \theta}\right)dx + f_1(\theta)$$

$$\qquad\qquad (13.64)$$

$$N_x = -\int\left(p_x + \frac{1}{r}\frac{\partial N_{x\theta}}{\partial \theta}\right)dx + f_2(\theta)$$

Here $f_1(\theta)$ and $f_2(\theta)$ represent arbitrary functions of integration to be evaluated on the basis of the boundary conditions. They arise, of course, as a result of the integration of partial derivatives.

EXAMPLE 13.8 Liquid Tank

Determine the stress resultants in a circular, simply supported tube of thickness t filled to capacity with a liquid of specific weight γ (Fig. 13.19a).

Solution The pressure at any point in the tube equals the weight of a column of unit cross-sectional area of the liquid at that point. At the arbitrary level mn (Fig. 13.19b), the outward pressure is $-\gamma a(1 - \cos\theta)$, where the pressure is positive radially inward; hence the minus sign. Then

$$p_r = -\gamma a(1 - \cos\theta), \quad p_\theta = p_x = 0 \qquad\qquad \textbf{(a)}$$

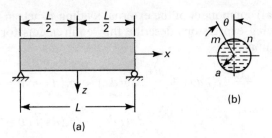

FIGURE 13.19. *Example 13.8. Cylindrical tube filled with liquid.*

Substituting the foregoing into Eqs. (13.64), we obtain

$$N_\theta = \gamma a^2 (1 - \cos\theta)$$

$$N_{x\theta} = -\int \gamma a \sin\theta \, dx + f_1(\theta) = -\gamma a x \sin\theta + f_1(\theta)$$

$$N_x = \int \gamma x \cos\theta \, dx - \frac{1}{a}\int \frac{df_1}{d\theta} dx + f_2(\theta)$$

$$= \frac{\gamma x^2}{2} \cos\theta - \frac{x \, df_1}{a \, d\theta} + f_2(\theta)$$

(b)

The boundary conditions are

$$N_x = 0, \qquad \text{for } x = L/2, \ -L/2 \qquad \textbf{(c)}$$

The introduction of Eq. (b) into Eq. (c) leads to

$$0 = \frac{\gamma L^2}{8} \cos\theta - \frac{L}{2a}\frac{df_1}{d\theta} + f_2(\theta)$$

$$0 = \frac{\gamma L^2}{8} \cos\theta + \frac{L}{2a}\frac{df_1}{d\theta} + f_2(\theta)$$

Addition and subtraction of these give, respectively,

$$f_2(\theta) = -\frac{\gamma L^2}{8} \cos\theta$$

$$\frac{df_1}{d\theta} = 0 \quad \text{or} \quad f_1(\theta) = 0 + c$$

(d)

We observe from the second equation of (b) that c in the second equation of (d) represents the value of the uniform shear load $N_{x\theta}$ at $x = 0$. This load is zero because the tube is subjected to no torque; thus $c = 0$. Then, Eq. (b), together with Eq. (d), provides the solution:

$$N_\theta = \gamma a^2 (1 - \cos\theta)$$

$$N_{x\theta} = -\gamma a x \sin\theta \qquad \textbf{(13.65)}$$

$$N_x = -\tfrac{1}{8}\gamma \, (L^2 - 4x^2) \cos\theta$$

The stresses are determined upon dividing these stress resultants by the shell thickness. It is observed that the shear $N_{x\theta}$ and the normal force N_x exhibit the same spanwise distribution as do the shear force and the bending moment of a beam. Their values, as may be readily verified, are identical with those obtained by application of the beam formulas, Eqs. (5.39) and (5.38), respectively.

It has been noted that membrane theory cannot, in all cases, provide solutions compatible with the actual conditions of deformation. This theory also fails to predict the state of stress in certain areas of the shell. To overcome these shortcomings, bending theory is applied in the case of cylindrical shells, taking into account the stress resultants such as the types shown in Fig. 13.3 and N_x, N_θ, and $N_{x\theta}$.

REFERENCES

13.1. UGURAL, A. C., *Stresses in Beams, Plates, and Shells*, 3rd ed., CRC Press, Taylor and Francis Group, Boca Raton, Fla., 2010.

13.2. TIMOSHENKO, S. P., and WOINOWSKY-KRIEGER, S., *Theory of Plates and Shells*, 2nd ed., McGraw-Hill, New York, 1959.

13.3. YANG, T. Y., *Finite Element Structural Analysis*, Prentice Hall, Upper Saddle River, N.J., 1986.

13.4. BATHE, K. I., *Finite Element Procedures in Engineering Analysis*, Prentice Hall, Upper Saddle River, N.J., 1996.

13.5. ZIENKIEWICZ, O. C., and TAYLOR, R. I., *The Finite Element Method*, 4th ed., Vol. 2 (Solid and Fluid Mechanics, Dynamics and Nonlinearity), McGraw-Hill, London, 1991.

13.6. BAKER, A. J., and PEPPER, D. W., *Finite Elements 1=2=3*, McGraw-Hill, New York, 1991.

13.7. ABRAHAM, L. H., *Structural Design of Missiles and Spacecraft*, McGraw-Hill, New York, 1963.

13.8. FAUPEL, J. H., and FISHER, F. E., *Engineering Design*, 2nd ed., Wiley, Hoboken, N. J., 1981.

PROBLEMS

Sections 13.1 through 13.10

13.1. A uniform load p_o acts on a long and narrow rectangular plate with edges at $y = 0$ and $y = b$ both clamped (Fig. 13.6). Determine (a) the equation of the surface deflection and (b) the maximum stress σ_y at the clamped edge. Use $\nu = \frac{1}{3}$.

13.2. An initially straight steel band 30-mm wide and $t = 0.3$-mm thick is used to lash together two drums of radii $r = 120$ mm (Fig. P13.2). Using $E = 200$ GPa and $\nu = 0.3$, calculate the maximum bending strain and maximum stress developed in the band.

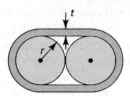

FIGURE P13.2.

13.3. For the plate described in Example 13.1, for $p_o = 20$ kN/m^2, $b = 0.6$ m, $t = 12$ mm, $E = 200$ GPa, and $\nu = \frac{1}{3}$, calculate the maximum deflection w_{max}, maximum strain $\varepsilon_{y,\,max}$, maximum stress $\sigma_{y,\,max}$, and the radius of the midsurface.

13.4. A thin rectangular plate is subjected to uniformly distributed bending moments M_a and M_b applied along edges a and b, respectively. Derive the equations governing the surface deflection for two cases: (a) $M_a \neq M_b$ and (b) $M_a = -M_b$.

13.5. The simply supported rectangular plate shown in Fig. 13.7a is subjected to a distributed load p given by

$$p = \frac{36P(a - x)(b - y)}{a^3 b^3}$$

Derive an expression for the deflection of the plate in terms of the constants P, a, b, and D.

13.6. A simply supported square plate $(a = b)$ carries the loading

$$p = p_o \sin\frac{m\pi x}{a} \sin\frac{n\pi y}{b}$$

(a) Determine the maximum deflection. (b) Retaining the first two terms of the series solutions, evaluate the value of p_o for which the maximum deflection is not to exceed 8 mm. The following data apply: $a = 3$ m, $t = 25$ mm, $E = 210$ GPa, and $\nu = 0.3$.

13.7. A clamped circular plate of radius a and thickness t is to close a circuit by deflecting 1.5 mm at the center at a pressure of $p = 10$ MPa. (Fig. 13.9). What is the required value of t? Take $a = 50$ mm, $E = 200$ GPa, and $\nu = 0.3$.

13.8. A simply supported circular plate of radius a and thickness t is deformed by a moment M_o uniformly distributed along the edge. Derive an expression for the deflection w as well as for the maximum radial and tangential stresses.

13.9. Given a simply supported circular plate containing a circular hole $(r = b)$, supported at its outer edge $(r = a)$ and subjected to uniformly distributed inner edge moments M_o, derive an expression for the plate deflection.

13.10. A brass flat clamped disk valve of 0.4-m diameter and 20-mm thickness is subjected to a liquid pressure of 400 kPa (Fig. 13.9). Determine the factor of safety, assuming that failure occurs in accordance with the maximum shear stress theory. The yield strength of the material is 100 MPa and $\nu = \frac{1}{3}$.

13.11. A circular plate of radius a is simply supported at its edge and subjected to uniform loading p_o. Determine the center point deflection.

13.12. A rectangular aluminum alloy sheet of 6-mm thickness is bent into a circular cylinder with radius r. Calculate the diameter of the cylinder and the maximum moment developed in the member if the allowable stress is not to exceed 120 MPa, $E = 70$ GPa, and $\nu = 0.3$.

13.13. The uniform load p_0 acts on a long and narrow rectangular plate with edge $y = 0$ simply supported and edge $y = b$ clamped, as depicted in Figure P13.13. Find (a) the equation of the surface deflection w; (b) the maximum bending stress.

13.14. An approximate *expression for the deflection* surface of a clamped triangular plate (Figure P13.14) is given as follows:

$$w = cx^2y^2\left(1 - \frac{x}{a} - \frac{y}{b}\right)^2 \tag{P13.14}$$

in which c is *a* constant. (a) Show that Equation P13.14 satisfies the boundary conditions. (b) Find the approximate maximum plane stress components at points A and B.

13.15. A square instrument panel (Figure P13.15) is under uniformly distributed twisting moment $M_{xy} = M_0$ at all four edges. Find an expression for the deflection surface w.

13.16. A cylindrical thick-walled vessel of 0.2-m radius and flat, thin plate head is subjected to all internal pressure of 1.5 MPa. Calculate (a) the thickness of the cylinder head if the allowable stress is limited to 135 MPa; (b) the maximum deflection of the cylinder head. *Given:* $E = 200$ GPa and $\nu = 0.3$.

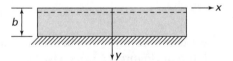

FIGURE P13.13.

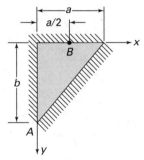

FIGURE P13.14.

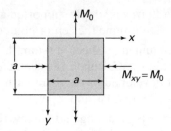

FIGURE P13.15.

13.17. A high-strength, ASTM-A242 steel plate covers a circular opening of diameter $d = 2a$. *Assumptions:* The plate is fixed at its edge and carries a uniform pressure p_o. *Given:* $E = 200$ GPa, $\sigma_{yp} = 345$ MPa (Table D.1), $v = 0.3$, $t = 10$ mm, $a = 125$ mm. Calculate (a) the pressure p_o, and maximum deflection at the onset of yield in the plate; (b) allowable pressure based on a safety factor of $n = 1.2$ with respect to yielding of the plate.

13.18. A 500-mm simply supported square aluminum panel of 20-mm thickness is under uniform pressure p_o. For $E = 70$ GPa, $v = 0.3$, and $\sigma_{yp} = 240$ MPa and taking into account only the first term of the series solution, calculate the limiting value of p_o that can be applied to the plate without causing yielding and the maximum deflection w that would be produced when p_o reaches its limiting value.

13.19. Verify that integrating Eq. (b) of Section 13.9 results in Eq. (c).

13.20. Apply the Rayleigh–Ritz method to determine the maximum deflection of a simply supported square plate subjected to a lateral load in the form of a triangular prism (Fig. P13.20).

Sections 13.11 through 13.16

13.21. Consider two pressurized vessels of cylindrical and spherical shapes of mean radius $r = 150$ mm (Figure 13.13), constructed of 6-mm-thick plastic material. Find the limiting value of the pressure differential the shells can resist given a maximum stress of 15 MPa.

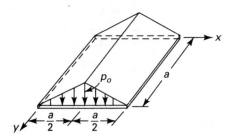

FIGURE P13.20.

13.22. A spherical vessel of radius $r = 1$ m and wall thickness $t = 50$ mm is submerged in water having density $\gamma = 9.81$ kN/m^3. On the basis of a factor of safety of $n = 2$, calculate the water depth at which the tangential stress σ_θ in the sphere would be 30 MPa.

13.23. Figure P13.23 shows a closed-ended cylindrical steel tank of a radius $r = 4$ m and a height $h = 18$ m. The vessel is completely filled with a liquid of density $\gamma = 15$ kN/m^3 and is subjected to an additional internal gas pressure of $p = 500$ kPa. Based on an allowable stress of 180 MPa, calculate the wall thickness needed (a) at the top of the tank; (b) at quarter-height of the tank; (c) at the bottom of the tank.

13.24. Resolve Prob. 13.23, assuming that the gas pressure is $p = 200$ kPa.

13.25. A cylinder of 1.2 m in diameter, made of steel with maximum strength $\sigma_{max} = 240$ MPa, is under an internal pressure of $p = 1.2$ kPa. Calculate the required thickness t of the vessel on the basis of a factor of safety $n = 2.4$.

13.26. A pipe for conveying water (density $\gamma = 9.81$ kN/m^3) to a turbine, called *penstock*, operates at a head of 150 m. The pipe has a 0.8-m diameter and a wall thickness of t. Find the minimum required value of t for a material strength of 120 MPa based on a safety factor of $n = 1.8$.

13.27. Resolve Prob. 13.26 if the allowable stress is 100 MPa.

13.28. A closed cylindrical tank fabricated of 12-mm-thick plate is under an internal pressure of $p = 10$ MPa. Calculate (a) the maximum diameter if the maximum shear stress is limited to 35 MPa; (b) the limiting value of tensile stress for the diameter found in part (a).

13.29. Given an internal pressure of 80 kPa, calculate the maximum stress at point A of a football of uniform skin thickness $t = 2$ mm (Fig. P13.29).

13.30. For the toroidal shell of Fig. P13.30 subjected to internal pressure p, determine the membrane forces N_ϕ and N_θ.

13.31. A toroidal pressure vessel of outer and inner diameters 1 and 0.7 m, respectively, is to be used to store air at a pressure of 2 MPa. Determine the required minimum thickness of the vessel if the allowable stress is not to exceed 210 MPa.

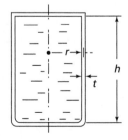

FIGURE P13.23.

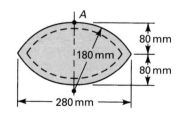

FIGURE P13.29.

13.32. Show that the tangential (circumferential) and longitudinal stresses in a simply supported conical tank filled with liquid of specific weight γ (Fig. P13.32) are given by

$$\sigma_\theta = \frac{\gamma(a-y)y}{t}\frac{\tan \alpha}{\cos \alpha}, \quad \sigma_\phi = \frac{\gamma(a - 2y/3)y}{2t}\frac{\tan \alpha}{\cos \alpha} \tag{P13.32}$$

13.33. A simply supported circular cylindrical shell of radius a and length L carries its own weight p per unit area (that is, $p_x = 0$, $p_\theta = -p\cos \theta$, and $p_r = p\sin \theta$). Determine the membrane forces. The angle θ is measured from the horizontal axis.

13.34. Redo Example 13.8 for the case in which the ends of the cylinder are fixed.

13.35. An edge-supported conical shell carries its own weight p per unit area and is subjected to an external pressure p_r (Fig. 13.13b). Determine the membrane forces and the maximum stresses in the shell.

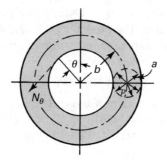

FIGURE P13.30.

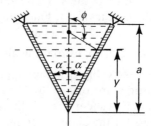

FIGURE P13.32.

Problem Formulation and Solution

A consistent, systematic procedure is required for solving problems in the mechanics of solids. A basic method of attack for any analysis problems is to define (or understand) the problem. Formulation of the problem requires consideration of the physical situations and an idealized description by the pertinent diagrams that approximate the actual component under consideration. The following five steps may be helpful in *formulation and solution* of a problem:

1. Define the problem and state briefly what is *given*.
2. State consistently what is to be *determined*.
3. List simplifying *assumptions* to be made.
4. *Apply* the appropriate *equations* to find the unknowns.
5. *Comment* on the results briefly.

Problem statement should indicate clearly what information is required. Free-body diagrams must be complete, showing all essential quantities involved. Assumptions or idealizations expand on the given information to further constrain the problem. For example, one might take the effects of friction to be disregarded or the weight of the member can be omitted in a particular situation. Solutions must be based on the principles of mechanics solids, formulas, tables, and diagrams. Comments present the key aspects of the solution.

Numerical Accuracy

In engineering problems of practical significance, the data are seldom known with an **accuracy** greater than 0.2%. Hence, the answers to such problems should not be written with an accuracy greater than 0.2%. Calculations are often performed by electronic calculators and computers, usually carrying eight or nine digits. So, the possibility exists that numerical result will be reported to an accuracy that has no physical meaning. Throughout this book, we usually follow a common engineering rule to report the *final results* of calculations:

- Numbers beginning with "1" are recorded to *four* significant digits.
- All other numbers (that begin with "2" through "9") are recorded to *three* significant digits.

Consequently, a force of 18 N, for example, should read 18.00 N, and a force of 56 N should read 56.0 N. *Intermediate results*, if retained for further calculations, are recorded to many additional digits to maintain the numerical accuracy. The values of π and *trigonometric functions* are calculated to several significant digits (10 or more) within the calculator or computer. We note that, in some cases, such as when data is read from a graph, fewer significant digits may be recorded. In these situations, given data are assumed to be accurate to the number of significant digits indicated in the preceding. Various computational tools that may be used to perform analysis calculations are discussed in Section 7.16.

Daily Planning

Learning to pay attention to our own internal *body clock*, or brain cells controlling the timing of our behavior, and *daily planning* can help us make the best of our time. A tentative schedule for the "morning person" who prefers to wake up early and go to sleep early is given in Table A.1 (Ref. A.1). The "evening person" works late and wakes up late. Most individuals may shift times from one to another and others combine some characteristics of both.

TABLE A.1. *The Best Time to Do Everything*

Time	Activity
6:00 a.m.	Wake
6:00–6:30 a.m.	Unsuitable to concentrate
6:30–8:30 a.m.	Suitable for creativity
8:30 a.m.–12:00 noon	***Suitable for problem solving***
12:00–2:30 p.m.	Unsuitable to concentrate
2:30–4:30 p.m.	***Suitable for problem solving***
4:30–8:00 p.m.	Rejuvenation
8:00–10:00 p.m.	Unsuitable for problem solving
10:00 (or 11:00) p.m.–6:00 a.m.	Sleep

We note that creativity serves well for open-ended thinking. Rejuvenation is renewing the mind with activities like reading, art work, and puzzle solving. It is unsuitable to concentrate when the body's biological clock changes. During times suitable for problem solving, concentration is the highest for analysis.

REFERENCE

A.1. UGURAL, A. C., *Living Better: A Guide to Health, Happiness, and Managing Stress*, Eloquent Books, New York, 2009, Sec. 20. (www.eloquentbooks.com or www.amazon.com)

APPENDIX B

Solution of the Stress Cubic Equation

B.1 PRINCIPAL STRESSES

There are many methods in common usage for solving a cubic equation. A simple approach for dealing with Eq. (1.33) is to find one root, say σ_1, by plotting it (σ as abscissa) or by trial and error. The cubic equation is then factored by dividing by $(\sigma_p - \sigma_1)$ to arrive at a quadratic equation. The remaining roots can be obtained by applying the familiar general solution of a quadratic equation. This process requires considerable time and algebraic work, however.

What follows is a practical approach for determining the roots-of-stress cubic equation (1.33):

$$\sigma_p^3 - I_1\sigma_p^2 + I_2\sigma_p - I_3 = 0 \tag{a}$$

where

$$
\begin{aligned}
I_1 &= \sigma_x + \sigma_y + \sigma_z \\
I_2 &= \sigma_x\sigma_y + \sigma_x\sigma_z + \sigma_y\sigma_z - \tau_{xy}^2 - \tau_{yz}^2 - \tau_{xz}^2 \\
I_3 &= \sigma_x\sigma_y\sigma_z + 2\tau_{xy}\tau_{yz}\tau_{xz} - \sigma_x\tau_{yz}^2 - \sigma_y\tau_{xz}^2 - \sigma_z\tau_{xy}^2
\end{aligned}
\tag{B.1}
$$

According to the method, expressions that provide direct means for solving both two- and three-dimensional stress problems are [Refs. B.1 and B.2]

$$
\begin{aligned}
\sigma_a &= 2S[\cos(\alpha/3)] + \tfrac{1}{3}I_1 \\
\sigma_b &= 2S\{\cos[(\alpha/3) + 120°]\} + \tfrac{1}{3}I_1 \\
\sigma_c &= 2S\{\cos[(\alpha/3) + 240°]\} + \tfrac{1}{3}I_1
\end{aligned}
\tag{B.2}
$$

Here the constants are given by

$$S = (\tfrac{1}{3}R)^{1/2}$$

$$\alpha = \cos^{-1}\left(-\frac{Q}{2T}\right)$$

$$R = \tfrac{1}{3}I_1^2 - I_2 \tag{B.3}$$

$$Q = \tfrac{1}{3}I_1 I_2 - I_3 - \tfrac{2}{27}I_1^3$$

$$T = (\tfrac{1}{27}R^3)^{1/2}$$

and invariants I_1, I_2, and I_3 are represented in terms of the given stress components by Eqs. (B.1).

The principal stresses found from Eqs. (B.2) are *redesignated* using numerical subscripts so that $\sigma_1 > \sigma_2 > \sigma_3$. This procedure is well adapted to a pocket calculator or digital computer.

B.2 DIRECTION COSINES

The values of the direction cosines of a principal stress are determined through the use of Eqs. (1.31) and (1.25), as discussed in Section 1.13. That is, substitution of a principal stress, say σ_1, into Eqs. (1.31) results in *two independent* equations in three unknown direction cosines. From these expressions, together with $l_1^2 + m_1^2 + n_1^2 = 1$, we obtain l_1, m_1, and n_1.

However, instead of solving one second-order and two linear equations simultaneously, the following simpler approach is preferred. Expressions (1.31) are expressed in matrix form as follows:

$$\begin{bmatrix} (\sigma_x - \sigma_i) & \tau_{xy} & \tau_{xz} \\ \tau_{xy} & (\sigma_y - \sigma_i) & \tau_{yz} \\ \tau_{xz} & \tau_{yz} & (\sigma_z - \sigma_i) \end{bmatrix} \begin{Bmatrix} l_i \\ m_i \\ n_i \end{Bmatrix} = 0, \qquad i = 1, 2, 3$$

The cofactors of the determinant of this matrix on the elements of the first row are

$$a_i = \begin{vmatrix} (\sigma_y - \sigma_i) & \tau_{yz} \\ \tau_{yz} & (\sigma_z - \sigma_i) \end{vmatrix}$$

$$b_i = -\begin{vmatrix} \tau_{xy} & \tau_{yz} \\ \tau_{xz} & (\sigma_z - \sigma_i) \end{vmatrix} \tag{B.4}$$

$$c_i = \begin{vmatrix} \tau_{xy} & (\sigma_y - \sigma_i) \\ \tau_{xz} & \tau_{yz} \end{vmatrix}$$

Upon introduction of the notation

$$k_i = \frac{1}{(a_i^2 + b_i^2 + c_i^2)^{1/2}} \qquad \textbf{(B.5)}$$

the direction cosines are then expressed as

$$l_i = a_i k_i, \qquad m_i = b_i k_i, \qquad n_i = c_i k_i \qquad \textbf{(B.6)}$$

It is clear that Eqs. (B.6) lead to $l_i^2 + m_i^2 + n_i^2 = 1$.

Application of Eqs. (B.2) and (B.6) to the sample problem described in Example 1.6 provides some algebraic exercise. Substitution of the given data into Eqs. (B.1) results in

$$I_1 = -22.7, \qquad I_2 = -170.8125, \qquad I_3 = 2647.521$$

We then have

$$R = \tfrac{1}{3}I_1^2 - I_2 = 342.5758$$
$$T = (\tfrac{1}{27}R^3)^{1/2} = 1220.2623$$
$$Q = \tfrac{1}{3}I_1 I_2 - I_3 - \tfrac{2}{27}I_1^3 = -488.5896$$
$$\alpha = \cos^{-1}(-Q/2T) = 78.4514°$$

Hence, Eqs. (B.2) give

$$\sigma_a = 11.618 \text{ MPa}, \qquad \sigma_b = -25.316 \text{ MPa}, \qquad \sigma_c = -9.001 \text{ MPa}$$

Reordering and redesignating these values,

$$\sigma_1 = 11.618 \text{ MPa}, \qquad \sigma_2 = -9.001 \text{ MPa}, \qquad \sigma_3 = -25.316 \text{ MPa}$$

from which it follows that

$$a_1 = \begin{vmatrix} (4.6 - 11.618) & 11.8 \\ 11.8 & (-8.3 - 11.618) \end{vmatrix} = 0.5445$$

$$b_1 = -\begin{vmatrix} -4.7 & 11.8 \\ 6.45 & (-8.3 - 11.618) \end{vmatrix} = -17.5046$$

$$c_1 = \begin{vmatrix} -4.7 & (4.6 - 11.618) \\ 6.45 & 11.8 \end{vmatrix} = -10.1939$$

and

$$k_1 = \frac{1}{(a_1^2 + b_1^2 + c_1^2)^{1/2}} = 0.0493$$

Thus, Eqs. (B.6) yield

TABLE B.1.　FORTRAN Program for Principal Stresses

```
        IMPLICIT REAL(A − Z)
        INTEGER J, D, X, Y
        DIMENSION A(3), B(3), C(3), S(3), L(3), M(3), N(3), K(3)
        WRITE(5, *) 'INPUT: SX, SY, SZ, TXY, TXZ, TYZ'
        WRITE(3, *) 'INPUT: '
        READ(5, *)SX, SY, SZ, TXY, TXZ, TYZ
        WRITE(5, 10)SX, SY, SZ
        WRITE(5, 20)TXY, TXZ, TYZ
        WRITE(3, 10)SX, SY, SZ
        WRITE(3, 20)TXY, TXZ, TYZ
10      FORMAT( /, 1X, 'SX = ', F7.3, 14X, 'SY = ', F7.3, 15X, 'SZ = ', F7.3)
20      FORMAT(1X, 'TXY = ', F7.3, 13X, 'TXZ = ', F7.3, 14X, 'TYZ = ', F7.3 //)
        I1 = SX + SY + SZ
        I2 = SX * SY + SX * SZ + SY * SZ − ((TXY) * * 2) − ((TXZ * * 2) − ((TYZ) * * 2)
        I3 = X * SY * SZ + 2. * TXY * TYZ * TXZ − SX * ((TYZ) * * 2) − SY * ((TXZ) * * 2)
    $   − SZ * ((TXY) * * 2)
        WRITE(5, *) 'OUTPUT: '
        WRITE (3, *) 'OUTPUT; '
        R = (1. /3. ) * ((I1) * * 2) − I2
        T = SQRT((1. /27. ) * R * * 3)
        Q = (1. /3. ) * I1 * I2 − I3 − (2. /27. ) * ((I1) * * 3)
        ST = SQRT((1. /3. ) * R)
        ALPHA = ACOS(−Q /(2. * T))
        S(1) = 2. * ST * COS(ALPHA /3. ) + (1. /3. ) * I1
        S(2) = 2. * ST * COS((ALPHA /3. ) + 2. 0944) + (1. /3. ) * I1
        S(3) = 2. * ST * COS((ALPHA /3. ) + 4. 1888) + (1. /3. ) * I1
        DO 50 X = 1, 2
        DO 40 Y = X, 3
        IF (S(X). LT. S(Y))GO TO 30
        GO TO 40
30      TEMP = S(X)
        S(X) = S(Y)
        S(Y) = TEMP
40      CONTINUE
50      CONTINUE
        WRITE(5, 60)S(1), S(2), S(3)
        WRITE(3, 60)S(1), S(2), S(3)
60      FORMAT( /, 1X, 'S1 = ', F7.3, 14X, 'S2 = ', F7.3, 15X, 'S3 = ', F7.3 /)
        DO 80 J = 1, 3
        A(J) = ((SY − S(J)) * (SZ − S(J)) − (TYZ) * * 2)
        B(J) = − ((TXY) * (SZ − S(J)) − (TXZ * TYZ))
        C(J) = ((TXY * TYZ) − TXZ * (SY − S(J)))
        K(J) = 1. /SQRT((A(J)) * * 2 + (B(J)) * * 2 + (C(J)) * * 2))
        L(J) = A(J) * K(J)
        M(J) = B(J) * K(J)
        N(J) = C(J) * K(J)
        WRITE(5, 70)J, L(J), J, M(J), J, N(J)
        WRITE(3, 70)J, L(J), J, M(J), J, N(J)
70      FORMAT(1X, 'L( ', I1, ') = ', F7.4, 12X, 'M(', I1, ') = ', F7.4, 13X,
    $   'N(', I1, ') = ', F7.4)
80      CONTINUE
        STOP
        END
```

INPUT:

SX = −19.000	SY = 4.600	SZ = −8.300
TXY = −4.700	TXZ = 6.450	TYZ = 11.800

OUTPUT;

S1 = 11.618	S2 = −9.001	S3 = −25.316
L(1) = 0.0266	M(1) = −0.8638	N(1) = −0.5031
L(2) = −0.620	M(2) = 0.3802	N(2) = −0.6855
L(3) = 0.7834	M(3) = 0.3306	N(3) = −0.5262

$$l_1 = 0.0266, \qquad m_1 = -0.8638, \qquad n_1 = -0.5031$$

As a check, $l_1^2 + m_1^2 + n_1^2 = 0.9999 \approx 1$. Repeating the same procedure for σ_2 and σ_3, we obtain the values of direction cosines given in Example 1.6.

A FORTRAN computer program is listed in Table B.1 to expedite the solution for the principal stresses and associated direction cosines. Input data and output values are also provided. The program was written and tested on a digital computer. Note that this listing may readily be extended to obtain the factors of safety according to the various theories of failure (Chap. 4).

REFERENCES

B.1. MESSAL, E. E., Finding true maximum shear stress, Mach. Des., December 7, 1978, pp. 166–169.

B.2. TERRY, E. S., *A Practical Guide to Computer Methods for Engineers*, Prentice Hall, Englewood Cliffs, N.J., 1979.

APPENDIX C

Moments of Composite Areas

C.1 CENTROID

This appendix is concerned with the geometric properties of cross sections of a member. These *plane area* characteristics have special significance in various relationships governing stress and deflection of beams, columns, and shafts. Geometric properties for most areas encountered in practice are listed in numerous reference works [Ref. C.1]. Table C.1 presents several typical cases.

The first step in evaluating the characteristics of a plane area is to locate the *centroid of the area*. The centroid is the point in the plane about which the area is equally distributed. For area A shown in Fig. C.1, the *first moments* about the x and y axes, respectively, are given by

$$Q_x = \int_A y\, dA, \qquad Q_y = \int_A x\, dA \qquad \text{(C.1)}$$

These properties are expressed in cubic meters or cubic millimeters in SI units and in cubic feet or cubic inches in U.S. Customary Units. The centroid of area A is denoted by point C, whose coordinates $\bar{x}$ and $\bar{y}$ satisfy the relations

$$\bar{x} = \frac{Q_y}{A} = \frac{\int_A x\, dA}{\int_A dA}, \qquad \bar{y} = \frac{Q_x}{A} = \frac{\int_A y\, dA}{\int_A dA} \qquad \text{(C.2)}$$

When an axis possesses *an axis of symmetry*, the centroid is located on that axis, as the first moment about an axis of symmetry equals zero. When there are *two axes of symmetry*, the centroid lies at the intersection of the two axes. If an area possesses no axes of symmetry but does have a *center of symmetry*, the centroid coincides with the center of symmetry.

TABLE C.1. *Properties of Some Plane Areas*

1. Rectangle

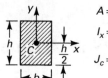

$$A = bh$$
$$I_x = \frac{bh^3}{12}$$
$$J_c = \frac{bh(b^2 + h^2)}{12}$$

2. Right triangle

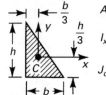

$$A = \frac{bh}{2}$$
$$I_x = \frac{bh^3}{36} \quad I_{xy} = -\frac{b^2h^2}{72}$$
$$J_c = \frac{bh(b^2 + h^2)}{36}$$

3. Ellipse

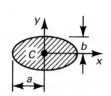

$$A = \pi ab$$
$$I_x = \frac{\pi ab^3}{4}$$
$$J_c = \frac{\pi ab(a^2 + b^2)}{4}$$

4. Isosceles triangle

$$A = \frac{bh}{2}$$
$$I_x = \frac{bh^3}{36} \quad I_y = \frac{hb^3}{36}$$
$$I_c = \frac{bh}{144}(4h^2 + 3b^2)$$

5. Circle

$$A = \pi r^2$$
$$I_x = \frac{\pi r^4}{4}$$
$$J_c = \frac{\pi r^2}{4}$$

6. Semicircle

$$A = \frac{\pi r^2}{2}$$
$$I_x = 0.110 r^4$$
$$I_y = \frac{\pi r^4}{8}$$

7. Thin tube

$$A = 2\pi rt$$
$$I_x = \pi r^3 t$$
$$J_c = 2\pi r^3 t$$

8. Half of thin tube

$$A = \pi rt$$
$$I_x = 0.095 \pi r^3 t$$
$$I_y = 0.5 \pi r^3 t$$

FIGURE C.1. *Plane area A with centroid C.*

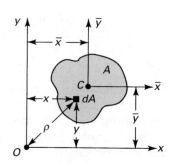

EXAMPLE C.1   Centroid of Triangular Area

Determine the ordinate $\bar{y}$ of the centroid of the triangular area shown in Fig. C.2.

Solution   A horizontal element with area of length x and height dy is selected (Fig. C.2). Considering similar triangles, $x = (h - y)b/h$, and

$$dA = \frac{b}{h}(h - y)\, dy$$

The first moment of the area with respect to the x axis is

$$Q_x = \int_A y\, dA = \frac{b}{h} \int_0^h (hy - y^2)\, dy = \tfrac{1}{6} bh^2$$

The second of Eqs. (C.2), with $A = bh/2$, then yields

$$\bar{y} = \frac{bh^2/6}{bh/2} = \tfrac{1}{3}h \tag{a}$$

Therefore, the centroidal axis $\bar{x}$ of the triangular area is located a distance of one-third the altitude from the base of the triangle.

Similarly, choosing the element of area dA as a vertical strip, it can be shown that the abscissa of the centroid is $\bar{x} = b/3$. The location of the centroid C is shown in the figure.

Frequently, an area can be divided into simple geometric shapes (for example, rectangles, circles, and triangles) whose areas and centroidal coordinates are known or easily determined. When a *composite area* is considered as an assemblage of n elementary shapes, the resultant or total area is the algebraic sum of the separate areas, and the resultant moment about any axis is the algebraic sum of the moments of the component areas. Thus, the centroid of a composite area has the coordinates

$$\bar{x} = \frac{\Sigma A_i \bar{x}_i}{\Sigma A_i}, \qquad \bar{y} = \frac{\Sigma A_i \bar{y}_i}{\Sigma A_i} \tag{C.3}$$

in which $\bar{x}_i$ and $\bar{y}_i$ represent the coordinates of the centroids of the component areas $A_i (i = 1, 2, \ldots, n)$.

In applying formulas (C.3), it is important to sketch the simple geometric forms into which the composite area is resolved, as shown next.

FIGURE C.2.   *Example C.1. Triangular area.*

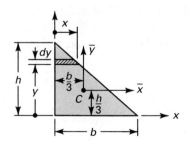

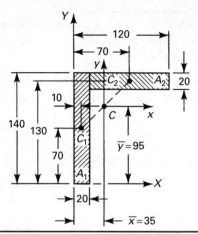

FIGURE C.3. *Example C.2. Area consisting of two parts.*

EXAMPLE C.2 Centroid of an Angle

Locate the centroid of the angle section depicted in Fig. C.3. The dimensions are given in millimeters.

Solution The composite area is divided into two rectangles, A_1 and A_2, for which the centroids are known (Fig. C.3). Taking the X, Y axes as reference, Eqs. (C.3) are applied to calculate the coordinates of the centroid. The computation is conveniently carried out in the following tabular form. Note that when an area is divided into *only two parts*, the centroid C of the entire area *always lies on the line* connecting the centroids C_1 and C_2 of the components, as indicated in Fig. C.3.

No.	A_i (mm^2)	$\bar{x}_i$ (mm)	$\bar{y}_i$ (mm)	$A_i \bar{x}_i$ (mm^3)	$A_i \bar{y}_i$ (mm^3)
1	$20(140) = 2800$	10	70	28×10^3	196×10^3
2	$20(100) = 2000$	70	130	140×10^3	260×10^3
	$\Sigma A_i = 4800$			$\Sigma A_i \bar{x}_i = 168 \times 10^3$	$\Sigma A_i \bar{y}_i = 456 \times 10^3$

$$\bar{x} = \frac{\Sigma A_i \bar{x}_i}{\Sigma A_i} = \frac{168 \times 10^3}{4800} = 35 \text{ mm}, \qquad \bar{y} = \frac{\Sigma A_i \bar{y}_i}{\Sigma A_i} = \frac{456 \times 10^3}{4800} = 95 \text{ mm}$$

C.2 MOMENTS OF INERTIA

We now consider the *second moment* or *moment of inertia* of an area (a relative measure of the manner in which the area is distributed about any axis of interest). The moments of inertia of a plane area A with respect to the x and y axes, respectively, are defined by the following integrals:

$$I_x = \int_A y^2 \, dA, \qquad I_y = \int_A x^2 \, dA \tag{C.4}$$

where x and y are the coordinates of the element of area dA (Fig. C.1).

Similarly, the polar moment of inertia of a plane area A with respect to an axis through O perpendicular to the area is given by

$$J_o = \int_A \rho^2 \, dA = I_x + I_y \tag{C.5}$$

Here ρ is the distance from point O to the element dA, and $\rho^2 = x^2 + y^2$. The *product of inertia* of a plane area A with respect to the x and y axes is defined as

$$I_{xy} = \int_A xy \, dA \tag{C.6}$$

In the foregoing, each element of area dA is multiplied by the product of its coordinates (Fig. C.1). The product of inertia of an area about any pair of axes is *zero* when either of the axes is an axis of *symmetry*.

From Eqs. (C.4) and (C.5), it is clear that the moments of inertia are always positive quantities because coordinates x and y are squared. Their dimensions are length raised to the fourth power; typical units are meters^4, millimeters^4, and inches^4. The dimensions of the product of inertia are the same as for the moments of inertia; however, the product of inertia can be positive, negative, or zero depending on the values of the product xy.

The *radius of gyration* is a distance from a reference axis or a point at which the entire area of a section may be considered to be concentrated and still possess the same moment of inertia as the original area. Therefore, the radii of gyration of an area A about the x and y axes and the origin O (Fig. C.1) are defined as the quantities r_x, r_y, and r_o, respectively:

$$r_x = \sqrt{\frac{I_x}{A}}, \qquad r_y = \sqrt{\frac{I_y}{A}}, \qquad r_o = \sqrt{\frac{J_o}{A}} \tag{C.7}$$

Substituting I_x, I_y, and J_o from Eqs. (C.7) into Eq. (C.5) results in

$$r_o^2 = r_x^2 + r_y^2 \tag{C.8}$$

The radius of gyration has the dimension of length, expressed in meters.

C.3 PARALLEL-AXIS THEOREM

The moment of inertia of an area with respect to any axis is related to the moment of inertia around a parallel axis through the centroid by the *parallel-axis theorem*, sometimes called the *transfer formula*. It is useful for determining the moment of inertia of an area composed of several simple shapes.

To develop the parallel-axis theorem, consider the area A depicted in Fig. C.4. Let $\bar{x}$ and $\bar{y}$ represent the centroidal axes of the area, parallel to the axes x and y, respectively. The distances between the two sets of axes and the origin are d_x, d_y, and d_o. The moment of inertia with respect to the x axis is

$$I_x = \int_A (y + \bar{y})^2 \, dA = \int_A y^2 \, dA + 2\bar{y} \int_A y \, dA + \bar{y}^2 \int_A dA$$

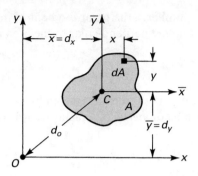

FIGURE C.4. *Plane area for deriving the parallel-axis theorem.*

The first integral on the right side equals the moment of inertia $I_{\bar{x}}$ about the $\bar{x}$ axis. As y is measured from the centroid axis $\bar{x}$, $\int_A y\,dA$ is zero. Hence,

$$I_x = I_{\bar{x}} + A\bar{y}^2 = I_{\bar{x}} + Ad_y^2 \qquad \textbf{(C.9a)}$$

Similarly,

$$I_y = I_{\bar{y}} + A\bar{x}^2 = I_{\bar{y}} + Ad_x^2 \qquad \textbf{(C.9b)}$$

The parallel-axis theorem is thus stated as follows: The moment of inertia of an area with respect to any axis is equal to the *moment of inertia around a parallel centroidal axis, plus the product of the area and the square of the distance between the two axes.*

In a like manner, a relationship may be developed connecting the polar moment of inertia J_o of an area about an arbitrary point O and the polar moment of inertia J_c about the centroid of the area (Fig. C.4):

$$J_o = J_c + Ad_o^2 \qquad \textbf{(C.10)}$$

It can be shown that the product of inertia of an area I_{xy} with respect to any set of axes is given by the transfer formula

$$I_{xy} = I_{\bar{x}\bar{y}} + A\bar{x}\,\bar{y} = I_{\bar{x}\bar{y}} + Ad_x d_y \qquad \textbf{(C.11)}$$

where $I_{\bar{x}\bar{y}}$ denotes the product of inertia around the centroidal axes. Note that the parallel-axis theorems, Eqs. (C.9) through (C.11), may be *employed only* if one of the two axes involved is a *centroidal* axis.

For elementary shapes, the integrals appearing in the equations of this and preceding sections can usually be evaluated easily and the geometric properties of the area thus obtained (Table C.1). Cross-sectional areas employed in practice can often be broken into a combination of these simple shapes.

EXAMPLE C.3 Moments of Inertia for a Triangular Area

For the triangular area shown in Fig. C.2, determine (a) the moment of inertia about the $\bar{x}$ and x axes and (b) the products of inertia with respect to the $\bar{x}\,\bar{y}$ and xy axes.

Solution   The area of a horizontal strip selected is

$$dA = \frac{b}{h}(h - y)\,dy \tag{a}$$

and the coordinates are related by

$$x = (h - y)\frac{b}{h} \tag{b}$$

as already found in Example C.1.

a. Then the moment of inertia about the centroidal $\bar{x}$ axis is (Fig. C.2)

$$I_{\bar{x}} = \int_A y^2\,dA = \int_{-h/3}^{2h/3} y^2 \frac{b}{h}(h - y)\,dy = \tfrac{1}{36}bh^3 \tag{c}$$

Similarly, the moment of inertia with respect to the x axis equals

$$I_x = \int_A y^2\,dA = \int_0^h y^2 \frac{b}{h}(h - y)\,dy = \tfrac{1}{12}bh^3 \tag{d}$$

This solution may also be obtained by applying the parallel-axis theorem:

$$I_x = I_{\bar{x}} + Ad_y^2 = \tfrac{1}{36}bh^3 + \tfrac{1}{2}bh\left(\frac{h}{3}\right)^2 = \tfrac{1}{12}bh^3$$

where $d_y = \bar{y}$ is the distance between the x and $\bar{x}$ axes.

b. From considerations of symmetry, the product of inertia of the horizontal strip with respect to the axes through its own centroid and parallel to the xy axes is zero. Its product of inertia about the xy axes, using Eq. (C.11), is then

$$I_{xy} = 0 + \int_A \bar{x}\,\bar{y}\,dA$$

Here $\bar{x}$ and $\bar{y}$ are the distances to the centroid of the strip. Referring to Fig. C.2, we have $\bar{x} = x/2$ and $\bar{y} = y$. Substituting these together with Eqs. (a) and (b) into the preceding equation, we have

$$I_{xy} = \int_0^h \frac{b^2}{2h^2}(h - y)^2 y\,dy = \tfrac{1}{24}b^2h^2 \tag{e}$$

The transfer formula now yields the product of inertia with respect to the centroidal axes:

$$I_{\bar{x}\bar{y}} = I_{xy} - A\bar{x}\,\bar{y} = \frac{b^2h^2}{24} - \frac{bh}{2}\left(\frac{b}{3}\right)\left(\frac{h}{3}\right) = -\tfrac{1}{72}b^2h^2$$

EXAMPLE C.4   Moment of Inertia for a *T* Section
Determine the moment of inertia of the *T* section shown in Fig. C.5a around the horizontal axis passing through its centroid. The dimensions are given in millimeters.

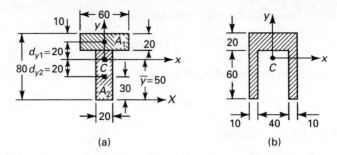

FIGURE C.5. *Example C.4. (a) T section; (b) channel section.*

Solution *Location of centroid:* The area is divided into component parts A_1 and A_2 for which the centroids are known (Fig. C.5a). Because the y axis is an axis of symmetry, $\bar{x} = 0$, and

$$\bar{y} = \frac{A_1\bar{y}_1 + A_2\bar{y}_2}{A_1 + A_2} = \frac{20(60)70 + 60(20)30}{20(60) + 60(20)} = 50 \text{ mm}$$

Centroidal moment of inertia: Application of the transfer formula to Fig. C.5a results in

$$I_x = \Sigma(I_{\bar{x}} + Ad_y^2)$$
$$= \tfrac{1}{12}(60)(20)^3 + 20(60)(20)^2 + \tfrac{1}{12}(20)(60)^3 + 20(60)(20)^2$$
$$= 136(10^4)\text{mm}^4$$

Interestingly, the properties of this T section about the centroidal x axis are the same as those of the channel (Fig. C.5b). Both sections possess an axis of symmetry.

C.4 PRINCIPAL MOMENTS OF INERTIA

The moments of inertia of a plane area depend not only on the location of the reference axis but also on the orientation of the axes about the origin. The variation of these properties with respect to axis location are governed by the parallel-axis theorem, as described in Section C.3. We now derive the equations for transformation of the moments and product of inertia *at any point* of a plane area.

The area shown in Fig C.6 has the moments and product of inertia I_x, I_y, I_{xy} with respect to the x and y axes defined by Eqs. (C.4) and (C.6). It is required to

FIGURE C.6. *Rotation of axes.*

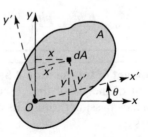

Appendix C   Moments of Composite Areas

determine the moments and product of inertia $I_{x'}$, $I_{y'}$, and $I_{x'y'}$ about axes x', y' making an angle θ with the original x, y axes. The new coordinates of an element dA can be expressed by projecting x and y onto the rotated axes (Fig. C.6):

$$x' = x \cos \theta + y \sin \theta, \qquad y' = y \cos \theta - x \sin \theta \tag{a}$$

Then, by definition

$$I_{x'} = \int_A y'^2 dA = \int_A (y \cos \theta - x \sin \theta)^2 dA$$

$$= \cos^2 \theta \int_A y^2 dA + \sin^2 \theta \int_A x^2 dA - 2 \sin \theta \cos \theta \int_A xy \, dA$$

Upon substituting Eqs. (C.4) and (C.6), the foregoing becomes

$$I_{x'} = I_x \cos^2 \theta + I_y \sin^2 \theta - 2I_{xy} \sin \theta \cos \theta \tag{b}$$

The moment of inertia $I_{y'}$ may be found readily by substituting $\theta + \pi/2$ for θ in the expression for $I_{x'}$. Similarly, using the definition $I_{x'y'} = \int_A x'y' \, dA$, we obtain

$$I_{x'y'} = (I_x - I_y) \sin \theta \cos \theta + I_{xy}(\cos^2 \theta - \sin^2 \theta) \tag{c}$$

The *transformation equations* for the moments and product of inertia may be rewritten by introducing double-angle trigonometric relations in the form

$$I_{x'} = \frac{I_x + I_y}{2} + \frac{I_x - I_y}{2} \cos 2\theta - I_{xy} \sin 2\theta \tag{C.12a}$$

$$I_{y'} = \frac{I_x + I_y}{2} - \frac{I_x - I_y}{2} \cos 2\theta + I_{xy} \sin 2\theta \tag{C.12b}$$

$$I_{x'y'} = \frac{I_x - I_y}{2} \sin 2\theta + I_{xy} \cos 2\theta \tag{C.12c}$$

In comparing the expressions here and in Chapter 1, it is observed that moments of inertia $(I_x, I_y, I_{x'}, I_{y'})$ correspond to the normal stresses $(\sigma_x, \sigma_y, \sigma_{x'}, \sigma_{y'})$, the *negative* of the products of inertia $(-I_{xy}, -I_{x'y'})$ correspond to the shear stresses $(\tau_{xy}, \tau_{x'y'})$, and the polar moment of inertia (J_o) corresponds to the sum of the normal stresses $(\sigma_x + \sigma_y)$. Thus, *Mohr's circle* analysis and the characteristics for stress apply to these properties of area.

The angle θ at which the moment of inertia $I_{x'}$ of Eq. (C.12a) assumes an extreme value may be obtained from the condition $dI_{x'}/d\theta = 0$:

$$(I_x - I_y) \sin 2\theta + 2I_{xy} \cos 2\theta = 0 \tag{d}$$

The foregoing yields

$$\tan 2\theta_p = -\frac{2I_{xy}}{I_x - I_y} \tag{C.13}$$

Here θ_p represents the two values of θ that locate the *principal axes* about which the *principal* or maximum and minimum *moments of inertia* occur.

C.4   *Principal Moments of Inertia*

653

When Eq. (C.12c) is compared with Eq. (d), it becomes clear that the product of inertia is *zero* for the principal axes. If the origin of axes is located at the centroid of the area, they are referred to as the centroidal principal axes. It was observed in Section 3.2 that the products of inertia relative to the axes of symmetry are zero. Thus, an *axis of symmetry coincides with a centroidal principal axis.*

The principal moments of inertia are determined by introducing the two values of θ_p into Eq. (C.12a). The sine and cosine of angles $2\theta_p$ defined by Eq. (C.13) are

$$\cos 2\theta_p = \frac{I_x - I_y}{2r}, \qquad \sin 2\theta_p = -\frac{I_{xy}}{r} \qquad \textbf{(e)}$$

where $r = \{[(I_x - I_y)/2]^2 + I_{xy}^2\}^{1/2}$. When these expressions are inserted into Eq. (C.12a), we obtain

$$I_{1,2} = \frac{I_x + I_y}{2} \pm \sqrt{\left(\frac{I_x - I_y}{2}\right)^2 + I_{xy}^2} \qquad \textbf{(C.14)}$$

wherein I_1 and I_2 denote the *maximum* and *minimum principal moments of inertia*, respectively.

EXAMPLE C.5 Principal Moments of Inertia for Angular Section
Calculate the centroidal principal moments of inertia for the angular section shown in Fig. C.7a. The dimensions are given in millimeters.

Solution *Location of centroid:* The x, y axes are the reference axes through the centroid C (Fig. C.7a), the location of which has already been determined in Example C.2. Again, the area is divided into rectangles A_1 and A_2.

Moments of inertia: Applying the parallel-axis theorem, with reference to the x and y axes, we have

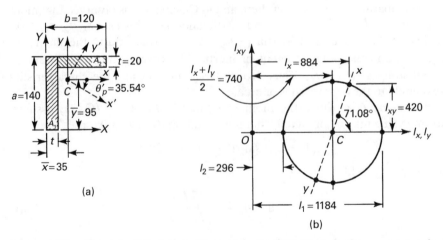

(a)

(b)

FIGURE C.7. *Example C.5. (a) Angle section; (b) Mohr's circle for moments of inertia.*

Appendix C   Moments of Composite Areas

$$I_x = \sum (I_{\bar{x}} + A d_y^2)$$

$$= \tfrac{1}{12}(20)(140)^3 + 20(140)(25)^2 + \tfrac{1}{12}(100)(20)^3 + 100(20)(35)^2$$

$$= 884(10^4) \text{ mm}^4$$

$$I_y = \sum (I_{\bar{y}} + A d_x^2)$$

$$= \tfrac{1}{12}(140)(20)^3 + 140(20)(25)^2 + \tfrac{1}{12}(20)(100)^3 + 20(100)(35)^2$$

$$= 596(10^4) \text{ mm}^4$$

The product of inertia about the xy axes is obtained as described in Section C.3:

$$I_{xy} = \sum (I_{\bar{x}\bar{y}} + A d_x d_y)$$

$$= 0 + 20(140)(-25)(-25) + 0 + 100(20)(35)(35)$$

$$= 420(10^4) \text{ mm}^4$$

Principal moments of inertia: Equation (C.13) yields

$$\tan 2\theta_p = -\frac{2(420)}{884 - 596} = -2.917, \qquad 2\theta_p = -71.08° \quad \text{or} \quad -251.08°$$

Thus, the two values of θ_p are $-35.54°$ and $-125.54°$. Using the first of these values, Eq. (C.12a) results in $I_{x'} = 1184(10^4)$ mm^4. The principal moments of inertia are, from Eq. (C.14),

$$I_{1,2} = \left[\frac{884 + 596}{2} \pm \sqrt{\left(\frac{884 - 596}{2} \right)^2 + 420^2} \right] 10^4 = [740 \pm 444] 10^4$$

or $I_1 = I_{x'} = 1184(10^4)$ mm^4 and $I_2 = I_{y'} = 296(10^4)$ mm^4. The principal axes are indicated in Fig. C.7a as the $x'y'$ axes.

The principal moments of inertia may also be determined readily by means of Mohr's circle, following a procedure similar to that described in Section 1.10, as shown in Fig. C.7b. Note that the quantities indicated are expressed in cm^4 and the results are obtained analytically from the geometry of the circle.

EXAMPLE C.6 General Formulation of Moments of Inertia for Angle Section

Derive general expressions for the centroidal moments and product of inertia for the angle section shown in Fig. C.7a. Write a computer program for the centroidal principal moments of inertia.

Solution Using the X, Y axes as reference (Fig. C.7a), Eqs. (C.3) yield

$$\bar{x} = \frac{a(t)(t/2) + t(b - t)(b + t)/2}{a(t) + (b - t)t} = \frac{b^2 + at - t^2}{2(a + b - t)} \qquad \textbf{(f)}$$

$$\bar{y} = \frac{a(t)(a/2) + (b-t)t(a-t/2)}{a(t) + (b-t)t} = \frac{a^2 + (b-t)(2a-t)}{2(a+b-t)} \qquad \text{(g)}$$

The transfer formula (C.9a) results in

$$I_x = \sum(I_{\bar{x}} + Ad_y^2)$$

$$= \tfrac{1}{12}ta^3 + at\left(\bar{y} - \frac{a}{2}\right)^2 + \tfrac{1}{12}(b-t)t^3 + (b-t)t\left[\bar{y} - \left(a - \frac{t}{2}\right)\right]^2$$

$$= \frac{t}{12}(a^3 + bt^2 - t^3) + (bt - t^2)\left[\bar{y} - \left(a - \frac{t}{2}\right)\right]^2 + at\left(\bar{y} - \frac{a}{2}\right)^2$$

$$\text{(h)}$$

Similarly,

$$I_y = \frac{t}{12}[(b-t)^3 + at^2] + (bt - t^2)[\bar{x} - \tfrac{1}{2}(b+t)]^2 + at\left(\bar{x} - \frac{t}{2}\right)^2$$

$$\text{(i)}$$

$$I_{xy} = (b-t)t[\bar{x} - \tfrac{1}{2}(b+t)]\left[\bar{y} - \left(a - \frac{t}{2}\right)\right] + at\left(\bar{x} - \frac{t}{2}\right)\left(\bar{y} - \frac{a}{2}\right)$$

$$\text{(j)}$$

For convenience in programming, we employ the notation

$$C_1 = \frac{t}{12}(a^3 + bt^2 - t^3), \qquad C_2 = bt - t^2$$

$$C_3 = \bar{y} - \left(a - \frac{t}{2}\right) \qquad\qquad C_4 = at$$

$$C_5 = \bar{y} - \frac{a}{2}, \qquad\qquad C_6 = \frac{t}{12}[(b-t)^3 + at^2]$$

$$C_7 = \bar{x} - \frac{t}{2}(b+t), \qquad C_8 = \bar{x} - \frac{t}{2}$$

and

$$D = \tfrac{1}{2}(I_x + I_y), \qquad E = \tfrac{1}{2}(I_x - I_y), \qquad F = -I_{xy}$$

Equations (h), (i), (j), (C.13), and (C.14) are thus

$$I_x = C_1 + C_2(C_3^2) + C_4(C_5^2)$$

$$I_y = C_6 + C_2(C_7^2) + C_4(C_8^2)$$

$$I_{xy} = C_2(C_3)C_7 + C_4(C_5)C_8$$

$$I_1 = D + (E^2 + F^2)^{1/2}$$

$$I_2 = D - (E^2 + F^2)^{1/2}$$

$$\theta P = \arctan\left(\frac{F}{E}\right)$$

TABLE C.2.   FORTRAN Program for Moments of Inertia

```
        REAL I1, I2, I, IX, IY, IXY, J, K, L, M, N
        PI = 3. 14159265
        WRITE(5, *) 'INPUT: '
        WRITE(3, *) 'INPUT: '
        READ(5, *)A, B, T
        WRITE(5, 10)A, B, T
        WRITE(3, 10)A, B, T
10      FORMAT( /, 11X, 'A = ', F10.4, 5X, 'B = ', F10.4, 5X, 'T = ', F10.5, //)
        WRITE(5, *) 'OUTPUT: '
        WRITE(3, *) 'OUTPUT: '
        XBAR = (B**2 − T**2 + T*A) /(2. *(A + B − T))
        YBAR = ((B − T)*(2. *A − T) + A**2) /(2. *(A + B − T))
        C1 = (T/12. )*(B*T**2 − T**3 + A**3)
        C2 = B*T − T**2
        C3 = YBAR − (A − T/2. )
        C4 = T*A
        C5 = YBAR − A /2.
        C6 = (T/12. )*((B − T)**3 + A*T**2)
        C7 = XBAR − (1/2. )*(B + T)
        C8 = XBAR − T/2.
        IX = C1 + C2*C3**2 + C4*C5**2
        IY = C6 + C2*C7**2 + C4*C8**2
        IXY = C2*C3*C7 + C4*C5*C8
        D = . 5*(IX + IY)
        E = . 5*(IX − IY)
        F = −IXY
        I1 = D + SQRT(E**2 + F**2)
        I2 = D − SQRT(E**2 + F**2)
        IF(IY. NE. IX) THETAP = (. 5*ATAN(F /E))*(180. /PI)
        IF(IY. EQ. IX) THETAP = 45.
        WRITE(5, 20)XBAR, YBAR, IX, IY, IXY, I1, I2, THETAP
        WRITE(3, 20)XBAR, YBAR, IX, IY, IXY, I1, I2, THETAP
20      FORMAT( /, 1X, 'XBAR = ', F18. 5, /, 1X, 'YBAR = ', F18. 5,
     +  /, 1X, 'IX = ', F18. 5, /, 1X, 'IY = ', F18. 5, /, 1X, 'IXY = ',
     $  F18. 5, /, 1X, 'I1 = ', F18. 5, /, 1X, 'I2 = ', F18. 5, /, 1X,
     +  'THETAP = ', F10. 5, /)
        STOP
        END

INPUT:

A =    140. 0000      B =    120. 0000      T =    20. 00000

OUTPUT:

XBAR =          35. 00000
YBAR =          95. 00000
IX =        8840000. 00000
IY =        5960000. 00000
IXY =       4200000. 00000
I1 =       11840000. 00000
I2 =        2960000. 00000
THETAP =   − 35. 53768
```

The required computer program, written in FORTRAN, is presented in Table C.2, along with input data (from Example C.5) and output values. The program was written and tested on a digital computer.

REFERENCE

C.1. YOUNG, W. C., *Roark's Formulas for Stress and Strain*, 6th ed., McGraw-Hill, New York, 1989, Chap. 5.

Tables and Charts

The properties of materials vary widely depending on numerous factors, including chemical composition, manufacturing processes, internal defects, temperature, and dimensions of test specimens. Hence, the approximate values presented in Table D.1 are not necessarily suitable for specific application. Tabulated data are for reference in solving problems in the text. For details, see, for example, Refs. D.1 and D.2.

The International System of Units (SI) replaces the U.S. Customary Units, which have long been used by engineers in this country. The basic quantities in the two systems are as follows:

Basic Units

| | SI Unit | | U.S. Unit | |
Quantity	Name	Symbol	Name	Symbol
Length	meter	m	foot	ft
Force*	newton	N*	pound force	lb
Time	second	s	second	s
Mass	kilogram	kg	slug	$lb \cdot s^2/ft$
Temperature	degree Celsius	C	degree Fahrenheit	F

*Dervied unit ($kg\ m/s^2$).

In the SI system, the acceleration due to gravity near Earth's surface equals approximately $9.81\ m/s^2$. A mass of 1 kilogram on Earth's surface will experience a gravitational force of 9.81 N. Thus, a mass of 1 kg has, owing to the gravitational force of Earth, a *weight* of 9.81 N. Interestingly, one *newton* is approximately the weight of (or Earth's gravitional force on) an average apple. Tables D.2 and D.3 contain conversion factors and SI unit prefixes in common usage.

TABLE D.1. Average Properties of Common Engineering Materials[a] (SI Units)

Material	Density, Mg/m³	Ultimate Strength, MPa			Yield Strength[c] MPa		Modulus of Elasticity, GPa	Modulus of Rigidity, GPa	Coefficient of Thermal Expansion, 10⁻⁶/°C	Ductility, Percent Elongation in 50 mm
		Tension	Compression[b]	Shear	Tension	Shear				
Steel										
Structural, ASTM-A36	7.86	400	—	—	250	145	200	79	11.7	30
High strength, ASTM-A242	7.86	480	—	—	345	210	200	79	11.7	21
Stainless (302), cold rolled	7.92	860	—	—	520	—	190	73	17.3	12
Cast iron										
Gray, ASTM A-48	7.2	170	650	240	—	—	70	28	12.1	0.5
Malleable, ASTM A-47	7.3	340	620	330	230	—	165	64	12.1	10
Wrought iron	7.7	350	—	240	210	130	190	70	12.1	35
Aluminum										
Alloy 2014-T6	2.8	480	—	290	410	220	72	28	23	13
Alloy 6061-T6	2.71	300	—	185	260	140	70	26	23.6	17
Brass, yellow										
Cold rolled	8.47	540	—	300	435	250	105	39	20	8
Annealed	8.47	330	—	220	105	65	105	39	20	60
Bronze, cold rolled (510)	8.86	560	—	—	520	275	110	41	17.8	10
Copper, hard drawn	8.86	380	—	—	260	160	120	40	16.8	4
Magnesium alloys	1.8	140–340	—	165	80–280	—	45	17	27	2–20
Concrete										
Medium strength	2.32	—	28	—	—	—	24	—	10	—
High strength	2.32	—	40	—	—	—	30	—	10	—
Timber[d] (air dry)										
Douglas fir	0.54	—	55	7.6	—	—	12	—	4	—
Southern pine	0.58	—	60	10	—	—	11	—	4	—
Glass, 98% silica	2.19	—	50	—	—	—	65	28	80	—
Nylon, molded	1.1	55	—	—	—	—	2	—	81	50
Polystyrene	1.05	48	90	55	—	—	3	—	72	4
Rubber	0.91	14	—	—	—	—	—	—	162	600

[a] Properties may vary widely with changes in composition, heat treatment, and method of manufacture.
[b] For ductile metals the compression strength is assumed to be the same as that in tension.
[c] Offset of 0.2%.
[d] Loaded parallel to the grain.

TABLE D.1. *Average Properties of Common Engineering Materials[a] (U.S. Customary Units)*

Material	Specific Weight, lb/in.³	Ultimate Strength, ksi			Yield Strength,[c] ksi		Modulus of Elasticity, 10⁶ psi	Modulus of Rigidity, 10⁶ psi	Coefficient of Thermal Expansion, 10⁻⁶/°F	Ductility, Percent Elongation in 2 in.
		Tension	Compression[b]	Shear	Tension	Shear				
Steel										
Structural, ASTM-A36	0.284	58	—	—	36	21	29	11.5	6.5	30
High strength, ASTM-A242	0.284	70	—	—	50	30	29	11.5	6.5	21
Stainless (302), cold rolled	0.286	125	—	—	75	—	28	10.6	9.6	12
Cast iron										
Gray, ASTM A-48	0.260	25	95	35	—	—	10	4.1	6.7	0.5
Malleable, ASTM A-47	0.264	50	90	48	33	—	24	9.3	6.7	10
Wrought iron	0.278	50	—	35	30	18	27	10	6.7	35
Aluminum										
Alloy 2014-T6	0.101	70	—	42	60	32	10.6	4.1	12.8	13
Alloy 6061-T6	0.098	43	—	27	38	20	10.0	3.8	13.1	17
Brass, yellow										
Cold rolled	0.306	78	—	43	63	36	15	5.6	11.3	8
Annealed	0.306	48	—	32	15	9	15	5.6	11.3	60
Bronze, cold rolled (510)	0.320	81	—	—	75	40	16	5.9	9.9	10
Magnesium alloys	0.065	20–49	—	24	11–40	—	6.5	2.4	15	2–20
Copper, hard drawn	0.320	55	—	—	38	23	17	6	9.3	4
Concrete										
Medium strength	0.084	—	4	—	—	—	3.5	—	5.5	—
High strength	0.084	—	6	—	—	—	4.3	—	5.5	—
Timber[d] (air dry)										
Douglas fir	0.020	—	7.9	1.1	—	—	1.7	—	2.2	—
Southern pine	0.021	—	8.6	1.4	—	—	1.6	—	2.2	—
Glass, 98% silica	0.079	—	7	—	—	—	9.6	4.1	44	—
Nylon, molded	0.040	8	—	—	—	—	0.3	—	45	50
Polystyrene	0.038	7	13	8	—	—	0.45	—	40	4
Rubber	0.033	2	—	—	—	—	—	—	90	600

[a] Properties may vary widely with changes in composition, heat treatment, and method of manufacture.
[b] For ductile metals the compression strength is assumed to be the same as that in tension.
[c] Offset of 0.2%.
[d] Loaded parallel to the grain.

TABLE D.2.   *Conversion Factors: SI Units to U.S. Customary Units*

Quantity	SI Unit	U.S. Equivalent
Acceleration	m/s^2 (meter per square meter)	3.2808 ft/s^2
Area	m^2 (square meter)	10.76 ft^2
Force	N (newton)	0.2248 lb
Intensity of force	N/m (newton per meter)	0.0685 lb/ft
Length	m (meter)	3.2808 ft
Mass	kg (kilogram)	2.2051 lb
Moment of a force	$N \cdot m$ (newton meter)	$0.7376 \text{ ft} \cdot \text{lb}$
Moment of inertia of a plane area	m^4 (meter to fourth power)	$2.4025 \times 10^6 \text{ in.}^4$
Power	W (watt)	$0.7376 \text{ ft} \cdot \text{lb/s}$
	kW (kilowatt)	1.3410 hp
Pressure or stress	Pa (pascal)	$0.145 \times 10^{-3} \text{ psi}$
Specific weight	kN/m^3 (kilonewton per cubic meter)	$3.684 \times 10^{-3} \text{ lb/in.}^3$
Velocity	m/s (meter per second)	3.2808 ft/s
Volume	m^3 (cubic meter)	35.3147 ft^3
Work or energy	J (joule, newton meter)	$0.7376 \text{ ft} \cdot \text{lb}$

TABLE D.3.   *SI Unit Prefixes*

Prefix	Symbol	Factor				
tera	T	$10^{12} =$	1 000	000 000	000	
giga	G	$10^9 =$		1 000 000	000	
mega	M	$10^6 =$		1 000	000	
kilo	k	$10^3 =$			1 000	
hecto	h	$10^2 =$			100	
deka	da	$10^1 =$			10	
deci	d	$10^{-1} =$			0.1	
centi	c	$10^{-2} =$			0.01	
milli	m	$10^{-3} =$			0.001	
micro	μ	$10^{-6} =$			0.000 001	
nano	n	$10^{-9} =$			0.000 000 001	
pico	p	$10^{-12} =$			0.000 000 000 001	

Note: The use of the prefixes hecto, deka, and centi is not recommended. However, they are sometimes encountered in practice.

The expressions for deflection and slope for selected members given in Tables D.4 and D.5 are representative of results found in a number of handbooks [Ref. D.3]. Restrictions on the application of these formulas include constancy of the flexural rigidity EI, symmetry of the cross section about the vertical y axis, and the magnitude of displacement v of the beam. In addition, the expressions apply to beams long in proportion to their depth and not disproportionally wide (see Secs. 5.4 and 5.6). The stress concentration factors K (Figures D.1 through D.8) were selected from extensive charts found in Refs. D.4 and D.5.

TABLE D.4. Deflections and Slopes of Beams

Loading	Deflection	Slope at
1.	$v = \dfrac{Px^2}{6EI}(3L - x)$ $v_{\max} = v_B = \dfrac{PL^3}{3EI}$	$\theta_B = \dfrac{PL^2}{2EI}$
2.	$v = \dfrac{Mx^3}{2EI}$ $v_{\max} = v_B = \dfrac{ML^2}{2EI}$	$\theta_B = \dfrac{ML}{EI}$
3.	$v = \dfrac{px^2}{24EI}(6L^2 - 4Lx + x^2)$ $v_{\max} = v_B = \dfrac{pL^4}{8EI}$	$\theta_B = \dfrac{pL^3}{6EI}$
4.	For $x \leq L/2$ $v = \dfrac{Px}{48EI}(3L^2 - 4x^2)$ $v_{\max} = v\left(\dfrac{L}{2}\right) = \dfrac{PL^3}{48EI}$	$\theta_A = -\theta_B = \dfrac{PL^2}{16EI}$
5.	$v = \dfrac{Mx}{6EIL}(L^2 - x^2)$ $v_{\max} = v\left(\dfrac{L}{\sqrt{3}}\right) = -\dfrac{ML^2}{9\sqrt{3}EI}$	$\theta_A = \dfrac{ML}{8EI}$ $\theta_B = -\dfrac{ML}{3EI}$
6.	$v = \dfrac{px}{24EI}(L^3 - 2Lx^2 + x^3)$ $v_{\max} = v\left(\dfrac{L}{2}\right) = \dfrac{5pL^4}{384EI}$	$\theta_A = -\theta_B = \dfrac{pL^3}{24EI}$

Load and Support	Reactions*	Deflection at Center

1.

$$R_A = R_B = \frac{P}{2}$$

$$M_A = M_B = \frac{PL}{8}$$

$$v_{max} = v_C = -\frac{PL^3}{192}$$

2.

$$R_A = \frac{Pb^2}{L^3}(3a + b) \quad R_B = \frac{Pa^2}{L^3}(a + 3b)$$

For $a > b$:

$$M_A = \frac{Pab^2}{L^2} \qquad M_B = \frac{Pa^2b}{L^2}$$

$$v_C = -\frac{Pb^2}{48EI}(3L - 4b)$$

3.

$$R_A = \frac{5}{16}P \qquad R_B = \frac{11}{16}P$$

$$M_B = \frac{3}{16}PL$$

$$v_C = -\frac{7PL^3}{768EI}$$

4.

$$R_A = R_B = \frac{pL}{2}$$

$$M_A = M_B = \frac{pL^2}{12}$$

$$v_{max} = v_C = -\frac{pL^4}{384EI}$$

5.

$$R_A = \frac{3}{32}pL \qquad R_B = \frac{13}{32}pL$$

$$M_A = \frac{5}{192}pL^2 \quad M_B = \frac{11}{192}pL^2$$

$$v_C = -\frac{pL^4}{768EI}$$

6.

$$R_A = \frac{3}{8}pL \qquad R_B = \frac{5}{8}pL$$

$$M_B = \frac{1}{8}pL^2$$

$$v_C = -\frac{pL^4}{192EI}$$

*For all the cases tabulated, the senses of the reactions and the notations are the same as those shown in case 1.
Note: The equivalent forces defined in Section 7.11 (F_{Ay}, F_{By}, M_A, M_B) have the *opposite sense* of the fixed-end reactions (cases 1, 2, 4, 5).

D.6. Stress Concentration Factors for Bars and Shafts with Fillets, Grooves, and Holes

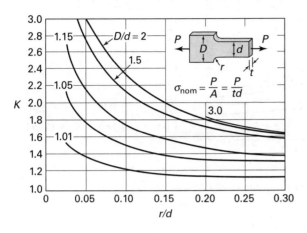

Approximate formula

$$K \approx B \left(\frac{r}{d} \right)^a, \text{ where:}$$

D/d	B	a
2.00	1.100	−0.321
1.50	1.077	−0.296
1.15	1.014	−0.239
1.05	0.998	−0.138
1.01	0.977	−0.107

FIGURE D.1. *Stress concentration factor K for a filleted bar in axial tension.*

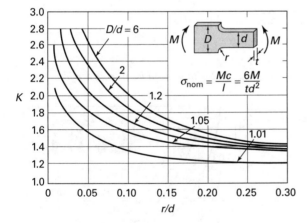

Approximate formula

$$K \approx B \left(\frac{r}{d} \right)^a, \text{ where:}$$

D/d	B	a
6.00	0.896	−0.358
2.00	0.932	−0.303
1.20	0.996	−0.238
1.05	1.023	−0.192
1.01	0.967	−0.154

FIGURE D.2. *Stress concentration factor K for a filleted bar in bending.*

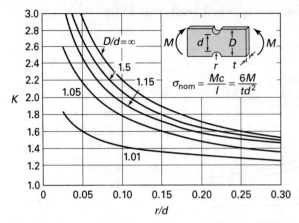

Approximate formula

$$K \approx B \left(\frac{r}{d} \right)^a, \text{ where:}$$

D/d	B	a
∞	0.971	−0.357
1.50	0.983	−0.334
1.15	0.993	−0.303
1.05	1.025	−0.240
1.01	1.061	−0.134

FIGURE D.3.   *Stress concentration factor K for a notched bar in bending.*

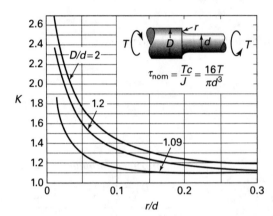

Approximate formula

$$K \approx B \left(\frac{r}{d} \right)^a, \text{ where:}$$

D/d	B	a
2.00	0.863	−0.239
1.20	0.833	−0.216
1.09	0.903	−0.127

FIGURE D.4.   *Stress concentration factor K for a shaft with a shoulder fillet in torsion.*

FIGURE D.5.   *Stress concentration factor K for a grooved shaft in axial tension.*

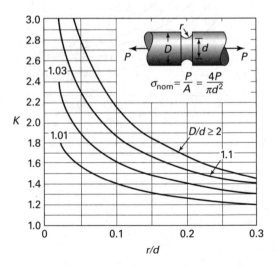

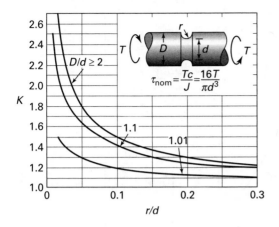

D/d	B	a
2.00	0.890	−0.241
1.10	0.923	−0.197
1.01	0.972	−0.102

Approximate formula

$$K \approx B \left(\frac{r}{d}\right)^a, \text{ where:}$$

FIGURE D.6. *Stress concentration factor K for a grooved shaft in torsion.*

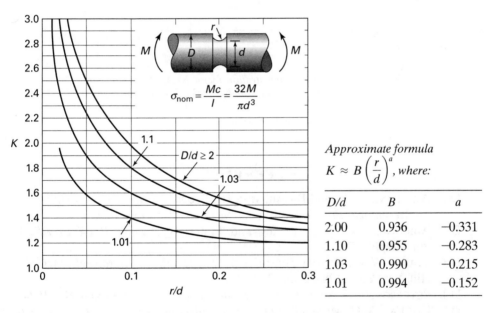

D/d	B	a
2.00	0.936	−0.331
1.10	0.955	−0.283
1.03	0.990	−0.215
1.01	0.994	−0.152

Approximate formula

$$K \approx B \left(\frac{r}{d}\right)^a, \text{ where:}$$

FIGURE D.7. *Stress concentration factor K for a grooved shaft in bending.*

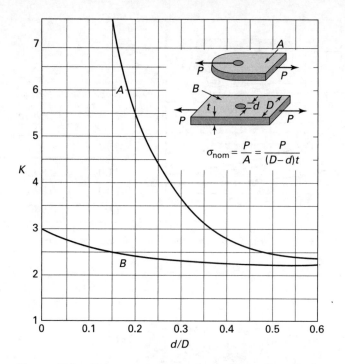

FIGURE D.8.   *Stress concentration factor K; A—for a flat bar loaded in tension by a pin through the transverse hole; B—for a flat bar with a transverse hole in axial tension.*

REFERENCES

D.1. AVALLONE, E. A., and BAUMEISTER III, T., eds., *Mark's Standard Handbook for Mechanical Engineers*, 10th ed., McGraw-Hill, New York, 1997.

D.2. American Society of Testing and Materials, *Annual Book of ASTM*, Philadelphia, Pa.

D.3. YOUNG, W. C., *Roark's Formulas for Stress and Strain*, 6th ed., McGraw-Hill, New York, 1989, Chap. 7.

D.4. PETERSON, R. E., *Stress Concentration Factors*, Wiley, Hoboken, N.J., 1974.

D.5. UGURAL, A. C., *Mechanical Design: An Integrated Approach*, McGraw-Hill, New York, 2004.

Answers to Selected Problems

CHAPTER 1 ANALYSIS OF STRESS

1.1. $P = 4.27$ kN

1.4. $P = 32.5$ kN, $\theta = 26.56°$

1.6. $P_{all} = 38.3$ kN

1.8. $\sigma_{x'} = 156$ MPa, $\tau_{x'y'} = -131$ MPa

1.10. $\sigma_{x'} = -13.15$ MPa, $\tau_{x'y'} = -15.67$ MPa

1.11. $F = 67.32$ kN/m³

1.14. $F_x = F_y = F_z = 0$

1.21. (a) $\sigma = 153.3$ MPa, (b) $\sigma = 23.1$ MPa

1.26. (a) $\sigma_1 = 121$ MPa, $\sigma = -71$ MPa, $\tau_{max} = 96$ MPa

     (b) $\sigma_1 = 200$ MPa, $\sigma_2 = -50$ MPa, $\tau_{max} = 125$ MPa

     $\theta'_p = 26.55°$, $\theta'_s = 71.55°$

1.31. (a) $\sigma_x = 38.68$ MPa, $\sigma_y = 12.12$ MPa, $\tau_{xy} = 7$ MPa

     (b) $\sigma_1 = 40.41$ MPa, $\sigma_2 = 10.39$ MPa, $\theta'_p = 13.9°$

1.34. $P = 3\pi pr^2$

1.37. $\tau_{max} = 48$ MPa, $\theta_p = 29°$

1.38. $\sigma_w = 61.5$ MPa, $\tau_w = 26.3$ MPa

1.42. (a) $\sigma = -0.237\tau_o$, $\tau = 0.347\tau_o$

     (b) $\sigma_1 = 3.732\tau_o$, $\sigma_2 = -0.268\tau_o$, $\theta'_p = 15°$

1.44. $\sigma_1 = -\sigma_2 = 51.96$ MPa, $\theta'_p = -30°$

1.46. $\sigma_1 = 46.17$ MPa, $\sigma_2 = 13.83$ MPa, $\theta'_p = -31.2°$

1.49. (a) $\sigma_x = 100$ MPa, $\tau_{xy} = -30$ MPa

(b) $\sigma_1 = 110$ MPa, $\theta'_p = -18.43°$

1.53. $P = 1069$ kN, $p = 467$ kPa

1.55. (a) $\sigma_x = 186$ MPa,   (b) $\sigma_1 = 188$ MPa, $\tau_{max} = 101$ MPa, $\theta'_p = 5.71°$

1.58. $p = 494$ kPa

1.62. $\sigma_1 = 66.06$ MPa, $l_1 = 0.956$, $m_1 = 0.169$, $n_1 = 0.242$

1.64. $\sigma_1 = 24.747$ MPa, $\sigma_2 = 8.48$ MPa, $\sigma_3 = 2.773$ MPa, $l_1 = 0.647$, $m_1 = 0.396$, $n_1 = 0.652$

1.69. (a) $\sigma_1 = 12.049$ MPa, $\sigma_2 = -1.521$ MPa, $\sigma_3 = -4.528$ MPa, $l_1 = 0.618$, $m_1 = 0.533$, $n_1 = 0.577$

1.73. $\sigma = 52.25$ MPa, $\tau = 36.56$ MPa

1.75. (a) $\tau_{13} = 8.288$ MPa, $\tau_{12} = 6.785$ MPa, $\tau_{23} = 1.503$ MPa

1.79. $\sigma_{oct} = 12$ MPa, $\tau_{oct} = 9.31$ MPa

1.82. (a) $\sigma_1 = 108.3$ MPa, $\sigma_2 = 51.7$ MPa, $\sigma_3 = -50$ MPa, $\theta'_p = 22.5°$

1.85. $\sigma = -12.39$ MPa, $\tau = 26.2$ MPa, $p_x = 16.81$ MPa, $p_y = -3.88$ MPa, $p_z = -23.30$ MPa

CHAPTER 2 STRAIN AND MATERIAL PROPERTIES

2.4. $c_0 = 2a_0, c_1 = 2(a_1 + b_0)$

2.9. (a) $\varepsilon_x = 667\,\mu$, $\varepsilon_y = 750\,\mu$, $\gamma_{xy} = -1250\,\mu$

(b) $\varepsilon_1 = 1335\,\mu$, $\varepsilon_2 = 82\,\mu$, $\theta''_p = 43.1°$

2.11. $\varepsilon_1 = 758\,\mu$, $\varepsilon_2 = -808\,\mu$, $\theta'_p = -36.65°$

2.13. $\Delta_{QB} = 0.008$ mm, $\Delta_{AC} = 0.016$ mm

2.15. (a) $\gamma_{max} = 200\,\mu$, $\theta_s = 45°$

(b) $\varepsilon_x = 350\,\mu$, $\varepsilon_y = 250\,\mu$, $\gamma_{xy} = -173\,\mu$

2.17. $\varepsilon_1 = -59\,\mu$, $\varepsilon_2 = -1141\,\mu$, $\theta'_p = 61.85°$

2.21. (a) $J_1 = -3 \times 10^{-4}, J_2 = -44 \times 10^{-8}, J_3 = 58 \times 10^{-12}$

(b) $\varepsilon_{x'} = 385\,\mu$

(c) $\varepsilon_1 = 598\,\mu$, $\varepsilon_2 = -126\,\mu$, $\varepsilon_3 = -772\,\mu$

(d) $\gamma_{max} = 1370\,\mu$

2.28. (a) $\Delta L = 5.98$ mm,   (b) $\Delta d = -0.014$ mm

2.33. (a) $\varepsilon_1 = 1222\,\mu$, $\varepsilon_2 = 58\,\mu$

(b) $\gamma_{max} = 1164\,\mu$, (c) $(\gamma_{max})_t = 1222\,\mu$

2.36. $\Delta_{BD} = 0.283/E$ m

2.39. $\varepsilon_x = 522\,\mu, \varepsilon_y = 678\,\mu, \gamma_{xy} = -1873\,\mu$

2.41. $\sigma_x = 72$ MPa, $\sigma_y = 88$ MPa, $\sigma_z = 40$ MPa, $\tau_{xy} = 16$ MPa, $\tau_{yz} = 64$ MPa, $\tau_{xz} = 0$

2.46. (b) $\Delta V = -2250$ mm^3

2.48. $\sigma_3:\sigma_2:\sigma_1 = 1:1.086:1.171$, $\sigma_1 = 139.947$ MPa, $\sigma_2 = 129.757$ MPa, $\sigma_3 = 119.513$ MPa

2.52. $\varepsilon_x = \gamma(a - x)/E, \varepsilon_y = -\nu\varepsilon_x, \sigma_x = \gamma(a - x), \gamma_{xy} = \tau_{xy} = \sigma_y = 0.$ Yes.

2.55. $U_1 = P^2L/2EA, U_2 = 5U_1/8, U_3 = 5U_1/12$

2.61. (a) $U = 60.981T^2a/\pi d^4G$, (b) $U = 2.831$ kN $\cdot$ m

2.67. $U_{ov} = 3.258$ kPa, $U_{od} = 38.01$ kPa

2.69. $U_v = \dfrac{1}{12}\dfrac{P^2L}{\pi r^2E}, \qquad U_d = \dfrac{5}{2}\dfrac{T^2L}{\pi r^4E} + \dfrac{5}{12}\dfrac{P^2L}{\pi r^2E}$

CHAPTER 3 PROBLEMS IN ELASTICITY

3.1. (b) $\Phi = \dfrac{px^3y^3}{6} - \dfrac{px^5y}{10} + (c_4y^3 + c_5y^2 + c_6y + c_7)x$
$$+ c_8y^3 + c_9y^2 + c_{10}y + c_{11}$$

3.2. $c_1 = \frac{3}{2}pa^2, c_2 = 2pa^3, c_3 = -\frac{1}{2}pa^4$

3.14. All conditions, except on edge $x = L$, are satisfied.

3.16. $\sigma_x = \dfrac{p}{a^2}(x^2 - 2y^2), \qquad \sigma_y = \dfrac{py^2}{a^2}, \qquad \tau_{xy} = -\dfrac{2pxy}{a^2}$

3.19. Yes, $\varepsilon_1 = 32.8\,\mu, \theta'_p = 5.65°$

3.20. $\sigma_x = \sigma_y = E\alpha T_1/(\nu - 1), \varepsilon_z = 2\nu\alpha T_1/(1 - \nu) + \alpha T_1$

3.24. $P_x = -161.3$ kN

3.31. (a) $(\sigma_x)_{\text{elast.}} = P/0.512L, (\sigma_x)_{\text{elem.}} = P/0.536L$
   (b) $(\sigma_x)_{\text{elast.}} = P/1.48L, (\sigma_x)_{\text{elem.}} = P/3.464L$

3.33. (a) $(\sigma)_{\text{elast.}} = 19.43F/L, (\sigma_x)_{\text{elem.}} = 20.89F/L$
   $(\tau_{xy})_{\text{elast.}} = 5.21F/L, (\tau_{xy})_{\text{elem.}} = 2.8F/L$

3.38. $r = 6.67$ mm, $d = 26.7$ mm

3.42. $T = 153.5$ N $\cdot$ m

3.47. (a) $\sigma_1 = 36.5$ MPa,   (b) $\tau_{\max} = 16.96$ MPa,
   (c) $\sigma_{\text{oct}} = 11.3$ MPa, $\tau_{\text{oct}} = 17.85$ MPa

3.49. (a) $\sigma_c = 1233$ MPa,   (b) $\sigma_c = 1959$ MPa

3.51. $\sigma_c = 418$ MPa, $b = 0.038$ mm

3.53. $a = 2.994$ mm, $b = 1.605$ mm, $\sigma_c = 505.2$ MPa

3.58. $\sigma_c = 1014.7$ MPa

CHAPTER 4 FAILURE CRITERIA

4.1. $P_{all} = 707$ kN

4.3. (a) Yes, (b) No

4.6. (a) $\sigma_{yp} = 152.6$ MPa, (b) $\sigma_{yp} = 134.9$ MPa

4.8. $t = 8.45$ mm

4.9. (a) $d = 27.95$ mm, (b) $d = 36.8$ mm

4.11. (a) $T = 31.74$ kN·m, (b) $T = 26.05$ kN·m

4.14. (a) $R = 932$ N, (b) $R = 959$ N

4.16. (a) $p = 6.466$ MPa, (b) $p = 5.6$ MPa

4.22. (b) $\sigma_1 = 75$ MPa, $\sigma_2 = -300$ MPa

4.25. (a) $\tau = \sigma_u/2$, (b) $\theta'_p = -26.57°$

4.30. $t = 9.27$ mm

4.36. $p = 11.11$ MPa

4.39. $t = 0.973$ mm

4.43. $P_{max} = 18.7$ kN

4.46. $\tau_{max} = 274.3$ MPa, $\phi_{max} = 4.76°$

CHAPTER 5 BENDING OF BEAMS

5.4. $M_o = 266.8$ N·m

5.6. (a) $P = 3.6$ kN, (b) $P = 3.36$ kN

5.9. (a) $\phi = -12.17°$, (b) $\sigma_A = 136.5$ MPa

5.10. (b) $\sigma_x = px^3/Lth^2$, (c) $(\sigma_x)_{elast.} = 0.998(\sigma_x)_{elem.}$

5.13. $p = 3.88$ kN/m

5.14. $P = 9320$ N

5.16. (a) $\sigma_A = 35.3$ MPa, (b) $\sigma_B = 23.5$ MPa
(b) $r_Z = -205.48$ m

5.19. (a) $\tau_{max}/\sigma_{max} = h/L$, (b) $p_{all} = 10.34$ kN/m

5.20. $P = 15.6$ kN

5.24. $e = 4R/\pi$

5.28. $R = -13pL/32$

5.30. $v = M_o x^2(x - L)/4EIL$, $R_B = 3M_o/2L$

5.34. $P = 1.614$ kN

5.35. (a) $\sigma_{max} = -177$ MPa, (b) $\sigma_{max} = -176.2$ MPa

5.45. (a) $\sigma_\theta = -182.3P$, (b) $\delta_p = 215.65P/E$ m

CHAPTER 6 TORSION OF PRISMATIC BARS

6.2. $D = 60.9$ mm

6.6. $(\tau_{max})_B = 130.4$ MPa

6.13. $T = 1.571$ kN $\cdot$ m

6.14. (a) $\tau_e > \tau_c$;   (b) $T_e > T_c$

6.16. $T = 256.5$ kN $\cdot$ m

6.18. $k = G\theta/2a^2(b-1)$

6.21. $\theta_A = aT/2r^4G, \theta_B = 2\theta_A$

6.23. $\tau_{max} = 15\sqrt{3}\, T/2h^3, \tau_{min} = 0, \theta = 15\sqrt{3}\, T/Gh^4$

6.26. $\tau_{max} = 76.8$ MPa, $\theta = 0.192$ rad/m

6.32. (a) $C = 2.1 \times 10^{-7}G, \tau_{max} = 112,860T$

6.33. $\theta = 0.1617$ rad/m

6.35. $\theta = 2T/9Ga^3t$

6.37. $\tau_{max} = 5.279$ MPa, $\theta = 0.0131$ rad/m

6.39. $\tau_2 = \tau_4 = \tau_{max} = 50.88$ MPa, $\theta = 0.01914$ rad/m

CHAPTER 7 NUMERICAL METHODS

7.3. $\tau_B = 0.0107G\theta$

7.6. $v(L) = 7PL^3/32EI$

7.10. $v_{max} = 0.01852pL^3/EI$

7.12. $v_B = -Pa^3/16EI, \theta_A = -Pa^2/16EI$

7.14. $v_{max} = 1.68182ph^4/EI, \theta_{max} = -0.02131pL^3/EI$

7.15. $R_A = R_B = P/2, M = PL/8$

7.18. (b) $u_2 = 2PL/11AE, u_3 = -3PL/11AE$
    (c) $R_1 = 2P/11, \ R_4 = 9P/11$

7.19. (b) $u_2 = PL/9AE, \ u_3 = PL/18AE$
    (c) $R_1 = 2P/3, \ R_4 = P/3$

7.24. (c) $u_2 = 0.9$ mm, $v_2 = -3.02$ mm
    (d) $R_{1x} = 4.5$ kN, $R_{1y} = 6$ kN, $R_{3y} = -4.5$ kN
    (e) $F_{12} = -7.5$ kN, $F_{23} = 36$ kN

7.27. $v_2 = -5pL^4/384EI$

7.29. $v_2 = -5pL^3/48EI, \ \theta_2 = PL^2/8EI$

7.32. (c) $v_3 = -7PL^3/12EI, \ \theta_3 = -3PL^2/4EI, \ \theta_2 = -PL^2/4EI$
    (d) $R_1 = -3P/2, \ R_2 = 5P/2, \ M_1 = PL/2, \ M_2 = -PL$

7.35. (c) $u_2 = 1.0$ mm, $v_2 = -3.9$ mm

(d) $R_{1x} = 7.63$ kN, $R_{1y} = 10.18$ kN, $R_{3x} = -7.5$ kN, $R_{3y} = 0$

7.42. $\{\sigma_x, \sigma_y, \tau_{xy}\}_a = \{66.46, 6.65, -92.12\}$ MPa

CHAPTER 8 AXISYMMETRICALLY LOADED MEMBERS

8.2. $p_i = 51.2$ MPa

8.3. (a) $r_x = \left[\dfrac{2n^2a^2\sigma_\theta + \Delta\sigma_\theta(n^2 + 1)n^2a^2}{\Delta\sigma_\theta(n^2 + 1) + 2\sigma_\theta n^2}\right]^{1/2}$, (b) $r_x = 27.12$ mm

8.5. (a) $p_i = 1.6p_o$, (b) $p_i = 1.16p_o$

8.7. $t = 0.59$ m

8.11. (a) $t = 0.825d_i$, (b) $\Delta d = 0.0074$ mm

8.13. (a) $\sigma_z = 2\nu\dfrac{a^2p_i - b^2p_o}{b^2 - a^2}$, (b) $\varepsilon_z = -\dfrac{2\nu}{E}\dfrac{a^2p_i - b^2p_o}{b^2 - a^2}$

8.16. $T = 5.073$ kN $\cdot$ m

8.21. $\Delta d_s = 0.23\delta_0$ m

8.23. $\sigma_{\theta,\text{max}} = 1.95E_bE_s(T_2 - T_1)/(E_s + 3E_b)10^5$

8.26. (a) $p_i = 155.6$ MPa, (b) $p_i = 179.3$ MPa

8.30. *Shaft*: $\sigma_\theta = \sigma_r = -49.4$ MPa

Cylinder: $\sigma_{\theta,\text{max}} = 63.8$ MPa, $\sigma_{r,\text{max}} = 49.4$ MPa

8.32. $T = 101.79$ N $\cdot$ m

8.36. $\omega = 36{,}726$ rpm

8.40. (a) $\sigma_{\theta,\text{max}} = 108.68$ MPa, (b) $\omega = 4300$ rpm

8.44. (a) $\sigma_{\theta,\text{max}} = 554.6$ MPa

CHAPTER 9 BEAMS ON ELASTIC FOUNDATIONS

9.1. $P = 51.18$ kN

9.3. $v = \dfrac{p_1}{k[1 + 4(\pi/\beta L)^4]}\sin\dfrac{2\pi x}{L}$

9.5. $v_{\text{max}} = 3.375$ mm, $\sigma_{\text{max}} = 103.02$ MPa

9.6. $P = 41.7$ kN

9.8. $\sigma_{\text{max}} = 196.3$ kN

9.11. $v_Q = \dfrac{p_o}{4\beta kL}[f_3(\beta a) - f_3(\beta b) - 2\beta Lf_4(\beta b) + 4\beta a]$

9.14. $v = -M_Lf_2(\beta x)/2\beta^2EI$

9.16. (a) $v = 10$ mm; (b) $v_L = 15$ mm, $v_R = 5$ mm

9.17. $v_C = 0.186$ mm

9.18. $v_C = 2.81 \times 10^{-8}$ m, $\theta_E = 5 \times 10^{-7}$ rad

CHAPTER 10 APPLICATIONS OF ENERGY METHODS

10.3. $v_p = (11Pc_1a^4/12E) + (7Pc_2a^3/3E) + (Pa^3/3EI_2)$

10.6. $\delta_A = 3PL^3/16EI, \theta_A = 5PL^2/16EI$

10.9. $\delta_D = Fab^2/8EI$

10.11. $\delta = \dfrac{PR^3}{r^4}\left(\dfrac{1}{E} + \dfrac{0.226}{G}\right)$

10.14. $\delta_D = 8.657WL/AE$

10.17. $v_c = pL^4/96EI$

10.18. $N_{BD} = 0.826P, \ N_{AB} = -0.131P, \ N_{BC} = 0.22P$

10.23. $R_A = -3M_o/2L, \ R_B = 3M_o/2L, \ M_A = M_o/2$

10.27. $R_A = 4p_oL/10, M_A = p_oL^2/15, R_B = p_oL/10$

10.29. $R_{Av} = 3(\lambda + 1)pL_1/2(3\lambda + 4), R_{Ah} = \lambda pL_1^2/4(3\lambda + 4)L_2, \lambda = E_2I_2L_1/E_1I_1L_2$

10.32. $\delta_E = 41.5(P/AK)^3$

10.37. $N_A = P/2\pi, M_A = PR/4$

10.43. $v = Px^2(3L - x)/6EI$

10.44. $v = Pc^2(L - c)^2/4EIL$

CHAPTER 11 STABILITY OF COLUMNS

11.1. $F = 357.5$ kN

11.3. $F_{all} = 12.93$ kN

11.5. $\sigma_{cr} = 183.2$ MPa

11.9. $L = 4.625$ m

11.12. (a) 6.25%, (b) $P_{all} = 17.99$ kN

11.15. (a) $\sigma_{cr} = 34$ MPa, (b) $\sigma_{all} = 35.25$ MPa

11.17. $L_e = L$

11.19. (a) $\Delta T = \delta/2\alpha L$, (b) $\Delta T = (\delta/2\alpha L) + (\pi^2I/4L^2A\alpha)$

11.21. Bar BC fails as a column; $P_{cr} = 1297$ N

11.23. $\sigma_{cr} = 59.46$ MPa

11.31. (a) $\sigma_{max} = 93.71$ MPa, $v_{max} = 150.6$ mm

11.33. $P = 0.89\pi^2EI/L^2$

11.35. $P_{cr} = 12EI/L^2$

11.37. $P_{cr} = 9EI_1/4L^2$

11.38. $v = \dfrac{4pL^4}{\pi^5IE} \displaystyle\sum_{n=1,3,5,\dots}^{\infty} \dfrac{1}{n^3(n^2 - b)} \sin\dfrac{n\pi x}{L}$

11.43. $P_{cr} = 16EI_1/L^2$

CHAPTER 12 PLASTIC BEHAVIOR OF MATERIALS

12.2. $\alpha = 42.68°$

12.5. $\sigma_{max} = 3Mh/4I$

12.7. (a) $P_{max} = 471$ kN, (b) $(\delta_{AB})_p = 0$, $(\delta_{BC})_p = 7.66$ mm

12.9. $P = 189.8$ kN

12.13. $M = 11ah^2\sigma_{yp}/54$

12.15. (a) $M_{yp} = 15.36$ kN $\cdot$ m, $M = 21.12$ kN $\cdot$ m

12.18. $f = 1.7$

12.19. $f = 1.4$

12.22. $P = 46.18$ kN

12.25. $P_u = 9M_u/2L$

12.28. (a) $M_u = 2M_{yp}$, (b) $M_u = 16b(b^3 - a^3)M_{yp}/3\pi(b^4 - a^4)$

12.36. $t_o = 6.3$ mm

12.41. (a) $p_u = 25.53$ MPa, (b) $p_u = 29.48$ MPa

CHAPTER 13 PLATES AND SHELLS

13.2. $\varepsilon_{max} = 1250\,\mu$, $\sigma_{max} = 274.7$ MPa

13.4. (b) $w = M_a(x^2 - y^2)/2D(1 - \nu)$

13.6. (b) $p_o = 12.05$ kPa

13.8. $w = M_o(a^2 - r^2)/2D(1 + \nu)$, $\sigma_{r,max} = \sigma_{\theta,max} = 6M_o/t^2$

13.10. $n = 3.33$

13.11. $w = \dfrac{p(a^2 - r^2)}{64D}\left(\dfrac{5 + \nu}{1 + \nu}a^2 - r^2\right)$

13.12. $d = 3.85$ mm, $M_{max} = 320$ N

13.16. (a) $t = 18.26$ mm, (b) $w_{max} = 0.336$ mm

13.21. $p = 1.2$ MPa

13.23. (a) $t = 11.11$ mm, (b) $t = 15.6$ mm, (c) $t = 17.1$ mm

13.29. $\sigma_{max} = 2.48$ MPa

13.30. $N_\phi = \dfrac{pa}{a \sin \phi + b}\left(\dfrac{a}{2}\sin \phi + b\right)$, $N_\theta = \dfrac{pa}{2}$

13.31. $t = 0.791$ mm

13.34. $N_x = \frac{1}{2}\gamma x^2\cos\theta + \nu\gamma a^2(1 - \cos\theta) - \frac{1}{24}\gamma L^2\cos\theta$

Index

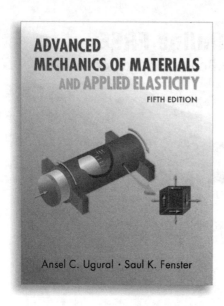